Regulating Audiovisual Services

Library of Essays in Media Law
Series Editors: Eric Barendt and Thomas Gibbons

Titles in the series:

Freedom of the Press
Eric Barendt

Media Freedom and Contempt of Court
Eric Barendt

Free Speech in the New Media
Thomas Gibbons

Regulating Audiovisual Services
Thomas Gibbons

Regulating Audiovisual Services

Edited by

Thomas Gibbons
University of Manchester, UK

ASHGATE

© Thomas Gibbons 2009. For copyright of individual articles please refer to the Acknowledgements.

All rights reserved. No part of this publication may be reproduced, stored in a retrieval system or transmitted in any form or by any means, electronic, mechanical, photocopying, recording or otherwise without the prior permission of the publisher.

Wherever possible, these reprints are made from a copy of the original printing, but these can themselves be of very variable quality. Whilst the publisher has made every effort to ensure the quality of the reprint, some variability may inevitably remain.

Published by
Ashgate Publishing Limited
Wey Court East
Union Road
Farnham
Surrey GU9 7PT
England

Ashgate Publishing Company
Suite 420
101 Cherry Street
Burlington, VT 05401-4405
USA

Ashgate website: http://www.ashgate.com

British Library Cataloguing in Publication Data
Regulating audiovisual services. - (Library of essays in media law)
  1. Television - Law and legislation 2. Webcasting - Law and legislation 3. Cellular telephone systems - Law and legislation
  I. Gibbons, Thomas
  343'.0994

Library of Congress Control Number: 2008939265

ISBN: 978 0 7546 2798 2

Printed and bound in Great Britain by
TJ International Ltd, Padstow, Cornwall

Contents

Acknowledgements vii
Series Preface ix
Introduction xi

PART I CONVERGENCE AND REGULATION

1 Wolfgang Hoffmann-Riem (1996), 'New Challenges for European Multimedia Policy: A German Perspective', *European Journal of Communication*, **11**, pp. 327–46. 3
2 Douglas W. Vick (2006), 'Regulatory Convergence?', *Legal Studies*, **26**, pp. 26–64. 23

PART II TECHNIQUES OF REGULATION

3 Cass R. Sunstein (2000), 'Television and the Public Interest', *California Law Review*, **88**, pp. 499–564. 65
4 Angela J. Campbell (1999), 'Self Regulation and the Media', *Federal Communications Law Journal*, **51**, pp. 711–71. 131
5 Andrew Murray and Colin Scott (2002), 'Controlling the New Media: Hybrid Responses to New Forms of Power', *Modern Law Review*, **65**, pp. 491–516. 193
6 Michael D. Birnhack and Jacob H. Rowbottom (2004), 'Shielding Children: the European Way', *Chicago-Kent Law Review*, **79**, pp. 175–227. 219

PART III STRUCTURAL REGULATION: MEDIA CONCENTRATION AND OWNERSHIP

7 Rachael Craufurd Smith (2004), 'Rethinking European Union Competence in the Field of Media Ownership: The Internal Market, Fundamental Rights and European Citizenship', *European Law Review*, **29**, pp. 652–72. 275
8 Peter Humphreys (1998), 'The Goal of Pluralism and the Ownership Rules for Private Broadcasting in Germany: Re-Regulation or De-Regulation?', *Cardozo Arts and Entertainment Law Journal*, **16**, pp. 527–55. 297
9 Christopher S. Yoo (2005), 'Architectural Censorship and the FCC', *South California Law Review*, **78**, pp. 669–731. 327
10 C. Edwin Baker (2005), 'Media Structure, Ownership Policy, and the First Amendment', *South California Law Review*, **78**, pp. 733–62. 391

11 Thomas Gibbons (2004), 'Control over Technical Bottlenecks – A Case for Media Ownership Law?', in *Regulating Access to Digital Television*, Strasbourg: European Audiovisual Observatory, pp. 59–67. 421

PART IV ISSUES IN REGULATING NEW MEDIA

12 Hernan Galperin and François Bar (2002), 'The Regulation of Interactive Television in the United States and the European Union', *Federal Communications Law Journal*, **55**, pp. 61–84. 433

13 Natali Helberger (2006), 'The "Right to Information" and Digital Broadcasting: About Monsters, Invisible Men and the Future of European Broadcasting Regulation', *Entertainment Law Review*, **17**, pp. 70–80. 457

14 Damien Geradin (2005), 'Access to Content by New Media Platforms: A Review of the Competition Law Problems', *European Law Review*, **30**, pp. 68–94. 469

15 Andrew T. Kenyon and Robin Wright (2006), 'Television as Something Special? Content Control Technologies and Free-to-air TV', *Melbourne Law Review*, **30**, pp. 338–69. 497

16 Horatia Muir Watt (2003), '*Yahoo!* Cyber-collision of Cultures: Who Regulates?', *Michigan Journal of International Law*, **24**, pp. 673–96. 529

17 Eli Noam (1998), 'Spectrum Auctions: Yesterday's Heresy, Today's Orthodoxy, Tomorrow's Anachronism. Taking the Next Step to Open Spectrum Access', *Journal of Law and Economics*, **41**, pp. 765–90. 553

18 Thomas W. Hazlett (1998), 'Spectrum Flash Dance: Eli Noam's Proposal for "Open Access" to Radio Waves', *Journal of Law and Economics*, **41**, pp. 805–20. 579

Name Index 595

Acknowledgements

The editor and publishers wish to thank the following for permission to use copyright material.

Cardozo Arts and Entertainment Law Journal for the essay: Peter Humphreys (1998), 'The Goal of Pluralism and the Ownership Rules for Private Broadcasting in Germany: Re-Regulation or De-Regulation?', *Cardozo Arts and Entertainment Law Journal*, **16**, pp. 527–55.

European Audiovisual Observatory/Council of Europe for the essay: Thomas Gibbons (2004), 'Control over Technical Bottlenecks – A Case for Media Ownership Law?', in *Regulating Access to Digital Television*, Strasbourg: European Audiovisual Observatory, pp. 59–67.

Melbourne University Law Review for the essay: Andrew T. Kenyon and Robin Wright (2006), 'Television as Something Special? Content Control Technologies and Free-to-air TV', *Melbourne Law Review*, **30**, pp. 338–69.

Michigan Journal of International Law for the essay: Horatia Muir Watt (2003), '*Yahoo*! Cyber-collision of Cultures: Who Regulates?', *Michigan Journal of International Law*, **24**, pp. 673–96.

Sage Publications for the essay: Wolfgang Hoffmann-Riem (1996), 'New Challenges for European Multimedia Policy: A German Perspective', *European Journal of Communication*, **11**, pp. 327–46. Copyright © 1996 Sage Publications.

Sweet and Maxwell for the essays: Rachael Craufurd Smith (2004), 'Rethinking European Union Competence in the Field of Media Ownership: The Internal Market, Fundamental Rights and European Citizenship', *European Law Review*, **29**, pp. 652–72. Copyright © 2004 Sweet and Maxwell; Natali Helberger (2006), 'The "Right to Information" and Digital Broadcasting: About Monsters, Invisible Men and the Future of European Broadcasting Regulation', *Entertainment Law Review*, **17**, pp. 70–80. Copyright © 2006 Sweet and Maxwell; Damien Geradin (2005), 'Access to Content by New Media Platforms: A Review of the Competition Law Problems', *European Law Review*, **30**, pp. 68–94. Copyright © 2005 Sweet and Maxwell.

University of California for the essay: Cass R. Sunstein (2000), 'Television and the Public Interest', *California Law Review*, **88**, pp. 499–564. Copyright © 2000 the California Law Review, Inc.

The University of Chicago Press for the essays: Eli Noam (1998), 'Spectrum Auctions: Yesterday's Heresy, Today's Orthodoxy, Tomorrow's Anachronism. Taking the Next Step to Open Spectrum Access', *Journal of Law and Economics*, **41**, pp. 765–90. Copyright © 1998 by the University of Chicago. All rights reserved; Thomas W. Hazlett (1998), 'Spectrum Flash Dance: Eli Noam's Proposal for "Open Access" to Radio Waves', *Journal of Law* *and Economics*, **41**, pp. 805–20. Copyright © 1998 by the University of Chicago. All rights reserved.

Wiley Publishers for the essays: Douglas W. Vick (2006), 'Regulatory Convergence?', *Legal Studies*, **26**, pp. 26–64. Copyright © 2006 Douglas W. Vick and © 2006 The Socierty of Legal Scholars; Andrew Murray and Colin Scott (2002), 'Controlling the New Media: Hybrid Responses to New Forms of Power', *Modern Law Review*, **65**, pp. 491–516. Copyright © 2002 the Modern Law Review.

Every effort has been made to trace all the copyright holders, but if any have been inadvertently overlooked the publishers will be pleased to make the necessary arrangement at the first opportunity.

Series Preface

Media law issues frequently dominate the news. A libel or privacy action by a politician or celebrity, an investigation into an alleged broadcasting scam, and the use of the Internet for downloading terrorist material or pornography are all stories which attract national, and increasingly international, publicity. Freedom of expression, whether on the traditional press and broadcasting media, or through the new electronic media, remains of fundamental importance to the workings of liberal democracies; indeed, it is impossible to see how a democracy could exist without a free, pluralist media for the dissemination of information and the discussion of political and social affairs. The media also provide us with celebrity gossip and popular entertainment.

But a free media does not entail the complete absence of law and regulation. Far from it. Laws are needed to balance the competing interests of the media, the public whom they inform and entertain and those individuals whose reputation, privacy, or even safety, might be endangered by newspapers, broadcasters and bloggers. All these branches of the media exercise considerable power and they can abuse it to distort the truth and harm individuals. Competition and other laws must be framed to prevent the emergence of media monopolies and oligopolies which are as incompatible with an effective democracy as is the domination of one political party. The Internet has been characterised by little or no regulation, beyond general criminal and civil laws, but it is legitimate to question whether this can remain the case given the ease with which, say, pornographic images can be circulated round the world in a moment. The globalisation of the media exacerbates legal problems, for a communication can be published more or less simultaneously in a number of different jurisdictions; some countries might, for example, protect privacy strongly, while others might not protect it at all because they consider privacy laws inimical to media freedom.

There is now a rich literature on many aspects of media law and regulation. The aim of these four volumes has been to present a sample of this literature, grouped round particular themes. Some of them concern topics which have been explored in legal periodicals for decades: freedom of the press, the balance between this freedom and reputation and privacy rights, media publicity prejudicing fair trials. Others deal with more modern aspects of the law, in particular whether and how the broadcasting and electronic media should be regulated. Inevitably, many essays are drawn from United States periodicals, as that country, with its strong attachment to freedom of speech and its powerful media industries, has produced an immense literature on all areas of media law. But we have also included articles from many Commonwealth countries. We have selected those which discuss issues of media law from a theoretical or comparative perspective. Lawyers in all jurisdictions can learn something from the treatment of common problems in other countries. The globalisation of the media means that knowledge of comparative law in this area is now of importance to practising lawyers.

ERIC BARENDT, *University College London*
THOMAS GIBBONS, *University of Manchester*

Introduction

The audiovisual industries have developed from terrestrial analogue radio and television broadcasting to encompass now a wide range of new media services based on mass communications to the general public by means of digital technology. The model of the vertically integrated broadcaster, dealing with all aspects of the process from the production of programmes to their transmission to the audience, has been replaced by more fragmented arrangements which have been made possible by convergence between different platforms. Digital technology means that the same content can be delivered by a range of methods, whether it be broadcasting, cable, satellite, the Internet or mobile telephony. Content can be sourced from a variety of creators, and packaged in a variety of ways, enabling users to exercise much greater choice over the kind of material that they receive and when they receive it. This volume of essays deals with the impact of the changing nature of audiovisual services on regulatory policy and design. Volume 3 in this series, *Free Speech in the New Media*, examines the issues of political and constitutional principle and theory that arise from the emergence of new media.

Convergence and Regulation

One effect of technological convergence is that the particular characteristics of the delivery mechanism – broadcasting, whether terrestrial or satellite, or telecommunications, whether mobile or cable – are much less relevant for regulatory purposes. There is little point in creating different regimes for each medium if the same content can be delivered across all of them. The general implication is that the focus of regulation will be on issues of content, and the appropriateness of imposing requirements that are common to all media.

At the outset, it should be cautioned that, although change is happening quickly, its pace and impact are rarely as radical as the industry predicts. Chapter 1, an essay by Wolfgang Hoffmann-Reim,[1] opens by noting that the advent of the multi-media age was being heralded in the mid-1970s. Yet it seems that traditional forms of programming still provide the core of audiences' media experiences and of business models.[2] Hoffman-Reim was writing soon after publication of the European Commission's Green Paper on Convergence(European Commission, 1997), which had invigorated policy debates about the implications of convergence (Goldberg, Prosser and Verhulst, 1998, Marsden and Verhulst, 1999; Marsden, 2000).

---

1 For decades, the leading expert on German media regulation, Hoffmann-Reim also served as Minister of Justice of the State of Hamburg, and was more recently a judge of the Federal Constitutional Court.

2 See, for example: Ofcom (2007); CRTC (2008) – a compilation of research and stakeholder views.

Hoffmann-Reim's perspicuous discussion of the challenges of new media will continue to have contemporary relevance. He argues that there is no reason to discard the ideas of responsibility and public interest, which are associated with notions of public service broadcasting, when considering the appropriateness of media policy and regulation for the future. Insofar as interventions in media activities are justified, in order to encourage individual self-development as well as the healthy functioning of democracy, thought needs to be given to the implications for the 'information' and 'knowledge' society. However, the issues will manifest themselves in different ways. Older forms of scarcity, in broadcasting frequencies, will be replaced by new scarcities created by technological and market barriers to access. In both cases, it is not the fact of scarcity in itself that justifies regulation,[3] but the excess of power by the media, over recipients and users of information, which it represents: 'scarcity constellations of all kinds are at the same time power constellations' (p. 9). As a response, Hoffmann-Reim does not advocate the mere continuance of traditional-style state regulation. He points out that the market and different regulatory techniques, such as self-regulation, will have important roles to play. However, he also maintains that a normative criterion – of fairness – should be applied in designing the best regulatory structures.

In Chapter 2, Douglas Vick provides a case study of developments in the United Kingdom, a jurisdiction where the implications of convergence generated an extensive policy discussion in the 1990s, leading to what was supposed to be an integrated and 'future-proof' scheme for regulating communications. Vick sets the debates in the context of the tensions between what he calls 'market liberalism' and 'social liberalism'. He argues that the differing approaches to regulating the media are explained less by perceptions of technological change than by shifts and tensions in political theory. In particular, he believes that they can account for variations in the level of official supervision (formal regulation or self-regulation), the amount of positive obligations, and the use of sector-specific regimes rather than general rules and principles. This use of 'models' of political thought is very helpful in highlighting the axes across which regulatory choices are made, provided that it is appreciated that there is a risk of oversimplification (as indeed does Vick).[4] With that caveat, the suggestion is that market liberalism, with its emphasis on minimal state interference, individual freedom of expression and the belief that market mechanisms can best allocate resources and satisfy readers' wants, explains the way that print and the press are treated. He suggests that telecommunications regulation is also explained by market liberalism, since regulation was implemented mainly to correct the market failure caused by the existence of what was regarded as a natural monopoly – the telecommunications network. By contrast, social liberalism accounts for the general approach to broadcasting policy, being concerned about the power of the media and the desirability of intervening to ensure that those who control that power will use it to benefit free democratic discussion. In the UK, this is manifested in the aspirations of public service broadcasting towards citizenship, universality and quality.

Vick offers a view of how these positions, and the compromises between them, appear to have influenced the passage of the UK Communications Act 2003. The Act was intended

---

3 For further discussion of the 'scarcity rationale' as a justification for media regulation, see Volume 3 of this series.

4 Not every theorist mentioned by Vick would endorse his classification. In particular, 'social liberal' covers many strands of what others might describe as democratic or republican theory.

to create a framework that would anticipate changes in media technology and business. It attempts to reconcile the need for competitiveness and innovation with the protection of basic social goals, adopting a 'light-touch' approach to regulation. A single regulator, Ofcom, has responsibility for all aspects of communications, including concurrent competition powers with the general competition regulator.[5] The Act has transposed the European Community's Electronic Communications Directives, which have liberalised telecommunications regulation and laid the foundation for a fully competitive market in electronic communications. In addition, it makes provision for spectrum use, including spectrum trading; it has simplified the licensing of digital channels; and it has eliminated controls on foreign ownership and on most concentrations of ownership. There is an increased emphasis on 'softer' forms of regulation such as co- and self-regulation. Nevertheless, an inclination towards economic regulation of communications markets is balanced by Ofcom's duties to promote public service television and its regulation of basic content standards in programming and in advertising.[6]

Techniques of Regulation

Although media convergence has encouraged a deregulatory mood, reasons for continuing at least some form of regulation continue to be advocated. The early part of the essay by Cass Sunstein (Chapter 3) is an example of the claim that there are justifications, based on democratic ideals, for regulating television programming in the public interest. Sunstein rejects the view that 'consumer sovereignty' is the appropriate benchmark. One set of reasons deals with the effects of market failure in broadcasting markets. The other set supports the overriding importance of certain non-market objectives: fostering the education of children; providing a substantive and diverse coverage of civic issues; and improving the viewing experience of the hearing impaired.[7] Throughout, he assumes that television programming has a distinctive social and economic role, although he recognizes that convergence may change that, albeit in the distant future. However, his piece is as interesting for its discussion of a range of mechanisms to promote the public interest, ones that are alternatives to 'command-and-control' forms of governmental intervention. The wish to regulate but to experiment with different techniques, is a feature of much contemporary writing about new media. In Sunstein's essay, the emphasis is on disclosure, economic incentives and voluntary self-regulation.

  The first proposal is that that every broadcaster should be required to disclose details of their public service and public interest activities. This is prompted by the apparent success of similar measures in the field of environmental regulation, where publicity about firms' environmentally negative activities has led to political and consumer resistance which has secured positive changes in behaviour. As Sunstein recognizes, the analogy with broadcasting is not quite apposite. Furthermore, a relatively similar approach in the UK, whereby commercial public service broadcasters are required to self-certify their positive, more-than-the-minimum contributions to quality programming, has not been sufficient to overcome

---

  5 Although Ofcom has no power to regulate Internet content, as such.

  6 For discussion of public service broadcasting, see Volume 3 of this series. For recent discussion of the further implications of Ofcom's remit, see Gardam and Levy (2008).

  7 Here, his arguments complement the discussion of justifications for regulation that are considered in Volume 3 of this series.

economic incentives to seek more popular scheduling.[8] Sunstein's second proposal is to provide economic incentives, again modelled on environmental regulation, by allowing a trade in public interest obligations. The idea of requiring the commercial sector to subsidize public interest media is not new[9] but, as he again acknowledges, the possibility of some firms' avoiding universal obligations may be more controversial. His third proposal is to introduce a self-regulatory code, although he is once more aware that there may be insufficient incentives or effective enforcement for that to make an impact.

Self-regulation has indeed been long regarded as a way of satisfying public interest concerns about an industry whilst minimising state interference in its activities.[10] It has been considered especially suitable for the media because self-restraint is consistent with freedom of speech.[11] However, as Angela Campbell shows in Chapter 4, there are a number of questions to be answered before self-regulation can be adopted as a policy option. First, the meaning of self-regulation is open to wide interpretation. It involves some form of regulation by the industry rather than government, but the exact relationship between the two is variable, with schemes having different levels of involvement in drafting a code and in enforcing it.[12] There are also nuances in the idea of 'self', with some industries genuinely subscribing to the public interest values at stake, while others are prompted to act when the threat of government intervention is considered imminent.[13] Secondly, the relative advantages and disadvantages of self-regulation are contestable. The industry will claim that it can achieve public interest objectives with flexibility and expertise at a reduced cost, whereas detractors will say that self-constraints are likely to be self-serving. Thirdly, and relatedly, the success of self-regulation may be difficult to measure.

Campbell discusses the problems in the light of the USA's experience in broadcasting and other media. As she notes, there is little reason to believe in the effectiveness of self-regulation, but it does have some political resonance. That may be why it appears attractive as a means of managing some aspects of new media, especially the Internet and mobile telephony, for which convergence may point in the direction of imposing content controls where previously there were none.[14] However, the realization that industry cannot generally be trusted to promote public policy through this means – because it has few incentives to do so – has prompted a more recent emphasis in Europe on 'co-regulation' This describes the situation where a regulator retains clear responsibility for imposing regulation, but sub-contracts one or more parts of its legislative brief to an industry body. The latter then implements the regulatory scheme but, importantly, the regulator retains 'back-stop' powers to intervene if the industry fails to be effective (Hans Bredow Institute, 2006).

---

8   In the UK, the problems are recognized in Ofcom (2008).
9   The Canadian radio and television regime is a good example.
10 For general discussion, see Baldwin and Cave (1999) chap.10; Ogus (1994), chap. 6; Black (1996), p. 24.
11 For example, the UK's Press Complaints Commission (at www.pcc.org.uk).
12 See also Hans Bredow Institute for Media Research and The Institute of European Media Law (2006). See links from the European Commission's Information Society Thematic Portal (at ec.europa.eu/information_society/activities/sip/index_en.htm.
13 On this and more generally, see Tambini, Leonardi and Marsden (2007).
14 For discussion in the context of European audiovisual regulation, see Volume 3 of this series.

Taking further the possibilities for alternative modes of regulation, in Chapter 5 Andrew Murray and Colin Scott offer a more theoretical analysis in the light of new media developments. Their essay offers a conceptual framework for designing regulation across all kinds of new media, including dimensions of Internet regulation not hitherto discussed. They concentrate on three problems associated with new media, which arise from opportunities that are presented by shifts in the power of those who conduct its operations, compared to traditional media. One is the problem of regulatory arbitrage, created when new media providers are sufficiently mobile to be able to choose the jurisdiction to govern their operations. Another is the problem of anonymity, which may prevent regulators from enforcing desired standards. The third problem identified is that of scarce resources. Here, Murray and Scott make the point that digital technology does not allow unlimited flows of information, since there are constraints on the amount of spectrum, bandwidth and network facilities that can be made available. Other problems, which are raised elsewhere in this volume, but which are equally compelling, are disparities in consumer access to digital media and the extent to which content should be controlled.

How, if at all, should regulators respond to these kinds of problem? Murray and Scott suggest that analysis should proceed in terms of Lessig's 'modalities of regulation'. For his terms – law, norms, markets and architecture – they substitute hierarchy, community, competition and design. Their aim is to capture a broader and more fluid set of control variables and, following Lessig, they envisage various combinations – 'hybrids' – as bases for regulation. Thus, self-regulation may have elements of hierarchy and community, combining official public standards with norms that are enforced socially, at the workplace or across an industry; an example would be the European Community's approach to preventing harm to children from Internet content. Of particular interest, for which Lessig's work is well known, is their discussion of design (architectural) elements, which include the organization of domain names on the Internet, content filtering and encryption. In each case, other modalities of trademarks, content rating and copyright may combine, respectively, with the design feature in order to create the requisite level of control. Murray and Scott's essay challenges us to think more creatively about regulation of audiovisual services and new media more generally. But it also reminds us that new forms of regulation are often evolutionary, building on traditional tools.

As Murray and Scott mention, an example of such a composite approach is the range of regulatory techniques adopted to prevent children from receiving harmful or illegal content. The essay by Michael Birnhack and Jacob Rowbottom (Chapter 6) provides a comparative analysis of the methods deployed in the USA and the European Community. In the latter, a broad set of 'Information Society' policies are intended to co-ordinate regulation of audiovisual services in a 'graduated' way in the light of media convergence. At one end of the spectrum, traditional television broadcasting ('linear services') should 'not include any programmes which might seriously impair the physical, mental or moral development of minors, in particular programmes that involve pornography or gratuitous violence'.[15] In respect of interactive and on-demand services ('non-linear services'), appropriate measures must be taken to ensure that similar material is only made available in such a way that minors will not normally hear or

---

15 Audiovisual Media Services Directive 2007, Art. 22.

see them.[16] Although the logic of convergence is that such non-linear services can be available on the Internet, the latter is regulated differently, through a combination of hard and soft law encompassed by the Safer Internet Programme 2009–2013.[17] This involves a number of initiatives, including the use of criminal law, technology, self-regulation and education.

Birnhack and Rowbottom point out that the European approach essentially leaves the regulation of material that is harmful to children to the market, but in the context of a governance framework that encourages self-regulation. One advantage of this apparently more liberal path is that it avoids the kind of constitutional challenges which have prevented legislative action from being successful in the USA. The latter has followed the route of 'direct public-ordering', which has led to open confrontation between the free speech principle and the child protection policy. Although the European Court of Human Rights has been willing to place more weight on the latter, in balancing the competing interests, European Community policy-making has followed the route of indirect or private ordering, whereby government promotes and supports self-regulation without explicit interference in media and Internet companies' decisions.[18] This route is not without its difficulties, as Birnhack and Rowbottom show: in particular, ratings and filtering are controversial; and the general problems of self-regulation, raised by Campbell in Chapter 4, are also apparent.

Structural Regulation: Media Concentration and Ownership

In economic terms, the structure of an industry is characterized by the number of firms operating in it and the degrees of power that they have over relevant markets. Where an industry shows tendencies to monopoly, market dominance or anti-competitive practices, a number of remedies are available under competition (anti-trust) law and regulation. The aim is to prevent too few firms from controlling too much of a market, both to prevent excessive prices being charged to consumers and to allow new products to be supplied. In general, the same problems and justifications for regulation will apply equally to media markets. However, is competition law sufficient to deal with those markets? Part III offers a selection of essays that explore the relevant arguments. Initially, they turn on whether media products are perceived to be the same as any other, or whether the association between the media and the political values of free speech and democracy renders at least some media outputs in some way special. But, even if they are special, it may still be the case that competition law is adequate to take account of that.

Two kinds of argument are typically advanced to show why structural regulation of media markets is desirable in order to prevent concentrations of control. One reflects the importance of securing a diversity of ideas and opinions to promote the interests that justify the priority given to free speech – the accumulation of knowledge and participation in a democracy. Here,

---

16 Audiovisual Media Services Directive 2007, Art. 3h. See the discussion by Craufurd Smith, in Volume 3 of this series.

17 This succeeds the Safer Internet Action Plan 1999–2004, and the Safer Internet *plus* Programme 2005–2008. See links from the European Commission's Information Society Thematic Portal.

18 The European Community's highest court, the European Court of Justice, follows the jurisprudence of the European Court of Human Rights (under the auspices of the Council of Europe), where appropriate.

creating an industry structure for a diversity of sources and platforms may be regarded as a necessary condition for pluralism – although it is important to appreciate that this will not in itself guarantee a plurality of viewpoints.[19] The other kind of argument is directed at the excess of power that may be wielded over public opinion and a broader understanding of the world, when only a few media firms (or individual proprietors) dominate public expression.

In both cases, the desire for a diversity of content might also be satisfied by directly imposing positive obligations to provide a range of perspectives, but that may be regarded as too much interference with freedom of the media. For the most part, such direct obligations may form part of a public service broadcasting remit but are rarely imposed on commercial media more generally (Craufurd Smith, 1997; Barendt, 1993).

In Chapter 7, Rachael Craufurd Smith discusses the issues in the context of the European Union, where debates about media concentration and ownership have been particularly active since the mid-1990s. National supervision of media ownership has long been a matter of peculiar political sensitivity, perhaps attributable to politicians' distrust of, yet respect for, perceived media power. Unsurprisingly, the details of national media ownership rules are usually context-specific. However, the European Commission floated the possibility of rationalizing the different approaches to be found in the European Community, at the same time questioning the technical basis for ownership regulation, through a draft directive on media ownership. The idea was to impose minimum requirements on all member states, based on a market share methodology for measuring concentrations of influence. Both the UK and Germany reformed their media ownership legislation in the light of the proposal, but the draft Directive was dropped, partly because of doubts about its legal standing, partly because competition law was thought by some to be a better tool, but mainly because of political resistance from the member states (Doyle, 2002; also Hitchens, 2006). Craufurd Smith's discussion covers the legal considerations in detail but she also explores various 'softer' techniques, developing some of the themes of the previous section. As she indicates, the expansion of new media does not, as some suppose, necessarily lead to greater diversity of content across the media, but it may require new responses to the problem.[20]

Whatever the solution, it is clear that media ownership regulation is highly sensitive to the politics of individual states. In Chapter 8, Peter Humphreys provides a case study of the policy process that laid the basis for the current position in Germany.[21] As he discusses, it is a dual system, which provides considerable scope for consolidation amongst private broadcasters, but on condition that the public service sector is maintained. Of particular interest is the role of the German constitutional court in mandating this approach and upholding the concept of broadcasting freedom as a buffer against state interference in media content. This reflects the fact that, while competition law will have an increasing role in dealing with concentrations of media power, the democratic functions of the media remain salient in German political discourse.

---

19 For a recent discussion, comparing pluralism and diversity in the UK, the USA and Australia, see Hitchens (2006).

20 In discussion leading to the revision of the Television without Frontiers Directive in 2007, the European Commission conceded that Community-wide co-ordination was not appropriate but it will continue to monitor developments: see European Commission (2007), p. 32.

21 Humphreys is a political scientist specializing in German media regulation.

Structural regulation of the audiovisual media has also been a feature of US communications regulation, albeit to a lesser extent compared to Europe. In the light of First Amendment resistance to direct content regulation, it has been the principal focus for regulation in the public interest to advance the democratic functions of freedom of speech.[22] The essay by Christopher Yoo (Chapter 9), and Edwin Baker's response to it (Chapter 10), represent major standpoints in relatively recent US debates on the topic.[23] Yoo argues that structural regulation actually has a negative impact on the content of speech, even though it has survived First Amendment scrutiny because it appears not to favour any particular speech. By way of example, he cites policies to promote free-to-air (advertising-supported) television, rate regulation of cable television, media ownership restrictions, and limits on vertical integration, policies which have been decided constitutionally not to implicate the content of expression. For Yoo, from an economics and law perspective, such 'architectural censorship' tends to diminish the overall quantity and quality of programming and also reduce the diversity of its content. However, he maintains that courts are unable to determine what should be the appropriate standard of scrutiny, for judging the status of these provisions, and are too deferential towards the policy objectives.

Baker's comments start by challenging the free market basis of Yoo's discussion. He focuses, instead, on the democratic significance of the media and the need to form arrangements that reflect democratic values. This entails the creation of structures to provide the greatest 'dispersal of power within public discourse' (p. 392). Baker's objections fall into three main categories. First, he rejects what he calls the 'reductionist commodification' underlying Yoo's analysis. He maintains that ownership rules are needed regardless of the empirical question of whether or not they actually produce greater diversity of content. For him, the presence of diversity of sources (voice) is the real issue. This perspective reflects only the expressive dimension of the free speech principle, however. A broader conception of media pluralism would also require that a diversity of information is indeed made available to audiences, but ownership regulation cannot do that without being supplemented by other measures to encourage a plurality of content and widespread access to it. Secondly, Baker takes issue with Yoo's 'enterprise-based' economics, arguing that they give insufficient weight to considerations of politics and welfare. Thirdly, he challenges the very idea of 'architectural censorship', claiming that the unintended consequences of structural design cannot be regarded as the suppression of opinion.

Structural regulation of ownership has been one of the principal ways of promoting the value of media pluralism in a democracy. But will it be appropriate for new media? The essay by Thomas Gibbons (Chapter 11) suggests that traditional, sector-specific approaches will become outmoded as media convergence occurs, but that other forms of regulation will be needed to deal with concentrations of power in the industry. The reason is that convergence between platforms for delivery will mean that the production and packaging (bundling) of content become more important in the media value chain than the means of dissemination. A particular issue is the control of 'bottlenecks', narrow points through which flows of content have to pass in the course of distribution. Competition law provides some redress for

---

22 See Sunstein, Chapter 3 in this volume. See also the essays in the Part I of Volume 3 of this series.

23 See also Baker (2007); Bagdikian (2004); Compaine and Gomery (2000). The FCC made a second attempt to relax local media ownership laws in December 2007.

economic bottlenecks, and the regulation to ensure access to services deals with some aspects of technical bottlenecks, but Gibbons argues that neither is sufficient to protect the interests of end-users in receiving information.

Issues in Regulating New Media

Part IV provides a selection of articles that deal in more detail with a range of significant issues raised by new media. The general theme is access and control over the dissemination of audiovisual material, and the ways that regulation interacts with technology, competition law, intellectual property rights, and contractual arrangements. Digital technology has the potential to confine, as well as to release, information flows.

A key feature of digital audiovisual media is their potential for interactive communication. In a comparison of the US and European approaches, the essay by Hernan Galperin and François Bar (Chapter 12) examines the regulatory implications of the development of interactive television services. Illustrating the kinds of questions that apply to new media more generally, it explores the relationship between content providers and network providers, interconnection between different services, consumer access to services, and barriers to competition. A particular concern is the possibility that, instead of allowing widespread access to information, there will be a tendency for services to depend on proprietary technologies and 'walled garden' business models. Anticipating what has come to be described as the 'digital divide', Galperin and Bar worry that this could 'leave second-class digital economy citizens with access to a limited array of entertainment, transaction and educational services' (p. 437). Their analysis is interesting because it suggests that, even if convergence leads to a deregulation of broadcasting in favour of the more market-oriented approach to be found in telecommunications, the retention of a broadcasting business model for new interactive services will inhibit the freer flow of information that might be anticipated. They show how there are opportunities and incentives for anti-competitive behaviour across the major components of a network: the transmission system, the return path (along which the user interacts with the provider) and the end-user's access (the set-top box). However, as they discuss, the US and European responses to the problems are somewhat different. The US regulator has been reluctant to intervene in an immature market, as the AOL/Time Warner case illustrated. By contrast, the European Commission has been willing to impose restrictions on dominant providers as a condition of approving joint ventures, as the BIB case illustrated.[24]

As Galperin and Bar observe, the EC's general competition rules are supplemented in this area by a co-ordinated framework of regulation for electronic communications,[25] which covers all aspects of the wireless and wired infrastructure. It is based on the premise that some *ex ante* regulation is required for enterprises with significant market power. However, that framework reflects the view, within the EC's broader Information Society policy, that infrastructure regulation can meaningfully be separated from content regulation. Natali Helberger challenges

---

24 Ofcom currently adopts broadly the same approach as that described for its telecommunications predecessor, Oftel.

25 For a comprehensive account, see Nihoul Rodford (2004). The Directives are under review in 2008; see links from the European Commission's Information Society Thematic Portal.

that view in Chapter 13,[26] in which she examines the connection between digital broadcasting and the right to receive information, in the light of the way that electronic access control changes the way that viewers access content. She observes that, under the jurisprudence of the European Court of Human Rights, the viewer's right is not to claim access to content that the media possess, but to be able to receive the content that the media choose to impart. This reinforces the idea that viewers are passive recipients of information, with the implication that media companies' control over access to content may have the effect of excluding members of the audience.

Helberger's focus is the European Community's review of the provision, in the 1997 amendment to the 'Television without Frontiers' Directive, for ensuring that major events of national importance can be broadcast free-to-air. That provision was a response to the growth of subscription television, both analogue and digital, and its effect on public service broadcasters' ability to secure the transmission rights for important occasions. However, because digital technology enables content to be put together into discrete digital packages, the distribution of material can be much more closely controlled and calibrated. Partly, the controls are technical, such as set-top boxes, and partly they are contractual, such as subscription management systems (tied to the technology) and the bundling of programming. In addition, intellectual property rights will interact with the technology to impose territorial restrictions on the circulation of content. As a response to those developments, as Helberger explains, a right to 'short reporting' became part of the reform agenda during the review of the Directive in 2004. That right has now been incorporated into the Audiovisual Media Services Directive 2007 (Articles 3j and 3k). However, she argues that the new provision does not take sufficient account of the way that digital television has altered the viewer's experience, from being a member of a large audience to being an individual user. She suggests that existing measures, to deal with what is broadcasters' access to infrastructure, needs to be supplemented with protection for individuals' access to specific content.

Nevertheless, without a vibrant market in content delivery, access rights may lack substance. As Damien Geradin explains, in his essay on the related competition law problems in the new media (Chapter 14), the ability to supply premium content, represented mainly by football matches and Hollywood blockbuster films, is critical to a media provider's business success and its capacity to supply other kinds of content. However, where dominant (generally pay-TV operators) control the distribution and showing of content, by means of contractual and intellectual property rights, there may be few opportunities for new firms to offer different packages. Geradin's essay deals with difficulties that are not usually discussed in the regulatory literature but which are immensely important for media practice. Indeed, one effect of convergence is to expose the sometimes complicated relationship between regulation and intellectual property rights. While technical convergence means that different forms of content delivery no longer need to be regulated differently, from a rights-holder's perspective, the ability to deliver the same content across a variety of platforms means that maximum economic value can be extracted by granting separate licences for each of them. Geradin provides a number of examples of such practices, reminding us that very large sums of money are at stake, which in itself will create an obstacle for new firms. The principal regulatory tool for dealing with the resulting problems is competition law and it is clear that, although the

---

26 The article draws on Helberger (2005).

general principles may appear straightforward, their application is highly context-specific. It requires empirical definition and analysis of the relevant markets, which may be frequently shifting, together with detailed analysis of the particular contractual arrangements. Geradin argues that it is these arrangements in the new media, especially those involving minimum guarantees and exclusivity agreements, which prevent new media platforms from gaining access to the market and which should be subject to greater regulatory intervention.

Interests in content may also be protected through new technology and this creates another area of interaction between regulatory policy and intellectual property law. Chapter 15, an essay by Andrew Kenyon and Robin Wright, explores some of the implications for free-to-air television, comparing US, European and Australian initiatives. Originally, one of the defining characteristics of broadcasting was that it was free-to-air – that is, capable of being received without a direct charge for the content. As mentioned earlier, digital technology allows direct charging to take place, providing a basis for subscription television. Will it make a difference to free-to-air programming? On one view, it will simply enable new forms of distribution across different platforms. But Kenyon and Wright suggest that the implications will be more significant if content providers are able to use the technology to restrict the use and reuse of material freely received. In practice, this could mean limiting the copying for the purposes of time-shifting or sharing with other users. Digital technology offers content providers the opportunity to enforce restrictions on usage that were not possible or practicable when material was freely available. From a copyright perspective, the policy issues turn on the balance to be struck between a fair reward for the content creator's efforts and the public's interest in the free flow of information. From a media-policy perspective, the issues involve the viability of the traditional, advertising-funded, mass audience approach to television viewing, and the desirability of content providers being able to exert direct influence media policy. Related, there is a public governance issue at stake, since copyright has traditionally been regarded as an area of private law-making, where wider democratic concerns have not been prominent. One of the key points made by Kenyon and Wright is that the evolution of digital technology requires the joint development of appropriate broadcasting and intellectual property policies.

The trans-national dimension of media transmission, whether it be by radio, satellite or the more recent Internet, has long given rise to problems of jurisdiction and enforcement. The European Community approach has already been mentioned, but there are broader questions posed by clashes between the fundamental values of states at an international level. These are considered by Horatia Muir Watt in Chapter 16, in which she examines a set of questions prompted by litigation where one state has attempted to impose local restrictions on material that can be accessed through the Internet. As Watt notes, problems typically arise because the libertarian perspective on freedom of expression, reflected in US constitutional law, comes into conflict with public values that deem certain content offensive. For example, in the Yahoo case, French regulators prohibited offensive content originating in California from being accessed in France. The conflict between jurisdictions is exacerbated in relation to the Internet because its history and development has been associated with the idea of 'cyberspace', a metaphor which connotes the last bastion of unrestricted flows of information, disconnected from geography and jurisdiction. How should such 'cyberconflicts' be resolved?

Watt argues that the design of the Internet does not determine the normative implications of such international conflicts. Perceiving the Internet as a norm-less space is an ideological choice. In fact, the architecture of the Internet has as much capacity to restrict information flows as it does to facilitate them, in particular, through filtering and zoning techniques (also known as 'geo-gating' or 'geo-blocking'). She suggests that the appropriate policy choice is to adopt the approach being adopted more generally in conflicts of law, which is to provide a remedy in the jurisdiction where the negative effects of conduct are experienced. Furthermore, she suggests that Internet technology will enable that to be done in a more calibrated way, which has minimal impact on more liberal jurisdictions. However, that will happen only if states show restraint in using such technology. But Watt's discussion makes it clear that it is political values, and not the technology, that are crucial.

One of the hazards of the discussion of new technology is that it can all too quickly become dated in a rapidly developing field. Predictions about the future are equally precarious but the discussion, in Chapter 17, of open spectrum access, by the respected commentator Eli Noam, offers a challenging speculation about what might be possible. Noam was writing soon after the US Federal Communications Commission had started to allocate electromagnetic spectrum by auction. His essay is partly a critique of the auction approach, his principal objection to it being that it establishes 'licensed exclusivity.' The alternative, made possible by new digital technology, is a model of open access to spectrum, which he sees as consistent with the constitutionally minimum abridgment of free electronic speech. Commenting on Noam's thesis, Chapter 18, an essay by notable media economist Thomas Hazlett, casts doubt on the rationale and the practicability of open access, and partly is a defence of spectrum auctioning and a rejection of the government intervention that would be needed to administer the system. Possibly, Noam was overoptimistic in believing that the wired Internet could provide a model for open wireless access; contemporary debates about net neutrality reflect the fact that the Internet's openness is by no means secure. Yet his argument is a refreshing counter to a seeming inevitability that the controlling, rather than liberating, potential of digitalization will dominate.

References

Baker, C.E. (2007), *Media Concentration and Democracy: Why Ownership Matters*, Cambridge: Cambridge University Press.
Bagdikian, B.H. (2004), *The New Media Monopoly*, Boston: Beacon Press.
Baldwin, R. and Cave, M. (1999), *Understanding Regulation: Theory, Strategy and Practice*, Oxford: Oxford University Press.
Barendt, E. (1993), *Broadcasting Law: A Comparative Study*, Oxford: Oxford University Press.
Black, J. (1996), 'Constitutionalising Self-Regulation', *Modern Law Review*, **59**, pp. 24–55.
Compaine, B.M. and Gomery, D. (2000), *Who Owns the Media? Competition and Concentration in the Media Industry* (2000) Mahwah, NJ: Lawrence Erlbaum Associates.
Craufurd Smith, R. (1997), *Broadcasting Law and Fundamental Rights*, Oxford: Oxford University Press;
CRTC (2008), *Perspectives on Canadian Broadcasting in New Media*, at www.crtc.gc.ca/eng/media/media.htm.
Doyle, G. (2002), *Media Ownership: The Economics and Politics of Convergence and Concentration in the UK and European Media*, London: Sage Publications.
European Commission (1997), *Green Paper on the Convergence of the Telecommunications, Media and Information Technology Sectors, and the Implications for Regulation*, COM (97) 623.

European Commission (2007), *Media Pluralism in the Member States of the European Union*, Brussels, SEC (2007).
European Commission's Information Society Thematic Portal, at ec.europa.eu/information_society/activities/sip/index_en.htm.
Gardam, T. and Levy, D.A. (eds) (2008), *The Price of Plurality: Choice, Diversity and Broadcasting Institutions in the Digital Age*, Oxford: Reuters Institute for the Study of Journalism.
Goldberg, D., Prosser, T. and Verhulst, S. (1998), *Regulating the Changing Media*, Oxford: Oxford University Press.
Hans Bredow Institute for Media Research and The Institute of European Media Law (2006), *Final Report Study on Co-Regulation Measures in the Media Sector*, A Study for the European Commission, Directorate Information Society and Media
Helberger, N. (2005), *Controlling Access to Content: Regulating Conditional Access in Digital Broadcastingi,* The Hague: Kluwer Law International.
Hitchens, L. (2006) *Broadcasting Pluralism and Diversity: A Comparative Study of Policy and Regulation*, Oxford: Hart Publishing.
Marsden, C. (ed.) (2000), *Regulating the Global Information Society*, London: Routledge.
Marsden, C. and Verhulst, S. (eds) (1999), *Convergence in European Digital TV Regulation*, London: Blackstone Press.
Nihoul, P. and Rodford, P. (2004), *EU Electronic Communications Law: Competition and Regulation in the European Telecommunications Market*, Oxford: Oxford University Press.
Ofcom (2007), *The Communications Market 2007*, at www.ofcom.org.uk.
Ofcom (2008), *Second Public Service Broadcasting Review – Phase 1*, at www.ofcom.org.uk .
Ogus, A. (1994), *Regulation: Legal Form and Economic Theory*, Oxford: Oxford University Press.
Tambini, D., Leonardi, D. and Marsden, C. (2007), *Codifying Cyberspace: Communications Self-Regulation in the Age of Internet Convergence*, London: Routledge.
UK Press Complaints Commission, at www.pcc.org.uk.

Part I
Convergence and Regulation

[1]

New Challenges for European Multimedia Policy
A German Perspective
■ *Wolfgang Hoffmann-Riem*

ABSTRACT

■ In order to ensure a certain quality of communicative achievements on behalf of the development of individuals and societies, media regulation will still be indispensable in the age of multimedia. The main question is not how to justify regulation but rather how to design it adequately. It will be a challenge to hold on to normative premises of political liberalism, of societal emancipation and of protection against manipulation as well as to stimulate the potential of differing values, information and ideas in Europe. The most promising form is regulation which concentrates on regulating structures which help improve the effectiveness of access at different levels and to different uses (especially fairness of communication, distribution and reception). ■

Key Words fairness of access, media responsibility, multimedia regulation, universal service

Innovation responsibility in the multimedia age

The advent of the multimedia age in Germany and in other European countries is accompanied by a host of visions and metaphors (cf. Schlosser and Bockholt, 1994: 122–4), however, this was not different 10 years ago, and a similar mood of new departure even prevailed 20 years ago. In 1975–6, for example, a Commission in Germany submitted a report

---

Wolfgang Hoffmann-Riem is a Professor at the University of Hamburg. He is currently working as the Head of the Justice Department for the state of Hamburg, Justizbehoerde, Drehbahn 36, 20354 Hamburg, Germany

(KtK, 1976) which irrigated the still fallow fields of telecommunications and also marked the start of the journey into the age of commercial broadcasting. Just under 10 years later, in 1984, observers in Ludwigshafen awaited the 'big bang' of the first pilot project (cf. Noelle-Neumann, 1985) — which, empirically speaking, then turned out to be more of a 'big sizzle', but which did usher in a new phase in the development of the broadcasting order: the end of the public broadcasting's monopoly and the starting phase of the dual (public and commercial) broadcasting system and of the further commercialization of the media. And at present, not only are new pilot projects being planned (see Stadek, 1994: 20, 24),[1] but efforts are in full swing, above all at countless conferences and symposia, in newspaper articles, at company meetings and in lobbyistic canvassing in state chancelleries, to enlist support for a new vision, the vision of the multimedia society.

As in former years, the comparisons drawn are not always apt. The new publicity-conscious terms 'Information Superhighway' and its German-language equivalent *Datenautobahn* are examples.[2] These lines of communication are depicted as the lifelines of the multimedia society; the terms, therefore, should stand for complex networking, for interactivity, for spontaneous and thus innovative communication, for undreamt-of speeds and for unrestricted access. Their promoters promise traffic without obstruction for all forms of communicative services — business and personal — from telebanking, homeshopping and telelearning to pay TV, video on demand and information on demand, from interactive services such as CD-ROM and other CD applications[3] to confession shows or pornographic pleasures on the Internet. And they emphasize that multimedia cannot develop in a straitjacket of state regulations, but that it requires entrepreneurial creativity which tolerates no patronization by the state.[4] Is the superhighway image appropriate in this context?

American superhighways and German *autobahns* channel traffic heading in one direction, grant access to motor vehicle owners only — not, for example, to cyclists and pedestrians — force all users to move in the same direction, have strict codes of conduct, even stipulate the speed by law or de facto through congestion, and do not always lead directly to the desired destination, but, as a rule, via secondary roads, traffic-abatement zones or country lanes. The motorist moves in a linear manner towards one destination; they can communicate with fellow motorists by flashing headlights or by making offensive gestures; of course, they can also drive back after reaching their destination, return to the superhighway via side-streets and country roads, and then travel on the opposite lane.

For the multimedia visions of networked computer worlds, virtual realities, facilities of interactivity and, above all, the wide range of unimagined information and communication services, the image of channelled and restricted traffic on the superhighway seems inappropriate. Or is it perhaps very appropriate indeed? Does interest focus in fact on spontaneous communication, individual interaction or communicative surprise? Or is it mainly concerned with commercial movements or even only with the really big flows of traffic and the rapid and cost-effective transportation of bulk goods? Do all persons involved perhaps regard strict traffic regulations as indispensable? Would some of them perhaps like to build the superhighway themselves, impose toll charges, and stipulate access and traffic regulations in line with commercial calculations? Perhaps they also hope that the hard shoulders and footpaths needed to reach the individual users will always remain an infrastructure task for the state if they cannot be refinanced via private toll charges? And what about the parallels to the external burdens of motorized traffic, such as the damage done to nerves and the mentally intoxicating emissions? Are cultural ozone holes to be expected? Does economic growth of multimedia require a back-up protection of the sociocultural multimedia ecology?

In Europe, such questions cannot, to remain in this metaphorical sphere, be simply 'bypassed'. We lack the 'just do it' mentality many Americans reputedly possess, although this is also embedded in a regulatory environment in the USA.[5] We also know from experience — worldwide experience — that effectively shaping media structures is only possible in the early stages. Once they have been established, these structures can hardly be meaningfully rearranged at a later date;[6] powerful actors have journalistic and political as well as economic power, which is generally superior to that of the state.

As always when it comes to innovations, however, the powers of imagination cannot suffice to forecast the course of future developments. This is inevitable: the interplay of media technology, media economics, media policy and media acceptance leads to complex networks and dynamic causal constellations with surprising feedback loops. Media developments are dovetailed with and influence general societal developments (cf. Saxer, 1994: 331). As in other problem areas of the risk society (cf. Beck, 1986), the implications cannot be exactly predicted and assessed and provisions against risks are fraught with uncertainty. The consequence should not be to abandon such provisions entirely for the future, in particular the safeguards of societal responsibility.

It should not, on the other hand, consist of recourse to outdated or even antiquated regulative mechanisms, especially whenever the latter primarily operate on the basis of behavioural prescripts and prohibitions. Such recourse should not only be ruled out because media supervision has been no success story over the past 10 years (cf. Hoffmann-Riem, 1996) not, first and foremost, because of the individual failure of the persons involved, but probably because of the lack of concordance between the goals and instruments of regulation. The regulatory philosophy of media laws to date will certainly come unstuck when faced by the globally structured multimedia and information society of the future. The actors in this environment can move with relative ease from one level and sphere of action to another, disguise themselves through structural interlocking, submerge in international networks or openly risk power struggles with supervisory bodies — and usually win on account of the numerous dependencies. What is more, as shown by the Internet (see Wisebrod, 1995; Grossman, 1994: 27), almost any regulation of behaviour by the state is virtually ruled out once a communication network sets up transnationally or even globally and trusts in a 'spontaneously' self-regulating structure. Users keen on evading state regulations here do not even need to be powerholders.

On the justification of regulation

Part of contemporary tradition on both sides of the Atlantic, however, is to ensure a certain quality of communicative achievements, achievements on behalf not only of the functioning of democracy, but, first and foremost, of the development of individuals and societies, i.e. on behalf of the communicative orientation, socialization and cultural development of citizens. Although the media are legally protected if they actually render these achievements, the communications order should be set up in such a way that these achievements are provided with a sufficient degree of probability and in line with certain rules. References to the public service orientation of broadcasting[7] or to the 'public task' of the press (cf. Rehbinder, 1962: 120) are the sediment of such thinking in terms of societal responsibility. The age of the information and knowledge society is now reputedly dawning particularly at a time when information and communication are poised to penetrate into all areas of life to a much greater extent than ever before. In other words, at a time of increasing universal dependency on the communications order, there is no reason to discard such normative expectations, i.e. media responsibility.

The objections to state or supranational media policies, therefore, are of a different nature. One of them runs to the effect that these positive goals can be achieved much more effectively by refraining from state intervention. This, for example, is the thesis forwarded by the 'Group of Personalities' — to be more precise, 'business personalities' — rallied around EC Commissioner Martin Bangemann (see Bangemann et al., 1994). The supporters of self-regulation by the market in its purest possible form admit that this is not faultless; nevertheless, they contend that it safeguards the public interest in the best possible way. This thesis is empirically verifiable and must consider the various findings on market failure and weigh them up against findings on the failure of the state — a comparison the Bangemann Group has not made.

Another associated objection disputes any legitimation whatsoever for state regulation. One correct aspect here is that the scarcity of transmission paths will soon no longer serve as justification for the need to regulate.[8] On closer examination, however, it becomes clear that scarcity was never a justification but merely a point of reference for regulation: scarcity demanded traffic regulations and allocation; the state reserved its right to organize both and linked both tasks to conditions whose content was not justified by scarcity but by the public interest goal. Consistent with this line of argument, the German Federal Constitutional Court did not justify the regulation of broadcasting by referring to the scarcity of frequencies alone, but, more generally, to existing barriers to access and to the power potential of the media.[9] The accumulation of power and barriers to access, however, will also exist in the age of abundant frequencies. To a certain extent, they will even increase, albeit at levels other than those of the technical availability of frequencies. Network operators, for example, can set up financial barriers, lay down technical standards and determine access conditions and traffic regulations. It is unclear whether fairness of access will be guaranteed for services and for suppliers or whether attempts will be made via technological, legal and financial access filters to create structurally anchored advantages for the marketing of certain services, such as those offered by affiliated companies. The interconnection between different media and varying levels of activity enables the application of subtle gains-oriented mechanisms. The more concentrated the media scene and the more pronounced the conglomerate character of the interest groups — a situation which is emerging in Germany (see EMI, 1995; Monopolkommission, 1994: 295) — the less capable a market is on its own to prevent such one-sided structures.

It is also unclear which access filters will exist for users, for example, *financial* — what kind of rates?; *technological* — in the design, say, of the set top box; *geographical* — will services also be provided to remote areas? or *media related* — how easily and how self-determinedly can navigational aids be used and how fair are they?

Personal media competence — 'media and computer literacy' — will be a further important filter to access. The information society can strengthen, indeed increase inequalities here which are anchored elsewhere in society, particularly in structurally disadvantaged life situations, which are formatively influenced, for example, by age, education, geographical location, poverty and ethnic origin. The market will not be able to guarantee protection against this, certainly insofar as the disadvantaged status also has economic causality.

The European Commission quite rightly addressed the risks of the emergence of a divided society which comprises information haves and have-nots (EC Commission, 1995: IX); it primarily considers the disparities regarding *technological* and *financial* access opportunities. The problem, however, is much broader: disparities are to be expected with respect to the sociocultural abilities to utilize communication, i.e. in the field of communicative competencies. A new social issue looms on the horizon: the 'social issue of the communications society'.

The information society will probably have an additional problem of new scarcities, in relative rather than absolute terms, in other words scarcities which are influenced by other framework conditions. The scarcity of available media content is already a huge problem, in the field of mass communications at any rate. The elimination of the scarcity of frequencies and the advent of new offers intensify the relative scarcity of programme software. Anyone who has a powerful or even dominant position at the upstream stages of distribution — especially in the field of production and copyright[10] — and anyone who, in addition, is powerful at downstream stages — resale in particular — can also control the organization of communication or the provision of services by others. The concrete marketing structures in the field of programme software will also decide whether the already substantial scarcity is artificially intensified or whether it is used also to create market power in other sectors. As, understandably, there is a call for powerful European players in global competition, the concentration trap threatens to snap shut: whoever holds power in global competition will find it easier to dominate local or regional, and certainly sectoral markets.

Another major scarcity problem exists. The extension of the range of offers and the increase in the transmission paths turn user attention into a particularly scarce resource. Competition for the scarce resource attention is in keeping with market requirements, but it can provoke aggressive supplier strategies in which media responsibility is no longer even administered in homoeopathic doses. Some of the excesses of reality TV or of similarly degrading shows already provide — *in statu nascendi* — illustrative material (cf. Wegener, 1994; Winterhoff-Spurk et al., 1994). Above all, however, there is a temptation at a time of media hyperabundance to exploit the lack of orientation of recipients when competing for attention as a scarce resource. Manipulative access to attention could be backed technologically, for example, through the design of the set top box or of the interface, or medially, through the design of special orientation aids.

This makes it clear that scarcity constellations of all kinds are at the same time power constellations. The crucial topic of regulation is the problem of use and abuse of power and the Federal Constitutional Court has always known this.[11] Up to now, it has linked the identification of the power risk with the scarcity of transmission paths and the high financial access barriers for broadcasting. The basic normative idea of necessary protection against the one-sided use of power, however, continues to apply, even if there is a shift in scarcity constellations and new abuse potentials become identifiable.

No new justification is required in the multimedia age for the aversion of traditional risks in fields in which the need for protection is generally accepted, for example, protection of the personality, such as the protection of honour or data privacy, but also the protection of children or the protection of property, such as copyright. State regulation as a setting of limitations, as specified in Article 5 II Basic Law, is not essentially questioned. A moot point, however, is whether the state should be allowed to regulate the media order as such, whether it assumes responsibility for its workability. This is a question which the Federal Constitutional Court customarily examines with reference to the norms relating to the structuring of the broadcasting order (cf. Ruck, 1992), a competence laid down in Article 5 I 2 Basic Law. The findings on scarcity and power abuse potentials outlined earlier suggest that this form of regulation will also remain fundamentally justifiable once the scarcity of transmission paths no longer exists.

Whether the market can stimulate and sustain sufficient countervailing forces from within to prevent abuse is, to put it mildly, not certain. Even the EC Commission has considerable doubts as reflected in the regulative proposals in various green papers and draft directives.[12]

Despite its rapid expiry period, the harmonization of regulation remains a constant leitmotif.

Even if the global, European, German and respective regional and local markets could be kept in balance by equipotent forces this alone would be unable to ensure a certain quality of content for communication offers and user possibilities. There are sufficient examples in the history of broadcasting which bear witness to this fact. The erosion of the public service idea (see Hoffmann-Riem, 1990) in the programmes offered by commercial broadcasting and its competition-induced consequences in public broadcasting too[13] are inevitable results of a marketing of mass communication whose inescapability is not disproved by selective examples to the contrary. Initially setting up the broadcasting order on a commercial basis and subsequently remonstrating with the broadcasters with a moral undertone for making use of their room for manoeuvre is hypocritical. Such hypocrisy can also be frequently found in the criticism that many programmes are characterized by popularization and trivialization, by sensationalism, by ritualization and stereotypification. To then bombard the broadcasters with appeals to mend their ways is ritual politics. No, the broadcasters behave in conformity with the market, i.e. usually in conformity with market ethics too. The question is whether this is enough.

There is no indication that the majority of programme offers will become fundamentally different in the age of programme multiplication, formatization, target group orientation or individual request. New types of offers will become possible, but presumably only for individual topic fields and users with sufficient financial resources, i.e. on a selective basis.

There is also no indication that the recipients will steer clear of the pleasure of the media world of illusion and turn away en masse from programmes without any special public service orientation. The pressure of competition will, if need be, induce broadcasters to launch counter-strategies. The technological possibilities will make it easier for them to identify user preferences, cater for these selectively, intensify them, and audiovisually take the recipients by the hand in their search for orientation. The technological potential, for example, for the creation of virtual realities, will also help. Subtle patronization by the private guardians of media access and the private moulders of media content will very probably increase in future. Nevertheless, this is no reason to call now for state patronization, irrespective of the fact that the still fertile humus for state patronization will dry up in the global information society.

The analysis so far has related to broadcasting. However, it is, in my opinion, applicable to all other fields in which communication is provided for the masses, regardless of how segmented and fragmented its manifestations and how individual its modes of operation may be. The pressure of competition and the fight for the distribution of scarce financial resources and for cost-effective channels of production will probably compound the generally observable trend towards standardized mass production (cf. Hoffmann-Riem and Vesting, 1995: 21) and the accompanying stereotypifications and simplifications (cf. Groebel et al., 1995: 98), even though this can be partly veiled by the clever use of technology, for example, through the creative application of the facilities of virtual reality. Multiple use in graduated cascades of utilization and virtually endless chains of repetition will also continue, although occasionally disguised, with technological assistance, by partial variations.

A similar trend may result for all standardized services, irrespective of whether they are legally classifiable as broadcasting, i.e. including those services for which the rigid role distribution between communicator and recipient is made flexible, through interactive technologies but prefabricated production modules are nevertheless used.

An increasingly frequent reaction to, in my opinion, the expected confirmation of further erosions of the public service orientation of mass-oriented communication, is to revalue the values. Supplier and consumer sovereignty had now, it is claimed, at long last been achieved; public service goals are no longer to be prescribed by the state and to be guaranteed by a 'positive order' of the field of broadcasting (along these lines, see, for example, Starck, 1995: 308). It is up to each society, of course, to decide whether the public service idea should be exhibited as a fossil in the media museum or entrusted to the custody of persons with a nostalgic fondness for certain values. It is fair to ask, however, whether the normative expectations so far regarding broadcasting, relating, for example, to the quality of content and plurality, to freedom from manipulation or to media responsibility, should now cease to be valid merely because technological and economic progress enables new and multiple forms of providing communication, whose social and individual function nevertheless remains, in many respects, the same as the one fulfilled by broadcasting so far.

I cannot say which way society will decide in the multimedia age. I would, however, like to discuss how the traditional value option could also be sustained in the future. This would be a real challenge for European media policy: to hold on to the normative premises of enlightened political liberalism, of societal emancipation and of protection against manipulation during periods of media transformation and to harness, indeed stimulate, the tremendous potential of differing values, information and ideas in Europe. The structures of the information society would then have to be set up in such a way as to guarantee that such goals stand a chance of realization.

Topics of regulated self-regulation

Such an effort does not mean a rejection of the market, and definitely not a rejection of other mechanisms which trust in a self-regulation of society. On the contrary, there is no indication whatsoever that the state or the EU could intervene in the complex, networked and highly sensitive communications sector with regulative-imperative success and without dysfunctional side-effects.[14] There is one thing both certainly cannot do and that is to affect the contents and quality of communication offers and user needs in such a way through prescripts and prohibitions that they are 'positively' influenced.

Rather, the discussion on regulation during recent years would suggest that the most promising form is regulation which concentrates on the creation of structures which help improve the effectiveness of societal self-regulation. Successful self-regulation is not only able to reduce state intervention. It can also contribute towards preventing or compensating for adverse societal effects, for example, by inducing the self-regulative internalization of such effects, i.e. wherever possible, their prevention, by those who cause them. The information and communications industry thus supplies society with important communication content. This supply, however, can lead to societally undesired sociocultural effects, whose 'disposal' must be organized, the discussion about the socialization effects of presentations of violence, about the media-intensified passivity of frequent viewers or about the consumption orientation and political apathy of the majority of recipients indicates possible after-effects, the management of which confronts society with difficult problems. Just as a cycle is envisaged in the industrial sector between supply and disposal with the aim of also reducing the disposal problem through changes in production and products, i.e. at the supply level, an attempt should be made to transfer as much of the responsibility for societally undesired sociocultural effects as possible to the suppliers of communication as well as to involve them in efforts to overcome remaining problems, in other words, to relieve the state. This is a difficult task, whose management needs to be structured by the state, but whose content has yet to be given adequate consideration.

There are also, however, less complicated approaches to the intensification of self-regulation. Insofar as self-regulation takes place via the market its efficiency needs to be guaranteed. This includes safeguards to reduce power asymmetries and to increase countervailing potentials. Power constellations, however, should not be exclusively measured in economic categories. The information and communication society will only function self-regulatively if the power problem is addressed in its various facets, i.e. also takes into account technological power, journalistic power and possible recipient power. Such power elements can be communicated by economic power. If, however, society allows economic power to become the main pillar of other elements of power it intensifies and concentrates this power, and, consequently, creates power imbalances. The communications sector is characterized by built-in mechanisms which disproportionately increase the power of these who are already powerful. One manifestation of this, which has been known for some time now, is the concentration effects of the advertisement financing of newspapers, which also has equivalents in the broadcasting sector (cf. Kruse, 1989; Owen et al., 1976). Power accumulations in just a few hands, or even in the hands of a single magnate, destroy self-regulation by society as a whole and replace it with self-regulation by the few, probably trans- and multinational enterprises.

Media policy, therefore, must seek to prepare various actors for the power struggle or to at least protect those from the use of power who are inevitably fighting a losing battle. Society alone cannot do this. The state and the EC as sovereign powers must set up structures for a socially compatible self-regulation of society. There are corresponding approaches in existing media and communications law, and also in the regulatory ideas of the EC. However, they require further elaboration. Only a few suggestions are possible here, limited to the field of mass communications.

Although the problem of access to the network has been realized it has so far almost always been set in relation to the field of telecommunications in the more narrow sense. The open, non-discriminatory access of various service providers is addressed in ONP (Open Network Provisions) directives and resolutions of the EC with respect to partial areas. In view of forthcoming liberalization and competing networks, the problem will increasingly surface for mass communication services too. 'Must carry obligations'[15] only help if the concrete conditions for equal access are laid down, i.e. not only to ensure access as such but also to provide for a fair rates structure which, for example, envisages preferential rates for services which are socially indispensable but which cannot be financed on a market basis. Above all, the access regulations must prevent subtle discriminations, regarding, for example, the allocation of services to the frequency spectrum or the linkage of various networks and, in general, the interaction of hardware and software. Or, to put it another way, the interoperationality of services and applications must be guaranteed in the field of mass communications too. Supervisory bodies and sovereign powers of control will probably remain essential here.

This fairness of access must continue at the level of distribution to users, for example, through fair access to the most common server and to the most commonly used set top box. In view of the host of offers, the nature of linkage with other offers is important for user access, i.e. the provision of attractive programme or service packages. Distribution fairness can also require 'must carry obligations' for societally important offers which would otherwise stand no chance on account of market structures.

At a final user level, the initial problem is fairness of reception, an aspect which is discussed in the field of telecommunications under the catchphrase 'universal service' (see EC Commission, 1995: 44; BMPT, 1995: 527). It deals with the infrastructure set-up, socially staggered rates and actual reception capabilities for major communicative services, in other words, not merely access to speech-only telephone or to minimal basic services, but to the entire range of vital media services in the information society. Up to now, the German Telekom functioned to a certain degree as a guarantor for fairness of access due to its monopoly position and its legally anchored public utility character. The kind of fairness of access familiar to the welfare state, however, is alien to a liberalized market. The Telekom, therefore, will also only be able to ensure fairness of access if structures exist which also impose similar obligations on private competitors in the field of telecommunication services.

Furthermore, fairness of operability is also important, i.e. the self-determined, especially manipulation-free utilizability of receivable information. This also relates to the protection of the consumer against being forced by new technological innovations constantly to update operating facilities. In addition, for the majority of users, for example, people aged 35 and above who have more or less grown up without computers, problems exist relating to the technical handling of the new technologies and, for members of this group and for others, problems of sociocultural management. The technology encompasses options in a spectrum between communication support and manipulation. The recipients must be enabled to handle competently the new media world and the media must be given incentives to do this themselves, i.e. not to delegate this task to media academics or even the state.

Fairness of both access and distribution for suppliers as well as fairness of recipience and handling for users are pressing challenges for the future. They cannot be guaranteed by the interaction of technology, journalism and economics alone, but require structural safeguards, i.e. a structuring media policy and, if need be, its supervision.

This analysis has still not addressed the aspect of whether suitable communication offers are made and where they come from. Up to now, German media policy and media law have misguidedly concentrated, as a rule, on the organization or provision of services. The contexts of distribution, origination and resale are at least equally important. Tremendous market power is currently evolving in the interaction frame of these various levels (cf. EMI, 1995), which cannot be effectively controlled at the distributional and organizational level alone. Media policy must learn to grasp the determinant factors of media production, of programme procurement and sale, of distribution and of financing (e.g. advertising or pay service rates) in their networked entirety, to examine feedback effects and to analyse, and if necessary curtail, the amplification of power by synergy effects. The extensive failure of anti-trust policies is also attributable to the lack of such an integral approach. The failure of the German Media Service GmbH[16] on account of anti-trust law provisions cannot guarantee that even more problematic interconnections at various levels of activity might not emerge in future and accumulate filtering power. Anti-trust law can only prevent market power in general, but it cannot do justice to the specific media-related problems from the perspective of a communicative orientation to the public interest (cf. Hoffmann-Riem, 1995, note 199).

The list of problems which need to be resolved could be extended, a move which would also increase the difficulties involved in finding the best solutions. Anyone seeking solutions must move away from a narrow sectoral focus on broadcasting law, telecommunications law, national law, European law or international law. Rather, the various legal orders could interact in such a way as to work as mutually supplementary orders, providing safety nets which offset functional inadequacies in one of the orders by correspondingly supplementing another. Anti-trust law, for example, could and must supplement communications law insofar as the communication order trusts in the workability of economic competition. This, however, cannot guarantee media-specific goals, such as the public service orientation, which means that other ways must be found for their realization.

The main problem, however, is finding appropriate instruments to attain acceptable goals. There should be no fear of mutual contacts when making the selection. The EC Commission's reservation, for example, expressing its reluctance to extend the regulation of telecommunications to other sectors (EC Commission, 1995: IX) or the fear of the public television networks in Germany that such an extension could encourage convergence hypotheses, are merely interim stages in a development marked by tremendous momentum.

The challenge for media policy begins at home and does not end at the EU level. Activities are required at many levels using varying approaches which supplement the essentially meaningful trust in market mechanisms. In the following, I assume the growing significance of the market as a control medium without closer analysis. Its power of control, however, is insufficient. As examples, I would like to outline five topics of regulation of societal self-regulation, excluding, however, traditional tasks such as the protection of the personality or of copyright — although I realize that their solution also has control effects for the media order as a whole.

1. Access to the market, distribution on the market and reception under market conditions should be improved through safeguards for fairness of access, distribution and reception. They must carry obligations and special rates, and other user arrangements could help; these should be differentiated according to the specific services concerned. Regulative safeguards should also selectively improve the opportunities with compensatory intention for those communication offers which are still viewed as societally important even if they cannot assert themselves solely on the basis of success on the market. The set-up of suitable structures and the issue of such regulations are initially the tasks of the communications industry itself. A state safety net, however, should guarantee minimum requirements, and incentives should be created to move beyond these minimum conditions wherever possible.
2. Fairness for the users with regard to reception of communication offers should be ensured through safeguards against the manipulative use of technology, for example, in the design of the user systems interface or in the electronically supported orientation aids and navigation guides. This is part of the more compre-

hensive task of protecting the recipients as 'consumers' of the services provided, including broadcasting programmes — not only, as has been the case so far, as consumers of the products promoted by such services. Once again, society itself should, wherever possible, afford this protection itself. This too, however, requires a regulatory setting, for example, through safeguards for the independent evaluation of offers and conditions of use and through consumer information and rights of intervention for consumer protection organizations. Furthermore, support is recommended in the setting up of systematic supervision and consumer information, for example, through a special foundation for media consumer information (cf. Groebel et al., 1995: 190).

3. The constantly growing horizontal, diagonal and vertical linkage between different media sectors corresponds to economic logic, but is inconsistent with the goal of limiting communicative power and preventing manipulation. A certain degree of concentration in the global information society, however, is inevitable and will probably be even more difficult to prevent in future than it has been so far. If it is to be accepted, although within limits, measures are needed to limit power, for example, through strict transparency safeguards and disclosure obligations for communication enterprises and effective instruments of control. Such safeguards should help involve public opinion as a watchdog (cf. Groebel et al., 1995: 187), including competing media enterprises. Anyone who makes the provision of communication the purpose of business activities and thus shapes the opinion of society as a whole must themselves become the object of this public opinion, with respect to internal structures too.

4. If major sectors of the communications order are jeopardized because of financial bottlenecks, their financing must be guaranteed. In view of the enormous earnings potential for communications markets (cf. Booz et al., 1994; Kessler, 1994: 89), I see no reason why the already overburdened state should help out financially. Businesses are not barred from increasing their profit-making opportunities by concentrating solely on lucrative activities. If, however, this leads to gaps in the reasonable provision of communication for all sections of the population it must be possible to partly skim off these profits and allocate the earnings compensationally to finance those communication services these firms neglect for reasons of

profit. The restructuring idea was a topic at the 1995 G7 summit and also characterizes reflections by the EC on fund financing in the telecommunications sector. It need not be limited to this. Compensatory responsibility can be understood in this context as a means of internalizing external effects: through the prospect of compensatory burdens media enterprises should be induced to fulfil the task of a comprehensive provision of communication themselves, thus eliminating any need for 'compensatory payments'.

5. The idea of the dual broadcasting order — a guaranteed mixture of commercial and public broadcasters — is also generalizable. Differently structured and financed partial media orders could be used to allocate the gains of one sector to offset the shortcomings of another and thus also help realize goals which are inevitably neglected through a pure market orientation. The maintenance of a viable pillar of the broadcasting order which is not financed by private enterprise, therefore, is important for the future. The idea could also be extended to upstream and downstream stages of the production and distribution of broadcasting programmes. Above all, it must be applied to new communicative services, especially if these are functional equivalents of previous broadcasting programmes. These new technological possibilities also create new scope for non-professional communication. Anyone observing developments in the Internet (cf. Grossmann, 1994) will already find numerous pointers to this trend. It should be examined whether an extension of the dual into a triple media order, whose third pillar shores up apartments for non-professionals, makes sense. As indicated by the current attempts to commercialize use within the frame of Internet, however, it must be clarified whether and to what extent the right of use for non-commercial features can be protected in the long term through safeguards.

This is just a selection of several possible approaches. None of them allow government interventions in the contents of communication; they merely protect and structure the scope of possibility. They also provide options for communicators and recipients insofar as they lack sufficient market power to ensure their development opportunities via the market alone.

A major challenge in all fields is to organize coexistence in such a way that all or as many persons as possible find a sociocultural environment with the option of appropriately satisfying their respective needs. In the knowledge and information society of the future, guaranteeing this for the communication sector will not necessarily become easier, in fact it will probably become more difficult than in the past.

Notes

1. The launch of a number of these pilot projects, however, is jeopardized.
2. The 'Information Superhighway' is a major integral part of the 'National Information Infrastructure', which is being intensely promoted in the USA by the Clinton Administration (see ITTF, 1993, and also Robinson, 1995: 36).
3. For details of the expected offers see Deutsches Video Institut (1995); Prognos (1995: 24).
4. Industry in particular calls for a 'liberal approach' which, on the one hand, enables entrepreneurial freedom, but, on the other hand, should guarantee planning certainty (see BDI/VPRT, 1995; Zmeck, 1994: 25).
5. A fact also emphasized by Noam (1995: 49) and Robinson (1995: 48).
6. A realization which the German Federal Constitutional Court has also been unable to ignore, see decisions of the Bundesverfassungsgericht, Vol. 31: 314, 325 (1971); Vol. 57: 295, 323 (1981).
7. For details on this concept see the contributions in Blumler (1992).
8. This is particularly favoured by the advocates of the model of 'pure' market broadcasting (e.g. Engel, 1995: 160).
9. Decisions of the Bundesverfassungsgericht, Vol. 57: 295, 322 (1981); Vol. 73, 118, 123 (1986).
10. On the level of 'vertical' concentration, see EMI (1995: 176).
11. See decisions of the Bundesverfassungsgericht, Vol. 57: 295, 322 (1981); Vol. 73: 118, 123 (1986).
12. On the host of regulative policy measures and proposals in the field of telecommunications alone, see EC Commission (1995: 162).
13. Some experts even diagnose the 'convergence' of the programmes of public and private broadcasters as one of the consequences; see Merten (1994), Schatz et al., (1989); critical on this thesis is Stock (1992: 745).
14. On the various possibilities and difficulties of state regulation, see Hoffmann-Riem (1996).
15. See Sections 4ff. US Cable Television Consumer Protection and Competition Act of 1992.
16. See the decision of the EC Commission, ABl. (EG) Nr. L 364/1ff. (1994) — Media Service GmbH.

References

Bangemann, Martin et al. (1994) 'Europa und die globale Informationsgesellschaft — Empfehlungen für den Europäischen Rat', Brussels.
BDI/VPRT (1995) Gemeinsame Stellungnahme Bundesverband der Deutschen Industrie e. V. (BDI), Verband Privater Rundfunk- und Telekommunikation e. V. (VPRT) und Verband der Telekommunikationsnetz- und Mehrwertdienstanbieter, Anforderungen der Industrie an die Informationsgesellschaft, 10 April 1995. Bonn: BDI/VPRT.
Beck, Ulrich (1986) *Risikogesellschaft — Auf dem Weg in eine andere Moderne*'. Frankfurt am Main: Suhrkamp.
Blumler, Jay G. (ed.) (1992) 'Multichannel TV and the Public Interest'. (Unpublished manuscript).
BMPT-Bundesministerium für Post und Telekommunikation (1995) Öffentliche Kommentierung zu den Eckpunkten für einen künftigen Regulierungsrahmen im Telekommunikationsbereich, ABI, (BMPT) pp. 527–48.
Booz, Allen and Hamiltion (eds) (1994) *Zukunft Multimedia*. Frankfurt: JMK.
Deutsches Video Institut (1995) *Multimedia-Shop*. Berlin: Deutsches Video Institute.
EC Commission (1995) Green Book ('über die Liberalisierung der Telekommunikationsinfrastruktur und der Kabelfernsehnetz'), Teil II, KOM (94) 682 final. Brussels: EC Commission.
EMI (Europäisches Medieninstitut) (1995) 'Bericht über die Entwicklung der Meinungsvielfalt und der Konzentration im privaten Rundfunk' pp. 127–220 in Die Landesmedienanstalten (eds) *Die Sicherung der Meinungsvielfalt*. Berlin: Vistas
Engel, Christoph (1995) 'Multimedien und das deutsche Verfassungsrecht', pp. 155–71 in Wolfgang Hoffmann-Riem and Thomas Vesting (eds) *Perspektiven der Informationsgesellschaft*. Baden-Baden: Nomos.
Groebel, Jo et al. (1995) *Bericht zur Lage des Fernsehens*. Gütersloh: Bertelsmann.
Grossman, Lawrence K. (1994) 'Reflections on Life along the Electronic Superhighway', *Media Studies Journal* 8: 27–48.
Hoffmann-Riem, Wolfgang (1990) *Erosionen des Rundfunkrechts*. Munich: C.H. Beck.
Hoffmann-Riem, Wolfgang (1995) 'Öffentliches Wirtschaftsrecht der Kommunikation und der Medien', Section 6 in Reiner Schmidt (ed.) *Öffentliches Wirtschaftsrecht. Besonderer Teil*. Berlin/Heidelberg: Springer.
Hoffmann-Riem, Wolfgang (1996) *Regulating Media*. New York: Guilford.
Hoffmann-Riem, Wolfgang and Thomas Vesting (1995) 'Ende der Massenkommunikation? Zum Strukturwandel der technischen Medien', pp. 11–30 in Wolfgang Hoffmann-Riem and Thomas Vesting (eds) *Perspektiven der Informationsgesellschaft*. Baden-Baden: Nomos.

ITTF (Information Infrastructure Task Force) (1993) *The National Information Infrastructure. Agenda for Action.* Washington: National Telecommunications and Information Administration.

Kessler, Martina (1994) 'Langfristiger Trend: Die wirthschaftlichen Chancen des digitalen Fernsehens', *tendenz* IV: 28-30.

Kruse, Jörn (1989) 'Ordnungspolitik im Rundfunk', pp. 77–111 in Michael Schenck and Joachim Donnerstag (eds) *Medienökonomie*. Munich: Reinhard/Fischer.

KtK (1976) 'Kommission für den Ausbau des technischen Kommunikationssystems, Telekommunikationsbericht', Bonn.

Merten, Klaus (1994) *Konvergenz der deutshen Fernsehprogramme*. Münster: Lit Verlag.

Monopolkommission (1994) 'Hauptgutachten 1992/1993 — Mehr Wettbewerb auf allen Märkten'. Baden-Baden: Nomos.

Noam, Eli M. (1995) 'Beyond Liberalization: From the Network of Networks to the System of Systems', pp. 49–59 in Wolfgang Hoffmann-Riem and Thomas Vesting (eds) *Perspektiven der Informationsgesellschaft*. Baden-Baden: Nomos.

Noelle-Neumann, Elisabeth (1985) 'Auswirkungen des Kabelfernsehens — 1. Bericht über Ergebnisse der Begleitforschung zum Kabel-Pilotprojekt Ludwigshafen/Vorderpfalz'. Berlin: VDE Verlag.

Owen, Bruce M., Jack H. Beebe and Willard Manning (1976) *Televisions Economics*. Toronto: Lexington Books.

Prognos AG (1995) 'Digitales Fernsehen — Marktchancen und ordnungspolitischer Regelungsbedarf ', Munich: Reinhard Fischer.

Rehbinder, Manfred (1962) *Die öffentliche Aufgabe und rechtliche Verantwortlichkeit der Presse*. Berlin: Duncker & Humblot.

Robinson, Kenneth (1995) 'Telekommukationspolitik der Clinton-Administration: Die ersten Jahre', pp. 36–53 in Herbert Kubicek et al. (eds) *Jahrbuch Telekommunikation und Gesellschaft 1995*. Heidelberg: R.v. Decker.

Ruck, Silke (1992) 'Zur Unterscheidung von Ausgestaltungs- und Schrankengesetzen im Bereich der Rundfunkfreiheit', *Archiv des öffentlichen Rechts* 117: 543–68.

Saxer, Ulrich (1994) 'Medien- und Gesellschaftswandel als publizistikwissenschaftlicher Forschungsgegenstand', pp. 331–54 in Otfried Jarren (ed.) *Medienwandel — Gesellschaftswandel*. Berlin: Vistas.

Schatz, Heribert, Nikolaus Immer and Frank Marcinkowski (1989) *Strukturen und Inhalte des Rundfunkprogramms der vier Kabelpilotprojekte*. Düsseldorf: Pressestelle der Landesregierung Nordstein-Westfallen.

Schlosser, Horst-Dieter and Andrea Bockholt (1994) 'Leitbilder und Metaphern im Diskurs über ISDN', pp. 117-34 in Herbert Kubicek et al. (eds) *Jahrbuch Telekommunikation und Gesellschaft 1994*. Heidelberg: R.v. Decker.

Stadek, Michael (1994) 'Digitales Fernsehen im Test', *tendenz* IV: 20-3.

Starck, Christian (1995) 'Grund- und Individualrechte als Mittel institutionellen Wandels in der Telekommunikation', pp. 291–3 in Ernst-Joachim Mestmäcker (ed.) *Kommunikation ohne Monopole II*. Baden-Baden: Nomos.
Stock, Martin (1990) 'Konvergenz im dualen Rundfunksystem', *Media Perspektiven* 745–54.
Wegener, Claudia (1994) *Reality TV. Fernsehen zwischen Emotion und Information*. Opladen: Westdeutscher Verlag.
Winterhoff-Spurk, Peter, Veronika Heidinger and Frank Schwab (1994) *Reality TV. Formate und Inhalte eines neuen Programmgenres*. Saarbrücken: Logos.
Wisebrod, Dov (1995) 'Controlling the Uncontrollable: Regulating the Internet', *Media and Communications Law Review* 4(3): 331–58.
Zmeck, Gottfried (1994) 'Gestalten statt behindern', epd/KiFu No. 24/25, 30, March, pp. 25–32.

[2]

Regulatory convergence?

Douglas W Vick*
Formerly Senior Lecturer, Department of Accounting, Finance and Law, University of Stirling

*PREFACE

This paper was accepted for publication prior to Doug Vick's premature and tragic death on 1 May 2004. It is being published in memory of Doug and is a fitting tribute to his scholarship. A US Attorney who had made his home in Scotland and was a highly respected and popular member of the Law staff at the University of Stirling, Doug had interests in a wide range of areas, but it was media law that was his main focus. This paper was part of a much wider project that Doug had been intending to complete on sabbatical. Sadly, this was not to be.

In highlighting the dominant themes that underscore recent dramatic changes in UK communications policy, Doug's analysis of the Communications Act 2003 (the Act) will have a lasting and profound resonance. The Act embodies some of the most sweeping reforms that have affected media and communications in decades. Doug's account of how the UK Government has interpreted and acted upon 'convergence' via the Act draws on a forensic and insightful analysis of the long-standing opposition between Market Liberal and Social Liberal influences affecting UK communications policy. The introduction of a converged regulatory body and conflict surrounding the precise nature of the responsibilities accorded to the Office of Communications (Ofcom) reflect what Doug compellingly argued is an uneasy accommodation between these competing ideologies. Tensions between these approaches will, of course, continue to be tested for many years to come as Ofcom begins to forge a lasting impression on the UK's telecommunications and broadcasting environment.

Gavin Little, Gillian Doyle and Nicole Busby, University of Stirling

The Communications Act 2003 can be seen as yet another attempt to reconcile the contradictions resulting from the great schism between Market and Social Liberalism that dominated twentieth-century political discourse in the West, in this case applied to the intricacies of media regulation. In this respect, the Act is simply the latest manifestation of an on-going process of philosophical accommodation that has been characteristic of British media policy, at least since World War II, if not before. This accommodation has always been imperfect, and the debates over the Act's more controversial provisions indicate that the tensions between the competing schools of liberalism will persist well into the twenty-first century.

INTRODUCTION

In media policy circles, 'convergence' was the catchword of the 1990s.[1] The term was originally used as a shorthand to describe the economic, technological and

1. See, eg, D McQuail and K Siune (eds) *Media Policy: Convergence, Concentration and Commerce* (London: Sage, 1998); TG Krattenmaker and LA Powe Jr 'Converging First Amendment principles for converging communications media' (1995) 104 Yale LJ 1719.

functional integration of the broadcast media with the telecommunications and computer industries. The emergence of digital and other technologies made it feasible for information in various forms – voice, text, data, sound or pictures – to be provided to a mass audience through a variety of means – television sets, radio receivers, personal computers or telephones.[2] These developments, in turn, drove the rapid vertical and horizontal integration of previously distinct communications industries, as large media conglomerates grew even larger in order to take full advantage of new media markets.[3]

But it did not take long for 'convergence' to take on broader meanings. The term acquired a near-utopian resonance for some, who prophesied that the physical characteristics and functions of different media fora would become indistinguishable, the variety of available programming and content unbounded, and special regulation of communications industries unnecessary.[4] The government's Communications White Paper (2000) claimed that 'convergence is not just a technology issue, but also an issue of culture and life style', blurring the boundaries between previously distinct media sectors and 'fuel[ling] a democratic revolution of knowledge and active citizenship' by making an unprecedented amount of information and analysis available to the general public.[5] Notwithstanding such claims, however, functional differences between media have persisted. Moreover, the converging sectors making a 'multi-media' world possible have been governed by very different regulatory regimes rooted in very different historical, economic and philosophical contexts. Maintaining these distinct regulatory regimes has become problematic: inconsistent regulatory approaches may hinder the development of multi-media services, but, at the same time, the multiplication of the means of delivering services provides opportunities for regulatory bypass. It is against this background that the Communications Act 2003 (the Act) emerged.

The Act, the most comprehensive legislation of its kind in British history, implements a sweeping programme of regulatory change. The cornerstone of New Labour's communications policy is the creation of the Office of Communications (Ofcom),[6] a statutory body that will be responsible for the regulation – whether economic, technical or cultural – of all sectors of the electronic communications industry, assuming the functions previously performed by five separate regulatory bodies. The government believes a single regulator will be better able to adapt to technological

---

2. See European Commission, Green Paper on the Convergence of the Telecommunications, Media and Information Technology Sectors, and the Implications for Regulation: Towards an Information Society Approach, COM (97) 623, 3 December 1997; Department of Trade and Industry and Department of Culture, Media and Sport *A New Future for Communications* Cm 5010, 2000 (Communications White Paper) paras 1.1.7 and 1.1.20, available at http://www.communicationswhitepaper.gov.uk/pdf/index.htm#top.
3. See, generally, McQuail and Siune, above n 1; ES Herman and RW McChesney *The Global Media: The New Missionaries of Global Capitalism* (London: Cassell, 1997); A Graham and G Davies *Broadcasting, Society and Policy in the Multimedia Age* (Luton: John Libbey, 1997); R Collins and C Murroni *New Media, New Policies: Media and Communications Strategies for the Future* (Cambridge: Polity Press, 1996).
4. See, eg, TG Krattenmaker and LA Powe Jr *Regulating Broadcast Programming* (Washington: AEI Press, 1994); JW Emord *Freedom, Technology, and the First Amendment* (San Francisco, CA: Pacific Research Institute, 1991); DE Lively 'The Information Superhighway: a First Amendment roadmap' (1994) 35 Boston College L Rev 1067.
5. Communications White Paper, above n 2, Annex D and para 1.1.15.
6. Established by the Office of Communications Act 2002.

convergence and consequent market changes than the fragmented regime it replaces, providing a 'comprehensive, coherent and joined up approach' to regulation, 'promot[ing] greater clarity and certainty', and avoiding the inefficiencies associated with the jurisdictional overlap of multiple regulators.[7] The Act also adopts significant reforms affecting the providers of electronic communications networks and delivery below-structures; promises a more flexible and decentralised approach to the regulation of broadcasting content; and dramatically liberalises the rules governing media and cross-media ownership. However, often conflicting policy objectives are pursued. On the one hand, the government has repeatedly stated that it wishes to deregulate communications industries wherever possible in order to 'unleash the potential' of technological convergence and make the UK 'home to the most dynamic and competitive communications and media market in the world' and to expand consumer choice in order to 'deepen democracy' and enhance the quality of entertainment and educational services. On the other hand, the government has indicated that it remains committed to ensuring that all members of society have access to 'diverse services of the highest quality' and that 'citizens and consumers' are protected both economically and from breaches of 'basic standards of decency and privacy'.[8]

One theme unifying the disparate parts of the government's communications strategy is New Labour's commitment to applying 'Third Way' ideology[9] to the regulation of electronic media. This occasionally vague and indeterminate political theory represents, at its core, an attempt to form a synthesis out of the contradictions of modern liberalism. These contradictions are largely the product of a rupture in liberal political philosophy that developed early in the twentieth century. Since then, most arguments made within Western democracies concerning rights, human nature, reason and the nature of truth – which in turn inform the arguments made about the legitimate scope of the state's regulatory power – can be roughly classified as belonging to one of two competing schools, Market Liberalism and Social Liberalism.[10] Market (or neo-classical) Liberalism is distinguished by its hostility toward the state and its strong commitment to free markets: economic libertarians freely associate personal autonomy with private property and often condemn state interference with the latter as an unwarranted intrusion upon the former.[11] In contrast, the Social Liberal (or social democratic) argument[12] is more optimistic that state power can be used for the public good and less convinced about the social and economic benefits of unfettered capitalism, actively promoting state intervention in the economy

---

7. Regulatory Impact Assessment *Setting up Ofcom as a Single Regulator* para 10, available at http://www.communicationsbill.gov.uk/pdf/OFCOM_assessment.pdf.
8. See Communications White paper, above n 2, paras 1.2.1–1.2.11.
9. The application of 'Third Way' politics in other policy areas is explored in M Powell (ed) *New Labour, New Welfare State?* (Bristol: Policy Press, 1999).
10. The framework used in this paper is suggested by the work of political scientist Richard Bellamy. See R Bellamy *Liberalism and Modern Society: An Historical Argument* (Cambridge: Polity Press, 1992); R Bellamy and J Zvesper 'The Liberal predicament: historical and logical' (1995) 15 Politics 1. See also RM Entman and SS Wildman 'Reconciling economic and non-economic perspectives on media policy: transcending the "Marketplace of Ideas" ' (1992) 42 J of Communication 1.
11. This school of thought is often associated with works such as FA Hayek *The Political Order of a Free People* (London: Routledge, 1979).
12. This school of thought is often associated with works such as J Maynard Keynes *The End of Laissez-Faire* (London: Hogarth Press, 1926).

through both publicly financed enterprises and the close regulation of privately owned businesses.[13]

The 'Third Way' (or 'new' social democracy) is a centre-left response to recent trends, in particular globalisation, that have weakened the state's ability to exercise control over the economy, especially through tax and spending policies.[14] It is less 'statist' than 'traditional' social democracy, conceding that state regulation sometimes can be contrary to the public interest and tends to view free markets more favourably than 'old way' social democrats, although insisting that private corporate power should be exercised in a 'socially responsible' manner.[15] At least in the context of media regulation, however, the Third Way is not a recent policy-making phenomenon: it is an approach that has broadly encapsulated the British attitude toward the media, particularly broadcasting, for much of the post-War period. At least since the introduction of commercial broadcasting to compete with the BBC in the mid-1950s, broadcast policy in the UK has represented a half-watermark between Market Liberalism, which has dominated media policy in the USA, and Social Liberalism, which supported the public monopoly approach favoured in much of continental Europe until the last decade of the twentieth century. But the accommodation of Market and Social Liberalism in the media context has a peculiar wrinkle that has not featured prominently in the Third Way literature: the conflicting perceptions of freedom of expression associated with the two schools and how those perceptions affect the role that should be played by the state vis-à-vis the media.

This paper examines the evolution of media policy in the UK in light of the tensions between Market and Social Liberalism. The paper first reviews the most commonly articulated rationales for treating different media differently, and concludes that all can be found wanting in some serious respect. The paper then posits that an alternative explanation for the disjointed regulatory environment of the media lies in the twentieth-century schism in liberal theory. The paper describes the Market Liberal and Social Liberal arguments as they pertain to media policy, particularly when freedom of expression concerns are at issue. Against this background, the paper then considers the changes wrought by the Act, and concludes that they are simply another step in the process of philosophical accommodation that has characterised British media policy for the past 50 years. Nonetheless, there are clear signals that the Market Liberal argument is gaining the upper hand.

---

13. See, eg, R Mullender 'Theorizing the Third Way: qualified consequentialism, the proportionality principle, and the new social democracy' (2000) 27 JLS 493 at 495. See also RG Picard *The Press and the Decline of Democracy: The Democratic Socialist Response in Public Policy* (Davenport, CT: Greenwood, 1985).
14. See W Hutton 'New Keynesianism and New Labour' in A Gamble and T Wright (eds) *The New Social Democracy* (Oxford: Blackwell, 1999) p 98. Leading 'Third Way' texts include A Blair *The Third Way: New Politics for the New Century* (London: Fabian Society, 1998); F Field *Making Welfare Work: Reconstructing Welfare for the Millennium* (London: Institute of Community Studies, 1995); F Field *Stakeholder Welfare* (London: Institute of Community Studies, 1996); A Giddens *The Third Way: The Renewal of Social Democracy* (Cambridge: Polity Press, 1998); A Giddens *The Third Way and its Critics* (Cambridge: Polity Press, 2000).
15. See Mullender, above n 13, at 495–500.

DISJOINTED REGULATION

Historically, different media have been distinguished by reference to their technological characteristics[16] and it is, thus, perhaps unsurprising that it became commonplace to attribute discrepancies in regulatory attitudes toward different media to technological distinctions. The most frequently cited example of divergent regulatory approaches is the contrast between the treatment of newspaper and magazine publishers – who are not hampered by a particularly intrusive regulatory regime, and usually are only required to comply with generally applicable legal rules[17] – and radio and television broadcasters, who have had a host of special affirmative obligations imposed on them, including the duties to be balanced and impartial in their political reporting, to present opposing points of view in their broadcasts, to provide educational services, and to air material that might appeal to minority audiences.[18] Until recently, the most common rationale for this difference in regulatory treatment was that spectrum scarcity – the presumed technical limitations of the radiomagnetic spectrum that prevented the broadcast of more than a few channels of programming without signal interference – meant that the number of potential providers of broadcasting services was too small to ensure that there was sufficient diversity in the programming offered to viewers and listeners.[19] Typically, the policy response to spectrum scarcity has been either to consign responsibility for all broadcasting to a publicly owned monopoly (as occurred in most European countries) or to regulate closely the commercial broadcasters to whom a limited number of broadcast frequencies would be allocated (as occurred in the USA).[20]

---

16. 'Broadcasting' (or terrestrial broadcasting) is a communications system whereby channels of audio or video programming are sent to radio and television sets on electromagnetic waves transmitted through the air by a ground-based network of antennas and relay stations; direct broadcast satellite transmits multi-channel video and audio programming to homes, pubs and hotels by first passing the signals through satellites in geo-stationery orbit before transmitting them to a receiving station or individual receiver; cable television systems transmit multi-channel video and audio programming electronically over coaxial or fibre-optic cables, either as analogue signals (electronic pulses) or as digital information; telephone systems allow interactive, real-time voice communication by using sophisticated switching technology to route analogue or digital signals over complex networks of cables and wires linking individual homes and businesses; and the Internet is, in essence, a set of technical rules and specifications ('protocols') for the transfer of data that allows the exchange of information between computers linked by wires, cables, telephone lines or satellites.
17. Although certainly some areas of the general law, such as libel, contempt of court and obscenity, have a disproportionate impact on the press.
18. See, eg, Broadcasting Act 1990, ss 1–12 and 83–96 (largely replaced by Communications Act 2003, ss 263–347); ITC Programme Code, available at http://www.itc.org.uk; Agreement between Secretary of State for National Heritage and the BBC, available at http://www.bbc.co.uk/info/BBCcharter/agreement/index.shtml.
19. See J Curran and J Seaton *Power Without Responsibility: The Press and Broadcasting in Britain* (London: Routledge, 5th edn, 1997) p 112.
20. It was argued that because the state was required to allocate broadcast frequencies to a small number of licensees, the programming offered by those licensees might be limited and biased unless special sector-specific rules were in place to guarantee diversity in programming and adequate and balanced coverage of issues of public importance. See, eg, *FCC v League of Women Voters* (1984) 468 US 364 at 376; *Red Lion Broadcasting Co v FCC* (1969) 395 US 367; *National Broadcasting Co v United States* (1943) 319 US 190 at 215–217.

While this and related technology-based rationales[21] withstood a hailstorm of criticism over the years,[22] the recent proliferation and convergence of electronic media have been their undoing. The emergence of multi-channel media like satellite and cable television, as well as the anticipated arrival of digital terrestrial broadcasting by the end of the decade, have undermined the scarcity rationale: the physical characteristics and limitations of the radio spectrum are no longer a decisive factor in explaining the structure of media markets, at least if audio-visual programming is the relevant point of reference. As was noted in the White Paper that laid the groundwork for the Act:

> 'public service television now includes the commercially funded ITV channels and Channels 4 and 5. There are 250 purely commercial channels available in the UK ... The internet already means we can listen to radio stations and receive watchable formats of television broadcasts from all around the world. Cable and satellite viewers can order films virtually on demand ... With the switchover to digital expected between 2006–2010, we are less than a decade away from every television household having access to dozens of channels.'[23]

The White Paper concluded that 'the era when the extent of broadcasting was determined by spectrum scarcity is drawing to a close',[24] and ignored other technology-based justifications for the continuation of the existing medium-specific regulatory regimes. The eclipse of the scarcity rationale has emboldened critics of the current regime of broadcasting regulation, with commercial broadcasters, in particular, questioning the special position held by the BBC and hoping to free themselves of the public service obligations imposed on them.[25] But the government has been unwilling to abandon regulatory distinctions between media altogether. Two alternative arguments for retaining these distinctions were offered in the White Paper, one economic and the other political.

The economic basis for retaining distinctions between media focused on the 'natural tendencies' of radio and television markets to 'concentrations of market power'.[26] This phenomenon is easily observed in the USA, where channel proliferation

---

21. Another frequently cited justification for a special regime of broadcast regulation was founded on the notion that broadcasting has a 'uniquely pervasive presence' in people's lives, with broadcast signals 'intruding' on the privacy of the homes of viewers and listeners and creating a substantial risk that children could be exposed to indecent material without parental supervision, and that sensitive adults would be exposed to offensive programming without adequate warning. See *FCC v Pacifica Foundation* (1978) 438 US 726 for the classic statement of this rationale.
22. See, eg, LA Powe *American Broadcasting and the First Amendment* (Chicago: University of Chicago Press, 1987); ML Spitzer *Seven Dirty Words and Six Other Stories* (New Haven: Yale University Press, 1986); MS Fowler and DL Brenner 'A marketplace approach to broadcast regulation' (1982) 60 Texas L Rev 207; R Coase 'The Federal Communications Commission' (1959) 2 J Law and Economics 1.
23. Communications White Paper, above n 2, paras 5.2.3–5.2.4.
24. Ibid, para 5.2.4. The government relied on the scarcity rationale only once during debates concerning the Act, in justifying rules prohibiting religious organisations from holding certain broadcasting licences. See 409 HC Official Report (6th series) cols 101–102, 14 July 2003.
25. See, eg, R Craufurd Smith *Broadcasting Law and Fundamental Rights* (Oxford: Clarendon Press, 1997) pp 43–63; G Born and A Prosser 'Culture and consumerism: citizenship, public service broadcasting and the BBC's fair trading obligations' (2001) 64 MLR 657 at 658–659.
26. Communications White Paper, above n 2, para 5.3.5.

began two decades ago and media markets have been largely deregulated.[27] In a country of roughly the same geographical size as Europe, four companies dominate terrestrial broadcasting and the top four cable system operators now control access to more than half of the cable television market.[28] This oligopolistic market structure is, at least partly, the consequence of the significant economies of scale in the mass broadcasting market attributable to the high fixed costs of establishing transmission networks and, especially, the high cost of producing quality programming.[29] As Chen observed, 'a very limited number of firms can amass sufficient capital to acquire, organize, deliver, and promote the constant stream of new programming needed to satisfy an easily bored public'.[30] The barriers to entry caused by high production costs may get worse if channel proliferation leads to audience fragmentation, because smaller per-channel audiences will likely increase the average costs of production.[31] Yet while the cost of programme production is high, the cost of programme reproduction and distribution is relatively low,[32] and getting lower. The White Paper noted that it is now possible for digital information to be 'endlessly edited, copied and merged with other information' and to be reproduced in many formats.[33] By thus increasing economies of scope in converging media industries, digital technology also increases the pressures towards economic concentration through multi-media mergers.[34] Further complicating matters is the tendency of communications industries to develop 'bottlenecks', where a company establishes monopolistic control over a scarce resource or facility (such as the set-top or built-in box necessary to provide access to digital broadcast signals, or the complex network of telephone cables, wires and switches that make interactive communications possible). If the facility is essential to the efficient provision of other products or services, the company can, by restricting access to the facility, establish monopoly control over the provision of those other products or services, even in markets that otherwise might be competitive.[35]

27. See especially Telecommunications Act 1996, Pub L No 104-104, 110 Stat 56.
28. See DJ Atkin 'Video dialtone reconsidered: prospects for competition in the wake of the Telecommunications Act of 1996' (1999) 4 Communications L and Policy 35 at 45. See, generally, BH Bagdikian *The Media Monopoly* (Boston: Beacon Press, 4th edn, 1992).
29. See Communications White Paper, above n 2, 50.
30. J Chen 'The Last Picture Show (on the twilight of federal mass communications regulation)' (1996) 80 Minnesota L Rev 1415 at 1489.
31. Communications White Paper, above n 2, 50.
32. Chen, above n 30, at 1425–1431; T Streeter *Selling the Air: A Critique of the Policy of Commercial Broadcasting in the United States* (Chicago: University of Chicago Press, 1996) pp 173–174.
33. Communications White Paper, above n 2, 50.
34. Ibid. The trend toward economic 'convergence', through which corporate giants in different media have merged into larger and larger transnational conglomerates, has not been confined by national borders. By the mid-1990s it was estimated that ten transnational media conglomerates dominated the global commercial media market, with an additional 30–40 media giants operating in niche or regional markets. See Herman and McChesney, above n 3. See also T Congdon et al *The Cross-Media Revolution: Ownership and Control* (London: John Libbey, 1995); PS Dunnett *The World Television Industry: An Economic Analysis* (London: Routledge, 1993).
35. The essential facilities doctrine was developed in several leading US anti-trust cases, including *United States v Terminal RR Association* (1912) 224 US 383; *Image Technical Services Inc v Eastman Kodak Co* 125 F2d 1195 (9th Cir 1997); and *MCI Communications Co v AT&T*, 708 F2d 1081 at 1132–1133 (7th Cir 1983). See also R Whish *Competition Law* (London: Butterworths, 4th edn, 2001) pp 611–624.

In short, the mere fact that it is technically feasible for more channels of programming to be delivered to consumers does not necessarily mean there will be more opportunities for greater competition in broadcast markets. This has been cited as justification for a relatively more intrusive regulatory regime for the electronic media than for the print media.[36] But the economic trends in electronic media industries are not unique. In many respects, the mass media are like any other segment of the economy, and many have argued that these media should be regulated in the same way as any other industry – subject to the strictures of general competition law, but little more – with the primary goal being the creation and maintenance of open and competitive markets that provide 'consumers' (viewers and listeners) with their choice of services and programmes at the lowest possible price.[37] In particular, it is difficult to explain the different regulatory treatment of the publishing and broadcasting industries by reference to the economic structure of those industries. The publishing industry in the UK became progressively more concentrated throughout the twentieth century,[38] and currently four companies control nearly 90% of the UK national newspaper market.[39] This pattern is observable in most Western democracies.[40] In economic terms, modern broadcasting markets are no more concentrated (or 'dysfunctional') than newspaper and magazine markets, and yet the print media have not been subjected to the same degree of sector-specific regulation as the electronic media.

If the discrepant treatment of print and broadcast media is not adequately explained by economics, the non-economic reasons offered by the government's Communications White Paper for retaining these distinctions must be considered more closely. The White Paper emphasises the 'democratic importance of public service broadcasting' that 'guarantees the availability of full and balanced information about the world at local, regional and global levels' and 'ensures that the interests of all viewers are taken into account'.[41] Retaining a system of public-service broadcasting is also justified on cultural grounds: 'The value of information, education . . . and entertainment is not limited to how much we are prepared to pay for them'.[42] These arguments hark back to what has been called the 'power' rationale,[43] which stresses that broadcasting has a unique capacity to influence the public and should be subjected to greater

---

36. Communications White Paper, above n 2, para 5.3.10.
37. See, eg, DL Brenner 'Ownership and content regulation in merging and emerging media' (1996) 45 Depaul L Rev 1009; JW Emord 'The First Amendment invalidity of FCC ownership regulations' (1989) 38 Catholic Univ L Rev 401.
38. See C Seymour-Ure *The British Press and Broadcasting Since 1945* (Oxford: Blackwell, 2nd edn, 1996) pp 118–137.
39. See, generally, G Doyle *Media Ownership* (London: Sage, 2002) p 88.
40. See, generally, WA Meier and J Trappel 'Media concentration and the public interest' in McQuail and Suine, above n 1, pp 38–59. In the USA, for example, most markets are served by fewer daily newspapers than broadcast outlets – see *Re Syracuse Peace Council* (1987) 2 FCCR 5043 – and, by 1990, 80% of daily newspapers were controlled by national chains, with 14 corporations controlling over half of the newspaper market, three corporations controlling over half of the magazine market and six controlling over half of the book publishing industry. See Bagdikian, above n 28, pp 18 and 22–24; S Lacy and TF Simon *The Economics and Regulation of United States Newspapers* (Norwood, NJ: Ablex, 1993).
41. Communications White Paper, above n 2, paras 5.3.9 and 5.3.10.
42. Ibid, para 5.3.11.
43. See MA Franklin and DA Anderson *Mass Media Law* (New York: Foundation Press, 5th edn, 1995) p 662.

regulatory oversight to ensure that this power is used for the public good. This argument has several strands.[44] One strand concerns the cultural dimension of broadcasting and its purported ability to create a collective consciousness of a common culture, even of a common group identity. A second strand focuses on the perceived moral dimension of broadcasting, including the need to provide 'quality' programming to a mass audience and to be sensitive to the effects of portrayals of sex and violence on society's vulnerable members.[45] A third strand focuses on the perceived power of the broadcast media to influence and potentially control political decision making. Linking all three strands is the fact that the activities of the electronic media uniquely implicate the fundamental right of free expression. Because of the social, cultural, moral and political facets of broadcasting, special regulatory measures are necessary to prevent abuse of power and to prevent the medium's dominance by either the state or commercial interests.[46]

But the power rationale is based on assumptions that are, at the least, contestable. The cultural strand of the argument often assumes that cultural or group identity is something that needs 'preservation' in the face of threats from media messages, rather than being the collective expression of individual choices that may shift and mutate over time; there is, it seems, an unspoken premise that culture is not resilient enough to withstand exposure to media content (especially 'foreign' content) that is not controlled to some extent.[47] The moral component of the argument is vulnerable to charges of paternalism, elitism and the perpetuation of a 'Nanny State' mentality.[48] And the third strand of the argument, particularly when applied to broadcasters but not publishers, rests uncomfortably on the unspoken assumptions that broadcasting is more powerful than print because a mass viewing audience is less sophisticated

---

44. See D Goldberg et al *EC Media Law and Policy* (London: Longman, 1998) pp 2–5.

45. This dimension of the 'power' argument overlaps with, but is distinct from, the technology-based 'intrusiveness' rationale discussed above n 21.

46. For variations and elaboration of these arguments, see JA Barron 'Access to the press – a new First Amendment right' (1967) 80 Harvard L Rev 1641; OM Fiss 'Why the state?' (1987) 100 Harvard L Rev 781; OM Fiss *Liberalism Divided: Freedom of Speech and the Many Uses of State Power* (Boulder, CO: Westview Press, 1996); J Keane *Media and Democracy* (Cambridge: Polity Press, 1991).

47. The notion that the consumption of too much imported media leads to the diminishment or disappearance of a people's collective (ethnic or national) identity can be traced to works on cultural imperialism. See, eg, C Hamelink *Cultural Autonomy in Global Communications* (New York: Longman, 1983); HI Schiller *Communication and Cultural Domination* (White Plains, NY: ME Sharpe, 1976). Other scholars have challenged this notion. Ferguson derided what she calls 'the myth of "cultural hegemony" ', which 'infers that the consumption of the same popular material and media products ... creates a metaculture whose collective identity is based on shared patterns of consumption'; M Ferguson 'The mythology about globalization' (1992) 7 European J of Communication 69 at 79–80. In her study of national identity in Puerto Rico, Morris demonstrated that the 'pressure on identity' caused by imported media can have the counter-intuitive effect of strengthening rather than diminishing group identities in receiving societies; N Morris *Puerto Rico: Culture, Politics, and Identity* (London: Praeger, 1995) pp 168–169.

48. For example, Rupert Murdoch argued that '[m]uch of what passes for quality on British television really is no more than a reflection of the values of a narrow élite which controls it and which has always thought that its tastes are synonymous with quality'; R Murdoch *Freedom in Broadcasting* Edinburgh Television Festival Lecture (1989), quoted in Craufurd Smith, above n 25, p 50.

than a literate, reading audience, and that the greater the likelihood that expression will be popular or persuasive, the greater is the need to control it.

In any event, the power rationale simply begs a larger question. The allocation and constraint of power is the dominion of politics and, ultimately, political theory. To appreciate why media policy is disjointed, it is important to appreciate the political theories underlying regulatory structures. The thesis advanced here is that variable regulatory regimes – in terms of the degree of official oversight, the imposition of legally enforceable affirmative obligations, and the application of sector-specific instead of general rules – are better explained by reference to the shifts and tensions within twentieth-century liberal thought than by the technological or economic characteristics of different media.

TWO MODELS OF MEDIA REGULATION

Political theory defines the parameters of regulatory decision making. This is not to trivialise the importance of immediate political concerns, such as whether giving Rupert Murdoch the opportunity to acquire a terrestrial television licence will secure his support in the next General Election. But the political options that can be justified – or even conceived – are bounded by overarching theoretical frameworks. At the foundation of many political debates in Western societies, including debates over the appropriate role of the state in regulating the mass media, are assumptions rooted in either Market Liberalism or Social Liberalism. These schools, as described here, are 'ideal types' – 'abstracted characterisations meant to serve as anchors on the continuum along which real-world analysts are arrayed'.[49] Inevitably, the dichotomy drawn between them is crude and does not do full justice to the subtleties of the work of twentieth-century liberal theorists, many of whom incorporate elements of both schools in their writing.[50] In fact, some question whether 'foundationalist' or 'universalist' claims about liberal theory are helpful, given the pluralism of values and perspectives in liberal societies.[51] There is some danger that the sort of broad theoretical classifications used here oversimplify debates over media policy, and obscure the contradictions and complexities of the views held by individual policy makers.[52] It is also tempting to interpret all real world events by reference to pre-defined theoretical models, however inapt they may be. While these problems must be acknowledged, broad theoretical classifications do serve the useful purpose of

---

49. Entman and Wildman, above n 10, at 6.
50. Nor is this simple dichotomy the only way to categorise theories of media regulation. Nordenstreng, for instance, offers five distinct paradigms of regulatory policy: the liberal-pluralist paradigm, the social responsibility paradigm, the critical paradigm, the administrative paradigm and the cultural negotiation paradigm. See K Nordenstreng 'Beyond the four theories of the press' in J Servaes and R Lie (eds) *Media and Politics in Transition* (Leuven: Acco, 1997).
51. See, eg, J Gray *Post-Liberalism: Studies in Political Thought* (London: Routledge, 1993). In particular, efforts to formulate coherent theories of media regulation are doomed to fail, critics claim, because they are contaminated by the unwitting biases of those who formulate them; because media systems are too complex and incoherent to be rationalised this way; and because theoretical structures are unable to accommodate the sheer diversity of media and the evolution of media technologies. See, eg, D McQuail *McQuail's Mass Communication Theory* (London: Sage, 4th edn, 2000) pp 155–156.
52. See Streeter, above n 32, p 11.

providing an analytical structure for organising and understanding an otherwise confusing mass of historical and social knowledge.[53] The categories of liberalism identified here provide a simple construction for making sense of competing media policy proposals.

Market and Social Liberalism are rooted in similar values, such as tolerance, equality, liberty, respect for individual choices and rationality.[54] Moreover, in the specific context of media regulation, both schools ground their theories by reference to a common set of arguments justifying the protection or promotion of freedom of expression. Most influential have been the arguments that conceptualise freedom of expression instrumentally, as an essential means to a number of beneficial ends: the advancement of knowledge and discovery of truth;[55] the realisation of an effective system of participatory democracy;[56] the erection of quasi-institutional curbs on the abuse of political or economic power;[57] and the achievement of non-violent political reform without undue social instability.[58] But the two schools diverge dramatically and often irreconcilably in their approaches to the accomplishment of these objectives.

The Market Liberal model

The Market Liberal argument[59] is a merger of *laissez faire* economics, which dominated Western thought by the turn of the twentieth century, with libertarian political theory. The economic theory is familiar enough: the laws of supply and demand as described by Adam Smith, if left undisturbed, will yield the products that people ('consumers' in today's parlance) desire at prices that provide a fair return for their producers, with an optimal outcome for society in terms of the allocation of its resources. The market is more efficient than the state in responding to, and thus enhancing, consumer choice: in a properly functioning market, if there is a significant unmet consumer demand for a product, someone will step in to meet that demand. The economic component of the Market Liberal argument holds that the media, like other industries, produce 'commodities' for the consumption of their audiences, and

---

53. Ibid.
54. Bellamy and Zvesper, above n 10, at 2.
55. See, generally, JS Mill *On Liberty and Other Writings* (Cambridge: Cambridge University Press, 1989). See also F Schauer *Free Speech: A Philosophical Enquiry* (Cambridge: Cambridge University Press, 1982) pp 15–34; K Greenawalt 'Free speech justifications' (1989) 89 Columbia L Rev 119 at 130–141.
56. See, eg, A Meiklejohn *Free Speech and its Relation to Self-Government* (New York: Harper, 1948); A Meiklejohn 'The First Amendment is an absolute' [1961] Supreme Court Rev 245.
57. See, eg, V Blasi 'The checking value in First Amendment theory' [1977] American Bar Foundation Research J 521.
58. See, eg, TI Emerson *The System of Freedom of Expression* (New York: Random House, 1970) p 7; Schauer, above n 55, pp 75–80; Greenawalt, ibid, at 141–142. A related argument is that respect for dissenting speech promotes tolerance. See, eg, L Bollinger *The Tolerant Society: Freedom of Speech and Extremist Speech in America* (New York: Oxford University Press, 1986).
59. Leading advocates of the argument in the media field include I de Sola Pool *The Technologies of Freedom* (Cambridge, MA: Belknap Press, 1983); Powe, above n 22; and Spitzer, ibid.

the forces of supply and demand will ensure that consumers have the widest choice of commodities of the highest possible quality available at the lowest possible price.

While most *laissez faire* arguments of the nineteenth century were restricted to property interests, at some point in the last century classical liberal economic theory coalesced with the libertarian conception of freedom of expression. The political component of the Market Liberal argument holds that the 'self-righting' mechanism of supply and demand not only is the best means of assuring the maximisation of a society's economic wealth, but is a generalisable principle describing the most efficient means of assuring that the political, cultural and educational needs of society are met. This notion is neatly expressed in the 'marketplace of ideas' metaphor made famous by US Supreme Court Justice Oliver Wendell Holmes in the aftermath of World War I: '[T]he ultimate good desired is better reached by free trade in ideas, [and] the best test of truth is the power of the thought to get itself accepted in the competition of the market'.[60] In a properly operating marketplace of ideas, people can choose for themselves what information and viewpoints they wish to receive. If certain messages are not being disseminated and there is an unmet 'demand' for such expression by a sufficient number of people, someone will come forward to meet the demand, just as a supplier of goods will enter markets where the demand for such goods exceeds supply.[61] Holmes's 'marketplace' analogy soon won acceptance beyond the confines of American constitutional law. In 1949, for example, the Royal Commission on the Press proclaimed that 'free enterprise is a prerequisite of a free Press'.[62]

This faith in market forces is complemented by a deep distrust of the state. The struggle for freedom of speech and freedom of the press was, for most of the Enlightenment, a story of conflict between the individual and an 'official' centralised authority – at first the Church and later the state.[63] As Sir Stephen Sedley observed, our conceptualisation of fundamental rights has derived 'from the historic paradigm, which has shaped our world, of the conscious human actor whose natural enemy is the state – a necessary evil – and in whose maximum personal liberty lies the maximum benefit for society'.[64] Consistent with the Enlightenment preoccupation with 'possessive individualism',[65] Market Liberals adopted a narrow view of the legitimate scope of state power. Official intervention affecting the content of media messages, however well meaning, at best distorts the 'marketplace of ideas' and, at worst, threatens basic liberties. Libertarians regard state regulation of the media as an 'unnatural' external impediment to rational communication among citizens; in contrast, the operation of the market (the 'invisible hand') is seen as a neutral, 'natural' force that allows individuals to exchange information and opinion freely and

---

60. *Abrams v United States* (1919) 250 US 616, 30 (Holmes J dissenting). See, generally, C Edwin Baker *Human Liberty and Freedom of Speech* (Oxford: Oxford University Press, 1989) pp 7–12.
61. See Keane, above n 46, pp 44–45.
62. *Report of the Royal Commission on the Press* Cmd 7700, 1949, para 682. Similarly, Richard Barbrook has observed that French courts have interpreted the free press rights conferred by the Law of the Press of 29 July 1881 as dependent on 'the natural right of the property ownership of printing presses'; R Barbrook *Media Freedom: The Contradictions of Communication in the Age of Modernity* (London: Pluto, 1995) p 17.
63. McQuail, above n 51, p 147.
64. Sir Stephen Sedley 'Human rights: a twenty-first century agenda' [1995] PL 386 at 386 (discussing the theoretical foundation of the European Convention for the Protection of Human Rights and Fundamental Freedoms 1950).
65. Ibid.

efficiently, enabling 'truth' to emerge. In its most extreme form, the Market Liberal argument concedes no distinction between the 'public interest' to be served by the media and the operation of media markets.

The profound misgivings about state power underlying the Market Liberal argument is reflected in the reluctance of libertarian theorists to make essential freedoms contingent upon corresponding public responsibilities, at least in any legal sense. This is largely because enforcement of such responsibilities would require an undesirable level of state interference with freedom of expression and with the property rights of the owners of media outlets.[66] The libertarian argument conceives rights, including the liberties of speech and press, as 'negative' freedoms, constraining the power of the legislative and executive branches of the state, but not imposing affirmative obligations that must be advanced by positive state action.[67] In other words, the state should neither infringe those rights nor promote them.[68] Market Liberals argue that it is best to let consumers exercise choice in an efficiently operating marketplace and the state should intervene (through application of general anti-monopoly laws) only when markets are non-competitive.

The Market Liberal argument was grounded on an idealised vision of a marketplace filled with numerous publishers fiercely competing for the hearts, minds and hard cash of the public. Arguably, this vision was not that far removed from reality around the middle of the nineteenth century: the economic and technological barriers to starting a newspaper or periodical were low, and, in theory, individuals who felt that their ideas and perspectives were not adequately represented in existing publications could start up competing presses comparatively easily.[69] But the print media of this time were not 'mass' media in any modern sense and newspapers tended to be sporadically produced publications with limited circulation.[70] Dramatic technological, industrial and legal developments of the late nineteenth and early twentieth centuries,[71] however, increased the size, speed and efficiency of existing media and spawned new ones – motion pictures, radio and television – which could reach audiences of unprecedented size. Media 'barons' at the forefront of this revolution were able to drive smaller competitors out of business and the escalating cost of competing with these larger media companies discouraged newcomers from entering the mass communications market.[72] As ownership of the commercial media became

---

66. See, eg, Emord, above n 37; Coase, above n 22; Fowler and Brenner, ibid.
67. See, eg, I Berlin 'Two concepts of liberty' in I Berlin *Four Essays on Liberty* (Oxford: Oxford University Press, 1969) pp 118 and 126–131; GC MacCallum Jr 'Negative and positive freedom' (1967) 76 Phil Rev 312 at 320–325; FS Siebert et al *Four Theories of the Press* (Urbana, IL: University of Illinios Press, 1963) p 93 ('Libertarian theory was born of a concept of negative liberty, which we can define loosely as "freedom from" and more precisely as "freedom from external constraint" ').
68. See, eg, D den Uyl 'Freedom and virtue' in TR Machan (ed) *The Libertarian Reader* (Totowa, NJ: Rowman and Littlefield, 1982) p 211.
69. See Commission on Freedom of the Press *A Free and Responsible Press* (Chicago: University of Chicago Press, 1947) p 14.
70. See, eg, de Sola Pool, above n 59, p 18.
71. For example, the invention of huge, hot-metal linotype machines in 1884 sped the printing process by fourfold, and nineteenth-century changes to company law in countries such as the UK and the USA facilitated the accumulation of capital for large-scale enterprises.
72. For a brief history of the British press in this period, see Curran and Seaton, above n 19, pp 10–57. Craufurd Smith observed that ideologically 'radical' publications were casualties of this trend; Craufurd Smith, above n 25, p 23.

concentrated in fewer and fewer hands, many questioned whether the economic marketplace was functioning in a way best suited to safeguard a healthy marketplace of ideas.[73]

Anxiety about the dangers of 'private censorship' by media corporations has been present almost from the time this consolidation process first became apparent, and was commonplace after World War I, when scepticism about the power of uninhibited markets to serve the public good reached new heights. Many doubted that media companies, left to their own devices, would provide the information and diversity of opinion required by democratic polities. Some attributed this to a common political and economic worldview shared by media moguls.[74] Others emphasised that, even if this consensus did not exist, and even if media markets were fully competitive, there was no guarantee that the public would be exposed to a diversity of views and opinions. In part, this is because the 'market' for much of the media operates differently from other markets. A fully functional market is one in which consumers exercise 'choice' by selecting the products they wish to purchase from the alternatives offered by competing producers. With the commercial media in particular, however, it is often the 'choice' of advertisers that matters, rather than that of the ultimate 'consumers' of media content.[75] Advertisers often fear having their products associated with controversial, disturbing or challenging content of the sort desired by substantial segments of the public, and advertiser-driven media often ignore the interests of groups with comparatively limited buying power (the poor, children, the elderly).[76] Since the interests of the ultimate consumers of media messages do not necessarily correspond with those of advertisers, media competition for advertisers is not necessarily a reliable way to deliver the benefits that competition is supposed to yield for those ultimate consumers.

This notion that media markets are in some way dysfunctional – that they cannot be relied upon to assure a true 'marketplace of ideas' – is an important aspect of the Social Liberal critique of the Market Liberal model of media regulation. More fundamentally, however, the political dimension of Social Liberalism challenges Market Liberal assumptions about the relative position of the individual and the state.

---

73. Fairly recent examples of this argument in the legal literature include Barron, above n 46; Fiss, *Liberalism Divided*, ibid; S Ingber 'The marketplace of ideas: a legitimizing myth' [1984] Duke LJ 1; KL Karst 'Equality as a central principle in the First Amendment' (1975) 43 U Chicago L Rev 20; J Weinberg 'Broadcasting and speech' (1993) 81 California L Rev 1101.
74. See Siebert et al, above n 67, pp 78–80; Commission on Freedom of the Press, above n 69, pp 59–62; WE Hocking *Freedom of the Press: A Framework of Principle* (Chicago: University of Chicago Press, 1947) pp 141–157; Keane, above n 46, p 46. While Market Liberals pointed out that media messages are not the product of single-minded monoliths, but of proprietors, reporters, editors, artists and writers who may hold differing viewpoints, critics have characterised the output of the commercial media as 'the diversity of a pack going essentially in one direction'; Fiss 'Why the State?', ibid, at 787 (quotation omitted). The beliefs of the journalists and artists responsible for media content may well differ from those of the owners of media outlets, but all of those involved in media production 'operate largely on the basis of a shared matrix of values'; Weinberg, above n 73, at 1152.
75. See ML Spitzer 'Justifying minority preferences in broadcasting' (1991) 64 So Cal L Rev 293 at 305; Keane, above n 46, p 83.
76. See, generally, T Gitlin *Inside Prime Time* (New York: Pantheon Books. 1983).

The Social Liberal model

The concentration of power in media industries around the turn of the twentieth century was but part of a broader trend toward the concentration of power in large governmental and extra-governmental collectives – multinational corporations, labour unions and bureaucracies – whose emergence challenged the individualistic assumptions of the classical Market Liberal model.[77] Curran recently observed that the classical liberal paradigm 'fail[ed] to recognize that people are represented primarily through political parties, interest groups and the myriad structures of civil society'.[78] The most trenchant critique of market ideology, of course, came from the socialist theorists, who stressed the corrosive effects of capitalism on human dignity and the values of equality and justice.[79] But socialist theorists were vulnerable to claims that they did not adequately address the dangers posed by twentieth-century collectivism to individual liberty. The Social Liberal (or social democratic) argument attempted to reconcile socialism's commitment to economic and social justice with liberalism's traditional commitment to individual liberty. Social Liberalism incorporated the socialist critique, but did not reject outright the basic structures of private ownership. Rather, it sought to manage rather than replace capitalism: 'Social democrats want to deploy the state to serve their ends, but do not see the state as so embodying those ends that it can replace the private sector'.[80]

Nonetheless, Social Liberals do not identify economic efficiency as the primary goal of media policy and maintain that unfettered media markets often have socially detrimental effects.[81] Indeed, they have tended to view private power and the operation of the economic marketplace as 'threats to a desirable communications order, as threats to diversity, and even as threats to freedom as freedom is normally understood'.[82] The modern commercial mass media are uniquely positioned to 'lay sentiments before the public, and it is they rather than the State who can most effectively abridge expression by nullifying the opportunity for an idea to win acceptance'.[83] Because of this unique power and because of the special privileges enjoyed by modern media conglomerates, it is reasonable to subject those conglomerates to certain obligations. Specifically, the media have an obligation to be 'socially responsible', assuring that diverse opinions, attitudes and perspectives are fairly presented to the public and that citizens are provided with sufficient information to make informed judgements about issues of public importance.

This emphasis on corporate social responsibility, now frequently associated with 'Third Way' politics, was prominent in media policy debates for much of the twentieth century. For example, this theme dominated the work of a private commission of

---

77. See Bellamy and Zvesper, above n 10, at 2.
78. J Curran 'Rethinking media and democracy' in J Curran and M Gurevitch (eds) *Mass Media and Society* (London: Edward Arnold, 3rd edn, 2000) p 120 at p 135.
79. Leading Marxist-influenced works addressing the social role of mass media and media policy interpret both as instruments of control for the dominant classes in society. See, eg, ES Herman and N Chomsky *Manufacturing Consent: The Political Economy of the Mass Media* (New York: Pantheon, 1988); H Marcuse 'Repressive tolerance' in RP Wolff et al (eds) *A Critique of Pure Tolerance* (Boston: Beacon, 1965) p 81.
80. Hutton, above n 14, p 97.
81. Entman and Wildman, above n 10, at 7.
82. C Edwin Baker 'Merging phone and cable' (1994) 17 Hastings Comm and Ent LJ 97 at 111.
83. Barron. above n 46. at 1655–1656.

inquiry, chaired by University of Chicago Chancellor Robert Hutchins, which issued several highly influential publications immediately after World War II.[84] The commission's main report criticised the press in the USA for its sensationalism, its commercialism, its conflation of factual reporting and opinion, and its failure to give voice to those outside society's elite classes.[85] A socially responsible press should present a 'truthful, comprehensive and intelligent account of the day's events in a context which gives them meaning', provide 'a forum for the exchange of comment and criticism', and give voice to the competing opinions and attitudes of the constituent groups in society.[86] The Hutchins Commission was but the first of several inquiries, some initiated by Western states, into the operation of the press in particular and the media's public purposes more generally.[87] In the UK, the first Royal Commission on the Press echoed many of the points made by the Hutchins commission, criticising, for example, the sensationalism of the tabloid press and the media's frequent lack of respect for individual privacy, and stressing the need for the media to allow the expression of a diversity of opinions.[88]

Many writers in the post-War period who urged large media organisations to exercise their power more responsibly ultimately advocated some form of industry self-regulation, a position that is easily reconciled with the anti-statist tradition of Market Liberalism.[89] But the logic of their arguments justified state intervention if the media failed to meet their public responsibilities. Contrary to the libertarian belief that any state regulation of speech and press imperils liberty, the Social Liberal argument posits that state power is often the only effective counterbalance to the power possessed by the media. It holds that a democratic society cannot function effectively if the media messages it receives lack diversity, and that sometimes diversity can be achieved only by making claims on the rights of the organisations that control media outlets.[90]

Under the social democratic theory of the press, freedom of expression is a positive freedom as well as a negative freedom. Social Liberals view the purely negative view of free expression promoted by Market Liberals as insufficient. The libertarian concern about freedom from external constraint ultimately focuses on the freedom of those who own or control media organisations at the expense of those who do not

---

84. See, eg, Commission on Freedom of the Press, above n 69; Hocking, above n 74; RA Inglis *Freedom of the Movies: A Report on Self-Regulation* (Chicago: University of Chicago Press, 1947).
85. See especially Commission on Freedom of the Press, ibid, pp 54–68.
86. Ibid, pp 20–21.
87. McQuail, above n 51, pp 148–150.
88. See, eg, *Report of the Royal Commission on the Press*, above n 62, paras 481–495 and 561–571.
89. See nn 96–103 below and accompanying text. A strong preference for industry-centred reform was apparent in the publications of the Hutchins Commission. For example, the Commission's main report concluded that the ideal of a socially responsible press was better achieved through industry self-regulation than government intervention. See, eg, Commission on Freedom of the Press, above n 69, ch 5. In a seminal treatise on the media's place in political theory first published in the mid-1950s, 'social responsibility' achieved through self-regulation was identified as a distinct model of media regulation. See Siebert et al, above n 67, pp 73–103.
90. See, eg, Barron, above n 46; Fiss *Liberalism Divided*, ibid; Fiss 'Why the State?', ibid; OM Fiss 'State activism and state censorship' (1991) 100 Yale LJ 2087; CR Sunstein *Democracy and the Problem of Free Speech* (New York: The Free Press, 1993); CR Sunstein 'Free speech now' (1992) 59 U Chicago L Rev 255.

(and, realistically, cannot). A concept of positive liberty imposes obligations on the state to guarantee affirmatively the presence of the conditions necessary for attaining desired societal goals. The proper ends of media regulation identified by Social Liberals – primarily those ends associated with the values underlying the freedoms of speech and press – are broadly similar to those goals identified by Market Liberals, but they are pursued through different means and with the rights of those without access to mass media outlets at the forefront.[91] To Social Liberals, the 'marketplace of ideas' is an idealised environment that the government has a positive obligation to promote and protect.

THE INFLUENCE OF THE TWO MODELS

To a significant extent, the inconsistencies of media regulation in Western democracies for much of the past century can be attributed to the sustained contest between Market and Social Liberalism. In the UK, the Market Liberal model best explains the regulatory environment of the print media, and accounts for the approach taken with the computer and telecommunications industries that are integral to the emerging new media technologies. On the other hand, the public-service broadcasting ideal associated with the BBC and, to a lesser extent, commercial television broadcasting is a product of Social Liberal theory.

The influence of Market Liberalism

The British attitude toward the regulation of publishers is the archetype of the Market Liberal approach to media regulation. It is now broadly agreed that a 'free' British press emerged in the mid-nineteenth century,[92] coinciding with (or, arguably, driving) publishing's emergence as a truly mass media. At that time, the assumptions of Market Liberalism were winning wide acceptance and 'freedom', in that theoretical construct, meant primarily freedom from the state. The origins of the modern press, then, can be traced to a time when *laissez faire* economics, social Darwinism and anti-statist liberalism were ascendant, and it was in this period that the regulatory model governing the press emerged. Media markets were perceived to be competitive, fluid and easy to enter. It was thought unnecessary for the state to devise special rules affecting those markets or the free exchange of ideas within them; adherence to the general law was all that was required of the press. This model has proved remarkably robust, surviving even the height of popularity of social democratic thought in the middle of the twentieth century. In fact, as far as the press has been concerned, the Social Liberal challenge to Market Liberal ideas has been answered in ways that can be reconciled with Market Liberal ideology, primarily through application of competition law principles and through the development of systems of industry self-regulation.[93]

---

91. See, generally, Hocking, above n 74.
92. See Curran and Seaton, above n 19, p 7. This has been the generally held view since the Victorian era.
93. This reaction to Social Liberal pressures on Market Liberal assumptions has not been universally observed throughout Europe. Some countries, for example, have subsidised minority publications and contemplated direct content and structural regulation of the press; McQuail, above n 51, p 150.

The concentration of economic power in the hands of relatively few companies played an important part in the emergence of Social Liberal opposition to Market Liberal ideology, and, thus, it is not surprising that competition law has been a flashpoint in the contest between Market and Social Liberalism. Those taking a 'pure' Market Liberal perspective argue that the media should be treated no differently than other industries, and the efficiency and competitiveness of media markets should be assessed under the same criteria applied to suppliers of other products and services and do not warrant industry-specific rules. Further, the goals of pluralism, diversity and quality are best achieved through the competition of the marketplace, as long as general competition law secures open markets; in other words, both economic and social goals are best achieved through the reactions of suppliers to consumer demand.[94] As in most Western countries, the ownership rules applicable to newspapers and periodicals in the UK largely follow this approach: checks on the concentration of ownership in the print media are imposed primarily by general competition law rather than a complex of media-specific rules.[95]

The most important influence of the Social Liberal critique on the Market Liberal regulatory model for the press has been the development of voluntary systems of industry self-regulation.[96] These schemes typically revolve around industry-drafted codes of practice, compliance with which is usually assured through non-coercive sanctions such as adverse publicity. In part, self-regulation schemes are defensive: the industry has a stake in their success because they dampen public disquiet and reduce the chances of the state imposing mandatory legislation. But Market Liberals also find industry self-regulation an attractive alternative to direct regulation as it is consistent with liberalism's traditional distrust of state interference with the press. The post-War Royal Commission on the Press, for example, was critical of the media's unbridled pursuit of commercial advantage,[97] but, nonetheless, rejected proposals for establishing state-subsidised newspapers or imposing state-enforced content regulations, insisting that free enterprise was the foundation of a free press.[98] Instead, the Royal Commission advocated the creation of an industry body that would adopt a code reflecting the 'highest professional standards',[99] and the Press Council was founded in response. Dissatisfaction with the performance of the press produced

94. See, eg, Fowler and Brenner, above n 22.
95. The main exceptions to this proposition are special cross-media ownership rules that limit the ability of newspaper companies to acquire interests in commercial broadcasting companies and vice versa (which are really aspects of broadcast regulation) – see Broadcasting Act 1990, Sch 2, part 4, as amended (replaced by Communications Act 2003, Sch 14) – and rules requiring that certain newspaper mergers receive the approval of the Department of Trade and Industry – see Fair Trading Act 1973, ss 57–62 and Enterprise Act 2002, s 69.
96. In the UK, industry-organised and -funded bodies that created and enforced voluntary codes of practice became particularly popular after World War II; McQuail, above n 51, pp 150–151. In fact, a rash of industry-drafted journalistic codes emerged in the USA and Europe in the wake of both world wars. See T Laitila 'Journalistic codes of ethics in Europe' (1995) 10 European J of Communication 527 at 530–531. This movement affected the press, broadcasting and the advertising industry in the UK, with industry-sponsored bodies such as the Press Council, the Advertising Standards Authority, the British Board of Film Censors and the Broadcasting Complaints Commission emerging over the years.
97. See, eg, *Report of the Royal Commission on the Press*, above n 62, paras 481–496 and 563–565.
98. Ibid, paras 682–683.
99. Ibid, paras 616–663 and 684.

two subsequent Royal Commission inquiries, in 1962 and 1977,[100] but both rejected the suggestion that coercive state sanctions should be used to force the press to act in a socially responsible way.[101] More recently, the press staved off the threat of legislation in the wake of the highly critical Calcutt Report[102] by replacing the Press Council with the somewhat more independent Press Complaints Commission and publishing a revised code of practice for editors and journalists.[103] Each time the press has been threatened with direct regulation by the state, it has rallied support through arguments centred on liberal conceptions of freedom of speech and press independence that favour self-regulation in lieu of legislation.

While the influence of Market Liberalism is most obvious when the regulation of the press is examined, its impact is also apparent when examining the regulatory environment of the telecommunications and computer industries, which play a central role in the emerging media. With regard to telecommunications, this may seem counter-intuitive, since historically that industry has been subjected to either very extensive state oversight or outright public ownership and control.[104] But these regulatory structures were the consequence of the long-held belief that telecommunications networks were natural monopolies,[105] meaning that competition over the provision of telecommunications services would be less efficient than simply granting private or public body monopoly rights in defined markets, with close oversight necessary to assure that these bodies did not abuse their monopoly power.[106] While achievement of certain social goals was thought to be facilitated by a monopoly – in particular, it allowed cross-subsidisation as a means of achieving universal service at an affordable price – the rationale for the traditional approach to telecommunications was almost entirely market-based. The market orientation of telecommunications

---

100. *Report of the Royal Commission on the Press* Cmnd 1811, 1962; *Final Report of the Royal Commission on the Press* Cmnd 6810, 1977.
101. See, eg, *Final Report of the Royal Commission on the Press*, ibid, para 2.12. Rather, less intrusive solutions such as improved training of journalists was advocated, despite the fact that both the 1962 and 1977 Commissions concluded that the greater concentration of power in the print media created by the operation of free markets had threatened the quality and diversity of public discourse and that there were no market-based solutions to this problem. See Curran and Seaton, above n 19, at 293.
102. *Report of the Committee on Privacy and Related Matters* Cm 1102, 1990.
103. The Calcutt Committee subsequently criticised the Press Complaints Commission, the code it applied and the reasoning of its determinations, and recommended its replacement with a statutory press tribunal; see *Review of Press Self-Regulation* Cm 2135, 1993, but, thus far, the system of self-regulation has not been abandoned.
104. Either through use of independent commissions to oversee private suppliers, as in North America, or a public body answerable to a government minister, as in most Western European countries, Australia and New Zealand.
105. A market is considered a natural monopoly if costs are minimised by relying on a single supplier rather than allowing free competition. Telecommunications markets were considered natural monopolies because of substantial economies of scale and scope attributable, in part, to engineering costs and, in part, to the peculiar features of the demand for telecommunications services. See LD Taylor *Telecommunications Demand: A Survey and Critique* (Cambridge: Ballinger, 1980); NW Sharkey *The Theory of Natural Monopoly* (Cambridge: Cambridge University Press, 1982) pp 182–187.
106. This conclusion has been questioned by critics of the 'natural monopoly' model for telecommunications. See, eg, GW Brock *The Telecommunications Industry: The Dynamics of Market Structure* (Cambridge, MA: Harvard University Press, 1981).

regulation has only been fortified by the liberalisation and privatisation policies now pursued in North America and Europe.[107] Technological developments in the post-War period made it possible for many services provided by the industry to be offered on a competitive basis[108] and, since the 1980s, several States, including the UK, have attempted to create the conditions that would make competition in certain sectors of the industry possible – in other words, to use regulatory power to mimic market conditions.[109] This approach reflects the regulatory culture of the Office of Telecommunications (Oftel), which was created to implement privatisation strategies during the height of the neo-liberal resurgence of the Thatcher years.[110]

Market Liberal assumptions are also discernable in connection with the regulation of the computer industry and computer-based communications systems. For example, a libertarian ethos monopolised the discourse about the Internet when it emerged as an important social and economic force in the 1990s, and that ethos, though no longer unchallenged, remains dominant.[111] The government's report on emerging multi-media technologies illustrates this:

> 'The Government's approach to Internet regulation is to encourage voluntary action backed up by the full force of the existing law, based on the application of general law on-line as off-line... Self-regulation is being taken further forward by the Internet Watch Foundation, funded by the United Kingdom Internet service provider industry.'[112]

This emphasis on industry self-regulation and resistance to medium-specific rules is consistent with Market Liberalism's insistence on minimum state interference with communications systems.

Social Liberalism and broadcasting policy

The economic considerations that for decades justified telecommunications monopolies can be contrasted with the justifications for the monopoly structure of

---

107. See, generally, I Walden and J Angel (eds) *Telecommunications Law* (London: Blackstone, 2001).
108. See, generally, HE Hudson *Global Connections: International Telecommunications Infrastructure and Policy* (New York: Van Nostrand Reinhold, 1997) pp 17–34 and 89–117.
109. See L Correa 'The economics of telecommunications regulation' in Walden and Angel, above n 107, pp 16–52. Among other things, liberalisation policies must give consideration to the unequal market power possessed by the incumbent supplier in a sector of the telecommunications industry and new entrants; the effects of prior investments on the incumbent's cost structures; and the problem of 'bottlenecks', where an incumbent's control of an 'essential facility' threatens the competitiveness of dependent services. See n 35 above.
110. Telecommunications Act 1984. See, generally, C Hall et al *Telecommunications Regulation: Culture, Chaos and Interdependence Inside the Regulatory Process* (London: Routledge, 1999).
111. See, eg, E Dyson *Release 2.0: A Design for Living in the Digital Age* (London: Viking, 1997); DR Johnson and DG Post 'Law and borders – the rise of law in cyberspace' (1996) 48 Stanford L Rev 1367. Part of the argument of the 'cyber-libertarians' is that this regulatory approach is compelled by the technological characteristics of the Internet. This contention has been refuted in the recent literature. See J Goldsmith 'Unilateral regulation of the Internet: a modest defence' (2000) 11 European J of Int'l Law 135.
112. Culture, Media and Sport Committee *The Multi-Media Revolution* 4th Report, HC 520-1, 1998, para 106.

broadcasting that prevailed in most European states for much of the twentieth century. Whatever the government's original rationale for establishing a broadcasting monopoly in 1922,[113] the perpetuation of this monopoly was grounded on social rather than economic considerations. John Reith, the first Director-General of the BBC, wanted a mass communications service that was 'moral in the broadest sense – intellectual and ethical; with determination that the greatest benefit would accrue from its output'[114] and believed that a market-based system was antithetical to these ends. As Craufurd Smith points out, Reith saw the 'brute force' of monopoly as central to his vision:

> 'Only by limiting audience choice could Reith aspire to introduce the public to a range of ideas and musical types which were alien or even disturbing. It was a paternalism encapsulated in the famous statement "[i]t is occasionally indicated to us that we are apparently setting out to give the public what we think they need and not what they want – but few know what they want and very few what they need".'[115]

Reith's patronising attitudes notwithstanding, the 'public-service broadcasting' (PSB) model he was instrumental in establishing was not simply the product of paternalism. It had deeper roots in the ideology of Social Liberalism. The dissatisfaction with the consequences of unregulated market capitalism that fuelled the rise of Social Liberalism was widespread when broadcasting emerged as a mass medium in the 1920s. Policy makers wished to insulate broadcasting from the 'corrupting' effects of economic competition and wealth maximisation, and in the immediate post-War period there was an unprecedented confidence in the potential of centralised management and distribution of public resources.[116] The BBC was the institution and PSB the regulatory model through which this interventionist approach to the utilisation of the public airways was to be achieved.

Established in 1927 by a Royal Charter made renewable every 10 years and revocable at any time, the BBC is overseen by a Board of Governors appointed by the Queen in Council. The Board are to act as trustees for the public interest and ensure that the terms of the Charter and other obligations are fulfilled. Currently, the main source of programming standards governing the BBC is set out in the form of a contract: the Agreement between the Secretary of State for National Heritage and the BBC.[117] These standards, however, have been strikingly vague and malleable, and, for the most part, the state has exercised control over the Corporation through the implied threat of non-renewal and through its influence over the BBC's funding and corporate governance rather than through detailed rules.[118] Potentially, this influence is great – most power within the BBC is exercised by a management board headed

---

113. The early history of the broadcasting monopoly is discussed in P Scannell and D Cardiff *A Social History of British Broadcasting, Volume 1: 1922–1939 Serving the Nation* (Oxford: Blackwell, 1991). See also T Burns *The BBC: Public Institution and Private World* (London: MacMillan, 1977) pp 1–11.
114. Quoted in Craufurd Smith, above n 25, p 45.
115. Ibid.
116. Curran and Seaton, above n 19, p 114.
117. See above n 18. The Royal Charter currently in force also requires the BBC to provide informational, educational and entertainment programming as part of its responsibility to provide broadcasting services 'as public services'. See the website available at http://www.bbc.co.uk/info/BBCcharter/charter/index.shtml.
118. See, generally, T Gibbons *Regulating the Media* (London: Sweet & Maxwell, 2nd edn, 1998) pp 243–250.

by a Director-General who effectively serves at the pleasure of the Prime Minister[119] – and the BBC's early history was marred by blatant instances of state censorship.[120]

Despite this, an embryonic version of PSB soon emerged that was grounded on the assumption that the public broadcaster must be given wide editorial independence from the state. As a concept, PSB has continually evolved since then, although its precise meaning has remained elusive.[121] There are probably as many definitions of PSB as there are people who attempt to define it and there has been little consensus about its precise contours.[122] At its core, however, the PSB ideal 'gives primacy to the needs of society or the collective needs of citizens rather than to individual rights, consumer freedom or market forces'[123] and PSB is directed at achieving certain 'public-interest' goals that 'the free market left to itself would fail to satisfy ... because it would not be profitable to do so'.[124] Barendt identified certain criteria as central to the PSB ideal: universal service, general geographical availability, concern for national identity and culture, diversity of programming, impartiality, independence from the state and independence from commercial interests.[125] Others have added to this list 'high quality programming in each genre, including innovation, originality and risk-taking, [and] a mission to inform, educate and entertain'.[126] These criteria are interrelated[127] and perceptions of the essential elements of PSB have shifted over time.[128]

119. For example, the Director-General was summarily dismissed by Margaret Thatcher in 1987 because of the government's displeasure with the BBC's coverage of Northern Ireland. See EM Barendt *Broadcasting Law: A Comparative Study* (Oxford: Clarendon Press, 1993) pp 67–68.
120. See, generally, WJ West *Truth Betrayed* (London: Duckworth, 1987). See also Craufurd Smith, above n 25, pp 31–32.
121. But see Communications Act 2003, s 264(4) and (6) (attempting to define PSB).
122. A representative sample of the voluminous literature addressing the subject would include M Raboy (ed) *Public Broadcasting for the 21st Century* (Luton: John Libbey, 1997); Broadcasting Research Unit *The Public Service Idea in British Broadcasting: Main Principles* (London: BRU, 1985); Barendt, above n 119, pp 50–74; M Feintuck *Media Regulation, Public Interest and the Law* (Edinburgh: Edinburgh University Press, 1999); the sources listed in Craufurd Smith, above n 25, p 46 n 5, and the sources listed in Born and Prosser, ibid, at 670 n 51.
123. McQuail, above n 51, p 156.
124. Ibid, p 157.
125. See Barendt, above n 119, p 52.
126. Born and Prosser, above n 25, at 671 n 32. A legislative definition of PSB was not attempted until s 264 was added to the text of the Act in the late stages of parliamentary debates over the legislation. The criteria identified in subss 4 and 6 largely reflect the elements discussed here.
127. Universal service and general geographical availability, for example, are founded on a certain conception of equality – all citizens have the equal right to receive essential broadcasting services, regardless of wealth or geographical location – that imposes affirmative obligations on the state. Universal availability of programming arguably fosters a common national culture, language and identity throughout the UK, and this was an original aim of the BBC (although it is now more common to see this objective referred to as the 'cultivation of social cohesion'). The goal of providing 'universalising' programming that 'invoke[s] commonality and enhances the creation of a "common culture" ', however, must be balanced against 'the key democratic function of staging for a society communicative and cultural (intersubjective) encounters between its plural communities and minority identities', with 'communicative tolerance' central to the task of 'mediating social and cultural differences'; Born and Prosser, ibid, at 676 and 672–673.
128. For example, 'diversity' emerged relatively late in the evolution of the PSB ideal. See especially *Report of the Committee on the Future of Broadcasting* Cmnd 6753, 1977 (Annan

Of all of the features of the PSB ideal, the most difficult to translate into reality is independence from the state and independence from commercial interests. The danger of domination by the state is particularly acute, since the government ultimately controls the BBC's purse-strings and its governing board. In the UK, several strategies have been employed to address this danger: one is to provide the BBC, through the licence fee, with a funding source separate from general tax revenue in order to minimise the uncertainties associated with competing for funding each year with other state-provided services; another is to impose content regulations requiring impartiality, mandating basic news and information programming, assuring coverage of certain political events and guaranteeing that the leading political parties have equal access to broadcast facilities to air their messages.

Containing the power of the government vis-à-vis the media is a traditional preoccupation of liberalism. In Reith's view, however, the BBC not only was to be above the political fray but also independent of commercial interests, and the Corporation was established as a non-profit organisation that was barred from airing advertisements.[129] A commercial system of broadcasting was rejected because it would merely appeal to the 'lowest common denominator' in order to maximise audience share and attract advertisers; the notion that the broadcaster should simply respond to 'consumer' tastes was inimical to the 'higher' purpose envisaged for broadcasting. This view is echoed in Born and Prosser's insistence that broadcast audiences not be reduced to an aggregate of individual consumers, and that the content of broadcasting should not be 'shaped simply by market signals from advertisers or (in subscription services) from viewers, but by an appeal to principles of citizenship, universality and quality'.[130] Similarly, Scannell has argued that equal access to PSB 'should be thought of as an important citizenship right in mass democratic societies'.[131] These contentions clearly echo Social Liberal ideology.

Report), but has taken an increasingly important place in the hierarchy of PSB objectives. 'Diversity' and 'pluralism' are terms that have been used loosely to describe a number of objectives, including the provision of a wide range of political opinions and arguments, the protection of local or regional broadcasting, the representation of minority cultures, and the provision of programming that caters to the needs of all groups in society. See, generally, Committee of Experts on Media Concentration and Pluralism *Consultant Study on 'Media Concentration in Europe: The Impact on Pluralism'* MM-CM (97) 12 rev (Strasbourg: Council of Europe, 1998); PM Napoli 'Deconstructing the diversity principle' (1999) 49 J of Communication 7; Entman and Wildman, above n 10, at 8.

129. See Curran and Seaton, above n 19, pp 114–116. Presently, the BBC does engage in some commercial activities, primarily through its wholly owned subsidiaries, BBC Worldwide and BBC Resources. These activities are regulated by general competition law as well as internal fair-trading rules mandated by an agreement between the BBC and the Secretary of State (see Cm 3152, 1996) that is designed to prevent cross-subsidies between the BBC's public and commercial services and to assure that the Corporation's commercial activities are not at odds with its public-service obligations.

130. Born and Prosser, above n 25, at 671 and 658. They argue 'that PSB is not purely responsible for filling gaps left in the market-place, but for aiding in the very definition and negotiation of social identities that are a core dimension of citizenship and, thereby, for establishing conditions of communication without which markets cannot work. In all the senses outlined, the constitutional role and the communicative functions of PSB are prior to the market, not simply part of it'; ibid, at 675.

131. P Scannell 'Public service broadcasting and modern public life' (1989) 11 Media, Culture and Society 135 at 164.

In fact, the PSB model is Social Liberalism's most lasting contribution to media policy debates. It is based on a positive conception of freedom of expression and arises as much from an ethic of obligation as one of individual rights; it places fear of corporate power on a more or less even footing with fear of state power; it rejects the assumptions of theories of 'consumer sovereignty' favoured by Market Liberals; and it contemplates a positive role for the state in mass communications, particularly in providing a forum for voices that might be marginalised in a communications marketplace.[132] But the hegemony of Social Liberal thought in constructing the regulatory environment for broadcasting ended with the introduction of a commercial television broadcaster in the mid-1950s. The BBC was widely perceived as being bureaucratic, complacent, unresponsive and elitist[133] when the Conservative Government of the 1950s decided to introduce market forces into the new medium of television broadcasting.[134] The Independent Television Authority (ITA)[135] was made responsible for licensing the use of broadcasting resources (through a system of franchising by competitive tender) and for the content broadcast by the licensees.

In the decades following the advent of commercial television, broadcasting policy became the primary battleground of proponents of the Market Liberal and Social Liberal models of media regulation, with efforts to reconcile these schools anticipating the 'Third Way' by nearly half a century. Despite the introduction of market forces into the broadcasting sector, however, the Social Liberal regulatory model for broadcasting has proved durable. Indeed, commercial broadcasting was quickly subsumed within the PSB ideal, which was reconceptualised as one 'essentially concerned with setting normative goals' that did not necessarily 'mandate any particular ownership structure'.[136] The criteria for allocating commercial television licences focused on the ability of applicants to provide a 'public service in broadcasting'[137] and, over the decades, the regulatory apparatus for tempering the influence of marketplace forces on programming decisions grew increasingly complex. Licensees had to comply with several wide-ranging compulsory codes governing programming standards, which, among other things, required commercial broadcasters to satisfy basic PSB requirements and serve minority audiences who might otherwise be ignored in an

---

132. See, eg, O O'Neill 'Practices of toleration' in J Lichtenberg (ed) *Democracy and the Mass Media* (Cambridge: Cambridge University Press, 1990) p 155.
133. See, eg, *Report of the Broadcasting Committee* Cmd 8116, 1951 (Beveridge Report) paras 182–186.
134. See Gibbons, above n 118, pp 150–151. See, generally, B Sendall *Independent Television in Britain: Volume 1 Origins and Foundations 1942–62* (London: Macmillan, 1982). Commercial radio was not officially sanctioned until the 1970s – see Sound Broadcasting Act 1972 – although the BBC faced competition from unauthorised sources such as Radio Luxembourg since the 1930s, and it was the popularity of 'pirate' stations in the 1960s that ultimately forced the creation of a 'legitimate' commercial radio sector.
135. Created by the Television Act 1954.
136. Craufurd Smith, above n 25, pp 46–47. Thus, while the regulatory apparatus for commercial operators is largely separate from that governing the BBC, the 'public' and 'commercial' broadcasters are 'unified under a common set of aspirations and goals'; ibid, p 38.
137. Gibbons, above n 118, p 152; see, generally, ibid, pp 150–178.

advertiser-driven system.[138] Not only is the content of commercial broadcasts regulated, so is the structure of the commercial broadcasting industry itself.[139] In stark opposition to the Market Liberal preference that the general principles of competition law govern the media, there have been many industry-specific rules governing who can own and control broadcasting companies that are meant to encourage competition, prevent the undue concentration of broadcast markets, avoid the anti-competitive effects of cross-subsidisation and ensure the diversification of the sources of programming content.[140]

A parallel between the evolution of broadcasting regulation and regulatory attitudes toward the print media can be drawn. The Social Liberal challenge to the market-orientated model applied to the print media was deflected by addressing many of the criticisms of the model without abandoning the model's base assumptions – primarily through the development of self-regulatory systems. Similarly, the Market Liberal challenge to the BBC monopoly, manifesting itself through the introduction of commercial broadcasting, simply meant that modifications to the PSB ideal were made, with the Social Liberal assumptions underlying broadcasting policy surviving intact. This suggests that, while the regulatory model governing a particular medium depends on which school of liberalism was in vogue when the medium emerged as a socially important force, once established it is difficult to dislodge; it is more likely that the model will adapt to pressures upon it rather than reject the theoretical assumptions that initially informed it.

This said, the past generation has seen an obvious preference for greater reliance on market forces to achieve policy goals in the broadcasting arena. For example, in 1980, the Thatcher Government established Channel 4 as a commissioning rather than production company in order to stimulate the growth of a competitive independent production sector.[141] The Cable Authority was established in 1984 to encourage exploitation of the economic potential of 'new' media such as cable and satellite television,[142] and it imposed comparatively undemanding programme standards on service providers. The Broadcasting Act 1990, although it effectively codified much of the PSB ideal and established a new regulatory body, the Independent Television Commission (ITC),[143] to regulate all television programme services other than the BBC and Welsh Authority, was nonetheless deregulatory in orientation, easing the burdens placed on the commercial television and radio sectors, promoting opportunities for independent producers and providing greater

---

138. Among other things, these codes require 'impartiality' on the part of licensees themselves; promote the dissemination of a wide range of opinion held by others; encourage diversity and balance in programming; prohibit offensive or 'immoral' programming; and generally insist upon the 'highest possible standards' for programme service and quality. See, eg, Broadcasting Act 1990, s 6 (replaced by Communications Act 2003, ss 264 and 319–320); ITC Programme Code, above n 18.
139. See, generally, Doyle, above n 39.
140. See, eg, Broadcasting Act 1990, Sch 12 (replaced by Communications Act 2003, Sch 14).
141. Broadcasting Act 1980.
142. Cable and Broadcasting Act 1984.
143. The ITC replaced the Independent Broadcasting Authority (the ITA's successor) and the Cable Authority. The Broadcasting Act 1990 also created the Radio Authority to oversee the independent radio industry.

flexibility in order to develop markets for new communications technologies.[144] Even as the cable and satellite industries were brought under the jurisdiction of the ITC, the minimalist regulatory regime established by the Cable Authority was maintained. The Broadcasting Act 1996 significantly liberalised media-ownership rules.[145] Meanwhile, the BBC responded to not-so-carefully veiled threats of privatisation from Conservative Governments by imposing 'quasi-market processes', both in conjunction with its internal decision making and its relationships with the independent production sector.[146]

But it has been the proliferation of forms and channels of mass communication made possible by digitalisation and the merger of telecommunication and audiovisual technologies that has posed the greatest threat to the Social Liberal model of broadcasting regulation.[147] Spectrum scarcity was an easy shorthand explanation for the special treatment of broadcasting that avoided more searching questions of political philosophy. The death of the scarcity rationale has raised questions that strike at the very foundations of broadcast policy. The commercial media stepped up pressures on policy makers to lighten the regulatory burdens imposed on them and attacked the special position held by the BBC under the present regulatory regime.[148] At one point, the government considered scrapping the licence fee[149] and many openly advocated a move to a wholly commercial broadcasting system akin to that of the USA. It was against this turbulent background that New Labour's communications policy was forged.

THE EVOLUTION OF NEW LABOUR'S COMMUNICATIONS POLICY

Since coming to power in 1997, the Blair Government has considered various policy responses to a perceived communications 'revolution', ranging from simply leaving the regulatory structure that New Labour inherited in place and encouraging greater cooperation between the various regulators, to creating a single body to oversee all

---

144. See, eg, Home Office *Broadcasting in the 90s: Competition, Choice and Quality* Cm 517, 1988; Home Affairs Committee *The Future of Broadcasting* Third Report, HC 262-I and II, 1987–1988.
145. See H Fleming 'Media ownership: in the public interest? The Broadcasting Act of 1996' (1997) 60 MLR 378.
146. See Born and Prosser, above n 25, at 666–670.
147. As the Communications White Paper stressed, technological developments have increased the amount and range of media content available to the public and created the potential for more direct competition between previously distinct and separate sectors of the communications industry. See Communications White Paper, above n 2, para 4.2.2. See also Department of Trade and Industry and Department of Culture, Media and Sport *Regulating Communications: Approaching Convergence in the Information Age* Cm 4022, 1998 (Communications Green Paper), paras 1.1–1.34.
148. Born and Prosser, above n 25, at 658. See also Independent Review Panel *Report, The Future Funding of the BBC* (1999) (Davies Report), available at http://news.bbc.co.uk/hi/english/static/bbc_funding_review/reviewco.pdf; Culture, Media and Sport Committee, above n 112.
149. See Davies Report, ibid (ultimately recommending that the licence fee should be retained).

communications industries (including the print media).[150] In a consultative document setting out the government's preliminary views on the implications of convergence published in 1998, a preference seemed to be expressed for an 'evolutionary approach to regulatory development', with the government reluctant to undertake 'major legislation so soon after the 1996 Broadcasting Act'.[151] Two years later, however, the government published its Communications White Paper, announcing its intention to abandon the existing regulatory structure, and with the subsequent publication of a draft Communications Bill,[152] the basic contours of New Labour's communications policy could be detected.

The strategy adopted was an aggressive one: the draft Bill proposed the most comprehensive and far-reaching reform of communications policy in the past 50 years and the environment in which decision making would take place was to be radically altered. A new regulator, Ofcom, would replace five independent bodies[153] responsible for implementing all aspects of media and telecommunications policy, except with respect to the print media and the BBC (later, even the BBC would be brought, at least partially, under Ofcom's jurisdiction). Ofcom would be charged with pursuing an agenda characterised by 'light touch' regulatory decision making that would eliminate 'unnecessary' regulatory burdens, encourage industry self-regulation and place greater reliance on market forces to achieve policy goals, and promote new media markets through measures designed to bolster consumer confidence.[154] The government proposed to eliminate many long-standing sector-specific rules designed to prevent the consolidation of ownership in media industries and to pave the way for spectrum trading. At the same time, however, calls for full deregulation of communications industries were rejected. The government reasserted its commitment to the principle of universal service in connection with traditional and emerging telecommunications services,[155] and seemingly embraced the fundamentals of PSB.[156] Nonetheless, the government insisted that 'the way in which that public service role is regulated and delivered by the broadcasters will have to change, to reflect the new

150. The latter option was favoured by Collins and Murroni, above n 3, pp 173–175. A Select Committee of the Department of Culture, Media and Sport advocated a single regulator for the entire electronic communications industry (excluding the print media). See Culture, Media and Sport Committee, above n 112. Among other options considered were the creation of two new regulators, one for the communications infrastructure and another for the 'content-providing' sectors of the industry; and the division of economic regulation and 'cultural and content' regulation between two regulators. See Communications Green Paper, above n 147, paras 5.7–5.17.
151. See Communications Green Paper, ibid, at 5–6 (Executive Summary). There was a similarly unenthusiastic response from European broadcasters and governments to a suggestion in a European Commission Green Paper (see above n 2) that there should be a single regulatory authority for broadcasting and telecommunications. See *Summary of the Results of the Public Consultation on the Convergence Green Paper* SEK(1998)1284, 29 July 1998.
152. Draft Communications Bill Cm 5508-I, 2002, available at http://www.commbill.net/thebill.htm.
153. Those bodies are the ITC, which regulated commercial television; the Radio Authority, which oversaw commercial radio; Oftel, the telecommunications regulator; the Broadcasting Standards Commission, the 'taste and decency' regulator; and the Radio Communications Agency, which managed radio spectrum allocation.
154. See Communications White Paper, above n 2, paras 8.5.1 and 8.10.1; Draft Communications Bill, above n 152, cls 3 and 5.
155. See, eg, Draft Communications Bill, ibid, cls 50–57.
156. Ibid, cls 3 and 181–230.

conditions in which they operate', and proposed to 'modernise' the regulatory framework to reflect better the particular missions and funding sources of different broadcasters; provide broadcasters with greater flexibility to 'innovate and respond to market and technological changes'; supplement the current reliance on programming codes with industry 'co- and self-regulation'; and generally move away from 'the detailed, prescriptive requirements – often dubbed "box ticking" – which are contained in present licences [and] may inhibit creative innovation, and thus harm both the public interest and the commercial success of companies'.[157]

The government's proposals were controversial. Some in the Conservative Party, for instance, complained that the government's liberalisation programme did not go far enough and that the demise of the scarcity rationale meant that detailed regulation of the communications sector could no longer be justified.[158] The more common criticism, however, was that the government's approach was too market-orientated and gave short shrift to the public's concerns as 'citizens'.[159] This theme informed the report published by a Parliamentary Joint Committee chaired by Labour Peer David Puttnam, which made 148 recommendations in response to the draft Bill, most of which were incorporated into the Bill ultimately placed before Parliament.[160] Even with these modifications, many felt that 'the Bill's central thrust of economic policy as the prime motivator' for regulatory reform would mean that 'public service broadcasting [would become] a secondary concern' and that 'the Bill's effect, by strengthening rather than containing market forces, could be to put public service broadcasting on the margins'.[161] Lord McNally, for example, decried the government's 'blind faith' that:

> 'deregulation to promote competitiveness and investment will produce the most dynamic and competitive communications industry in the world. Maybe so, or it may result in concentrations of power in the hands of multi-media global conglomerates that will pose a threat to programme quality, cultural diversity, regional identity and the effective workings of our democracy.'[162]

157. Communications White Paper, above n 2, paras 5.2.6, 5.4.5, 6.1 and 5.4.3.
158. See, eg, 646 HL Official Report (5th series) col 661, 25 March 2003 (Baroness Buscombe stated that 'although we welcome the move to liberalise existing restrictions prohibiting media convergence, we believe that the Government's approach to ownership is prescriptive and timorous ... In essence, content is driven by consumer demand, not by ownership. The two issues should not be confused. We should have more confidence in our culture and allow the consumer to choose ... [A]lthough the restrictions governing ownership of different media could previously be justified with reference to the scarcity of available spectrum, the media have changed exponentially in the past decade, with the advance of new technology and communications').
159. In fact, one suspects that if a thorough content analysis of the draft Bill and accompanying documentation were conducted, references to 'consumers' would far out number references to 'citizens' and often the two terms were conflated. See, eg, Communications White Paper, above n 2, paras 1.2.10–1.2.11.
160. See Report of the Joint Committee on the Draft Communications Bill, HL 169-I and HC 876-I, 31 July 2002, available at http://www.parliament.the-stationery-office.co.uk/pa/jt200102/jtselect/jtcom/169/16901.htm, and *Government's Response to the Report of the Joint Committee on the Draft Communications Bill* Cm 5646, 2002, available at http://www.communicationsbill.gov.uk/pdf/Joint_cttee_CBill.pdf.
161. 646 HL Official Report (5th series) col 670, 25 March 2003 (speech by Lord Bishop of Manchester).
162. Ibid, col 666. Along similar lines, Lord Puttnam dismissed the notion that plurality and diversity can be stimulated by encouraging 'market dominance' as 'frankly, risible. Recent history points only one way – towards the inevitable consolidation of conformity and power':

Constant pressure from a cross-party coalition led by Lord Puttnam in the House of Lords won some concessions from the government that arguably softened the impact of the Act ultimately adopted. For example, a significant proportion of parliamentary debate over the Communications Bill focused on how Ofcom's general duties would be defined. When first laid before Parliament, the Bill did not prioritise Ofcom's often-conflicting general duties, leaving questions about which duties should take precedence in particular situations to Ofcom's discretion. Many feared that this meant that immediate, tangible economic concerns would inevitably prevail over less quantifiable cultural and civic considerations. A last-minute compromise potentially strengthened the public-service dimension of the Act, which, as enacted, provides that:

> 'it shall be the principal duty of Ofcom, in carrying out their functions, (a) to further the interests of citizens in relation to communications matters; and (b) to further the interests of consumers in relevant markets, where appropriate by promoting competition.'[163]

Proposals to elevate the cultural and political interests of 'citizens' *over* the economic interests of 'consumers', however, were rejected,[164] and it will be left to Ofcom to reconcile these interests in the vast array of situations in which they are implicated.

In fact, the Act is an uneasy accommodation of ideas derived from both the Market Liberal and Social Liberal schools. The influence of the former can be seen in provisions establishing the regulatory framework for the networks and services that form the below-structure for the transmission of audio, video and other digital content to the public.[165] In addition to implementing a number of liberalising EC Directives, these provisions allow Ofcom to extend the market-based strategies employed by Oftel, the old telecommunications regulator, for promoting competition and encouraging access to communications markets for new entrants, and contemplate the creation of a secondary market in UK spectrum rights. Similarly, the Act significantly liberalises the competition and ownership rules applicable to the commercial media. The government's strategy in this area derives from three basic principles: that the commercial media in the UK need new sources of investment; that the source of this investment is unimportant so long as effective content regulation assuring quality and diversity is in place; and 'ownership rules must reflect the reality of a global marketplace'.[166] Pursuant to these principles, the Act abolishes rules that prevented non-EU media companies from holding UK broadcasting

---

ibid, col 674. Lord Puttnam stressed that in an industry 'in which the cost of entry has become all but entirely prohibitive, [it is] absolutely certain [that] plurality and diversity are not a natural by-product of unregulated market forces'; ibid, col 673.
163. Communications Act 2003, s 3(1).
164. The government feared that these proposals could expose Ofcom to multiple judicial review proceedings and undermine the regulator's effectiveness. See 409 HC Official Report (6th series) cols 44–45, 14 July 2003.
165. See Communications Act 2003, ss 32–184.
166. 395 HC Official Report (6th series) col 793, 3 December 2002 (speech by Culture Minister Tessa Jowell). See also Department of Trade and Industry and Department of Culture, Media and Sport *Consultation on Media Ownership Rules* November 2001, para 1.8, available at http://www.dtg.org.uk/news/archive/ownershi.pdf.

licences (raising concerns in some quarters that large US media conglomerates might enter UK markets and threaten the financial stability and independence of European broadcasting companies);[167] significantly weakens sector-specific media-ownership restrictions,[168] giving major commercial radio and television broadcasters unprecedented opportunities to expand their market share in the UK and potentially allowing a single company to acquire most or all of the 16 national and regional ITV licences;[169] and repeals most of the cross-ownership rules that have long been in force, including rules that have prevented national newspaper proprietors with a market share in excess of 20% from also owning Channel 5, opening the door to Rupert Murdoch's possible entry into terrestrial broadcasting in the UK.[170] Instead of sector-specific ownership restrictions, the government clearly prefers to rely on general competition law, supplemented by content regulation, to safeguard media plurality and diversity.[171]

On the other hand, a residue of Social Liberal ideology remains. The government insisted that the Act does not abandon socially orientated objectives, particularly those associated with PSB, and rejected suggestions that market forces will always guarantee the diverse, high-quality programming currently available to viewers and listeners in the UK, or promote the democratic, educational and cultural ends served by PSB.[172] Nonetheless, the Act attempts to renovate the

---

167. Communications Act 2003, s 348. Critics of this provision worry that British radio and television may become a dumping ground for cheap, inferior-quality programming produced in the USA. The government believes this can be avoided through appropriate content regulation and, in any event, see no difference between UK broadcast companies being controlled by a company such as Vivendi Universal (which cannot be barred from ownership under EC law) and multinationals such as Sony or Time Warner.
168. For example, the Act removed the 15% upper audience limit for ownership of UK television broadcasting, revoked the rule that banned single ownership of the two London ITV licences and abandoned the 'points' system that limited UK-wide ownership of radio broadcasting to a 15% share of commercial audiences. See Communications Act 2003, s 350 and Sch 14.
169. In addition, the Act permits consolidation of ownership in the radio industry that could result in many regions of the country receiving radio broadcasts from stations owned by only two companies other than the BBC.
170. In addition, the Act eliminates the ban on common ownership of national television and national radio licences and will, for the first time, allow joint ownership of a national ITV licence and the Channel 5 licence, and joint ownership of both a regional ITV licence plus a local radio licence for the same service area. Only a few core cross-media restrictions deemed necessary to ensure pluralism remain: no one controlling over 20% of the national newspaper market (ie Murdoch) will be permitted to hold an ITV licence, and no one controlling more than 20% of a regional newspaper market will be allowed to own the regional ITV licence for the same area; Communications Act 2003, s 350 and Sch 14.
171. This is consistent with the government's stated strategy to 're-base' structural regulation as far as possible on the principles of general competition law and retain only those industry-specific rules deemed 'essential'. See Communications White Paper, above n 2, paras 8.9.1 and 4.2.8. The Competition Commission and Office of Fair Trading will retain primary responsibility for monitoring communications markets under the Competition Act 1998 and the Enterprise Act 2002, but they will be expected to consult with Ofcom whenever mergers involving communications companies are at issue. See, eg, Communications Act 2003, ss 369–372 and 381–386.
172. See, eg, Communications White Paper, above n 2, paras 5.2.5, 5.2.6 and 5.3.5–5.3.12.

regulation of broadcast content through the adoption of a 'three-tier' approach.[173] Ofcom will issue codes of practice establishing a basic tier of regulation, applicable to all broadcasters, directed at ensuring that all UK television and radio programming complies with basic domestic, European and international obligations.[174] Under the second tier of content regulation, the commercial PSBs – the ITV licensees, Channel 4, Channel 5 and the public teletext service – and the BBC will be required to meet quantifiable and measurable obligations imposed on them by Ofcom.[175] The third tier concerns the qualitative aspects of public-service television broadcasting provided by the commercial PSBs, a category including numerous quality and diversity obligations, such as requirements to air educational, arts or religious programmes and the duty to serve minority audiences. The government having concluded that requirements of this sort are 'less easy to quantify without

---

173. Like many three-tier regulatory structures, the government's policy concerning media content has four tiers. What the Communications White Paper called 'tier zero' contemplates an unobtrusive role for the state in connection with the dissemination of video and audio material over the Internet and via telephony. See ibid, para 5.9.1. The government has taken the view that 'people make a clear distinction between their expectations of broadcasting and of the internet'; ibid, para 6.6.7, and that material disseminated through the new media should only be subject to the general law and industry self-regulation. Although the government 'expect[s] to see public service broadcasters applying the same high standards and high quality in their services on the internet and via telephony as they do on their traditional broadcast businesses'; ibid, para 5.9.1, it did not bring these media within the Act's framework for media content regulation.

174. This category includes matters such as taste and decency; accuracy and impartiality; the protection of minors; advertising and sponsorship; rules governing access for disabled people; and the provision of training and rules on equal opportunities. See, eg, Communications Act 2003, ss 303–308, 319–328 and 337. While in the abstract the same standards apply to all licensed broadcasters, Ofcom will likely take into account differences in the sorts of services being provided to viewers and listeners in drafting its codes. For example, Ofcom might consider issues such as the likely degree of harm that would be caused by including particular material in specific types of programming; the size and composition of the audience that would be likely to see or hear this material; the extent to which audiences can be warned about broadcast content in advance; and the likelihood of accidental exposure to potentially harmful material; ibid, s 319(4). Ofcom may also consider audience expectations 'as to the nature of a programme's content', a factor that could depend, in part, on expectations about the content typically available through different delivery platforms (terrestrial, cable, satellite, etc); ibid, s 319(4)(c). See also Communications White Paper, above n 2, para 6.3.7.

175. Quantifiable obligations include quotas for original or independent productions, targets for regional programming, targets for scheduling of news during peak time and rules governing party political broadcasts. See Communications Act 2003, ss 277–289. These obligations will not be the same for every broadcaster: Channel 5, for example, is subject to fewer public-service programming obligations than ITV (Channel 3). Ofcom will also adopt regulations governing the independent radio sector analogous to the quantifiable regulation characteristic of Tier 2 television regulation, even though commercial radio stations are not regarded as 'public-service broadcasters' in the UK. Prominent among these regulations will be rules designed to assure that the range of programming available in any locality in the UK is not narrowed and that the 'localness' of local radio is preserved. See ibid, ss 312–315.

tipping back into box-ticking',[176] the Act envisages that they will be delivered and monitored through a system of self-regulation.[177]

In sum, the Act radically alters the environment in which regulatory decision making affecting communications industries will take place. Many long-standing rules affecting media and communications have been eliminated and the stage has been set for the elimination of many more. The BBC, which heretofore has stood largely outside more formal regulatory structures, will find itself subject to Ofcom's rule-making power.[178] Most fundamentally, industry sectors that have grown up under very different regulatory regimes will be brought together under a common regulatory umbrella. The next section of this paper addresses the implications of these changes for the future of communications regulation.

---

176. Communications White Paper, above n 2, para 5.8.1.

177. The Act requires each commercial public-service broadcaster to develop detailed statements of programming policy consistent with its own particular public-service remit and to provide a self-evaluation of its performance in light of its stated policy. Ofcom is then required to review and report on the state of PSB periodically. See Communications Act 2003, s 264. Reserve powers are given to Ofcom in the event that self-regulation fails, but those powers can only be exercised if that failure is not attributable to economic factors; ibid, s 270.

178. As originally devised, the regulatory structure proposed by the government only indirectly affected the BBC. Criticism of the government's failure to place the BBC under Ofcom's jurisdiction, however, dominated early debates over the Communications Bill. Among other things, the BBC was accused of arrogance and secretiveness; of abusing an unfair competitive advantage given it by the licence fee; and of being 'judge and jury in its own court' when resolving complaints made about its programmes. See, eg, 646 HL Official Report (5th series) cols 662–663, 25 March 2003 (speech by Baroness Buscombe); ibid, col 731 (speech by Lord Harris of Highcross); 400 HC Official Report (6th series) col 190, 25 February 2003 (speech by Andrew Mitchell); ibid, col 191 (speech by Michale Fabricant); ibid, col 210 (speech by John Whittingdale). Calls for reconsideration of the licence fee and greater regulatory and financial oversight of the BBC's activities were made. See, eg, ibid, col 166. In response to these complaints, the Bill was amended, subjecting the BBC to the first two tiers of regulation and giving Ofcom unprecedented powers in relation to the BBC. Ofcom's relationship with the BBC will be defined primarily in amendments to the Agreement between the Secretary of State and the BBC, which will provide that, for the first time, a media regulator will have the power to impose financial penalties or take other remedial actions against the BBC in the event specified standards are contravened. The BBC is made subject to all of Ofcom's codes of practice issued under Tier 1 of the new regulatory scheme, except to the extent they concern accuracy and impartiality, which will remain the exclusive concern of the BBC's Board of Governors. The quantifiable Tier 2 rules will also apply to the BBC. For now, the BBC is not, strictly speaking, subject to the third tier of content regulation, as Ofcom has not been given the backstop powers it can theoretically exercise against the commercial PSBs. Nevertheless, the BBC will be required to publish an annual statement of programming policy and must consider Ofcom's guidance and feedback in formulating that policy. See, generally, *Proposed Amendments to the Main BBC Agreement dated 25 January 1996* (20 March 2003), available at http://www.communicationsbill.gov.uk/pdf/proposed_amendments_to_bbc_march.pdf. Moreover, the government will consider further expansion of Ofcom's powers when the renewal of the BBC's Charter is considered in 2005–2006; 400 HC Official Report (6th series) col 212, 25 February 2003.

REGULATORY CONVERGENCE OR MEDIA COMMODIFICATION?

A review of the history of the communications industries in the UK indicates that the regulatory model governing a medium of communication is determined, in large part, by whether the Market Liberal or Social Liberal school was in ascendancy at the time that medium first attracted the attention of policy makers. The regulatory model governing the press reflects the assumptions of early liberal theory and cemented itself into place when *laissez faire* economic and social theory was at its height: broadcasting emerged during the Social Liberal backlash against the excesses of marketplace ideology; the new media became important during the neo-liberal revival of the 1980s and 1990s. Once established, regulatory models are difficult to dislodge, surviving subsequent shifts in the fortunes of the political theories underlying them. Thus, the Market Liberal model for press regulation has accommodated the Social Liberal critique by adopting systems of self-regulation; the PSB model has, until now, withstood commercial challenges without losing its core identity. Sometimes the arguments used to justify the status quo will shift: the disappearance of the scarcity rationale, for example, has forced defenders of PSB to focus more on arguments derived from the perceived power of broadcasting media or the alternatives 'neutral' broadcasters provide to the partisan reporting of the press.[179] But while underlying rationales may be reassessed, the basic regulatory framework remains. No doubt there are many reasons for the persistence of regulatory models. Simple inertia probably explains a lot: once a system is established, it is easier to make modifications at the margins than to discard the template and start over with a structure founded on different theoretical assumptions. The status quo is reinforced by insulated and entrenched bureaucratic structures, which typically are resistant to change. Moreover, while the different schools of political thought have enjoyed pre-eminence at different points in time, neither has ever enjoyed unchallenged hegemony, and any effort to reform existing regulatory structures in a truly fundamental way (as opposed to attempting reforms at the periphery) will attract a discouraging level of opposition.

Does the Act represent a break in this historical pattern? At first blush, it might seem that it does not. As the discussion above demonstrates, the influence of both Market Liberal and Social Liberal thought can be discerned in various provisions of the Act, and a dichotomous approach to regulation will likely persist for the foreseeable future under the new regime. While the Act, on the whole, seems to propel media policy toward a more market-orientated approach and away from the moderating influence of Social Liberal ideas, the sheer scope of the Act is evidence that the regulatory environment of electronic media will not resemble the paradigm for print media anytime soon.

Nonetheless, Lord Currie of Marylebone, the economist and former business school dean named the first chairman of Ofcom, has suggested that 'we are creating a new regulator, with a new culture and a new approach to effecting regulation'.[180] Indeed, it is possible that the decision to create a single regulator whose authority will extend to the regulation of all aspects – economic, technical and socio-political – of telecommunications, television, radio and the new media could mark a sea change in communications regulation. Historically, the UK has been hostile to the idea that a single body should establish and pursue a single set of strategic goals for

---

179. See, eg, Communications White Paper, above n 2, paras 6.6.1–6.6.2.
180. 646 HL Official Report (5th series) col 684, 25 March 2003.

communications industries, favouring instead a 'fragmented and compartmentalised' approach to policy making.[181] However inefficient this may seem, there are advantages to dispersing policy-making power across different regulatory bodies. Gibbons argued that with a single regulator disagreements over regulatory objectives could 'become a matter of office politics rather than democratic debate', while having separate regulators helps make the inevitable conflicts between telecommunications and broadcasting specialists, for example, more transparent.[182] It also may be easier for a smaller regulator with a comparatively limited brief to accomplish its goals than it is for a larger regulator pursuing a more burdensome set of competing objectives. Perhaps most importantly, a system of multiple regulators and dispersed decision making reduces the risk that social and cultural goals will take a back seat to the economic imperatives of the industry being regulated. By bringing under one roof regulatory traditions for different communications sectors that are often at odds with one another, there is a danger of regulatory incoherence or, alternatively, the domination of one tradition over the other.

It is not inevitable that replacing multiple regulators of various traditions with Ofcom will alter the fundamental approach taken in regulating the distinct sectors of the communications industry. For one thing, the government's refusal to give Ofcom regulatory power over the press or the Internet will militate against full regulatory convergence – that is, the emergence of a homogenous approach to all communications sectors. Further, a smaller-scale regulatory merger of the sort mandated by the Act has occurred before, with no significant change in the regulatory environment of the industry sectors affected: after the Independent Broadcasting Authority and Cable Authority were merged in 1990 to form the ITC, the regulation of commercial terrestrial broadcasting retained its Social Liberal orientation, while the regulation of cable and satellite continued to reflect the more *laissez faire* approach of the Cable Authority. In addition, while the government has attempted to avoid strict divisions between the regulation of communications below-structures and content providers, or between economic and 'cultural and content' regulation, the Act's provisions on the ownership and management of communications below-structures are clearly separated from provisions on regulating content, and this division ultimately may be reflected in the formal or informal internal organisation of Ofcom. If there are distinct divisions or units for broadcast and telecommunications regulation, or for below-structure and content regulation, the distinct regulatory approaches that can be observed today may remain largely unchanged. For example, the Act itself requires Ofcom to establish a subcommittee, called the Content Board, which will have significant influence in connection with the regulation of the content of television and radio broadcasts transmitted through terrestrial, cable and satellite platforms;[183] a prominent role for a largely

---

181. See Seymour-Ure, above n 38, p 232; J Tunstall *The Media in Britain* (London: Constable, 1983) p 238. Seymour-Ure observed that British policies concerning the media have been 'generally uncoordinated, reactive, expediential, partial and indirect; a matter of broad objectives and attitudes ("freedom of the press", "an independent British film industry", "public service broadcasting", "deregulation") rather than of detailed programmes and plans'; Seymour-Ure, ibid, p 228.
182. See Gibbons, above n 118, p 304.
183. Communications Act 2003, ss 12–13.

independent Content Board may lead to a bifurcation of content and economic regulation within Ofcom.[184]

But if regulatory convergence in fact occurs – if experts pulled together from different regulatory cultures and professional backgrounds actually exchange ideas and learn to look at problems in new ways and are not isolated from one another within Ofcom – a new regulatory environment, and a 'new breed' of regulators, could emerge. If so, it is likely to be an environment more sympathetic to the assumptions of Market Liberalism than Social Liberalism. Experience indicates that regulatory models are shaped by the assumptions of the dominant philosophical school at the time the model first comes into existence. For much of the past generation neo-liberalism has enjoyed, both nationally and globally, something of a Renaissance, notwithstanding occasional setbacks. Taking a broad perspective of the past 20 years, these setbacks have tended to be temporary, probably because a coherent and persuasive alternative to market-based regulation has rarely been offered by the UK's leading political parties. New Labour's Third Way ideology, the centre-left's response to the pressures of economic and social globalisation, offers little in resistance to these liberalising trends. Hutton has argued that because New Labour has failed to embrace a coherent economic theory that combines a non-Marxist critique of capitalism with a workable alternative economic and social programme, it is prone to associate reform with market imperatives because 'there is no intellectual framework that can offer any contrary ballast'.[185] The tendency is to equate the public interest with 'the interplay of private interests in a free market' because Third Way ideology offers 'no other reference point'.[186]

Consistent with this analysis, the Act provides many signs of movement in a Market Liberal direction. First, it is worth noting that Ofcom's very existence is the consequence of economic considerations. The primary reason given for Ofcom's creation was that there was a 'real risk' that the existing regulatory environment was too complex to respond effectively to rapid technological and economic changes, failing to 'meet the needs' of communications industries.[187] A single regulator was deemed necessary to assure that 'consumer interests' are served 'in the current and future markets'.[188] Ofcom's *raison d'être* is to help unleash the economic potential of convergence and promote 'a dynamic, growing consumer market' for the information and entertainment services converging technologies can provide.[189] The government could not have expressed the importance of communications industries to the UK's economy more graphically than when it observed that more money is spent on communications services 'than is spent on beer'.[190] And it was revealing when Ofcom's first chairman described 'plurality, impartiality, high quality, diversity, and

---

184. See, eg, 646 HL Official Report (5th series) col 683, 25 March 2003 (Lord Currie states that the Content Board 'will manage the high-profile content issues within a strategic context set by the main Ofcom board, so that the main board will not be diverted from its central role as a strategic competition authority for the communications sector').
185. Hutton, above n 14, p 99.
186. Ibid.
187. See, eg, Communications White Paper, above n 2, para 8.2.1; Draft Regulatory Impact Assessment, para E3, available at http://www.communicationsbill.gov.uk/pdf/RIA.pdf.
188. Communications White Paper, ibid, para 8.2.3.
189. Ibid, para 1.2.2.
190. Ibid, para 1.1.18.

effective support for democratic discourse' as 'critical outputs'.[191] In light of this background, it would be surprising if market considerations did not take precedence over, and sometimes displace, other regulatory objectives in Ofcom's decision making.

This conclusion is reinforced by the government's repeated insistence that regulation must 'be kept at the minimum necessary level to deliver our goals for consumers and society';[192] that greater emphasis should be placed on industry self-regulation, even in connection with the fulfilment of PSB remits; that the Act is essentially deregulatory; and that Ofcom must 'roll back regulation promptly . . . when the market is sufficiently strong that competition will flourish on its own accord'.[193] One of the justifications the government offers for retaining PSB is that it helps drive the domestic programme production market,[194] and the Act excuses failure to meet certain PSB obligations if that failure is due to 'economic or market conditions'.[195] While, for now, the government accepts that reliance on general competition law alone will not be sufficient to guarantee plurality and diversity in commercial broadcasting,[196] this conviction is explained in quasi-market language that cites the under-provision of 'merit goods' like plurality by 'market mechanisms'.[197] The government insists that the system of self-regulation introduced for the qualitative aspects of PSB does not signify a wavering commitment to PSB ideals, but its White Paper openly acknowledged the possibility that, sometime after the digital switchover expected by the end of the decade, 'competitive pressure' on advertising revenue may adversely affect the ability of the commercial sector 'to deliver all the current public service broadcasting obligations in the same way as at present', and the new regulatory framework is designed to 'leave flexibility to review' how PSB obligations will be fulfilled in the future.[198]

Two themes in particular run through the documents setting out the government's communications policy that challenge the long-term survival of the Social Liberal influence on media regulation. The first is a tendency to justify PSB, not by reference to its philosophical roots, but by reference to audience expectations – the notion that PSB should be retained because PSB is what audiences have come to expect.[199] But reliance on audience expectations not only makes PSB vulnerable to changes in those expectations, it exposes a certain lack of faith in the continued persuasiveness of the socio-political belief system that gave birth to PSB in the first place. The second theme casting doubt on the commitment of policy makers to that belief system is the disproportionate weight given to 'market failure' as the basis for special rules for the communications industry. For instance, the Davies Committee, convened by the

---

191. 649 HL Official Reports (5th series) col 17, 23 June 2003 (speech by Lord Currie of Marylebone).
192. Communications White Paper, above n 2, para 8.11.2.
193. Draft Regulatory Impact Assessment, above n 187, para 25.
194. See Communications White Paper, above n 2, para 5.3.3.
195. Communications Act 2003, s 270(2) (concerning Tier 3 regulation).
196. See Communications White Paper, above n 2, paras 4.2.1–4.2.8.
197. See Draft Regulatory Impact Assessment, above n 187, para D1.
198. Communications White Paper, above n 2, para 5.6.10.
199. See, eg, Communications Act 2003, s 319(4) (factors to be considered in setting or revising standards objectives); Communications White Paper, above n 2, para 6.3.1 ('viewers and listeners have different expectations about acceptability of content provided to them in different ways or circumstances'); ibid, para 6.4.1 (stressing the need for Ofcom to conduct research to keep abreast of audience expectations and attitudes).

Labour Government to re-examine the funding of the BBC, concluded that the needs of British audiences made 'a strong case for a comprehensive public service broadcaster like the BBC', but, in doing so, declared that 'some form of market failure must lie at the heart of any concept of public service broadcasting'.[200] The Davies Report indicated that an 'expensive organisation dedicated to public service television' could only be justified if 'a combination of the private sector's profit motive, plus regulation, is insufficient to repair the market failure'.[201] Similarly, the Regulatory Impact Assessment prepared for the draft Bill emphasised that 'in a fully competitive market there should be no need for regulation, as the market would regulate itself'.[202] While perceptions of market failure certainly spurred the early development of the Social Liberal argument, the Social Liberal model of regulation has come to incorporate deeper philosophical considerations – assumptions about power, rights and the proper role of the state – that go far beyond immediate questions about the competitiveness of markets.

This weak conception of the philosophical foundations of PSB gives rise to a related problem: it does not provide a counterweight to the natural tendency of policy makers to give way to the perceived economic exigencies of the day. The immediate economic concerns of an industry are frequently thought to be more urgent and, in the short term, more compelling than most non-economic considerations. In part, this may be because it is possible to quantify economic phenomena and explain them in well-developed models of cause and effect, lending economics an aura (or, perhaps, illusion) of objectivity. It is relatively easier to formulate a policy response to economic phenomena than it is to address the more ambiguous social and cultural implications of mass communications. The 'box-ticking' associated with the previous regulatory regime, so lamented by the government, might have provided some insulation for PSB values against the pressures of perceived market imperatives; but 'box-ticking' apparently is to be abandoned by Ofcom in favour of a more 'flexible' approach to decision making.

Born and Prosser argue that the normative basis for PSB in the UK is, ultimately, 'of uncertain legal status':[203]

> 'In other European nations the concept of public service, especially that of PSB, has entered fully into constitutional culture and expectations . . . It does not mean, of course, that there is a core and universally agreed conception of public service; this continues to be the subject of heated debate, like many other constitutional concepts. It does mean, however, that there is a definitive source of principle to which appeal can be made in the search for the meaning of the concept, and one that stands above everyday politics. By contrast, in the UK the key characteristic has been the constitutional and judicial exclusion of the audio-visual field.'[204]

PSB, the authors conclude, is an example of 'soft' law[205] – in that its content is insufficiently precise, its scope is uncertain and it consists of divergent norms that do

---

200. Davies Report, above n 148, at 10.
201. Ibid. See also Communications White Paper, above n 2, para 7.4.1; Communications Green Paper, above n 147, paras 4.38–4.40.
202. Draft Regulatory Impact Assessment, above n 187, para E4.
203. Born and Prosser, above n 25, at 664.
204. Ibid, at 663.
205. A concept borrowed from public international law. See KC Wellens and GM Borchardt 'Soft law in European Community law' (1989) 14 European L Rev 267.

not always create rights and obligations in a strictly legal sense – and this conclusion is only reinforced by the decision to leave the realisation of many core PSB principles to a system of self-regulation. A vague conception of PSB without clear legal foundation may be acceptable in times of broad cultural consensus about such broadcasting, but whatever consensus did exist before technological convergence has broken down in the past decade.[206] In this milieu, 'soft' values like the cultural and citizenship purposes that are the bases for PSB are easily subordinated to the 'hard' values underlying competition and fair-trading law, for example.[207] Merging multiple regulators with divergent regulatory traditions may dilute the influence of the tradition that has promoted PSB, and the 'soft' values of PSB may be neglected in favour of more immediately tangible economic concerns.

CONCLUSION

The conventional explanations for the divergence of regulatory approaches in the communications industry refer to the technological or economic characteristics of different sectors of that industry. The thesis advanced here is that this divergence owes more to the shifts and tensions within twentieth-century liberal thought, and particularly the historical accident of which school of liberalism was in fashion at the time the medium emerged as a socially important force. The philosophical assumptions underlying regulatory models are not so much a response to technological or economic 'reality', but rather the lens that shapes perceptions of reality itself at the time those models were established. Once established, however, regulatory paradigms have proven resilient, withstanding challenges resulting from the changing fortunes of the political philosophies that gave rise to them.

The central ideological struggle underlying political conflicts in many Western countries in the twentieth century has been that between Market and Social Liberalism. Inevitably, this struggle has made itself felt in the debates over one of the most important social and political developments of the twentieth century, the emergence of truly *mass* media communications. It may well be that the tension between Market and Social Liberalism will continue to influence media policy well into the twenty-first century; certainly, the nature of the debates over the Act is evidence that these tensions are still very much present. But the creation of Ofcom may mark the beginning of a new regulatory paradigm, one that reflects the hazy ideological groundings of Third Way politics. The Third Way is, in essence, an attempt to reconcile – or converge – the competing schools of liberalism. It is, in many ways, a reflection of the insecurity of progressive politics in Britain, battered as it has been by the seemingly unappeasable forces of social and economic globalisation. Hutton observes that Third Way politics is an effort to redefine social democracy by dumping the baggage of its historic socialist component and reaching an 'intellectual accommodation and political alliance with [market] liberalism to form a stronger counterbalance to contemporary conservatism'.[208] But it also seems to lack faith in the philosophical alternatives to marketplace ideology and, because of this, accommodation can start to look like capitulation.

---

206. Born and Prosser, above n 25, at 663.
207. Ibid, at 659–660.
208. Hutton, above n 14. p 98.

In the context of media policy, the Third Way may be more retrogressive than progressive. PSB is Social Liberalism's most enduring contribution to British media policy. It is based on a positive conception of freedom of expression, reflects an ethic of obligation, recognises that the excesses of corporate activities can be just as dangerous as the abuse of state power and is suspicious of the rampant consumerism informing so much of contemporary policy making. The Third Way, however, seems to offer a miserly interpretation of PSB that relies almost entirely on perceptions of market failure and consumer expectations. New Labour proposes the creation of a single regulatory body that, given the circumstances of its creation, is more likely to be sensitive to the pressures of short-term industry interests than the socio-political interests that traditionally have informed PSB. By converging regulatory structures for the communications industry, the long-term survival of PSB will be endangered, and the continued relevance of Social Liberalism to media in the twenty-first century is put in doubt.

Part II
Techniques of Regulation

[3]

Television and the Public Interest

Cass R. Sunstein†

TABLE OF CONTENTS

Introduction	501
I. History, Puzzles, Problems	506
A. A Brief Historical Overview	506
B. Two Puzzles	508
C. Identifying the Problem	509
II. Preferences and Audiences	511
A. Three Market Failures	514
1. Eyeballs as the Commodity	514
2. Informational Cascades and Broadcaster Homogeneity	515
3. Externalities and Collective Action Problems	516
B. Problems on Nonmarket Criteria: Children, Deliberative Democracy, and Related Issues	518
1. Children and the Hearing Impaired	518
2. Balkanization	519

---

† Karl N. Llewellyn Distinguished Service Professor, University of Chicago, Law School and Department of Political Science. From 1997 to 1998, the author served on the Advisory Committee on the Public Service Obligations of Digital Television Broadcasters, which produced a report in December 1998 on which this Article draws. *See* ADVISORY COMMITTEE ON PUBLIC INTEREST OBLIGATIONS OF DIGITAL TELEVISION BROADCASTERS, CHARTING THE DIGITAL BROADCASTING FUTURE: FINAL REPORT OF THE ADVISORY COMMITTEE ON PUBLIC INTEREST OBLIGATIONS OF DIGITAL TELEVISION BROADCASTERS (1998). Many of the arguments in this Article were framed through the opportunity to see, through work on the Advisory Committee, two groups close up. The first was a large number of broadcasters—most of them public-spirited, most of them extremely nervous about growing competition from cable and elsewhere, most of them sharply opposed to government mandates, few of them unwilling to accept governmental help. The second was the National Association of Broadcasters (NAB)—at the time an extraordinarily defensive, fearful, uncooperative, aggressive, and self-protective organization, consisting of many honorable people, but often unwilling to compromise or even to reflect. In this way, the NAB was reminiscent of the tobacco industry in, say, the 1960s, and was enthusiastic about using the First Amendment in the same way that the National Rifle Association uses the Second Amendment, that is, as an all-purpose shield against any action adverse to their interests. The author is grateful to other members of the Advisory Committee for many helpful discussions of these problems and, in particular, to the two co-chairs of the Committee, Norman Ornstein and Leslie Moonves. The Committee's final report has influenced the treatment here, not least when there are disagreements. He is also grateful, for helpful comments, to Douglas Lichtman, Eric Posner, and Richard Posner, and to participants in a conference held in honor of Jürgen Habermas in Frankfurt, Germany in July 1999, especially Clause Offe, Bernhard Peters, and Jürgen Habermas.

3. Citizens, Consumers, and Precommitment
   Strategies 520
4. Endogenous Preferences 521
5. Paternalism? Elitism? 522
6. Constitutional Notes 524

III. Principle, Policy, Technology 525
   A. Practice 525
   B. Communications Past, Present, and Future: Planned
      Obsolescence and Beyond 526
      1. Predicting the Future 527
      2. Regulatory Options and Technological Change 529

IV. Disclosure 531
   A. Precursors 531
   B. Rationale 533
   C. The Minimal Proposal 535
   D. Of Realism and Ineffectiveness 536

V. Economic Incentives 538
   A. Of Nature and Coase 538
   B. Taxes, Public Bads, Hot Potatoes, and Cold Spots 539
   C. Economic Incentives and the Constitution 542
   D. Expanding the Viewscreen: A Glance at the Cathedral 543
   E. A Brief Note on Cultural Policies and
      Cultural Subsidies 549

VI. Voluntary Self-Regulation: Aspirations, Trustees, and
   "Winner-Take-Less" Codes 549
   A. The Problem and a Recently Emerging Strategy 551
   B. A Code: Sample Provisions 553
   C. A Code: Problems and Prospects 555
   D. Notes on the First Amendment and Antitrust Law 557
   F. Less Puzzling Puzzles 558

VII. A Summary 559
Conclusion 563

Television and the Public Interest

Cass R. Sunstein

The communications revolution has thrown into question the value of imposing public interest obligations on television broadcasters. But the distinctive nature of this unusual market—with "winner-take-all" features, with viewers as a commodity, with pervasive externalities from private choices, and with market effects on preferences as well as the other way around—justifies a continuing role for government regulation in the public interest. At the same time, regulation best takes the form, not of anachronistic command-and-control regulation, but of (1) disclosure requirements, (2) economic incentives ("pay or play"), and (3) voluntary self-regulation through a privately administered code. Some discussion is devoted to free speech and antitrust issues, and to the different possible shapes of liability and property rules in this context, treating certain programming as a public "good" akin to pollution as a public bad.

INTRODUCTION

There is a large difference between the public interest and what interests the public. This is so especially in light of the character and consequences of the communications market. One of the central goals of the system of broadcasting, private as well as public, should be to promote the American aspiration to deliberative democracy,[1] a system in which citizens are informed about public issues and able to make judgments on the basis of reasons. Both norms and law should be enlisted in this endeavor; if one fails, the other becomes all the more important. These are the claims that I attempt to bring to bear on the so-called communications revolution.

This revolution has been driven by extraordinary technological change.[2] The rise of cable television, the Internet, satellite television, direcTV, and digital television has confounded ordinary understandings of "television."[3] Before long, digital television may enable viewers to choose

---

1. *See, e.g.*, AMY GUTMANN & DENNIS THOMPSON, DEMOCRACY AND DISAGREEMENT 52-94 (1996) (discussing ideals of deliberative democracy); Joshua Cohen, *Democracy and Liberty*, in DELIBERATIVE DEMOCRACY 185, 185-231 (Jon Elster ed., 1998) (discussing foundations of deliberative democracy).

2. A valuable general discussion can be found in ANDREW L. SHAPIRO, THE CONTROL REVOLUTION (1999).

3. *See* BRUCE M. OWEN, THE INTERNET CHALLENGE TO TELEVISION 311-26 (1999) (dealing with other dramatic technological developments).

among over a thousand programs.[4] The possible combination of television and the Internet, a combination now in its early stages, may prove an equally dramatic development; the fact that the Internet is a partial substitute for television has already introduced a measure of competition between the two.[5]

Law has responded to these developments in fits and starts, largely by attempting to engraft legal requirements designed for the old environment onto an altogether new communications market. The result is a high degree of anachronism, misfit, and drift, and in the view of many observers, a series of constitutional violations.[6] Most of the modern debate involves a vigorous but increasingly tired contest between those defending the old regulatory order[7] and those urging rapid movement toward "simple rules" for government control of television, above all well-defined property rights and freedom of contract.[8] Strikingly similar debates, about the value of "simple rules," the place of regulatory safeguards, and the role of television in a democracy, can be found in many nations.[9]

My aim in this Article is to discuss an important part of the intersection between the emerging communications market and law: public interest obligations imposed on television broadcasters.[10] Since the initial rise of broadcasting in the United States, government has treated the license as a kind of "grant" that is legitimately accompanied by duties.[11] Congress and the FCC have required broadcasters to follow a range of requirements—a form of old-style "command-and-control" regulation, growing out of an

---

4. *See* Lawrie Mifflin, *As Band of Channels Grows, Niche Programs Will Boom*, N. Y. TIMES, Dec. 28, 1998, at A1.

5. *See* OWEN, *supra* note 3, at 311-26. *But see* David Goldberg et al., *Conclusions, in* REGULATING THE CHANGING MEDIA: A COMPARATIVE STUDY 295, 297 (David Goldberg et al. eds., 1998) ("[C]onvergence has been unequal between different nations, has been slower than expected and will not break down distinctions between different markets as rapidly as is claimed.").

6. *See* RATIONALES & RATIONALIZATIONS: REGULATING THE ELECTRONIC MEDIA (Robert Corn-Revere ed., 1997); Thomas G. Krattenmaker & L.A. Powe, Jr., *Converging First Amendment Principles for Converging Communications Media*, 104 YALE L.J. 1719, 1725 (1995); Mark S. Fowler & Daniel L. Brenner, *A Marketplace Approach to Broadcast Regulation*, 60 TEX. L. REV. 207, 209-10 (1982).

7. This seems to me the general thrust of LEE C. BOLLINGER, IMAGES OF A FREE PRESS (1991) and OWEN M. FISS, THE IRONY OF FREE SPEECH (1996).

8. *See* RICHARD A. EPSTEIN, SIMPLE RULES FOR A COMPLEX WORLD 275-306 (1995) (discussing environmental protection and presenting a general account of the role of government in a way that is easily adapted to the area of communications); RATIONALES AND RATIONALIZATIONS, *supra* note 6 (offering a number of essays challenging any role for government aside from the definition and enforcement of property rights).

9. *See* MONROE E. PRICE, TELEVISION: THE PUBLIC SPHERE AND NATIONAL IDENTITY (1995); Goldberg et al., *supra* note 5.

10. I use this term to refer to literal broadcasters, and thus not to include cable providers. As we will see, however, the distinction between the two seems increasingly (though not yet entirely) artificial, and much of the discussion will bear on the appropriate regulatory stance toward television in general.

11. *See* discussion *infra* Part I.A.

understanding that there would be three, and only three, private broadcasting stations. Much, though far from all, of this regulation was eliminated in the 1980s.[12] A large question is the extent to which public interest requirements continue to make sense, or even to survive constitutional scrutiny, in an entirely different communications market where broadcasters occupy a decreasingly distinctive position.

The question was posed starkly with the enactment of the Telecommunications Act of 1996,[13] one of whose central concerns involved the rise of digital television. The Act had to deal with two issues. First, who would have the right to broadcast digital television? Should the licenses be sold, or auctioned, or given outright to existing broadcasters? Second, what public interest obligations, if any, should attach to the ownership of a right to broadcast digital television? The Act squarely answered the first question,[14] but was inconclusive on the second. In an extremely controversial step, Congress did not sell or auction the right to broadcast digital television, but basically gave the right to existing broadcasters for free.[15] This has been described, and reasonably so, as a "huge giveaway" of "a $70 billion national asset."[16] At the same time, Congress refused to eliminate public interest obligations, delegating to the FCC the power to decide whether such obligations should be imposed on digital television broadcasters, and if so, in what form.[17] The FCC has not yet made that decision or even commenced formal proceedings.

In this Article, I offer two basic claims, one involving ends, the other involving means. The first is that in view of the character and consequences of television programming, any system for the regulation of television should be evaluated in democratic as well as economic terms. The economic ideal of "consumer sovereignty" is ill-suited to the communications market. It follows that, at least in the near term, the changes introduced by the emerging communications system do not justify abandoning the idea that broadcasters should be required to promote public interest goals. Educational programming and programming that deals with civic questions can promote the aspiration to deliberative democracy; reliance

---

12. *See id.*
13. Pub. L. No. 104-104, 110 Stat. 56 (codified in scattered sections of 47 U.S.C. §§ 151-613 (Supp. III 1997)) (amending the Communications Act of 1934).
14. *See* 47 U.S.C. § 336(a).
15. *See id.*
16. *What Price Digital Television?*, N.Y. TIMES, Dec. 26, 1998, at A26; *see also Federal Management of the Radio Spectrum: Advanced Television Services: Hearing Before the Subcomm. on Telecomm. and Fin. of the House Comm. on Commerce*, 104th Cong. 82 (1996) (statement of Robert M. Pepper, Chief, Office of Plans and Policy, FCC) (stating that an auction would bring between $11 to $70 billion in revenue). For general criticism, see Thomas G. Krattenmaker, *The Telecommunications Act of 1996*, 29 CONN. L. REV. 123, 163-64 (1996), and Matthew Spitzer, *Dean Krattenmaker's Road Not Taken: The Political Economy of Broadcasting in the Telecommunications Act of 1996*, 29 CONN. L. REV. 353 (1996).
17. *See* 47 U.S.C. § 336(d).

on an unregulated market may not. There are also legitimate grounds for encouraging broadcasters to make programming accessible to people with disabilities, above all the hearing impaired. I emphasize in this connection some special characteristics of the broadcasting market, characteristics that make it hazardous to rely on "consumer sovereignty" as the exclusive basis for regulatory policy. Instead communications policy should be assessed, at least in part, by reference to its effects on the public sphere.[18]

My second claim is that in order to promote the relevant goals, government should decreasingly rely on command-and-control regulation,[19] and should consider instead three less intrusive and more flexible instruments, each of which is well-adapted to a period of rapid technological change. The instruments are: (1) mandatory public disclosure of information about public interest broadcasting, unaccompanied by content regulation; (2) economic incentives, above all subsidies and "play or pay"; and (3) voluntary self-regulation, as through a "code" of appropriate conduct, to be created and operated by the industry itself. These instruments have played an increasing role in regulatory policy in general, especially in the environmental arena.[20] But they have rarely been discussed in the area of communications, where they have a natural place;[21] and despite its growing importance, the general topic of industry self-regulation has received little academic attention.[22]

By requiring broadcasters to disclose information about their public interest activities, the government might be able to enlist public pressure and social norms so as to create a kind of competition to do more and better. This is the simplest and least intrusive of regulatory instruments. By allowing broadcasters to buy their way out of certain public interest obligations, the government should be able to ensure that those with an

---

18. See PRICE, supra note 9, at 194-246. See generally JÜRGEN HABERMAS, THE STRUCTURAL TRANSFORMATION OF THE PUBLIC SPHERE (Thomas Burger trans., 1989) (offering an extended historical discussion).

19. A question not addressed here is the content of any minimal requirements; I emphasize the more flexible alternatives as the instruments of choice, without denying the need for some minima as a "backstop." In the current system, for example, it may well make sense to require a degree of children's programming and also free air time for candidates. See ADVISORY COMMITTEE ON PUBLIC INTEREST OBLIGATIONS OF DIGITAL TELEVISION BROADCASTERS, CHARTING THE DIGITAL BROADCASTING FUTURE: FINAL REPORT OF THE ADVISORY COMMITTEE ON PUBLIC INTEREST OBLIGATIONS OF DIGITAL TELEVISION BROADCASTERS 45-64 (1998) [hereinafter FINAL REPORT]. The precise extent of mandatory programming is beyond the scope of the present discussion, though I do refer to mandates at several points below.

20. See generally NEIL GUNNINGHAM ET AL., SMART REGULATION: DESIGNING ENVIRONMENTAL POLICY 422-48 (1999); NATIONAL ACAD. OF PUB. ADMIN., THE ENVIRONMENT GOES TO MARKET 9-20 (1994); ANTHONY OGUS, REGULATION 121-49, 245-56 (1998).

21. The best discussion can be found in Angela J. Campbell, *Self-Regulation and the Media*, 51 FED. COMM. L.J. 711 (1999).

22. The principal exception can be found in an illuminating symposium issue in an Australian law review. See Symposium, *Special Issue on Self-Regulation*, 19 LAW & POL'Y 363 (1997) [hereinafter *Special Issue on Self-Regulation*].

incentive to produce good programming are actually doing so, while also producing the lowest-cost means of promoting public interest programming. And by encouraging (not mandating) voluntary self-regulation, the government can help overcome a kind of prisoner's dilemma faced by participants in a "winner-take-all" market,[23] a prisoner's dilemma that contributes to a range of social problems, often stemming from a kind of "race to the bottom" with respect to programming quality.

It should be clear that this basic approach combines a recognition of the serious limits of unrestrained communications markets in promoting social goals with a plea for rejecting traditional regulation and for enlisting more flexible, market-oriented instruments in the service of those goals. This approach is consistent with some incipient but quite general trends in regulatory law.[24] If the approach is sound, it is well-suited to the emerging communications market; but it is easily adapted to other areas as well, including environmental degradation, occupational safety and health, and other social problems. It is much too soon to say whether there is a "third way" between traditional command-and-control regulation and reliance on free trade and well-defined property rights.[25] But if there is indeed a "third way," it is likely to be found in proposals of this kind.

A general theme of this Article is that disclosure, economic incentives, and voluntary self-regulation might displace government command-and-control in a variety of areas of regulatory law. Specific themes include requiring producers simply to disclose goods and bads; relaxing antitrust law so as to permit cooperation designed to reduce some of the problems associated with "races to the bottom"; and building on emerging developments in environmental protection so as to allow far more imaginative "trades" among producers. In short, it is time to move beyond the view that market ordering and content regulation are the only two possibilities for communications law. There are many alternatives, and real progress can come only from exploring the choices among them.

The Article comes in seven parts. Part I sets the stage, outlining the history of regulation, identifying some relevant puzzles, and exploring some diverse problems with television in its current form. Part II, the theoretical heart of the Article, evaluates and rejects the claim that in the emerging media market there is no longer room for public interest regulation of any kind. I suggest that television is no ordinary commodity, partly

---

23. *See* ROBERT H. FRANK & PHILLIP J. COOK, THE WINNER-TAKE-ALL SOCIETY 189-209 (1995). The most important distinguishing feature of "winner-take-all" markets is that rewards are based on "relative rather than (or in addition to) absolute performance." *Id.* at 24. In such markets, "rewards tend to be concentrated in the hands of a few top performers." *Id.*

24. *See, e.g.*, GUNNINGHAM ET AL., *supra* note 20, at 37-91; NATIONAL ACAD. OF PUB. ADMIN., *supra* note 20; Cass R. Sunstein, *Informational Regulation and Informational Standing: Akins and Beyond*, 147 U. PA. L. REV. 613, 618-33 (1999).

25. For a general discussion, see ANTHONY GIDDENS, THE THIRD WAY (1998).

because of the collective benefits of good programming, partly because of the link between television and democracy, and partly because viewers are more like products offered to advertisers than consumers paying for entertainment on their own. Part III discusses the relation between principle and practice; it traces likely stages of the emerging market, with broadcast programming becoming increasingly like general-interest magazines. Part IV deals with disclosure, exploring the possibility that relevant private groups, invoking widespread social norms, can interact to produce improvements in the broadcasting market without compulsory programming of any kind. Part V deals with economic incentives, beginning with the idea of "play or pay," and then adapting some ideas from the law of tort to the law of broadcasting. Part VI examines whether a code of broadcasting might operate as a kind of positional arms control agreement, helping to counteract a situation in which broadcasters compete to the detriment of collective goals. Part VII is a brief summary of regulatory options.

I
HISTORY, PUZZLES, PROBLEMS

A. *A Brief Historical Overview*[26]

Broadcast licenses have never been treated like ordinary property rights, open for sale on the free market.[27] Since the initial enactment of the Communications Act of 1934, the government has awarded licenses to broadcasters in accordance with "convenience, public interest, [and] necessity."[28] The Federal Radio Commission early described the system as one in which broadcasters "must be operated as if owned by the public. . . . It is as if a community should own a station and turn it over to the best man in sight with this injunction: 'Manage this station in our interest. . . .'"[29] Under this "public trustee" standard, the FCC has imposed a range of obligations on broadcasters—an idea that perhaps made special sense when a small number of companies dominated the television market.

In its initial set of guidelines, the FCC required stations to meet the "tastes, needs, and desires of all substantial groups among the listening public . . . ."[30] This required "a well-rounded program, in which

---

26. This Section draws on the first section of the FINAL REPORT, *supra* note 19, at 3-16. See also Campbell, *supra* note 21, for detailed discussion of many relevant developments.
27. See the colorful, skeptical presentation in PETER HUBER, LAW AND DISORDER IN CYBERSPACE: ABOLISH THE FCC AND LET COMMON LAW RULE THE TELECOSM 3-9 (1997).
28. 47 U.S.C. § 303 (1994).
29. *The Federal Radio Commission and the Public Service Responsibility of Broadcast Licensees*, 11 FED. COMM. BAR J. 5, 14 (1950) (quoting *Schaeffer Radio Co.*, an unpublished 1930 Federal Radio Commission decision).
30. Great Lakes Broadcasting Co., 3 Fed. Radio Comm'n, Ann. Rep. 32, 34 (1929), *modified on other grounds*, 37 F.2d 993 (D.C. Cir. 1930).

entertainment, consisting of music of both classical and lighter grades, religion, education and instruction, important public events, discussions of public questions, weather, market reports, and news, and matters of interest to all members of the family, find a place."[31]

Often this kind of guidance operated as a general plea, with little systematic enforcement. In 1960, however, the FCC went so far as to outline fourteen of the "major elements usually necessary to meet the public interest."[32] These included: religious programming, programs for children, political broadcasts, news programs, sports programs, weather and market services, and development and use of local talent.[33] The FCC eventually specified its general guidelines, which were merely indicia of the types and areas of appropriate service. The specifications included minimum amounts for news, public affairs, and other nonentertainment programming, including the controversial "fairness doctrine"[34] and also access rules for prime-time.[35]

Substantial changes occurred in the 1980s, a period of significant deregulation. The head of the FCC, Mark Fowler, declared (in a kind of soundbite, or bumper-sticker, for the market approach to the topic) that television is "'just another appliance,'" a "'toaster with pictures.'"[36] The fairness doctrine was largely eliminated, and many of the more particular public interest requirements were removed.[37] Nonetheless, a number of such requirements remain. For example, the FCC continues to say that if a broadcaster sells airtime to one candidate, it must sell similar time to opposing candidates as well. Congress itself has codified a right of this kind.[38] A long-standing statutory provision, in the obvious self-interest of law makers, requires that if a broadcaster offers to sell time, it must do so at the "lowest unit rate of the station" during the forty-five days before a

---

31. *Id.*
32. Report and Statement of Policy Res: Commission en banc Programming Inquiry, 44 F.C.C. 2303, 2314 (1960).
33. *See id.*
34. In brief, the fairness doctrine required broadcasters to attend to public issues and to ensure a diversity of views. On the doctrine, see discussion *infra* Part III.A.
35. *See* Amendment to Section 0.281 of the Commission's Rules: Delegations of Authority to the Chief, Broadcast Bureau, 59 F.C.C.2d 491, 493 (1976); Amendment of Part 73 of the Commission's Rules and Regulations with Respect to Competition and Responsibility in Network Television Broadcasting, 23 F.C.C.2d 382, 385-88 (1970).
36. Bernard D. Nossiter, *Licenses to Coin Money: The F.C.C.'s Big Giveaway Show*, 241 NATION 402 (1985) (quoting radio address given by Mark Fowler).
37. *See* The Revision of Programming and Commercialization Policies, Ascertainment Requirements, and Program Log Requirements for Commercial Television Stations, 98 F.C.C.2d 1076 (1984); *see also* Revision of Applications for Renewals, 49 R.R.2d 470 (1981) (postcard renewals); Syracuse Peace Council v. Television Station WTVH, 2 F.C.C. Rcd. 5043, 5054-55 (1987) (repealing most of fairness doctrine), *aff'd sub nom.* Syracuse Peace Council v. FCC, 867 F.2d 654 (D.C. Cir. 1989).
38. *See* 47 U.S.C. § 315 (1994).

primary election and during the sixty days before a general or special election.[39]

In recent years Congress has devoted special attention to children's programming and to television access for the hearing impaired. In 1990, Congress enacted the Children's Television Act of 1990,[40] limiting the advertising on children's programming (twelve minutes per hour during weekdays and ten and a half minutes per hour on weekends).[41] Under this statute, the FCC has further required broadcasters to provide three hours of children's programming per week.[42] The Television Decoder Circuitry Act of 1990[43] requires new television sets to have special decoder chips, allowing them to display closed-captioned television transmissions for the hearing impaired. The Telecommunications Act of 1996[44] requires use of "v-chip" technology, designed to facilitate parental control over what enters the home; it also contains ancillary requirements intended to ensure "ratings" of programming content.[45]

B. Two Puzzles

Turn now to the present, or at least to the more recent past. From 1997 to 1998, a presidential advisory committee met to discuss the public interest obligation of television broadcasters.[46] Several of the broadcasters on the Committee were quite skeptical about governmental mandates, but highly receptive to the idea of adopting some kind of broadcasting "code," akin to the kind approved and administered by the National Association of Broadcasters (NAB) between 1928 and 1979.[47] The Committee eventually moved toward endorsing the notion of a code, and the idea received considerable attention in the trade press.[48]

In its annual meeting, however, the NAB signaled skepticism about the idea and came very close to saying "no" and "never." A large part of the broadcasters' objection was that any "code" would violate the antitrust laws. This was very odd because in their discussions, members of the NAB

---

39. *Id.* § 315(b).
40. Pub. L. No. 101-437, 104 Stat. 996 (codified in scattered sections of 47 U.S.C.).
41. *See* 47 U.S.C. § 303(a) (1994).
42. *See* Broadcast Services: Children's Television, 61 C.F.R. § 43981, 43988 (1996).
43. 47 U.S.C. § 303(u).
44. Pub. L. No. 104-104, 110 Stat. 56 (codified in scattered sections of 47 U.S.C. §§ 151-613 (Supp. III 1997)).
45. See the general discussion in JAMES T. HAMILTON, CHANNELING VIOLENCE: THE ECONOMIC MARKET FOR VIOLENT TELEVISION PROGRAMMING 302-11 (1998).
46. I draw on personal recollections here. *See* FINAL REPORT, *supra* note 19. Transcripts of the relevant meetings can be found on the web site of the Advisory Committee on the Public Interest Obligations of Digital Television Broadcasters at <http://www.ntia.doc.gov/pubintadvcom/piacreport.pdf>.
47. *See* discussion *infra* Part V.B; *see also* Campbell, *supra* note 21, at 720-35.
48. *See, e.g., Gore Proposals Go to White House*, TELEVISION DIGEST, Dec. 21, 1998.; *Gore Recommendations on Digital Standards Go to White House*, COMM. DAILY, Dec. 21, 1998.

treated the possibility of an antitrust violation as extremely good news. Take this as the first puzzle; it is not often that high-level corporate officials are smiling when they discuss the possibility that their own action would be found unlawful.

Consider a second puzzle. During the committee's deliberations, some people argued on behalf of a "play or pay" system, in which broadcasters would be relieved of public interest obligations (to "play") if they agree instead to "pay" someone else—another broadcaster—to do so. But many of the broadcasters on the committee were quite skeptical of this approach, arguing that public interest obligations were part of the (sacred?) duty of every broadcaster, and that no one should be exempted for a price. This was also very odd. It is not often that high-level corporate officials prefer rigid government mandates to more flexible approaches. What explains these puzzles? The answers—offered in closing here[49]—reveal a great deal about the emerging market.

C. Identifying the Problem

To evaluate particular proposals, it is necessary to have a concrete sense of why some people think that even well-functioning television markets are inadequate. Consider the following possibilities, each of which has produced public concern in the last decade:[50]

1. There may be insufficient educational programming for children. The existing fare may be insufficient because there is too little simply in terms of amount (for example, for people who lack cable), or because children do not watch the stations on which it is available, or because the quality is too low.

2. Programming may not be sufficiently accessible to people who are hearing impaired; this may be a particular problem if citizens are unable to find out about emergencies, or if they are unable to understand programming that bears on central public issues.[51] The exclusion may have practical consequences; it may even produce a form of humiliation.[52]

3. At least on the major networks, programming may be too homogenous, in a form of "blind-leading-the-blind" programming. Since a significant percentage of Americans do not receive

---

49. *See infra* Part IV.F.
50. For different angles, see, for example, FINAL REPORT, *supra* note 19; HAMILTON, *supra* note 45, at 3-50 (discussing television violence as a public policy issue); NEWTON N. MINOW & CRAIG L. LAMAY, ABANDONED IN THE WASTELAND: CHILDREN, TELEVISION, AND THE FIRST AMENDMENT 10-45 (1995) (arguing that the system serves the best interests of advertisers rather than children); DANNY SCHECHTER, THE MORE YOU WATCH, THE LESS YOU KNOW (1997) (examining the content of news).
51. *See* FINAL REPORT, *supra* note 19, at 30-31.
52. On the notion of humiliation, see AVISHAI MARGALIT, THE DECENT SOCIETY (Naomi Goldblum trans., 1996).

cable television and depend on broadcasters, the result may be insufficient variety in programming.[53]

4. Some programming may be affirmatively bad for children if, for example, it contains excessive violence, or otherwise encourages behavior that is dangerous to self and others. The result of such programming may be to produce violent or otherwise dangerous behavior in the real world.[54]

5. There may be too much violent programming in general, with adverse consequences for adults, not only children.[55] The adverse consequences may include an increase in violence (because of changes in social norms or "copycat" effects[56]), general demoralization and fear, or a misperception of reality.

6. News coverage may be a form of "infotainment," dealing not with real issues, but with gossip about celebrities and unsubstantiated charges of various kinds.[57]

7. There may be too little coverage of serious questions, especially during political campaigns. The relevant coverage may involve sensationalism and "sound bites," or attention to who is ahead ("horse-race issues") rather than who thinks what and why. The result may be an insufficiently informed citizenry.[58]

8. Stations too rarely cover international issues or developments in other nations. The result is that people are extremely ill-informed about the global background for national events, including proposed financial assistance and possibly even war, and also about practices other than their own. This ignorance makes it difficult

---

53. *See* discussion *infra* Part II.A.2.
54. See HAMILTON, *supra* note 45, at 20-30, for evidence.
55. *See id.* (providing a detailed discussion).
56. For evidence, see ELLIOT ARONSON, THE SOCIAL ANIMAL 62-64, 263-67 (6th ed. 1992).
57. *See generally* SCHECHTER, *supra* note 50.
58. See KIKU ADATTO, SOUND-BITE DEMOCRACY (1990), for general discussion; and JOHN DEWEY, THE PUBLIC AND ITS PROBLEMS: AN ESSAY IN POLITICAL INQUIRY 179-80 (Gateway Books 1946) (1927):

> A glance at the situation shows that the physical and external means of collecting information in regard to what is happening in the world have far outrun the intellectual phase of inquiry and organization of its results. Telegraph, telephone, and now the radio, cheap and quick mails ... have attained a remarkable development. But when we ask what sort of material is recorded and how it is organized, when we ask about the intellectual form in which the material is presented, the tale to be told is very different. "News" signifies something which has just happened, and which is new just because it deviates from the old and regular.... [W]e have here an explanation of the triviality and "sensational" quality of so much of what passes as news. The catastrophic, namely, crime, accidents, family rows, personal clashes and conflicts, are the most obvious forms of breaches of continuity ... they are the *new* par excellence....

for people to deliberate well about important questions and evaluate purely national practices.[59]

9. It may be too expensive for candidates to reach the electorate via television. The result may be excessive competition to accumnlate funds simply in order to have access to television; this competition may have corrosive effects on the electoral process. Free air time would be a possible response, perhaps qualified by an obligation, on the part of the candidate, to speak for at least fifty percent of the time or to refrain from negative campaigning.

10. There may be too little substantive diversity of view—too little debate among people with genuinely different perspectives about issues of policy and fact. Here too, the result may be an insufficiently informed citizenry.[60]

11. The problem may be not homogeneity but heterogeneity, which may result in a highly balkanized viewing public, in which many or most people lack shared viewing experiences, or in which people view programming that largely reinforces their own convictions and prejudices.[61] The result can be extremism and fragmentation.[62]

To be sure, some of these problems cannot be corrected through regulation that is either feasible or constitutional. Moreover, these various conceptions of the relevant problem point toward diverse solutions, some of which would raise serious First Amendment problems, as discussed below.[63] A particular challenge is to develop approaches that would allow a high degree of flexibility, minimize government involvement in programming content, and also do some good.

II

PREFERENCES AND AUDIENCES

It has increasingly been urged that any objections to existing television are elitist or outmoded.[64] On one view, public interest obligations have no place in modern law, particularly because the "scarcity" rationale for

---

59. *See generally* SUSAN D. MUELLER, LUMPUSSION FATIGUE (1999); MARTHA C. NUSSBAUM, CULTIVATING HUMANITY (1997).
60. *See* C. Edwin Baker, *The Media that Citizens Need,* 147 U. PA. L. REV. 317, 383-408 (1998).
61. See the discussion of oversteering and balkanization in SHAPIRO, *supra* note 2, at 105-32.
62. *See id.* See also the discussion of group polarization in PATRICIA WALLACE, THE PSYCHOLOGY OF THE INTERNET 73-78 (1999); David Schkade et al., *Are Juries More Erratic than Individuals?*, COLUM. L. REV. (forthcoming 2000); and Cass D. Snnstein, The Law of Group Polarization (Jan. 3, 2000) (unpublished manuscript) (on file with author).
63. *See* discussion *infra* Part II.B.6.
64. *See generally* HUBER, *supra* note 27; Fowler & Brenner, *supra* note 6.

regulation grows weaker every day;[65] it has even been urged that the FCC no longer has any appropriate role.[66] Once the problem of scarcity has been eliminated, individual consumers can design their own preferred communications package, at least if the government permits them to do so. Consider this utopian picture of a system of unrestricted markets in communications:

> There will be room enough for every sight and sound, every thought and expression that any human mind will ever wish to convey. It will be a place where young minds can wander in adventurous, irresponsible, ungenteel ways. It will contain not innocence but a sort of native gaiety, a buoyant, carefree feeling, filled with confidence in the future and an unquenchable sense of freedom and opportunity.[67]

A conceptual point first: though many people claim to argue for "deregulation," that route is not in fact an option, or at least not a reasonable one. What "deregulation" really means is a shift from the status quo to a system of different but emphatically legal regulation, more specifically one of property, tort, and contract rights, in which government does not impose specific public interest obligations but instead sets up initial entitlements and then permits trades among owners and producers. This is a regulatory system as much as any other. If it seems close to the current system for newspapers and magazines, it is no less a regulatory system for that reason; a great deal of law (inevitably) governs the rights and duties of newspapers and magazines.[68] Such law imposes rights and duties,

---

65. As noted below, the rationale has not disappeared; well over one-third of American households continue to depend on free, over-the-air broadcasting. *See* U.S. BUREAU OF THE CENSUS, STATISTICAL ABSTRACT OF THE UNITED STATES: 1998 at 573 (118th ed. 1998) (Table No. 915, Utilization of Selected Media: 1970 to 1996). On some of the difficulties with the whole notion of scarcity, see R.H. Coase, *The Federal Communications Commission*, 2 J.L. & ECON. 1 (1959). The defect of Coase's analysis is his (remarkable) lack of self-consciousness about the idea that consumer sovereignty is the appropriate ideal for broadcasting; it is as if that idea is so self-evidently correct that it need not even be defended.

66. *See* HUBER, *supra* note 27, at 3-9. See also, in a different vein, LAWRENCE LESSIG, CODE AND OTHER LAWS OF CYBERSPACE 188-90 (1999) (arguing that a growing body of technical research suggests the FCC is unnecessary); Yochai Benkler & Lawrence Lessig, *Net Gains*, THE NEW REPUBLIC, Dec. 14, 1998, at 15 (same).

67. HUBER, *supra* note 27, at 206.

68. I do not deal here with the question why broadcasters and newspapers should be subject to different legal regimes. As a matter of fact, the difference seems to be a historical accident, associated with the particular form of regulation chosen for broadcasting. *See id.* at 4-9. As a matter of principle, the difference has been justified as a way of ensuring two competing regulatory regimes, each well-designed to combat the vices associated with the other. *See generally* BOLLINGER, *supra* note 7. In my view, this justification is serious but not convincing, and some measures designed to promote a well-functioning democratic culture might well be justified as applied to newspapers too. *See* CASS R. SUNSTEIN, DEMOCRACY AND THE PROBLEM OF FREE SPEECH 107-08 (1993). For example, it would not be unconstitutional, in my view, for government to require large metropolitan newspapers to have a "letters to the editor" page, or to require such newspapers to disclose their public service activities, or to require such newspapers to publish, on a nondiscriminatory basis, paid political advertising. For the

permissions and prohibitions; among other things, it ensures, via the law of property, that some people, and not others, will have access to the public sphere. The issue is thus not whether to "deregulate," but whether one or another regulatory system is better than imaginable alternatives.

Notice that this is a purely conceptual claim; it is not a normative argument of any kind. Any market system necessarily depends on regulatory controls, in the form of an assignment (by law) of property rights and (legal) rules of contract.[69] A system of television is hard to imagine without ownership rights; in the absence of ownership rights, who could use whose spectrum for what purpose? It is no answer to point to voluntary arrangements. Such arrangements are likely to break down without rules of law allowing some people to exclude others. If the Columbia Broadcasting System does not have a legal right to own spectrum and to enter into binding agreements with others, it will not be able to provide television as we know it. This point should be a familiar one for land and other "tangible" property; it is no less true for the services provided by television. Indeed, it holds, though to a lesser extent, for those who have web sites and provide services over the Internet; without legal protection against trespasses, and without a right to enter into legally enforceable agreements, web sites would be a modern version of the state of nature—a battleground rather than a framework for productive relationships.[70]

I therefore turn to the general question whether there remains any reason for government to regulate television in the "public interest." My concern here is both theoretical and empirical. The question is whether, in the current market, broadcasters are likely to provide viewers what they would like to see, and if so, whether that point is decisive on the question whether public interest obligations should be imposed.[71] The brief answer is that the idea that broadcasters show "what viewers want" is a quite inadequate response to the argument for public interest obligations.[72] The discussion

---

most part, however, such requirements do not seem necessary. But I cannot discuss these issues in detail here.

69. In certain circumstances, norms may successfully do the work of law. *See* ROBERT C. ELLICKSON, ORDER WITHOUT LAW 4-6 (1991). But even in such circumstances, legal rules of property and contract generally loom in the background.

70. *See* Neil W. Netanel, *Cyberspace Self-Governance: A Skeptical View from Liberal Democratic Theory*, 88 CALIF. L. REV. 395 (2000). Of course it is possible to imagine the Internet as a form of genuine anarchy, unregulated by legal rules, with norms and self-help (in the form of code) doing the work ordinarily done by the law of contract and property. Interestingly, the Internet is not subject to the usual legal realist claim that private ordering is dependent on law and that law is inevitable. *See* CASS R. SUNSTEIN, THE PARTIAL CONSTITUTION 51-54 (1993). But it is very hard to imagine this state of affairs for television, and those most critical of the existing legal structure seek only to replace one legal regime with another. *See* HUBER, *supra* note 27, at 5-30.

71. At several points I draw on the superb discussion found in C. Edwin Baker, *Giving the Audience What It Wants*, 58 OHIO ST. L.J. 311 (1997).

72. *See id.*

here deals with the technological present and the short-term future; later I introduce complications from emerging technological developments.

A. Three Market Failures

According to the economic model, a well-functioning television market would promote the ideal of consumer sovereignty. On this view, the point of markets is to satisfy consumer preferences. In this system, people would satisfy their "preferences," as these are measured via the criterion of private willingness to pay. People would be able to choose from a range of options, and suppliers would cater to their tastes. To a considerable extent, of course, the existing system already approaches this ideal, and this is increasingly the case. But there are three serious problems, each suggesting that the economic ideal of consumer sovereignty is not in fact served by free markets in programming. These are market failures if it is assumed that the purpose of a well-functioning television market is to ensure that programming is well-matched to viewer preferences. In the next Part, I question the market ideal itself.

1. Eyeballs as the Commodity

The first point is the simplest. Currently television is not an ordinary product, for broadcasters do not sell programming to viewers in return for cash. A system of "pay-per-view" would indeed fit the usual commodity model; but "pay-per-view" continues to be a relatively rare practice. The difference between the existing broadcasting market and "pay-per-view" is quite important. The key problem here is that viewers do not pay a price, market or otherwise, for television. As C. Edwin Baker has shown, it is more accurate to say that viewers are a commodity, or a product, that broadcasters deliver to the people who actually pay them: advertisers.[73]

This phenomenon introduces some serious distortions, at least if we understand an ideal broadcasting market as one in which viewers receive what they want. From the standpoint of consumer sovereignty, the role of advertisers creates market failures. Of course broadcasters seek, other things being equal, to deliver more rather than fewer viewers because advertisers seek, other things being equal, more rather than fewer viewers. But advertisers have issues and agendas of their own, and the interests of advertisers can push broadcasters in, or away from, directions that viewers, or substantial numbers of them, would like.

This is a substantial difference from the ordinary marketplace. Advertisers like certain demographic groups and dislike others, even when the numbers are equal; they pay extra amounts in order to attract groups that are likely to purchase the relevant products, and this affects programming

---

73. *See* C. EDWIN BAKER, ADVERTISING AND A DEMOCRATIC PRESS 25-87 (1994).

content.[74] Advertisers do not want programming that draws product safety into question, particularly if it concerns their own products and sometimes even more generally.[75] In addition, advertisers want programming that will put viewers in a receptive purchasing mood, and hence not be too "depressing."[76] Advertisers also tend to dislike programming that is highly controversial or that is too serious, and hence avoid sponsoring shows that take stands on public issues.[77] In these ways, the fact that broadcasters are delivering viewers to advertisers—this is largely their charge, under existing arrangements—can produce offerings that diverge considerably from what would emerge if viewers were paying directly for programming. To this extent the notion of consumer sovereignty is seriously compromised whenever programming decisions are a product of advertiser wishes.

2. Informational Cascades and Broadcaster Homogeneity

A second problem is that it is not clear whether broadcasters are now engaged, in anything like a systematic or scientific way, in catering to public tastes. At first glance it would seem obvious that broadcasters must be engaged in this endeavor (subject to the qualification just stated, involving the role of advertisers); if broadcasters are maximizing anything, they must be maximizing viewers (subject to the same qualification). In general, attracting viewers is their job.[78] But there is reason to question this judgment, at least in its simplest form. Sometimes rational people make decisions not on the basis of a full inspection of the alternatives, but on the basis of an understanding of what other people are doing.[79] Because people obtain information from other people's actions, individual actions carry with them one or more "informational externalities," which potentially affect the decisions of others. Thus, rational and boundedly rational people, in business as elsewhere, rely on the signals provided by the words and deeds of others.[80] This reliance can produce cascade effects, as B follows A, and C follows B and A, and D, as a rational agent, follows the collected wisdom embodied in the actions of A, B, and C. Informational cascades often produce unfortunate outcomes, in fact outcomes far worse than those that would result if individuals accumulated information on their own. Sometimes, moreover, people use the "availability" heuristic, deeming an event more probable if an instance of its occurrence can be readily brought

---

74. See id. at 66-70.
75. See id. at 50-56.
76. See id. at 62-66.
77. See id.
78. See generally SCHECHTER, supra note 50.
79. See Sushil Bikhchandani et al., Learning from the Behavior of Others: Conformity, Fads, and Informational Cascades, 12 J. ECON. PERSP. 151, 164 (1998).
80. In a related vein, see Andrew Caplin & John Leahy, Miracle on Sixth Avenue: Information Externalities and Search, 108 ECON. J. 60 (1998).

to mind.[81] Using the availability heuristic, broadcasters might reason that if one show or another has attracted substantial viewers in the past, they should copy it. The result would be "fads" and "fashions" in programming.[82]

In theory, then, broadcasters might be building on the programming judgments of other broadcasters, often, perhaps, reacting to the "availability" of salient recent instances in which a particular program was especially popular (dealing, let us suppose, with the Clinton-Lewinsky scandal or the O.J. Simpson trial) or especially unpopular (dealing, let us suppose, with South Africa). If this is true, private decisions by broadcasters may produce both mistakes and homogeneity—mistakes, in the form of programming that is not what viewers want, and homogeneity, in the form of the "blind leading the blind."[83]

Recent evidence suggests that this theoretical account has considerable truth. A careful study shows that there is a good deal of simple imitation, as networks provide a certain kind of programming simply by imitating whatever other networks are doing.[84] Recently popular shows tend to create cascade effects. This imitative behavior is not in the interest of viewers. On the contrary, it creates a kind of homogeneity and uniformity in the broadcasting market, and thus makes for problems in terms of providing what viewers "want."[85] This is not a conventional market failure, but it suggests that existing decisions are unlikely to promote consumer sovereignty. The problem is rapidly diminishing with an increase in available programming options, but to the extent that substantial numbers of people continue to depend on a small number of broadcasters, the existence of informational cascades suggests that the market does not entirely promote consumer sovereignty.

3. Externalities and Collective Action Problems

Even if broadcasters did provide each viewer with what he or she wanted, a significant problem would remain, and from the economic point of view, this is probably the most serious of all. Information is a public good, and once one person knows something (about, for example, product hazards, asthma, official misconduct, poverty, welfare reform, or abuse of

---

81. *See* Timur Kuran & Cass R. Sunstein, *Availability Cascades and Risk Regulation*, 51 STAN. L. REV. 683, 711-14 (1999).
82. *See* Bikhchandani et al., *supra* note 79, at 161-62.
83. *See* David Hirshleifer, *The Blind Leading the Blind: Social Influence, Fads, and Informational Cascades*, *in* THE NEW ECONOMICS OF HUMAN BEHAVIOR 188 (Mariano Tommasi & Kathryn Ierulli eds., 1995).
84. *See* Robert E. Kennedy, Strategy Fads and Competitive Convergence: An Empirical Test for Herd Behavior in Prime-Time Television Programming (1999) (Harvard Business School Working Paper 96-025, on file with the *California Law Review*).
85. *See id.*

power), the benefits of that knowledge will probably accrue to others.[86] Note in this regard Amartya Sen's remarkable observation that no famine has ever occurred in a democratic country with a free press.[87] This observation is complemented by a series of less dramatic ones, showing the substantial benefits for individual citizens of a media that is willing and able to devote attention to public concerns, including the plight of the disadvantaged.[88] Individual choices by individual viewers are highly likely to produce too little public interest programming in light of the fact that the benefits of viewing such programming are not fully "internalized" by individual viewers. Thus, individually rational decisions may inflict costs on others at the same time that they fail to confer benefits on others. In this respect, the problem "is not that people choose unwisely as individuals, but that the collective consequences of their choices often turn out to be very different from what they desire or anticipate."[89]

Most generally, there are multiple external effects in the broadcasting area; some of these are positive, but unlikely to be generated sufficiently by individual choices, while others are negative, and likely to be excessively produced by individual choices. Consider a decision to watch violent programming.[90] In short, the effects of broadcasting depend on social interactions. Many of the resulting problems are connected with democratic ideals. A culture in which each person sees a high degree of serious programming may well lead to better political judgments; greater knowledge on the part of one person often leads to more knowledge on the part of others with whom she interacts.[91] Perhaps most important, a degree of serious attention to public issues can lead to improved governance through deterring abuses and encouraging governmental response to glaring problems. In these various ways, public interest programming can produce social benefits that will not be adequately captured by the individual choices of individual citizens; the same is true for programming that produces social costs, including apathy, fear, and increased criminal activity.[92] Because of the collective action problem, an unregulated market will underproduce public goods and overproduce public bads.

---

86. An illuminating and detailed discussion is Baker, *supra* note 71, at 350-85.
87. *See* JEAN DREZE & AMARTYA SEN, INDIA 76 (1995).
88. *See id.* at 75-76, 173, 191.
89. FRANK & COOK, *supra* note 23, at 191; *see also* PIERRE BOURDIEU, ON TELEVISION 9-29 (Priscilla Parkhurst Ferguson, trans., The New Press 1998); Baker, *supra* note 71, at 350-85.
90. *See* HAMILTON, *supra* note 45, at 3-49 (discussing the market for violence).
91. *See* Baker, *supra* note 71, at 350-67.
92. *See id.* at 355-56; *see also* HAMILTON, *supra* note 45, at 20-30, 285-322.

B. Problems on Nonmarket Criteria: Children, Deliberative Democracy, and Related Issues

Thus far the discussion has emphasized difficulties with television markets on conventional economic grounds; but it is wrong to endorse purely market approaches to television.[93] Television is not best understood as an ordinary commodity, subject to the forces of supply and demand. There are several reasons why this is so. The unifying theme is that the American political tradition is committed to the ideal of deliberative democracy,[94] an ideal that has animated much First Amendment doctrine and media regulation in general.[95] Even if the media market were well-functioning from the economic point of view, there would be room for measures designed to promote a well-functioning system of democratic deliberation, especially in view of the importance of television to people's judgments about what issues are important and about what it is reasonable to think.[96] Consider, by way of general orientation, John Dewey's suggestion:

> [W]hat is more significant is that counting of heads compels prior recourse to methods of discussion, consultation and persuasion, while the essence of appeal to force is to cut short resort to such methods. Majority rule, just as majority rule, is as foolish as its critics charge it with being. But it never is *merely* majority rule.... The important consideration is that opportunity be given that idea to spread and to become the possession of the multitude.... The essential need, in other words, is the improvement of the methods and conditions of debate, discussion and persuasion. That is *the* problem of the public.[97]

1. Children and the Hearing Impaired

A well-functioning market may fail to serve certain categories of viewers. Of these the most obvious is children, who may be poorly served by an absence of educational programming[98] or adversely affected by violent programming.[99] It is reasonable to treat the resulting problems as "externalities," but the more natural conclusion is that the television market is creating difficulties even in the absence of a market failure. Because television has a significant role as an educational instrument, the failure to

---

93. Here, too, Baker has provided very illuminating discussions, and I draw on his account. *See* Baker, *supra* note 71, at 355-65.

94. *See* JOSEPH M. BESSETTE, THE MILD VOICE OF REASON 6-39 (1994).

95. *See* MINOW & LAMAY, *supra* note 50, at 15-65; SUNSTEIN, *supra* note 68, at 53-92.

96. *See generally* SHANTO IYENGAR, IS ANYONE RESPONSIBLE? HOW TELEVISION FRAMES PUBLIC ISSUES 127-44 (1991); SHANTO IYENGAR & DONALD R. KINDAR, NEWS THAT MATTERS (1987).

97. DEWEY, *supra* note 58, at 207-08.

98. *See* MINOW & LAMAY, *supra* note 50, at 10-65.

99. *See* HAMILTON, *supra* note 45, at 76-128.

serve children is a significant problem,[100] for children lack much ability and much willingness to pay, and the result can be inadequate attention to their needs.[101] To be sure, some children do have willingness to pay in the sense that they can pressure their parents to purchase certain programming and the products that support it. But it is implausible to say that market criteria exhaust the goals of a system of broadcasting with respect to the interests of children.

A well-functioning market may also disserve people who are hearing impaired, if they are deprived of access to television by the existing use of technology. This is a particular problem if they are unable to watch the news or to understand descriptions of emergency conditions. Here too there is potential room for a regulatory response, partly in order to include the hearing impaired in civic activities by informing them of electoral issues and news in general. To be sure, some people who are hearing impaired are willing to pay for closed captioning. But such people face serious collective action problems in making their needs and wishes clear to advertisers and producers. In any case, there is an important democratic interest in ensuring that certain programming is available to the hearing impaired, quite apart from their willingness to pay. There is thus strong reason to support communications policies that promote the enfranchisement of people with disabilities.

2. Balkanization

Imagine that in a technological future, each person could devise her own preferred communications "menu"; imagine, in other words, that programming could be fully, and not just partially, individuated. On economic grounds, this would seem to be a striking advance, a guarantee of a kind of optimality, a victory for both freedom and welfare. But from the democratic standpoint, it is the stuff of science fiction, and it contains serious risks, above all because it may well result in a situation in which many or most are not exposed to diverse views, but instead hear louder and louder echoes of their own preexisting convictions.[102] One of the advantages of a well-functioning system of freedom of expression is that it supplies one or more genuinely public spheres, in which diverse points of view are presented and confront one another, and are exposed to people who have a willingness to learn. General-interest newspapers and magazines often do precisely this, and it is important to make provision for multiple forums of

---

100. *See* MINOW & LAMAY, *supra* note 50, at 100-26.
101. *See id.*; *see also* PRICE, *supra* note 9, at 192, 246.
102. *See* SHAPIRO, *supra* note 2, at 105-32. See also the discussion of group polarization in Schkade et al., *supra* note 62; Sunstein, *supra* note 62.

this kind. Insofar as they disregard this point, market conceptions of communications miss a central matter.[103]

3. *Citizens, Consumers, and Precommitment Strategies*

The most important point is that a market system may fail to provide a system of communication that is well-adapted to a democratic social order. People are quite aware of this fact, whatever they may choose in their capacity as viewers and listeners; and they may, and often do, seek collective corrections. The problem with the economic approach is that it makes private preferences normative, or decisive, for purposes of policy.

In short, there is a pervasive difference between what people want in their capacity as viewers (or "consumers of broadcasting") and what they want in their capacity as citizens.[104] Both preferences and values are a function of the setting in which people find themselves; they are emphatically a product of social role.[105] In these circumstances, it would be wrong to think that the choices of individual viewers are definitive, or definitional, with respect to the question of what individuals really prefer. On the contrary, a democratic public, engaged in deliberation about the world of telecommunications, may legitimately seek regulations embodying aspirations that diverge from their consumption choices.

Participants in politics may be attempting to promote their meta-preferences, or their preferences about their own preferences. They may be attempting to carry out a precommitment strategy of some kind. They may be more altruistic or other-regarding in their capacity as citizens, perhaps because of the nature of the goods involved. They may be more optimistic about the prospects for change when acting collectively, and therefore able to solve a collective action problem faced in their individual capacities.[106] In this last respect, the democratic argument for departing from private consumption choices converges with the argument emphasizing the character of information as a public good.

When participants in a democracy attempt to make things better and do not simply track their consumption choices, it is not helpful to disparage

---

103. *See* LESSIG, *supra* note 66, at 185-86. See also JÜRGEN HABERMAS, BETWEEN FACTS AND NORMS 362 (1996), emphasizing that

> [t]he diffusion of information and points of view via effective broadcasting media is not the only thing that matters in public processes of commuuication, nor it is the most important.... [T]he rules of a *shared* practice of communication are of greater significance for structuring public opinion. Agreement on issues and contributions *develops* only as the result of more or less exhaustive controversy in which proposals, information, and reasons can be more or less rationally dealt with.

104. A good discussion is Daphna Lewinsohn-Zamir, *Consumer Preferences, Citizen Preferences, and the Provision of Public Goods*, 108 YALE L.J. 377 (1998).

105. *See generally* Cass R. Sunstein, *Social Norms and Social Roles*, 96 COLUM. L. REV. 903, 906-17 (1996).

106. The latter two points are emphasized in *id.* at 944-46; in the particular context of broadcasting, see Baker, *supra* note 71, at 401-04.

their efforts as "paternalism" or as "meddling." Their efforts at reform represent democracy in action.[107] It is entirely appropriate for government to respond to people's aspirations and commitments as expressed in the public realm. This is especially so when a democratic polity is itself attempting to ensure more in the way of democratic deliberation.

4. Endogenous Preferences

On the market view, freedom consists in the satisfaction of viewer preferences, whatever their content. But this is an inadequate conception of freedom.[108] It is important to ensure a degree of freedom in the formation of preferences, and not only in preference satisfaction. If people's preferences are formed as a result of the existing arrangement, including limitations in available opportunities, or of exposure to a limited kind of television, then it makes no sense to say that the existing arrangement can be justified by reference to their preferences.

It seems clear that the public's "tastes," with respect to television programming, do not come from nature or from the sky. They are partly a product of current and recent practices by broadcasters and other programmers. They are often generated by the market.[109] What people want, in short, is partly a product of what they are accustomed to seeing. It is also a product of existing social norms, which can change over time, and which are themselves responsive to existing commercial fare. Tastes are formed, not just served, by broadcasters.

This point raises doubts about the idea that government policy should simply take viewers' tastes as given. In an era in which broadcasters are providing a good deal of public interest programming, dealing with serious issues in a serious way, many members of the public will cultivate a taste for that kind of programming. This effect would promote democratic ideals by disseminating information and helping to increase deliberation.[110] In an era in which broadcasters are carrying sensationalistic or violent material, members of the public may well cultivate a taste for more of the same. "Free marketeers have little to cheer about if all they can claim is that the market is efficient at filling desires that the market itself creates. . . . Just as culture affects preferences, so also do markets influence culture."[111] If this is so, the ideal of consumer sovereignty is placed under some pressure; market activities cannot easily be justified by reference to tastes that they themselves generate.

---

107. *See* Baker, *supra* note 71, at 401-11.
108. *See generally* JOSEPH RAZ, THE MORALITY OF FREEDOM 369-78 (1986) (discussing social preconditions for autonomy); AMARTYA SEN, COMMODITIES AND CAPABILITIES (1985).
109. *See* Baker, *supra* note 71, at 404-10.
110. For evidence that the effects of television on this count are far from fanciful, see IYENGAR, *supra* note 96, at 26-116.
111. FRANK & COOK, *supra* note 23, at 201.

This point should not be overstated. Probably broadcasters have limited power to push tastes very dramatically in one direction or another. But at a minimum, the idea that viewers' tastes are endogenous to existing fare should be taken as a cautionary note about treating consumption choices as decisive for purposes of policy. There is nothing illegitimate about policies that depart from consumption choices in favor of widely held social aspirations. But there is reason for broader concern about the adverse effects of certain kinds of programming—including a failure to cover serious issues in a serious way—on democratic judgments.[112]

5. Paternalism? Elitism?

It might be tempting to respond that the arguments thus far are unacceptably paternalistic, indeed elitist. If individual listeners and viewers prefer fare of a certain kind, how can there be any ground for legitimate complaint? Perhaps children pose a special case, but even here parental guidance is far from unusual. Why should government displace the choices of adults, including parents, and substitute choices of its own?

Let us take the charge of paternalism first. Notice that insofar as the argument stresses a collective action problem faced by individual consumers, paternalism is not at work at all. Notice too that insofar as the argument centers on people's desires in their capacity as citizens, no paternalism is involved; the claim is that (a majority of) the people seek to push consumption patterns in certain directions. This form of precommitment strategy, or autopaternalism, should not be confused with paternalism of any objectionable kind.[113] To be sure, this argument depends on an empirical proposition to the effect that in their capacity as citizens, people would like a communications market of a certain kind.[114] But the proposition seems at the very least highly plausible.

To the extent that I have emphasized the endogeneity of preferences, my argument might seem to verge on objectionable paternalism. Certainly preferences are not being taken as given. But there is nothing objectionable about insisting that in a democracy, free and equal citizens are entitled to a public culture that will promote their freedom and their equality. It is one thing to say that a government should not be authorized to overcome people's judgments, when those people are armed with adequate information. It is quite another to say that government should be permitted to take modest, viewpoint-neutral steps to promote the operation of a democratic order by, for example, ensuring free air time for candidates, or subsidizing

---

112. *See* IYENGAR, *supra* note 96, at 46-68 (discussing television news' framing of poverty, unemployment, and racial inequality).

113. *See* JON ELSTER, ULYSSES AND THE SIRENS 36-47 (1979) (discussing precommitment strategies).

114. Of course some such judgments would run afoul of the First Amendment—if, for example, they involved a form of viewpoint discrimination.

certain kinds of fare, or promoting substantive discussion of substantive questions. And the charge, indeed the very notion, of paternalism becomes harder to understand when the preferences involved are a product of the very system whose legitimacy is at issue.

Nor is there anything unacceptably "elitist" about a communications policy that fosters education of children, and more substantive and diverse coverage of civic issues. To the extent that substantive programming is said to have special appeal for "elites" (a vague and empirically uncertain claim), the problem lies not in a policy that encourages such programming, but in unjust background conditions, and in particular, in unjust inequalities in education. A communications policy that attempts to promote more discussion and understanding of public issues is a partial way of overcoming those unjust inequalities. It is not a way of catering to them.

To be sure, the charge of elitism would have force if programming content were dictated by a political elite, promoting its own preferred fare free from effective electoral control. It is crucial to the argument offered here that regulatory strategies—from the FCC or from Congress—are subject to democratic supervision. It is also crucial that any effort to promote programming of a certain kind is defended, not by the preferences of the regulators, but by democratic values that should, at least in principle, meet with widespread public approval. The judgment on behalf of deliberative democracy (and corresponding regulatory strategies) does not itself come from the sky. Any view on its behalf depends, not on a claim that it is extracultural, but on the arguments that are made on its behalf.

There is no argument here for any particular conception of public interest broadcasting. A talk-show on racial violence may well be at least as desirable, for democratic purposes, as a public debate between candidates for public office. Staged and self-serving statements by politicians on C-Span may not add more, and may add less, to public understanding than (easily imaginable) rock or rap music videos. The point of a system of public service broadcasting would be to encourage those who produce television programming to use their own creativity to promote deliberation about public issues, not to force programming into any particular mode.

Of course public interest programming will do little good if people simply change the channel. No one urges that the government should require people to watch governmentally preferred programming. The only suggestion is that if the government, responsive as it is to citizen aspirations, seeks to ensure more public interest programming than the market does, there is no principled ground for complaint. In any case it is likely that some people would watch the resulting programming and develop a taste for it; that empirical probability is all that is necessary to vindicate the suggestions made here.

6. Constitutional Notes

This is not an Article about the First Amendment or about constitutional law; but it will be useful to conclude this Section with some brief notations on those subjects. From what has been said thus far, it should be clear that a central purpose of the First Amendment is to ensure a well-functioning democratic order.[115] A system of free expression is designed in large part to protect the preconditions for a form of sovereignty that is suited to genuine self-governance. It is not designed to protect whatever happens to come out of the mouths, or pens, or word-processors of those who are attempting to speak. This view receives support on both historical and philosophical grounds.[116]

One implication is that government may well be permitted to regulate speakers who are not contributing to democratic deliberation—for example, those who advertise cigarettes. Another implication is that government efforts to promote a well-functioning democratic order should not be invalidated even if they involve content regulation, so long as there is no discrimination against any point of view. If government seeks to ensure a certain level of educational programming, or if it allows free air time for candidates, or if it provides a right of access for those who attempt to speak on political issues, it is not violating the free speech guarantee merely by virtue of the fact that it is intruding on the discretion of those who own stations. A conclusion to the contrary would convert the First Amendment into a form of Herbert Spencer's *Social Statics*,[117] or at least Richard Epstein's *Simple Rules for A Complex World*;[118] it would tear the First Amendment from its theoretical underpinnings.[119]

On this view of the First Amendment, there is no tension between constitutionalism and democracy, or between individual rights and majority rule, properly understood; robust rights of free expression are a precondition for both democracy and majority rule, properly understood. In this way, private autonomy is in no tension with, but is on the contrary inextricably intertwined with, the notion of popular sovereignty.[120] The point is not limited to freedom of speech. Protection of private property, for example, can be seen as a precondition of the status of citizenship; those whose holdings depend on the beneficent exercise of government

---

115. *See* ALEXANDER MEIKLEJOHN, FREE SPEECH AND ITS RELATION TO SELF-GOVERNMENT (1948); SUNSTEIN, *supra* note 68, at 121-66.

116. *See generally* HABERMAS, *supra* note 18, at 283-328 (discussing theory of deliberative democracy); SUNSTEIN, *supra* note 68, at 241-50 (same).

117. HERBERT SPENCER, SOCIAL STATICS (1913).

118. EPSTEIN, *supra* note 8 (arguing for six simple rules of governance).

119. None of this means that the First Amendment is only about democratic self-government. It has other purposes as well. *See generally* SUNSTEIN, *supra* note 68, at 137-48.

120. *See* JÜRGEN HABERMAS, THE INCLUSION OF THE OTHER 258-60 (Ciaran Cronin & Pablo De Greiff eds., 1998).

discretion are hardly in a position to operate as independent citizens in the public domain. And the protection against unreasonable searches and seizures, and other abuses of authority by the police, are safeguards of individual liberty that simultaneously prevent attacks on popular sovereignty. The account offered here thus provides a basis for understanding many individual rights in a way that fuses the "freedom of the ancients" and the "freedom of the moderns."[121]

III
PRINCIPLE, POLICY, TECHNOLOGY

These points suggest that there is good reason, in principle, for some kind of regulatory response to existing markets in television. But nothing said thus far argues for any particular governmental initiative. There is no simple "match" between the identifiable market and nonmarket failures and public interest requirements in general. For example, it is hard to imagine a legitimate governmental response to the problem of excessive homogeneity on the major networks. We have also seen that existing public interest obligations are extremely varied; each of them must be assessed on its own.

At first glance, policies that attempt to promote better programming for children are most securely supported by the arguments made thus far (as a response to what is reasonably classified as a market failure, and also as a way of increasing positive externalities and of promoting social aspirations). Efforts to ensure that hearing impaired people are able to enjoy television, through closed captioning, are justifiable on similar grounds. There is room also for efforts to ensure better coverage of electoral campaigns—perhaps through a requirement of free air time for candidates, perhaps through a private "code" designed to ensure more substantive discussion. Disclosure requirements, allowing the public to have a general sense of broadcaster performance, seem to be justified as a nonintrusive method for allowing civic aspirations to help influence future programming. As a response to the possibly unfortunate effects of advertiser pressures, and also as a way of ensuring against a destructive "race to the bottom" with respect to programming content, a general code of broadcaster behavior seems appealing.[122]

A. Practice

There are important pragmatic questions here. To say that a response is justified in principle is not at all to say that it will succeed in practice. Just as in the environmental area, where command-and-control regulation

---

121. *Id.* at 258.
122. *See* discussion *infra* Part VI.

has produced unintended adverse consequences,[123] many problems have emerged with command-and-control regulation of the kind that has typified FCC regulation for most of its history. Consider, for example, the fairness doctrine, designed to ensure exposure to public issues and to allow diverse voices to have access to the airwaves. A serious problem with the fairness doctrine is that it appears to have discouraged stations from covering controversial issues at all, thus ensuring a kind of bland uniformity that disserves democratic goals.[124] A uniform set of mandates may also produce waste and poor programming; if a network is especially bad at generating good shows for children and has a hard time attracting a children's audience, is it so clear that that network should be faced with the same obligations as everyone else? In view of the great diversity of the broadcasting market, a "one-size-fits-all" approach may be far more costly, and less effective, than creative alternatives.

It is useful to distinguish here between approaches that suppress markets and approaches that supplement markets.[125] Market-suppressing approaches include minimum wage and maximum hour laws, or price and wage controls. Market-supplementing approaches include job-training programs and the earned income tax credit. In telecommunications policy, the fairness doctrine was a market-suppressing remedy; so too are requirements that broadcasters provide three hours of educational programming per week, or a certain amount of time for candidates for public office. By contrast, the grant of public funds to the Public Broadcasting System (PBS) is a market-supplementing approach; so too is a subsidy granted to each of the networks, designed to ensure a certain amount of public interest programming. To be sure, the line between the two approaches can be thin when the market-supplementing approach ends up displacing material that would otherwise be supplied in accordance with forces of supply and demand.

B. Communications Past, Present, and Future: Planned Obsolescence and Beyond

The arguments offered thus far have not specifically addressed the new market in communications. An especially important question is whether emerging changes in television technology should strengthen satisfaction with market outcomes.

---

123. *See* CASS R. SUNSTEIN, FREE MARKETS AND SOCIAL JUSTICE 245-283 (1997).
124. *See* Thomas W. Hazlett & David W. Sosa, *Was the Fairness Doctrine a "Chilling Effect"? Evidence from the Postderegulation Radio Market*, 26 J. LEGAL STUD. 279 (1997) (offering an affirmative answer to the question in the title).
125. *See* DREZE & SEN, *supra* note 87, at 21-24 (discussing market supplementing strategies).

1. Predicting the Future

Of course the most striking feature of the emerging communications market is a dramatic increase in the number of available stations and programming options. The existing regulatory regime was designed for a system with three private broadcasting networks and PBS. In 1996, about thirty-five percent of people who had television remained dependent on broadcasters, which results in access to five or six stations.[126] This means that about two-thirds of viewers had access to between fifty and one hundred stations, including the all-news stations C-SPAN and CNN, and a range of "soft news" stations such as MSNBC. By itself this is an extremely significant change, and the shift from this situation to one in which most people have access to (say) 500 stations may be only one of degree.

More dramatic innovations are coming in the future, with the possible ultimate "convergence" of various television sources, including digital television and the Internet.[127] If a "television set" becomes akin to a computer monitor that provides access to the full range of American magazines, would not the case for public interest regulation be substantially weakened? The foregoing discussion offers an ambivalent answer. Some of the problems with the market status quo would dissipate, but others would remain. Let us explore these questions in more detail.

For purposes of analysis, we might separate the market for television into four rough stages. The first is the period between the 1940s and 1970s—the market for which the existing regulatory system was designed; call this the old regime. As noted, the old regime included three large networks and also PBS; the four stations provided all of what Americans knew as television. The second market is one of the 1990s; call this the transitional state. Here the most dramatic change is in the number of available options. This is a system in which a substantial percentage (about thirty-five percent) of the viewing public relies on five broadcasters and PBS, but in which sixty-five percent of the public has access to cable television, and thus is able to choose among fifty or more options.[128] Of that sixty-five percent, a growing segment is able to see well over one hundred stations. But in the transitional state, broadcasters continue to have a special role, both because a substantial number of people do not have access to cable at all, and because even cable viewers watch the major networks disproportionately.

The third stage, likely to begin shortly, is continuous with the transitional state; call this the stage of multiple options. This will be a market in which broadcasters will continue to be seen by more people than other

---

126. *See* U.S. BUREAU OF THE CENSUS, *supra* note 65.
127. *See* OWEN, *supra* note 3, at 327-33.
128. *See* U.S. BUREAU OF THE CENSUS, *supra* note 65.

providers, but they will become decreasingly distinctive in terms of both the size and the nature of their audience. Even more people will have access to cable or other options, and many of those who rely on broadcasting will be able to have more options too. But broadcasters will continue to be seen by a disproportionate number of people, if only because a shrinking but still substantial percentage of viewers will continue to have access only to broadcasters; this significant subgroup is important partly because of its sheer size and partly because it includes an especially high percentage of people, including children, who are poor and poorly educated. In this stage, broadcasters will in some ways be akin to *Newsweek*, *Time*, and *U.S. News & World Report*, in the sense that they will have a relatively dominant role in terms of sheer numbers. But they will have to compete with other programmers, some general, some specialized, with analogues (most of them now in place on cable or even the Internet) to *Sports Illustrated*, *The Economist*, *Dog Fancy*, *National Review*, *National Geographic*, *The New Republic*, *Consumer Reports*, *Playboy*, and many more. This third stage will be marked by the rise of digital television, which is allowing broadcasters to "multiplex," that is, to provide two, three, four, or even five programs where they could previously provide only one. The result may be to make broadcasters themselves more specialized.[129] It is hard to speculate about the future, but the best prediction is that a situation of this kind will prevail for the next decade and more.

The final stage—call it one of technological convergence—is one of substantial or possibly even complete shrinkage in the distinctive role of broadcasters.[130] This will be a stage in which television programming can be provided via the Internet, over telephone lines, or both; a television, or one kind of television, may itself be a simple computer monitor, connected to various programming sources from which viewers may make selections. If it is economically feasible for broadcasters to continue as such, they are unlikely to have a special role and will be among a large number of providers. At most, and extending the analogy to *Time* or *Newsweek*, they will have a somewhat larger and more general audience than most of their competitors; perhaps they will be entertainment conglomerates with multiple stations and programs. Perhaps they will not be distinctive at all.[131] The most extreme version of this final stage would be akin to the market for books, in which people make individual choices, usually not filtered by an intermediary offering packages.[132]

---

129. *See id.* at 100.
130. This shrinkage may or may not occur as a result of convergence. For discussion, see *id.* at 329.
131. This speculation is questioned in OWEN, *supra* note 3, at 328-33, and Goldberg et al., *supra* note 5, at 296.
132. I put to one side the complexities introduced by Amazon.com and various book clubs.

2. Regulatory Options and Technological Change

All of the arguments offered thus far—about market and nonmarket failures—make sense for the old regime and the transitional state. They also seem to make sense for the emerging third stage of multiple options; recall that even here, a substantial segment of the public will depend on broadcasters only, and people with access to cable and other alternatives continue, statistically speaking, to watch a disproportionate amount of broadcast fare. But arguments for public interest obligations would be less sensible as applied to a market in which broadcasters occupy no special role, partly because some of the relevant problems would be diminished, partly because in such a market it seems peculiar to impose on broadcasters, and no one else, a special duty to protect public interest goals. In that market, perhaps every station should be faced with some of the requirements discussed below (in particular, the disclosure requirement). But the case for other requirements (snch as uniform mandates, a broadcaster-only code, or "play or pay") would be greatly weakened, not least since it wonld seem arbitrary to single out broadcasters for such requirements.

Let us examine, in more detail, the force of the particular arguments as the market changes over time. Even if informational influences produce a degree of homogeneity among broadcasters, and even if broadcasters tend to follow one another, the increasing number of channels means that for most Americans, there is far more heterogeneity now than there was a decade ago, and a great deal more heterogeneity is likely in the near future, probably dramatically increasing heterogeneity. At the same time, advertisers are likely to have an increasingly weak role in determining overall programming content. When a few broadcasters exhaust the market, advertiser preferences could have a more substantial effect than they now do. Thus, two of the arguments made above—involving the market failures resulting from advertiser pressures and informational cascades—are significantly weakened.[133]

But some of the arguments offered above—especially those focused on democratic ideals—retain considerable force. Even in the very long term, there will continue to be substantial external benefits from public interest programming, benefits that are not adequately captured by individual viewer choices. And to the extent that citizens seek to push communications policy toward (for example) more and better programming for children, or free air time for candidates, or greater access for the hearing impaired, changes in the evolving market offer only partial answers. Heterogeneity may be an inadequate solution here. It

---

133. Of course these changes should not be overstated. Recall that 32 percent of American households with televisions still lack cable television, and that a substantial number of Americans are likely to depend on over-the-air programming for the not-too-distant future; hence, the increasing heterogeneity is not quite as dramatic as it might seem.

is reasonable for citizens to believe that there should be very general public exposure to public issues, and that it is not sufficient to have one, or two, or three, or even more stations (CNN, C-SPAN, MSNBC) that take such issues seriously. Indeed, citizens may favor a kind of general precommitment strategy—operating against their own particular viewing choices—through which broadcasters, at least, are required to devote some time to educational or civic programming. Thus, the presence of news-only stations, especially on cable, is not a sufficient response to those who want broadcasters to do more and better. It is insufficient partly because a significant segment of the population will have no access to cable at all, and partly because in their capacity as citizens, people may favor a precommitment strategy that favors certain public interest goals.

In the very long run, this argument too will be weakened. As we have seen, the strongest objection would come when broadcasters are no longer genuinely distinctive; when this is the case, it will seem arbitrary to encourage broadcasters, but not others, to provide certain kinds of programming. In the face of such changes, it will indeed make sense to adapt the proposals discussed below to this dramatically altered market. But it is hard to explore this question in the abstract; everything turns on the particular regulatory proposal that is at issue.

The question of adaptation will arise at several points below. For now let us observe only that in the extreme situation—when broadcasters are not in any sense distinctive—the case for regulation limited to broadcasting would be very weak, and alternative strategies, involving funding of public interest programming, would be better. Thus, my emphasis here—on disclosure, economic incentives, and voluntary self-regulation—is designed for a (likely not inconsiderable) period in which broadcasters continue to occupy a special role. To simplify a complex story, the very long-term may call for a combination of public subsidies for high-quality programming and disclosure requirements for general-interest stations. But in the shorter term, there is a great deal more to consider. And it is important to see that in the long-term much may be lost, as well as gained, with a highly balkanized communications system in which many people are not exposed to serious programming, many others simply hear echoes of their own voices, and widely shared viewing experiences become rarer.[134]

There is a large issue in the background here: how to define the market from which to assess proposals for public interest requirements. Traditionally, "television broadcasting" has been viewed as the relevant category, but skeptics might well ask why the real category is not "sources of information and entertainment." If that category were the relevant one, there might seem to be no problem calling for requirements at all—not in the 1930s, not in the 1950s, not in the 1970s, not in the 1990s, and not in

---

134. *See* Lessig, *supra* note 66, at 185-86.

the foreseeable future. The traditional approach seems best defended on the ground that television has a special role and salience; if this is no longer true, a particular regulatory regime, limited to television, loses its rationale.[135] The best argument for the future is that even if television as a whole is far from the only source of information and entertainment, it continues to have a distinctive social role, and efforts to promote the goals associated with public interest programming might do considerable good.

IV
DISCLOSURE

Consider a simple proposal: broadcasters should be required to disclose, in some detail and on a quarterly basis, all of their public service and public interest activities. The disclosure might include an accounting of any free air time provided to candidates, educational programming, charitable activities, programming designed for traditionally under-served communities, closed captioning for the hearing impaired, local programming, and public service announcements.[136] The hope, vindicated by experience with similar approaches in environmental law,[137] is that a disclosure requirement will by itself trigger improved performance, by creating a kind of competition to do better, and by enlisting various social pressures in the direction of improved performance. A requirement of this sort would be part of a general trend in federal regulation, one with considerable promise.[138]

A. Precursors

Many statutes and regulations now require the disclosure of information. Some of these are designed to assist consumers in making informed choices; such statutes are meant to be market-enhancing. By contrast, others are designed to trigger political rather than market safeguards; such statutes are meant to enhance democratic processes. The most famous of these is the National Environmental Policy Act (NEPA).[139] Enacted in 1972, the principal goal of NEPA is to require the government to compile and disclose environmentally-related information before the government goes forward with any projects having a major effect on the

---

135. But see BOLLINGER, *supra* note 7, defending two competing regulatory regimes, one for television and one for the print media.
136. See FINAL REPORT, *supra* note 19, at 104-05, for an illustration.
137. *See* Madhu Khanna et al., *Toxics Release Information: A Policy Tool for Environmental Protection*, 36 J. ENVTL. ECON. & MGMT. 243 (1998).
138. *See* WESLEY A. MAGAT & W. KIP VISCUSI, INFORMATIONAL APPROACHES TO REGULATION (1992) (discussing possible effectiveness of disclosure); Sunstein, *supra* note 24, at 618-29 (discussing this trend).
139. 42 U.S.C. §§ 4321-4370 (1994).

environment.[140] NEPA does not require the government to give environmental effects any particular weight, nor is there judicial review of the substance of agency decisions.[141] The purpose of disclosure is principally to trigger political safeguards, coming from the government's own judgments or from external pressure. Governmental indifference to adverse environmental effects is perfectly acceptable under NEPA: the idea behind the statute is that if the public is not indifferent, the government will have to give some weight to environmental effects.

Probably the most successful experiment in information disclosure is the Emergency Planning and Community Right-to-Know Act (EPCRA).[142] Under this statute, firms and individuals must report to state and local government the quantities of potentially hazardous chemicals that have been stored or released into the environment.[143] On the basis of the relevant results, the EPA publishes pollution data about the releases of over 300 chemicals from over 20,000 facilities.[144] This has been an exceptional success story, one that has well exceeded expectations at the time of enactment.[145] A detailed report suggests that EPCRA has had important beneficial effects, spurring innovative, cost-effective programs from the EPA and from state and local government.[146]

Many other statutes involving health, safety, and the environment fall into the category of information disclosure measures. The Animal Welfare Act[147] is designed partly to ensure publicity about the treatment of animals; thus covered research facilities and dealers are required to file reports with the government about their conduct, with the apparent goal that the reports will deter noncompliance and also allow continuing monitoring.[148] In addition to its various command-and-control provisions, the Clean Air Act[149] requires companies to create and disclose "risk management plans" involving accidental releases of chemicals; the plans must include a worst-case scenario.[150] The Safe Drinking Water Act[151] was amended in 1996 to

---

140. *See* Calvert Cliffs' Coordinating Comm., Inc. v. United States Atomic Energy Comm'n, 449 F.2d 1109 (D.C. Cir. 1971).
141. *See* Strycker's Bay Neighborhood Council, Inc. v. Karlen, 444 U.S. 223, 227-28 (1980).
142. 42 U.S.C. §§ 11001-11050 (1994).
143. *See id.* § 11023.
144. *See* HAMILTON, *supra* note 45, at 302.
145. *See* ROBERT V. PERCIVAL ET AL., ENVIRONMENTAL REGULATION: LAW, SCIENCE, AND POLICY 612-16 (2d ed. 1996) (discussing informational approaches); James T. Hamilton, *Exercising Property Rights to Pollute: Do Cancer Risks and Politics Affect Plant Emission Reductions*, 18 J. RISK & UNCERTAINTY 105, 106-08 (1999) (discussing success of toxic release inventory); Khanna et al., *supra* note 137, at 243 (discussing success of toxic release inventory and stockholder responses to release of relevant information).
146. *See* U.S. GENERAL ACCOUNTING OFFICE, TOXIC CHEMICALS (1991) (report to Congress).
147. 7 U.S.C. §§ 2131-2159 (1994).
148. *See id.* §§ 2140, 2142.
149. 42 U.S.C. §§ 7401-7671 (1994).
150. *Id.* § 7412(r).

require annual "consumer confidence reports" to be developed and disseminated by community water suppliers.[152] Statutes governing discrimination and medical care also seem committed partly to the idea that "sunlight is the best of disinfectants";[153] thus they require covered institutions to compile reports about their conduct and compliance with applicable law. The Federal Election Campaign Act[154] requires political committees to disclose a great deal of information about their activities.

Of course there is an overlap between informational regulation designed to assist consumers and informational regulation designed to trigger political checks. A statute that requires companies to place "eco-labels" on their products may produce little in the way of consumer response, but shareholders and participants in the democratic process may attempt to sanction those companies whose labels reveal environmentally destructive behavior. Companies will know this in advance, with likely behavioral consequences. The risk of sanctions from shareholders and state legislatures may well produce environmental improvement even without regulation.

A great deal of recent attention has been given to informational regulation in the particular context of the communications industry. As an alternative to direct regulation, which raises especially severe First Amendment problems, the government might attempt to increase information instead. Thus, the mandatory "v-chip" is intended to permit parents to block programming that they want to exclude from their homes; the v-chip is supposed to work hand-in-hand with a ratings system.[155] Similarly, a provision of the 1996 Telecommunications Act requires television manufacturers to include technology capable of reading a program rating mechanism; requires the FCC to create a ratings methodology if the industry does not produce an acceptable ratings plan within a year; and requires that broadcasters include a rating in their signals if the relevant program is rated.[156] Spurred by this statute, the networks have generated a system for television ratings, which is now in place.[157] The question is whether disclosure requirements might be enlisted more generally.

B. Rationale

Why has information disclosure become such a popular regulatory tool? There are several answers. For various reasons, a market failure may

---

151. 42 U.S.C. § 300g (1994).
152. 42 U.S.C. § 300g-3(c)(4) (Supp. II 1996).
153. The phrase comes from LOUIS D. BRANDEIS, OTHER PEOPLE'S MONEY 92 (1914).
154. 2 U.S.C. § 441a-411h (1994).
155. *See* HAMILTON, *supra* note 45, at 289-92.
156. *See* Telecommunications Act of 1996, Pub. L. No. 104-104, § 551, 110 Stat. 56; *see also* HAMILTON, *supra* note 45, at 302.
157. *See* § 551, 110 Stat. at 141-42.

come in the form of an inadequate supply of information.[158] Because information is generally[159] a public good—something that if provided to one is also provided to all or many—workers and consumers may attempt to free ride on the efforts of others, with the result that too little information is provided. For this reason, compulsory disclosure of information can provide the simplest and most direct response to the relevant market failure.

Information disclosure is often a far less expensive and more efficient strategy than command-and-control, which consists of rigid mandates about regulatory ends (a certain percentage reduction in sulfur dioxide emissions, for example), regulatory means (a technological mandate, for example, for cars), or both.[160] A chief advantage of informational regulation is its comparative flexibility. If consumers are informed of the salt and sugar content of foods, they can proceed as they wish, trading off various product characteristics however they see fit. If workers are given information about the risks posed by their workplace, they can trade safety against other possible variables (such as salary, investments for children or retirement, and leisure).[161] If viewers know the content of television programs in advance, they can use market methods (by refusing to watch) or political methods (by complaining to stations) to induce changes. From the standpoint of efficiency, information remedies can be better than either command-and-control regulation or reliance on markets alone.

From the democratic point of view, informational regulation also has substantial advantages. A well-functioning system of deliberative democracy requires a certain degree of information so that citizens can engage in their monitoring and deliberative tasks. Subject as they are to parochial pressures, segments of the government may have insufficient incentives to disclose information on their own; consider the Freedom of Information Act (FOIA)[162] or the Federal Election Campaign Act (FECA),[163] designed to counteract the self-interest of government or private groups, which may press in the direction of too little disclosure. A good way to enable citizens to oversee private or public action, and also to assess the need for less, more, or different regulation, is to inform them of both private and public

---

158. *See* ANTHONY OGUS, REGULATION: LEGAL FORM AND ECONOMIC THEORY 121-25 (1994).
159. Of course it is possible to give information more "private good" characteristics, and innovative approaches can be expected in the next decade. Consider, for example, fees for access to information on the Internet, or the subscription-based *Consumer Reports*; neither of these approaches converts information into a private good, but both would reduce the range of people who may, without high cost, have access to it. It is possible to imagine a range of approaches that would diminish the cost of access for some while increasing it, or holding it constant, for others.
160. *See* OGUS, *supra* note 158, at 121-49.
161. *See* SUNSTEIN, *supra* note 123, at 329-30.
162. 5 U.S.C. § 552 (1994).
163. 2 U.S.C. § 441a-441h (1994).

activity. The very fact that the public will be in a position to engage in general monitoring may well be a spur to desirable outcomes.

EPCRA is the most obvious example here. Sharp, cost-effective, and largely unanticipated reductions in toxic releases have come about without anything in the way of direct regulation.[164] One of the causes appears to be adverse effects on stock prices from repeated disclosure of high levels of toxic releases.[165] In the area of broadcasting, it is possible to hope that disclosure of public interest programming, and the mere need to compile the information each year, can increase educational and public affairs programming without involving government mandates at all. A primary virtue of informational regulation is that it triggers political safeguards and allows citizens a continuing oversight role, one that is, in the best cases, largely self-enforcing.

None of this is to say that informational regulation is always effective or desirable. Under imaginable assumptions, such regulation would be much less effective than command-and-control regulation and much more expensive than reliance on markets unaccompanied by disclosure requirements. Sometimes informational strategies cost more than they are worth, and may be ineffectual or even counterproductive. Whether these are convincing objections depends on the incentives faced by those who disclose, and these incentives are likely to differ with context. Undoubtedly, the most successful cases of disclosure involve well-organized groups able to impose reputational and financial harm on those engaged in harmful activity.[166]

C. *The Minimal Proposal*

We are now in a position to discuss a disclosure requirement for public interest programming in somewhat more detail. On a quarterly basis, every broadcaster should be required to make public the full range of public interest and public service activities in which it has engaged. The relevant activities might involve free air time for candidates, educational programming, public service announcements, access for disabled viewers (as through closed captioning or video descriptions), charitable activities, emergency warnings and services, and the like. The FCC should require completion of a relatively simple form to ensure accurate and uniform accounting, and FCC staff should sanction those stations that have failed to disclose, or that have done so inaccurately.

A special advantage of disclosure requirements is that they appear to fit well with the emerging communications market insofar as they allow maximum flexibility and do not impose requirements that may be rapidly

---

164. *See* PERCIVAL ET AL., *supra* note 145, at 611-16.
165. *See* Khanna et al., *supra* note 137, at 243.
166. *See* GUNNINGHAM ET AL., *supra* note 20, at 296-300.

outrun by changing technologies. Even in a period in which broadcasters are akin to *Time* and *Newsweek*, such requirements would make a good deal of sense as a means of creating some democratic pressure for improvement.[167] Of course it is reasonable to think that, as the market evolves, disclosure requirements should be placed on all programmers, and not be limited to broadcasters. The hope—based on good results in the environmental context—would be that such requirements would produce a kind of "race," at least in some markets, to do more and better.

D. *Of Realism and Ineffectiveness*

Is the hope realistic? People did not anticipate that the Toxic Release Inventory (TRI)[168] would by itself spur behavioral changes; the question is whether the same forces might operate here. The answer depends on whether the mechanisms that have produced significant voluntary changes in the environmental arena will also be triggered in this setting.

In order for voluntary improvements to occur, the disclosure requirements must be accompanied by political activity or existing norms that will increase public interest programming. With respect to the TRI, well-organized groups have been able to threaten, or to use, publicity so as to induce companies to undertake voluntary reductions. Environmental groups have mobilized when disclosure shows high levels of toxic emissions; anticipating this, companies have reduced emissions voluntarily.[169] Thus, the effect of the TRI has been to draw private and perhaps governmental attention to the most serious polluters, who have an incentive to reduce on their own. Once this process is underway, there has been a kind of competition to produce further reductions, as each polluter seeks to be substantially below the group of most serious polluters.

The question is whether the same might happen here. The answer depends first on the existence of external monitoring and second on the

---

167. An obvious question is whether, if the case for disclosure has been made out, similar requirements ought to be imposed on magazines and even newspapers. Indeed, the same question might be asked about economic incentives and voluntary self-regulation, as discussed below. I do not discuss these questions here. For those who believe that there is increasingly little difference between television and print media, it might seem that if requirements of this sort are not desirable for the latter, they are also undesirable for the former. The best response to this argument is in BOLLINGER, *supra* note 7, at 85, with the suggestion that the different regulatory regimes for the broadcast and print media are well-suited to two different images of press freedom, one image involving democratic self-government, the other involving a form of economic laissez-faire. Bollinger believes that the existence of two parallel regulatory regimes makes appropriate space for each of these images, and that if one regulatory regime goes wrong (through, for example, excessive regulation of television, or an excessive "race to the bottom" in magazines), the other can serve as a corrective. *See id.* at 128-32. This Article is in the general spirit of Bollinger's approach, but it attempts to develop more flexible tools for implementing it, tools that are better adapted to the emerging television market.

168. 42 U.S.C. § 11023 (Supp. III 1997).

169. *See* PERCIVAL ET. AL., *supra* note 145, at 612-16; Hamilton, *supra* note 145, at 106-19; Khanna et. al., *supra* note 137, at 243-44.

power of the monitors to impose reputational or financial harm on broadcasters with poor records.[170] The external monitors may include public interest groups seeking to "shame" badly performing broadcasters; they may include rivals who seek to create a kind of "race to the top." From the disclosures, it should be clear which broadcasters are doing least to promote the public interest, and perhaps those broadcasters will be specially targeted by private groups and competitors. The ultimate effect cannot be known a priori. If public interest organizations, and viewers who favor certain programming, are able to mobilize, perhaps in concert with certain members of the mass media, substantial behavioral effects might be expected. It is even possible that a disclosure requirement would help create its own monitors.

These are analogies between environmental disclosure and disclosure of public interest activities, but there are important differences as well. It is possible that toxic releases are such a salient and easily quantified public "bad" that a political response is quite likely; perhaps a failure to provide public interest programming is a far less salient "bad." An announcement that a certain company has emitted a certain level of toxic pollutants may produce a rapid public outcry and considerable media attention; an announcement that a certain station has failed to provide free air time for candidates may be met with public indifference, even a yawn. Partly this is because acts seem worse than omissions (whatever the conceptual difficulties with making the distinction). Partly this is because the harmful effects of toxic pollution seem serious and real in the abstract, a point that is far less clear with a particular station's refusal to provide programming for children. The harmful effects of that refusal might be mitigated by the fact that programming for children is available on other stations, and indeed any problems caused, or not solved, by profit-seeking television companies might well be remedied by other sources (newspapers, magazines, the Internet). These points suggest that information disclosure may well work better in the environmental context, in which the nature and extent of the problem make public concern far more likely.

In short, companies are responsive to economic incentives, as well as to existing social norms. Information disclosure works best when market pressures, or political pressures, are likely to result in significant costs for those whose performance is poor. In the environmental context, disclosure strategies have worked well when companies have feared their consequences.[171] In the context of television, the risk is that disclosure will have no effects at all—a purely symbolic measure. Indeed, the very collective action problems that argue for public service obligations raise the possibility that an information system will be quite ineffectual.

---

170. *See* GUNNINGHAM ET AL., *supra* note 20, at 296-300; Khanna et al., *supra* note 137, at 243.
171. *See* GUNNINGHAM ET AL., *supra* note 20, at 68-69.

But in view of the relative unintrusiveness of a disclosure requirement, and the flexibility of any private responses, this approach is certainly worth trying. At worst, little will be lost. At most, something will be gained, probably in the form of better programming and greater information about the actual performance of the broadcasting industry—and also about the circumstances in which disclosure requirements will be effective on their own. In light of the aspirations of most viewers, the possible result of disclosure will be to improve the quality and quantity of both educational and civic programming in a way that promotes the goals of a well-functioning deliberative democracy. The most effective system of disclosure would work in concert with well-organized advocacy groups willing to publicize poor performance and to bring general attention to those who do both worst and best.

V
ECONOMIC INCENTIVES

In this Part, I explore the possibility that broadcasters might meet their public interest responsibilities, not through a set of uniform requirements, but through economic incentives. As we will see, the most creative and promising approach, modeled on recent environmental reforms, involves "play or pay," in which broadcasters are given a choice between complying with public interest requirements or paying someone else to put public interest programming on the air. I begin with a discussion of "play or pay," and then move to a more ambitious treatment of the alternatives, growing out of the law of tort, and posing a debate between market-suppressing and market-supplementing approaches.

A. *Of Nature and Coase*

Ronald Coase's work on efficiency, free trades, and transaction costs originated in the area of communications, and in particular, in an attack on the FCC; but it has been most influential in the environmental arena.[172] In that area, there has been a great deal of dissatisfaction with rigid governmental commands, and there has also been an unmistakable movement in the direction of more flexible economic instruments, which are likely to be far more efficient.[173] A command might say, for example, that every coal-fired power plant must reduce its sulfur dioxide emissions by fifty percent, or that it must use technology of a governmentally specified kind. With respect to environmental protection, incentives typically come in two different forms: pollution fees, imposed on those who produce environmental

---

172. *See* R.H. Coase, *The Market for Goods and the Market for Ideas*, 64 AM. ECON. REV. PAPERS & PROC. 384 (1974), *reprinted in* R.H. COASE, ESSAYS ON ECONOMICS AND ECONOMISTS 64 (1994); *see also* R.H. Coase, *supra* note 65.
173. *See* GUNNINGHAM ET AL., *supra* note 20, at 391-421.

harm, and tradeable pollution right or "licenses," given to those who produce pollution. Under the pollution fee model, the government might say that companies must pay a certain amount per unit of sulfur dioxide emission. Under the tradeable pollution right model, the government might say that each company is permitted to emit a certain specified amount of sulfur dioxide, but that its permission, or right, can be bought and sold on the free market. In the former model, it pays to reduce pollution simply in order to reduce the level of the tax. In the latter model, it also pays to reduce pollution because the reduction can be used to engage in more of the relevant activity or in order to obtain money from another who cannot reduce so cheaply. Fees or tradeable licenses should create good dynamic incentives for pollution reduction, and also move environmental protection in the direction of greater cost-effectiveness.[174]

There is a complex literature on the choice between pollution fees and tradeable emission rights.[175] The solution depends largely on an inquiry into what the government knows and does not know.[176] In general, a fee is better if the government is able to calculate the damage done per unit of pollution but has difficulty in calculating the appropriate aggregate pollution level. In those circumstances, a fee is better because the government is unlikely to err in setting it, whereas a system of tradeable permits will produce mistakes. By contrast, a tradeable permit is better if the government knows the appropriate aggregate level, but is unable to calculate the damage done per unit of pollution. In either case, the government can capitalize on the informational advantage held by private businesses participating in pollution control, so as to allow them to decide on the most effective, least expensive method of achieving any particular pollution reduction. If, under a system of pollution fees, it is extremely expensive for Company A to reduce its current level, it may choose to pay a high tax. If the system is one of tradeable pollution rights, it may simply pay someone else, capable of reducing pollution more cheaply, to produce the relevant reduction instead. For any desired level of reduction, a system of economic incentives should produce the right result at a lower cost, by allocating burdens to those most able to bear them.

B. *Taxes, Public Bads, Hot Potatoes, and Cold Spots*

Although the FCC has experimented with allocating communications rights via auction, little thought has been given to the possibility of using economic incentives to promote public interest goals in the

---

174. *See* WILLIAM J. BAUMOL & WALLACE E. OATES, THE THEORY OF ENVIRONMENTAL POLICY 177 (2d ed. 1988).
175. For an overview, see Richard H. Pildes & Cass R. Sunstein, *Reinventing the Regulatory State*, 62 U. CHI. L. REV. 1, 72-85 (1995).
176. *See* STEPHEN BREYER, REGULATION AND ITS REFORM 271-83 (1982).

communications market. In principle, however, both the fees approach and the license approach may well be preferable to government commands. At least this is so if we think of public interest programming as a "good," which people should pay for failing to produce, just as pollution is a "bad," which people should pay for producing. Consider educational programming and free time for presidential elections. Suppose, for example, that ABC is in an especially good position to produce high-quality programming for children, whereas CBS is in an especially good position to promote high-quality programming involving presidential elections. Rather than requiring both ABC and CBS to produce educational programming and programming involving presidential elections, the government might allow each to pay a fee, or a tax, if it is to be relieved of the requirement of providing one or the other. This is the "tax" model of public interest programming. Alternatively, the government might adopt the tradeable emission right model, and allow CBS to sell ABC its obligation with respect to educational programming, while permitting ABC to sell CBS its obligation with respect to presidential elections.

A large problem with a tax is that it is very hard to calculate. Should the government use market measures of some kind, or attempt to measure the public loss, or lost public gain, from the broadcasters' behavior? Either approach would be quite difficult. In these circumstances, the simplest approach would be for the government to experiment with "play or pay" approaches, in which broadcasters have a presumptive obligation to provide public service programming but can buy their way out by paying someone else to provide that programming instead. Such approaches have had considerable success in the environmental area, despite a number of familiar reservations.[177] People have objected, for example, that emissions trading will make an unfortunate "statement" about pollution, thus legitimizing it,[178] or that trading will result in the concentration of pollution in dangerous "hot spots,"[179] or that the administrative burdens of a trading system are overwhelming.[180] Practice has generally shown these objections to be unconvincing.[181] If there is an analogy between environmental protection and broadcasting regulation,[182] a system in which those who do not provide public interest programming must pay a kind of "fee" has an important advantage, because it is so much more flexible than one in which the government imposes uniform obligations on everyone. In this respect, a

---

177. *See* Robert N. Stavins, *What Can We Learn from the Grand Policy Experiment? Lessons from SO_2 Allowance Trading*, J. ECON. PERSP., Summer 1998, at 69.
178. *See* STEVEN KELMAN, WHAT PRICE INCENTIVES?: ECONOMISTS AND THE ENVIRONMENT 110-65 (1981).
179. *See* Stavins, *supra* note 177, at 82.
180. *See id.* at 80.
181. *See id.* at 79-82.
182. See HAMILTON, *supra* note 45, at 285-322, for an instructive discussion.

system of "play or pay" seems to be the most cost-effective means of promoting public interest goals, just as emissions trading is the most cost-effective means of reducing pollution. For those who dislike it, a public interest obligation can be treated as a kind of "hot potato"; fortunately, from their point of view, it is one that they can transfer to others, as a gift accompanied by cash.

It is possible to respond, as has been conventionally thought, that public interest responsibilities are a general part of the public trust and not alienable, and that broadcasters should not be permitted to "buy their way" out of those obligations. But it is unclear what content to give to this statement; the question is what concrete harm would be created by a right to "pay" rather than "play." If the "play or pay" option had corrosive effects on the norms of the broadcasting industry, by making people take public responsibilities less seriously, that would indeed be a problem; but there is little reason to believe that the option would have this effect. The simple question is this: What if a broadcaster were willing to give ten million dollars to PBS in return for every minute, or every thirty seconds, of relief from a public interest responsibility? At first glance, the nation would be better off as a result, simply because the result would be to provide the same level of public interest broadcasting at lower cost. Any objection to a system of tradeable rights would have to be more subtle.

A conceivable problem with an economic incentive in this context is that it may undermine the general purpose of public interest programming by producing a situation in which that programming is confined to a small subset of stations—"cold spots"—in a kind of communications equivalent, or converse, of the "hot spot" problem in the environmental area.[183] The "hot spot" problem arises when trades result in a concentration of pollution in a single area, with serious adverse health effects; it is generally agreed that steps must be taken to ensure that this does not happen.[184] In the communications context, the problem will arise if all of the widely viewed broadcasters end up selling their obligations to a single station or set of stations. This is undesirable if it results in a kind of "ghettoization" of public interest programming and if it is believed, as seems quite sensible, that all or most viewers ought to have access to some public interest programming.

A second problem is both conceptual and administrative. When a trade is made, what is being traded? Perhaps it seems simplest, and most sensible, to trade minutes for minutes. But all broadcast minutes are not the same. An "internal" trade could be one in which ABC (for example) trades an hour of prime-time programming for an hour of 3:00 A.M. programming; an external trade could involve a transfer of one hour of ABC's

---

183. *See* Pildes & Sunstein, *supra* note 175, at 24.
184. *See, e.g.*, Stavins, *supra* note 177, at 82.

highly popular evening hours to (say) FOX's far less popular show in the same period. Steps must be taken to ensure that in any trade, there is an equal public interest benefit for public interest loss. Perhaps a test of audience shares—"viewer-per-viewer" trades—is the best way to start.

An approach of this kind would have the fortunate consequence of helping to handle the "cold spot" problem as well. Part of the problem can be handled by monitoring the sales to make sure that a high-viewer broadcaster is trading to other high-viewer stations. If the "minute-for-minute" trades were adjusted to take account of the number of viewers, a trade to a low-viewer station would be especially expensive. Demographic considerations could play a role as well. The details are less important than the suggestion that a creative administrative solution could reduce the relevant problems, just as these have been handled in the environmental arena.

C. Economic Incentives and the Constitution

What is the relationship between economic incentives and the First Amendment? A direct tax on undesirable programming, or on the failure to provide desirable programming, would raise serious constitutional questions. This is because the tax would be a regulation of speech on the basis of content. Whether such a regulation would be unconstitutional should turn on many of the questions raised in debates over the legitimacy of the "fairness doctrine," designed to compel coverage of serious issues and an opportunity to speak for opposing views.[185] Many people have argued that with decreasing scarcity, the fairness doctrine is no longer legitimate, if it ever was.[186] If this objection is correct, an economic incentive in the form of a tax would be questionable too.

This point raises the question why, if a tax would be constitutionally problematic, a system of "play or pay," which has similar motivations and consequences, would not be constitutionally problematic as well. The intuition might be that a tax is a direct penalty on a certain programming content, whereas "play or pay" simply provides an alternative ("pay") to a legitimate mandate. But this seems to be a form of wordplay. If a tax is questionable, "play or pay" should be questionable as well.

My suggestion here is that there should be no constitutional objection to the extent that the government is acting, in a viewpoint-neutral fashion, to promote educational goals and attention to civic affairs. Current law gives no clear answer to that question.[187] For those who believe that the

---

185. See *Syracuse Peace Council v. FCC*, 867 F.2d 654 (D.C. Cir. 1989), for an outline of the issues.
186. This is the FCC's current position. *See id.* at 656.
187. *See* Turner Broad. Sys., Inc. v. FCC, 520 U.S. 180, 191 (1997) (upholding "must carry" rules as effort to promote widespread dissemination of information from a multiplicity of sources, but emphasizing that the rules are content-neutral). For a general discussion, see CASS R. SUNSTEIN, ONE CASE AT A TIME 172-207 (1999).

government is prohibited from favoring programming of a particular content, "play or pay" should be unacceptable. The problem with this view is that it seems to convert the First Amendment into a species of Herbert Spencer's *Social Statics*,[188] in a way that loosens the connection between the free speech principle and underlying democratic goals. If the First Amendment is associated with democratic self-government and with deliberative democracy, "play or pay," of the sort suggested here, would be perfectly consistent with the free speech guarantee.[189]

D. Expanding the Viewscreen: A Glance at the Cathedral

Thus far I have been exploring economic incentives by contrasting taxes and tradeable rights with command-and-control regulation. But if we wanted a more complete picture, we would widen the viewscreen a bit. In a classic article, Guido Calabresi and A. Douglas Melamed proposed four "rules" that courts might adopt for nuisance suits.[190] Two of the rules come from a situation in which either the plaintiff or the defendant is given the relevant entitlement, and it is protected via a "property rule," in which case the entitlement could be reallocated only through a trade.[191] The other two rules come from a situation in which either is given an entitlement protected by a "liability rule," in which case the entitlement could be reallocated through a legally forced exchange, at a price determined through the legal system (assumed to be the market price).[192] Calabresi and Melamed also discuss "inalienability rules," in which no exchanges are permitted, either voluntarily or through the legal system.[193]

There is a great deal of room for exploring, through this lens, the system of public interest regulation. If public interest programming is desirable, and if certain programming is undesirable, it makes sense to think of ways of requiring broadcasters to pay "damages" or instead requiring the taxpaying public to pay for better programming. Valuation, of course, is a serious problem. Suppose that certain programming (educational or civic, for example) is a public good, producing positive externalities, and that certain programming (violent material, for example) is a public bad, producing negative externalities. How can the government assign monetary values to the desirable and undesirable effects? Is it constitutional for the government to do so? These questions are hard enough in the area of torts;

---

188. SPENCER, *supra* note 117.
189. I discuss the relationship between democratic goals and regulation of speech in SUNSTEIN, *supra* note 68, at 17-51.
190. *See* Guido Calabresi & A. Douglas Melamed, *Property Rules, Liability Rules, and Inalienability: One View of the Cathedral*, 85 HARV. L. REV. 1089 (1972).
191. *See id.* at 1106.
192. *See id.* at 1107.
193. *See id.* at 1111-15.

they are far harder in the context of broadcasting. I restrict myself here to a comparison of some leading alternatives.

Rule 1. The government requires all broadcasters to provide public interest programming; no bargaining is allowed.

Comment: This is the traditional model, with the debate being about its scope; the number of obligations was sharply reduced in the 1980s, but without rethinking the basic model. Under this system, the public's interest in the relevant programming[194] is protected by an inalienable property rule. The entitlement is granted to the government, and it is entitled to mandate broadcaster performance. Broadcasters have a kind of "split" property right; they own the right to broadcast as they choose (in general), but the public has a kind of lien on the property, giving it ownership rights over certain areas. Those who like the traditional approach appear to think that it has good social consequences, by, for example, ensuring that public interest programming is not relegated to unpopular times and channels, and also that it has desirable "expressive" effects, by, for example, affirming the status of broadcasters as public trustees. As we have seen, it also has several problems; its rigidity is likely to lead to inefficiency, and it may well produce unintended adverse consequences, as in the case of the fairness doctrine.

Rule 2a. The government requires broadcasters who do not provide public interest programming to pay a kind of "damage award" to be determined by the government and then used to fund public interest programming by others, such as PBS.

Comment: Under this approach, the public continues to have the relevant entitlement, which is protected by an unusual liability rule. The broadcasters' failure to provide educational programming for children or free air time for candidates would count as a kind of social harm for which broadcasters would have to pay. (The same might be said of the provision of violent or sexually explicit programming, though here the First Amendment problems would be quite serious.) One problem with this approach is the need to calculate the level of the "damage award." There are no clear market measures for this amount, which will therefore have a level of arbitrariness.

Rule 2b. The government requires each broadcaster to provide a certain level of public interest programming, but permits broadcasters to transfer their obligations (accompanied by money) to others, at a market-determined rate.

---

194. I am so describing it for convenience, without making any normative judgment.

Comment: This is akin to Rule 2a, in the sense that the public has the relevant entitlement, which is protected by a kind of liability rule, but here the market, rather than the legal system, determines the value of not playing the public interest programming. One station might sell to another its obligation to provide, say, one hour of educational programming; the selling station would pay the market-determined amount to ensure that the buying station will find it worthwhile to take on the new duty. This kind of market determination could be a substantial advantage in light of limited information on the government's part, as the government is in an extremely poor position to calculate any such "damage award."

As compared with Rule 1, a potential problem with this approach is that some people may avoid the stations that "play" and may not see the relevant programming at all. On the other hand, the empirical question remains whether under Rule 1 most members of the viewing audience will see and benefit from the mandated programming.

Rule 2c. The government establishes a minimum total content of public interest broadcasting on broadcast networks each year (for example, 6 hours of free air time for candidates, 150 hours of educational programming for children); assigns initial obligations to each broadcaster; and then permits broadcasters to trade the obligations at market-determined prices.

Comment: This is very close to Rule 2b; the only difference is that it is a precise analogy to certain initiatives in environmental law, in which the government establishes a maximum level of pollution in the relevant area, provides pollution permits, and then allows trades among polluters.[195] A disadvantage of this approach, as compared to Rule 2a, is that it may be harder to calculate the total level of appropriate programming than to decide on the appropriate tax for those who do not "play." On the other hand, the opposite may be true.[196] An additional difficulty, as discussed above, is that broadcasting hours are not fungible. An hour of children's programming is much less valuable at 3:00 A.M. on Monday than at 9:00 A.M. on Saturday.

Rule 3a. The government pays broadcasters (at reasonable but government-determined rates) to provide public interest programming.

Comment: Under this approach, the broadcasters have the relevant entitlement, which is protected by a liability rule. The broadcaster owns the entitlement, but the government is permitted to obtain a forced exchange,

---

195. This is the acid deposition program under the Clean Air Act, 42 U.S.C. § 7651 (1994).
196. These are brief notes on a complex issue. For details in the environmental context, see Jonathan Baert Wiener, *Global Environmental Regulation: Instrument Choice in the Legal Context*, 108 YALE L.J. 677, 706-35 (1999).

just as it is in the general law of eminent domain. In fact it is possible to see some kind of payment as the constitutionally compelled solution, at least if the right to provide such programming as broadcasters choose is taken to be, by constitutional decree, an entitlement in broadcasters. For example, the government might be able to compel coverage of important issues, or attention to the needs of children, or programming involving emergencies, but the public has to pay.

Interestingly, the government appears not to have tried this approach, at least not as a general rule. An advantage of this approach is that it should not be difficult to calculate the value of the exchange; the market will help answer that question. This is an important advantage over Rule 2a. On the other hand, Rule 3a has probably been resisted, as compared with Rules 1, 2a, and 2b, with the thought that because broadcasters are beneficiaries of public largesse, they should not be paid to promote public interest goals. This is of course a distributional concern, and the underlying judgment—that "broadcasters," rather than "taxpayers," should pay—is not clearly correct in light of the complexity of the incidence of the burden imposed under Rules 1, 2a, or 2b. The burden under the latter rules does not simply fall on "broadcasters," but more likely on advertisers and ultimately on consumers—perhaps to the benefit of those who advertise in newspapers and on cable. Rule 3a might be preferable if it is amended as suggested in Rule 5a, which would require broadcasters to buy spectrum rights.

An important difference between Rule 3a and Rules 2a, 2b, and 2c is that the latter rules give the entitlement to the public, in a way that may have important psychological effects on any trades. The initial allocation of the entitlement creates an endowment effect and tends to "stick,"[197] in part because of its legitimating function. Those who favor public interest obligations might be skeptical of Rule 3a for this reason alone.

An additional problem with this kind of system is that it may provide broadcasters with an incentive to produce less public interest broadcasting on their own than they otherwise would, or at least to understate the amount that they would voluntarily provide. In an unrestricted market, political and shareholder pressures, conscience, and advertiser and viewer demand will result in a nontrivial amount of public interest programming.[198] But if the government proposes to pay broadcasters for whatever public interest programming they provide, voluntary service may be substantially reduced. This is a pervasive problem with paying people to do good or not to do bad; the payment may induce less of the good or more of

---

197. *See* RICHARD H. THALER, THE WINNER'S CURSE: PARADOXES AND ANOMALIES OF ECONOMIC LIFE 63-66 (1992).

198. *See* NATIONAL ASSOCIATION OF BROADCASTERS, BROADCASTERS: BRINGING COMMUNITY SERVICE HOME 2-4 (1998).

the bad. The question for Rule 3 is whether it is possible to generate a "baseline production level" from which any subsidy could be calculated.

Rule 3b. The government must—and does—buy the right to ensure public interest programming, at market-determined rates.

Comment: This is a variation on 3a in the sense that it transforms the broadcasters' entitlement into one protected by a property rule. It is akin to a system of free markets in broadcasting, but with two qualifications: Broadcasters are not required to pay for their entitlement in the first instance, and government stands ready to compete with others who seek to obtain access to viewers. One advantage of Rule 3b over 3a is that it operates entirely on the basis of market-determined prices; no one has to calculate a special government rate. But if one believes that existing fare is not problematic, the government's purchases will be wasteful. And those committed to public interest programming may object that the government will not purchase enough (unless this is specified in some way in advance); they may also object that so long as licenses are being given away free, other rules, not giving a "windfall" to broadcasters, are better.

Rule 4. Broadcasters provide such public interest broadcasting as they choose, including none at all.

Comment: Under this approach, the broadcasters' interest is protected by a property rule. Broadcasters own the relevant entitlement. Just as occurs in an ordinary market, people can pay broadcasters to provide public interest programming, at market-determined prices. The problem with this approach is that the market price might be too high, for all of the reasons discussed in Part I.B of this Article. This approach is similar to Rule 3b, except that the government does not stand ready to ensure a certain level of public interest programming.

Rule 5a. Broadcasters must buy, via auction, the right to broadcast, and once they do that, they can provide such public interest broadcasting as they choose, including none at all.

Comment: This is a genuine market solution. It does not involve a governmental "giveaway" of a scarce resource, and after the valuable commodity has been purchased, free trades are allowed. It seems to have all the advantages of Rule 4, with the further advantage that the valuable property right is purchased rather than simply conferred. The problems with this approach should be easy to identify from Part I.B of this Article. In short, the result may be insufficient public interest programming, partly because of the endowment effect, partly because of collective action problems, partly because of the limitations of the market model in the communications industry, and partly because of public aspirations diverging from private consumption choices.

Rule 5b. Broadcasters must buy, via auction, the right to broadcast, but once they do that, they may be asked or (if for some reason necessary) compelled to provide public interest broadcasting at market-determined prices.

Comment: This is quite similar to Rule 5a. The difference is that the government stands ready to pay for public interest programming on broadcast stations, at market prices; and if broadcasters for some reason refuse, the government can force an exchange. Under Rule 5b, part of the entitlement is owned, but protected only by a liability rule. This approach is in one sense a cousin of Rule 2b. The major difference is that here the taxpayers are paying for public interest programming (at the same time that they receive money from the sale of the spectrum), rather than "broadcasters." The most important difference is therefore distributional. As noted above, that difference is more complex than it seems in light of the fact that when a burden is imposed on "broadcasters," their advertisers are likely to be paying much of the bill, and the result will be complex effects on consumers, on other communications outlets, and on advertising choices. There is no simple redistribution, in Rule 2b, from "broadcasters" to the "public."

Rule 5c. Broadcasters must buy, via auction, the right to broadcast; once they do so, they may be subject to public interest obligations, but they can pay a "damage award," to be used by some other station to support public interest programming, if they fail to fulfill their obligations themselves.

Comment: This should also decrease the amount paid for the entitlement, as compared with Rule 5a. This is a "play or pay" version of Rules 5a and 5b. The problem here lies in determining the level of any such "damage award." This approach is in one sense a cousin of Rule 2a, as a "play or pay" system.

***

A full understanding of these possibilities would greatly facilitate both conceptual and empirical inquiry. Undoubtedly a choice among the various options should depend partly on the particular public interest obligation involved; for example, a requirement of emergency warnings might take the form of Rule 1, whereas a requirement of free air time for candidates might be some combination of Rules 1, 2b, and 5b. The analysis thus far suggests that Rules 2b and 5b have special advantages over the alternatives. Rule 2b, the basic "play or pay" system, seems well-suited to the current period, as a kind of interim improvement over the regulatory status quo. In some areas, Rule 5b may well be better in the longer term, as broadcasters come to resemble general-interest magazines. Movement in the direction of selling the spectrum, rather than giving it away to preselected owners, would also be highly desirable.

E. A Brief Note on Cultural Policies and Cultural Subsidies

The immediately preceding discussion raises obvious questions about when and whether it is appropriate for the government to devote collective resources to the promotion of a better culture, and in particular, to promotion of a better democratic order. A "cultural policy" might, for example, involve the use of taxpayer funds to promote opera or high art, perhaps by subsidizing musicians and artists, perhaps by making it easier for poor people and children to afford to go, perhaps by creating a situation in which people can attend at vastly reduced rates or even for free. In Washington, D.C., for example, people can attend public museums, including those devoted to historical and democratic issues, for free, a practice that creates a distinctive atmosphere for people who visit the nation's capitol. Such visits would have a different tone if, for example, it were necessary to pay three dollars to attend the National Archives, which house the original Declaration of Independence, Constitution, and Bill of Rights.

The general topic of public funding of cultural endeavors is highly controversial and has received considerable attention.[199] The discussion thus far suggests, at a minimum, that such policies should be less controversial when individual consumption choices involve a collective action problem, or when in their capacity as citizens, most people urge their government to promote some activity in order to promote genuinely public aspirations. The argument is most secure when those aspirations involve democracy itself. There is reason for the government to support programming that promotes public education about civic affairs, whether or not that programming is provided by markets themselves; and at a minimum, government subsidies of this kind are firmly supported on theoretical grounds.[200]

VI
VOLUNTARY SELF-REGULATION: ASPIRATIONS, TRUSTEES, AND "WINNER-TAKE-LESS" CODES

In this Part, I discuss the possibility of promoting public interest goals through voluntary self-regulation, as through a "code" of conduct to be issued and enforced by the National Association of Broadcasters (NAB), or

---

199. *See, e.g.*, Ronald Dworkin, *Can A Liberal State Support Art?*, *in* A MATTER OF PRINCIPLE 221, 221-35 (1985).
200. *See* Baker, *supra* note 60, at 383-408.

perhaps by a wider range of those who produce television for the American public. The idea of voluntary self-regulation—of television and content on the Internet—has received growing attention in many nations.[201] For many decades, in fact, the NAB did indeed impose a code, partly to promote its economic interest (by raising the price of advertising), partly to fend off regulation (by showing that the industry was engaged in self-regulation), and partly to carry out the moral commitments of broadcasters themselves.[202] And voluntary self-regulation has played a role in numerous areas of media policy, including, for example, cigarette advertising, children's advertising, family viewing, advertising of hard liquor, and fairness in news reporting.[203] A code has a great deal of potential. Above all, it could address a far greater number of problems than could an economic incentive (for First Amendment reasons), and it appears to have far more potential for producing good, and reducing bad, than a disclosure requirement.

For example, a code could address all public interest obligations mentioned thus far, but also attempt to protect against sexually violent material, against subliminal advertising, against sensationalistic treatment of politics, and against a wide range of other problems with television. The question is whether it is possible, in the current era, for broadcasters to overcome some of the unfortunate effects of the marketplace with voluntary measures. An underlying question, likely to be faced in many areas of regulatory policy both domestically and internationally, is whether a code would work as a kind of unfortunate cartelization or instead as protection against an undesirable "race to the bottom."

A code might do a great deal of good, partly because of the likely existence of external monitors, partly because of a code's capacity to help develop a kind of internal morality likely to affect many of its signatories. A general lesson is that the antitrust laws ought not to be invoked too readily to prevent producers from undertaking cooperative action in circumstances in which competition is producing palpable social harms. In such contexts, a code can provide some of the advantages of government regulation, but do so in a more flexible and better-informed fashion. It is hazardous to invoke the antitrust laws to prevent an industry from providing the kinds of benefits that might be provided, more crudely and expensively, by direct regulation.[204]

---

201. *See* Goldberg et al., *supra* note 5, at 312. For a general discussion, see IAN AYRES & JOHN BRAITHWAITE, RESPONSIVE REGULATION: TRANSCENDING THE DEREGULATION DEBATE 103-16 (1992).
202. *See infra* Part VI.B.3.
203. *See* Campbell, *supra* note 21, at 715-37.
204. *See* FRANK & COOK, *supra* note 23, at 225-27.

A. The Problem and a Recently Emerging Strategy

Notwithstanding the qualifications described above, competitive pressures often can and do provide programming that people would like to see. In an era of cable and satellite television, and an increasingly large range of options, competitive pressures will be especially important in producing "niche" programming for people who have a particular interest in serious programming. The communications market increasingly resembles the market for magazines; recall the possibility, in a digital market, of over one thousand stations. But competitive pressures have a downside. They can lead to sensationalistic, prurient, or violent programming, and to a failure to provide sufficient attention to educational values, or to the kind of programming that is indispensable to a well-functioning democracy.[205] This is so especially in light of the fact that a small relative advantage can lead to huge increases in viewers, a fact that presses television in tabloid-like directions. As Robert Frank and Philip Cook have suggested,

> [i]ncreasingly impoverished political debate is yet another cost of our current cultural trajectory. Complex modern societies generate complex economic and social problems, and the task of choosing the best course is difficult under the best of circumstances. And yet, as in-depth analysis and commentary give way to sound bites in which rival journalists and politicians mercilessly ravage one another, we become an increasingly ill-informed and ill-tempered electorate.[206]

It would be possible to respond to the harmful effects of competitive pressures in various ways. Probably the simplest response would take the form of voluntary self-regulation, through some kind of "code" of good programming; this approach is specifically designed to respond to the problems that can be introduced by market pressures. In various nations, including the United States, cooperative action has played a constructive role in situations of this kind.[207] Though it has yet to receive much academic commentary,[208] voluntary self-regulation via industry agreements is emerging as a regulatory strategy of choice, especially in the environmental arena.[209] The EPA, for example, has encouraged companies that produce pesticides to agree on pesticide reduction strategies, and here

---

205. *See* HAMILTON, *supra* note 45, at 129-284 (discussing television violence).
206. FRANK & COOK, *supra* note 23, at 203.
207. See the discussion of ethical codes in Israel in MOSHE NEGBI, THE ENEMY WITHIN 14 (1998) (discussion paper published by the Joan Shorenstein Center at Harvard University's John F. Kennedy School of Government). Note also that many international bodies attempt to certify quality in a kind of cooperative action designed to reduce adverse effects of market pressures. *See infra* note 218.
208. An exception is Neil Gunningham & Joseph Rees, *Industry Self-Regulation: An Institutional Perspective*, 19 LAW & POL'Y 363 (1997). *See also* AYRES & BRAITHWAITE, *supra* note 201, at 101-32; Campbell, *supra* note 21, at 715-45.
209. *See* GUNNINGHAM ET AL., *supra* note 20, at 50-56, 300-10.

the fact of broad agreement is crucial.[210] California has attempted to deal with the problem of workplace accidents via a "cooperative compliance program" involving self-enforced safety plans on large construction projects. The result has been a significant drop in accident rates.[211] Self-regulating agreements are now in place in Canada's system for forest management; in the Responsible Care program of the chemical industry, now operating in more than forty countries; and in national regulation of nuclear power plants.[212]

What accounts for the increasing popularity of industry self-regulation, as part of the general project of "reinventing government"?[213] From the industry's standpoint, self-regulation allows far more flexibility than government mandates. From the standpoint of government, a special advantage of codes is that they avoid the kind of informational overload that comes from government prescriptions.[214] It is partly for this reason that voluntary agreements among companies have had good effects in the area of occupational safety and health.[215] In a point of special relevance to television, codes also have been found to have the power to influence community attitudes in a way that tends to contribute to the development of a custodial ethic.[216] Thus, codes have helped to develop an institutional morality that brings the behavior of industry members within a normative framework.[217]

Such cooperative action raises concerns about antitrust violations and self-interested profit-seeking under a public-spirited guise. This is of course a risk, but the antitrust law can go wrong when it prevents cooperative action that overcomes palpably adverse effects of market pressures. Indeed, the International Standards Organization is designed specifically to ensure a form of cooperation to overcome those adverse effects;[218] the question is whether that experience has communications analogues.

---

210. *See id.* at 300.
211. *See* Gunningham & Rees, *supra* note 208, at 369.
212. *See* GUNNINGHAM ET AL., *supra* note 20, at 51.
213. *See Special Issue on Self-Regulation, supra* note 22.
214. *See* AYRES & BRAITHWAITE, *supra* note 201, at 110-12.
215. *See* JOHN MENDENHALL, REGULATING OCCUPATIONAL SAFETY AND HEALTH 41-48 (1989); JOSEPH V. REES, REFORMING THE WORKPLACE: A STUDY OF SELF-REGULATION IN OCCUPATIONAL SAFETY 80-82 (1988).
216. *See* GUNNINGHAM ET AL., *supra* note 20, at 162.
217. *See* Gunningham & Rees, *supra* note 208, at 371-72.
218. See, for example, Michael Prest, *Profit Bows to Ethics*, INDEPENDENT (London), Oct. 26, 1997, at 3, stating that:

> Some of the world's biggest companies are putting their weight behind a new, verifiable code of conduct intended to answer mounting consumer criticism of the exploitative conditions under which the goods they sell are produced in poor countries.... The code, called SA8000 (Social Accountability 8000), is the brainchild of the Council on Economic Priorities, an American public interest group, which tries to improve corporate responsibility. It has been drawn up by companies, non-governmental organisations, trade unions, and other interested groups, and is due to start operating next year. The code covers the basic issues of child

My emphasis here is on allowing programmers and journalists to do what, in an important sense, they would actually prefer to do. It is worth underlining this point. Many journalists in the world of broadcasting would very much like to do better;[219] competitive pressures are the problem, not the solution, and a voluntary code could help them and the public as well.

B. A Code: Sample Provisions

The question is whether it might be possible to adopt a new code for broadcasting, specifically designed for the new communications market. There have been many precursors for voluntary self-regulation via codes.[220] A new code might update the old NAB code, and help overcome current problems, without having the degree of tepidness of the existing "standards." A code might even promote some of the goals associated with deliberative democracy.

What provisions might a new code include? The appropriate level of specificity is an important concern; especially clear provisions ("three hours of educational programming per week") risk excessive rigidity, whereas vague provisions ("reasonable efforts to provide educational programming for children") risk meaninglessness. Discussions of code-making in general have stressed the need for "the public announcement of the principles and practices that the industry presumptively accepts as a guide to appropriate conduct and also as a basis for evaluating and criticizing performance."[221] This point argues in favor of a degree of specificity.

Consider the following possible code provisions,[222] simply for the sake of illustration:

1. Each broadcaster shall provide three hours of free air time for candidates during the two-month period preceding the election. In return for free air time, candidates shall discuss substantive

---

labour, forced labour, health and safety, trade union rights, discrimination, discipline, working hours, and pay.... As the name SA8000 suggests, it is the first to be modelled on existing and widely accepted commercial standards such as ISO9000, drawn up by the International Standards Organisation in Geneva, which is used to determine whether companies have the management systems to meet required product quality. But the real strength of the new approach is commercial sanctions. A company which adopts the code also agrees to be independently inspected to see whether it is abiding by the conditions laid down. It will be able to attract customers and gain a competitive advantage by advertising the fact that its factories and suppliers meet the standard.

219. *See* SCHECHTER, *supra* note 50, at 455-58.
220. *See* FINAL REPORT, *supra* note 19, at 114-16.
221. Gunningham & Rees, *supra* note 208, at 383.
222. These provisions are adapted but substantially revised from the more detailed code suggested by the Advisory Committee's FINAL REPORT, *supra* note 19. I believe that the provisions described there are too vague and tepid to be useful (I can attest that this was a quid pro quo for Committee agreement); for reasons discussed in the text, vague and tepid provisions create a high probability of futility and ineffectiveness.

issues in a substantive way and must provide something other than short "sound bites."

2. Each broadcaster shall provide one hour of educational programming for children each day. Broadcasters shall attempt to ensure that children are not exposed to excessively violent programming or programming that is otherwise harmful to, or inappropriate for, children. Broadcasters shall avoid programming that encourages criminal or self-destructive behavior; they should also be sensitive in presenting sexual material that children might encounter.

3. News coverage shall be substantive and issue-oriented. It should not emphasize the sensational and the prurient. It should concern itself with claims and disagreements on matters of substance. Consistent with the exercise of legitimate station discretion, stations should not give excessive or undue attention to sensational accusations, or to reports of "who is ahead in the polls," at the expense of other issues.

4. Morbid, sensationalistic, or alarming details not essential to a factual report, especially in connection with stories of crime or sex, should be avoided. News should be broadcast in such a manner as to avoid panic and unnecessary alarm. News programming should attempt to avoid prurience, sensationalism, and gossip. Stations should make an effort to devote enough time to public issues to permit genuine understanding of problems and disagreements.

5. Violence should be portrayed responsibly and not exploitatively. Presentation of violence should avoid the excessive, the gratuitous, the humiliating, and the instructional. The use of violence for its own sake and the detailed dwelling upon brntality or physical agony, by sight or sound, should be avoided. Programs involving violence should venture to present the consequences to its victims and perpetrators. Particular care should be exercised where children may see, or are involved in, the depiction of violent behavior. Programs should not present rape, sexual assault, or sexual violence in an attractive or exploitative light.

6. Broadcasters shall ensure that their programming is responsive to the needs of citizens with disabilities. To this end, broadcasters shall ensure that programming is accessible, through the provision of closed captioning and other means, to the extent that doing so does not impose an undue burden on the broadcaster. Particular efforts should be made to provide full access to news and public affairs programming. Hearing impaired citizens are

sometimes at risk of a form of disenfranchisement, or even physical danger, because steps are not taken to ensure that television broadcasting is available to them. Stations should take special steps to ensure that information about disasters and emergencies is fully accessible to those who are hearing impaired, ideally in "real time."

7. Broadcasters shall cover international as well as domestic questions and give appropriate coverage to important events in other nations. They should recognize that purely national questions are often hard to evaluate without an understanding of the practices of others, and also that many questions, including those of war and peace, cannot be well-understood without the kind of background that comes from suitable attention to developments and events abroad.

C. A Code: Problems and Prospects

Such a code would of course raise many questions. The first would involve the problem of enforcement. Without an enforcement mechanism, a code might have no effect at all, indeed it might be a form of public deception.[223] The enforcement question is a central part of the general inquiry into the preconditions for effective self-regulation.[224]

There are several obvious possibilities. The simplest would be for the NAB to undertake enforcement on its own, just as it did under the old code. It might, for example, give a seal of approval to those who are shown to comply with its provisions and deny a seal of approval to those who have been shown not to have complied. The NAB might also give special public recognition to those stations that have compiled an excellent public service record in the past year. Such recognition might be awarded for, among other things, meeting the needs of children in a sustained and creative way; offering substantive and extended coverage of elections, including interviews, free air time, and debates; offering substantive and extended coverage of public issues; and providing opportunities for discussion of problems facing the local community. At the time of license renewal, a notation might be given to the FCC that there has been compliance or continuing or egregious noncompliance with the code. If the NAB is unwilling to enforce a code of this kind,[225] perhaps a private group

---

223. *See* John Braithwaite, *Responsive Business Regulatory Institutions*, *in* BUSINESS, ETHICS, AND LAW 91 (C. Cody & C. Sampford eds., 1993) (discussing need for enforcement mechanisms); Campbell, *supra* note 21, at 756-69 (offering a skeptical view of the likelihood of success from any code). For a helpful overview of the preconditions for success, see Douglas C. Michael, *Federal Agency Use of Audited Self-Regulation as a Regulatory Technique*, 47 ADMIN. L. REV. 171 (1995).

224. *See generally* GUNNIGHAM ET AL., *supra* note 20, at 155-72; Michael, *supra* note 223.

225. In the summer of 1998 and January 1999, the NAB decided not to oppose the idea of a code publicly and officially, but did suggest "serious concern" about any government effort to interfere with

could take the initiative, both promulgating the code and publicizing it, and in that (modest) sense sanctioning violations.[226] A special problem here is that in light of increasing competition from nonbroadcast programming sources, a code would not be in the economic interest of broadcasters even if generally adopted, and this is an unpromising fact for a code's effectiveness.[227] Perhaps supplemental enforcement will come from rivals of those who defect from the agreement and violate the code; this is a reasonable prediction in theory, and something similar has been found in analogous areas.[228]

Any enforcement by the NAB or even a private monitoring group would be most likely to succeed if accompanied by external pressures of one sort or another. As in the case of disclosure requirements, the most promising possibilities include public interest groups able to mobilize relevant social norms and to focus media attention on derelict actors.[229] Perhaps such activity would be accompanied by market pressures of various sorts, as consumer action has had significant effects on code enforcement in related areas.[230] A degree of FCC interest in the existence of code violations would also help.

These points raise a related question: the appropriate scope of any such code. Undoubtedly such a code was less painful and easier to operate when three broadcasters exhausted the universe of television. Broadcasters now find themselves in competition with many other entertainment sources, including cable and the Internet. In these circumstances, broadcasters are not likely to constrain themselves if their competitors are not similarly constrained; the competition for an audience for news is much affected by the existence of "tabloid television," and a rule limited to broadcasters would raise an obvious question about fairness. A broadcaster who ties himself to the mast may find himself with a significantly reduced audience. This point suggests that in the development of a code, broadcasters should perhaps be joined by the National Association of Cable Television.

By itself, however, a code limited to broadcasters should do considerable good, even if some broadcasters are reluctant to subscribe to it. When the market reaches the stage in which broadcasters are merely some of a

---

editorial freedom. *See* Paige Albiniak, *Preparing for Battle*, BROADCASTING & CABLE, July 6, 1998, at 22; *Fox Still Undecided on Leaving NAB, Expresses Concerns*, COMM. DAILY, Jan. 14, 1999, at 6.

226. This approach would have special advantages in light of the fact that some stations are not members of the NAB. For NAB stations that must compete with nonmember stations, a code creates an obvious competitive risk, and there is a continuing question whether it is possible to design strategies, public or private, to combat this risk.

227. *See* GUNNINGHAM ET AL., *supra* note 20, at 53-54 (emphasizing relation between economic self-interest and self-regulation).

228. *See* Gunningham & Recs, *supra* note 208, at 403 (finding enforcement from competitors).

229. *See id.* at 390-92.

230. *See id.* at 391 (discussing hostile consumer action).

large number of providers, with no distinctive status, it might make sense to think of a more general code (with suitable adjustments for particular kinds of programmers).[231] It is worth underlining the point that, as in the context of disclosure, the likelihood of success will increase with the existence of third-party monitors, both public and private, and also with a threat of more intrusive action should it prove necessary.[232]

D. Notes on the First Amendment and Antitrust Law

A code for television broadcasters might be thought to raise issues of both constitutional and antitrust law. The constitutional issues are relatively straightforward; the antitrust issues are a bit more complex. I offer a brief discussion here.

There is essentially no risk that a code of the sort suggested here would create serious First Amendment problems. By itself, a code is a private set of guidelines, and private guidelines by themselves raise no First Amendment issue.[233] If a private group decides to impose restrictions on the speech of its members, and government is not involved, the First Amendment is irrelevant.[234] Of course things would be different if government mandated any such code.

Nor would provisions like those described above be likely to violate the antitrust laws. The Department of Justice has so concluded, as have district courts in two important cases in which the private antitrust actions at issue involved parts of previous codes. The district court in *American Brands, Inc. v. National Ass'n of Broadcasters*[235] examined a claim that code standards forbidding cigarette advertising violated antitrust laws. That court refused to issue an injunction against such standards. In *American Federation of Television and Radio Artists v. National Ass'n of Broadcasters*,[236] the district court upheld code standards regarding advertising aimed at children. The courts in both of these cases found the code standards reasonable and in the public interest. As I have noted, the most

---

231. For example, it is hardly clear that a station devoted to children should be required to provide free air time for political candidates.
232. *See* GUNNINGHAM ET AL., *supra* note 20, at 55-56.
233. *See* San Francisco Arts & Athletics, Inc. v. United States Olympic Comm., 483 U.S. 522 (1987) (holding that the Constitution is not applicable to Olympic Committee); Rendell-Baker v. Kohn, 457 U.S. 830 (1982) (holding that the Constitution is inapplicable to private actors); Flagg Bros. v. Brooks, 436 U.S. 149 (1978) (noting that a private contract raises no First Amendment issue). If a code is a product of government threat, and is effectively required by government, the First Amendment comes into play. There can be no question that a governmentally mandated code, not voluntary but taking the form outlined here, would raise legitimate constitutional problems. This does not necessarily mean that the First Amendment would be violated, but it does mean that the code would have to be tested for compliance with First Amendment principles, including constitutional limits on content regulation.
234. *See* cases discussed *supra* note 233.
235. 308 F. Supp. 1166 (D.D.C. 1969).
236. 407 F. Supp. 900 (S.D.N.Y. 1976).

recent code's disintegration came at the hand of a 1979 antitrust action, in which the Department of Justice alleged that certain Code provisions violated the Sherman Act. But the Justice Department's complaint was quite narrow, and the court's decision[237] would not invalidate a code of the sort suggested here.[238] The television market's unique characteristics would allow a code of the sort porposed here to survive a "rule of reason" inquiry, which requires balancing of the relevant factors. It would be most reasonable to hold that any restrictions contained in the code would promote competition as well as various public interests goals, such as education of children, access for the handicapped, and other civic functions.[239]

E. Less Puzzling Puzzles

We are now in a position to disentangle the two puzzles discussed in Part I.B above. Why were broadcasters pleased that the proposed NAB code would violate antitrust laws? Why did they prefer rigid government mandates to a flexible, "play or pay" approach? The broadcasters on the Committee favored a code partly because they thought it a good idea in principle and partly because they had little to lose from it. Though generally "winners," they were selected for the Committee because of their commitment, through both words and deeds, to moderating some of the adverse effects of competition. They were vulnerable to "winner-take-all" effects insofar as they were reluctant to engage in certain competitive practices. In this way, a code might even help them. But the NAB would not like a meaningful code at all. The broadcasting industry as a whole would be hurt by such a code, especially because cable television would not be bound by it. Why should broadcasters, in an intensely competitive market, give a significant edge to cable? Especially if the result would be that cable could take a lot more, of the viewing audience than it now does? It is not surprising if broadcasters who supply a large degree of public interest programming believe that they would be net winners with a code— and if the broadcasting industry as a whole believes that it would be a net loser. Such an industry, cautious about invoking its own economic interest alone, is all too likely to invoke the antitrust laws (or the First Amendment) for purely strategic and self-interested reasons.

This point also helps explain the Committee broadcasters' skepticism about "play or pay" alternatives. A set of rigid public interest requirements does not hurt them and may even help them, insofar as it places their competitors under legal duties that they would themselves meet voluntarily (because of their aspirations or because of the particular demands of their audience and their advertisers). A system of "play or pay" would mean that

---

237. *See* United States v. National Ass'n of Broadcasters, 536 F. Supp. 149 (1982).
238. *See* FINAL REPORT, *supra* note 19.
239. *See id.* at 120-21.

these broadcasters would be undercut by competitors who, unwilling to play, would pay—and capture a large audience share, in a version of "winner-take-most." But this is not a convincing objection to a system of "play or pay." It is true that some of those who "play" will be at a competitive disadvantage with respect to some of those who "pay" others to play instead. But by itself this competitive disadvantage is not worthy of concern, any more than we should be concerned when, in the environmental context, some of those who reduce pollution are at a competitive disadvantage with respect to those who, instead of reducing pollution, pay a substantial fee to third parties who have reduced pollution. The question is what approach yields the best outcome for the public. If there is too little public interest broadcasting, or too much pollution, the solution is not the simple command to "play," or to "reduce," but to increase the price for failing to play or for failing to reduce.[240]

The best defense of a code of the sort I have discussed is that it would produce "winner-take-less" outcomes, in a way that would provide significant benefits for the public by diminishing some of the adverse effects of market competition and by strengthening broadcaster norms in favor of obligations to children and to democratic values. And if this is so, it provides a general lesson about how voluntary private action might sometimes handle problems usually dealt with by direct regulation—and a lesson about the reflexive use of the antitrust laws to prevent producer cooperation. A key question is whether mechanisms might be created to ensure compliance with any such code.

VII
A SUMMARY

This discussion has ranged over a number of regulatory tools, and it may be helpful, by way of summary, to discuss them all briefly and at once. We have seen that both disclosure and codes have the advantage of ensuring a minimal government role, in a way that reduces constitutional concerns and also allows a high degree of flexibility. The danger is that these remedies will have little effect; here the key question is whether there are good external monitors, able to impose reputational or other costs on those who do poorly. We have also seen that "taxes" on programming that does not serve public interest goals, and subsidies to programming that does serve such goals, can have similar effects. A principal problem with subsidies is that they create an incentive not to provide such programming voluntarily. The following table charts the basic territory:

---

240. With the qualification that this could involve hot spots and cold spots, discussed *supra* at Part V.B.

TABLE

	Examples	Potential Virtues	Potential Problems	Most Appropriate Context and Overall Evaluation
1. Mandate	Fairness doctrine; mandatory three hours of children's educational programming; Clean Air Act and Clean Water Act "technology" approaches	Effectiveness; simplicity	Rigidity; high expense; futility (if viewers fail to watch), even counter-productive (if producers are able to circumvent the system and to do worse than they would have done)	Three or four broadcasters and no other television sources; generally ill-suited to current situation, except as "backstops" against egregious behavior
2. Disclosure	Toxic Release Inventory; warning labels for foods and drugs; television ratings system	Flexibility, low administrative costs	Futility (if disclosure does no good), a particular problem iu the event of market failure	Some external monitoring by private groups; well-suited to the likely future
3. Economic incentive—"tax"	"Green taxes"; polluters pay	Efficiency and hence a good programming "mix"; correction of any market failure and serving nonmarket goals	Difficulty of calculating the relevant amount; putting good broadcasters at a competitive disadvantage	Ability to calculate the tax along with market failure or use of nonmarket criteria; well-suited to the short term at least

Table (continued)

	Examples	Potential Virtues	Potential Problems	Most Appropriate Context and Overall Evaluation
4. Economic incentive—subsidy	Public Broadcasting System; National Endowment for the Arts; "cash for clunkers" in environmental policy	Efficiency and hence a good programming "mix"; correction of any market failure and serving nonmarket goals (same as 3)	Difficulty of calculating the subsidy; harmful secondary effects, as in a decrease in voluntary provision of public interest programming (why play when you can be paid?)	Ability to calculate the subsidy along with market failure or use of nonmarket criteria; also well-suited to the short term, in the form of subsidies to PBS
5. Voluntary self-regulation	Old National Association of Broadcasters Code; Current "Responsible Care" program for chemical industry; International Standards Organization	Flexibility, use of industry knowledge, and effectiveness; creation of a kind of "trustee culture"	Lack of enforcement; a charade designed to fend off regulation; an effort by some producers to prevent competition from others; possible unfairness limited to broadcasters	External monitoring and/or industry bona fides and/or threat of mandates if this approach fails; sensible until the distinction crumbles between broadcasters and others
6. Market ordering	Ordinary commodities	Efficiency; consumer sovereignty	Market failure, e.g., externalities, public good problems; use of noneconomic criteria, e.g., democratic ideals	Brave new world (where broadcasters have no special role); perhaps appropriate, probably alongside subsidies and perhaps disclosure, in the very long term

Some of these tools could serve as complements rather than as alternatives. Disclosure makes sense with or without additional strategies. It should be the least controversial item on the list; the only real question is what else should accompany it. Disclosure is likely to work especially well in tandem with voluntary self-regulation, indeed the two tools are natural allies. By contrast, mandates and economic incentives are genuine competitors, and in the current context, economic incentives generally seem best, with mandates operating as a "backstop," probably to be eliminated in the long term.[241]

As I have emphasized, these recommendations are designed for the current stage of telecommunications technology; they are also likely to make sense for the near term. For the next decade, the key question is whether initiatives designed for broadcasters should be applied to cable programmers as well (especially disclosure and compliance with a code). In the very long term, when broadcasters occupy no special role, it may be best to impose disclosure requirements on general-interest stations, and also to subsidize high-quality programming of various sorts. A reasonable conclusion would be that in the long-term, public interest programming will be best promoted via subsidies, not through regulation of any kind. In such an era, much will have been gained and much will also have been lost, with the fragmentation of the television market, with less in the way of common experiences, and, possibly, with less frequent exposure to serious coverage of serious issues.[242] It is too soon to know whether the very long term will come in the next decade or long thereafter.[243]

One final note: The discussion here has been focused on law, not on norms or culture; but the argument for a certain conception of the social role of television bears at least as much on norms and culture as on law. It is possible to think that before long, it will not make sense to impose regulatory requirements on CBS, NBC, ABC, and Fox, any more than it makes sense to impose regulatory requirements on *Newsweek, Time, The New York Times, USA Today, National Review,* or *The New Republic*. I have argued against this conclusion, but it must be taken seriously. Even if it is accepted, it is crucial to say that those who provide television, like those who produce newspapers and magazines, have a distinctive cultural responsibility, associated with the promotion of a democratic culture. Many people not subject to regulation take this responsibility seriously; many do not. If law diminishes as a constraining force, it is all the more important to speak on behalf of the basic norms—protection of children,

---

241. For an overlapping discussion, emphasizing the value of "industry mixes" in the context of environmental law, see GUNNINGHAM ET AL., *supra* note 20, at 422-48.
242. *See* LESSIG, *supra* note 66, at 185-86.
243. *See* OWEN, *supra* note 3, at 311-26 (predicting that convergence will take a long time, with the suggestion that the Internet may never become an important means of delivering television).

serious attention to serious issues, inclusion of a wide variety of social groups, including the hearing impaired—that have undergirded regulatory law in the past. If it flourishes, a "trustee culture" can do much of the work of the law.

Conclusion

Much of the discussion here has involved appropriate regulatory tools. For the most part, the policy instruments of choice should not involve rigid dictates or commands, which are expensive and potentially counterproductive, and in any case ill-suited to an era of rapidly changing technology. I have suggested a strong preference for the less intrusive options of disclosure, economic incentives, and voluntary self-regulation. Disclosure has been a surprisingly successful, low-cost strategy in other areas of regulatory law. If certain broadcasting is seen as a public good, analogous to clean air, economic incentives may be able to accomplish a good deal, and to do so at relatively low cost. Because competitive pressures are frequently the engine behind poor broadcaster performance, voluntary self-regulation may turn out to be a desirable kind of "cartel," helping to counteract short-term interests. Through this route it may be possible to develop an intermediate system of controls, responding to the market and nonmarket failures of current markets, but without introducing the rigidity and inefficiencies of command-and-control regulation. These points hold notwithstanding current and anticipated technological developments.[244] Of course, the appropriate attitude toward such instruments is pragmatic, empirical, and experimental, rather than dogmatic or theological. If these instruments fail to work, it will be worthwhile to consider alternatives.

If measures of this kind have promise in the areas of environmental protection and public interest programming, there is every reason to explore them in other, less familiar contexts as well. In many areas of law, command-and-control regulation has proved a partial or complete failure, and the natural alternative—a system of well-defined property rights and freedom of contract—may produce serious problems of its own. In such circumstances, any "third way," if it is ultimately to develop for the modern regulatory state, is likely to place heavy reliance on disclosure, economic incentives, and voluntary self-regulation.

---

244. See Goldberg et al., *supra* note 5, at 296 (footnote omitted), arguing that:
> A further point of some importance is that there is no single new media form or market, and there is never likely to be such uniformity. Markets remain distinct; for example there is still a clear distinction between television-type services and on-line services. Technological convergence may be imminent in the form of television Internet access (or Web TV) becoming cheaply available, but the cultures remain radically different. Indeed, in the context of television, it seems likely that, though delivery forms may change, the culture may not, and that new types of media may supplement rather than replace existing ones. Again the message is one of caution before we scrap existing regulatory arrangements.

The most fundamental points involve the appropriate understanding of a system of free expression—an understanding of what gives such a system its motivation and point. I have emphasized the external benefits that come from public interest programming and also the peculiar characteristics of the television market, where viewers, or eyeballs, are a commodity provided to advertisers. Because of the collective action problem, regulatory efforts that attempt to promote democratic goals, or that provide captioning for the hearing impaired, are easily defended in principle. But I have also argued that purely economic principles should be rejected as the foundation for communications policy. With respect to public interest programming, viewers' tastes may be a product of an undesirable set of communications options, and in their capacity as citizens, people may well want to make things better rather than worse. Especially in light of the role of the communications media in the production of culture—and on both preferences and values—it is entirely legitimate for a democratic government to refuse to make "consumption choices" the exclusive basis for policy design. I have emphasized that a public committed to deliberative democracy might well support initiatives designed to provide better programming for children and better coverage of public issues.[245] So long as they are subject to democratic control, such initiatives should be regarded not as a paternalistic interference with private choices by a regulatory elite, but as an effort by a self-governing public to promote a political culture that is consistent with its own highest aspirations. Above all, this is the sense in which there is a difference between the public interest and what interests the public.

---

245. See PRICE, *supra* note 9, at 245-46, for related conclusions.

[4]

Self-Regulation and the Media

Angela J. Campbell*

I.   INTRODUCTION 712
II. SELF-REGULATION 714
     A. The Definition of Self-Regulation 714
     B. Arguments in Favor of Self-Regulation 715
     C. Arguments Against Self-Regulation 717
     D. Conditions for Successful Use of Self-Regulation 719
III. USE OF SELF-REGULATION BY THE MEDIA 720
     A. The NAB Code 720
         1. Radio Code 721
         2. Television Code 722
         3. Effectiveness of the NAB Code 725
     B. Children's Advertising Review Unit 735
         1. Effectiveness of CARU's Advertising Guidelines 737
         2. Children's Advertising Review Unit's Online Privacy
            Guidelines 740
     C. Other Self-Regulatory Efforts Involving Media 744
         1. Advertising of Hard Liquor 744
         2. Self-Regulation of News: The Press Councils 746
         3. Comic Books 749

---

* Professor, Georgetown University Law Center. The Author wishes to thank Randi Albert, Michael Kautsch, Joel Reidenberg, Mark Tushnet, and Stefaan Verhulst, as well as the participants at the University of Kansas's Legal Theory Workshop and at the Georgetown Faculty Research Workshop, for their helpful comments and suggestions. The Author is also indebted to Nicole Lemire for her valuable research assistance. The views expressed here are solely those of the Author. However, as Associate Director of the Institute for Public Representation, a clinical program at Georgetown University Law Center, Professor Campbell serves as counsel to some of the parties involved in some of the proceedings discussed in this Article, and her experience informs her perspective. For example, Professor Campbell has worked closely with the Center for Media Education and other advocacy groups to protect the online privacy of children and to implement the V-Chip legislation.

  4. Motion Pictures 750
  5. Video Games 752
  6. Television Violence: Ratings and the V-Chip 753
IV. ANALYSIS.. 755
 A. *The Advantages and Disadvantages of Self-Regulation* 756
 B. *Conditions in Which Self-Regulation Can Work* 757
  1. Industry Incentives and Expertise 758
  2. Agency Ability to Audit 760
  3. Objective v. Subjective Standards.................................. 761
  4. Fair Process and Public Participation 761
  5. Other Conditions.. 763
 C. *Implications for Digital Television* 763
 D. *Implications for Privacy on the Internet* 768
V. CONCLUSION.. 772

I. INTRODUCTION

Calls for self-regulation of electronic media have recently been heard in Washington, D.C. In December 1998, a Presidential Advisory Committee recommended that digital television broadcasters adopt a voluntary Code of Conduct highlighting their public interest commitments.[1] The Advisory Committee Report even included a "Model Voluntary Code" drafted by a subcommittee headed by Professor Cass Sunstein.[2]

Similarly, President Clinton has called for industry self-regulation to address consumer privacy concerns on the Internet.[3] This theme has been

---

  1. NTIA ADVISORY COMMITTEE ON PUBLIC INTEREST OBLIGATIONS OF DIGITAL TELEVISION BROADCASTERS, CHARTING THE DIGITAL BROADCASTING FUTURE, FINAL REPORT OF THE ADVISORY COMMITTEE ON PUBLIC INTEREST OBLIGATIONS OF DIGITAL TELEVISION BROADCASTERS 46 (1998), *available at* <http://www.ntia.doc.gov/pubintadvcom/piacreport.pdf> [hereinafter PIAC REPORT]. As described in this Report, "[d]igital television is a new technology for transmitting and receiving broadcast television signals. It delivers better pictures and sound, uses the broadcast spectrum more efficiently, and adds versatility to the range of applications." *Id.* at xi. The Advisory Committee was charged with developing recommendations concerning public interest obligations to be imposed on television stations as they convert to digital television. Exec. Order No. 13,038, 62 Fed. Reg. 12,063 (1997). It consisted of 22 members representing both the broadcast industry and the public. *See* NTIA, *Advisory Committee on Public Interest: Obligations of Digital Television Broadcasters* (visited Mar. 15, 1999) <http://www.ntia.doc.gov/pubintadvcom/pubint.htm>. The Advisory Committee met over a period of 15 months. The topic of self-regulation was discussed at the Advisory Committee meetings held on June 8, 1998, and September 9, 1998. Transcripts of these meetings may be found at <http://www.ntia.doc.gov/pubintadvcom>.
  2. PIAC REPORT, *supra* note 1, at 106-26.
  3. PRESIDENT WILLIAM J. CLINTON & VICE PRESIDENT ALBERT GORE, JR., A FRAMEWORK FOR GLOBAL ELECTRONIC COMMERCE 18 (1997).

echoed by the Department of Commerce's National Telecommunications and Information Administration (NTIA)[4] and the Federal Trade Commission (FTC).[5] However, the FTC released a report in June 1998, finding that self-regulation had not been successful thus far in protecting consumer privacy.[6] It recommended congressional action specifically to address the issue of children's privacy online, while giving more time to industry to prove it can regulate itself before calling for general privacy legislation.[7]

In the last few days of the 1998 session, Congress in fact passed legislation designed to protect the privacy of children online. Yet, the regulatory scheme established by this legislation contains specific incentives for industry to self-regulate.[8]

Why are the Advisory Committee, the Clinton Administration, and the FTC all calling for self-regulation? It is most likely that they see self-regulation as superior to (or at least more politically acceptable than) government regulation. Many scholars have also touted the benefits of self-regulation, but self-regulation is not without its critics.

The Administration supports private sector efforts now underway to implement meaningful, consumer-friendly, self-regulatory privacy regimes. These include mechanisms for facilitating awareness and the exercise of choice online, evaluating private sector adoption of and adherence to fair information practices, and dispute resolution.

. . . If privacy concerns are not addressed by industry through self-regulation and technology, the Administration will face increasing pressure to play a more direct role in safeguarding consumer choice regarding privacy online.

Id.

4. Elements of Effective Self Regulation for the Protection of Privacy and Questions Related to Online Privacy, *Notice and Request for Public Comment*, 63 Fed. Reg. 30,729 (NTIA 1998).

5. *See, e.g.*, FTC, PRIVACY ONLINE: A REPORT TO CONGRESS i-ii (1998) (noting that the FTC's "goal has been to encourage and facilitate effective self-regulation as the preferred approach to protecting consumer privacy online") [hereinafter PRIVACY REPORT].

6. *Id.* at 41.

7. *Id.* at 42. *See also Electronic Commerce: Hearings on H.R. 2368 Before the Subcomm. on Telecommunications, Trade, and Consumer Protection of the House Comm. on Commerce*, 105th Cong. 307 (1998) [hereinafter *Electronic Commerce Hearings*] (statement of Robert Pitofsky, Chairman, FTC).

8. Children's Online Privacy Protection Act of 1998, Pub. L. No. 105-277, §§ 1301-1308, 112 Stat. 2681, 2728-35 (to be codified at 15 U.S.C. §§ 6501-6506). Section 1303 directs the FTC to promulgate regulations that generally require commercial Web site operators to give notice and obtain consent from the parent of a child under the age of 13 before collecting personally identifiable information from the child. Section 1304 directs the FTC to "provide incentives for self-regulation" by offering a "safe harbor" from prosecution for companies that comply with self-regulatory guidelines issued by industry that have been found, after notice and comment, by the FTC to meet the requirements of the law. *See also Electronic Commerce Hearings*, *supra* note 7, at 309-10 (statement of Robert Pitofsky, Chairman, FTC) (describing how a safe harbor would work).

Part II of this Article reviews the literature on self-regulation to define what is meant by the term, to identify the purported advantages and disadvantages of self-regulation, and to identify the conditions needed for its success. Part III examines situations involving media where self-regulation has been utilized to determine why it was undertaken and whether is has been successful.[9] Based on these examples, Part IV suggests some tentative conclusions about the circumstances under which self-regulation works. Finally, it applies these findings to recent proposals to utilize self-regulation in connection with digital television and privacy on the Internet.

II. SELF-REGULATION

A. *The Definition of Self-Regulation*

The term self-regulation means different things to different people. In introducing a collection of papers analyzing the prospects of self-regulation for protecting privacy on the Internet, Assistant Secretary of Commerce Larry Irving observed:

> Most basically, we need to define what we mean, as the term "self-regulation" itself has a range of definitions. At one end of the spectrum, the term is used quite narrowly, to refer only to those instances where the government has formally delegated the power to regulate, as in the delegation of securities industry oversight to the stock exchanges. At the other end of the spectrum, the term is used when the private sector perceives the need to regulate itself for whatever reason—to respond to consumer demand, to carry out its ethical beliefs, to enhance industry reputation, or to level the market playing field—and does so.[10]

To devise a definition for purposes of this Article, it is useful to break apart the term "self-regulation." The word "self" refers to the actor. It could mean a single company. More commonly, however, and for purposes of this Article, it is used to refer to a group of companies acting collectively, for example, through a trade association.[11] The word "regulation" refers to what the actor is doing. Regulation has three components: (1) legislation, that is,

---

9. In analyzing success, the Author primarily considers whether self-regulatory codes have been successful in achieving their stated purposes. The separate, but important question of whether the stated purposes are in the public interest is beyond the scope of this Article.

10. Larry Irving, *Introduction* to PRIVACY AND SELF-REGULATION IN THE INFORMATION AGE (NTIA 1997), *available at* <http://www.ntia.doc.gov/reports/privacy/privacy_rpt.htm>.

11. Other writers may define self-regulatory activities more broadly. *See, e.g.*, Everette E. Dennis, *Internal Examination: Self-Regulation and the American Media*, 13 CARDOZO ARTS & ENT. L.J. 697 (1995) (including public opinion and media critics in survey of self-regulatory efforts of print media).

defining appropriate rules; (2) enforcement, such as initiating actions against violators; and (3) adjudication, that is, deciding whether a violation has taken place and imposing an appropriate sanction.[12]

Thus, the term "self-regulation" means that the industry or profession rather than the government is doing the regulation. However, it is not necessarily the case that government involvement is entirely lacking.[13] Instead of taking over all three components of regulation, industry may be involved in only one or two. For example, an industry may be involved at the legislation stage by developing a code of practice, while leaving enforcement to the government, or the government may establish regulations, but delegate enforcement to the private sector. Sometimes government will mandate that an industry adopt and enforce a code of self-regulation.[14] Often times, an industry will engage in self-regulation in an attempt to stave off government regulation. Alternatively, self-regulation may be undertaken to implement or supplement legislation.[15]

B. Arguments in Favor of Self-Regulation

The claimed advantages of self-regulation over governmental regulation include efficiency, increased flexibility, increased incentives for compliance, and reduced cost. For example, it is argued that industry participants are likely to have "superior knowledge of the subject compared to [a] government agency."[16] Therefore, it is more efficient for government to rely on

---

12. Peter P. Swire, *Markets, Self-Regulation, and Government Enforcement in the Protection of Personal Information*, in PRIVACY AND SELF-REGULATION IN THE INFORMATION AGE, *supra* note 10, at 9.

13. *But see* Robert Corn-Revere, *Self-Regulation and the Public Interest*, in DIGITAL BROADCASTING AND THE PUBLIC INTEREST: REPORTS AND PAPERS OF THE ASPEN INSTITUTE COMMUNICATIONS AND SOCIETY PROGRAM 63 (Charles M. Firestone & Amy Korzick Garmer eds., 1998), *available at* <http://www.aspeninst.org/dir/polpro/CSP/DBPI/dbpi14.html> (arguing that self-regulation is best promoted by ending all direct and indirect government content control and that efforts to promote government policies by means of threat, indirect pressure, or suggested industry codes are not true self-regulation).

14. IAN AYRES & JOHN BRAITHWAITE, RESPONSIVE REGULATION: TRANSCENDING THE DEREGULATION DEBATE 103 (1992) (viewing self-regulation as a form of subcontracting regulatory functions to private actors). In some countries, laws require industries to adopt codes of practice. For example, Australia requires broadcasting industry groups to develop codes of practice, in consultation with the regulatory authority, concerning such topics as preventing the broadcast of unsuitable programs, promoting accuracy and fairness in news and current affairs, and protecting children from harmful program material. Broadcasting Services Act, 1992, § 123 (Austl.).

15. Frank Kuitenbrouwer, *Self-Regulation: Some Dutch Experiences*, in PRIVACY AND SELF-REGULATION IN THE INFORMATION AGE, *supra* note 10, at 113.

16. Douglas C. Michael, *Federal Agency Use of Audited Self-Regulation as a Regulatory Technique*, 47 ADMIN. L. REV. 171, 181-82 (1995); AYRES & BRAITHWAITE, *supra* note 14, at 110-12.

the industry's collective expertise than to reproduce it at the agency level. This factor may be particularly important where technical knowledge is needed to develop appropriate rules and determine whether they have been violated.

Second, it is argued that self-regulation is more flexible than government regulation.[17] It is easier for a trade association to modify rules in response to changing circumstances than for a government agency to amend its rules. Not only are government agencies bound to follow the notice and comment procedures of the Administrative Procedure Act, but it is often difficult for an agency to obtain the political support and consensus needed to act. It is argued that industry is better able to determine when a rule may be changed to result in better compliance. Moreover, self-regulation can be more tailored to the particular industry than government regulation. While "command and control" regulation may have worked well in the past when addressing near monopolies, it does not work well with different types of market failures.[18] Given the sheer magnitude of individual problems, general rules may lead to absurd results.

Another argument in support of self-regulation is that it provides greater incentives for compliance.[19] It is thought that if rules are developed by the industry, industry participants are more likely to perceive them as reasonable. Companies may be more willing to comply with rules developed by their peers rather than those coming from the outside.[20]

Fourth, it is argued that self-regulation is less costly to the government because it shifts the cost of developing and enforcing rules to the industry.[21] Of course, the government may still be involved in supervision, but supervision requires fewer resources than direct regulation. Indeed, Ian Ayres and John Braithwaite argue that self-regulation is an attractive alternative to direct government regulation because the state "cannot afford to do an adequate job on its own."[22] They acknowledge, however, that self-regulation will only result in a net reduction of cost if the costs to industry are lower than the government's cost savings.[23]

---

17. Michael, *supra* note 16, at 181-82; AYRES & BRAITHWAITE, *supra* note 14, at 110-12.
18. Michael, *supra* note 16, at 186-88.
19. *Id.* at 181, 183-84; AYRES & BRAITHWAITE, *supra* note 14, at 115.
20. Swire, *supra* note 12, at 4; AYRES & BRAITHWAITE, *supra* note 14, at 115-16.
21. *See, e.g.*, Michael, *supra* note 16, at 184; AYRES & BRAITHWAITE, *supra* note 14, at 114.
22. AYRES & BRAITHWAITE, *supra* note 14, at 103.
23. *Id.* at 120-21. If industry expertise is important, it may be the case that the costs to industry are lower.

Self-regulation may also be justified where the rules or adjudicatory procedures differ from the surrounding community or the rules of the surrounding community are inapplicable. Specifically, the argument is sometimes made with respect to the Internet, where jurisdictional and sovereignty issues make it difficult for nations to enforce their laws.[24]

Finally, self-regulation may be used instead of governmental regulation to avoid constitutional issues.[25] For example, it is doubtful under the First Amendment whether government can prohibit the advertising of alcoholic beverages.[26] However, no constitutional question arises if a station or group of stations independently decides not to accept alcohol advertising.

C. Arguments Against Self-Regulation

Critics of self-regulation question the basis for the arguments in favor of self-regulation. For example, while acknowledging that industry may possess greater technical expertise than government, Professor Peter Swire questions whether companies will use that expertise to the benefit of the public, suggesting instead that they are more likely to employ their expertise to maximize the industry's profits.[27] Similarly, the idea that industry will comply more willingly with its own regulations than those imposed from the outside seems somewhat weak where industry is actively involved in developing regulations at the agency.[28]

Other criticisms are directed against self-regulation itself. Leaving regulation to the industry creates the possibility that industry may subvert regulatory goals to its own business goals; or as one article put it, "self-regulators often combine—and sometimes confuse—self-regulation with self-service."[29] Self-regulatory groups may be more subject to industry pressure than government agencies. Moreover, the private nature of self-

---

24. *See, e.g.*, David R. Johnson & David Post, *Law and Borders—the Rise of Law in Cyberspace*, 48 STAN. L. REV. 1367, 1370-76 (1996).

25. *See generally* Duncan A. MacDonald, *Privacy, Self-Regulation, and the Contractual Model: A Report from Citicorp Credit Services, Inc.*, in PRIVACY AND SELF-REGULATION IN THE INFORMATION AGE, *supra* note 10, at 133, 134 (arguing that self-regulatory measures can avoid constitutional issues arising under the First, Fourth, and Fifth Amendments).

26. *See generally* 44 Liquormart, Inc. v. Rhode Island, 517 U.S. 484 (1996).

27. Swire, *supra* note 12, at 13.

28. In the Author's experience in rule makings at the FCC, industry tends to dominate the process. An alternative process that may offer similar opportunities for industry involvement and commitment to regulations is the negotiated rulemaking process.

29. Donald I. Baker & W. Todd Miller, *Privacy, Antitrust and the National Information Infrastructure: Is Self-Regulation of Telecommunications-Related Personal Information a Workable Tool?*, in PRIVACY AND SELF-REGULATION IN THE INFORMATION AGE, *supra* note 10, at 93-94.

regulation may fail to give adequate attention to the needs of the public or the views of affected parties outside the industry.

Many question the adequacy of enforcement in self-regulatory regimes.[30] Industry may be unwilling to commit the resources needed for vigorous self-enforcement.[31] It is also unclear whether industry has the power to enforce adequate sanctions. At most, a trade association may punish noncompliance with expulsion. Whether expulsion is an effective deterrent depends on whether the benefits of membership are important.[32] In many cases, expulsion or other sanctions, such as denial of the right to display a seal, are insufficient.[33]

Without adequate incentives to comply, "bad actors" will be unlikely to comply, and the "good actors" that do comply will be placed at a competitive disadvantage.[34] Where a company can make greater profit by ignoring self-regulation than complying, it is likely to do so, especially where noncompliance is not easily detected by the consumer or likely to harm the particular company's reputation.[35] Like cartels, self-regulatory frameworks may unravel because of cheaters.[36] On the other hand, when enforcement actions are taken, concerns are raised about the exercise of unreviewable discretion.[37]

Another problem with self-regulation is that it can facilitate anticompetitive conduct.[38] Self-regulation, as that term is used in this Article, involves competitors getting together to agree on how they will conduct their

---

30. *See, e.g.*, Michael, *supra* note 16, at 189; Mark E. Budnitz, *Privacy Protection for Consumer Transactions in Electronic Commerce: Why Self-Regulation Is Inadequate*, 49 S.C. L. REV. 847, 874-77 (1998); Deidre K. Mulligan & Janlori Goldman, *The Limits and the Necessity of Self-Regulation: The Case for Both*, *in* PRIVACY AND SELF-REGULATION IN THE INFORMATION AGE, *supra* note 10, at 67-68.

31. Stephen Balkam, *Content Ratings for the Internet and Recreational Software*, *in* PRIVACY AND SELF-REGULATION IN THE INFORMATION AGE, *supra* note 10, at 145 (pointing out that self-regulation requires considerable effort, time, resources, good judgment, and honesty).

32. Henry H. Perritt, Jr., *Regulatory Models for Protecting Privacy in the Internet*, *in* PRIVACY AND SELF-REGULATION IN THE INFORMATION AGE, *supra* note 10, at 110.

33. Such sanctions may be ineffective where consumers lack the knowledge of how a company is viewed by its peers. Moreover, trade associations generally are reluctant to expel their members, especially when the members pay dues to support the association's activities.

34. *Electronic Commerce Hearings*, *supra* note 7, at 356 (testimony of Kathryn Montgomery, President, Center for Media Education).

35. Swire, *supra* note 12, at 6.

36. Perritt, *supra* note 32, at 109-10.

37. Michael, *supra* note 16, at 190; AYRES & BRAITHWAITE, *supra* note 14, at 124-25.

38. *See, e.g.*, Baker & Miller, *supra* note 29, at 93; Joseph Kattan & Carl Shapiro, *Privacy, Self-Regulation, and Antitrust*, *in* PRIVACY AND SELF-REGULATION IN THE INFORMATION AGE, *supra* note 10, at 99.

business. As one article points out, this type of agreement inherently raises antitrust issues, and agreements by professional organizations have sometimes been challenged by the government under antitrust laws.[39]

D. Conditions for Successful Use of Self-Regulation

Professor Douglas C. Michael has surveyed the use of "audited self-regulation" by federal agencies.[40] This term refers to the delegation of power to implement laws or agency regulations to a nongovernmental entity where the federal agency is involved in verifying the soundness of rules, checking compliance, and spot-checking the accuracy of information supplied to it.

Although audited self-regulation is somewhat more narrow than the examples of self-regulation discussed below, Michael's observations about the conditions for successful self-regulation might have broader application. Michael reviewed the literature and hypothesized that audited self-regulation would work best where certain conditions were met:

> First, the private entity to which self-regulatory authority is granted must have both the expertise and motivation to perform the delegated task. Second, the agency staff must possess the expertise to "audit" the self-regulatory activity, which includes independent plenary authority to enforce rules or to review decisions of the delegated authority. Third, the statute must consist of relatively narrow rules related output-based standards. . . . Finally, the agency's and delegated authority's decision must observe rules for notice, hearing, impartiality, and written records of proceedings and decisions.[41]

Michael examined twelve different self-regulatory programs of seven different agencies in the areas of financial institutions, services and products, government benefit programs, nuclear power, and agricultural marketing.[42] His survey confirmed the importance of the industry organization having both expertise and the incentive to self-regulate.[43] Where industry expertise and incentives were missing, self-regulatory programs were abandoned or modified.[44] However, it was not essential to have a preexisting industry organization; rather, "successful regulatory organizations [could] be established contemporaneously with the regulation."[45] The lack of agency expertise also turned out not to be an obstacle because agencies were able to develop the required expertise.[46] Confirming his third hypothesis, he found

---

39. Kattan & Shapiro, *supra* note 38, at 99.
40. Michael, *supra* note 16.
41. *Id.* at 192.
42. *Id.* at 203-41. Some of these programs were successful, while others were not.
43. *Id.* at 241.
44. *Id.* at 242.
45. *Id.* at 241-42.
46. *Id.* at 242.

that the programs with the most subjective standards experienced the most difficulty in implementation.[47] Finally, he found that self-regulatory programs failed where procedural fairness, through such means as rule making on the record with notice and opportunity for comment from all affected groups, was lacking.[48]

III. USE OF SELF-REGULATION BY THE MEDIA

Self-regulation has been tried since the earliest days of electronic media, beginning with radio in the 1920s. The National Association of Broadcasters' (NAB) Radio Code existed for over fifty years, while the Television Code was in effect for about thirty years. Advertisers also have their own codes applicable to broadcast advertising. Examples of the latter include the Guidelines of the Children's Advertising Review Unit (CARU) and the Code of Practice of the Distilled Spirits Industry (DISCUS). Self-regulation has been employed with respect to news reporting, comic books, motion pictures, video games, and the Internet as well. This Part examines several of these examples of self-regulation. It focuses primarily on the experience with the NAB Code and the CARU Guidelines since they are most closely related to the proposals to use self-regulation for digital television (DTV) and privacy on the Internet.

A. *The NAB Code*

The NAB's efforts at self-regulation date back to its beginnings. Founded in 1923,[49] the NAB sought to address the problem of interference by asking radio stations to operate voluntarily only on their assigned wavelengths at assigned hours.[50] This effort failed, however, because too many stations refused to go along. As one commentator notes, "[t]he problem was enforcement of a self-regulatory plan in a circumstance where the natural incentives to break the rules were overwhelming. It might be in the industry's best interest for a particular broadcaster to leave the airwaves, but why would any particular broadcaster voluntarily do this?"[51] The recognition that

---

47. *Id.*
48. *Id.* at 245.
49. The NAB was formed originally in an attempt to resolve disputes over copyright issues between broadcasters and the American Society of Composers, Authors, and Publishers (ASCAP). David R. Mackey, *The Development of the National Association of Broadcasters*, 1 J. BRDCST. 307, 307 (1957).
50. Mark M. MacCarthy, *Broadcast Self-Regulation: The NAB Codes, Family Viewing Hour, and Television Violence*, 13 CARDOZO ARTS & ENT. L.J. 667, 668-69 (1995).
51. *Id.* at 669.

self-regulation was not working to eliminate the chaos on the airwaves was one of the factors leading to the passage of the Radio Act of 1927.[52]

1. Radio Code

Passage of the 1927 Radio Act made clear that if broadcasters failed to control their activities, the government would do it for them.[53] So, after more than a year of discussion and drafting, the NAB adopted its first Radio Code in 1929. It contained two sections—a code of ethics and standards of commercial practices. However, it lacked any enforcement provisions. Once the document was passed, it was largely forgotten.[54]

In 1933, the NAB submitted the Radio Code to the National Recovery Administration, and it was signed by President Roosevelt. This had the effect of turning the voluntary code into a federal law with a federally appointed authority to supervise compliance.[55] However, in 1935, the Supreme Court found such codes unconstitutional in *Schechter Poultry Corp. v. United States.*[56] The NAB thereafter adopted a voluntary code that was, according to one commentator, "placed . . . in an obscure office file . . . not to be dusted off again until 1939."[57] In that year, the Code was amended in response to strong criticism of the industry to make it more specific and to create an enforcement group called the Compliance Committee.[58]

For the first time, the Code was enforced. A provision that prohibited the sale of time for the airing of controversial views was applied to stations that sold time to the popular anticommunist, anticapitalist, anti-Semitic demagogue Father Charles E. Coughlin. Enough stations complied with the NAB Code that Father Coughlin found it difficult to find outlets and eventu-

---

52. *Id.*; Mackey, *supra* note 49, at 320-21.
53. BRUCE A. LINTON, SELF-REGULATION IN BROADCASTING 12 (1967). *See also* Lynda M. Maddox & Eric J. Zanot, *Suspension of the NAB Code and Its Effect on the Regulation of Advertising*, 61 JOURNALISM Q. 125, 126 (1984) (noting that after the Federal Radio Commission called for regulation of the amount and character of advertising, the NAB adopted the Radio Code calling for the voluntary elimination of all commercials between 7 P.M. and 11 P.M.).
54. LINTON, SELF-REGULATION IN BROADCASTING, *supra* note 53, at 12.
55. *Id.*
56. *Schechter Poultry Corp.*, 295 U.S. 495 (1935).
57. LINTON, SELF-REGULATION IN BROADCASTING, *supra* note 53, at 13.
58. *Id.* These changes were adopted in part to respond to the perception that the FCC was prepared to regulate content in wake of the broadcast of *War of the Worlds*. MacCarthy, *supra* note 50, at 672. Among other things, the 1938 Code called for close supervision of children's programs, required fair discussion of controversial issues, banned the sale of time for controversial issues, required news to be fair and accurate, limited the amount and type of commercials, and prohibited advertising of hard liquor and fortune telling. *Id.* at 672-73.

ally went off the air in 1940.[59] A few years later, the NAB revised the Code to conform to a Federal Communications Commission (FCC or Commission) ruling that airtime should be sold for the airing of controversial views[60] and made it clear that the Code provisions were meant merely to guide broadcasters and would not be enforced.[61]

2. Television Code

The first Television Code was adopted at the end of 1951.[62] Based on both the Radio Code and the Motion Picture Code,[63] it was drafted to head off proposed legislation that would have created a citizens advisory board for radio and television.[64]

The Television Code was amended at various points.[65] It probably had its greatest effect in the 1960s and 1970s. In 1961, the NAB expanded its budget and staff to include an overall Code Authority Director for both the Television and Radio Codes.[66] Writing in 1967, Professor Bruce Linton describes the activities of the Code Authority.[67] One function was to interpret the Code by providing advice, publishing guidelines and amendments to clarify Code provisions, and issuing rulings on specific programs or commercials. He notes that most cases were brought to successful conclusion through negotiation rather than issuing a ruling.[68] Most of the daily work of the Code staff concerned commercials rather than program content.[69]

---

59. MacCarthy, *supra* note 50, at 673.

60. United Brdcst. Co., *Decision and Order*, 10 F.C.C. 515 (1945).

61. MacCarthy, *supra* note 50, at 673; LINTON, SELF-REGULATION IN BROADCASTING, *supra* note 53, at 13-14.

62. LINTON, SELF-REGULATION IN BROADCASTING, *supra* note 53, at 15-16; MacCarthy, *supra* note 50, at 674.

63. The Motion Picture Code is discussed *infra* Part III.C.4.

64. MacCarthy, *supra* note 50, at 674 (describing Senator William Benton's bill to establish a National Citizens Advisory Board for Radio and Television to oversee programming and to submit a yearly report to Congress on the extent to which broadcasting served the public interest). Brenner attributes the Television Code "partly as a reaction to congressional ire over hard liquor advertisements on TV." Daniel Brenner, Note, *The Limits of Broadcast Self-Regulation Under the First Amendment*, 27 STAN. L. REV. 1527, 1529 (1975). *See also Liquor Advertising over Radio and Television: Hearings on S. 2444 Before the Senate Comm. on Interstate and Foreign Commerce*, 82d Cong. 227-28 (1952).

65. In the 1950s, changes were made in the aftermath of the quiz show scandals. MacCarthy, *supra* note 50, at 675.

66. LINTON, SELF-REGULATION IN BROADCASTING, *supra* note 53, at 18.

67. *Id.* at 34-46.

68. In 1966, for example, the New York Code Office dealt with 115 advertising agencies and acted on over 1,640 commercials for over 800 different products. *Id.* at 37.

69. LINTON, SELF-REGULATION IN BROADCASTING, *supra* note 53, at 39.

Another function of the Code Authority was that of enforcement. Linton notes that there were

> two areas of compliance involved—one with the [advertising] agencies and producers, and the other with the Code subscribers. The agencies and producers must comply with Code rulings or risk not having their material scheduled on Code stations. The Code subscribers must conform to the Code and comply with Code rulings or run the risk of losing the Code Seal and Code membership.[70]

He found a "considerable amount of 'due process'" for advertisers.[71] All negotiations were confidential unless a finding of noncompliance was made. While confidentiality protected advertisers from criticism, it also "serve[d] to mask from the public any real appreciation of the work of the Code Authority."[72]

Linton observed that determining whether stations were in compliance with the NAB Code was "a fantastically difficult job."[73] Compliance was checked through a system of twice-per-year monitoring backed up by letters of complaint, which came mostly from competing stations. The only sanction for violation was the revocation of subscription, which meant that the station no longer had the right to display the Seal of Good Practice. He noted that the process for removal was "complex and full of 'due process.'"[74]

Writing in the mid-1970s, Daniel Brenner provides a similar account of the Code administration bureaucracy.[75] The Code Authority handled the Code's day-to-day operations.[76] The Code Authority had three offices—one each in New York, Hollywood, and Washington, D.C.[77] It was involved in "clearing every new network series and individual controversial episodes, consulting with advertising agencies and independent program producers on standards of taste, evaluating commercials for Code compliance, receiving complaints, and monitoring subscribers."[78] The Code Authority even published *Code News*, a monthly newsletter.[79]

---

70. *Id.* at 40. Other functions of the Code Authority included acting as a liaison with government and other organizations and publicizing the Code's work. *Id.* at 43-46.
71. *Id.* at 40.
72. *Id.* at 41.
73. *Id.*
74. *Id.* at 42.
75. This description of the Code enforcement bureaucracy is from Brenner, *supra* note 64, at 1530-33.
76. *Id.* at 1530.
77. *Id.* at 1531 n.20.
78. *Id.* at 1530 n.15.
79. *Id.* at 1531 n.20.

Decisions of the Code Authority were reviewed by the Television Code Review Board.[80] This Board consisted of nine members representing subscribing stations, including one member for each of three major networks.[81] It reported directly to the Television Board of Directors at the NAB. The Television Code Review Board also considered revisions to the Code.[82]

Both the Radio and Television Codes were repealed after the Department of Justice (DOJ) filed suit against the NAB alleging that the advertising provisions of the Code violated the antitrust laws.[83] The DOJ argued that provisions limiting the number of minutes per hour of commercials, the number of commercials per hour, and the number of products advertised in a commercial, had the actual purpose and effect of manipulating the supply of commercial television time, with the result that the price of the time was raised to advertisers in violation of section 1 of the Sherman Act.[84] After the district court granted partial summary judgment for the DOJ, the NAB entered into a consent decree in which the NAB agreed to cease enforcing the advertising guidelines.[85] Although only certain advertising provisions had been challenged, the NAB abandoned the Code in its entirety in January 1983.[86]

At the time the NAB decided to abandon the Code, it had become clear that the FCC, then under Republican Chairman Mark Fowler, was abandoning the public trustee concept and dismantling the system of broadcast regulation that had grown up around it.[87] In 1981, the FCC had "deregulated" radio by repealing ascertainment requirements,[88] repealing processing

---

80. *Id.* at 1530.
81. *Id.*
82. *Id.*
83. United States v. NAB, 536 F. Supp. 149 (D.D.C. 1982).
84. The NAB argued that the limits were voluntary guidelines enacted in the public interest. *See* Patricia Brosterhous, United States v. National Association of Broadcasters: *The Deregulation of Self-Regulation*, 35 FED. COMM. L.J. 313 (1983).
85. United States v. NAB, *Proposed Final Judgment and Competitive Impact Statement*, 47 Fed. Reg. 32,810 (DOJ 1982).
86. A few years after the Code was eliminated, Professor Linton found that each of the three major networks had adopted parts of the NAB Code, which were implemented through the networks' Standards and Practices Divisions. He found that the absence of the Code had the greatest impact at the station level. Bruce A. Linton, *Self-Regulation in Broadcasting Revisited*, 64 JOURNALISM Q. 483 (1987).
87. Instead of treating broadcasters as public trustees, Fowler and Brenner believed that the public interest was best served by leaving broadcasters to respond solely to market forces. *See, e.g.*, Mark S. Fowler & Daniel L. Brenner, *A Marketplace Approach to Broadcast Regulation*, 60 TEX. L. REV. 207 (1982); Brosterhous, *supra* note 84, at 342-43.
88. Under ascertainment, stations were required to interview representatives of important groups within their community, determine their concerns, and provide programming responsive to those concerns.

guidelines that essentially required stations to broadcast news, public affairs, and other informational programming, eliminating limits on the amount of time devoted to commercials, and eliminating program reporting requirements.[89] With the deregulation of radio and the FCC's fundamental questioning of the public trustee concept, television deregulation was sure to follow.[90] Because the NAB Code had been adopted to stave off government regulation, once the threat of government regulation was removed, the industry saw no reason to retain the Code.[91]

3. Effectiveness of the NAB Code

It is difficult to evaluate the effectiveness of the NAB Code because there has been little academic study of it.[92] Moreover, the NAB Code covered many different aspects of broadcasting and was frequently amended. In evaluating effectiveness, this Article first examines some general limitations on the Code's effectiveness and then looks at the effectiveness of specific provisions.

---

89. Deregulation of Radio (Part 1 of 2), *Report and Order (Proceeding Terminated)*, 84 F.C.C.2d 968, 49 Rad. Reg. 2d (P & F) 1 (1981), *recons.*, *Memorandum Opinion and Order*, 87 F.C.C.2d 797, 50 Rad. Reg. 2d (P & F) 93 (1981), *aff'd in part and remanded in part*, Office of Comm. of United Church of Christ v. FCC, 911 F.2d 803 (D.C. Cir. 1990) [hereinafter *Deregulation of Radio Report and Order*]. That same year, the FCC eliminated the requirement that stations include in their renewal applications any information about their program efforts. Radio Broadcast Services: Revision of Applications for Renewal of License of Commercial and Noncommercial AM, FM, and Television Licensees, *Report and Order*, 49 Rad. Reg. 2d (P & F) 740 (1981), *recons.*, *Memorandum Opinion and Order*, 87 F.C.C.2d 1127, 50 Rad. Reg. 2d (P & F) 704 (1982), *aff'd sub nom.* Black Citizens for a Fair Media v. FCC, 719 F.2d 407 (D.C. Cir. 1983).

90. In fact, television was deregulated in 1984. Revision of Programming and Commercialization Policies, Ascertainment Requirements, and Program Log Requirements for Commercial Television Stations, *Report and Order*, 98 F.C.C.2d 1076, 56 Rad. Reg. 2d (P & F) 1005 (1984), *recons.*, *Memorandum Opinion and Order*, 104 F.C.C.2d 358, 60 Rad. Reg. 2d (P & F) 526 (1986), *aff'd in part and remanded in part*, Action for Children's Television v. FCC, 821 F.2d 741 (D.C. Cir. 1987) [hereinafter *Deregulation of Television Report and Order*].

91. In 1990, the NAB did issue "voluntary programming principles" concerning children's television, indecency and obscenity, drugs and substance abuse, and violence. MacCarthy, *supra* note 50, at 687. However, it made clear that the application, interpretation, and enforcement of these principles remained at the sole discretion of individual licensees. *Id.* at 688. The NAB's Statement of Principles of Radio and Television Broadcasting are included as Appendix C to the PIAC REPORT, *supra* note 1, at 127.

92. Professor Bruce A. Linton, the Chairman of the Radio-TV-Film Department of the University of Kansas, published a useful outline in 1967. LINTON, SELF-REGULATION IN BROADCASTING, *supra* note 53, at 12. It summarizes the history, current activities, and some of the problems with the NAB Code, but it cites no studies analyzing its effectiveness. Most law review writing about the NAB Code focuses, not surprisingly, on the antitrust suit and the Writers Guild's challenge to the Family Viewing Hour.

a. General Limitations on Effectiveness

Linton, writing in 1967, found that the biggest obstacle to the NAB's self-regulation was the extent of subscribership.[93] At that time, 396 out of 631 television stations (63 percent) subscribed, and the percentage for radio stations was even lower.[94] The most common reason for a station's refusal to subscribe was that it could not comply with the advertising restrictions and still make a profit.[95] Some stations objected more on grounds of ideology or fear of government regulation: They viewed the Code of Practice as providing *"a clear blueprint for increased government control of broadcasting."*[96] Yet, other stations declined to subscribe because they found the standards too low.[97]

In a 1975 article, Brenner characterized the "[a]pprehension of violations by the Code Authority [as] spotty at best."[98] Code monitoring was done by a randomly rotated, quantitative review of subscribers' program logs conducted twice yearly. Because the staff could not be expected to cover member stations with any regularity, it was forced to "rely mainly on complaints by those few viewers aware of [the Code Authority's] existence or by subscribers themselves, who request Code review."[99]

Where violations were found, the Code provided for "a 21-part suspension procedure with final review residing with the Television Board of Directors."[100] Suspension was rare.[101] As Brenner points out, the NAB was reluctant to suspend stations because it did not want to lose dues-paying

---

93. *Id.* at 41-42.
94. These figures are consistent with Brenner's statement that as of January 1, 1975, 60% of television stations subscribed. Brenner, *supra* note 64, at 1530 n.17. In 1979, 65% of television stations were members of the NAB, but the Television Code's influence may have been greater than these numbers suggest. These 65% of stations accounted for 85% of all viewing. Moreover, as far as commercials were concerned, since all of the three major networks were members of the NAB and most stations at that time were network affiliates, the network commercials presumably complied with the Code even if the affiliate was not a member. Maddox & Zanot, *supra* note 53, at 125.
95. LINTON, SELF-REGULATION IN BROADCASTING, *supra* note 53, at 50.
96. *Id.* at 52.
97. Brenner, *supra* note 64, at 1531 n.21.
98. *Id.* at 1531 n.22.
99. *Id.* Linton notes that stations listened to each other and often complained to the Code Authority if a competitor was not in compliance. LINTON, SELF-REGULATION IN BROADCASTING, *supra* note 53, at 41.
100. Brenner, *supra* note 64, at 1532 n.22.
101. Indeed, the only case of suspension the Author found, involved the suspension or withdrawal of 39 stations in the early 1960s for airing a commercial for "Preparation H," which the Code Review Board found unsuitable for television. However, several years later, the Television Board of Directors reversed the Board's position, and 23 of the stations rejoined. *Id.*

members.[102] Moreover, the NAB's ultimate sanction was ineffective. Broadcast historian Erik Barnouw describes the Television Code's "enforcement machinery" as

> among its most absurd features. If a subscribing station was charged with violating the code and found guilty by an NAB review board, the station (according to the rules) would lose the right to display on the screen the NAB "seal of good practice." Since the seal meant nothing to viewers and its absence would be virtually impossible to notice, the machinery meant nothing.[103]

Thus in general, effective enforcement of the Television Code was hampered by the less-than-universal industry participation, limited resources, and inadequate enforcement incentives.

b. *Commercial Provisions*

Most enforcement actions involving the NAB Code concerned the commercial restrictions.[104] The NAB Code contained a variety of restrictions on the amount and type of commercial matter that could be broadcast. For example, the amount of nonprogram material (including advertising) was limited to 9.5 minutes per hour in prime time and to sixteen minutes per hour at all other times.[105] Advertising of hard liquor, fortune telling, and fireworks was prohibited, while commercials for beer, wine, and products of a personal nature were permitted so long as they were in "good taste."[106] In addition, broadcasters were responsible for making available documentation to support the truthfulness of claims, demonstrations, and testimonials contained in commercials.[107]

In 1963, the FCC sought to transform the commercial time restrictions contained in the Code into regulations,[108] but this effort was abandoned due to the objections of the broadcasters.[109] Nonetheless, during the 1960s and

---

102. *Id.*
103. 3 ERIK BARNOUW, THE IMAGE EMPIRE: A HISTORY OF BROADCASTING IN THE UNITED STATES FROM 1953, at 251 (1970).
104. LINTON, SELF-REGULATION IN BROADCASTING, *supra* note 53, at 39.
105. NAB, THE TELEVISION CODE 16-20 (19th ed. June 1976) [hereinafter TELEVISION CODE]. Different advertising limits applied to children's programs. *See infra* Part III.A.3.b(ii).
106. TELEVISION CODE, *supra* note 105, at 10-12.
107. *Id.* at 12-14. *See generally* Herbert J. Rotfeld et al., *Self-Regulation and Television Advertising*, 19 J. ADVER. 18, 19 (1990) (describing the advertising clearance procedures when the NAB Code was in effect and how they changed after repeal of the Code).
108. Amendment of Part 3 of the Comm'n's Rules and Regulations with Respect to Advertising on Standard, FM, and Television Broadcasting Stations, *Report and Order*, 36 F.C.C. 45 (1964).
109. Broadcasters overwhelmingly opposed the FCC's proposal. *Id.* para. 2. In addition to opposing the FCC's proposals in their comments, broadcasters complained to Congress

1970s, the FCC staff utilized the NAB Code as a processing guideline in reviewing station license renewals.[110] These processing guidelines, in effect, established NAB Code compliance as a "safe harbor" with detailed government review of the record of stations not meeting the Code. Few stations were found to exceed the guidelines.[111] Thus, at least during the 1960s and 1970s, the NAB Code was generally successful in limiting the amount of commercials. Next, this Article examines some specific types of advertising—cigarette advertising and advertising to children.

(i) Cigarette Advertising

One type of advertising that the NAB Code was unsuccessful in restricting was that of cigarette advertising. In the 1960s, there was a great deal of public concern about the impact of smoking on health.[112] Much of this concern focused on cigarette advertising. At that time, cigarette advertising accounted for 10 percent of all broadcasting advertising revenues.[113]

According to the congressional testimony of Warren Braren, who was the manager of the New York Code Authority office from 1960 until he resigned in 1969, the NAB Code Authority staff proposed strict guidelines for cigarette advertising in the summer of 1966.[114] The proposed guidelines

---

and influenced the House of Representative to pass a bill that would have prevented the FCC from adopting rules limiting the number of commercials. ERWIN G. KRASNOW ET AL., THE POLITICS OF BROADCAST REGULATION 193-96 (3d ed. 1982).

110. KRASNOW ET AL., *supra* note 109, at 196-97. *See, e.g.*, Amendment of Part O of the Comm'n's Rules, *Order*, 43 F.C.C.2d 638, 640 (1973) (delegating authority to Chief of the Broadcast Bureau and directing that television stations proposing more than 16 minutes per hour of commercial time be referred to the full Commission). These "delegations of authority" to the staff, which permitted the staff to renew radio and television licenses so long as they did not exceed certain advertising limits based on the NAB Code, were repealed in 1981 for radio and 1984 for television. *See Deregulation of Radio Report and Order*, *supra* note 89; *Deregulation of Television Report and Order*, *supra* note 90.

111. *Deregulation of Radio Report and Order*, *supra* note 89, para. 83; *Deregulation of Television Report and Order*, *supra* note 90, para. 59. In the antitrust litigation, however, the court found the extent to which the NAB Code influenced the supply and price of commercial time to be "a disputed issue of material fact." United States v. NAB, 536 F. Supp. 149, 158 (D.D.C. 1982).

112. *See generally* RICHARD KLUGER, ASHES TO ASHES: AMERICA'S HUNDRED-YEAR CIGARETTE WAR, THE PUBLIC HEALTH, AND THE UNABASHED TRIUMPH OF PHILIP MORRIS 221-62 (1996).

113. MacCarthy, *supra* note 50, at 676. Moreover, broadcast advertising accounted for approximately 70% of the cigarette makers' advertising budgets. KLUGER, *supra* note 112, at 302.

114. *Regulation of Radio and Television Cigarette Advertisements: Hearing Before the House Comm. on Interstate and Foreign Commerce*, 91st Cong. 5-10 (1969) [hereinafter *Cigarette Regulation Hearing*] (statement of Warren Braren, Former Manager, New York Office, National Association of Broadcasters Code Authority). *See also* KLUGER, *supra* note 112, at 331-32 (describing how Braren's testimony came about).

would have prohibited the use of heroes to promote smoking, banned depictions of smoking in advertisements, and ruled out cigarette advertisements in sports settings. Yet, the Television Code Review Board rejected these proposals, instead adopting weaker guidelines that required few or no changes in then-existing advertising practices.

Even these limited guidelines were resisted by the tobacco companies and television networks. Braren describes how the attempts at self-regulation were rendered ineffectual due to the actions of some networks and tobacco companies.[115] He attributed the network subscribers' lack of support for the Code Authority to the impact on their advertising revenues.[116] Braren asserted that "the code authority is little more than a 'step-child' of the NAB. The autonomy and power needed in order to become a truly professional body acting objectively to serve the public interest is absent."[117] The NAB's failure to address the concerns about cigarette advertising was one of the factors responsible for Congress's adoption of an outright prohibition on broadcast advertising of cigarettes.[118]

(ii) Children's Advertising

The Television Code met greater success, at least for a time, in limiting the amount of advertising on children's programs as well as preventing direct governmental regulation of advertising on children's television. The NAB first adopted toy advertising guidelines in 1961. These guidelines were modest in scope and substance. At that time, there was little advertising directed at children and little research about children's understanding of advertising.[119] By the 1970s, however, the nature of advertising to children had changed. The concentration of children's programming on Saturday mornings led to more focused advertising.[120] Concerns about advertising on chil-

---

115. For example, he describes how the American Tobacco Co., which had withdrawn from the tobacco industry's own code in 1967, repeatedly violated the broadcasting guidelines by airing commercials that had never been submitted to the Code Authority. American Tobacco took the view that if a commercial was approved by television networks, there was no reason for the Code Authority to raise any questions. *Cigarette Regulation Hearing, supra* note 114, at 7.

116. *Id.* at 9.

117. *Id.* at 10.

118. Federal Cigarette Labeling and Advertising Act, 15 U.S.C. § 1335 (1994). Interestingly, it was broadcasters, rather than the tobacco companies that unsuccessfully challenged the constitutionality of the advertising prohibition. *See* Capital Brdcst. Co. v. Mitchell, 405 U.S. 1000 (1972).

119. Dale Kunkel & Walter Gantz, *Assessing Compliance with Industry Self-Regulation of Television Advertising to Children*, 21 J. APPLIED COMM. RESEARCH 148, 151 (1993) [hereinafter Kunkel & Gantz, *Assessing Compliance*].

120. *Id.*

dren's television were raised in the early 1970s by groups such as Action for Children's Television (ACT). Action for Children's Television called on the FCC, as well as the FTC, to prohibit or limit advertising directed at children.

While the FCC deliberated on the ACT proposal, the NAB amended the Television Code to diminish the number of children's advertisements.[121] In 1974,[122] the FCC declined to adopt limits on children's advertising, noting that the NAB had proposed limits to go into effect in 1976 of 9.5 minutes on weekend children's programs and twelve minutes for children's programs shown during the week.[123]

In 1979, a task force established by the FCC found that broadcasters had in general complied with the NAB's advertising guidelines.[124] However, with the subsequent abandonment of the NAB Code in 1983 and the deregulation of television in 1984, all limits on advertising on children's shows were removed and advertising increased.[125] Eventually, Congress directed the FCC to adopt numerical limits in the Children's Television Act of

---

121. KRASNOW ET AL., *supra* note 109, at 198. In response to the FCC's urging, the NAB also amended its Code to prohibit host selling on children's programs. *See* American Federation of Television and Radio Artists v. NAB, 407 F. Supp. 900, 901-02 (S.D.N.Y. 1976).

122. Also in 1974, the National Association of Advertisers created the Children's Advertising Review Unit (CARU), which is discussed *infra* Part III.B.

123. Petition of Action for Children's Television (ACT) for Rulemaking Looking Toward the Elimination of Sponsorship and Commercial Content in Children's Programming and the Establishment of a Weekly 14-Hour Quota of Children's Television Programs, *Children's Television Report and Policy Statement*, 50 F.C.C.2d 1, para. 43, 31 Rad. Reg. 2d (P & F) 1228 (1974) [hereinafter *Children's Television Report and Policy Statement*]. The FCC also relied on the fact that the Association of Independent Television Stations (INTV) had also proposed limits. Moreover, the NAB had recently amended its advertising Code to require a separation device between programming and commercials and to prohibit host selling. *Id.* paras. 49, 52. The FCC acknowledged that some stations were not members of either the NAB or INTV, but warned they too would be expected to bring their advertising practices into conformance with the Code requirements. *Id.* para. 43 n.13. Action for Children's Television appealed the FCC's decision to rely on industry self-regulation, and the court found the FCC's actions reasonable. *See* Action for Children's Television v. FCC, 564 F.2d 458, 481 (D.C. Cir. 1977).

124. Children's Television Programming and Advertising Practices, *Noticed of Proposed Rulemaking*, 75 F.C.C.2d 138, para. 5 (1979) (citing 2 FCC, TELEVISION PROGRAMMING FOR CHILDREN: A REPORT OF THE CHILDREN'S TELEVISION TASK FORCE 47-60 (Oct. 1979)).

125. *See, e.g.*, Dale Kunkel & Donald Roberts, *Young Minds and Marketplace Values: Issues in Children's Television Advertising*, 47 J. SOC. ISSUES 57, 68 (1991); Dale Kunkel & Walter Gantz, *Children's Television Advertising in the Multichannel Environment*, 42 J. COMM. 134, 143-44, 147 (1992) (finding children's advertising on networks in 1990 averaged 10:05 minutes per hour compared to eight minutes in 1983; also noting the development of a new type of advertising/program combinations).

1990.[126] The numerical limits in the Act—10.5 minutes per hour on weekends and twelve minutes per hour on weekdays[127]—were slightly higher than those in the former NAB Code. Thus, with the exception of cigarette advertising, the NAB Code seems to have enjoyed relative success in limiting advertising.

c. Programming Provisions

In contrast to advertising, the NAB's Television Code seems to have had less effect in the programming area. The Program Standards of the Television Code contained both general and specific prescriptions and proscriptions. For example, Part I, Principles Governing Program Content, advised broadcasters to "be conversant with the general and specific needs, interests and aspirations . . . of the communities they serve. They should affirmatively seek out responsible representatives of all parts of their communities so that they may structure a broad range of programs that will inform, enlighten, and entertain the total audience."[128]

Other sections dealt with responsibility toward children, community responsibility, special program standards, treatment of news and public events, controversial public issues, political telecasts, and religious programs. As one writer put it, the Code served as "a readymade articulation of . . . professional conscience that may be exhibited in the event of inquiry by the FCC or community groups."[129] It was closely tied to and represented a further articulation of the public trustee concept in broadcasting.[130]

Few studies analyzing compliance with programming provisions exist, and they are limited to children's programming.[131] Such studies would have been difficult to conduct because many of the program areas covered by the Television Code were also subject to FCC regulation,[132] and thus, it would

---

126. Children's Television Act of 1990, Pub. L. No. 101-437, 104 Stat. 996 (codified as amended at 47 U.S.C. § 303a (1994)).

127. 47 U.S.C. § 303a(b) (1994).

128. TELEVISION CODE, *supra* note 105, at 2. This principle essentially embodied the FCC's then-existing requirement that broadcast stations conduct ascertainment. *See supra* note 88.

129. Brenner, *supra* note 64, at 1531.

130. Brosterhous, *supra* note 84, at 314.

131. There is some anecdotal evidence that Code provisions were effective in keeping some types of programming off the air. For example, in the *Mark* case, NBC cited the NAB Code as the reason for declining to allow an astrologist to appear as a guest on the Tonight Show. *Mark v. FCC*, 468 F.2d 266 (1st Cir. 1972).

132. Brosterhous, *supra* note 84, at 314.

be difficult if not impossible to attribute findings of compliance to the Television Code rather than to the FCC's regulation.[133]

The study by the FCC Task Force suggests that self-regulation of children's programming was generally ineffective. The one program provision in the NAB Code that seemed to be effective—the so-called "Family Viewing Policy"—was subject to litigation and eventually repealed.

(i) Children's Program Provisions

Part II of the Television Code, "Responsibility Toward Children," addressed affirmative programming responsibilities:

> Broadcasters have a special responsibility to children. Programs designed primarily for children should take into account the range of interests and needs of children from instructional and cultural material to a wide variety of entertainment material. In their totality, programs should contribute to the sound, balanced development of children to help them achieve a sense of the world at large and informed adjustments to their society.[134]

This Part appears to have had little effect. Peggy Charren, the founder of Action for Children's Television and a member of the Advisory Committee, observed at the meeting discussing the proposed voluntary code:

> There are aspects of the old [NAB] code, like the description of how to serve children, that were never paid any attention to. They actually sounded like I wrote them, and if they were happening during the 30 years I was in business, I never would have started [Action for Children's Television].[135]

Indeed, the Children's Television Task Force, which was directed by the FCC to "determine the effectiveness of industry self-regulation," found that broadcasters had not complied with programming guidelines.[136] Specifically, it found that broadcasters had failed to provide a reasonable amount of educational and informational programming.[137] Public and congressional

---

133. For example, the provisions concerning controversial public issues (Part VI) track the FCC's Fairness Doctrine. Indeed, the section on Political Telecasts (Part VII) refers to the Communications Act and FCC regulations on political broadcasting. TELEVISION CODE, *supra* note 105.

134. *Id.* at 3.

135. *Open Meeting of Advisory Committee on Public Interest Obligations of Digital Television Broadcasters, June 8, 1998*, at 24-25 (visited Mar. 15, 1999) <http://www.ntia.doc.gov/pubintadvcom/junemtg/transcript-am.htm>.

136. 1 FCC, TELEVISION PROGRAMMING FOR CHILDREN: A REPORT OF THE CHILDREN'S TELEVISION TASK FORCE REPORT 2 (Oct. 1979) [hereinafter TASK FORCE REPORT VOL. 1].

137. 2 FCC, TELEVISION PROGRAMMING FOR CHILDREN: A REPORT OF THE CHILDREN'S TELEVISION TASK FORCE REPORT 25 (Oct. 1979) [hereinafter TASK FORCE REPORT VOL. 2]. It also found that television stations had failed to increase the overall amount of children's programming, *id.* at 17, had failed to air age-specific programs for children, *id.* at 28, and

dissatisfaction with broadcasters' program offerings for children eventually led to the passage of the Children's Television Act of 1990,[138] and ultimately, the guideline that each station offer a minimum of three hours per week of programming specifically designed to serve the educational and informational needs of children.[139]

(ii) The Family Viewing Policy

In addition to imposing a duty to affirmatively provide programming for children, the Television Code was amended in 1975 to limit children's exposure to programming containing violence, sex, or offensive language. Under what came to be known as the Family Viewing Policy, programs deemed inappropriate for general family audiences could not be shown during the first two hours of network programming in prime time.[140] The practical effect of this policy was to require that such programs be scheduled after 9 P.M. This action came after a great deal of pressure from Congress and the FCC to address the problem of televised violence.[141] The Writers Guild, which represented program producers, challenged the Family Viewing Policy in district court. It argued that the NAB's adoption of the Family Viewing Policy was not the result of voluntary industry self-regulation, but had been coerced by the government, and as a result, it constituted state action in violation of the First Amendment.

The district court allowed extensive discovery, and some of its findings provide interesting insights into the operation of the NAB Code. For example, the court found that a high level official at CBS was genuinely concerned with the level of violence on television, but

> feared that if CBS publicly committed itself to [the family viewing policy] that the commitment would work to CBS's competitive disadvantage in the absence of a binding enforcement mechanism applicable to the industry at large. Past experience in children's programming had led him to the conviction that broadcasters, more interested in dollars than in the public interest, would use violence as a tool to hike program ratings if they were left free to program in their own discretion. CBS was thus prepared to delegate its program discretion

---

that many stations were continuing to schedule children's programming only on weekends, *id.* at 42.

138. Pub. L. No. 101-437, 104 Stat. 996-1000 (codified as amended at 47 U.S.C. §§ 303a, 303b, and 394 (1994)).

139. Policy and Rules Concerning Children's Television Programming, *Report and Order*, 11 F.C.C.R. 10,660, paras. 159-63, 3 Comm. Reg. (P & F) 1385 (1996).

140. Writers Guild of Am., W., Inc. v. FCC, 423 F. Supp. 1064, 1072 (C.D. Cal. 1976).

141. *Id.* at 1094-95.

to the NAB, but only if its major competitors could be persuaded to do so as well.[142] The court found that at least as to the Family Viewing Policy, "the NAB function[ed] as an effective enforcement mechanism."[143]

The district court agreed with the Writers Guild on the state action claim and found that the FCC had violated the First Amendment.[144] Although the court conceded that there would be no First Amendment issue if the networks had truly decided to regulate themselves, or even if the regulation was the product of government "encouragement," it found that the type of indirect coercion engaged in by the government raised censorship issues under the First Amendment and the Communications Act.[145]

The district court's decision was overturned by the court of appeals for lack of jurisdiction.[146] The court of appeals also questioned the existence of state action and remanded the case to the FCC for determination. Not surprisingly, the FCC concluded that no improper coercion had occurred.[147] Yet, the Family Viewing Policy was never restored. In the meantime, the DOJ had brought an antitrust suit against the NAB. Although nothing in the antitrust suit required abandonment of the parts of the Code concerning violence and other programming inappropriate for children, the NAB decided to abandon the Code in its entirety.[148]

A few years later, Congress passed a law exempting broadcasters from the antitrust laws so that they could take collective action to reduce violent programming.[149] They did nothing to take advantage of this legislation, however. The broadcasters' failure to do so once the risk of antitrust liability was removed suggests that they had instituted the Family Viewing Policy

---

142. *Id.* at 1094.

143. *Id.* at 1123-34.

144. *Id.* at 1151. The court also found that the FCC violated the Administrative Procedure Act because it had adopted a new policy without conducting a notice and comment rule making as required by section 553. *Id.*

145. *Id.* The court made detailed factual findings about the role of the FCC, and particularly its Chairman, in pressuring the networks and the NAB into adopting the family viewing hour policy. *Id.* at 1092-122.

146. Writers Guild of Am., W., Inc. v. American Brdcst. Cos., 609 F.2d 355 (9th Cir. 1979).

147. Primary Jurisdiction Referral of Claims Against Government Defendant Arising from the Inclusion in the NAB Television Code of the "Family Viewing Policy," *Report*, 95 F.C.C.2d 700 (1983).

148. *See supra* Part III.A.2.

149. Television Program Improvement Act of 1990, Pub. L. No. 101-650, 104 Stat. 5127 (codified at 47 U.S.C. § 303c (1994)). *See also* MacCarthy, *supra* note 50, at 686-93 (describing the Television Program Improvement Act of 1990, sponsored by Senator Paul Simon, that granted broadcasters a three-year exemption from the antitrust laws to permit the industry to draft joint standards to reduce violence on television).

only to respond to government pressure and that they believed the First Amendment insulated them from such pressure.[150]

Thus, in general, during the thirty years in which the Television Code was in effect, it had its greatest effect on advertising. However, some of the advertising provisions were found to violate the antitrust laws. The programming provisions were generally ignored or at least not enforced. The one programming provision that actually had an impact—the Family Viewing Hour—was found to violate the First Amendment.

B. *Children's Advertising Review Unit*

Although not as old as the NAB Code, the Guidelines of the Children's Advertising Review Unit (CARU) have been in operation for almost twenty-five years. The National Advertising Division (NAD) of the Council of Better Business Bureaus created CARU in 1974, the same year that the FCC issued its *Children's Television Report and Policy Statement*.[151] At that time, Action for Children's Television had not only asked the FCC to eliminate or restrict advertising on children's programs, but it and other consumer groups had petitioned the FTC to take action with respect to children's advertising. In response, the FTC proposed to prohibit all advertising to children under age eight and to limit other types of advertising directed at children, particularly sugared cereals.[152]

The CARU was "established to forestall efforts by groups outside the industry which would severely restrict or even ban advertising to children."[153] Yet, its creation was met with considerable skepticism by the FTC and consumer groups that questioned whether an organization funded by industry, especially one with so little consumer representation, could objectively regulate advertising practices.[154]

---

150. Eventually, of course, most of the industry agreed to provide program ratings that were designed to allow parents to block programming they thought unsuitable for their children. *See infra* Part III.C.6.

151. The NAD also administers a code for advertising generally. *See* NAD, *Procedures* (visited Mar. 15, 1999) <http://www.bbb.com/advertising/nadproc.html>. This Article does not attempt to assess the effectiveness of that effort. Self-regulation of advertising is fairly common throughout the world. *See generally* JEAN J. BODDEWYN, GLOBAL PERSPECTIVES ON ADVERTISING SELF-REGULATION (1992) (surveying self-regulation of advertising in 38 countries).

152. FTC, FTC STAFF REPORT ON TELEVISION ADVERTISING TO CHILDREN 320-24, 328-41 (Feb. 1978). The FTC never adopted these rules, however, as Congress terminated its authority to do so. *See* Federal Trade Commission Improvements Act of 1980, Pub. L. No. 96-252, § 11, 94 Stat. 374 (amending 15 U.S.C. § 57a (1994)).

153. Gary M. Armstrong, *An Evaluation of the Children's Advertising Review Unit*, 3 J. PUB. POL'Y & MKTG. 38, 40 (1984).

154. *Id.* at 38.

The CARU promulgated guidelines that apply to all forms of children's advertising. The CARU Guidelines list six basic principles for advertising directed to children.[155] For example, advertisers are advised to "always take into account the level of knowledge, sophistication and maturity of the audience" and that they "have a special responsibility to protect children from their own susceptibilities."[156] Likewise, they should "exercise care not to exploit unfairly the imaginative quality of children."[157] In addition, separate sections address specific problem areas: product presentation and claims; sales pressure; comparative claims; endorsement and promotion by program or editorial characters; premiums, promotions, and sweepstakes; and safety.[158]

Some provisions duplicate FCC policies, while others go beyond them. For example, the prohibition on the use of program personalities to sell products is similar to the FCC's policy against "host selling,"[159] while the prohibition on urging children to ask parents or others to buy products is not an FCC policy. Other provisions delineate deceptive or unfair advertising practices that might run afoul of section 5 of the Federal Trade Commission Act.[160] The Guidelines have been amended from time to time. Most significantly, CARU issued special guidelines in 1994 concerning recorded telephone messages and again in 1997 concerning online advertising. With the demise of the NAB Code in 1982, the CARU Guidelines were left as the only remaining industry-wide policy for children's advertising.[161]

1. Effectiveness of CARU's Advertising Guidelines

Several studies have evaluated CARU's effectiveness.[162] Armstrong reviewed CARU's activities during its first ten years.[163] He found that a

---

155. *The Children's Advertising Review Unit 1997 Self Regulatory Guidelines for Children's Advertising* (visited Mar. 15, 1999) <http://www.bbb.org/advertising/caruguid.html> [hereinafter *CARU Guidelines*].
156. *Id.* Principle 1.
157. *Id.* Principle 2.
158. *Id.*
159. *Children's Television Report and Policy Statement, supra* note 123, paras. 49-52.
160. For example, in Azrak-Hamway International, Inc., *Consent Agreement*, 61 Fed. Reg. 6841 (FTC 1996), the FTC alleged that certain toy advertisements on television depicting toys doing things that they could not do in actual use were false and misleading under section 5 of the FTC Act. As part of the proposed settlement, the company agreed to send letters to the stations running the spots advising them of the CARU Guidelines. The letter notes that in adopting the CARU Guidelines, "the advertising industry has undertaken various self-regulatory efforts to assist companies to comply with the law and to promote other industry goals." *Id.* at 6846.
161. Kunkel & Gantz, *Assessing Compliance, supra* note 119, at 148.
162. One study by Gary M. Armstrong was published in 1984 and evaluated the first 10 years of CARU's existence. *See supra* note 153. The other study was conducted by Dale

large proportion of CARU's activities involved casework.[164] The CARU staff (which consisted of four persons including secretarial personnel) monitored children's advertising. It also received complaints from outside sources, but Armstrong found that only 14 percent of cases came from sources other than monitoring.[165] The staff would evaluate advertisements against the Guidelines. The CARU could address problems informally[166] or by opening a formal investigation. After an investigation, which may involve internal review, product testing, or consultation with outside experts, the CARU had three options. First, it could find the advertisement acceptable and dismiss the case. Second, it could find the advertisement unacceptable and request that the advertiser discontinue or modify the advertisement. Third, it could find the advertisement questionable and request substantiation.[167] Armstrong found that CARU dismissed 7 percent of the cases, requested advertisers to discontinue or modify the advertisement in 71 percent of the cases, and sought substantiation in 22 percent of the cases.[168]

Compliance with CARU's rulings is purely voluntary. The CARU has no sanctions or enforcement power. Where an advertiser is noncooperative, CARU may refer the case to regulatory agencies such as the FTC and FCC.[169] Armstrong observes that "CARU's most powerful sanction is pub-

---

Kunkel and Walter Gantz in 1990. The results of this study are presented in two different papers: DALE KUNKEL & WALTER GANTZ, TELEVISION ADVERTISING TO CHILDREN: MESSAGE CONTENT IN 1990, REPORT TO THE CHILDREN'S ADVERTISING REVIEW UNIT (Jan. 1991) [hereinafter KUNKEL & GANTZ, CARU REPORT] and Kunkel & Gantz, *Assessing Compliance*, *supra* note 119.

163. The following description is from Armstrong, *supra* note 153, at 41.

164. Other activities include a voluntary film submission program in which advertisers may submit review copies before running a commercial, a Clearinghouse for Research on Children's Advertising, and occasional "outreach" activities. Armstrong notes that CARU's activities in these areas had been substantially cut back. *Id.*

165. Specifically, CARU staff monitors network, independent, and cable television programming at "child viewing times," as well as child-directed advertisements at nonchild viewing times. It also monitors children's magazines and comic books. *Id.*

166. *Id.* This behind-the-scenes activity was not reported. The practice of dealing with problems informally continues today. The CARU Web page reports that in 1994 while it monitored more than 17,000 television commercials, it initiated only 55 informal inquiries and five formal cases. *Children's Advertising Review Unit* (visited Mar. 15, 1999) <http://www.bbb.org/advertising/childrensMonitor.html>.

167. Armstrong, *supra* note 153, at 47-49.

168. *Id.*

169. *See, e.g.*, Dr. Frederick Beitenfeld, Jr., President WHYY, Inc., *Letter*, 7 F.C.C.R. 7123, 71 Rad. Reg. 2d (P & F) 1014 (1992). CARU filed a complaint alleging that a public television station in Philadelphia violated the FCC's policy against host selling. *Self-Reg Unit Refers Toy Ad to FTC*, FTC: WATCH, Sept. 18, 1996 (noting that CARU referred a television advertisement to the FTC alleging possible deception and safety concerns about a television advertisement portraying a doll whose eyelids, lips, and legs are shown changing color and that is shown with a razor resembling a real razor, and the advertiser

lic notice."[170] Case summaries are reported by the National Advertising Division and are typically covered by the press.[171]

Armstrong analyzed all reported cases from the creation of CARU through October 1982.[172] There were 147 cases, 63 percent of which dealt with television advertisements. The cases involved eighty-five different advertisers, the majority of whom were selling toys or food products. He found that deception was the most common violation.

Armstrong found that CARU had a small case load, averaging fifteen cases per year and falling. He notes that "[t]hough casework may provide input for the development and testing of the [CARU's] guidelines, this volume of cases is probably too small to enforce the guidelines effectively or to set meaningful precedents for the improvement of advertising to children."[173]

In addition, he found that "[t]he level of case activity appears to fluctuate substantially with external pressures on children's advertising. For example, efforts peaked in 1978-79 when the industry was threatened by the FTC trade regulation rule on children's advertising. When the threat had passed, the caseload decreased considerably."[174] He concludes that "[i]n the face of harsh constraints, [CARU] has made notable contributions toward improvement of child-directed advertising, especially in its earlier years. However, CARU has been given too much to do with too little means."[175]

Another study by Kunkel and Gantz was designed to assess: "How effective are [the CARU] self-regulatory guidelines at generating compliance with their stated policies?"[176] They identified thirty-nine measurable criteria

---

refused to participate in self-regulatory process). Referral to federal agencies is rare. According to CARU Director Elizabeth Lascoutx, in it first 23 years, CARU only referred complaints to federal agencies three times—two to the FTC and one to the FCC. Michael Hartnett, *Hold on There, Soupy! Advertising Guidelines by the Children's Advertising Review Unit*, 16 FOOD & BEVERAGE MKTG. 24 (1997). Recently, at least one additional referral has been made. *See* FTC: WATCH, Jan. 26, 1998 (reporting that CARU referred complaint to FTC where advertisement showed baby doll shedding tears without any method of operation and producer promised but failed to make recommended changes).

170. Armstrong, *supra* note 153, at 41.

171. According to Armstrong, the summaries were reported regularly in *Advertising Age* and the *New York Times*. In addition, they were published in a Council of Better Business Bureaus (CBBB) volume titled NAD/NARB Decisions. *Id.* at 42. That practice continues today.

172. Decisions may be appealed through the National Advertising Review Board (NARB) appeals procedure. However, at the time Armstrong was writing, no appeals had been made. *Id.* at 41.

173. *Id.* at 51.

174. *Id.* at 51-52.

175. *Id.* at 52.

176. Kunkel & Gantz, *Assessing Compliance*, *supra* note 119, at 152.

in the guidelines,[177] and using samples of programs recorded in 1990, they analyzed the commercials using the identified criteria.[178]

Out of 10,329 commercials reviewed, they found 385 (3.7 percent) violated the CARU Guidelines in one or more respects.[179] They found violations across all channels, with cable networks more likely to have violations than independent stations, and independent stations more likely to have violations than major network (ABC, CBS, and NBC) affiliates. They also found violations across product types. Of six basic product groups, advertisements for toys and breakfast cereals were the least likely to have violations, while advertisements for fast foods and recorded telephone message service were the most likely to violate the CARU Guidelines.[180]

Given the limits on CARU's operation and the burgeoning marketplace of advertising to children, Kunkel and Gantz found it "impressive" that 96 percent of all commercials observed in the study were found to comply; and that even greater compliance was found for the two types of products most frequently advertised on children's programs.[181] At the same time, they caution that their study does not provide an overall assessment of the CARU Guidelines. They cite research indicating that some problems with children's advertising are not addressed by the Guidelines.[182]

It is difficult to draw any firm conclusions about the effectiveness of the CARU Guidelines from these two studies conducted at different times looking at different data. While Kunkel and Gantz suggest that CARU has been largely successful in implementing the stated goals of the Guidelines, another interpretation is possible. One might wonder why if almost 4 percent of children's commercials shown in one week violate the CARU Guidelines,

---

177. *Id.* Two guidelines were excluded from the study because the researchers were not able to measure compliance: The guideline that states that "personal endorsements should reflect the actual experiences and beliefs of the endorser" was excluded because the researchers lacked knowledge of the endorser's experience. *Id.* at 153. Similarly, the guideline that states that "care should be taken not to exploit a child's imagination" was excluded as too subjective to evaluate. *Id.*

178. They recorded a total of 604 hours of children's programming, which included 10,329 commercials (many were repetitions). Samples were recorded in seven medium to large television markets around the country. The sample included programming from broadcast networks, independent stations, and basic cable channels. They used a technique called the composite week. *Id.* at 152-53.

179. *Id.* at 154. Of the 385 advertisements found in violation, 85 contained multiple violations, for a total of 492 violations. *Id.*

180. *Id.* at 155.

181. *Id.* at 159.

182. *Id.* at 160. For example, while cereal advertisers adhere to the guidelines that require sugared cereals to be portrayed as only a part of a healthy diet, research showed that disclaimers accomplished little in terms of improving children's understanding about the product. *Id.*

CARU has only brought an average of fifteen cases per year. Moreover, the number of cases in recent years seems to have declined. During the first eight months of 1995, CARU reportedly monitored more than 10,000 commercials, yet initiated only thirty-five informal inquiries and three formal cases.[183] Children's Advertising Review Unit Director Elizabeth Lascoutx asserts that more cases are not filed because advertisers have learned to follow the rules, and "'[t]hey know that the alternative to this kind of self-regulation is government regulation that could be much more intrusive.'"[184] But, the low number of cases could also mean that CARU is not doing its job.[185]

2. Children's Advertising Review Unit's Online Privacy Guidelines

In 1997, with the rapid growth of the Internet and other online services, CARU amended its Guidelines to take into account a new form of advertising to children—online advertising. The Children's Advertising Review Unit's action was prompted by concerns raised about the collection of information from children on Web sites by the FTC. This issue was first discussed at the FTC's roundtable on Consumer Privacy on the Global Information Infrastructure held in June 1996.[186] After the FTC announced that it was going to hold a second workshop on privacy issues, CARU revised its Guidelines to include online advertising in April 1997.[187]

In the section labeled "Guidelines for Interactive Electronic Media (e.g. Internet and Online Services)," CARU made it clear that the general principles regarding children's advertising applied to online and Internet advertising. It also adopted some specific guidelines governing online sales to children and the collection of data from children online. For example, in the section on data collection, the Guidelines state that advertisers should: (1) remind children to obtain parental permission before asking them to supply information; (2) disclose why information is being requested and whether it will be shared with others; (3) disclose the passive collection of information; and (4) make "reasonable efforts" to obtain parental permission if they col-

---

183. Rick Montgomery, *Wouldn't It Be Cool If . . .* , KANSAS CITY STAR, Nov. 5, 1995, at A21.
184. *Id.* (quoting Elizabeth Lascoutx, Director, CARU).
185. *Id. See also* Ray Richmond, *Singing? Under Water? With the Holidays Coming, It's Time to Beware of Deceptive Toy Ads*, CHI. TRIB., Nov. 23, 1992, at E3 (noting complaints by consumer representatives that many television commercials targeting children exaggerate, deceive, and mislead and that effective policing is lacking).
186. FTC, *Workshop on Consumer Privacy on the Global Information Infrastructure* (visited Mar. 15, 1999) <http://www.ftc.gov/bcp/privacy/wkshp96/privacy.htm> [hereinafter *Privacy Workshop*].
187. *CARU Guidelines, supra* note 155.

lect identifiable information. When CARU finds that an advertisement or Web site is inconsistent with the Guidelines, it "seeks changes through the voluntary cooperation of advertisers."[188]

Since adopting these new Guidelines, CARU has brought a few formal cases involving Web sites as well as undertaken some informal reviews. The Children's Advertising Review Unit's first case was against the Beanie Babies Web site. This site came to CARU's attention because of a report issued by the Center for Media Education. The Children's Advertising Review Unit found that although the site contained several areas where visitors could enter information about themselves and communicate directly with each other, it had no notice of its information collection or privacy policies.[189] In 1998, CARU took formal action involving at least two other Web sites.[190] In addition, CARU undertook informal investigations of Web sites and conducted compliance reviews on request.[191]

The FTC examined the adequacy of the CARU Guidelines in its *Report to Congress* in June 1998. It found that the CARU Guidelines were consistent with the principles outlined earlier by the FTC staff in an opinion letter regarding the KidsCom Web site.[192] Moreover, it noted that CARU has an enforcement mechanism in place and has

---

188. *Id.*
189. Ty, Inc., operator of the Beanie Babies Web site, responded by making changes to the site. *Beanie Babies Website*, 27 NAD Case Reports (CBBB) 194 (Nov. 1997). *See also* Stuart Elliott, *Self-Regulation in Cyberspace: The Web Site for Beanie Babies Undergoes Several Changes*, N.Y. TIMES, Dec. 8, 1997, at D3.
190. *Lisa Frank Website*, 28 NAD Case Reports (CBBB) 225 (Sept. 1998) (finding failure to give notice regarding information collection policy or privacy policy and inadequate labeling of advertising material); *Trendmasters, Inc. Website*, 28 NAD Case Reports (CBBB) 244 (Sept. 1998) (finding Web site featuring information about toy product lines, downloadable games, and survey failed to include privacy policy and did not clearly identify advertising).
191. The NAD Case Reports notes that when advertising is "immediately modified or discontinued, CARU will not open a formal case in the matter" but does report the results. *Inquiries*, 28 NAD Case Reports (CBBB) 247 (Sept. 1998). The NAD Case Reports discloses informal investigations of the Nickelodeon and Dole Web sites and review of the Chevron site. *Id. See also The Jupiter Interview: Elizabeth Lascoutx, CARU*, DIGITAL KIDS REPORT, Aug. 1996, at 12 (describing how General Mills' You Rule School advertising site went through approval process with CARU).
192. PRIVACY REPORT, *supra* note 5, at 17. In contrast to the CARU Guidelines, the FTC found that the Children's Guidelines of the Direct Marketing Association did not conform to the Staff Opinion Letter. *Id.* at 18. The Staff Opinion Letter was issued in July 1997, in response to a complaint filed by the Center for Media Education, for which the Author served as counsel. The complaint alleged that KidsCom, a Web site directed at children ages four to fifteen, violated section 5 of the FTC Act. Since KidsCom had changed its practices, the FTC declined to take enforcement action against it. However, the FTC letter set forth the staff's analysis of KidsCom's past practices "[t]o provide guidance in this area" and to offer "several broad principles [which] apply generally to online infor-

achieved a remarkably high level of compliance under this mechanism in the offline media over a long period of time. While CARU has worked to encourage Web sites to adhere to its privacy guidelines with respect to the collection of personal information from children online, to date it has not achieved the same widespread adherence it has achieved in other media.[193]

Indeed, the FTC found that as of March 1998, nearly a year after issuance of the CARU Guidelines, serious problems still remained.[194] The FTC surveyed 212 Web sites directed to children. It found that 89 percent of those sites collected one or more types of personal information from children,[195] and that "[o]ften the sites that collect personal identifying information also collect several other types of information, enabling them to form a detailed profile of a child."[196] The survey revealed that sites used a variety of techniques to solicit personal information, including registration, contests, imaginary characters, guest books, pen pal programs, and prizes.[197]

Using a very broad definition of disclosure, the FTC found that 54 percent of children's sites had some kind of disclosure.[198] However, no site's information practice disclosure statement discussed the full range of fair information practice principles. Moreover, many sites disclosed children's personal information to third parties, thereby creating a risk of injury or exploitation. Only 23 percent of Web sites told children to ask parents for

mation collection from children." Letter from Jodie Bernstein, Director, FTC's Bureau of Consumer Protection, to Kathryn C. Mongtomery, President, and Jeffrey A. Chester, Executive Director, Center for Media Education 2 (July 15, 1997) (on file with author).

The FTC found that KidsCom's information collection practices had likely violated section 5 in two ways. First, the FTC found that it was
> a deceptive practice to represent that a Web site is collecting personally identifiable information from a child for a particular purpose (*e.g.*, to earn points to redeem a premium), when the information will also be used for another purpose which parents would find material, in the absence of a clear and prominent disclosure to that effect.

Id. at 3-4 (citation omitted). Second, the FTC found it was an unfair practice "to collect personally identifiable information . . . from children and sell or otherwise disclose such identifiable information to third parties without providing parents with adequate notice . . . and an opportunity to control the collection and use of the information." *Id.* at 5.

193. PRIVACY REPORT, *supra* note 5, at 17 (footnotes omitted).

194. *Id.*

195. *Id.* at 31. Personal information collected included name, e-mail address, postal address, telephone number, social security number, age, date of birth, gender, education, interests, and hobbies. *Id.* at 31-32.

196. *Id.* at 32.

197. *Id.* at 33.

198. *Id.* at 34. Of sites that collect personal information and have at least one information practice disclosure, 43% say they provide children or parents choice about how information is used; 12% say they offer access or an opportunity to correct information; 8% say they provide security; and 12% say they will notify parents. Only 24% of sites that collected personal information had posted a privacy policy notice. *Id.*

permission, while only 1 percent required parental consent before personal information was collected or used.[199] The FTC concluded that its survey showed "a very low level of compliance with the basic parental control principles contained in the staff opinion letter and the CARU guidelines more than seven months after these documents were released."[200]

Finding its own authority insufficient to address the problems, the FTC recommended that Congress adopt legislation. Even in calling for legislation, however, the FTC emphasized its preference for self-regulation:

> The Commission has encouraged industry to address consumer concerns regarding online privacy through self-regulation. The Internet is a rapidly changing marketplace. Effective self-regulation remains desirable because it allows firms to respond quickly to technological changes and employ new technologies to protect consumer privacy. Accordingly, a private-sector response to consumer concerns that incorporates widely-accepted fair information practices and provides for effective enforcement mechanisms could afford consumers adequate privacy protection. To date, however, the Commission has not seen an effective self-regulatory system emerge.[201]

Nonetheless, the FTC's legislative proposal closely tracked CARU's Guidelines.[202]

The legislation passed by Congress authorizes the FTC to promulgate regulations that generally require commercial Web site operators to give notice and obtain consent from the parent of a child under the age of thirteen before collecting personally identifiable information from the child. It also directs the FTC to "provide incentives for self-regulation" by offering a "safe harbor" from prosecution for companies that comply with self-regulatory guidelines issued by industry that have been found by the FTC, after notice and comment, to meet the requirements of the law.[203]

In sum, CARU has had moderate to good success in ensuring that television advertisers comply with its advertising guidelines. It has had little success, however, with the children's online privacy guidelines. Several factors might explain the different levels of success. First, the number of products advertised to children on television is fairly limited, consisting pri-

---

199. *Id.* at 37. Only 8% say that parents can ask for a child's information to be deleted or not used ("opt-out").

200. *Id.* at 38.

201. *Id.* at 41.

202. *Compare id.* at 43, *with CARU's Reasonable Efforts Standard* (visited Mar. 15, 1999) <http://www.bbb.org/advertising/caruefforts.html> (noting that both would require prior parental consent where personally identifiable information would allow someone to contact the child offline or where it is publicly posted or disclosed to third parties).

203. Children's Online Privacy Protection Act of 1998, Pub. L. No. 105-277, § 1304, 112 Stat. 2681, 2732-33 (to be codified at 15 U.S.C. § 6503).

marily of toys, breakfast foods, and fast foods.[204] However, the number of companies offering Web sites to children, while including these same advertisers, appears to be much larger and diverse.[205] Moreover, in the case of television advertising, if CARU was unable to get voluntary cooperation from an advertiser, it could always file a complaint with the FCC or FTC. In the case of children's privacy, however, the FTC acknowledged that its ability to act in this area was limited.[206] Finally, the privacy guidelines have been in effect for a much shorter period of time.

C. *Other Self-Regulatory Efforts Involving Media*

1. Advertising of Hard Liquor

While the NAB Code prohibited the advertising of hard liquor,[207] broadcast advertising of hard liquor was also prohibited by the "Code of Good Practice" of DISCUS.[208] Distilled Spirits Council of the United States is the national trade association of producers and marketers of distilled spirits. Thus, even after the NAB Code was repealed in 1983, the DISCUS Code prohibited advertising on radio and television stations as well as on cable and satellite services.[209]

In March 1996, Seagram, the second largest marketer of distilled spirits, violated the Code of Practice by airing a liquor advertisement on a small sports cable network. A few months later, it violated the ban again by airing an advertisement on an ABC affiliate in Corpus Christi, Texas. Instead of imposing sanctions, however, DISCUS voted in November 1996 to repeal

---

204. KUNKEL & GANTZ, CARU REPORT, *supra* note 162, at 23.

205. For example, the FTC Staff Survey of Child-Oriented Commercial Web Sites included sites for television networks, crafts, magazines, and books, as well as a variety of sites oriented to young people such as KidsCom and Yahooligans Club.

206. PRIVACY REPORT, *supra* note 5, at 40-41.

207. *See* TELEVISION CODE, *supra* note 105, at 10-12 ("The advertising of hard liquor (distilled spirits) is not acceptable.").

208. The radio ban was adopted in 1936 and the television ban in 1948. The 1995 DISCUS Code of Good Practice defined the broadcast media to include cable and satellite. Separate provisions prohibited advertising on the screen of motion picture theaters or videotapes and prohibited paying compensation for advertising "plugs" on the broadcast media.

209. CODE OF GOOD PRACTICE FOR DISTILLED SPIRITS ADVERTISING AND MARKETING (Distilled Spirits Council, 1995) [hereinafter DISCUS CODE OF GOOD PRACTICE]. Despite these voluntary restrictions, some companies did seek to advertise hard liquor on television once the NAB Code was repealed. However, these advertisements were soon stopped in response to public and congressional pressure. Linton, *Self-Regulation in Broadcasting Revisited*, *supra* note 86, at 484-85.

the voluntary prohibition.[210] Competitive concerns as well as changes in technologies had undermined industry support for the voluntary ban.[211] According to DISCUS's President, the association saw no basis for allowing the broadcast advertising of beer and wine and not other alcoholic beverages.[212] A Seagram's executive also pointed out that the ban on television advertising no longer made sense when distilled spirits could be advertised on the Internet.[213]

The members of DISCUS were undoubtedly aware of the Supreme Court's decision in *44 Liquormart, Inc. v. Rhode Island* announced in May 1996.[214] That decision struck down a state law prohibiting the advertisement of liquor prices. Because the law banned truthful commercial speech about a lawful product, the Court reviewed it with "'special care.'"[215] This decision effectively removed the credible threat of government regulation.[216]

Although DISCUS repealed the ban on broadcast advertising, other provisions of the DISCUS Code of Practice remained in effect. For example, the Code cautioned that distilled spirits should be portrayed "in a responsible manner" and "should not be advertised or marketed in any manner directed or primarily intended to appeal to persons below the legal purchase age."[217] The Codes of Practice of the beer and wine industries have similar provisions. Recently, however, the Federal Trade Commission has questioned the efficacy of some of these provisions. In August 1998, the FTC began an inquiry into the advertising practices of eight of the nation's top marketers of

---

210. Stuart Elliott, *Liquor Industry Ends Its Ad Ban in Broadcasting*, N.Y. TIMES, Nov. 8, 1996, at A1. Repeal of the voluntary ban was protested by various public health groups. Two FCC Chairmen attempted to start proceedings to consider imposing a ban or other restrictions, but lacked the votes to proceed. A bill to prohibit the broadcasting of hard liquor advertisements was introduced in Congress, but did not pass. *Id.*

211. "Liquor consumption [declined] 40% since its peak in 1979 as drinkers have shifted to wine and beer." Denise Gellene, *Seagram Bucks Voluntary Ban on TV Advertising with Spot on Cable*, L.A. TIMES, May 1, 1996, at D3.

212. Elliott, *supra* note 210. Added another industry executive, "'The members of the distilled-spirits industry have felt for many years that their competitive position has been with one hand tied behind their back . . . because they too would like access to a medium they think would be very efficient for them.'" *Id.*

213. *Id. See generally* CENTER FOR MEDIA EDUCATION, ALCOHOL & TOBACCO ON THE WEB: NEW THREATS TO YOUTH (Mar. 1997) (describing Web sites promoting alcoholic beverages) (on file with author).

214. *44 Liquormart, Inc.*, 517 U.S. 484 (1996).

215. *Id.* at 504. Although they employed different reasoning, all nine Justices found the Rhode Island statute unconstitutional. *Id.*

216. *See generally* Claudia MacLachlan, *Law Murky on Stopping Liquor Ads*, NAT'L L.J., Nov. 25, 1996, at A1. ("[M]any lawyers believe a ban on liquor ads would be deemed unconstitutional in light of the Supreme Court's decision last May in *44 Liquormart, Inc. v. Rhode Island*.").

217. DISCUS CODE OF GOOD PRACTICE, *supra* note 209.

beer, wine, and liquor.[218] It specifically sought information about how the companies had implemented Code provisions that prohibited advertising intended to appeal to or reach persons below the legal drinking age.[219]

On the same date, the FTC filed a complaint and proposed consent decree charging that advertisements for Beck's beer that depicted young adults partying and drinking beer on a sail boat were "unfair acts or practices" in violation of section 5(a) of the Federal Trade Commission Act. The complaint noted that the advertisements were inconsistent with the Beer Institute's Code because they portrayed boating passengers drinking beer "while engaged in activities that require a high degree of alertness and coordination to avoid falling overboard."[220] These recent actions by the FTC suggest that the self-regulatory codes of the alcoholic beverages industry are not being effectively enforced.

2. Self-Regulation of News: The Press Councils

One type of media self-regulation that has clearly been unsuccessful in the United States is the attempt to promote public accountability and fairness in news reporting by the use of a press council. Modeled after the British Press Council, press councils were established in several states in the late 1960s, and in 1973, the National News Council (NNC) was established. However, the NNC closed in 1984, and by that time, most of the state press councils had closed as well.[221]

---

218. Sally Beatty & John Simons, *FTC Eyes Liquor Ads' Kid-Appeal*, WALL ST. J., Aug. 7, 1998, at B1.

219. *Order to File Special Report* (visited Mar. 15, 1999) <http://www.ftc.gov/os/1998/9808/6(b)sptr.fin.htm> and <http://www.ftc.gov/os/1998/9808/6(b)brre.fin.htm >.

220. Beck's North America, Inc., *Complaint, available at* <http://www.ftc.gov/os/1998/9808/9823092.cmp.htm>. *See also* Beck's North America, Inc., *Agreement Containing Consent Order*, File No. 982-3092, *available at* <http://www.ftc.gov/os/1998/9808/9823092.agr.htm>; *Alcohol Companies to Supply Data on Their Self-Regulatory Activities to FTC* (visited Mar. 15, 1999) <http://www.ftc. gov/opa/1998/9808/alcohol.htm>.

221. The only surviving press council in the United States is in Minnesota. *Minnesota News Council* (visited Mar. 15, 1999) <http://www.mtn.org/newscouncil>. However, press councils are found in many other countries. One study reports that 48 countries have press councils. K. TRIKHA, THE PRESS COUNCIL: A SELF-REGULATORY MECHANISM FOR THE PRESS 2 (Bombay, Somaiya Publications Pvt. 1986). The first press council was established in Sweden in 1916. Today, 12 countries in the European Union have press councils. *See* EMMANUEL E. PARASCHOS, MEDIA LAW AND REGULATION IN THE EUROPEAN UNION: NATIONAL, TRANSNATIONAL AND U.S. PERSPECTIVES ch. 9 (Ames: Iowa State University Press 1998) (describing press councils in Austria, Belgium, Denmark, Finland, Germany, Greece, Italy, Luxembourg, the Netherlands, Portugal, Sweden, and the United Kingdom).

Patrick Brogan has done a detailed study of the NNC and the reasons for its failure.[222] The idea of an NNC came out of a task force put together by the Twentieth Century Fund in 1971. At the time, the press had been under a great deal of attack, particularly by the Nixon Administration.[223] According to Brogan, those who set up the NNC were concerned that if the press failed to establish its own standards, public criticism would increase to the point where pressure to change the freedoms enjoyed by the press under the First Amendment could not be resisted.[224]

The NNC consisted of fifteen members representing both the public and the media. It considered complaints against national newspapers, news agencies, magazines, and television networks. Complainants would waive their right to use any of the council's proceedings as evidence in court.[225] The NNC staff analyzed complaints and made initial judgments about their merit. If a complaint was found to have merit, it was sent to a grievance committee composed of members of the Council, which in turn, made recommendations to the full Council.[226] The Council judged the cases and issued verdicts. Over the decade of its existence, the NNC dealt with 227 complaints.[227] Its decisions were made public, although not widely reported.[228] It had no power of enforcement, but relied on publicity to encourage the press to mend its ways.

Brogan found that

[d]espite all its good intentions and ten years of strenuous endeavor, the council was spurned by the press and neglected by the public. Without press or public support, it could win no publicity. Without that, it could not raise the money it needed to carry on operations—and earn the support of press or public.[229]

---

222. *See generally* PATRICK BROGAN, SPIKED: THE SHORT LIFE AND DEATH OF THE NATIONAL NEWS COUNCIL (Priority Press Publications, N.Y. 1985) (A Twentieth Century Fund Paper).

223. *Id.* at 10-12.

224. *Id.* at 4.

225. *Id.* at 6.

226. The NNC also considered matters of general journalistic ethics and attacks upon the press from government, private interests, and individuals. *Id.* at 49-50, 53-54.

227. Brogan describes many of the cases as trivial, but asserts that over time, "the council's rulings established a body of case law that made a useful contribution to journalistic ethics and practice." *Id.* at 38.

228. For a time, its decisions were published in the *Columbia Journalism Review*. *Id.* at 47, 61. However, the failure to obtain wide dissemination of its operations is one of the reasons contributing to the demise of the NNC. GANNETT CENTER FOR MEDIA STUDIES AND SILHA CENTER FOR THE STUDY OF MEDIA ETHICS, MEDIA FREEDOM AND ACCOUNTABILITY: A CONFERENCE REPORT 27 (1988) [hereinafter GANNETT REPORT].

229. BROGAN, *supra* note 222, at 7.

From the beginning, the NNC faced vigorous opposition from a large segment of the press, including the *New York Times*, which viewed the NNC as a threat to its First Amendment freedoms.[230] Some press organizations declined to cooperate with the Council. While over time the NNC gained more support among the press, the industry was never willing to provide financial support for its operations.[231]

Most of the funding for the NNC came from two foundations: the Twentieth Century Fund and the Markle Foundation.[232] Lack of sufficient funding contributed to the problems faced by the NNC. To effectively review whether a story had been reported fairly, the Council had to in effect report a story, which required a large amount of resources and a staff of experienced journalists capable of retracing and assessing the steps taken by the original reporting team. Lacking funding for such an effort, the NNC tended to rely upon graduate students to do its investigations.[233] Lack of funding also made it difficult for the NNC to pay high enough salaries to attract high quality staff.[234]

In analyzing why the NNC has failed but the British Press Council has succeeded, Brogan notes that "the situation is different in Britain. The British Bill of Rights does not mention the press, and there is nothing to stop Parliament from imposing statutory control over the press should the public ever insistently demand it."[235] Moreover, Brogan observes that

> [t]he council was conceived in a period when the press was under attack and feared that its enemies might carry the day and seriously restrict its freedom. In the event, the fears proved exaggerated. The press easily survived Richard Nixon and Spiro Agnew, and therefore

---

230. *Id.* at 27-29.

231. *Id.* at 28.

232. *Id.* at 21-22 & app. E. The failure of the NNC to obtain funding from the Ford Foundation is also cited as one of the reasons contributing to its demise. *Id.* at 21. *See also* GANNETT REPORT, *supra* note 228, at 24.

233. GANNETT REPORT, *supra* note 228, at 28.

234. BROGAN, *supra* note 222, at 45, 95.

235. *Id.* at 6. Of course, there are other differences as well. Brogan notes that the British Press Council started small and took a long time to become accepted. *Id.* at 90-91. More-over, according to its Director, the British Press Council

> enjoys two significant advantages that an American equivalent would be unlikely to obtain. First, the targets of the Council's censure willingly publish the results of the proceeding in full, prominently displayed, . . . —something that few U.S. publications, and probably no network news departments, would allow. . . . Second, complainants to the Council are obliged to treat its proceedings as an alternative to libel litigation rather than a preliminary skirmish, and thus persuade news organizations to open their books and cooperate, whereas in the litigious United States a potential plaintiff would not . . . sign away his rights, and therefore a potential defendant would likely resist all inquiries.

GANNETT REPORT, *supra* note 228, at 27.

never saw any need for the dubious protection of the fledgling council.[236]

Thus, without the threat of government regulation, effectively barred in the United States by the First Amendment, members of the press saw no reason to pay for or submit to outside reviews of their fairness.[237]

3. Comic Books

The comic book industry engaged in self-regulation after crime and horror comics became popular in the 1950s, and many states passed laws making it unlawful to distribute such comic books to minors.[238] In 1954, Congress held hearings about the effects of comic books on youth evidencing concern about violence. While earlier attempts at self-regulation had failed, the Association of Comic Magazine Publishers promised renewed action when the Senate began investigating. It adopted the Comics Code and provided for the display of its seal of approval on comics that met Code requirements.[239]

This action had the desired effect, at least in the short to intermediate term. The subcommittee decided not to recommend regulation of comic books, but instead recommended reliance on industry self-regulation. Many publishers of crime and horror comics went out of business because wholesalers refused to distribute comics without a seal.[240] But after twenty years, comic book violence began to make a comeback. Publishers found they could avoid complying with the Code by shipping directly to specialty stores.[241] Even major publishers began to produce non-Code compliant editions. Thus, over time, self-regulation of violence in comic books lost its effectiveness.[242]

---

236. BROGAN, *supra* note 222, at 92.

237. *See also* Dennis, *supra* note 11. In this survey of self-regulation of the print media, Dennis notes that because of the First Amendment, the print media have been virtually exempt from regulation, and accountability has been a strictly voluntary affair.

238. Kevin W. Saunders, *Media Self-Regulation of Depictions of Violence: A Last Opportunity*, 47 OKLA. L. REV. 445, 446-47 (1994). Some of these statutes were found unconstitutional.

239. *Id.* at 452.

240. *Id.* (citing Margaret A. Blanchard, *The American Urge to Censor: Freedom of Expression Versus the Desire to Sanitize Society—from Anthony Comstock to 2 Live Crew*, 33 WM. & MARY L. REV. 741, 793 (1992)).

241. *Id.*

242. *Id.* at 452-53. Saunders notes that even though "the level of crime, horror and violence in present day comics is as bad today as it was in 1955," and some public concern has been expressed, the issue appears to be of less concern than in the 1950s. *Id.* at 452. He attributes the lesser concern to "the availability of violent images in so many other media" that "make comic book violence not seem so bad." *Id.* at 453.

4. Motion Pictures

The 1930 Production Code of the Motion Picture Association of America (MPAA) resulted from public pressure, largely organized through the Roman Catholic Church, to "clean up" movies.[243] Like many other media industry codes, the Production Code contained both general provisions and specific affirmative and negative provisions.[244] The Production Code proved quite successful because "[f]ilms without the Code Seal of Approval were doomed to failure" since the theaters, which were mostly owned by major studios, would not exhibit films without the MPAA seal.[245]

Several factors have been identified as contributing to the success of the Production Code in dictating content during the 1930s and 1940s.[246] First, there was little competition from other forms of entertainment.[247] Second, the oligopolistic structure of the movie industry enabled the MPAA-member companies to enforce the Code.[248] Third, the Code was insulated from constitutional challenge by the position of the Supreme Court that movies were not entitled to First Amendment protection.[249]

After World War II, circumstances changed. The public demanded more realistic films. Television started competing with the movies. These

---

243. Roy Eugene Bates, *Private Censorship of Movies*, 22 STAN. L. REV. 618, 619 (1970).

244. For example, under the first section, General Principles, it is stated that the "motion picture has special *Moral obligations*." The Motion Picture Code of 1930, *reprinted in* HOLLYWOOD'S AMERICA: UNITED STATES HISTORY THROUGH ITS FILMS 142, 144 (Steven Mintz & Randy Roberts eds., 1993). The second section, Working Principles, states that "[n]o picture should lower the moral standards of those who see it." *Id.* at 145. Specific provisions address such topics as the portrayal of sin and evil, adultery, vulgarity, dance, and religion. For example, the subject of adultery should be avoided and is never a fit subject for comedy. When it is portrayed in serious drama, it should not appear justified or presented as attractive or alluring. *Id.* at 147.

245. Bates, *supra* note 243, at 619. In its first 30 years, 25,000 films, including 12,000 full-length features were reviewed by the Code Office. Producers would submit scripts, which would be read by at least two members of the staff to determine whether they met the Code. The staff met daily to discuss problems. They would send written decisions to the producer, including suggestions as to how any problems might be solved. Finished films were also reviewed by the Code staff. Appeals could be made to a review board. It appears that most difficulties were worked out by making changes. *See generally Self-Policing of the Movie and Publishing Industry: Hearing Before the Subcomm. on Postal Operations of the House Comm. on Post Office and Civil Service*, 86th Cong. 28-30 (1960) (statement of Geoffrey M. Shurlock, Director, Production Code Administration, Motion Picture Association of America).

246. Bates, *supra* note 243, at 619-20.

247. *Id.* at 619.

248. *Id.* at 619-20.

249. *Id.* at 620; Mutual Film Corp. v. Ohio Indus. Comm'n, 236 U.S. 230, 243-45 (1915), *overruled in part by* Joseph Burstyn, Inc. v. Wilson, 343 U.S. 495 (1952).

developments created pressure to break away from Code-imposed standards. In addition, the *Paramount* case altered the industry structure and loosened the hold of the studios, which made enforcement difficult.[250] The number of independently produced movies increased, and starting in 1952, the Supreme Court came to recognize motion pictures as a form of expression protected by the First Amendment.[251] With this decision, as one commentator noted, "the main reason for movie industry self-regulation—fear of governmental censorship—almost disappeared."[252]

As the Code lost most of its effectiveness in limiting film content and movies became more explicit in the treatment of nudity and sex in the 1960s, the public demanded government control. In 1965, the City of Dallas enacted the first movie classification ordinance designed solely to protect children. Although the Supreme Court overturned the ordinance, *dictum* in its decision supported the use of an age-classification scheme.[253] These developments convinced MPAA officials that industry self-regulation would have to take the form of an age-classification system to prevent a flood of new censorial statutes. Motion Picture Association of America President Jack Valenti met with leaders of industry and outside parties to revise the Code.[254] The rating system that grew out of this process took effect in 1968, and remains substantially unchanged today.

Under the MPAA rating systems all films produced or distributed by MPAA members are submitted to the ratings board for rating.[255] Nonmembers may also submit films to be rated. Preliminary ratings are based on the script, while final judgment is reserved until the film is viewed. Producers unhappy with their rating may appeal. While no one is required to obtain a rating, most producers do so because approximately 85 percent of theaters cooperate with the MPAA.[256] Films that are not rated or are rated with an X rating (now called NC-17) find their opportunities for distribution limited.[257]

---

250. United States v. Paramount Pictures, 334 U.S. 131 (1948). In this case, the Supreme Court found that the ownership of the majority of movie theaters by the major studios violated the Sherman Act, and it ordered the dissolution of this monopoly.

251. *Joseph Burstyn, Inc.*, 343 U.S. 495.

252. Bates, *supra* note 243, at 621.

253. Interstate Cir., Inc. v. City of Dallas, 390 U.S. 676, 682 (1968).

254. For Jack Valenti's description of how the ratings system came about and how it works, see *Swope v. Lubbers*, 560 F. Supp. 1328, 1335-38 (W.D. Mich. 1983).

255. For a description of the MPAA rating scheme and how films are rated, see Richard P. Salgado, *Regulating a Video Revolution*, 7 YALE L. & POL'Y REV. 516, 519-20 (1989).

256. *Swope*, 560 F. Supp. at 1338.

257. Salgado, *supra* note 255, at 523-25 (describing obstacles faced by X-rated films). *See generally* Jacob Septimus, Note, *The MPAA Ratings System: A Regime of Private Censorship and Cultural Manipulation*, 21 COLUM.-VLA J.L. & ARTS 69 (1996).

5. Video Games

Another medium where a rating scheme has been utilized is video games.[258] Senator Lieberman became concerned about violent video games, such as Mortal Kombat. In December 1993, Senators Lieberman and Kohl convened hearings where they proposed legislation establishing an independent agency to oversee development of voluntary industry standards.[259] On the day of the hearings, the Software Publishers Association, the largest trade association in the computer software sector, announced its intent to create its own rating and warning system that would meet the elements specified by Senator Lieberman.[260]

The industry commissioned the development of a self-rating system.[261] The nonprofit Recreational Software Advisory Council (RSAC) was formed to administer the rating system. Its bylaws require that a majority of its board members come from outside the industry.[262] Software makers using RSAC's self-rating scheme must sign a contract that permits, among other things, RSAC to require corrective labeling, consumer and press advisories, product recalls, and monetary fines.[263] Spot checks and audits are performed by the Psychology Department at Yale.[264] Disputes over ratings may be addressed by the RSAC Appeals Committee.[265]

Although use of the RSAC rating system is voluntary, Balkam suggests that video game companies feel compelled to rate or face limited opportunities for distribution. Under pressure from members of Congress, WalMart, Toys R Us, and other retailers announced they would only stock rated games.[266] Balkam concludes that "[t]hrough a process of carrot and stick, the government has ensured that the industry has 'voluntarily' imposed a regulatory rating scheme upon itself without the need of a dedicated government department and all the expenditure required to bring one into place."[267]

---

258. Balkam, *supra* note 31, at 139. *See also* Saunders, *supra* note 238, at 458.
259. Balkam, *supra* note 31, at 139.
260. *Id.*
261. *Id.* at 140 (describing how self-rating works).
262. According to Stephen Balkam, Executive Director of RSAC, outside board members were "a vital part of its early success that the organization could be seen to be fair, balanced and not unduly influenced by game makers." *Id.* at 140.
263. *Id.*
264. *Id.*
265. *Id.*
266. *Id.* at 141. Balkam notes that Senator Lieberman and Senator Kohl wrote to major retail outlets and held a press conference praising those who agreed and criticizing others.
267. *Id.*

6. Television Violence: Ratings and the V-Chip

A somewhat similar approach—that of ratings—was recently implemented to address violent and sexual content on television. After the failure of the Family Viewing Policy,[268] many efforts were made to reduce violent and sexual content on television.[269] Unlike video games, however, Congress eventually did pass legislation that provided for a voluntary rating scheme. The Telecommunications Act of 1996 mandates that television sets be equipped with a chip that will permit programs with certain ratings to be blocked.[270] It gave the industry one year to come up with "*voluntary* rules for rating video programming that contains sexual, violent, or other indecent materials about which parents should be informed before it is displayed to children" and to agree "*voluntarily* to broadcast signals that contain ratings of such programming."[271] If the industry failed to develop rules acceptable to the FCC, the FCC was required to establish an advisory committee to recommend a rating system; to prescribe guidelines and procedures for rating video programs; and to require stations to include the ratings on any program that is rated.[272]

Regarding the statutory language, Professor J.M. Balkin notes:

> The Act's "fail-safe" provision deliberately stops short of requiring that broadcasters accept the ratings system devised by the advisory committee. It requires only that, if video programming already is rated by the broadcaster, the rating must also be encoded so that it can be read by a V-Chip system.[273]

The Act's

> fail-safe provision is left deliberately toothless to avoid constitutional problems of prior restraint and compelled speech. Instead, the true goal of the legislation is to present broadcasters with a set of unpalatable alternatives. If they do nothing, they risk the appointment of an advisory committee telling them how to rate their programs. Even if the FCC cannot constitutionally require that they accept the ratings system as a condition of broadcasting, there will be enormous public pressure on broadcasters to accept a system that has already been worked out with attendant public fanfare. Faced with this possibility, broadcasters and distributors will instead choose to create their own ratings system.[274]

---

268. *See supra* Part III.A.3.c(ii).
269. *See* MacCarthy, *supra* note 50, at 685-95.
270. Pub. L. No. 104-104, § 551(d), 110 Stat. 56, 141 (codified at 47 U.S.C. § 330(c)(1)-(4) (Supp. II 1996)).
271. *Id.* § 551(e)(1)(A), (B), 110 Stat. at 142 (emphasis added).
272. *Id.*
273. J.M. Balkin, *Media Filters, the V-Chip, and the Foundations of Broadcast Regulation*, 45 DUKE L.J. 1131, 1157 (1996).
274. *Id.* at 1158.

In fact, they did.[275] As the one-year deadline approached, the movie, broadcast, and cable industries—as represented by the MPAA, NAB, and National Cable Television Association (NCTA)—jointly developed a rating system based in large part upon the existing system for rating motion pictures.[276] After public and congressional opposition,[277] the proposal was revised. The FCC found the revised rating scheme to be acceptable, and thus there was no need to convene an advisory committee.[278]

It is too early to assess whether this system is going to be successful in meeting the stated goal of providing parents with effective tools to supervise their children's viewing of inappropriate content.[279] However, it is an inter-

---

275. Although television executives initially threatened to challenge the V-Chip legislation in court, they agreed to develop an industry rating system after a White House summit in February 1996. Paul Farhi & John F. Harris, *TV Industry Agrees to Use Rating System*, WASH. POST, Feb. 29, 1996, at A1. Noting that the industry had been forced to act at the White House Summit because of the V-Chip legislation, Ted Turner reportedly said, "'Let's be honest; this is not voluntary.'" Richard Zoglin, *Prime-Time Summit*, TIME, Mar. 11, 1996, at 64, 66.

276. Letter from Jack Valenti, President, Motion Picture Association of America, et al., to William F. Caton, Secretary, FCC (Jan. 17, 1997), *available at* <http://www.fcc.gov/Bureaus/Cable/Public_Notices/1997/fcc97034.txt>.

277. For example, Senator Hollings introduced the Children's Protection from Violent Programming Act on February 26, 1997, to make it unlawful "to distribute . . . violent video programming not blockable by electronic means *specifically on the basis of its violent content* during hours when children are reasonably likely to comprise a substantial portion of the audience." 143 CONG. REC. S1670, S1671 (daily ed. Feb. 26, 1997) (emphasis added). This bill would also have given the FCC the right to revoke a license for failure to rate based on violent content. *Id.* at S1670-71 (statement of Sen. Hollings). Senator Coats also introduced legislation to authorize the FCC not to grant or renew a license unless the network used a content descriptive rating system. 143 CONG. REC. S4015, S4016 (daily ed. May 6, 1997) (statement of Sen. Coats). To fend off legislation of this type, the industry, except for NBC, agreed to changes in the industry-proposed rating system. Regarding the changes, MPAA President Jack Valenti said, "'[t]his is not something we celebrated as a great victory . . . [t]his is something we did because we had to do it.'" Paige Albiniak, *Ratings Get Revamped*, BRDCST. & CABLE, July 14, 1997, at 4.

278. Implementation of Section 551 of the Telecommunications Act of 1996; Video Programming Ratings, *Report and Order*, 13 F.C.C.R. 8232, paras. 18-26, 11 Comm. Reg. (P & F) 934 (1998) [hereinafter *Implementation Report and Order*]. The proposal was revised in response to public protest that the original proposal failed to provide adequate information to parents about the type of objectionable content. *Id.* paras. 12-17. The compromise reached after negotiations was to retain the age-based categories but to add content descriptions. Thus, for example, instead of rating a program simply TV-PG, the rating might include V indicating moderate violence, S indicating some sexual situation, L indicating infrequent coarse language, or D for some suggestive dialog. *Id.* para. 7. Most of the major industry players except NBC and Black Entertainment Television (BET) agreed to go along with the revised rating system. *Id.* para. 30.

279. The full effect cannot properly be assessed until a significant number of television sets equipped with the V-Chip are in use, which will not occur for some time. The FCC has required television manufacturers to include blocking technology on at least half of their sets with a screen 13 inches or larger by July 1999, and the rest by January 2000.

esting example of where self-regulation may be preferable to government regulation because government regulation would raise constitutional difficulties. Indeed, some have questioned whether the industry decision to utilize ratings is sufficiently voluntary to avoid constitutional problems.[280]

IV. ANALYSIS

The examples discussed above include a broad range of self-regulatory efforts involving the media. They provide some support for the claimed advantages and disadvantages of self-regulation as well as general support for Michael's hypotheses about the conditions needed for effective self-regulation. In addition, they provide a basis for assessing the proposed use of self-regulation for digital television and online privacy.

A. *The Advantages and Disadvantages of Self-Regulation*

The examples discussed above do not provide a great deal of support for the claimed advantages of self-regulation. At best, some, such as the video games example, illustrate the ability of self-regulatory organizations to act more quickly than government. But on the other hand, in the case of protecting children's privacy on the Internet, it has taken some time for the industry to act, and the government presumably could have acted more quickly but preferred to wait to give industry a chance to fix the problem first.

Some examples also suggest that self-regulation can result in the costs of regulation being borne by the industry instead of the government. Examples might include the NAB Code[281] and the CARU Advertising Guidelines. The examples also supply limited support for the claim that self-regulation

---

Technical Requirements to Enable Blocking of Video Programming Based on Program Ratings, *Report and Order*, 13 F.C.C.R. 11,248, para. 23, 11 Comm. Reg. (P & F) 907 (1998). Once television sets with V-Chips are available, it will still take years for people to replace their old sets. Initial sales have been slow. Marta W. Aldrich, *Parents Slow to Embrace V-Chip*, ASSOCIATED PRESS, Feb. 4, 1999, *available at* 1999 WL 11924613. However, a recent study by the Kaiser Family Foundation does suggest some problems. DALE KUNKEL ET AL., RATING THE TV RATINGS: ONE YEAR OUT, AN ASSESSMENT OF THE TELEVISION INDUSTRY'S USE OF V-CHIP RATINGS (Kaiser Family Foundation 1998). For example, it found that "[p]arents cannot rely on the content descriptors, as currently employed, to effectively block all shows containing violence, sexual material, or adult language." *Id.* at 89.

280. *See, e.g.*, Corn-Revere, *supra* note 13, at 64-65 (arguing that industry acceptance of ratings is not voluntary due to congressional threats to adopt even less palatable legislation and because broadcasters must seek license renewal from the FCC).

281. Running the Code Authority accounted for approximately 14% of the NAB's budget. Maddox & Zanot, *supra* note 53, at 130. When the Code was abandoned, however, it appears that most of the enforcement costs were shifted to the networks and stations rather than to the government. *Id.* at 128-30.

can be used where direct government regulation would raise constitutional problems. For example, it has been argued that the MPAA rating scheme would be unconstitutional if imposed by the government.[282] However, the experience with the Family Viewing Policy suggests limits to what the government can do to encourage self-regulation without turning voluntary action into state action.[283] The V-Chip example presents a case where the voluntary nature of the self-regulation is questionable.[284]

Another benefit of self-regulation is that it can provide a forum for testing rules that may ultimately become regulations.[285] For example, the NAB Code's limits on children's advertising formed the basis for the Children's Television Act's limit on advertising.[286] Similarly, the FTC's legislative proposals regarding children's privacy seemed to draw upon the CARU self-regulatory guidelines.[287]

The examples also support some arguments against self-regulation. Some examples suggest that inadequate enforcement may be a problem under self-regulation. There are few examples of the NAB taking any action against television stations that violated the Code. The Children's Advertising Review Unit brings only a small number of cases each year against television advertisers, even though evidence suggests that hundreds of noncompliant advertisements are broadcast each week. And more recently, CARU has only concluded a few cases involving Web sites that collect information from children, even though the FTC has documented that a large number of Web sites are not in compliance with the CARU Guidelines.

Inadequate sanctions also presented a problem in some cases. For example, denial of the right to display the NAB seal did not provide a meaningful sanction for broadcast stations. Finally, the DOJ's antitrust suit against the NAB illustrates how self-regulation can result in anticompetitive conduct.[288]

B. Conditions in Which Self-Regulation Can Work

There are many ways in which one might measure the "success" of self-regulatory schemes. It might be measured in terms of whether the self-regulation meets the stated goals or whether the stated goals are the correct or best goals. Likewise, self-regulation could be considered successful when

---

282. Bates, *supra* note 243, at 625; Septimus, *supra* note 257, at 86-87.
283. *See supra* Part III.A.3.c(ii).
284. *See supra* Part III.C.6.
285. Mulligan & Goldman, *supra* note 30, at 65-67.
286. *See supra* Part III.A.3.c.
287. *See supra* note 202.
288. *See supra* Part III.A.2.

it meets other (perhaps unstated) industry objectives, such as "avoiding intrusive government regulation" or restricting competition, even when those goals may not benefit the public.

For purposes of this Article, the focus is on whether self-regulatory codes have been successful in achieving their stated purposes. At least some of the above examples of self-regulatory schemes involving the media have enjoyed success in that way. Indeed, in some cases, voluntary codes have been in effect for a long period of time. Among the successful (or at least partially successful) examples of self-regulation are the NAB commercial time limits, including the NAB's children's advertising limits, the NAB's Family Viewing Policy, CARU's Advertising Guidelines, DISCUS's prohibition on broadcast advertising, the MPAA Production Code, the MPAA rating scheme, and the Comic Book Code. Yet, of these arguably successful schemes, only two—the CARU advertising limits and the MPAA rating scheme—remain in effect. The NAB commercial time limits were found to have raised antitrust problems; the Family Viewing Policy was found unconstitutional; the NAB's children's advertising limits were repealed and ultimately replaced by a statute; DISCUS has rescinded its ban on broadcast advertising; the MPAA Production Code has been replaced by a rating system; and the Comic Book Code seems to have lost its effectiveness.

Among the unsuccessful self-regulatory schemes are the NAB's early attempt to regulate frequency assignments, the NAB's attempt to limit cigarette advertising, the NAB's attempt to encourage diverse cultural and educational programming for children, and the NNC's attempt to promote fairness in news coverage. The Children's Advertising Review Unit's attempt to protect children's online privacy has largely been ineffective as well; however, it may achieve greater success over time.[289]

What are some of the factors that may account for these successes and failures? First, these examples are used to test the four criteria identified by Michael.[290] Second, they suggest an additional factor that may influence the success or failure of self-regulatory schemes.

1. Industry Incentives and Expertise

Michael hypothesized that for self-regulation to be successful, the self-regulatory body must have both the expertise and motivation to perform the self-regulation. The media examples support this hypothesis. In most exam-

---

289. Several other schemes—video games and the V-Chip—are too new to assess. Although the content provisions of the alcoholic beverages industry codes have been in existence for some time, the Author is not aware of any studies of their effectiveness. The FTC's current investigation may shed some light on this question.

290. *See supra* Part II.D.

ples, a major motivating factor was fear that if the industry failed to act on its own, the government would regulate.[291] Where the threat of government regulation receded—as in the case of the National News Council—self-regulation failed. Further, in cases where the credible threat of governmental regulation disappeared, so did the self-regulation. For example, the NAB decided to abandon the entire Code instead of simply eliminating the sections challenged in the antitrust suit because it was clear that the FCC was no longer interested in regulating broadcast content.[292] Likewise, CARU's efforts to police children's advertising have varied depending on the intensity of governmental interest.[293]

In the above examples, the threat of government regulation was due to a change in the political climate. In other cases, legal constraints have mitigated the possibility of government regulation. For example, following the Supreme Court decision granting First Amendment protection to motion pictures, the Movie Production Code fell apart.[294] Similarly, DISCUS's decision to eliminate the prohibition on advertising distilled spirits followed the Supreme Court's decision in *44 Liquormart*.[295]

Economics is another important incentive. One might explain the effect of the NAB's limitation on cigarette advertising by looking at broadcasters' advertising revenues. The limitations on the amount of advertising per hour and the number of products advertised in a single spot increased broadcasters' revenues by raising the price of advertising time, while limits on cigarette advertising would have cut into advertising revenues. Yet, economic incentives cannot provide the sole explanation, for surely the NAB's and DISCUS's prohibition on advertising distilled spirits also had the effect of reducing advertising revenues.[296] Ultimately, the distilled spirits industry's

---

291. Some of the media examples demonstrate that the failure of self-regulation does in fact lead to government regulation. For example, the NAB's inability to voluntarily work out frequency assignments led to the Radio Act of 1927, under which the FCC awarded licenses for specific frequencies. Likewise, the failure of the NAB and the tobacco industry to restrict cigarette advertising contributed to the passage of a law prohibiting broadcast advertising. Repeal of the NAB Code provisions limiting the amount of children's advertising led to a law limiting the amount of advertising. However, the converse is not necessarily true—that is, that successful self-regulation will obviate the need for government regulation. *See infra* Part IV.B.2.

292. *See supra* Part III.A.2.

293. Not only did Armstrong's study indicate that CARU's efforts correlated with the degree of governmental interest, but CARU increased its activities regarding online advertising when it became clear that the FTC was considering action in this area. *See supra* Part III.B.1-2.

294. *See supra* Part III.A.2.

295. *See supra* Part III.C.1.

296. The different treatment of alcohol and tobacco advertising may be attributed to historical differences in how these products have been perceived by the public. The expe-

perceived need to advertise in order to regain market share lost to beer and wine resulted in the failure of self-regulation.

Economic incentives can sometimes combine with altruism as a motive for self-regulation. For example, in the case of the Family Viewing Hour, the district court found that CBS genuinely wished to reduce violence on its network, but did not want to act alone because it would suffer competitive harm.[297]

Michael argues that industry must not only be willing to self-regulate, but it must possess the requisite expertise. Lack of expertise presented a problem in only one of the media examples—the National News Council. One of the factors contributing to the NNC's demise was its reliance on inexperienced graduate students to conduct investigations. Of course, this lack of expertise was directly related to the lack of funds, which in turn was due to the lack of motivation within the industry.[298]

The NNC also provides the only media example where there was no preexisting organization willing to take on the self-regulation. The lack of a preexisting organization may have been a contributing factor to its failure since the new organization was unable to establish a sufficient source of funding.[299] In the other media examples, whether successful or not, a trade association was already in existence, although, in some cases—CARU and RSAC—the trade association created a separate unit or organization to actually carry out the self-regulation.

Thus, the examples of self-regulation of the media support the hypothesis that industry motivation is essential to successful self-regulation. If the source of that motivation is removed or weakened, then self-regulation is likely to falter.

2. Agency Ability to Audit

Michael's second hypothesis—that agency staff must have the authority and expertise to audit the self-regulatory activity—applies specifically to the type of self-regulation he was interested in, that is, audited self-regulation by federal agencies.[300] While none of the media examples in-

---

rience of Prohibition suggests that large segments of the public have traditionally disapproved of alcohol consumption, while smoking was not generally viewed as undesirable until the Surgeon General's Report publicized the health risks in the 1960s.

297. *See supra* Part III.A.3.c(ii). In advocating for protection of children online, the Author sometimes heard similar arguments from industry representatives. They would say that they wanted to protect children, but were afraid that their competition would not, and the "good guys" would suffer in the marketplace.

298. *See supra* Part III.C.2.

299. *See supra* Part III.C.2.

300. *See supra* Part II.D.

volved audited self-regulation, they suggest that even where self-regulation is not required by a federal agency, it is more effective where a federal agency has authority to regulate and is available to enforce rules against the noncompliant.

The "successful" instances of self-regulation generally involved situations where there was some government regulation. For example, NAB's implementation of advertising limits was supported by the fact that the FCC looked at the number of commercials in renewing licenses.[301] Similarly, the Children's Advertising Review Unit's efforts to deter unfair and deceptive advertising to children were backed up by FTC (and sometimes FCC) enforcement actions.[302] The recent FTC case against Beck's similarly suggests that the FTC is willing to enforce restrictions on alcohol advertising where the industry fails to adequately police itself.[303] Thus, having a government agency with authority to regulate as a backup appears to be an important factor in successful self-regulation.

3. Objective v. Subjective Standards

Michael hypothesized greater success where rules were relatively narrow and susceptible to output-based standards and found that programs with the most subjective standards experienced the most difficulty in implementation. Again, the media examples confirm this finding. When Code requirements were vague and subjective, compliance was less likely than when they were concrete and measurable. For example, the FCC's Children's Television Task Force found that most stations complied with the NAB's children's advertising limits, which could be measured in terms of minutes per hour, but did not comply with the more subjective obligation to provide a variety of educational and cultural programming.[304] Similarly, compliance with an outright ban on the broadcast of advertising distilled spirits is easily determined. These examples generally support the conclusion that self-regulation is more successful when the regulation is susceptible to output-based standards.

4. Fair Process and Public Participation

Michael argued that to enhance the likelihood of success,

---

301. *See supra* Part III.A.3.b.
302. *See supra* Part III.B.1.
303. *See supra* Part III.C.1.
304. TASK FORCE REPORT, VOL. 1, *supra* note 136, at 4. *See also* Policies and Rules Concerning Children's Television Programming, *Notice of Inquiry*, 8 F.C.C.R. 1841, para. 7, 3 Comm. Reg. (P & F) 2291 (1993) (noting compliance with "specific, palpable performance standards" for children's advertising but little improvement concerning vague program obligations).

[t]he self-regulatory organization should engage in its rulemaking on the record, with notice and opportunity for comment given to all affected groups to the extent possible, with particular emphasis on notice to nonmembers who might be adversely affected by the proposed rule, and responses to all significant comments required in the rulemaking record.[305]

Most of the media examples did not engage in "rule making" on the record with notice and opportunity to affected groups. Indeed, in many media examples, there appears to be little public awareness of the self-regulation, much less public involvement in the rulemaking and enforcement processes. For example, the NAB does not appear to have consulted with viewers or consumers in developing or amending the Code nor to have encouraged or accepted complaints from the public.[306] Indeed, most of its enforcement was done behind closed doors. Moreover, the public was largely unaware of the NAB seal. It is impossible to know whether this lack of public awareness and participation affected the NAB's effectiveness.

The Children's Advertising Review Unit's mode of operation is similar to the NAB's, although it occasionally acts on complaints from the public, makes some effort to publicize its actions, has an academic advisory board, and consults with nonindustry groups from time to time.[307] The industry groups that developed the V-Chip rating scheme—MPAA, NAB, and NCTA—were criticized for developing the initial rating scheme without sufficient input from the public,[308] but did eventually meet with the groups and added representatives of the public to their advisory board.[309]

Other self-regulatory groups have included nonindustry representation. A majority of the members on the RSAC, which developed video game ratings, comes from outside the video game industry.[310] Moreover, the NNC

---

305. Michael, *supra* note 16, at 245. He also argues that in enforcement activities, the self-regulatory organization should provide notice and opportunity for a hearing before an impartial decision maker who is required to decide on the record. *Id.* The media examples, such as the NAB Code, CARU Guidelines, and the movie ratings, seemed to comply with these procedural safeguards.

306. Several articles commented on the public's lack of awareness of the NAB Code and the muted public response to its repeal. *See, e.g.*, Maddox & Zanot, *supra* note 53, at 125.

307. *See supra* Part III.B.1. Members of CARU's academic advisory group are listed on its Web page. Some of the organizations the Author represents met with CARU representatives to comment on the draft guidelines for online media.

308. *See, e.g.*, Jube Shiver, Jr., *TV Industry to Use Ratings Before Regulatory Review*, L.A. TIMES, Dec. 19, 1996, at A1; Jenny Hontz & Christopher Stern, *D.C. Goes Rating-Baiting*, VARIETY, Feb. 24-Mar. 2, 1997.

309. *Implementation Report and Order*, *supra* note 278, paras. 12-16.

310. Balkam, *supra* note 31, at 140. Whether RSAC is ultimately successful, will, as in the case of the V-Chip, depend on whether it can earn the public's knowledge and trust.

consisted of members drawn from both the public and the journalism profession and was chaired by a member of the public. Yet, despite this public involvement, the NNC's failure was attributed in large part to its failure to obtain widespread public awareness and support.[311] On the other hand, the success of movie ratings has depended upon widespread public awareness and support.[312]

In sum, the media examples are inconclusive on whether public participation in rule making and enforcement is important for effective self-regulation. Some have enjoyed success without public participation while others have failed even with public participation.

5. Other Conditions

The media examples also suggest that the size and structure of the industry are important factors in the success of self-regulation. Logic suggests that the fewer industry participants, the easier it would be to self-regulate. The media examples also indicate that the existence of market power may play a role in being able to effectively enforce industry self-regulation.

The NAB enjoyed success in limiting the amount and kind of advertising, even though there was a large number of advertisers and advertising agencies, because the number of television stations was limited, and most belonged to one of three major networks. Since access to the network affiliates, most of whom complied with the NAB Code, was essential to reach the majority of markets, advertisers had a strong incentive to comply with the Code.[313] Similarly, there are only a small number of movie studios. When they controlled the majority of movie theaters, they could ensure that the theaters showed only films meeting the Production Code.[314]

The success CARU has achieved with television commercials compared to Web sites may also be due to the structure of those industries. The number of companies that advertise to children on television is fairly limited. However, the number of companies offering Web sites to children is quite large and includes many new entrants as well as the traditional cereal, toy,

---

311. *See supra* Part III.C.2. Moreover, Brogan attributes some of the NNC's problems to the requirement that it be chaired by a member of the public. BROGAN, *supra* note 222, at 17. He found that the only effective chairman was one that came from a journalism background, who, because of his background, was able to command the respect of the media and be effective in fund-raising. *Id.* at 47.

312. *See* Swope v. Lubbers, 560 F. Supp. 1328, 1340-41 (W.D. Mich. 1983) (describing public awareness and use of movie ratings based on surveys).

313. While advertisers could always find non-Code stations to carry their commercials, "faced with the choice of abiding by the Code or incurring the expense and possible public relations headaches of making two sets of ads for Code and non-Code stations, advertisers generally chose the former." Rotfeld et al., *supra* note 107, at 19.

314. *See supra* Part III.C.4.

and fast food companies.[315] Thus, additional factors affecting the success of self-regulation are the number of industry participants and whether there are dominant players that can use their market power to enforce self-regulatory provisions. With these factors in mind, this Article now analyzes the two recent proposals for self-regulatory initiatives—digital television and online privacy.

C. *Implications for Digital Television*

The Advisory Committee has recommended adoption of a Model Voluntary Code of Conduct for Digital Television Broadcasters. The Advisory Committee explains the reasons for its recommendation:

> A new industry statement of principles updating the 1952 Code would have many virtues. The most significant one is that it would enable the broadcasting industry to identify the high standards of public service that most stations follow and that represent the ideals and historic traditions of the industry. A new set of standards can help counteract short-term pressures that have been exacerbated by the incredibly competitive landscape broadcasters now face, particularly when compared to the first 30-some years of the television era. Those competitive pressures can lead to less attention to public issues and community concerns. A renewed statement of principles can make salient and keep fresh general aspirations that can easily be lost in the hectic atmosphere and pressures of day-to-day operations.[316]

Unlike the old NAB Code, the Model Voluntary Code does not address advertising.[317] Apart from this change, the Model Voluntary Code closely tracks the old NAB Television Code. Both include sections addressing responsibility toward children, the treatment of news and public events, com-

---

315. *See supra* Part III.B.2.

316. PIAC REPORT, *supra* note 1, at 46. In a draft of this section of the report, Professor Sunstein makes similar arguments in support of a voluntary code:

> The principle virtue is that it enables the broadcasting industry to identify and to adhere to high standards of public service, standards that are already followed by many (though not all) stations, and that are consistent with the best historical traditions of the industry. A code can help counteract short-term pressures that can lead to less attention to public issues, less and worse programming for children, and more sensationalism and prurience than is desirable. A code can also identify and help promote general aspirations that can sometimes be lost in day-to-day operations. Because a code involves self-regulation, it has the distinctive advantage of not permitting government officials to be in the business of making decisions about television content.

CHARTING THE DIGITAL BROADCASTING FUTURE, APPENDIX B: VOLUNTARY CODE OF CONDUCT 2-3 (draft Sept. 4, 1998) (citation omitted) (on file with author) [hereinafter PIAC DRAFT CODE OF CONDUCT].

317. Presumably, this was done in part to avoid any antitrust issues. *See* PIAC REPORT, *supra* note 1, at 120-21 (explaining why the Model Code is unlikely to be found to violate antitrust laws).

munity responsibility, controversial public issues, and special program standards (violence, drugs, gambling, etc.).[318] Unlike the old NAB Television Code, the Model Voluntary Code does not contain a section on religious programming, but has new sections on covering elections and responsibility toward individuals who are deaf or hard of hearing.[319]

Like the old NAB Television Code, the Model Voluntary Code provides for the position of Code Authority Director and a Television Code Review Board. While the NAB's work in the past was largely conducted in private, the Advisory Committee has made several recommendations designed to invite public and governmental awareness and participation. It urges the NAB to draft the Code with input from community and public interest leaders.[320] Moreover, the proposed Model Voluntary Code provides for "special public recognition" to stations with an excellent public service record.[321] It also would require that the FCC be informed at license renewal time whether or not a station is in compliance with the Code, although this notation is to "lack any legal force or effect."[322] In addition, the Television Code Review Board is to "report to the public the names of complying, noncomplying, and specially commended stations" and "report continuing or egregious violations of the code to Congress, the public, and FCC on an ongoing basis."[323]

There are three reasons to be skeptical about the Advisory Committee's recommendation for a voluntary code. First, it is unclear whether the NAB will follow it. Second, even if the NAB does adopt a voluntary code along the lines suggested by the Advisory Committee, it is doubtful that the code will be effective in achieving the stated goals. Finally, the Model Voluntary Code raises similar questions regarding voluntariness that could cause it to be subject to constitutional challenge.

---

318. Within these sections, some changes have been made. For example, the section on Responsibilities Toward Children in the Model Voluntary Code, which states that "[e]ach broadcaster should endeavor to provide a reasonable amount of educational programming for children each week," PIAC REPORT, *supra* note 1, at 107, is somewhat more specific than the old NAB Code's statement that "[b]roadcasters have a special responsibility to children. Programs designed primarily for children should take into account the range of interests and needs of children from instructional and cultural material to a wide variety of entertainment material." TELEVISION CODE, *supra* note 105, at 3.
319. PIAC REPORT, *supra* note 1, at 107-08, 112.
320. *Id.* at 47.
321. *Id.* at 113.
322. *Id.* at 114.
323. *Id.*

The initial reaction of many broadcasters was to oppose adoption of a voluntary code.[324] After the Advisory Committee discussed the proposed voluntary code in the summer of 1998, *Broadcasting & Cable* reported that "[w]hile broadcasters hate the idea of any so-called voluntary code of conduct, the NAB board decided . . . to proceed with caution—for now."[325] After a "robust" discussion on whether to strongly oppose such a plan, the Board yielded to arguments not to oppose the Code before it was even recommended. Instead, the Board passed a resolution expressing "serious concern" regarding any government efforts to impose limits on broadcasters' editorial freedom.[326] At its January 1999 board meeting, the Board took no action regarding the Advisory Committee Report except to decide to file comments.[327] Before the meeting, an NAB staffer was quoted as saying, "'The board may well decide that it's better off to say nothing now,'" and to wait for the FCC and Congress to act.[328]

As shown *supra*, having an organization that is willing to commit adequate resources to self-regulation is essential to a successful self-regulatory scheme. Here, the incentive for self-regulation, if any, comes from the threat of government regulation. So, the question is whether the broadcast industry's fear of government regulation provides sufficient incentive for it to engage in self-regulation.

Government regulation is certainly a real possibility here. The statute permits broadcasters to receive, free of charge, the exclusive use of extremely valuable electromagnetic spectrum. Instead of being required to bid for the spectrum in an auction, as many other spectrum licensees are, the broadcasters are expected to serve the public interest. Thus, as a condition of using the spectrum, the broadcasters are required to do something in the public interest. The only question is what that something is.

The Advisory Committee recommended (over the dissents of some broadcast members) that the FCC impose some minimal public interest

---

324. At the Advisory Committee meeting on September 9, 1998, some members sought unsuccessfully to get the NAB's reactions to the proposed voluntary code on record. *Advisory Committee on Public Interest Obligations of Digital Television Broadcasters, Morning Session* 38-43 (visited Mar. 15, 1999) <http://www.ntia.doc.gov/pubintadvcom/sepmtg/> [hereinafter September Transcript]. However, one member from the broadcast industry did state his understanding that the NAB hierarchy opposed voluntary standards out of fear that they would evolve into mandatory minimum standards. *Id.* at 47-49 (comments of Paul La Camera).

325. Paige Albiniak, *Preparing for Battle*, BRDCST. & CABLE, July 6, 1998, at 22.

326. *Id.*

327. *Fox Still Undecided on Leaving NAB, Expresses Concerns*, COMM. DAILY, Jan. 14, 1999, at 6.

328. *Networks Seek Repeal of NAB Endorsement of 35% Station Cap*, COMM. DAILY, Jan. 6, 1999, at 2.

standards, but it left the content of those standards up to the FCC.[329] The FCC is expected to initiate a rule making to determine what those public interest standards will be in the spring of 1999.

It is not clear whether the FCC will propose, much less adopt, significant public interest requirements for digital broadcasters.[330] If the FCC proposes to adopt serious public interest requirements, the NAB may decide that while it opposes a voluntary code, it opposes government-mandated standards even more. From the broadcast industry's point of view, the lesser of the two evils may well be the voluntary code.

On the other hand, the NAB is likely to opt for self-regulation only if it can avoid or render insubstantial the FCC rules. If the FCC adopts serious public interest requirements for digital broadcasters, broadcasters will have little incentive to engage in voluntary self-regulation. Thus, although having both government regulation and industry self-regulation may provide the best conditions for successful self-regulation, such a scenario seems an unlikely outcome in this case.

Another reason why broadcasters may be less than enthusiastic about embracing self-regulation relates to the structure of the video industry. Since the period in which the old NAB Code was in effect, competition from outside the broadcast industry has increased. Competition comes primarily from cable television and somewhat from satellite television and the Internet. The proposed code, however, would apply only to broadcast digital television and not to other television delivery systems. To address this problem, the group drafting the Model Voluntary Code suggested that the Code should be applied to television programmers that are not broadcasters.[331] However, it seems doubtful that these other industries, which unlike the broadcasters are not getting the benefit of valuable, free spectrum, have any incentive to adopt a similar code of conduct.

While the NAB may agree to self-regulation to avoid government regulation, it is unlikely to commit the necessary resources to make self-regulation effective. As demonstrated above, when the former Television Code existed, little attention and few resources were devoted to enforcing the program provisions compared to the advertising provisions. This may have been in part due to the fact that program provisions were vague and thus it was more difficult to evaluate compliance. Similarly, many of the proposed

---

329. PIAC REPORT, *supra* note 1, at 47-48.

330. In an earlier rule making, the FCC put off this question. Advanced Television Systems and Their Impact upon the Existing Television Broadcast Service, *Fifth Report & Order*, 12 F.C.C.R. 12,809, paras. 3-5, 7 Comm. Reg. (P & F) 863 (1997). Since that time, the composition of the FCC has changed.

331. PIAC DRAFT CODE OF CONDUCT, *supra* note 316, at 3. *See also* September Transcript, *supra* note 324, at 41. However, this proposal does not appear in the final report.

provisions in the Model Code are vague and not easily measured or enforced.[332]

Finally, even if the NAB adopts a code of conduct, its action may be subject to constitutional challenge.[333] To be sure, just because adoption of the voluntary code is not truly voluntary does not necessarily mean that the First Amendment would be violated.[334] But it could cause delay and, as a practical matter, the ultimate abandonment of the Code, as happened with the Family Viewing Hour.

To avoid First Amendment difficulties, the Advisory Committee notes that "it is extremely important that we are arguing on behalf of a code as a simple recommendation to private organizations, above all the NAB, and *not* as a proposed mandate from the government, either the FCC or Congress."[335] The Report explains that the First Amendment is irrelevant where the industry as a whole decides what to broadcast without government involvement. However, "if a code were a product of government threat, and were effectively required by government," the First Amendment would apply, and the content regulation would be subject to scrutiny by the courts.[336]

It is difficult to take seriously the Advisory Committee's claim that the code is purely voluntary. As discussed *supra*, digital broadcasters are getting free use of valuable spectrum in return for serving the public interest. Moreover, on its face, the proposed code would require the FCC to be notified at license renewal time whether a station is in compliance with the voluntary code. Broadcast licensees must seek renewal from the FCC every

---

332. For example, the Model Voluntary Code states that "news programming should be both substantive and well-balanced" and should "provide appropriate coverage to topics of particular concern to the local community." PIAC REPORT, *supra* note 1, at 108, 110. However, some provisions are more objective and quantifiable. For example, one provision states that broadcasters should provide well over 75 public service spots per week. *Id.* at 110.

333. While presumably the NAB could not bring such a challenge, perhaps dissident members or third parties might have standing to bring a constitutional challenge.

334. A First Amendment violation would only arise if government coerced industry to do something indirectly that it could not require directly. While some of the proposed provisions, if they were direct government regulations, might violate the First Amendment, many would not. For example, the proposed children's television provision, *see supra* note 318, is similar to the Children's Television Act's requirement that each broadcaster provide some programming serving the educational and informational needs, but is less specific than the FCC's interpretation of that Act, which established a guideline of three hours of children's educational programming per week. *See supra* Part III.A.3.c(ii). Since neither the Children's Television Act nor the FCC rules have been found unconstitutional, presumably the broadcasters' decision to provide a reasonable amount of educational programming for children would not violate the First Amendment, even if it were the result of government coercion.

335. PIAC REPORT, *supra* note 1, at 117.

336. *Id.*

eight years, and even though nonrenewal is extremely rare, no broadcaster wants to risk a challenge to its license. Although the Model Voluntary Code states that the notation as to compliance lacks any legal force or effect, it is hard to imagine that the FCC would feel free to ignore a finding of "continuing or egregious noncompliance."[337] In sum, past experience with self-regulation of the media provides little hope that the Advisory Committee's recommended voluntary code for digital television will be successful.

D. Implications for Privacy on the Internet

There is also reason to be skeptical about the ability of self-regulation to protect consumer privacy on the Internet. Despite numerous calls for self-regulation, industry appears to be dragging its feet. Although the White House first called for industry self-regulation in the *Framework for Global Electronic Commerce* issued in July 1997, by November 1998, only limited progress had been made. In its First Annual Report, the U.S. Government Working Group on Electronic Commerce, found that "[i]ndustry was slow to respond to the President's call in the *Framework* for the development of effective self-regulation."[338] It noted that since the FTC published its *Report to Congress* in June 1998, however, serious efforts to protect privacy through self-regulation had begun, citing as an example, the efforts of the Online Privacy Alliance.[339] But it warned that "if self-regulation is to work, these efforts must expand over the next year."[340]

Likewise, the FTC began calling for industry self-regulation at least as early as the June 1996 workshop.[341] In congressional testimony in July 1998, the Chairman of the FTC gave industry until the end of the year to

---

337. *Id.* at 114.

338. U.S. GOVERNMENT WORKING GROUP ON ELECTRONIC COMMERCE, FIRST ANNUAL REPORT 16 (Nov. 1998), *available at* <http://www.ecommerce.gov> [hereinafter FIRST ANNUAL REPORT]. *See also* Elizabeth deGrazia Blumenfeld, *Privacy Please: Will the Internet Industry Act to Protect Consumer Privacy Before the Government Steps In?*, 54 BUS. LAW. 349, 382 (1998).

339. FIRST ANNUAL REPORT, *supra* note 338, at 16. According to its Web page, www.privacyalliance.org, the Online Privacy Alliance is a coalition of over 60 global corporations and associations, formed to "lead and support self-regulatory initiatives that create an environment of trust and that foster the protection of individuals' privacy online and in electronic commerce." Online Privacy Alliance, *Mission* (visited Mar. 15, 1999) <http://www.privacyalliance.org/mission/>. The Alliance has proposed privacy principles and advocates an enforcement system based on a seal. *Id.*

340. FIRST ANNUAL REPORT, *supra* note 338, at 17.

341. *See Privacy Workshop*, *supra* note 186, at 436 (closing statement of Chairman Pitofsky), *available at* <http://www.ftc.gov/bcp/privacy/wkshp96/pw960605.pdf>.

come up with effective self-regulation, or the FTC would seek legislation.[342] It does not appear that industry has met this challenge.[343]

Although the government argues that it is in the economic interests of business to develop effective self-regulation because "it is essential to assure personal privacy in the networked environment if people are to feel comfortable doing business online,"[344] on balance, the economic incentives probably run the other way.[345] It is quite profitable for companies to collect personal information. It costs companies little to collect personal information, and they can sell it or use it to better target their sales efforts.[346] Self-regulation that requires them to disclose their information practices and allows the public to opt-out (or even worse from the industry point of view, having to get them to opt-in) will increase the costs of information collection.[347]

Thus, the major incentive for industry to self-regulate is to avoid the threat of government regulation. Since there is no government agency that currently has sufficient authority to regulate privacy, legislation would be required. The possibility of legislation is not entirely remote. Both the Administration and the FTC have threatened to seek legislation if the industry fails to self-regulate.[348] Moreover, Congress did pass legislation to protect children's privacy online.[349] Nonetheless, the threat of legislation may not be sufficiently realistic to overcome the obstacles to effective self-regulation.

---

342. *Electronic Commerce Hearings, supra* note 7, at 303 (statement of Robert Pitofsky, Chairman, FTC); Jeri Clausing, *Group Proposes Voluntary Guidelines for Internet Privacy*, N.Y. TIMES, July 21, 1998, at D4.

343. In February 1999, an FTC spokeswoman stated that the Commission would still "'rather have industry regulate this than government.'" Courtney Macavinta, *Government Delivers Privacy Ultimatum* (visited Mar. 15, 1999) <http://www.news.com/News/Item/Textonly/0,25,31822,00.html>. The FTC plans a survey for March 1999 to determine the industry's progress. *Id.*

344. Elements of Effective Self Regulation for the Protection of Privacy and Questions Related to Online Privacy, *Notice and Request for Public Comment*, 63 Fed. Reg. 30,729 (NTIA 1998).

345. It has been reported that companies selling personal information had gross annual revenues of $1.5 billion. Trans Union Corp., *Initial Decision*, FTC Docket No. 9255, at 53 n.354 (July 31, 1998), *available at* <http://www.ftc.gov/os/1998/ 9808/d9255pub.id.pdf>.

346. *See, e.g.*, Budnitz, *supra* note 30, at 853; Blumenfeld, *supra* note 338, at 351.

347. *See, e.g.*, Perritt, *supra* note 32, at 108 (describing transaction costs involved in opt-out); Jerry Kang, *Information Privacy in Cyberspace Transactions*, 50 STAN. L. REV. 1193, 1255 (1998) (noting that even disclosing the fact of collection is costly to firms).

348. Several academics have also called for legislative action to address privacy concerns. *See, e.g.*, Kang, *supra* note 347, at 1193; Budnitz, *supra* note 30. Numerous bills have been introduced in Congress that are related to privacy on the Internet. *See generally Electronic Privacy Information Center* (visited Mar. 15, 1999) <http://www.epic.org>.

349. *See supra* note 8. However, it may be easier to get the political support needed for legislation when the subject of the legislation is children and the number of companies affected is smaller.

Even if the companies were to agree to a self-regulatory regime, it may be difficult to enforce.[350] All of the self-regulatory schemes being discussed rely on "notice and choice"; that is, the Web site would disclose its privacy practices, and the consumer could exercise choice by declining or continuing to do business with the company. One problem, however, is that it is difficult for consumers to verify whether a company in fact complies with its stated policies.[351] Thus, it would be easy for companies to cheat and difficult for consumers to confirm compliance. Although the Better Business Bureau is developing an enforcement mechanism that would award a "privacy seal" to sites that meet its standards,[352] there is no reason to believe this enforcement mechanism would be any more effective than the old NAB seal.[353] These problems would be aggravated by the lack of government oversight since there is no government agency that clearly has authority to oversee privacy protection.[354]

As in the case of digital television, self-regulation as an adjunct to government regulation seems more promising. The safe harbor provisions of the children's privacy legislation provide a useful model in this regard.[355] The

---

350. Professor Reidenberg has similarly argued that self-regulation has not been nor is it likely to become a successful way to protect the privacy of U.S. citizens. *See* Joel R. Reidenberg, *Restoring Americans' Privacy in Electronic Commerce*, 14 BERKELEY TECH. L.J. (forthcoming Apr. 1999).

351. *See, e.g.*, Joel R. Reidenberg & Paul M. Schwartz, Legal Perspectives on Privacy 27-28 (Oct. 29, 1998) (paper presented as part of the Information Privacy Seminar Series, Georgetown University, January 1998) (discussing how actual information practices are largely hidden from public view, and barriers for individuals to discover how businesses use personal information are often insurmountable; at the same time, businesses profit enormously from trade in personal information); Swire, *supra* note 12, at 6; Mary J. Culnan, A Methodology to Assess the Implementation of the Elements of Effective Self-Regulation for Protection of Privacy 11-12 (discussion draft June 1, 1998), *available at* <http://www.georgetown.edu/culnan/>.

352. Macavinta, *supra* note 343.

353. Indeed, a recent press report illustrates one of the problems with seals. Trust-E is an organization, funded by Microsoft and nine other companies, that monitors the Internet privacy policies of about 500 companies. Companies that meet the criteria are awarded a seal of approval designed to assure consumers that their privacy will be protected. Although a customer complained to Trust-E that Microsoft was collecting personally identifiable information even when customers explicitly indicated they did not want this information collected, Trust-E declined to deny Microsoft the use of its seal or even to audit the company. *Watchdog Group Won't Pursue Microsoft*, SAN JOSE MERCURY NEWS, Mar. 23, 1999 (visited Mar. 15, 1999) <http://www.mercurycenter.com/svtech/news/breaking/merc/docs/ 009780.htm>.

354. It has been suggested that the FTC could initiate enforcement actions against companies that post privacy disclosure policies yet fail to comply with them. However, apart from the difficulties in determining whether companies are complying when they alone know what they do with the information, the remedies are also inadequate for consumers. Reidenberg & Schwartz, *supra* note 351, at 26-27.

355. *See supra* note 8.

safe-harbor approach would appear to shift some of the costs of regulation to the private sector, while ensuring that all industry participants are subject to minimum standards. Likewise, this approach can allow flexibility and take advantage of industry's superior knowledge, without having to rely solely on industry self-interested choices. The ability of the public to comment on the FTC rules and on the adequacy of industry guidelines provides an additional safeguard against industry subversion of self-regulation to its own ends.

Finally, the industry structure militates against effective self-regulation. As Professor Budnitz notes,

> meaningful regulation requires participation by the entire electronic commerce industry. Unfortunately, the presence of great diversity in this industry makes universal participation unlikely. In fact, in this context, it is probably inaccurate to talk about the electronic commerce industry in the singular, for several industries are involved.[356]

No single industry organization comparable to the NAB exists that could undertake self-regulation. Although the Online Privacy Alliance "represents significant online players, . . . the group's membership is only a drop in the bucket given the seemingly infinite number of sites on the Web."[357] Given the large number and diversity of parties involved, it is difficult to see how self-regulation could work.

V. CONCLUSION

Self-regulation has been portrayed as superior to government regulation for addressing problems of new media such as digital television and the Internet. This Article has analyzed the effectiveness of self-regulation by looking at the track record of self-regulation in other media. After describing and analyzing past uses of self-regulation in broadcasting, children's advertising, news, alcohol advertising, comic books, movies, and video games, it concludes that self-regulation rarely lives up to its claims, although in some cases, it has been useful as a supplement to government regulation. It then identifies five factors that may account for the success or failure of self-regulation. These include the industry incentives, the ability of government to regulate, the use of measurable standards, public participation, and industry structure. Applying these five factors to digital television public interest responsibilities and privacy on the Internet, it concludes that self-regulation is not likely to be successful in these contexts.

---

356. Budnitz, *supra* note 30, at 874.
357. Macavinta, *supra* note 343.

[5]

Controlling the New Media: Hybrid Responses to New Forms of Power

Andrew Murray and Colin Scott***

The development of new media industries, stimulated by the technology of digitalisation, has thrown up an important literature on mechanisms for regulation and control. In this article we elaborate on and develop Lawrence Lessig's 'modalities of regulation' analysis. As we reconceive them the four basic control forms are premised upon hierarchy, competition, community and design and can be deployed in fifteen pure and hybrid forms. This analysis is enriched through elaborating on the essential elements of control systems (standard-setting, monitoring and behaviour modification) to demonstrate the importance and variety of hybrid forms that real-world control systems take in the new media domains. Although the article does not provide any universal prescriptions as to which control forms are likely to be most appropriate in particular domains, it does provide a richer analytical base both for understanding existing control mechanisms and the potential for using greater variety. The development of regulatory regimes which are both legitimate and effective in any given domain is likely to require sensitivity to the particular context and culture of both the domain and the jurisdiction within which it is located.

Introduction

The emergence and identification of the new media, premised upon the development and application of digital technologies, has created new sources and locations of power, many not fully documented or understood. Those new configurations of power which have been identified have stimulated distinctive literatures about the most appropriate mechanisms of control. With much of the literature classical or 'command and control' regulation is held either to be undesirable or unfeasible in the face of the new policy challenges. For one school of thought the changing market structures associated with the new media indicate a reduced role for classical regulation and its virtually total displacement by

---

* Law Department, London School of Economics.
** Research School of Social Sciences, Australian National University and Centre for Analysis of Risk and Regulation, London School of Economics.
Though we accept full responsibility for errors and infelicities we are grateful to the following for comments on an earlier draft: Julia Black, John Braithwaite, Neil Duxbury, Peter Grabosky, Mathias Klang, Robin Mansell, David Post, and participants in the Competition Law and the New Economy Workshop, University of Leicester, July 2001.

competition law.[1] For another school the emergence of the Internet presents insuperable problems for classical regulation and alternative mechanisms of control based on self-regulation and architecture are more likely to be effective.

In this article we draw together some of the regulatory problems presented by the new media and apply a developed and modified version of Lawrence Lessig's 'modalities of regulation'[2] analysis to thinking about the range of mechanisms which have been developed to address these problems. Accordingly we first provide a description of some of the key problems identified with controlling the new media. Our modified version of Lessig's analysis claims that there are four bases of regulation – hierarchy, competition, community and design. We set the analysis to work demonstrating that these four bases of regulation are observable as means of addressing the range of regulatory problems of the new media. The tendency to privilege one basis for regulation over others appears to us to be consistent neither with empirical observation nor with the normative considerations of institutional design for good regulation. What we observe is the prevalence of hybrid forms of control which, when better understood, could provide the basis for a better informed policy debate about the control of the new media.

Differences in approach may partly be explained by reference to the cultures and preoccupations within different jurisdictions. The UK, and many European Union states, have a strong tradition of self-regulation in the media generally and the legitimacy of this form of governance is widely accepted.[3] Private governance forms are generally less well recognised and accepted in the United States and have been the subject matter of fierce debate over their legitimacy.[4] A related bias in the US literature is the very high value placed on the constitutional ideal of freedom of speech which feeds into a strong libertarian underpinning to much discussion of regulation of new media.[5] Though freedom of speech may have some constitutional protection in EU states, the extent to which such a right is qualified by other collective considerations is quite pronounced. Thirdly American scholarship on new media issues is dominated by the 'legal centralist' perspective of law and economics which accords less recognition to the potential for pluralism in the generation of norms than is true of some European scholarship.[6]

---

1 The Chicago School of Law and Economics supports market control where markets are competitive. If the market is uncompetitive, competition law provides an adequate remedy. This premise has been attacked in relation to layered communications networks. See L. Lessig, *The Future of Ideas: The Fate of the Commons in a Connected Worlds* (New York: Random House, 2001) 110; C. Salop & R.C. Romaine, 'Preserving Monopoly: Economic Analysis, Legal Standards and Microsoft' (1999) 7 *George Mason Law Review* 617.
2 L. Lessig, *Code and Other Laws of Cyberspace* (New York: Basic Books, 1999) 88.
3 J. Black, 'Constitutionalising Self-Regulation' (1996) *Modern Law Review* 24.
4 The suspicion of private governance institutions in American legal scholarship is forcefully represented by Michael Froomkin's critique of the Internet Corporation for Assigned Names and Numbers (ICANN): M. Froomkin, 'Wrong Turn in Cyberspace: Using ICANN to Route Around the APA and the Constitution' (2000) 50 *Duke LJ* 17; Compare the (European) views of W. Kleinwächter, 'The Silent Subversive: ICANN and the New Global Governance' (2001) 3 *Info* 259 which are largely approving of the innovation in governance created by ICANN.
5 M. Castells, *The Internet Galaxy* (Oxford: Oxford University Press, 2001) 33. S. Venturelli, 'Inventing E-Regulation in the US, EU and East Asia: Conflicting Social Visions of the Internet and the Information Society' paper presented to 29th Research Conference on Communication, Information and Internet Policy, October 2001, Alexandria, Virginia available at <http://www.arxiv.org/ftp/cs/papers/0110/0110002.pdf> (visited 19 December 2001).
6 The analysis is succinctly made by R. Ellickson, 'The Aim of Order Without Law' (1994) 150 *Journal of Institutional and Theoretical Economics* 97, which provides a summary of the fuller treatment in R. Ellickson, *Order Without Law* (Cambridge, MA: Harvard University Press, 1991). See also R. Cooter, 'Against Legal Centrism' (1993) 81 *California Law Review* 417; L. Lessig, 'The Regulation of Social Meaning' (1995) *University of Chicago Law Review* 943.

Whatever the effects of intellectual biases we suggest that research and thinking on control of new media sectors has generated novel insights on regulation which are of wider application. In particular the current debate on how forces of control may be used to shape the future development of networks is of wider interest to researchers in the fields of law, economics and social policy. The debate is centred upon the role of the commons in the fledgling third generation Internet. Sunstein's claim that 'there is no avoiding "regulation" of the communications market'[7] has been met by an equally forceful counterclaim by Lessig that '[t]he issue for us will not be which system of exclusive control – the government or the market – should govern a given resource. The question for us comes before: not whether the market or the state but, for any given resource, whether that resource should be controlled or free'.[8] Lessig's call for a debate on this issue provides a powerful rallying call to those lobbying for the deregulation of, in the sense of making free, all layers of the Internet infrastructure. The debate called for by Lessig is not new. The open source movement led by Richard Stallman and the Free Software Foundation has lobbied for deregulation of the code level since the mid 1980s.[9] Deregulation at the content level was built into the original Internet infrastructure by network designers such as Paul Baran, Jerome Saltzer, David Clark and David Reed.[10] This has since been substantially eroded by the development of intelligent networks such as Resource Reservation Protocol (RSVP).[11] There is no doubt many in the new media sector will respond to Lessig's analysis and over the next five years the wider dialogue of the role of regulation within political science, media and economics will be strongly influenced by this currently narrow legal debate. Thus, while we make extensive use of examples drawn from new media, we suggest that the developed models of control which we discuss are of interest to policy makers and researchers with interests in governance and regulation generally.

New media and the problems of effective control

Processes of digitalisation associated with the development of new media have brought about important reconfigurations of power. The Internet, for example, provides widespread access to technology based on a network of networks and addressing systems which connect computers globally.[12] It is said to create a space where users can engage in a variety of activities with a substantial autonomy from state power which does not exist in non-digital media.[13] Digitalisation of broadcasting and mobile telecommunications create niches for new forms of service provider, shifting power away both from those who own the physical infrastructure of networks and from those who own content. We identify in this section three general problems of new media (that is problems which apply

---

7 C. Sunstein, *Republic.Com* (Princeton NJ: Princeton University Press, 2001) 128.
8 Lessig, n 1 above, 12.
9 *ibid* 52–61.
10 *ibid* 34–44; J. Saltzer, D. Reed and D. Clark, 'End-to-End Arguments in System Design' (1984) 2 *ACM Transactions in Computer Systems* 277. Online version at <http://web.mit.edu/Saltzer/www/publications/endtoend/endtoend.pdf> (visited 4 January 2002).
11 Discussed below, 499.
12 B. Leiner et al, 'A Brief History of the Internet' *Internet Society* available at <http://www.isoc.org/internet/history/brief.shtml> (visited 7 January 2002); Castells, n 5 above, ch 1.
13 S. Sassen, 'Digital Networks and the State: Some Governance Questions' (2000) 17 *Theory, Culture and Society* 19, 20. According to Lawrence Lessig much of this autonomy is hard-wired into the network by its end-to-end architecture, Lessig, n 1 above, 26–41.

generally or to more than one medium) which arise from shifts in power. None of these problems is exclusive of the new media, though each emerges with interesting new features in this context. They are the problems of regulatory arbitrage, anonymity and scarcity of resources. In each case once prevalent governance forms based on public ownership are no longer fashionable (and for some no longer feasible) enhancing the urgency of investigating other forms of control. We should be clear that these are not the only problems associated with the new media. Among the other pressing policy problems are the issues relating to accessibility of digital broadcasting and communications services to less advantaged consumers (which can be defined both in economic and social terms)[14] and the extent to which content of digital broadcasting should be controlled (in the manner that both negative and positive content controls apply to analogue broadcasting).[15] Discussion of these issues is precluded for reasons of space and in the belief that the theoretical frame developed is sufficiently addressed by the policy problems which we do discuss.

The regulatory arbitrage problem

The problem of regulatory arbitrage emerges wherever subjects of regulation have sufficient mobility in their operations or activities that they can choose to be regulated by one regime rather than another. The effect is to create a form of market for regulation within which dissatisfied subjects can 'exit' one regime in favour of another. Regulatory arbitrage, seen as a problem for authorities attempting to capture activities within their web, can also be seen as a solution to problems of excessive or inappropriate regulation as it limits the capacities of authorities.[16] The problem has an interesting double-edged character in the new media, since options to relocate to avoid particular regulatory regimes may be available both to service providers and consumers. Thus broadcasters can relocate their operations to different jurisdictions to evade national regulation (and this predates digitalisation) while listeners and viewers can relocate from the more controllable forms of delivery to satellite and Internet. One of the problems raised by regulatory arbitrage is the risk that competing standards for the new digital broadcasting transmission services might develop. This is squarely addressed with harmonised rules requiring all member states to legislate for common standards in the EU, notably in respect of consumer equipment for conditional access to services.[17] Under the terms of European Union legislation the EU rules on broadcasting regulation apply only to broadcasters established in a state to which the applicable directive applies.[18] The directive's requirements that member states apply their domestic broadcasting rules to all broadcasters established within the

---

14 M. Lemley and D. McGowan, 'Legal Implications of Network Economic Effects' (1998) 86 *California Law Review* 479; P. David, 'The Evolving Accidental Information Super-Highway' (2001) 17 *Oxford Review of Economic Policy* 159; M. Cave and R. Mason, 'The Economics of the Internet: Infrastructure and Regulation' (2001) 17 *Oxford Review of Economic Policy* 188.
15 See. C. Sunstein, 'Television and the Public Interest' (2000) 88 *California Law Review* 499; D. Goldberg, T. Prosser and S. Verhulst, *Regulating the Changing Media: A Comparative Study* (Oxford: Oxford University Press, 1998).
16 See W. Bratton, J. McCahery, S. Picciotto and C. Scott (eds), *International Regulatory Competition and Coordination* (Oxford: Oxford University Press, 1996).
17 Directive 95/47/EC OJ 1995 L281, 23.11.95 p 51, Art 4; Broadcasting Act 1996.
18 Council Directive 89/552/EEC OJ 1989, L298 p 23 as amended by Directive 97/36/EC, OJ 1997 L202 p 60, Art 2(1); B. Drijber, 'The Revised Television Without Frontiers Directive: Is it Fit for the Next Century' (1999) 36 *Common Market Law Review* 87, 92.

state has been interpreted so as to require member states to apply their rules as intensely to broadcasters directing their programming at other member states.[19] This interpretation is intended to preclude countries like the UK establishing themselves as attractive locations for establishment of overseas broadcasters through the application of a more liberal regime than would apply to domestic broadcasters.[20] This is a particular issue with broadcasters seeking to evade what they regard as overly restrictive domestic rules, for example on advertising to children or transmission of pornography.

Regulatory arbitrage in Cyberspace (that is applying to the Internet) is a focal point for two opposing schools of thought, the Cyberlibertarians and the Cyberpaternalists. The primary argument of the Cyberlibertarians is that Cyberspace is unregulable due to its design. Cyberspace is a unique jurisdiction as it has no physicality or real-world existence. It is possible to conceive of Internet users simultaneously in Cyberspace and in a grounded, real-world jurisdiction.[21] It is this duality and the non-physicality of Cyberspace which allows for regulatory arbitrage. In the physical world sovereignty is exercised by governments over defined physical territories. A user who wishes to be regulated by a different regulatory structure may take steps to relocate either themselves or their activities. In Cyberspace users may transcend physical borders with ease and may choose to take on any guise or form desired (see below 'The Anonymity Problem'). Users who prefer a regulated environment where there are structured discussions on carefully selected topics, and where content is closely monitored and censored may choose to join a regulated and monitored cybercommunity such as America Online (AOL). Users seeking uncensored discussion and complete freedom of speech may make use of a virtual chat room on the USENET system or may use an Internet Service Provider (ISP) to enter unmonitored discussion boards on the Web. These freedoms allow users to choose freely the regulatory structure they wish to follow while in Cyberspace. Thus a citizen of Germany can enter a USENET discussion group on the Holocaust and post denial messages, something he or she would be unable to do freely in their home state. Similarly a UK citizen may post information which is in breach of the Official Secrets Acts. Although strictly speaking these citizens are still committing offences within their physical jurisdiction, they can do so without fear of prosecution as in Cyberspace they have taken on a different personality and thus are unlikely to be traced and prosecuted.[22] These citizens have effectively removed themselves from the regulatory control of their sovereign government and have chosen to be regulated by another set of regulatory values and norms. This is because, as dramatically put by David Post, 'Cyberspace ... does not merely weaken the significance of physical location it destroys it ... they do not cross geographical boundaries (in the way that say environmental pollution crosses geographical boundaries), they ignore the existence of boundaries altogether'.[23]

---

19 *Commission* v *United Kingdom* Case C-222/94, [1996] ECR I-4025.
20 Broadcasting Act 1990, s 43.
21 n 2 above, 190.
22 With a degree of computer literacy they can ensure that it would be almost impossible for law enforcement agencies in the physical world to track them down and prosecute. This is discussed further below at 496–497.
23 D. Post, 'Governing Cyberspace' (1997) 43 *Wayne Law Review* 155. Online version at <http://www.temple.edu/lawschool/dpost/Governing.html> (visited 4 January 2002).

The anonymity problem

The non-physicality of Cyberspace allows Internet users to choose to adopt a different persona from their real-world personality (pseudonymity) or to hide all details of their personality (anonymity). Pseudonymity and anonymity provide a further set of problems for regulators. As well as facilitating regulatory arbitrage by allowing citizens to conceal their identity, thereby inhibiting the application of civil, administrative and criminal regimes while in Cyberspace, pseudonymity and anonymity also allow Netizens to carry out transactions in an unregulated manner.[24] Two examples which may be given are the distribution of hate or defamatory speech, and access to regulated content.

To begin with the latter, there are certain areas in our physical societies where we regulate access to certain persons. Children are not permitted access to public bars or licensed sex shops. In addition there are activities that are restricted to certain persons. Only those with driving licenses may legally drive and only those who are members of the appropriate professional society may practice as a lawyer. A lack of physical persona makes the regulation of such simple activities much more complex in Cyberspace. A child may take on an adult personality and gain access to pornographic content.[25] In the physical world a child entering a licensed sex shop would be removed by the manager, whereas in Cyberspace the elements of physicality are lost and the ability to regulate is impaired. This is not to say the anonymity problem renders regulation of access impossible. Community-based control structures, supported by design-based elements have met with a high degree of success.[26] More worryingly, the access control problem allows for the potentially more harmful conduct of adults passing themselves off as children. In the same way that children are prevented from accessing certain adult areas of the physical world, there are areas where unauthorised adults are kept out to protect children.[27] Children nowadays are educated to keep away from strangers and to be wary of any unusual adult contact. Again the lack of physicality in Cyberspace raises problems. Users cannot discern the age of others in the chatroom intended for children. As it is at the user's discretion how much information he wishes to reveal about himself there is no practical method of ensuring that adults do not pose as minors for as long as Cyberspace supports an anonymous culture. And given that any attempt to remove the currently available culture of pseudonymity/anonymity would probably lead to a high level of regulatory arbitrage there is no apparent means to deal with such problems.

Further, the easy availability of anonymous messaging allows individuals to take part in activities without being required to meet usual societal norms. Individuals may make antisocial comments without fear of being ostracised by society at large. The technology of anonymous remailers when coupled with encryption technology can ensure an untraceable message source.[28] This may be used to distribute comments about an individual or organisation without fear of prosecution or social

---

24 Netizen is the universally accepted term for a 'citizen of the Internet'.
25 n 2 above, 174.
26 See below 'Other Forms of Control'.
27 Examples would be schools, children's playgrounds, nurseries and other controlled environments.
28 The technology is described in some detail by Michael Froomkin in 'The Internet as a Source of Regulatory Arbitrage', in B. Kahin and C. Nesson (eds), *Borders in Cyberspace* (Cambridge, MA: MIT Press, 1997). Online version available at <http://www.law.miami.edu/~froomkin/articles/arbitr.htm> (visited 4 January 2002).

exclusion.[29] Anonymity in Cyberspace creates a unique culture where expression free from the normal constraints of legal and social control is common. Even the United States with its particular emphasis on the right to free speech cannot allow completely unfettered or unrestricted freedom of expression.[30] Cyberspace uniquely offers a forum for unfettered free expression.[31] Although it may be argued that ISPs or other moderators of discussion groups may remove offending messages, they may be reposted somewhere else in Cyberspace almost immediately. Also Netizens may directly address others via e-mail. Again although this practice, known as spamming is regulated in Europe by the Distance Selling and E-commerce Directives[32] and by other enactments worldwide, the availability of anonymous communications renders such enactments impotent within Cyberspace. He who cannot be caught cannot be punished. Anonymity therefore allows for perfect freedom of expression, which in the physical world has been tempered by even the most liberal of regimes.

The scarce resources problem

Regulators in the new media are called upon to oversee systems of allocation of scarce resources. All new media sectors draw heavily on limited resources, whether these be natural resources such as spectrum for the telecommunications or broadcasting sectors or man-made resources such as domain names in relation to Cyberspace. Digital developments do, in some respects reduce existing scarcity problems. Thus digital broadcasting uses spectrum more efficiently and thus enhances capacity.[33] This may in turn create a problem for regulators seeking to maintain controls designed to ensure pluralism in the broadcasting sector.[34]

The spectrum scarcity problem is exemplified by the emergent market for third generation (3G) mobile communications.[35] 3G mobile will make multimedia

---

29 Following the enactment of the Communications Decency Act 1996 a US-based ISP has no third party liability for any libellous messages carried on their system (s 230). In the UK and the European Union ISPs may have third party liability if they fail to act once the nature of a libellous message is drawn to their attention. See *Godfrey* v *Demon Internet* [1999] 4 All ER 342 and the E-Commerce Directive (Directive2000/31/EC OJL 178 , 17/07/2000, 1–16) Art 12.
30 n 7 above, 151–153. For a fuller account of the philosophical foundations upon which restrictions on the first amendment are justified see F. Schauer, 'The Aim and Target in Free Speech Methodology' (1989) 83 *Northwestern University Law Review* 562; R.K. Greenawalt, 'Free Speech Justification' (1989) 89 Col. LR 119.
31 Several commentators cited the success of the complainers in *UEJF* (*L'Association Union des Etudiats Juifs de France) et Licra (La Ligue Contre le Racisme et l'Antisémitisme)* v *Yahoo! Inc*, L'ordonnance du Tribunal de Grande Instance, 20 November 2000 as evidence of the ability of courts to regulate expression in Cyberspace. This confidence has been substantially eroded following the finding of Judge Fogel in *Yahoo! Inc* v *Licra* ND Cal Filed 7 November 2001, that the French order is not enforceable in the United States as '[it] chills Yahoo's First Amendment Rights ... and that the threat to its constitutional rights is real and immediate.' (at 23) Decision available at <http://www.cand.uscourts.gov/cand/tentrule.nsf/4f9d4c4a03b0cf70882567980073b2e4/daaf80f58b9f-b3e188256b060081288f/$FILE/yahoo%20sj%20%5Bconst%5D.PDF> (visited 20 December 2001).
32 Distance Selling Directive, Directive 97/7/EC OJ L 144, 04/06/1997, pp 19–27; E-Commerce Directive, Directive 2000/31/EC, OJ L 178, 17/07/2000 pp. 1–16. The Distance Selling Directive was implemented in the UK through the Consumer Protection (Distance Selling) Regulations 2000 SI 2334. At the date of writing the E-Commerce Directive awaits implementation.
33 R. Collins, 'Back to the Future: Digital Television and Convergence in the United Kingdom' (1998) 22 *Telecommunications Policy*, 383, 384–385.
34 M. Cave, 'Regulating Digital Television in a Convergent World' (1997) 21 *Telecommunications Policy* 575, 590.
35 Sometimes known as Universal Mobile Telecommunications System (UMTS).

services available to mobile phone users anywhere in the world, combining satellite and terrestrial digital capacities. This development has the potential both substantially to displace a number of current communications technologies, notably second generation mobile and fixed link telephony, and to grow new markets in mobile communications. Most EU Member States have concluded that spectrum scarcity permits them to licence between 4 and 6 network operators for 3G mobile.[36] The objectives of the licence allocation processes have been to promote the development of competitive markets, to allocate the spectrum to those best placed to use it, and in many cases to secure windfall fee-income to the finance ministry. Further policy making will be necessary to determine the terms on which service providers who do not have network operators licences can have access to the networks for the provision of services.

Scarce resources are also a problem in Cyberspace. The Internet is often seen as a network without resource constraints. If more resources are needed more computers can be added to the network. This though only increases the available processing power of the net, there are other key areas where resources remain scarce. One area is bandwidth.[37] Modern telecommunications networks rely on the ability to transmit data from one source to another and in this respect the Internet is no different from mobile telecommunications networks. Network content is increasingly sophisticated. Consumers are demanding faster and more stable access to the network, to allow them to listen to real time audio transmissions and to view streaming video transmissions. These additional network demands are putting the current network protocols under strain and commercial providers of such services are calling for the current protocols to be substantially overhauled to provide for the flow of such services free from the current problems of latency (delays in transmission) and jitter (variations in delays).[38] These problems are caused by the current network protocol, Internet Protocol version 4 (IPv4) which employs a 'best effort' quality of service.[39] The best effort service is simply an onward transmission service which routes packets of information based upon information on congestion given to the sender from the next point or node in the network. This means packets of information relating to a single transmission can become separated and can arrive with delay variation causing jitter. Simple Internet applications such as e-mail or web-browsing can tolerate these delays and differentials, but streaming audio and video cannot: Internet telephony for example cannot tolerate a delay of more than 250 milliseconds.[40] To deal with these problems network designers have suggested the creation of an intelligent network which would allow for quality of service (QoS) solutions.[41] The implementation of QoS systems involve either the implementation of a complex virtual overlay network (VON) which would allow traffic from a single network flow to pass

---

36 P. Curwen, 'Next Generation Mobile: 2.5G or 3G?' (2000) 2 *Info* 455, 461.
37 See J. Glasner, 'Move Over, Pork Bellies' *Wired News* May 20 1999 available at <http://www.wired.com/news/business/0,1367,19796,00.html> (visited 4 January 2002); Lessig, n 1 above, 47.
38 See for example C. Huitema (Microsoft Corporation), 'How Will IPv6 Change the World?' paper presented to IPv6 2000, October 19–20, 2000 Washington DC available at <http://www.ipv6forum.com/navbar/events/xiwt00/presentations/html/huitema/> (visited 20 December 2001); Y. Pouffary (Compaq), 'The IPv6 Advantage' paper presented to IPv6 2000, October 19–20, 2000 Washington DC available at <http://www.ipv6forum.com/navbar/events/xiwt00/presentations/html/pouffary/> (visited 20 December 2001).
39 David, n 14 above, 173.
40 Lessig, n 1 above, 46.
41 Lessig, n 1 above, 46–47. Generally Lessig is wary of such solutions as adding intelligence to the network allows for control in the content layer.

through routers without competing with traffic from other network flows[42] or as seems more likely the implementation of a new network protocol, Internet Protocol version 6 (IPv6).[43] IPv6 offers many advances over IPv4. It allows for better homogeneity of transmission. In the event of network queuing it allows for streaming transmissions to be packaged together. This means time critical transmissions such as streaming audio and video may be prioritised over less time sensitive transmissions such as e-mails. Also it crucially supports the Resource Reservation Protocol (RSVP) developed by Cisco Systems and MCI WorldCom which allows service providers to sell bandwidth to users allowing them to prioritise their transmissions over other traffic using the same routers.[44] This functionality comes at a cost. These developments will almost certainly lead to the development of fragmented proprietary networks within the wider network structure and an end to the current end-to-end infrastructure of the Internet.[45]

Although bandwidth scarcity is not unique to Cyberspace the scarcity of domain names is.[46] It may seem bizarre to claim domain names are a scarce resource. The permutations of domain names seem almost limitless. They may be made up of a string of up to 61 characters[47] in any permutation and a top level domain of which there are more than 250.[48] Despite this there is a scarcity of usable domain names. Usable domain names reside almost exclusively in the .com top level domain and are made up of recognisable terms in major languages.[49] There is a paucity of such names as usable domain names are of a one mark one owner architecture whereas previous trade mark systems had been of a one mark many owners architecture.[50] Competing demands for usable domain names quickly arose and the bodies charged with overseeing the domain name system (initially the Internet Assigned

---

42 David, n 14 above, 173; Computer Science and Telecommunications Board, National Research Council, *The Internet's Coming of Age* (Washington DC: National Academy Press, 2001) 102–103. Available at < http://bob.nap.edu/html/coming_of_age/ > (visited 20 December 2001).

43 IPv6 also solved the problem of a scarcity of Internet Protocol (IP) addresses. Currently there are just under 4 billion available Ipv4 addresses. Although this seems a healthy figure large organisations such as AT&T and MIT hold up to 16 million addresses each. Currently if you use dial-up access you will be allocated a temporary IP address while connected. This allows several users to share the same IP address. With new networked tools some analysts suggested the total fund of IP addresses would be exhausted by 2004. IPv6 allows for 10^{38} IP addresses more than enough for the foreseeable future.

44 'Traffic to Take the High Road – for a Price' *Wired News* March 27 1997. Available at < http://search.hotwired.com/search97/s97.vts?Action=FilterSearch&Filter=docs_filter.hts&ResultTemplate=news.hts&Collection=news&QueryMode=Internet&Query=IPv6 > (visited 20 December 2001).

45 David, n 14 above, 174–178. A. Odlyzko, *The Economics of the Internet: Utility, Utilization, Pricing, and Quality of Service* AT&T Labs-Research 1998, 27–28. Available at < http://www.dtc.umn.edu/~odlyzko/doc/internet.economics.pdf > (visited 20 December 2001).

46 M. Mueller, 'Competing DNS Roots: Creative Destruction or Just Plain Destruction?' Paper presented to 29th Research Conference on Communication, Information and the Internet, October 2001, Alexandria, Virginia, available at < http://www.arxiv.org/ftp/cs/papers/0109/0109021.pdf > (visited 19 December 2001).

47 The total length of a domain name (excluding root) may be up to 63 characters. As the shortest top level domains are the two letter country code domains, this means the longest lower level domain possible is 61 characters. See M. Galperin and I. Gordin, 'The Domain Name System.' Available at < http://www.rad.com/networks/1995/dns/dns.htm > (visited 4 January 2002).

48 Currently there are 239 Country Code top level domains (ccTLDs) detailed in ISO-3166, 12 generic top level domains (gTLDs) and two US Federal TLDs (.gov & .mil).

49 The latest Network Wizards Domain Name Survey (July 2001) records 37,502,747 .com domains. By comparison the Oxford English Dictionary only contains 'over half a million words'. (Source: About the *Oxford English Dictionary* < http://www.oed.com/public/inside/ > (visited 4 January 2002)).

50 See for example A. Brunel and M. Laing, 'Trademark Troubles with Internet Domain Names and Commercial Online Service Screen Names' [1997] 5 *International Journal of Law and Information Technology* 1; A. Murray, 'Internet Domain Names: The Trade Mark Challenge' [1998] 6 *International Journal of Law and Information Technology* 285.

Numbers Authority (IANA) and Network Solutions Inc, and more latterly the Internet Corporation for Assigned Names and Numbers (ICANN))[51] were required to develop a policy to deal with these competing claims. This policy, the Uniform Domain-Name Dispute-Resolution Policy, attempts to balance the rights of trade mark holders against the first-user policy previously applied. It is an extremely controversial policy and will be examined in depth below when we analyse the effectiveness of control mechanisms in the new media.

Extending the 'modalities of regulation' analysis

Lawrence Lessig's *Code and Other Laws of Cyberspace* is widely regarded as one of the most complete analytical attempts to capture the variety of forms which regulation of new media does or may take.[52] Lessig contends that there are four distinct modalities of regulation. He attaches to these the labels law, markets, norms and architecture. He thinks of these in terms of constraints on action.[53] Thus law constrains through the threat of punishment, social norms constrain through the application of societal sanctions such as criticism or ostracism, the market constrains through price and price-related signals, and architecture physically constrains (examples include the locked door and the concrete parking bollard).

Lessig's work is of great value for reminding us of the importance of architecture as a basis for regulation. The potential for controls to be built into architecture have long been recognised, as exemplified by Jeremy Bentham's design for a prison in the form of a panopticon (within which the architecture permitted the guards to monitor all the prisoners) and the more recent observations of the way in which visitors to Disney World are controlled by an architecture in which nearly every aspect of the design has a disciplinary function.[54] Lessig observed the various constraints that are built into software by their designers. Such architectural constraints in software code are chiefly used for commercial purposes (such as restricting the user's use to what they have paid for or segmenting the market so as to charge higher prices in some segments without the risk of arbitrage) but may also be used for other regulatory purposes (as with the controls placed on users by Filterware).[55] Lessig suggests that as a means of regulation architecture is self-executing and thus different at least from norms and

---

51 IANA and ICANN are non-governmental not for profit agencies. Network Solutions Inc is a subsidiary of Verisign Inc a for-profit publicly listed company.
52 See n 2 above; cf the five way analysis of controllers in ordering society put forward by Ellickson, n 6 above, 131. Ellickson sees order as a product of first party (or self-control) second party (or contractual) control, third party control (based on social forces and norms), organisation (with associated institutional apparatus) and, lastly (significantly) government with the laws. There is a substantial political science literature on alternative instruments of governance see J. Kooiman (ed), *Modern Governance* (London: Sage, 1993) and C. Hood, *The Tools of Government* (London: Macmillan, 1983).
53 n 2 above, 235–239.
54 M. Foucault's research on the history of the prison has been responsible for generating new interest in surveillance generally and Bentham's panopticon in particular. *Discipline and Punish: The Birth of the Prison* (Harmondsworth: Penguin, trans A. Sheridan, 1977) ch 3. C. Shearing and P. Stenning, 'From the Panopticon to Disney World: The Development of Discipline' in A. Doob and E. Greenspan (eds), *Perspectives in Criminal Law* (Aurora: Canada Law Book Co, 1984). Crime control through design is exemplified by A. Lester, *Crime Reduction through Product Design* (Australian Institute of Criminology, Trends and Issues in Criminal Justice no 206, 2001); N. Kaytal, 'Architecture as Crime Control' (2002) 111 *Yale Law Journal* (forthcoming).
55 Filterware is discussed further below. See 'Other Forms of Control'.

law.[56] This claim appears correct up to a point. However the analysis which separates the functions of a control system shows that the standard-setting element of architecture is not self-executing but is, by definition, designed by human hands. Some architecture-based regimes may be self-executing as to monitoring and behaviour modification. A parking bollard, for example, requires no further agency on the part of a regulator to control parking. Other architectural controls do rely on actions by the controller. For example, Bentham's panopticon requires that prison guards actively monitor prisoners and intervene to control deviance. The panopticon can thus be seen as a hybrid of hierarchy and architecture.

The importance of Lessig's analysis is to draw attention to the variety of bases for control which can be deployed in the face of anxiety that technological change (such as the Internet) and economic change (such as globalisation) tends to make a variety of different forms of conduct unregulatable. The argument that variety in forms of activity requires an equal or greater variety of bases for control if regulation is to be effective has found formal expression in the cybernetics 'law of requisite variety'. It is expressed in other terms as the principle that 'only variety can destroy variety'.[57] The sceptical position which Lessig challenges is premised in part upon a myth that social and economic activity has traditionally been highly amenable to regulation, conventionally defined. Recent scholarship on the limits to control has emphasised the problems of trying to regulate social and economic activity.[58] This work has emphasised the importance of developing regulatory regimes which seek to steer or stimulate activities within the target system indirectly as an alternative to external command and control.[59] Lessig's work has the potential to support efforts to reconceive regulation in a sense that is both more modest in its claims and ambitions and more useful in providing mechanisms not only, or perhaps mainly, of direct control but also of indirect control. A key method of this new approach, which we deploy in this article, is to identify effective regulation in whatever form it takes and to seek to support it, develop it or extend it by analogy to other domains in which there are problems of regulation.

The concept of regulation deployed in Lessig's analysis is a broad one, extending beyond the narrowly defined 'systematic oversight by reference to rules' to encompass four 'modalities of regulation' which have the object or effect of holding behaviour within one state among all the possible states which the behaviour might take. Lessig refers to the '"net regulation" of any particular policy...' domain as the 'sum of the regulatory effects of the four modalities together'.[60] Regulation in this expansive sense is conceptually closer to the usage

---

56 n 2 above, 236–237. Lessig claims that markets have in common with norms and law the fact that they require human agency and are not self-executing. This claim is contentious (though Lessig does not recognise this) as the control exerted by a market does not operate at the level of the individual seller and buyer, but rather in an aggregate. In the perfectly competitive market model the decisions of no individual buyer or seller can affect the operation of the market.
57 S. Beer, *Decision and Control* (London: Wiley, 1966) 279–280.
58 P. Grabosky, R. Smith and G. Dempsey, *Electronic Theft: Unlawful Acquisition in Cyberspace* (Cambridge: Cambridge University Press, 2001) 5–11; P. Nonet and P. Selznick, *Law and Society in Transition* (New York: Harper & Row, 1978); I. Ayres and J. Braithwaite, *Responsive Regulation: Transcending the Deregulation Debate* (New York: Oxford University Press, 1992); N. Gunningham and P. Grabosky *Smart Regulation: Designing Environmental Policy* (Oxford: Oxford University Press, 1998).
59 This prescription has its origins in systems theory: n 57 above; and is found in both the political science literature: A. Dunsire, 'Tipping the Balance: Autopoiesis and Governance' (1996) 28 *Administration and Society* 299, and legal literature: G. Teubner, 'Juridification: Concepts, Aspects, Limits, Solutions' in G. Teubner (ed), *Juridification of Social Spheres* (Berlin: De Gruyter, 1987).
60 L. Lessig 'The Law of the Horse: What Cyberlaw Might Teach' (1999) 113 Harv L Rev 501, 508.

of biologists and sociologists than to that of lawyers.[61] It refers to any control system. To be viable, within the terms of control theory, a control system must have some standard-setting element, some means by which information about the operation of the system can be gathered, and some provision for modifying behaviour to bring it back within the acceptable limits of the system's standards.[62] With regulation information gathering is usually achieved through monitoring by an agency, department or self-regulatory body and deviations addressed by application of formal and informal sanctions (See Figure 1 below).

When locating Lessig's description within the stricter analysis of control theory some problems emerge both with the labels and the concepts which they describe. Put simply the conceptual schema, drawn from Lessig's work in law and economics, needs enriching if it is to capture the institutional variety in control. Our earlier discussion of control theory suggests that the appropriate schema involves not only a four way division between different bases of control, but also a further fine grained analysis of the three different elements necessary to generate a control system (standard-setting, information gathering and behaviour modification). This development of the analysis provides a clearer descriptive framework for understanding how control is or can be achieved and opens up the possibility for identifying the wide range of control systems which appear as hybrids of two or more modalities of regulation. To develop this analysis we draw not only on Lessig's work, but also on attempts to deploy cultural theory to identify variety in control systems.[63] This analytical frame has recently been put to work in analysing variety in risk regulation regimes.[64] The term 'regime' is apt to capture variety not only in standards and standard-setting (which represents the bias in Lessig's analysis) but also in the institutional dimensions of information gathering and behaviour modification. The regime analysis makes it transparent that the various functions which contribute to viable control systems can be widely dispersed among state and non-state actors, even within a single regime, and can be assembled in mixed or hybrid forms.

Lessig's conceptualisation of 'law as command'[65] suffers from a weakness in that it fails to capture all of the control systems which are within the set of command based or, as we label it, hierarchical control. Law, in this conception, refers only to state law (whether made by judges, or, more commonly in this context, legislatures)[66] and neglects the plurality of forms which hierarchical control structures may take. The richer conception of hierarchy looks to the form of control rather than its source. Thus the regime for developing Internet domain names has important elements which are non-state in character and yet which are distinctly hierarchical (and are discussed further below). The term law also suffers from the difficulty that it is often deployed in a way which infers only standards and not the institutional elements of a control system (*viz* information gathering and behaviour modification). Law in Lessig's terms is merely the constraint placed upon the individual. Accordingly hierarchical control provides both a better label and a substantively enriched conception of this modality of regulation.

---

61 R. Baldwin, C. Scott and C. Hood (eds), 'Introduction' in *Socio-Legal Reader on Regulation* (Oxford: Oxford University Press, 1998); M. Clarke, *Regulation* (London: Macmillan, 2000).
62 C. Hood, H. Rothstein, and R. Baldwin, *The Government of Risk: Understanding Risk Regulation Regimes* (Oxford: Oxford University Press, 2001) 21–27.
63 C. Hood, 'Control Over Bureaucracy: Cultural Theory and Institutional Variety' (1996) 15 *Journal of Public Policy* 207.
64 n 62 above, 9–14.
65 n 2 above, 235.
66 n 60 above, 507.

The concept of norms as it is deployed in Lessig's analysis follows a usage developed in the social psychological literature – referring to shared patterns of behaviour – but which is unconventional and unhelpful in the study of law. Even in its psychological usage the term norm does not describe the institutional dimensions of a control system, but rather a set of standards which exist between a particular social group for the time being. We argue that the preferred meaning of the word norm is as the generic term for standards, guidelines and legal and non-legal rules.[67] The control form which involves societal or group standards, peer-based information gathering and behaviour modification based on social sanctions such as ostracisation or disapproval, we refer to as community-based control. This category includes not only the social norms which exist generally or between particular groups, but also some elements of more formalised regimes, as where self-regulatory standards are socially generated and written down and then combined in a hybrid form with hierarchical elements to create a self-regulatory control system which is a hybrid between community and hierarchical bases.

The concepts of markets and architecture as they are deployed by Lessig are each under-inclusive. Rivalry and competition provide a form of control in environments where there is no identifiable market. Indeed recent public sector reforms have made widespread use of what we will call competition-based controls in non-market situations.[68] Additionally there is a marked element of regulatory competition applying to the development of regulatory standards in some domains both in the US and the EU.[69] Where the conditions for such regulatory competition exist (a topic of hot debate), and states are permitted to develop their own rules, competition for client businesses is said to create a check on any tendency to 'over-regulate'.[70]

The concept of architecture, referring in Lessig's terms to the whole built environment with and without intended effects,[71] does not capture the whole set of control mechanisms which are premised upon design as a basis of control. Thus there are social and administrative systems which have design features which create control in a way in which the regulatee cannot affect. A key example is the deployment of 'contrived randomness' in the oversight of taxpayers or employees so as to reduce the scope of these groups to exploit a wholly predictable system of opportunities and pay-offs.[72] Accordingly we re-label this fourth modality of regulation as design.[73] The different elements of each of the four types of regulation are illustrated in Figure 1.

---

67 P. Drahos and J. Braithwaite, *Global Business Regulation* (Oxford: Oxford University Press, 2000) 20.
68 P. Self, *Government by the Market* (London: Macmillan, 1993).
69 D. Esty and D. Gerardin (eds), *Regulatory Competition and Economic Integration* (Oxford: Oxford University Press, 2001).
70 W. Bratton, J. McCahery, S. Picciotto and C. Scott, 'Introduction: Regulatory Competition and Institutional Evolution' in Bratton, McCahery, Picciotto and Scott (eds), *International Regulatory Competition and Coordination* (Oxford: Oxford University Press, 1996).
71 n 60 above, 507–508.
72 n 63 above, 211–214.
73 This concept of design has a loose affinity with the deployment of the term 'technologies' in the Foucauldian literature on governmentality. It is possible that the term technologies 'linking together forms of judgement, modes of perception, practices of calculation, types of authority, architectural forms, machinery and all manner of technical devices with the aspiration of producing certain outcomes in terms of the conduct of the governed' (N. Rose, 'Government and Control' (2000) 40 *British Journal of Criminology* 321, 323) infers a rather wider range of instrumentalities than are inferred by the concept of design in this article. For deployment of the concept of technologies in regulatory theory see J. Black, 'Decentring Regulation: Understanding the Role of Regulation and Self-Regulation in a 'Post-Regulatory' World' (2001) 54 *Current Legal Problems* 103.

Element of a Control System	Hierarchical Control	Community-Based Control	Competition-Based Control	Design-Based Control
Standard Setting	Law or Other Formalised Rules	Social Norms	Price/Quality Ratio (and equivalents with non-market decisions)	Inbuilt design features and social and administrative systems
Information Gathering	Monitoring (by agencies or third parties)	Social Interaction	Monitoring by dispersed buyers, clients, etc	Interaction of design features with environment
Behaviour Modification	Enforcement	Social Sanctions (eg ostracism, disapproval)	Aggregate of decisions by buyers, clients, etc on purchase, take-up, location etc	As for information gathering (self-executing)

Figure 1: Elements of Control Systems

It is part of Lessig's argument that there is scope for the use of hybrid forms of regulation which link two or more of the 'pure' modalities of regulation noted above.[74] In particular he suggests there is scope to link what are in his terms law and architecture, for example by mandating software designers to build certain elements into software code in pursuit of public regulatory objectives.[75] However we think he underplays the extent to which contemporary control is already based on hybrid regulatory forms and the extent to which a wide variety of regulatory hybrids may be useful in developing regulatory control. Indeed, underlying Lessig's argument is a claim that there is considerable novelty to the nature of law in Cyberspace, a view seemingly accepted by those Cyberlibertarians who contest the normative dimension to Lessig's work.[76] Nowhere in the work of Lessig or his critics is this claim substantiated. As Lessig himself recognises, features which we might call design or architecture have long been fundamental to the way we are governed, whether by features of the built environment (such as the Parisian boulevard system) or the Byzantine systems of an obscure public bureaucracy or of commercial actors such as banks and insurance companies. It is not clear that design of software is fundamentally different from design in other aspects of social and economic activity. Wherever it is deployed it has controlling effects and a potential for those controlling effects to be turned towards different or modified effects.

If each of the four pure bases of regulation is theoretically capable of being deployed on its own and with each of the other three bases (giving four single bases, six pairings, four threesomes and one foursome) then there are fifteen forms of regulation in total. There is no empty set since all domains are subject to some form of regulation (or else, by definition, they could not be a domain since they would not hold a recognisable shape). Even regimes which apparently exhibit a pure basis of regulation may have the dominant form tempered by another. For

---

[74] n 60 above, 511–514.
[75] *ibid*, 514–522.
[76] n 2 above, 5–6; D. Post, 'What Larry Doesn't Get: Code, Law and Liberty in Cyberspace' (2000) 52 Stan L R 1439, 1443.

example much hierarchical regulatory enforcement is tempered by more co-operative relationships more characteristic of community, and where there is a proliferation of hierarchical regulators in a particular domain (telecommunications and competition authorities in the communications domain for example)[77] then hierarchy may be tempered by a form of institutional competition as regulators jockey for position and custom.

Among the widely observed hybrid forms are competition law and co-regulation and enforced self-regulation. Though competition law is often equated with competition in its control dimensions competition law exemplifies hierarchical control, with elements of competition possible where third party actions are widely deployed. Co-regulation and enforced self-regulation each link some of the strengths of community-based control (notably within self-regulatory regimes) with the use of hierarchy, for example by state approval of standards set by industry groups (co-regulation) or mandating firms to establish and sometimes enforce their own standards (enforced self-regulation). Other less prevalent forms are observable but do not have widely accepted labels. Thus mandatory design features (for example in product design) are hierarchy/design hybrids which we could refer to as 'enforced design'. The form taken by some self-regulatory efforts to inhibit access to undesirable websites is a community/design hybrid.

One further set of remarks is necessary concerning the bases of control. Different forms of control work differently in different contexts. Markets, hierarchies, communities and design are each embedded in wider social practices.[78] Key social networks may be a factor in explaining relations of interdependence and thus how power is played out in particular social settings.[79] Similarly the effects of controls may vary depending on how they are perceived in the cognition of those whom they affect. Thus some individuals or societies may respond with resistance to controls which are met with compliance by others or at other times. Thus an analysis of modalities of regulation does not, by itself, provide a toolkit for decisions on the design of controls, but rather a more limited analytical understanding of controls which have been observed and might be deployed in certain environments and which might be expected to be effective under appropriate conditions.[80]

Putting controls to work

The importance of the reconfiguring and development of the modalities of regulation argument further extends to institutional choices for seeking to use controls for public policy objectives. Whereas Lessig places greater emphasis on top-down institutional approaches, of which regulatory agency forms represent the leading example, we contend that an emphasis on hybrid forms of control will tend to lead to the deployment of hierarchical controls as instruments to steer organic or bottom up developments, whether in the form of competition, community or

---

77 C. Scott, 'Institutional Competition and Coordination in the Process of Telecommunications Liberalization' in McCahery, Bratton, Picciotto and Scott (eds), n 70 above.
78 J. Rogers Hollingsworth and R. Boyer (eds), *Contemporary Capitalism: The Embeddedness of Institutions* (Cambridge: Cambridge University Press, 1997).
79 R. Rhodes, *Understanding Governance* (Buckingham: Open University Press, 1997) especially ch 3. See also P. Drahos and J. Braithwaite, *Global Business Regulation* (Oxford: Oxford University Press, 2000) (discussion of 'regulatory webs', ch 23); C. Scott, 'Analysing Regulatory Space: Fragmented Resources and Institutional Design' [2001] *Public Law* 329.
80 We are grateful to Julia Black for this point.

design-based control. In some instances successful regimes have combined three or even all four of the bases for regulation.

Hierarchy/community

Hierarchy and community-based controls are often combined either to ensure that industries effectively collaborate on controlling their sector or to give sectoral self-regulation greater authority. The hierarchy/community hybrid bases of regulation are exemplified by the structures established to address scarcity in domain names. By regulating the domain name system ICANN plays a key role in the regulation of Cyberspace.[81] ICANN and its predecessors, IANA and Network Solutions Inc, have long provided regulatory control over the domain name system but have done so not as a function of hierarchical control, but rather to assist in the development of the domain name system as required by the community and to ensure the system design remained intact.

A simple example of the deployment of hierarchical controls to assist in the development of community based controls may be seen in the promulgation by both Network Solutions and ICANN of Domain-Name Dispute-Resolution Policies.[82] These procedures are used to counteract the primary problem of misappropriation of scarce resources. The procedure appears to have been extremely successful in countering the problem of 'cybersquatting'. The practice of cybersquatting was recognised at an early stage of development of the Web. In its simplest form it is the ability of unscrupulous individuals to register valuable domains such as Disney.com and then to offer them on at a profit to the rightful holder of the trademark in question. Individuals who entered into such practices were quickly dubbed 'cybersquatters' by the Web community, a reflection of their standing within the community as equivalent to persons who unlawfully misappropriate physical property in the real world. Community opinion was brought to bear. These people were acting antisocially but social sanctions failed to affect their actions; being ostracised in Cyberspace did not affect their everyday lives. Their actions were, though more than socially unacceptable, they were also a threat to the developing architecture of the domain name system. By controlling domain names which reflected well known identifiers from the real world they posed a threat to the system. How could people navigate the Web if they couldn't rely on the knowledge they had developed in the physical world?[83] Although courts could intervene in cases where cybersquatters had misappropriated another's trademark[84] regulatory arbitrage meant enforcement of orders could sometimes prove problematic.

What was required was a regulatory regime which would apply to all registrations and could be applied whatever the jurisdiction of the parties. This led directly to the first Network Solutions Inc Domain-Name Dispute-Resolution Policy, a policy which has now been adopted and refined by ICANN. The policy has proven successful as it treats the domain name space as a separate jurisdiction, thus preventing regulatory arbitrage. Anyone who resides in the ICANN domain name space must contractually agree to be bound by the policy, and must agree to

---

81 As discussed above ICANN controls the allocation of a scarce resource and therefore plays an important regulatory role.
82 Kleinwächter, n 4 above, 271–272.
83 For example, if cybersquatters controlled domains such as disney.com, mcdonalds.com and microsoft.com how would users navigate their way to the sites of these well known companies?
84 See eg *Panavision* v *Toeppen* 945 F.Supp. 1296 (1996); *British Telecommunications plc and others* v *One in a Million Ltd* [1999] RPC 1.

the arbitration procedure contained therein. Thus the values of the cybercommunity may be upheld by ICANN through the arbitration process. Secondly, the ICANN policy of using low-cost online arbitration at the expense of court proceedings meets the needs of the community. One of the key problems with usable domain names was they were unusually an inexpensive scarce commodity. Scarce commodities often carry a proportionately high price tag, as demonstrated by the UK and German 3G mobile spectrum licence auctions.[85] This is a simple application of the economic model of demand, supply and equilibrium pricing. Domain names though do not fit the economic model particularly well as the market as a whole is oversupplied while a small percentage of that market is undersupplied or scarce. As registrars cannot differentiate useful (and therefore scarce) domain names from the majority it means market-based controls may be circumvented and a scarce and therefore valuable domain name may be had for as little as $25. This allows for a high degree of speculation in domain names.

The previous Network Solutions Domain-Name Dispute-Resolution Policy required the complainer to obtain a court order. This meant it was in many cases cheaper to buy the disputed domain name from the defender than to pursue an action to recover the name, especially if the dispute had an international element. The present ICANN Uniform Domain-Name Dispute-Resolution Policy, through its use of inexpensive arbitration procedures provides a regulatory process which takes account of market conditions. This is not to say that the policy is not without its critics. There is strong criticism of the ICANN policy on the grounds that it now favours trademark holders over domain name holders who fail, for whatever reason to comply with US trademark law.[86] This has led to a practice known as 'Reverse Domain Name Hijacking' occurring.[87] This is a potential flaw in the ICANN policy. As discussed the policy was originally introduced to deal with cybersquatters who were perceived as socially unacceptable and a potential threat to continued utility of the architecture of the domain name system. The policy now needs to develop to provide a more balanced approach between the competing interests of parties. Fortunately there is evidence that the arbiters under the policy may be developing such a mature and balanced approach. There were some initial claims that the policy was being used to restrict free speech.[88] Recently though, decisions of the arbitration panels have shown the policy has a degree of flexibility which may allow them to develop the policy to meet the demands of the community at large.[89] Clearly the regulatory authority was implementing a hierarchical control system to support the development of community-based and design-based controls.

---

85 The UK raised US$35.4bn by auctioning 5 UMTS spectrum licences, while Germany raised $46.1 bn by auctioning twelve spectrum blocks. In both cases the number of interested bidders exceeded the number of licences available creating a scarcity of resources. This may be contrasted with the position in the Netherlands where the auctioning of five licences was met with five serious bidders and raised only $2.5 bn or in Italy where a similar situation to the Netherlands saw the Italian Government raise only $10 bn.
86 See eg Froomkin, n 4 above, 96–101; C. Perry, 'Trademarks as Commodities: The Famous Roadblock to Applying Trademark Dilution Law in Cyberspace' (2000) 32 Conn L Rev, 1127, 1155–1157.
87 Examples involving American Express and QVC may be found at <http://www.ejacking.com/> (visited 4 January 2002).
88 These claims are based in the so-called 'sucks' cases. Domains such as directlinesucks.com (D2000-0583) and freeservesucks.com (D2000-0585) were transferred to the trademarks holders following arbitration. Claims followed that decisions such as these were restricting free speech.
89 Three recent decisions wallmartcanadasucks.com (D2000-1104), lockheedmartinsucks.com (D2000-1015) and michaelbloombergsucks.com (FA0097077) have all found in favour of the respondent. These cases may signal a new approach in relation to such free speech cases.

Hierarchy/competition

The combination of hierarchical with competition-based controls is well established in the media and communications sectors. Thus regimes which apply economic or content controls more intensely to some firms than to others effectively create a continuum within which firms exerting dominance are often located closer to the hierarchy end while smaller and/or less powerful firms are located towards the market end. Within the 'responsive regulation' theory this approach is labelled 'partial industry regulation'.[90] The logic of the approach is that the benefits sought for regulation may be secured less intrusively by applying regulation only to a proportion of the firms, whilst creating space for other firms to be controlled more by market elements. Typical patterns of more intense regulation of broadcast over print media are said to have reduced risks of censorship and promoted pluralism.[91] In the telecommunications sector 'asymmetric regulation' has been deployed to provide tighter controls over dominant incumbents both to maintain service levels and to promote access to the market by new entrants.[92] With the new media other forms of control which mix hierarchy and competition have been developed.

With the scarcity issue related to spectrum, conventional hierarchical controls have been displaced by a hierarchy/competition hybrid in some domains. With 3G mobile governments have attempted to use spectrum allocation mechanisms to promote competitive markets, to promote efficient allocation of resources and in some cases to secure fee-income windfalls for finance ministries. Attempting to set policies that were friendly to the development of advanced infrastructure the European Commission initially recommended that Member States should allocate licences to 3G mobile operators free of charge.[93] Only Finland and Sweden, among the first movers on Universal Mobile Telecommunications System (UMTS) licensing, followed this policy course. All the other Member States decided to charge for the licences. Cynical accounts claim that the decision to charge was premised upon the greed of finance ministries. But there is a more principled explanation for the policy which is posited as a solution to one of the key problems of scarcity – that governments may fail to allocate scarce resources to those who are best able to exploit them to the general benefit.

The conventional instrument for the allocation of scarce spectrum is the exercise of government's hierarchical authority to examine potential applicants and make a decision along the lines of a 'beauty contest'.[94] This method was used in eight of

---

90 I. Ayres and J. Braithwaite, *Responsive Regulation* (Oxford: Oxford University Press, 1992), ch 5.
91 L. Bollinger, 'Freedom of the Press and Public Access: Towards a Theory of Partial Regulation of the Mass Media' (1976) 75 *Michigan Law Review* 1.
92 A. Perucci and M. Cimatoribus, 'Competition, Convergence and Asymmetry in Telecommunications Regulation' (1997) 21 *Telecommunications Policy* 493. Partial industry regulation in telecommunications is exemplified by US rules which apply greater restrictions to the commercial packaging of digital subscriber lines (DSL) provided by telecommunications companies than apply to functionally equivalent cable modems provided by cable communications (formerly tv) companies: M. Lemley and L. Lessig, 'The End of End-to-End: Preserving the Architecture of the Internet in the Broadband Era' (2001) 48 *UCLA Law Review* 925.
93 European Commission *Communication from the Commission to the Council, the European Parliament, the Economic and Social Committee and the Committee of the Regions: Strategy and Policy Orientations with Regard to the Further Development of Mobile and Wireless Communications (UMTS)* COM (97) 513 Final.
94 The theoretical basis of the shift towards auctions, which lies in game theoretic approaches is described in D. Salant, 'Auctions and Regulation: Reengineering of Regulatory Mechanisms' (2000) 17 *Journal of Regulatory Economics* 195.

the Member States.⁹⁵ The weakness of this method is said to lie in its dependence on the knowledge and judgement of the applicable state bureaucracy both to guess the appropriate fee to charge successful applicants and which applicants are best placed to exploit the spectrum. This 'limited knowledge' problem is perhaps more acute in the 3G mobile sector where there is little consensus on the commercial prospects for services which are made possible in the digital environment but which have not yet been tested in the market place.

The alternative method for allocating spectrum used in the remainder of the member states was to auction the licences, combining hierarchy with competition as the basis of control. Deviating from the sealed bid method used in previous spectrum auctions, the UK government and others decided to use a transparent (ie no sealed bids) simultaneous multi-round ascending auction under which bidders' offers would be revealed at the end of each round and whoever held the highest bid when the number of bidders was reduced to equal the number of licences would win the particular licence. In this way the price mechanism is used to determine which firms should have access to the scare resource controlled by government. The outcome of the UK auction was that payments for licences totalling 22 billion pounds were much higher than was expected by commentators and government.⁹⁶ Details of auction rules and incentives resulted in less successful outcomes in some other member states.⁹⁷ The UK experience initially suggested the auction had been successful in revealing a true value of the licences well above government estimates. Commentators still do not agree on whether the high cost of licences, particularly in the UK and Germany, will stifle the market as operators struggle to repay the cost.⁹⁸ The German regulator has already indicated that it may allow the operators to share infrastructure costs and the same thing may happen in the UK.⁹⁹ This divergence between the actual operating conditions (and reduction in costs) over those projected at the time of the auctions suggests that the injection of competition in the licence allocation process has generally been less than successful.

With the problem of regulatory arbitrage the solutions are often put in terms of regulatory competition or coordination. In other words arbitrage may be overcome by providing coordinated or harmonised rules across jurisdictions or arbitrage itself may seen as a solution to the problem of excessive regulation. Regulatory harmonisation was for a long time the favoured way of providing a level playing field for competition in the internal market of the EU. However, this exercise of hierarchical authority raises practical difficulties in terms of the scale of resource necessary to achieve it, and is said to risk stultifying the very markets which are to be liberalised. A partial response to the practical problems of harmonisation was the decision of the European Court of Justice in the *Cassis de Dijon* case which gave judicial authority to a principle of mutual recognition.¹⁰⁰ Regulatory competition is said to provide the flexibility for jurisdictions to develop standards to match the local requirements (whether technical or political), the capacity to innovate in regulation while

---

95 n 36 above, 461.
96 M. Cave and T. Valletti, 'Are Spectrum Auctions Ruining Our Grandchildren's Future?' (2000) 2 *Info* 347.
97 n 36 above, 474–475.
98 n 96 above; J. Bauer, 'Spectrum Auctions, Pricing and Network Expansion in Wireless Telecommunications' paper presented to 29th Research Conference on Communication, Information and Internet Policy, October 2001, Alexandria, Virginia available at <http://www.arxiv.org/ftp/cs/papers/0109/0109108.pdf> (visited 19 December 2001).
99 'MMO2 und T-Mobile schliesen UMTS-Kooperationsvertrag' Frankfurter Allgemeine Zeitung, 22/09/01.
100 *Rewe-Zentral AG v Bundemonopolverwaltung für Branntwein* [1979] ECR 649.

encouraging states to adopt rules of minimum necessary burden on business or others (because of the threat that such regulatory clients might shift their business elsewhere). A recent analysis suggests that the choice between competition and coordination is a false one both in practice and normatively and that what we are likely to see is elements of competition (for example between institutions) emerging in domains that are notionally coordinated and vice-versa. Thus it is better to talk of 'regulatory co-opetition', a hierarchy/community hybrid form of control, both as description of the phenomena and as normative aspiration.[101]

Regulatory arbitrage is a well recognised phenomenon of Cyberspace, though commentators reach different conclusions as to its significance.[102] Cyberlibertarians argue that regulatory arbitrage prevents hierarchical regulation of Cyberspace. This is most clearly and famously put in David Johnson and David Post's seminal article, *Law and Borders – The Rise of Law in Cyberspace*.[103] For Johnson and Post the practical effect of regulatory arbitrage is that hierarchical controls are rendered impotent. Netizens may choose to reject hierarchical controls they find unpalatable by moving to another part of Cyberspace. As previously outlined Netizens may choose *how* they wish to be regulated much more freely than citizens of physical jurisdictions. The only effective regulatory system according to Cyberlibertarian theory is therefore one which is acceptable to all (or the vast majority of) Netizens. Johnson and Post therefore suggest a bottom-up or organic regulatory model. They envisage a self-regulatory governance system along similar lines to that developed to regulate the domain name system. Lessig disagrees with their conclusion. He agrees that Cyberspace is a separate space and can be seen as a distinct jurisdiction. He disagrees though with the conclusion that it is a jurisdiction which requires the organic development of regulatory regimes. For Lessig, once you isolate Cyberspace as a distinct space you may use its unique architecture to establish a hierarchical regulatory structure. The argument of the Cyber-paternalists is therefore that once a recognised regulator emerges in any given activity they may impose regulatory regimes on Netizens through the unique man-made architecture of the Web, its code.[104]

To the extent that regulatory arbitrage is a problem with new media generally, and usage of the Internet in particular, it remains an open question to what extent the balance between competition and coordination might be deployed to resolve issues. For many commentators the nature of Internet technology makes regulatory arbitrage inevitable and difficult to forestall, whatever may be desirable from a policy point of view. It is the high mobility both of providers and users within

---

101 D. Esty and D. Geradin, 'Regulatory Co-Opetition' in Esty and Geradin (eds), *Regulatory Competition and Economic Integration* (Oxford: Oxford University Press, 2001).
102 See for example, M. Froomkin, 'The Internet as a Source of Regulatory Arbitrage' n 28 above; D. Johnson and D. Post, 'Law and Borders – The Rise of Law in Cyberspace' (1996) 48 Stan L Rev 1367; L. Lessig, 'Zones in Cyberspace' (1996) 48 Stan L Rev 1403; P. Samuelson, 'Five Challenges for Regulating the Global Information Society' in C. Marsden (ed), *Regulating the Global Information Society* (London: Routledge, 2001). An interesting side effect of regulatory arbitrage is a regression to the least interventionist standard in a given area. The can most clearly be seen in relation to freedom of speech following the decision of the US Supreme Court in *ACLU* v *Reno* 177 S. Ct. 2329 (1997) where an online movement towards the US free speech standard may be detected. For further discussion on this see D. Vick, 'Exporting the First Amendment to Cyberspace: The Internet and State Sovereignty' in N. Morris and S. Waisbord (eds), *Media and Globalisation: Why the State Matters* (Lanham: Rowman & Littlefield, 2001).
103 *ibid*. Online version available at <http://www.temple.edu/lawschool/dpost/Borders.html> (visited 4 January 2002).
104 n 2 above, *passim*. See also L. Eko, 'Many Spiders, One Worldwide Web: Towards a Typology of Internet Regulation' (2001) 6 *Communications Law and Policy* 445.

Cyberspace which makes it difficult to envisage coordinative solutions. For some this is a strength militating against excessive control of Cyberspace. It was argued that regulatory arbitrage acted as a (limited) check on stringent UK legislation governing state monitoring of electronic communications generally in the Regulation of Investigatory Powers Act 2000.[105] Arguably any solutions here are likely to be a product of cooperation and community-based controls involving both governments and businesses rather than of co-ordination between governments, as through the EU or the World Trade Organisation (WTO).

Hierarchy, competition and design

In addition to the use of hierarchical/community controls discussed earlier, ICANN is also applying a design/competition-based hybrid in an attempt to alleviate the pressure on the domain name system. As domain names are a man made rather than natural phenomenon they do not have to be rationed in the manner of natural resources such as bandwidth. Whereas governments cannot simply create additional bandwidth to meet the demand of mobile phone operators,[106] ICANN hopes to solve the domain name problem by creating additional resources. To this end on 16 November 2000 ICANN announced seven new top level domains.[107] It is the hope of ICANN that by creating competition in new, more specialised domains, demand will be lowered in the oversubscribed .com domain and a solution will be found to the scarcity problem. There has been profound disquiet about allegedly anti-competitive outcomes from ICANN's allocation of new top-level domain (TLD) names. Thus the allocation of some of these new resources (notably .pro and .info) has been made to organisations already controlling other key TLDs such as .com, .net and .org. The refusal to create other new top-level domain names (for example .xxx for pornography) has been criticised for inhibiting design-based controls over access or exploitation of particular sites. The solution to these problems posited by one key critique is to open the domain name market to greater competition between assignment organisations and use competition as a key form of control.[108] It may, though, already be too late for any competition-based approach to work in relation to domain names. The scarcity of resource problem in relation to domain names appears to be restricted to the .com TLD. As discussed previously there are already a large number of alternative top-level domains available.

Attempts previously to turn country code TLDs into generic TLDs have not released useful domain names. The most concerted effort has been in relation to the .ws (Western Samoa) domain, which is being promoted as a 'World Site' domain.

---

105 The House of Lords debated this possibility at some length at the Report Stage of the Bill (HL Deb vol 615 cols 381–388; cols 400–452, 13 July 2000). See in particular the debate on Amendment No 64 at col 408–418.
106 Lessig offers an alternative solution to the problem of undersupply of bandwidth. Although acknowledging supply of radiocommunications bandwidth is naturally limited he suggests we are extremely wasteful of the resource available. He rejects the use of beauty contests or auctions to propertise bandwidth as outlined above and suggests instead a design solution allowing a more efficient use of bandwidth as a free or common resource. Lessig, n 1 above, ch 12.
107 They are .aero, .biz, .coop, .info, .museum, .name and .pro. For further details on these names including who may apply for a name within these new domains see < http://www.icann.org/tlds/ >.
108 M. Mueller, 'Domains Without Frontiers' (2001) *Info* 97, 99. See also M. Froomkin, 'Is ICANN's New Generation of Internet Domain Name Selection Process Thwarting Competition?' Presentation to US House of Representatives Committee on Energy & Commerce, Subcommittee on Telecommunications, February 8, 2001. Available online at < http://personal.law.miami.edu/~froomkin/articles/commerce8Jan2001.htm > (visited 4 January 2002).

In many instances holders of current generic TLDs simply replicated their registration in the new domain. There is little evidence that the creation of manufactured additional resources deals with this particular scarce resources problem. The availability of these alternatives has not encouraged sufficient competition to effect the base of the .com domain. The relevant market, appears therefore not to be the market in TLDs as a whole, or even generic TLDs, but is restricted to the .com TLD. The .com TLD is, it appears, too well established to be affected by the creation of alternative domains. The creation of such alternatives does not appear to introduce competition within the relevant market it merely creates alternative markets in which mere replication of registration occurs.

The only possible methodology which would appear to provide for a functioning competition-based solution to the .com problem would be to increase the marketability of competing TLDs. The current ICANN policy is for the creation of alternative TLDs which they expect will increase in marketability through the efforts of the registrars who deal in such names. They are relying upon a free market rhetoric which states that those with saleable assets will work to increase the marketability of their asset through advertising and marketing. ICANN believes that the domain name system is thus a free market in which demand may be created in new products through advertising and marketing. Unfortunately the free market rhetoric does not apply to domain names in this manner. They are more than simply saleable assets. Firstly valuable domain names are, in many cases, a reflection of currently held trademarks. As has been previously alluded to, the creation of alternative TLDs fails to release alternative resources due to replication of registrations by current holders of trademarks and valuable .com domain names to prevent any risk of cybersquatting. Secondly all domain names are a method of indexing information and navigation. Thus they are streetnames not just marketable assets. And as with all other communities the Web has its desirable areas and its undesirable areas. In this virtual community .com is the business and financial district. It is the Web's equivalent to the City of London, Wall Street or Rodeo Drive. And just as businesses in the real world will pay a premium for such addresses so the focal point for competition in relation to domain names will remain in the .com domain. Due to these problems the scarcity issue in relation to domain names may be as ingrained as the bandwidth problem in relation to telecommunications and a more radical solution may be required in the future.

Other forms of control

The emphasis of current thinking on alternatives to hierarchical control is largely focused on linking hierarchy to competition or to community-based methods of control. This focus largely excludes two major classes of forms of control, one defined in terms of excluding hierarchy and the other defined in terms of including design.

Design-based regulation

A key example which is located in both sets (employing design and excluding hierarchy) is the use of regional management codes by DVD producers and equipment manufacturers. Producers and equipment manufacturers have collaborated in a regional coding system which allows for market segmentation within the DVD industry. Regional coding was developed to permit studios to control the home release of movies within different geographical regions allowing the staggering of cinematic releases.[109] The studios required that DVD software codes included a simple code that could be used to prevent playback of certain discs in certain geographical regions. The equipment manufacturers assisted by producing region specific DVD players, each player being given a code for the region in which it is sold. The player will refuse to play discs that are not encoded for that region. This means that discs bought in one country may not play on players bought in another country. The addition of regional management codes are entirely optional for the maker of a disc, discs without codes will play on any player in any country. These codes should not be confused with the DVD Content Scramble System, discussed below, which acts as a copy-control measure. Regional management codes are not an encryption system, they are merely one byte of information on the disc, which denotes one of eight different DVD regions.[110] Thus an encoded DVD bought in the US will not be viewable on a European DVD player. There is no hierarchical element to this. Customers are not prevented by contract or any other laws from buying DVDs in other countries. The control is effected by features of the diverse product standards which make a DVD useless when paired with a player with a different coding.

Including design

A related example is the use of a hierarchy/design hybrid in an attempt to manage the high levels of digital piracy which occur on the Web. Copy-control devices have been employed by almost all copyright holders who trade in digital media. These controls have met with varied degrees of success, but are supported by not only industry groups such as the Motion Picture Association of America (MPAA) and the Recording Industry Association of America (RIAA), but also have been given the force of law through the actions of the World Intellectual Property Organisation (WIPO)[111] as enacted within the European Union by the Directive on Certain Aspects of Copyright and Related Rights in the Information Society,[112] and in the United States through the Digital Millennium Copyright Act 1998.[113] With the legal support offered by these enactments several copy-control systems have been developed and implemented by bodies representing copyright holders mostly against the wishes of the community at large. One such standard developed by the MPAA for use on DVD releases is the Content Scramble System (CSS). CSS was

---

109 The coding also allows studios to sell exclusive distribution rights to a variety of foreign distributors. See D. Marks and B. Turnbull, 'Technical Protection Measures: The Intersection of Technology, Law and Commercial Licences' paper presented to World Intellectual Property Organisation Workshop on Implementation Issues of the WIPO Copyright Treaty (WCT) and the WIPO Performances and Phonograms Treaty (WPPT), December 6–7 1999, Geneva available at <http://www.wipo.org/eng/meetings/1999/wct_wppt/pdf/imp99_3.pdf> (visited 7 January 2002). The European Commission is currently investigating whether DVD producers are using the technology to illegally partition the market. See *Financial Times* 11 June 2001.
110 These are as follows: Region 1 USA and Canada; Region 2 Japan, Europe and Middle East; Region 3 Southeast and East Asia; Region 4 Australasia, Central and South America and Caribbean; Region 5 Eastern Europe, India and Africa; Region 6 China ; Region 7 Reserved and currently unused; Region 8 Special Venues (Planes, Cruise Ships etc.)
111 Article 11 of the WIPO Copyright Treaty requires contracting parties to, 'provide adequate legal protection and effective legal remedies against the circumvention of effective technological measures that are used by authors in connection with their exercise of rights under this Treaty or the Berne Convention and that restrict acts, in respect of their works, which are not authorised by the authors concerned or permitted by law.'
112 Directive 2001/29/EC, Art. 6.
113 §§ 1201(a)(1) and 1201 (a)(2).

developed by two hardware companies, Matsushita Electric and Toshiba, for the motion picture industry and was adopted as industry standard in 1996. The system involves a dual key encryption system which encrypts all sound and graphic files contained on a DVD release. The files may be decrypted by the appropriate decryption algorithm which is made up of a series of keys stored on both the DVD and the DVD player. This means that only players and discs containing the appropriate keys may decrypt the necessary files and play the movies stored on the DVDs.[114] The CSS system did not prevent direct copying of DVD discs, the contents of a DVD (while encrypted) could be copied directly from one DVD to another. CSS did though prevent the uploading of the contents of a DVD on to hard disc or a web server. The concern of some users was that CSS systems were only licensed for use on Macintosh and Windows based operating systems (and for dedicated DVD players). Users of open source operating systems such as GNU/Linux could not play a CSS encoded DVD on their system. This led to a campaign of civil disobedience leading to the development of a decryption code for CSS which would allow the playing of CSS encrypted DVDs on any platform. The CSS code was a quite weak 40 bit encryption system and in September 1999 it was successfully hacked independently by an anonymous German hacker and a member of the 'Drink or Die' cracking community.[115] This development meant that CSS encrypted DVDs could now be used on unlicensed DVD players and that DVD material could be placed directly onto the Web. Such a development was an obvious threat to the continued use of CSS by DVD producers. Action was taken immediately in Norway where Jon Johansen who had been erroneously identified as the author of DeCSS was prosecuted and in the United States where Universal Studios successfully obtained injunctions under the Digital Millennium Copyright Act against several individuals who were distributing the DeCSS code from US-based websites.[116] The decision in this case has been extensively criticised by many commentators, including Lessig who argues that 'DeCSS didn't increase the likelihood of piracy. All DeCSS did was (1) reveal how bad an existing encryption system was; and (2) enable disks presumptively legally purchased to be played on Linux (and other) computers'.[117] Lessig is extremely critical of the use of law to support these design controls arguing that they create an 'imbalance where traditional rights are lost in the name of perfect control by content holders.'[118]

This view taken by Lessig in his new book *The Future of Ideas* may though prove to be unduly pessimistic. There is as yet no evidence of content holders attaining the perfect control he fears in Cyberspace. Indeed the victory of Universal Studios and the MPAA has proved to date to be pyrrhic. As is often the case in Cyberspace when hierarchical/design controls are used to regulate the community at large the community will rally in an attempt to defeat the regulatory control mechanisms. The DeCSS code may currently be obtained from any one of hundreds of websites which

---

114 For more detail on CSS see the opinion of Judge Kaplan in *Universal Studios Inc v Reimerdes et al* 111 F.Supp 2d 294 (2000). Affirmed *Universal Studios Inc v Corley et al* 28 November 2001, Second Circuit Court of Appeals Docket No. 00-9185 Available at <http://eon.law.harvard.edu/openlaw/DVD/NY/appeals/opinion.pdf> (visited 4 January 2002).
115 The media wrongly attributed the development of DeCSS to a fifteen year old Norwegian Jon Johansen. Although Mr. Johansen was a member of the 'Masters of Reverse Engineering' community which released DeCSS he was not the author of the program. This is made clear in a text file which accompanied the release of the program. The text file is available at <http://www.lemuria.org/DeCSS/dvdtruth.txt> (visited 4 January 2002).
116 *Universal Studios Inc v Reimerdes et al*, n 114 above.
117 Lessig, n 1 above, 189. See further 187–190.
118 *ibid* 200.

remain out of the reach of the US authorities.[119] Currently the producers of DVD titles and the hacking community are involved in a war of code. The motion picture industry has updated the CSS code which means the DeCSS code no longer decrypts the latest DVD releases. This has simply encouraged hackers to produce new, more powerful, second generation decryption codes such as DVD-Decrypter. Both parties continue to battle for the control of DVD encryption/decryption codes. The producers of DVD titles and the community at large are both using design tools to attempt to protect their position. The producers presently have the advantage, due primarily to a weakness of current technology. At the moment the lack of widely available broadband technology prevents distribution of decrypted movie data over the Web: the producers hold the upper hand. As distribution technology improves the movie industry may find that their design solutions cannot effectively function without either the support of the community at large or far greater reliance upon the hierarchical control elements introduced by the Digital Millennium Copyright Act and the Directive on Copyright and Related Rights in the Information Society. Producers of DVDs will need to decide within the next few years whether they wish to rely on a hierarchy/design hybrid or a community/design hybrid.[120]

Excluding hierarchy

A successful example of a community using design tools to effect a regulatory scheme is the community-based approach to protecting children in Cyberspace. As discussed above the anonymity problem raises two distinct dangers for minors in Cyberspace. One is that they gain access to materials which are unsuitable for minors and the other is that adults take advantage of anonymity to forge improper relationships with minors. Hierarchical controls fail to remedy these problems but a community-based solution has proved extremely successful, especially when linked with design-based solutions. Within organised cybercommunities children may be supervised by the community. Communities such as AOL encourage family membership where parents register the details of the family as a whole and each individual member has their own password. Unless the child were to compromise an adult password, their status can therefore be made known to the community and the community can supervise and protect the child while he is online. Children cannot be watched all the time and the community cannot take over all parenting responsibilities. To assist, additional design-based tools may be used. In addition to the community supervision, parents may employ software solutions such as CYBERsitter and Net Nanny. These products allow parents to set acceptable parameters for their children when in Cyberspace.[121] Combined, the

---

119 The website operated by Shawn Reimerdes (one of the defendants in the MPAA action) contains the following advice: 'A Federal Judge removes this link by court order! We are fighting for the right to put this link back up for you! I am not allowed to have this decryption information anymore, so I will just tell you the obvious: Go to your favorite search engine and enter 'DeCSS'. You will find one of thousands of websites that has decided to post this information'. Doing so will allow you to locate sites such as the DeCSS mirror site at <http://heavymusic.8m.com/> (visited 20 December 2001); Download.com <http://www.download.cnet.com/> (visited 4 January 2002) and <http://www.lemuria.org> (visited 4 January 2002) all of whom currently have the DeCSS program available for downloading.
120 ee further Lessig, n 1 above, chapter 11.
121 Such software programs are called alternatively Filterware or Censorware (depending very much upon your political viewpoint). Many programs such as the ones listed use stand-alone value judgements to categorise websites based on their content. The software provider will review sites and will put them on either an 'allowed' or a 'not allowed' list. Other programs rely upon the Platform for Internet Content Selection (PICS) a standardised industry system which allows content to be rated in various categories including: topics such as 'sexual content', 'race', and 'privacy', under the control of the user.

515

role of the community and the security provided by these products appear to provide a relatively successful solution to the access problem.

Conclusions

New and unpredictable configurations of power are among the hallmarks of the new media. It is not surprising that the problem of control has attracted such a high degree of interest among scholars. Not only are there interesting problems of designing regimes to provide appropriate constraints on undesirable activities, there are also challenges in securing the maximum benefit to the community of new technologies such as the Internet and 3G mobile (each of which is said to be subject to 'network effects' such that the more users there the greater the benefit to the community generally). The new media phenomena present scholars with at least two temptations. One is to overstate the novelty of the problems presented, with a consequent tendency to reject 'old' forms of control.[122] The second is to overstate the extent to which the media themselves 'hardwire' or constrain the possible means to addressing the problems. Both tendencies are prevalent in analyses of the control problem as it applies to the Internet.

The alternative, which we have argued for, is to locate problems of controlling the new media squarely within well established analyses of problems of regulatory control. Such analysis encourages us to look at the mechanisms of control which already subsist within the target system and to find ways to stimulate or steer those indigenous mechanisms towards meeting the public interest objectives of regulation. Thus a central role for hierarchy is to steer systems which involve other forms of control based in community, market or design (or combination thereof). This does not exclude the possibility that effective control may occur through competition, design or community, together or separately, without hierarchical involvement.

A key challenge presented by such novel governance mechanisms is how to deploy them in such a way that are perceived as legitimate. The legitimacy of democratic government is linked to processes of representation and open decision making. Though other governance mechanisms may be legitmated in similar ways, in many cases it will either be alternative process elements and/or outcomes which are more important in generating legitimacy. Judgements on the appropriate balance between democratic and other forms of legitimation are likely to differ within different political cultures. This is evidenced in markedly different responses in Europe and the United States to the creation of ICANN. For some it represents an unacceptable delegation of government authority to a private body.[123] For others it is an efficient technical solution to a pressing problem, even if its decision making is not wholly technical. A key challenge in deploying ideas about the mixture of control forms advanced in this article is to balance these twin concerns about efficiency and legitimacy. The conditions for achieving an acceptable balance are likely to vary in different places and different times.

---

122 A useful early consideration of the 'newness' issue in Cyberspace is I. Trotter Hardy, 'The Proper Legal Regime for "Cyberspace"' (1994) 55 *University of Pittsburgh Law Review* 993. More recently see M. Price, 'The Newness of New Technology' (2001) 22 *Cardozo Law Review* 1885.
123 Froomkin 'Wrong Turn in Cyberspace' n 4 above; J. Wienberg 'Geeks and Greeks' (2001) 3 *Info* 313; cf R. Marlin Bennett 'Icann and Democracy: Contradictions and Possibilities' (2001) 3 *Info* 299.

[6]

SHIELDING CHILDREN: THE EUROPEAN WAY

MICHAEL D. BIRNHACK* & JACOB H. ROWBOTTOM**

INTRODUCTION

At the time of writing, the dangers posed by the Internet to children are making regular headlines in the United Kingdom and elsewhere in Europe. In *Operation Ore*, British police have been investigating a reported seven thousand credit card subscribers to a single child pornography web site based in the US. With this come reports that suspects include judges, lawyers, teachers, university lecturers, policemen, and a few celebrities. Some have argued that this is creating a moral panic.[1] The controversy has sparked many difficult questions as to how many such users may be based in the UK, and whether there is anything wrong in looking at pictures (as opposed to actual child abuse).[2] In these cases, the issue concerns material viewed by adult Internet users and whether that material is linked to the actual abuse of children by encouraging such pictures to be made and by fuelling the viewers' fantasies that may turn to action. While the harm caused to children using the Internet has not been overlooked, it has again been concerned with a link to actual child abuse, especially through the use of chat rooms. Stories have been reported of adults arranging to meet children after posing as teenagers in chat rooms.[3]

---

* Lecturer, Faculty of Law, University of Haifa, Israel; J.S.D., New York University School of Law, 2000; LL.M., New York University School of Law, 1998; LL.B., Tel Aviv University, 1996.
** Fellow, King's College, Cambridge University; UK. Barrister; LL.M., New York University School of Law, 2000; B.A., Oxford University, 1996.
We wish to thank Nick Barber and Guy Harpaz for helpful comments, and Avihay Dorfman for able research assistance.
  1. See *Calm the Witch-Hunt: Even Child Porn Suspects Have Rights*, GUARDIAN, Jan. 18, 2003, at 21, *available at* http://www.guardian.co.uk/leaders/story/0,3604,877205,00.html.
  2. See Philip Jenkins, *Cut Child Porn Link to Abusers*, GUARDIAN, Jan. 23, 2003, *available at* http://www.guardian.co.uk/online/story/0,3605,879877,00.html; Matthew Parris, *Child Abuse, or a Crime in the Eye of the Beholder?* TIMES (London), Jan. 18, 2003, at 24; *Networks of Trust: The Internet and the Abuse of Innocence*, Editorial, TIMES (London), Jan. 15, 2003, at 21.
  3. *Father Rescues Naked Girl After Net Rendezvous*, TIMES (London), Jan. 28, 2003, at 8.

Less sensational are instances where children access material on the Internet that may not put them at risk of abuse, but that may still be harmful. Such material may include sexual content or scenes of violence, material that poses little threat to adults and which adults should be free to read. This paper investigates the approach taken to the problem of Internet material that is harmful to children in Europe and the UK, and locates the discussion within the emerging constitutional jurisprudence in Europe.[4]

In a nutshell, and inasmuch one can generalize, the current European solution, unlike the mostly unsuccessful legislative attempts in the US, tends to leave the regulation of material that is harmful to children to the market. However, this is not necessarily a civil libertarian heaven. Rather, it is a guided, or directed, legal framework which actively fosters and encourages self-regulation. In this, it is closer to—though not exactly the same as—Amitai Etzioni's suggestion that the legal response should first aim at separating children and adults so to minimize the "spillover" onto the rights of adults, and alternatively, if the first avenue is ineffective, proposing that limitations on adults are justified when the harm to children is substantial.[5]

We begin in Part I by drawing the contours of the issue at stake. We propose an intuitive metaphoric framework to examine the issues at stake by thinking of the producer of the harmful material and the child as two ends of a chain, which we call the *"pornography chain."* In between there are various other links. We set out several baselines of the discussion. Firstly, we distinguish between material that should be put out of reach of both adults and children, such as child pornography, and that which is harmful to children but not to adults (sometimes referred to as an illegal/harmful distinction).[6] Secondly, we assume that there is such harm, and thirdly, we assume that adults do

---

4. "European Law" is a rather broad term, as there are several levels of legal systems in Europe; first, each country has its own legal system, second, countries which are members of the Council of Europe are bound by the European Convention for the Protection of Human Rights and Fundamental Freedoms ("ECHR"), and third, there is the European Union in which fifteen states are members at this point. We will discuss the different layers of legal systems in Europe, especially that of the ECHR. The legal response in the United Kingdom will serve as a leading example throughout the discussion.

5. *See* Amitai Etzioni, *On Protecting Children from Speech*, 79 CHI.-KENT L. REV. 3 (2004).

6. We adopt Dr. Etzioni's distinction between "children" (twelve years and under) and "teenagers" (thirteen to eighteen years old), and the generic term "minors," which refers to both groups together. *See id.* at 43.

have a constitutional right to access free content online.⁷ We then turn, in Part II, to set out the European constitutional background, wherein free expression is recognized as a human right, and is defined both in a wider and a narrower manner than the American First Amendment, in that it explicitly covers the right to receive information, but it includes built-in limitations.

In Part III, we survey various possible legal responses to the issue. Firstly, a "direct public-ordering approach" in which the State, through a statute, administrative act, or judicial decision, announces what is prohibited and what is permitted. Thus far, European legal systems have not chosen this approach, but nevertheless, we assess the constitutional meaning of such a response. Secondly, an "indirect public-ordering approach," in which the State does not interfere in as blunt of a manner in the digital environment, but is a player in the field; it creates various incentives for the players to act in a publicly desired manner. Thirdly, a "private-ordering approach," where the State refrains from any kind of interference with the digital arena and leaves the playground to self-regulation. The approach opted for in European legal systems seems, at least at this point, a combination of the latter two.

This is evident, for instance, in the topic of filtering software and rating programs (private-ordering), public programs of hotlines for reporting illegal material, encouraging the adoption of codes of conduct (public support for private-ordering), rules that impose liability on Internet Service Providers ("ISP") (indirect public ordering), and education (public involvement). We demonstrate the legislative approaches by analyzing some of these rich regulatory tools. In way of conclusion, we raise a few thoughts as to why it is these approaches that were preferred in Europe.

I. SETTING THE PROBLEM

A. *The Interest in Protecting Children and the Pornography Chain*

For the purpose of this Article, we do not quarrel with the assumption that pornography does indeed harm children who are ex-

---

7. In this case, "free" is used both as in "free speech" and as in "free beer." The distinction was made in the context of the free software movement by Richard Stallman. *See* SAM WILLIAMS, FREE AS IN FREEDOM: RICHARD STALLMAN'S CRUSADE FOR FREE SOFTWARE ch.9 (2002), *available at* http://www.faifzilla.org/ch09.html.

posed to it. The evidence discussed in Etzioni's article suffices to establish that there is a public interest in protecting children from pornography, and more so, from violent material.[8] This interest will later be phrased in constitutional terms as "necessary in a democratic society," which is one of the conditions upon which the ECHR allows restricting freedom of speech.[9]

Fulfilling this interest does not come without a cost. The cost is one of limiting the freedom of consenting adults to access these materials. Before considering the direct clash between the public interest and the freedom to access online available material, it is first necessary to identify the *chain of pornography*.[10] There are several links in the chain of pornography, from production to consumption: the producer of the material, the web site operator who offers it, the ISP who provides access to the site or service, the institution through which access is offered, parents, and, finally, the child end-user. Not all links appear in all situations: for instance, when we surf from the privacy of our home, the institutional link drops out of the picture. One strategy to protect the child end-user might be to impose liability on one of the links in the *pornography chain*. Another strategy would be to focus not on *who* can prevent the harm, but on the *content* that passes through the *pornography chain*. We first briefly examine the various links of the chain, and examine whether we can curtail the *pornography chain* there, and whether it is a good solution. We then examine the second strategy.

1. The Producer or Web Site Operator

Regulation at this point tackles the problem at the source of the material and thereby restricts every individual's access to the material regardless of age. But directly imposed limitations on the producer of the material or the web site operator might run into both serious constitutional and technological difficulties. An adult has the right to produce certain content, as long as it in itself does not harm others (as is the case with child pornography) or where there is a constitutionally valid limitation on this right. This issue raises the need to distinguish the illegal from the legal, an issue which we will address shortly. Furthermore, due to the architecture of the Internet, especially its

---

   8. *See* Etzioni, *supra* note 5, at 33–40.
   9. In the US, this interest can be phrased as a "compelling state interest."
   10. In the discussion to follow, we focus mostly on pornography, but the arguments apply to violent content as well.

borderless character, it might be inefficient to try to block the pornography at its source; end-users will access the same harmful material, now relocated on web sites operated from other countries, where the legal standard is more permissive.

2. The ISP

Perhaps we can aim at the next link in the chain: the ISP could be required to block children from accessing the harmful material. But current technology does not permit an ISP to easily identify child users, and therefore any restrictions on content are likely to apply to adults as much as children. A ten-year-old child who will seek the services of an ISP might be denied access, most likely because of her inability to provide the ISP with assurance that she can pay for the services. But once the service to a home or a public library is established, the ISP cannot effectively know whether it is an adult or a child who uses it at any given minute.

Furthermore, imposing a duty on the ISPs to block the harmful material raises a host of complex questions, as to the effect on the rights of the ISPs themselves (their right to property and contract), the effects of imposing such liability on the development of the Internet in general and of e-commerce in particular, the "chilling effect" on the ISPs and thus the speech-effects on end-users, effects on the costs which are associated with imposing liability, and much more. We will address some of these issues later on in our discussion of the *indirect public-ordering approach*.[11]

3. The Facility

Some institutions have the ability to control access to the physical facility where the computers are located, and we can assume that some of these institutions adopt a clear policy as to who may have access to the location and use of the computers. An Internet café, for example, is more likely to refuse entrance to children, perhaps because children are less likely to be able to pay for the services. In any case, the operator of the small Internet café directly faces the patron. Just like the seller at the newsstand can recognize that it is an eight-year-old who wishes to buy *Playboy*, and hence refuse to sell it to the child, so can the operator of the café refuse access. Other institutions have the power to adopt and implement clear and enforceable rules

---

11. *See infra*, Part III.C.

as to the access and use of the Internet within their physical boundaries. An elementary school, for example, in which there are computers and access to the Internet, is likely to prohibit access to pornographic web sites. The physical presence of both the operator and the child enables control over access.

The difficulty identified by Etzioni arises in situations where both adults and minors use the same physical location to access the Internet, such as many public libraries. Indeed, much of the legal debate has evolved around libraries.[12]

4. The Parents

The parents have a place in the *pornography chain* by providing access to the Internet in the home, or by providing the parental consent that some public facilities require before granting access. By deciding where and when the child can have access to the Internet, the parent can determine what content the child views. This approach creates the impression that the rights of a child are an adjunct to and subordinate to those of a parent. The ECHR, discussed below, gives little guidance on resolving conflicts between the rights of the child and the parent that may arise in this situation,[13] and views the family as a zone of *laissez-faire,* trusting parents to be able to make the best choices for the child. This may explain the general preference in Europe for parental restrictions on access to the Internet, rather than externally imposed limitations. While both the conservatives and civil libertarians prefer to trust the parental choices, the premise is questionable given that many parents lack the understanding to restrict what children access on the Internet and may not know much about the level of harm that may be caused.[14] Consequently, such an approach must be supported with sufficient resources and support to allow parents to make an informed choice.

---

12. In the US, see *Kathleen R. v. City of Livermore*, 104 Cal. Rptr. 2d 772, 777 (Ct. App. 2001); *Mainstream Loudoun v. Bd. of Trs. of the Loudoun County Library*, 24 F. Supp. 2d 552 (E.D. Va. 1998); *Am. Library Ass'n v. United States*, 201 F. Supp. 2d 401 (E.D. Pa. 2002), *rev'd,* 123 S. Ct. 2297 (2003); Mark S. Nadel, *The First Amendment's Limitations on the Use of Internet Filtering in Public and School Libraries: What Content Can Librarians Exclude?*, 78 TEX. L. REV. 1117 (2000); and Junichi P. Semitsu, Note, *Burning Cyberbooks in Public Libraries: Internet Filtering Software Vs. The First Amendment*, 52 STAN. L. REV. 509 (2000). In the UK, see *infra*, text accompanying note 147.

13. *See* Jane Fortin, *Rights Brought Home for Children*, 62 MOD. L. REV. 350, 354, 357 (1999).

14. *See* Lilian Edwards, *Pornography and the Internet, in* LAW AND THE INTERNET: A FRAMEWORK FOR ELECTRONIC COMMERCE 307 (Lilian Edwards & Charlotte Waelde eds., 2000).

5. Users

Perhaps we should turn to the last link in the chain—the minor consumers. Etzioni suggests that we distinguish between minors of various ages, which he roughly divides into two groups: children and teenagers.[15] This is a much-needed distinction, but the difficulty of distinguishing between these two groups is the same that drives and underlies the entire problem discussed here: the current architecture of the Internet lacks the ability to recognize the user. A famous *New Yorker* cartoon features a dog sitting by a computer, accompanied by the caption, "On the Internet, nobody knows you're a dog."[16] A web site operator cannot know who the end-user is. At most, the operator can recognize the Internet Protocol (IP) address of the user. The IP address can be analyzed, but the information recovered will only indicate the ISP used by the user to connect to the Internet. This might indicate a rough geographical location, but usually not more than that.[17] The ISP, as discussed above, is also limited in its ability to recognize the user. Hence, the way to recognize the end-user depends on the minor user's own cooperation.

But minor users cannot be trusted to identify themselves as minors or as adults. In the absence of strong social condemnation against surfing web sites with "adult content," and as long as democratic societies value the privacy of users, including children, then counting on the subjects of the public interest will not be an efficient solution. A requirement to "click here if you are under 18" is unlikely to deter many minors. Hence, using the law to curtail the *pornography chain* at the minor-user's link is unlikely to succeed. Of course, ultimately, it is all a matter of education, and the question addressed here is whether the law or technology should—or could—replace education or aid it.

6. Technology

There has been an attempt to develop and utilize technological measures to differentiate the end-users, including various age-verification measures, which ask the end-user to prove his or her age by providing a driver's license number, credit card number, and the

---

15. *See* Etzioni, *supra* note 5, at 43.
16. The author of the cartoon is Peter Steiner. *See* NEW YORKER, July 5, 1993, at 61.
17. Several sites offer an analysis of users' privacy in order to demonstrate the ease with which information can be retrieved. *See, e.g.*, http://privacy.net/.

like, or by using authentication certificates.[18] However, these can be easily bypassed, either technologically or by providing false information, or simply by using an adult's documents. These measures have the further unintended effect of deterring adults from accessing legal web sites and imposing heavy costs on various service providers. In addition, as the U.S. Supreme Court noted, requiring web-site operators to install age verification measures imposes heavy costs on non-commercial speakers.[19]

The intermediate conclusion is that given the practical problems in regulating the various links of the chain described above, the current solutions using technology to differentiate between children and adults are likely to be only partially successful at best. In this context, it is important to note that Etzioni's child-adult separation approach refers to the physical, off-line links in the *pornography chain*, such as the facility, rather than to the on-line links, such as the ISPs.[20] However, the physical, institutional, educational, and technological barriers are not impassable.

A different strategy to prevent harm might be to target the *content* that passes through the *pornography chain*, rather than to target the links thereof. This requires that we are able to define "good" content, or at least "harmful" content. This distinction is crucial for another basic assumption which accompanies the debate, namely, that adults have a right to produce and/or to access pornography or violent material, even if the same material is harmful to children. It also affects the scope of the rights of web site operators.[21]

B. Illegal and Legal Content

Various European institutions have explicitly made the distinction between the legal and the illegal and treated them in two different ways.[22] This brings to mind the American distinction between

---

18. For a discussion of authentication, see LAWRENCE LESSIG, CODE AND OTHER LAWS OF CYBERSPACE 30–36 (1999) (arguing that "the absence of self-authenticating facts in cyberspace reduces its regulability").
19. Reno v. ACLU, 521 U.S. 844, 880 (1997).
20. *See* Etzioni, *supra* note 5, at 29–30. Etzioni proposes that libraries allocate separate computers to children and to adults. This proposal sets out a simple and easy solution for libraries. However, it does not resolve the broader problem of children accessing harmful material in other situations.
21. Later on, we explain how this distinction between adults and children relates to another distinction we make, regarding the various kinds of regulation. *See infra*, Part III.A.
22. *See* European Commission, *Green Paper on the Protection of Minors and Human Dignity in Audiovisual and Information Services*, COM(96)483 final at 6 (recognizing a category

"obscene" and "indecent." While the First Amendment does not cover the former, the latter enjoys constitutional protection.[23] Obviously, the difficulty lies in drawing the line between the two kinds of content—a problem with which American courts struggle.[24] This difficulty in itself has a price—the unclear boundaries of the "illegal" might deter not only illegal speech, but also legitimate content. The laws determining what content is illegal in Europe are drawn up by each Member State, and different countries will draw the balance differently. In this section, we consider the English attempt to define the line between the legal and the illegal, and examine its applicability to the Internet.

It is obvious that it would not be satisfactory to make all material harmful to children illegal. The question of illegality raises the constitutional issue of determining what types of material no one should have access to and that deserve no protection. Powerful reasons exist to make some types of speech illegal, as, for example child pornography has been made under the Protection of Children Act of 1978 in the UK.[25] Other types of material are harmful to some parts of society but not others, thereby deserving of at least some constitutional protection. This may include written words that have some sexual or adult themes, descriptions of violence, or strong language that may be

---

of material that violates human dignity and that should be banned for everyone regardless of age) [hereinafter Green Paper]; *see also* Council Recommendation 98/560/EC of 24 Sept. 1998 on the Development of the Competitiveness of the European Audiovisual and Information Services Industry by Promoting National Frameworks Aimed at Achieving a Comparable and Effective Level of Protection of Minors and Human Dignity, art. 17, 1998 O.J. (L 270) 48 (noting that the distinction between materials that are offensive to human dignity and those that are harmful to minors is vital, and that the two types of problems require a different approach) [hereinafter Council Recommendation].

23. *See* Chaplinsky v. New Hampshire, 315 U.S. 568, 571–572 (1942). For what it means to be "covered," see FREDERICK SCHAUER, FREE SPEECH: A PHILOSOPHICAL ENQUIRY 89 (1982).

24. The definitive test was set forth in *Miller v. California*, 413 U.S. 15 (1973). It states the basic guidelines:

(a) whether 'the average person, applying contemporary community standards' would find that the work, taken as a whole, appeals to the prurient interest; (b) whether the work depicts or describes, in a patently offensive way, sexual conduct specifically defined by the applicable state law; and (c) whether the work, taken as a whole, lacks serious literary, artistic, political, or scientific value.

Id. at 24 (citations omitted). The digital environment raises some challenges to this test. For example, what is the "community" and according to whose standards is the decision made? The Supreme Court Justices have expressed various opinions in this regard. *See* Ashcroft v. ACLU, 535 U.S. 564 (2002). The Supreme Court remanded the case to the Third Circuit, which once again ruled that COPA is unconstitutional, albeit on different grounds than the previous holding. *See* ACLU v. Ashcroft, 322 F.3d 240 (3rd Cir. 2003), *cert. granted*, 124 S. Ct. 399 (2003).

25. Protection of Children Act, 1978, c.37.

unsuitable for a child. The difficulty with this type of material is determining the balance between the two competing groups.

The difficulties in striking this balance and the application to the Internet are illustrated by the British obscenity laws. In England and Wales, the line between illegal and harmful material is blurred by the way illegal *obscene* speech is defined. Under the obscenity laws, it is a criminal offense to publish an obscene article and to possess an obscene article with the intent to publish it for gain.[26] By focusing on the publisher, the Act tackles the dissemination of such material at the source, although the Act is enforced against those involved in dissemination lower down the chain, such as the seller. Obscene material was first defined in the common law by the courts, in the Victorian case of *Hicklin*,[27] as material that tended to "deprave and corrupt" those into whose hands the publication *may* fall. If the *Hicklin* test were to be applied "as is" to the Internet, it would have a far-reaching effect, as most material has the potential to be accessed by at least a small number of minors, on whom it may have greater corrupting effect.

The common law test has since been replaced by the 1959 and 1964 Obscene Publications Acts that retain the "deprave and corrupt" test, but provide that it is to be applied to persons who are *likely* to read, see, or hear the matter contained or embodied in it.[28] "Persons" has been held to mean both a "significant"[29] and "more than negligible"[30] proportion of those likely to read the material, and the test varies according to the circumstances of each case.[31] The Act therefore does not impose liability if the material will "deprave and corrupt" only a small number of incidental viewers. Nor does the threshold for "deprave and corrupt" assume some standard of purity in most readers. For example, if the readers were already "corrupted" and familiar with pornographic material, it can still be obscene in so far as it feeds an existing habit or makes it worse. Consequently, a different standard applies to material likely to be read by adults, as opposed to teenagers. As Lord Wilberforce stated in *DPP v. Whyte*:

---

26. *See* Obscene Publications Act, 1959, c.66; Obscene Publications Act, 1964, c.74. The Acts do not prohibit the possession of an obscene article for private use.
27. The Queen v. Hicklin, 3 L.R. 360, 371 (1868). The *Hicklin* test was used by the US courts, but was rejected by the Supreme Court in *Roth v. United States*, 354 U.S. 476, 489 (1957). *See also ACLU*, 535 U.S. at 574–75.
28. Obscene Publications Act of 1959 § 1(1).
29. R. v. Calder & Boyars Ltd., [1969] 1 Q.B. 151 (C.A.).
30. Dir. of Pub. Prosecutions v. Whyte, [1972] A.C. 849, 864–66 (H.L.).
31. R. v. Perrin [2002] EWCA Crim 747, ¶ 30 (C.A.).

the tendency to deprave and corrupt is not to be estimated in relation to some assumed standard of purity of some reasonable average man. It is the likely reader. And to apply different tests to teenagers, members of men's clubs or men in various occupations or localities would be a matter of common sense.[32]

While this test creates a flexible approach that does not reduce all permissible speech to that suitable for a child, problems in controlling access could lead to greater liability for publications via the Internet. Materials that are legal in other media, such as non-hardcore pornography,[33] may be more easily accessible by children, for example, where no password or fee is required, and would have a corrupting effect on those children.[34] Consequently, a significant proportion of the likely readership of Internet material may be children, giving the Obscene Publications Act a further reach on the Internet than with traditional media.[35]

Whether such a broad application of the Act would arise in relation to the Internet is questionable, given the more liberal approach of English juries in recent decades. Section 3 of the Human Rights Act of 1998 also works against such an interpretation, as legislation has to be interpreted to give effect to the ECHR, including the right to free expression.[36] Furthermore, the police and prosecutors practice tolerance[37] and do not seek to enforce the laws on pornography in traditional formats, such as magazines, that could be accessed by children. Such tolerance prevents the Act from being used as an instrument of moral paternalism in practice, even though that is the

---

32. *Whyte*, [1972] A.C. at 863.

33. While the Obscene Publications Act is most frequently invoked against materials with sexual content, it can apply to any material thought to deprave and corrupt, such as materials encouraging drug use or depicting violence. See R. v. Skirving, [1985] 1 Q.B. 819 (C.A.); *see also* Dir. of Pub. Prosecutions v. A. & B.C. Chewing Gum Ltd., [1968] 1 Q.B. 159.

34. *See Perrin*, [2002] EWCA Crim at ¶11–12. In *Perrin*, the trial jury convicted the defendant of publishing obscene material that was featured in a trailer free of charge to anyone with access to the Internet, but acquitted for materials that required name, address, and credit card details.

35. Even when considering material that only adults can purchase, courts should still consider the likelihood of that material falling into the hands of a child. The British Video Appeals Committee thought videos sold at specialty adult stores would be accessed by children infrequently. *See* R. v. Video Appeals Committee of the British Board of Films Classification, [2000] E.M.L.R. 850, ¶ 24 (Q.B.).

36. The Human Rights Act, 1998, c. 42, has a significant impact on UK constitutional law, and we shall return to it later on.

37. Prosecutors have been reported to have policies regarding which material deserves prosecution. For example, prosecutors will tolerate material with nudity, but draw the line at images of an erect penis. *See* GEOFFREY ROBERTSON, FREEDOM, THE INDIVIDUAL AND THE LAW 190 (6th ed. 1989). Prosecutors may now demonstrate greater tolerance since those reports, given that hardcore pornography can be legally sold at some licensed stores.

rationale behind the wording of the statute, as discussed below. Such an approach has led to criticisms that the Obscene Publications Act is inconsistently applied and does not represent a clear principle.[38] The scope of the Act is further limited by the statutory defense that the article is in the public good on the grounds that it is in the interests of science, literature, art, or learning, or other objects of general concern.[39] This helps address the concern that the Act could restrict information that would be essential for children, for example, information on family planning or safe-sex education.

The focus on the likely reader contrasts with the US test for obscenity, which refers to the average person,[40] and unlike the US test, the Obscene Publications Act makes no reference to the offensiveness of the material.[41] In England and Wales, if the material is so offensive that it would repulse and thereby avert any corrupting influence, it will remain legal. The question as to what material would "deprave and corrupt" is an issue of fact to be decided by the jury and means more than just material that is loathsome or lewd.[42] The application of the standard varies according to the composition, background, and values of the jury, and will be assessed in the light of contemporary standards.[43] The question is not determined by looking at the *content* of the material, but rather on its *effect* on the mind of the reader. Consequently, demonstrating that reading the material will lead to a specific harmful activity is unnecessary.[44] In this, the Act takes a paternalistic approach to the harm; it does not aim to prevent individuals from being confronted with publications that they do not want to see, but stops readers from seeing material that they may well

---

38. Yaman Akdeniz & Nadine Strossen, *Sexually Oriented Expression*, in THE INTERNET, LAW AND SOCIETY 207, 211 (Yaman Akdeniz et al. eds., 2000) (*citing* David Pannick, *Question: When Is Disgusting Not Obscene?*, TIMES (London), Sept. 8, 1998, at 39).

39. Obscene Publications Act of 1959 §§ 4(1), (2). This defense balances the interest of the community in receiving the material against the harm to the individual identified in the first part of the offense. Compare this defense to the third prong of the *Miller* test, applied in the US. See Miller v. California, 413 U.S. 15, 24 (1973).

40. *See Miller*, 413 U.S. at 24; ERIC BARENDT, FREEDOM OF SPEECH 264 (1985).

41. *See Miller*, 413 U.S. at 24; BARENDT, *supra* note 40, at 264. However, offensiveness of content is still relevant to common law offenses, such as outraging public decency. *See* R. v. Gibson, [1990] 2 Q.B. 619, 622–24 (C.A.).

42. *See* R. v. Anderson, [1972] 1 Q.B. 304, 305, 311–15 (C.A. 1971).

43. *See* BARENDT, *supra* note 40, at 256–57. Compare this to the first prong of the *Miller* test, applied in the US, where the standard is that of the "average person" in the community. *See Miller*, 413 U.S. at 24.

44. For a discussion of the link between the two, see REPORT OF THE COMMITTEE ON OBSCENITY AND FILM CENSORSHIP [Cmnd. 7772], 61–95 (1979).

enjoy for fear that it undermines their moral state.[45] While the restriction may seem contrary to the principles of a liberal account of free speech that stress individual autonomy and the freedom to choose lifestyle and moral actions,[46] it may seem more suitable for children that are not yet deemed responsible enough to make their own choices as to what materials are suitable to read.[47] However, this comes at a high price if it requires restricting the choices of adults and older minors.

By defining illegal speech by reference to its audience, the Act may encourage publishers to restrict access to potentially corrupting material. However, the Act is distinct from the approach taken by Etzioni, the focus of which is to *restrict* access to the material rather than to *suppress* it.[48] By contrast, the Act is a blunt instrument in that material cannot be published at all if thought to "deprave and corrupt" the likely reader, and thereby cannot be viewed by those potential readers on whom it has little chance of harming at all. However, the types of media used at the time of enactment may have influenced the strategy employed by the Act. Suppression of the source has traditionally been thought to be easier than regulating access. However, when addressing pornography on the Internet, unlike in print media, suppressing the first link in the *pornography chain* is not always easy, to say the least. The source can hide behind anonymous names, use technical means to disguise his or her identity, and can be outside the jurisdiction. Despite these difficulties, the Act has been applied to the Internet, though the practical issues concerning enforcement in this context will be considered below.

The Obscene Publications Act provides a theoretical route for preventing the publication of materials deemed harmful to children by making such publications illegal. However, such a route is unlikely, given the more relaxed approach to the application of the Act in which only publications with an extreme sexual content are targeted for prosecution and likely to secure a jury conviction. Furthermore, the features of the Internet make this type of regulation inappropri-

---

45. For a criticism of such paternalism in relation to *Hicklin*, and comparable problems in the US law, see 2 JOEL FEINBERG, OFFENSE TO OTHERS 165–89 (1985).

46. *See* John Gardner, *Freedom of Expression, in* INDIVIDUAL RIGHTS AND THE LAW IN BRITAIN 209 (Christopher McCrudden & Gerald Chambers eds., 1994).

47. While the problems of an overbroad interpretation have been considered, the Act may offer too little protection for children from material that is thought to be harmful to them, but not enough to "deprave and corrupt."

48. *See* Etzioni, *supra* note 5, at 42–47.

ate. By making the standard dependent on the likely audience, the Act could suppress a wider range of material and thereby spill over into adult rights of expression. Furthermore, it is especially difficult to bring prosecutions that suppress the source of Internet material. Instead, the approach that is promoted by the UK government in limiting the harm caused to children on the Internet is through self-regulation and the promotion of responsible use.

C. Adults' Rights

The discussion thus far has assumed that adults have a different stake than children and that material that is likely to harm the latter is unlikely to harm the former. But the argument is stronger than this. It is that adults have a *right* to access this material, and that this right is part of, or derivative of, their freedom of expression. Given the much-debated nature of pornography, this assumption is not trivial and requires some elaboration. The discussion is limited to the material that is not deemed harmful to adults. In regard to such material, and given the notorious effects of pornography, what is the free-speech interest in consuming it?[49] The position described above reflects the civil libertarian view. However, under "European law" this is only the beginning of the constitutional scrutiny. Acknowledging "rights" does not necessarily infer that they are trumps. As we will discuss in the next part, there is room for balancing.

The American discourse regarding freedom of speech stems from the First Amendment, which reads in its relevant part, "Congress shall make no law ... abridging the freedom of speech, or of the press."[50] This language seems to cover the rights of speakers, but does not extend to listeners. Indeed, listeners' rights were recognized only in an indirect manner. Courts acknowledged that under one of the theories of free speech—the one introduced by Alexander Meiklejohn half a century ago—protecting free speech is the means to achieve a public goal, which is the self-government of the people.[51] Under this instrumental view of the First Amendment, listeners have an important *interest* in receiving information, but not necessarily a

---

49. For a feminist critique of pornography, see CATHARINE A. MACKINNON, FEMINISM UNMODIFIED: DISCOURSES ON LIFE AND LAW (1987).
50. U.S. CONST. amend. I.
51. *See* ALEXANDER MEIKLEJOHN, FREE SPEECH AND ITS RELATION TO SELF-GOVERNMENT (1948); William J. Brennan, Jr., *The Supreme Court and the Meiklejohn Interpretation of the First Amendment*, 79 HARV. L. REV. 1, 18 (1965).

right.[52] In the liberal rights-talk[53] which dominates American legal discourse, this is an important distinction, for it means that no one has a duty to provide listeners with information.[54] The only limit imposed on the government under this liberal view is that *speakers* should not be limited. Under this rights talk, accessing information produced by another is at most a "negative liberty," *i.e.*, it implies only that the government should not interfere with the activity.[55]

The European response, under ECHR jurisprudence, is easier than the one given in the US. Article 10(1) of the ECHR instructs that

> [e]veryone has the right to freedom of expression. This right shall include freedom to hold opinions *and to receive and impart information* and ideas without interference by public authority and regardless of frontiers. This Article shall not prevent States from requiring the licensing of broadcasting, television or cinema enterprises.[56]

Under these plain words, adults have the right to receive information. Indeed, the European Court of Human Rights ("the Court") interpreted the term "information" to also include information and ideas that "offend, shock or disturb."[57] The interpretation of the term "information" is informed by the underlying rationale of freedom of expression, as the Court stated:

> Freedom of expression is one of the essential foundations of a democratic society, one of the key requirements for progress and for the development of every individual. Subject to paragraph 2 of Article 10, it applies not only to "information" and "ideas" that are viewed favourably or regarded as inoffensive or immaterial, but also to those that are conflicting, shocking or disturbing: this is the meaning of pluralism, tolerance and the spirit of openness, without which there is no "democratic society."[58]

---

52. In *Board of Education v. Pico*, 457 U.S. 853, 866–68 (1982), Justice Brennan recognized "the right to receive ideas," basing this conclusion on the Meiklejohnian theory of the First Amendment. For a theoretical discussion, see SCHAUER, *supra* note 23, at 35–56.

53. The term is borrowed from MARY ANN GLENDON, RIGHTS TALK: THE IMPOVERISHMENT OF POLITICAL DISCOURSE (1991).

54. As Professor Ronald Dworkin explains, the public has an important interest in receiving information, but not a "right." See RONALD DWORKIN, A MATTER OF PRINCIPLE 76 (1985).

55. For the meaning of "negative liberty" vis-à-vis positive liberty, see ISAIAH BERLIN, *Two Concepts of Liberty*, in FOUR ESSAYS ON LIBERTY 118, 122 (1969).

56. ECHR, Art. 10(1) (emphasis added).

57. *See* Handyside v. United Kingdom, 24 Eur. Ct. H.R. (Ser. A) (1976). See also the opinion of the ECHR Commission, as incorporated in the Court's judgment in *Jersild v. Denmark*, 19 Eur. H.R. Rep. 1, 14 (1995) (in regard to racist speech) (quoted in Fressoz v. France, 31 Eur. H.R. Rep. 2, 56 (1999)).

58. Aksoy v. Turkey, 34 Eur. H.R. Rep. 57, ¶ 51 (2002). *See also* Da Silva v. Portugal, 34 Eur. H.R. Rep. 56, ¶ 30 (2002).

The right to receive information, whether we like the information or not, and as long as it is not illegal, is thus considered to be an inseparable part of freedom of speech.[59] The emerging European jurisprudence of freedom of expression focused mostly on political speech, but the same reasoning applies to the harmful material discussed here: it is offensive to children and many adults, and it might be shocking and disturbing, but for consenting adults it is nevertheless "information."

We are now back to the fundamental conflict: on the one hand we have a valid public interest in protecting children from harmful material, and on the other hand we have an important freedom of consenting adults to access the very same content, a freedom which enjoys a constitutional anchor. Various mechanisms to differentiate adults from minors are either not satisfactory (as in the case of public libraries), or impractical (as in limiting the speakers), or partial (as in using age verification mechanisms), or have negative unintended consequences (as in imposing liability on ISPs). What then should be the legal response in the face of a frontal conflict of this kind? This is the question addressed in the next part.

II. CONSTITUTIONAL PROTECTION OF EXPRESSION UNDER THE EUROPEAN CONVENTION ON HUMAN RIGHTS AND IN THE UK

Before considering actual examples of the attempts to control the various links in the *pornography chain*, it is first necessary to outline the scheme for protecting speech rights. The hurdles that must be overcome to survive such scrutiny will make certain types of control more desirable. In this section, the differences between the US, European, and UK approaches will be examined. The constitutional methodology is important here, for if our starting point is one that does not value free speech as one of the most important human rights, if not the most important one, we might slip down the slippery slope.[60] But even those who believe it to be a fundamental human

---

59. Thus, in *Jersild*, 19 Eur. H.R. Rep. at 25–26, where the Court interpreted the free speech rights of a journalist, the Court stated that "the public also has a right to receive [information and ideas]. Were it otherwise, the media would be unable to play their vital role of 'public watchdog.'" *See also Fressoz*, 31 Eur. H.R. Rep. at 59.

60. In fact, Dr. Etzioni argues that "free speech can be highly valued even if one ranks it somewhat lower than it has been recently held and that children are now to be more highly regarded." Etzioni, *supra* note 5, at 41. This attitude allows him to trade-off free speech with the interest of protecting children. While we fully accept Etzioni's ultimate conclusion, we disagree with this constitutional methodology. Categorical balancing can take place, but at the same time, we should explain and justify the compromise of free speech rights, which in itself should

right do not argue that it is an absolute imperative.[61] There are countervailing interests (whether just "interests" in liberal rights talk, or "rights"), important in themselves; protecting children from harmful material is one of them.

A. United States

The American approach is to examine the clash by way of *categories*: first, is the content at stake covered by the First Amendment?[62] If it is, the next step would inquire what sort of regulation is at stake: is it content-based or content neutral? And if the former, is it viewpoint neutral?[63] "Balancing" is a foreign concept in the official methodology of American constitutional law.[64] The categorical constitutional methodology has the effect of channeling the complex picture of rights, freedoms, and interests into a binary juxtaposition: if there is a First Amendment right at stake, it is almost certain to overcome the opposite interest.[65] Hence, it is surprising that Congress opted first for a rather blunt direct regulation in order to protect children,[66] and later for a narrower regulation,[67] but nevertheless, one of

---

come with an acknowledgment of the "price" society pays—a moral regret—when compromising speech. This argument requires much elaboration, which we cannot conduct here. For more on the meaning of constitutional tests and the notion of moral regret, see Lawrence G. Sager, *Some Observations About Race, Sex, and Equal Protection*, 59 TUL. L. REV. 928 (1985).

61. In the US, the view that the First Amendment is an absolute has not gained support beyond Justice Black's positions. *See* Hugo L. Black & Edmond Cahn, *Justice Black and First Amendment "Absolutes": A Public Interview*, 37 N.Y.U. L. REV. 549 (1962).

62. *See* Miller v. California, 413 U.S. 15, 25–26 (1973).

63. Compare the majority's opinion in *Turner Broadcasting System, Inc. v. FCC*, 512 U.S. 622, 642–677 (concluding that the must-carry rules imposed on cable television operators are content neutral), with Justice O'Connor's opinion, *id.* at 677 (concluding that the same rules are content-based). *See also* Thomas v. Chi. Park Dist., 534 U.S. 316 (2002) (content neutral scheme for issuing permits for rallies).

64. The term, though, is more complex. *See* T. Alexander Aleinikoff, *Constitutional Law in the Age of Balancing*, 96 YALE L.J. 943 (1987). Not all constitutional systems shy away from balancing, though. *See, e.g.*, Aharon Barak, *A Judge on Judging: The Role of a Supreme Court in a Democracy*, 116 HARV. L. REV. 16, 93–97 (2002) (President of the Israeli Supreme Court outlining the constitutional methodology of balancing).

65. Once the examined regulation is identified as content-based, it takes a compelling state interest to overcome it. One of the rare cases in which this has happened is *Burson v. Freeman*, 504 U.S. 191 (1992) (holding that a statute limiting political canvassing near polling places was constitutional). However, in contemporary terms, this regulation will be considered content-based, but view-point neutral.

66. Communications Decency Act ("CDA"), Pub. L. 104-104, 110 Stat. 133 (1996). The CDA was declared unconstitutional in *Reno v. ACLU*, 521 U.S. 844 (1997).

67. Congress next enacted the Child Online Protection Act ("COPA"), Pub. L. 105-227, 112 Stat. 2681-736 (1998) (codified as amended at 47 U.S.C. § 231 (2000)). COPA was subsequently declared unconstitutional for the purposes of granting a preliminary injunction in *ACLU v. Reno*, 217 F.3d 162 (3d Cir. 2000). The Supreme Court reversed and remanded in

direct public-ordering. It is less of a surprise that, thus far, this direct public-ordering approach has failed in the courts and the First Amendment has prevailed.

B. Europe

The emerging constitutional and Human Rights jurisprudence in Europe does not shy away from explicit balancing.[68] In fact, balancing is a concept embedded in Article 10 of the ECHR itself. Contracting States to the ECHR enjoy a margin of appreciation in striking the balance,[69] but the article spells out the guidelines for doing so, though the Court repeatedly held that this does not exclude European supervision.[70] After outlining the contours of the right,[71] Article 10(2) states:

> The exercise of these freedoms, since it carries with it duties and responsibilities, may be subject to such formalities, conditions, restrictions or penalties as are *prescribed by law* and *are necessary in a democratic society*, in the interests of national security, territorial integrity or public safety, for the prevention of disorder or crime, for the protection of health or *morals*, for the protection of the reputation or rights of others, for preventing the disclosure of information received in confidence, or for maintaining the authority and impartiality of the judiciary.[72]

This structure of the limitation allows overriding freedom of expression if, and only if, several conditions are met, and these exceptions should be interpreted narrowly, as follows:[73]

Ashcroft v. ACLU, 535 U.S. 564 (2002). The Third Circuit again held that COPA was unconstitutional, albeit on different grounds, for purposes of granting a preliminary injunction, 322 F.3d 240 (2003), and the Supreme Court has recently granted certiorari to hear the case again. 124 S. Ct. 399 (2003).

68. For an overview of European constitutionalism, see J.H.H. WEILER, THE CONSTITUTION OF EUROPE: "DO THE NEW CLOTHES HAVE AN EMPEROR?" AND OTHER ESSAYS ON EUROPEAN INTEGRATION (1999).

69. The margin might change according to the kind of speech being regulated and the goal the restrictions aims at. Thus, for instance, when political speech is at stake, the margin is rather narrow, whereas in the "sphere of morals," such as blasphemy, member states enjoy a wider margin of appreciation. *See, e.g.*, Wingrove v. United Kingdom, 24 Eur. H.R. Rep. 1, 30 (1997). For elaboration on the concept of "margin of appreciation," see *Handyside v. United Kingdom*, 1 Eur. H.R. Rep. 737, 754 (1976). For application in the context of Article 10(2), see *Bowman v. United Kingdom*, 26 Eur. H.R. Rep. 1 (1998).

70. *See, e.g.*, *Wingrove*, 24 Eur. H.R. Rep. at 3.

71. ECHR, Art. 10(1).

72. ECHR, Art. 10(2) (emphasis added).

73. For the judicial instruction to interpret the exceptions narrowly, see *Da Silva v. Portugal*, 34 Eur. H.R. Rep. 56, ¶¶ 30, 33 (2000).

Restriction of Speech: First, there is a dual preliminary condition: that the regulated act is considered "expression," and that the regulation restricts it.[74] It does not matter whether the restriction is direct or indirect. In a leading case on this issue, the European Court found that an English statute that limited the expenditures of people who were not standing for election, in connection with the elections, was to be considered a "restriction."[75] In another case, the Court found that a sanction imposed after a defamatory publication took place "hampers" the press.[76] Closer to the subject discussed here, the European Commission[77] found that the screening of a gay porn movie in a back room of a sex shop was within the realm of freedom of expression.[78]

"Prescribed by Law": Secondly, Article 10(2) sets a formal condition: that the limitation of the right should be "prescribed by law."[79] This term was interpreted to include not only statutes and constitutions,[80] but also unwritten law like the English Common Law.[81] In any case, the law should be formulated with sufficient clarity to enable foreseeability.[82] Thus, when interpreting the Swiss Criminal Code's prohibition of making or possessing *obscene* material—a term which is not defined in the Swiss Code—the Court pointed to several consistent decisions by Swiss courts and found them to supplement the let-

---

74. This is parallel to the US requirements which trigger the First Amendment—that the speech is "covered" by the First Amendment and that the regulation abridges the First Amendment right.

75. *See Bowman*, 26 Eur. H.R. Rep. at 9–10. For a discussion of the impact of free expression rights on political funding in the United Kingdom, see Jacob Rowbottom, *Political Donations and the Democratic Process: Rationales for Reform*, 2002 Pub. L. 758, 771.

76. *See* Lingens v. Austria, 8 Eur. H.R. Rep. 407, 420 (1986).

77. The ECHR initially provided a procedure for individuals to complain of breaches by the states that are a party to it. The European Commission on Human Rights received and investigated initial complaints of a breach of the Convention. If the dispute was not settled, a report was provided to the state involved and the Committee of Ministers. The state concerned or the Commission could then bring the complaint before the European Court of Human Rights. The Commission has since been abolished and a full time court established. *See* A.W. BRADLEY & K.D. EWING, CONSTITUTIONAL AND ADMINISTRATIVE LAW 417–418 (11th ed. 1993).

78. *See* Scherer v. Switzerland, 18 Eur. H.R. Rep. 276, 284–285 (1994). The Commission subsequently concluded that the applicant's conviction for showing the film violated Article 10 of the Convention. However, the applicant died before the Court reached a decision and the Court thus struck the case out of its list.

79. *See Ek v. Turkey*, 35 Eur. H.R. Rep. 41 (2002), in which the Court found that the law according to which the applicant was convicted was no longer in force, and hence was a breach of Article 10(2).

80. *See e.g.*, Refah Partisi (Welfare Party) v. Turkey, 35 Eur. H.R. Rep. 3, 67 (2002).

81. *See* Wingrove v. United Kingdom, 24 Eur. H.R. Rep. 1, 26–27 (1996) (finding that the English Common Law of blasphemy is "law" within the meaning of Article 10(2)).

82. *See id.* at 26.

ter of the Code, and thus to meet the condition of "prescribed by law."[83] As for administrative decisions, the Court ruled that as long as the discretion is conferred by law, and the scope of the discretion and manner of exercise are clear, the condition is satisfied.[84]

Legitimate Aim: The restriction on expression must fulfill a "legitimate aim." This requirement, though not explicit in the ECHR, refers to the list of enumerated causes which allow the restriction of freedom of expression.[85] The list includes "morals," a term which was applied in several obscenity cases, where the Court found that, given the margin of appreciation accorded to Member States, regulation which targets pornography aims at protecting the (public) morals, and thus satisfies this condition.[86]

Proportionality: The European judiciary added one more important condition not found in the text of Article 10(2), that of proportionality:[87] the restriction on freedom of expression should be proportionate to the legitimate aim pursued.[88] The principle of proportionality is the heart of the balancing of freedom of expression with the opposing interests. It was developed and elaborated by European courts[89] to include several prongs. *Firstly*, a connection be-

---

83. *See* Müller v. Switzerland, 13 Eur. H.R. Rep. 212, 226 (1988).
84. *Wingrove*, 24 Eur. H.R. Rep. at 26–28 (discussing the authority of the British Board of Film Classification, which derives from the Video Recordings Act 1984).
85. *See* Bowman v. United Kingdom, 26 Eur. H.R. Rep. 1, 10 (1998); Aksoy v. Turkey, 34 Eur. H.R. Rep. 57, ¶47 (2000).
86. *See, e.g., Müller*, 13 Eur. H.R. Rep. at 230; Scherer v. Switzerland, 18 Eur. H.R. Rep. 276, 285–87 (1994) (Commission's position).
87. See, *e.g., Bowman*, 26 Eur. H.R. Rep. at 13. For discussion of the principle of proportionality in the UK, see Richard Clayton, *Regaining A Sense of Proportion: The Human Rights Act and the Proportionality Principle*, 2001 EUR. HUM. RTS. L. REV. 504.
88. *Bowman*, 26 Eur. H.R. Rep. at 13. The principle of proportionality has also been applied in regard to other rights which are enumerated in the ECHR. For further discussion, see Michael Supperstone & Jason Coppel, *Judicial Review After the Human Rights Act*, 1999 EUR. HUM. RTS. L. REV. 301, 312–13. For a thorough discussion of the principle of proportionality in the European Community ("EC"), see TAKIS TRIDIMAS, THE GENERAL PRINCIPLES OF EC LAW 89–94 (1999).
89. The principle was especially developed by the European Court of Justice (of the EC) and the European Court of Human Rights (of the Council of Europe). For discussion of the relationship between EC law and the ECHR, see TRIDIMAS, *supra* note 88, at 236–43 (explaining that the EC is not formally bound by the ECHR, though the Treaty on the European Union refers to the ECHR and commits the EU to respect basic rights, and concluding that the "two jurisdictions are in a relationship of co-operation and not one of confrontation"). Indeed, the Treaty reads, in Article 6(2) (formerly article F(2)): "The Union shall respect fundamental rights, as guaranteed by the European Convention for the Protection of Human Rights and Fundamental Freedoms signed in Rome on 4 November 1950 and as they result from the constitutional traditions common to the Member States, as general principles of Community Law." In the preamble of its 2000 Charter of Fundamental Rights of the European Union, the EU reaffirmed its commitment to the ECHR. Article 11(1) of the Charter reads: "Everyone has the right to freedom of expression. This right shall include freedom to hold opinions and to receive

tween the restriction and the legitimate aim should be shown. This prong requires more of a general observation: Are the means suitable to the aim?[90] *Secondly*, a direct connection and proportion between the goal and the means is required: Are the means applied to achieve the interest necessary to the restriction of freedom of expression? This is the "necessity" prong, and the question asked is often: Are there alternative, less restrictive means that could be applied to achieve the same goal? *Thirdly*, though this is sometimes a neglected prong, the question to be asked is: Has the measure chosen had an excessive effect?[91]

Proportionality requires examining the facts of the legislation at stake.[92] Any law that will attempt to protect children in a manner that will restrict the freedom of expression of adults would have to pass this barrier. It is our opinion that any restriction that limits adults' opportunities to those that are permissible for children fails at least the second prong: there should be less restrictive means applied. We will return to this point later on.

C. United Kingdom

The protection of expression has not traditionally played as fundamental role in the UK Constitution as the First Amendment has in the US. Traditionally, under the principle of Parliamentary sovereignty, the legislature is free to pass whatever laws it likes regardless of whether it violates fundamental rights. In this sense, rights to speech are merely residual, as people are free to say whatever they want in so far as it is not otherwise restricted. This did not mean that rights were ignored under the traditional approach, but rather that

---

and impart information and ideas without interference by public authority and regardless of frontiers." Article 52(3) of the Charter instructs that the meaning and scope of the rights recognized in the Charter are the same as in the ECHR. *See also* COUNCIL OF THE EUROPEAN UNION, CHARTER OF FUNDAMENTAL RIGHTS OF THE EUROPEAN UNION: EXPLANATIONS RELATING TO THE COMPLETE TEXT OF THE CHARTER (2000).

90. *See* Supperstone & Coppel, *supra* note 88, at 313.

91. See TRIDIMAS, *supra* note 88, at 91–93, for a discussion of the three-part test. An ancillary test looks at the reasons given by the state (or its relevant authority) for the measure it chose to apply, and queries whether they are "relevant and sufficient." *See, e.g.*, Worm v. Austria, 25 Eur. H.R. Rep. 454, 455–456 (1997); Tidende v. Norway, 31 Eur. H.R. Rep. 16, 434 (2000). Various scholars present these three prongs in slightly different terms. *See, e.g.*, Supperstone & Coppel, *supra* note 88, at 313–14 (defining the three prongs as "suitability," "necessity," and "balance").

92. For example, see the factual analysis in *Jersild v. Denmark*, 19 Eur. H.R. Rep. 1, ¶¶ 33–45 (1994). Many cases in various contexts were decided on this point. *See, e.g.*, Lingens v. Austria, 8 Eur. H.R. Rep. 407 (1986); *Tidende*, 31 Eur. H.R. Rep. 16; Aksoy v. Turkey, 34 Eur. H.R. Rep. 57 (2000); Da Silva v. Portugal, 34 Eur. H.R. Rep. 56 (2000).

they were protected by individual members of Parliament committed to a culture of liberty who were democratically accountable. By the early 1980s, when the political consensus broke down and the Thatcher government was able to dominate the House of Commons, greater concern arose that individual rights were too easily bypassed. By this time, the ECHR had given the legal protection of rights, such as expression, a more prominent role in the UK. In 1991, the House of Lords recognized that ambiguous legislation should be construed to be in conformity with the ECHR.[93] However, the Lords noted that the ECHR was not incorporated into UK law and held that there was no presumption that a statutory discretion should be exercised in conformity with the ECHR.[94] Consequently, while rights gained some recognition, UK law was still nowhere near the rights talk of the US. During this time, a similar channel of protection found favor in the courts, in which rights were not embodied in a constitutional provision, but rather were found in the principles of the Common Law. This meant that it was open to the legislature to restrict rights, but an interpretive presumption existed that they would not intend to do so. Consequently, restrictions imposed on prisoners' rights to free speech could not be imposed using powers granted under an ambiguous or generally worded statute.[95] This seemed to go beyond the approach in *Brind*, and required a presumption that discretionary powers would be exercised in conformity with basic rights, to be displaced only through express words or by necessary implication of an Act of Parliament.

A more active approach from the courts emerged as greater concern arose in relation to the orthodox constitutional theory. The huge majorities of the Thatcher and Blair governments meant Parliament gave little chance for political channels to protect these rights, creating a dissatisfaction that led to calls for the ECHR to be incorporated into domestic legislation. These calls were met in 1998 when the Human Rights Act ("HRA") was passed by the UK Parliament and came into effect in October 2000. The operation of the Act is devel-

---

93. Brind v. Sec'y of State for the Home Dept., [1991] 1 A.C. 696, 703 (1990).
94. *Id.* at 708–09. In *Brind*, the House of Lords upheld an executive order banning the broadcast of statements of certain organizations using the actual voice of the speaker. As no ambiguity or uncertainty existed in the statutory provision granting the Secretary of State's power to issue the ban, the discretion conferred by that provision did not have to be exercised in accordance with the ECHR.
95. *See* R. v. Sec'y of State for the Home Dept., ex. parte. Simms, [1999] E.M.L.R. 689 (H.L.).

oping fast, and space precludes a detailed discussion.[96] By incorporating the ECHR, the scheme for protecting rights is similar to that of the Strasbourg court outlined above.[97] The Act does not adopt an absolutist approach to speech rights, and the limitations in Article 10(2) are generic, leaving it for the courts to determine the scope and extent of protection.[98]

The HRA is designed to leave the last word on the meaning of rights with Parliament. The courts cannot strike down legislation as invalid, but under Section 4, they can make a declaration of incompatibility. The Act envisages a political pressure on the government to change the terms of legislation in the event of a declaration, or at least to justify the restriction. However, Section 3 places a significant obligation on the courts to read legislation as compatible, even if this means departing from the literal meaning of the text. The House of Lords' decision in *Regina v. A* demonstrates the importance of this power, arguably giving the court a role in re-writing legislation to make it compatible with the ECHR.[99] Such a power may be equal to, or even greater than, the power to strike down legislation.[100]

Under Section 6 of the HRA, it is unlawful for a public authority to act in a way incompatible with an ECHR right, unless required by an act of Parliament. The term "public authority" clearly includes local authorities responsible for public libraries and schools that provide Internet access.[101] If incompatibility is found in the public authority's act, the court can award whatever remedy it thinks is necessary, as opposed to the declaration in the case of primary legislation.[102] The influence of local level public authority in drawing the balance be-

---

96. For an overview, see K.D. Ewing, *The Human Rights Act and Parliamentary Democracy*, 62 Mod. L. Rev. 79 (1999). For a discussion on some of the major decisions in the first two years of the Act, see Francesca Klug & Claire O'Brien, *The First Two Years of the Human Rights Act*, 2002 Pub. L 649; Keir Starmer, *Two Years of the Human Rights Act*, 2003 EUR. HUM. RTS. L. REV. 14.

97. *See, e.g.*, O'Shea v. MGN, [2001] E.M.L.R. 943 (holding that a pornographic advertisement was protected expression).

98. In contrast to other rights that do not have such a provision.

99. R. v. A. *(No. 2)*, [2001] 1 A.C. 45, 67–68.

100. Of course, striking down legislation is a powerful tool, but it is a binary decision. Rewriting legislation, on the other hand, is a creative task that goes beyond judicial review, in that it is not just stating what the legislature can do, but de facto, replacing it.

101. *See* Poplar Housing & Regeneration Cmty. Assoc. Ltd. v. Donoghue, [2002] Q.B. 48, 66 (C.A. 2001) (providing that "public authority" includes persons, even private persons, performing functions of a public nature); Heather v. Leonard Cheshire Foundation, [2002] H.R.L.R. 30, 838 (C.A.).

102. Human Rights Act, 1998, c. 42, § 8 (providing that "in relation to any act (or proposed act) of a public authority which the court finds is (or would be) unlawful, it may grant such relief or remedy, or make such order, within its powers as it considers just and appropriate.").

tween the right and the competing aim depends on the level of deference the courts accord to the primary decision maker when applying the proportionality standard. While accepting proportionality as more intense than the traditional standard of review,[103] the appropriate level of deference to be shown to the primary decision maker is still debated. A more stringent and less deferential approach will be granted depending on the nature of the speech in question. For example, political speech is granted a higher level of protection than most other types of expression.[104] By contrast, sexually oriented speech would receive a much weaker standard of protection, thereby rendering the state controls discussed in this paper more likely to survive scrutiny. A further question of deference depends on the nature of the primary decision maker. At the European level, the European Court of Human Rights applies the doctrine of margin of appreciation, in which the court shows deference to the different cultures that might result in different balances in signatory states.[105] While there is no reason this should apply in the UK, a similar doctrine of deference may apply to show respect to the competence or democratic legitimacy of other institutions.[106] The approach taken by the courts to this question will determine whether the restriction of expression on the Internet is compatible with the Act. However, under the approach developing under the HRA, and following the Strasbourg jurisprudence, attempts to prevent harmful material from

---

103. R. v. Sec'y of State for the Home Dept., ex parte Daly, [2001] 2 A.C. 532, 547. For a discussion of the standard of review before and after the Human Rights Act, see Mark Elliott, *Scrutiny of Executive Decisions Under the Human Rights Act 1998: Exactly How "Anxious"?*, 2001 J.R. 166; Clayton, *supra* note 87, at 507.

104. R. (ProLife Alliance) v. British Broad. Corp., [2002] E.M.L.R. 41, 921–22; *see also* N.W. Barber, Note, *A Question of Taste*, 118 L.Q. REV. 530, 531 (2002).

105. Handyside v. United Kingdom, 1 Eur. H.R. Rep. 737, 754 (1976).

106. The deference can arise at different stages of the review, such as the consideration of whether the subject matter precludes judicial intervention, and the determination as to whether the limitation is unlawful. For a discussion of deference, see *R. v. Dir. of Pub. Prosecutions, ex parte Kebilene*, [2000] 2 A.C. 326, 380–81; *R. (Alconbury Developments Ltd) v. Secretary of State for Transport the Environment and the Regions*, [2003] 2 A.C. 295, 320–22 (2001); *Brown v. Stott*, [2003] 1 A.C. 681, 694–95, 703 (P.C. 2000); *R. v. Lambert*, [2002] 2 A.C. 545, 623 (2001); and *R. v. Sec'y of State for the Home Dept.*, [2001] 2 AC 532, 535–36. For a discussion of the case law, see Paul Craig, *The Courts, The Human Rights Act and Judicial Review*, 117 L.Q. REV. 589 (2001); Richard A. Edwards, *Judicial Deference Under the Human Rights Act*, 65 MOD. L. REV. 859 (2002); Ian Leigh, *Taking Rights Proportionately: Judicial Review, the Human Rights Act and Strasbourg*, 2002 PUB. L 265; Nicholas Blake, *Importing Proportionality: Clarification or Confusion*, 2002 EUR. HUM. RTS. L. REV 19. The approach has been criticized for lacking a coherent principle and that the courts have acted pragmatically. For example, Richard Edwards argues that the courts should not shy away from morally complex decisions as the purpose of the Human Rights Act is to create a culture of justification from the decision making bodies and that deference should not simply grant a license to the legislatures. *See* Edwards, *supra*, at 878.

being accessed on the Internet by children are less likely to fall foul of the courts than in the US, even if they entail some degree of overspill into the protected speech of adults. The way in which the courts approach this issue will be examined in the next section.

III. LEGAL RESPONSES AND THEIR CONSTITUTIONAL MEANING

A. The Various Kinds of Regulation

Regulation of expression is a tricky task. Almost any interference in the "marketplace of ideas" has a negative effect on the free expression rights of some participants in the market. Hence, direct regulation, which interrupts "normal" market behavior and rules out some of the activities going on within it, needs to pass a rather high threshold. But there are other narrowly tailored, less intrusive ways to achieve more of the public interest while causing less harm on the free speech side. For example, opting for a shift in the regulatory mode—from a *direct public-ordering approach* to an *indirect public-ordering approach*—is one such way. Instead of the government explicitly declaring which activities are allowed and which are prohibited, the government creates a mechanism that provides the players within the market with incentives to reevaluate their behavior, and adapt it toward the public interest. They are not obliged to comply, but are encouraged to do so.

Yet another way is even less intrusive, and it is one which leaves the market (or field, for those who prefer a less capitalistic metaphor) entirely in the hands of the players: if they wish, they can undertake self-regulation. The motivation to do so might be the players' own sense of responsibility, or their commercial fear of the market's reaction, or the political fear that if they stay idle, government will eventually interfere. This is an approach of *private-ordering*.

The legal experience thus far in the area of protecting children has provided us with examples of all three possible avenues and various combinations thereof. The US first attempted a direct public-ordering approach and, later, an indirect one, whereas the European way tended from the very start towards a *private-ordering approach*. The kinds of regulation described above will generally focus on different stages of the *pornography chain*.[107] Direct controls on content,

---

107. *See supra* Part I.A.

in aiming to suppress the dissemination of material, will be most effective against the producer/speaker (the first link of the *pornography chain*). Indirect public-ordering and private-ordering are similar in that they attempt to create self-regulatory, flexible, or individualized controls, and will often be targeted at the later links of the chain, such as the ISP, facility, or user. These later stages allow for greater variation in regulation according to the potential recipient, whereas targeting the producer of material will result in a uniform restriction of the material.

At first sight, then, the different regulatory strategies might seem to parallel the links in the chain of pornography; direct public-ordering aims at the producer of the material, whereas indirect public-ordering and private-ordering aim at other links in the chain. However, the parallel between strategy and link is not inevitable. For example, restraints on content may be applied against not only the producer, but also the individual user or library to prevent the downloading of unlawful material. The discussion below will illustrate that the parallel is kept in place by problems of enforcing such restraints at the lower levels.[108]

B. *Direct Public-Ordering*

Direct public-ordering poses the greatest difficulty in terms of free expression, in that it is a state intervention that rules out certain types of activities. The approach taken may be to prohibit certain activities or types of speech, thereby rendering speech that was previously thought of as harmful to be unlawful. The methods and problems of this type of control have been discussed in relation to the illegal/harmful distinction above. For such reasons, direct public-ordering is the bluntest method and has the greatest incidental impact. An example of this would arise if a country within Europe attempted to pass its own version of the American Communications Decency Act ("CDA") or the Child Online Protection Act ("COPA"). The question therefore arises as to whether this legislation would survive the ECHR or domestic protection of rights. It seems that the production of the material would be considered "expression," and that such a measure restricts it. In order to survive, the "Europeanized CDA" would have to be prescribed in a law, or at

---

108. Enforcing content restraints on individual users, while possible, would require extensive policing in order to suppress the material. It is likely to be found only in combination with attempts to retrain the higher levels of the *pornography chain*.

least an anchor in a law, that outlines the contours of discretion of an administrative body, and it would have to address the definitions of the prescribed material in as clear of a manner as possible. Such legislation would be likely to be considered as furthering a legitimate aim, and "necessary in a democratic society." The focus would be on the proportionality of such a measure, and especially, we anticipate the question to be whether there are less restrictive means to achieve the legitimate goal without impacting the freedoms of adults. Etzioni's proposal to adopt an adult-child separation approach, *i.e.*, allocating separate computers to children and adults in public libraries, perfectly fits this test. This would be a completely different way of looking at things, compared to the US; the legislation is not thought of in terms of one legal regime (*i.e.*, the First Amendment), but rather as a *balance* between the competing social interests of protecting free speech *and* protecting children. At the end of the day, the details of each legislation will determine its fate.

Direct measures that proscribe expression harmful to children have already reached the European Court, for example in *Handyside*, a pre-Internet case. In that case, the publishers of *The Little Red Schoolbook* claimed that their expression rights were infringed by the seizure of books and prosecution under the UK Obscene Publications Acts. The book contained a twenty-six-page section on sex, including such topics as masturbation, pornography, venereal disease, and abortion—subjects that would not be thought of as harmful to an adult reader. Although the book was distributed through ordinary channels, it was aimed at school children age twelve and older. The European Court held the protection of morals and of young people to be a legitimate aim under Article 10(2).[109]

The Court placed importance on the fact that the book was aimed at young persons and had a factual style that would be easily understandable, even by readers younger than the targeted group, and that the book was to be distributed widely.[110] The Court stated that this case was not analogous to pornographic publications, sex shops, and adult entertainers that might be exposed to young people in some circumstances, because the *Little Red Book* was aimed at and easily accessible to young people.[111] Consequently, on the necessity of the measures, the Court in *Handyside* rejected the argument that a

---

109. *Handyside*, 1 Eur. H.R. Rep. at 753, 755.
110. *Id.* at 755.
111. *Id.* at 758.

restriction on the sale of the book (for example, to adults only) would suffice,[112] as there was no sense in restricting to adults the sale of a work destined for the young.[113]

As noted in earlier sections, similar considerations apply to the Internet where minors can easily access material aimed at adults, especially if a password or a fee is not required. Such an approach has been followed in relation to the Internet in *Perrin*, where the UK Court of Appeal rejected the argument that self-regulation or blocking software was the only proportionate way to pursue the aim of protecting children.[114] The restriction of content at the source may be suitable where the material is aimed at or easily accessed by minors. However, where the harmful material is produced with adults as the intended consumers and is not targeted to minors, a direct restriction that limits adult access is more likely to fall foul of the ECHR. The possibility of restricting access to such sites may be thought sufficient to protect children when the material is aimed at adults and access by minors is incidental.

It is interesting to note that the approach in *Handyside* balanced the adult's right to produce and impart expression with the interest in protecting the child's welfare. Even though children contributed to *The Little Red Book*, the rights of children to produce and receive information were not considered.[115] An approach based on the child's right to produce would have similarities with Etzioni's distinction between the First Amendment rights of children and teenagers, although it remains unclear whether such a framework would have made a difference to the outcome of the case.[116] *Handyside* demonstrates that in some circumstances, the ECHR permits direct re-

---

112. *Id.*
113. *Id.* at 759.
114. R. v. Perrin, [2002] EWCA Crim. 747 (C.A.). In *Perrin*, the Court noted that impact on freedom of expression was limited, as the prosecution had only been successful in relation to a trailer that did not require a password or credit card details. The jury did not convict on material that was accessible only through membership. *See id.* at ¶ 50.
115. Fortin, *supra* note 13, at 353.
116. Geraldine Van Bueren argues that if the action was framed in terms of an older child's right to receive information, the Court may have considered the total prohibition of the book as disproportionate to the aim, thereby focusing more on proportionality than margin of appreciation. GERALDINE VAN BUEREN, THE INTERNATIONAL LAW ON THE RIGHTS OF THE CHILD 135 (1995). Ursula Kilkelly notes that the Commission on Human Rights dismissed a complaint brought by a mother and her thirteen- and seventeen-year-old children, alleging that the ban infringed their right to receive information. However, the application failed as a second, modified, edition of the book was freely available. URSULA KILKELLY, THE CHILD AND THE EUROPEAN CONVENTION ON HUMAN RIGHTS 131 (1999).

straints on adults' expression rights in relation to materials that would not harm adults in order to protect children.

This jurisprudence appears to leave open the option of suppressing material on the Internet in some circumstances to protect the welfare of children without falling foul of the ECHR. Given this, it may come as a surprise that the principal means of combating harmful material in Europe has not been through direct restrictions on permissible content. One reason may be that European countries have benefited from the American experience of the CDA and do not want to be vulnerable to such challenges. Another factor is the sheer difficulty in enforcing such laws.

Several problems have already emerged when attempting to enforce the Obscene Publications Act in the UK in relation to the Internet. Initial uncertainty over the applicability of the law to the Internet was resolved in 1994, when the Act was amended to include data stored on a computer disc as an "article,"[117] and covered the transmission of data as publication. However, even if the Act does apply to the Internet, the problem remains in deciding *whom* to prosecute. The global nature of the Internet means obscene material can be downloaded from sites outside the UK. The legal response has been to take a broad interpretation of the Act and define "publication" as occurring when the information is downloaded in the UK, as the electronic data stored overseas is transmitted into the UK.[118] As a

---

117. Criminal Justice and Public Order Act, 1994, c.33, Sched 9 ¶ 3. For a discussion of these issues, see Colin Manchester, *Computer Pornography*, 1995 CRIM. L. REV. 546, 548–52. The Act has been applied to activity on the Internet on a few occasions, mainly in relation to hardcore pornography that is corrupting to adults. See Akdeniz & Strossen, *supra* note 38, at 210 (referring to *R. v. Jack (Colin Mason)*, Norwich C.C., 4 July 1994).

118. In *R. v. Waddon*, [2000] All E.R. 502, a businessman was prosecuted even though the websites concerned were based in the US. The materials were, however, prepared and uploaded to the website in the UK. The Court of Appeal suggested that publication can occur when images are downloaded from a foreign website. Rose, L.J., stated,

> As it seems to us, there can be publication on a Web site abroad, when images are there uploaded; and there can be further publication when those images are downloaded elsewhere. That approach is, as it seems to us, underlined by the provisions of s.1(3)(b) as to what is capable of giving rise to publication where matter has been electronically transmitted.

Id. at ¶ 12. That approach was confirmed by the Court of Appeal in *Perrin*, [2002] E.W.C.A. Crim 747, in which a man was prosecuted for publishing an obscene article when the material was prepared and uploaded abroad into a foreign website. The Court rejected a parallel with *ACLU v. Reno*, 521 U.S. 844 (1997), given the difference in constitutional protection and relations between the states, and decided there was no need to show that major steps in relation to publication were taken within the jurisdiction of the court. The Court of Appeal held that the images were published in England when downloaded by the police in the UK. For a criticism of the decision in *Perrin*, see Michael Hirst, *Cyberobscenity and the Ambit of English Criminal Law*, 13 COMP. & L. 25, 28 (2002).

consequence, producers of obscene content based overseas can be liable in the UK for publishing materials that are downloaded in the UK. This does not, however, resolve the problem of enforcing this provision where the publisher remains based overseas. Consequently, strategies seeking the cooperation of those at every stage in the chain have more chance of being effective and practically implemented, even if direct ordering may be constitutionally permitted.

C. Indirect Public-Ordering

A less restrictive measure to achieve the goal of protecting children from harmful material is to offer a set of incentives to various players in the field: if they act in a certain manner, they will enjoy some benefits, and if they behave in a different manner, they might suffer some losses. In the digital environment, the prominent example of such a regulatory approach is the liability of Internet Service Providers ("ISP") for third-party content.[119]

There are various kinds of ISPs. Some offer only access to the Internet and, in this respect, are similar to telephone companies ("common carriers"). Others offer content produced by users, such as web sites which allow the posting of comments, including links and files. These are either supervised or unsupervised, and operate either synchronically (chat rooms) or a-synchronically ("forums" or bulletin boards). Accordingly, the services offered range from a mere platform to a more active role. Yet other services offer location tools, such as search engines or indexes, or technology which enables users to communicate directly, such as Instant Messaging or peer-to-peer systems like Kazaa.

Imposing liability on ISPs seems, at least at first, to be a reasonable measure; if aiming at the speakers might limit their freedom of expression, why not aim at another link in the chain, which is, in many cases, also a technological bottleneck, since much of the communication passes through its services. Since it is not the ISPs who produce the speech, the argument continues, imposing liability does not raise

---

119. The discussion that follows does not purport to be exhaustive. There is abundant literature on this topic. *See, e.g.*, Assaf Hamdani, *Who's Liable for Cyberwrongs?*, 87 CORNELL L. REV. 901 (2002). The problem has arisen in the context of copyright infringement by third parties, with some similar considerations. *See, e.g.*, Niva Elkin-Koren, *Copyright Law and Social Dialogue on the Information Superhighway: The Case Against Copyright Liability of Bulletin Board Operators*, 13 CARDOZO ARTS & ENT. L.J. 345, 372–380 (1995); Alfred C. Yen, *Internet Service Provider Liability for Subscriber Copyright Infringement, Enterprise Liability, and the First Amendment*, 88 GEO. L.J. 1833, 1840–44 (2000).

any particular free-expression difficulties. The latter argument depends, of course, on the kind of service imposed: a common carrier or a search engine does not have a free expression interest in the communication it carries or points to, just like the telephone company does not have such an interest in the communication of the users, but a web site that offers editorial services might have such an interest, and thus its freedom of expression will be triggered. However, given the application of Article 10 discussed above, and the more limited nature of the restraint here, a challenge asserting the expression rights of the ISP seems less likely to succeed.

Imposing liability on ISPs would cause them to undertake measures to avoid the risk of liability. An ISP may adopt a reviewing procedure to evaluate the material users wish to post on-line, either before the message is posted, or to adopt a "take down" policy, where the ISP reviews the material *after* it is posted, and then deletes whatever it believes is harmful (or for that matter, any illegal material). Another possible policy would be to "take down" content only after a specific complaint is made ("notice"), accompanied with a procedure to inform the person who posted the material in the first place. These policies can be accompanied with specific terms embodied in the contract between the ISP and the user.

It is easy to see that this system is much more attractive than direct public-ordering. However, it also has quite a few apparent unintended consequences. Firstly, it imposes costs on ISPs. Establishing and operating a "notice and take down" policy, for example, requires substantial amounts of money, and a review mechanism requires even higher amounts. There is also a question of the technological possibility: can there be a meaningful review process in a system where millions of messages are exchanged every minute? If we do impose liability, the result would be that fewer operators will be able to undertake their operations. This would result in concentration of ownership,[120] impediment to competition, and raising prices. The users, needless to say, would be those who would eventually bear the costs.

Secondly, since the ISP would naturally wish to avoid liability, it would have to make decisions regarding the legality of the content. This task is almost impossible, given the global nature of the Internet:

---

120. This concentration of private power might later on become attractive for the state, for example, in pursuing its battle against terror. This raises another host of questions. For discussion, see Michael D. Birnhack & Niva Elkin-Koren, *The Invisible Handshake: The Reemergence of the State in the Digital Environment*, 8 VA. J.L. & TECH. (forthcoming) (on file with authors).

what is legal in one place might be illegal elsewhere.[121] Such liability will cause ISPs to remove any material about which a complaint is made, even where the complaint lacks merit.[122] Hence, we can expect that the ISP will act in a censorial spirit; whenever there would be a doubt as to whether some material is harmful to children, the ISP would be quick to click on the "delete" key. In other words, imposing liability on ISPs results in a chilling effect on the ISPs, and it would affect quite immediately the rights of users—the potential speakers and receivers of information.[123]

The fact that the legal environment would now be that of private law—the question being whether the ISP violated the contract with the user—further limits the rights of users. The ISP is not obliged to provide service; it is not accountable and does not have to explain its decision (especially in light of the broad disclaimers which are often included in "terms of use" contracts). In other words, replacing direct public-ordering with indirect public-ordering has the effect of privatizing the enforcement of the protection of freedom of expression, in a manner that expression is likely to lose. A challenge by a user would be dependent on some horizontal effect of the rights.[124]

Thirdly, in light of rules of liability, some ISPs might be drawn to change their business model: they will reduce their involvement and strive to offer less editorial services and more of a "common carrier" service. Liability will actively discourage ISPs from taking responsibility, and attempting to monitor content, as ignorance could act as a

---

121. The Yahoo! controversy in France is a clear example. Whereas trading Nazi memorabilia is illegal in France, it is "covered" by the First Amendment. This did not deter a French court from ruling that the French subsidiary of the American Yahoo! is bound by French law when operating in France. *See* LICRA (League Against Racism and Antisemitism) v. Yahoo! Inc., (County Court, Paris, Nov. 20, 2000), *available at* http://www.cdt.org/speech/international/001120yahoo.france.pdf. Following this decision, an American court declared the French ruling to be unenforceable in the US. *See* Yahoo!, Inc. v. La Ligue Contre le Racisme et L'Antisemitisme, 169 F. Supp. 2d 1181, 1194 (N.D. Cal. 2001).

122. *See* Press Release, Cyber-Rights & Cyber-Liberties (UK), *U.K. ISP Found Liable for Defamation*, http://www.cyber-rights.org/press/1999.htm (Mar. 26, 1999); Yaman Akdeniz, *Case Analysis of* Laurence Godfrey v. Demon Internet Limited, 1999 J. CIV. LIBERTIES 260, 260–67, *available at* http://www.cyber-rights.org/reports/demon.htm; Kit Burden, *Damned for Defamation*, 15 COMP. L. & SECURITY REP. 260 (1999); Lillian Edwards, *Defamation and the Internet*, *in* LAW AND THE INTERNET: A FRAMEWORK FOR ELECTRONIC COMMERCE, *supra* note 14, at 267.

123. This seems to be the main consideration which led the U.S. Congress to accord ISPs strong immunity in many situations. *See* 47 U.S.C. § 230(c)(2) (2000); Zeran v. Am. Online, Inc., 129 F.3d 327, 330–31 (4th Cir. 1997).

124. Space precludes a detailed discussion, but in the context of the UK Human Rights Act, see Murray Hunt, *The "Horizontal Effect" of the Human Rights Act*, 1998 PUB. L. 423; Richard Buxton, *The Human Rights Act and Private Law*, 116 L.Q. REV. 48 (2000); H.W.R. Wade, *Horizons of Horizontality*, 116 L.Q. REV. 217 (2000).

shield for liability.[125] If this is the effect, then society loses twice: it has less valuable services to choose from and the content is not reviewed by anyone. Thus, the harmful material will find its way to the "market."

After some experiments, the European Community opted for a rather general structure of liability imposed on ISPs. Before we look into this scheme, it might be a good idea to look at a couple of prior events. One is the *Somm* case in Germany, the other being the English case of *Godfrey v. Demon*.

The *Somm* case illustrates some of the difficulties of imposing liability on ISPs.[126] Felix Somm was the chief executive of CompuServe, Germany, a subsidiary of CompuServe USA. The German company provided access services to the Internet, including access to material stored on the servers of the American company. Some users posted material such as child pornography, bestiality, and violent games on various Usenets. The distribution of these was illegal in Germany (and probably in other jurisdictions as well).[127] In 1995, the German police acted and informed Somm of the illegal material. Immediately thereafter, CompuServe blocked access to all Usenet newsgroups, all over the world.[128] That was an unprecedented response (which has apparently not since repeated itself). The access remained blocked for two months.

Somm was charged and, in May 1998, convicted by the Munich Local Court of assisting the dissemination of material harmful to minors.[129] The case raises many interesting and important issues, such as conflict of laws and choice of law, but for the current purpose we should note the ability of Somm to control the material and delete it: he had no such control, since the material was stored on servers owned by another company (though not a stranger to the German subsidiary) in another country. All Somm did was to provide access.

---

125. *See* Andrew Joint, *Paedophiles and Their Use of Online Chat Facilities*, 152 NEW L.J. 1602, 1602 (2002).
126. *See* People v. Somm, Amtsgericht, File No. 8340 Ds 465 Js 173158/95 (1998) (English translation by Christopher Kuner available at http://www.cyber-rights.org/isps/somm-dec.htm (Sept. 1998)).
127. Thus, for example, the German Federal Review Board listed the games at stake as morally harmful to minors. *See id.* at § II.2.
128. *See* Mark Konkel, Comment, *Internet Indecency, International Censorship, and Service Providers' Liability*, 19 N.Y.L. SCH. J. INT'L & COMP. L. 453, 454–55 (2000). The number of users affected was reported to be as high as 4.3 million. *See* MARGARET JANE RADIN ET AL., INTERNET COMMERCE: THE EMERGING LEGAL FRAMEWORK 1058 (2002).
129. *See Somm, supra* note 126.

Even if Somm did have the technical ability, what should he have done, given that 99 percent of the material available in the Usenets was legal?[130] Furthermore, it is quite likely that some of the material was illegal in Germany but legal elsewhere (we can assume that at least the games were legal in the US).

The events that accompanied the case were no less interesting. The German legislature amended the relevant statutes so as to clarify that ISPs are immune from liability for third-party content, and the prosecutors changed their minds. They themselves appealed the conviction, and in 1999, the appellate court acquitted Somm.[131]

Although the case dealt with illegal material, it illustrates the difficulties of imposing liability on ISPs. Not all ISPs are technical bottlenecks. In addition, it is difficult, if not impossible, to separate the legal from the illegal. Furthermore, the potential liability has a "freezing" effect on speech. The difficulty to identify the unwanted content is even more acute when the legality turns not only on the content, but on the user—that is, whether she is an adult or a minor.

In the English case of *Godfrey v. Demon Internet Ltd.*,[132] the ISP (Demon) offered a Usenet service in which users' postings were stored for two weeks. An unknown person made an obscene and defamatory posting purporting to be written by Godfrey. Godfrey informed Demon that the posting was a forgery and requested that it be removed. Demon failed to do this, and the message remained until its expiry. The libel proceedings subsequently brought by Godfrey tested the issue of ISP liability in the UK. Under the Common Law standard, Mr. Justice Morland held that Demon was the publisher and that the transmission of such a posting from the service provider to any person accessing it was publication:

> the defendants, whenever they transmit and whenever there is transmitted from the storage of their news server a defamatory posting, publish that posting to any subscriber to their ISP who accesses the newsgroup containing that posting.[133]

Demon was not just the mere passive conduit of information, but chose to receive the postings, to store them, to make them available to users, and could have removed them. Demon's defense of "innocent dissemination" under Section 1(1) of the Defamation Act of

---

130. *See Somm, supra* note 126, at § V.
131. For a discussion of these events, see Konkel, *supra* note 128, at 463–65.
132. [2001] Q.B. 201.
133. *Id.* at 208–09.

1996 was rejected as Demon had been given notice of the posting by Godfrey.[134] The effect of the decision is to require ISPs providing that type of service to remove the material once they acquire knowledge of the unlawful content.

This approach indeed fits the requirements of the EC E-Commerce Directive[135] that was implemented in the UK in August 2002.[136] Unlike the US, which has two separate legal regimes that govern liability of ISPs—one in regard to copyright law and one for all other kinds of content—the European Directive determines a unified system.[137] The European regime differentiates between the kinds of service provided. Firstly, it assures immunity to ISPs who are mere conduits. Article 12 sets a few conditions which ensure that the ISP is indeed a conduit and is not involved in initiating or editing the message transmitted, or editing it, and does not determine the parties of the transmission.[138] Secondly, it provides the ISP with immunity for caching, i.e., automatic, intermediate, and temporary storage. Caching is a vital technological step for transmission of information under the current architecture of the Internet, in that it eases the traffic over the Internet and enables a rather quick and accurate transmission.[139] This immunity is subject to the ISP not interfering with the transmissions, and to its removal of material upon requiring knowledge that the original material is no longer available. Thirdly, the Directive requires that ISPs be awarded immunity for content stored on their servers by users.[140] This immunity is subject to lack of knowledge on

---

134. Section 1(1) of the Defamation Act, 1996, c.31, provides:
   In defamation proceedings a person has a defence if he shows that—
   (a) he was not the author, editor or publisher of the statement complained of,
   (b) he took reasonable care in relation to its publication, and
   (c) he did not know, and had no reason to believe, that what he did caused or contributed to the publication of a defamatory statement.

135. *See* Council Directive 2000/31/EC of 8 June 2000 on Certain Legal Aspects of Information Society Services, in Particular Electronic Commerce, in the Internal Market, 2000 O.J. (L 178) 1–16 [hereinafter E-Commerce Directive].

136. *See id.*; Electronic Commerce (EC Directive) Regulations 2002, No 2013; Graham Smith & Alex Hand, *Implementing the E-Commerce Directive*, 152 NEW L.J. INFO. TECH. SUPPLEMENT 1597, 1597–99 (2002). Under European Community law, a directive requires the Member States to adopt their local law so as to comply with the general principles enumerated in the Directive.

137. The liability of ISPs for copyright infringement is defined in 17 U.S.C. § 512 (2000), and for any other content, in 47 U.S.C. § 230 (2000), notwithstanding the effect of criminal law, intellectual property law, state law, and the Electronic Communications Privacy Act of 1986. *See* 47 U.S.C. § 230(e) (2000).

138. E-Commerce Directive, *supra* note 135, Art. 12. Article 12 further clarifies that it refers to "automatic, intermediate and transient storage of information."

139. E-Commerce Directive, *supra* note 135, Art. 13.

140. E-Commerce Directive, *supra* note 135, Art. 14.

behalf of the ISP, and to the condition that once such knowledge is acquired, the content is removed. This is usually referred to as the "notice and take-down" principle. Whether an ISP can be deemed to have such constructive knowledge will depend on the circumstances; it may in some cases be blatantly obvious from the name of the Usenet group that it is to be used for obscenity or defamatory purposes.[141] Finally, the Directive requires that "Member States shall not impose a general obligation on providers... to monitor the information which they transmit or store, nor a general obligation actively to seek facts or circumstances indicating illegal activity."[142]

This is not a perfect system either, and it does impose costs on the ISPs, as well as risks. If the ISPs, upon receiving a notice of harmful content, take it down, and it later turns out that the material was perfectly legal, they face the risk of being sued by the user who posted the content at stake.[143] In the UK, the police had initially considered pursuing ISPs to combat the problems of pornography.[144] However, a balance seems to have been struck between increasing the expense of monitoring and risking liability, since now the Internet Watch Foundation ("IWF") notifies the ISPs of any unlawful material, and the ISP can then take the necessary steps to avoid liability.[145]

The overall result is one of a careful regulation that adopts an indirect public-ordering approach. It provides the ISPs with an option to choose immunity, so as not to act in a censorial mode, and thus protects both the ISPs' commercial and proprietary interests and, more importantly, does not interfere with the "marketplace of ideas" since the users are not affected. But, unlike the US regime regarding defamation, violations of privacy, or even negligence to remove child pornography,[146] this immunity is not absolute. Upon notice, the ISP

---

141. Gavin Sutter, *'Nothing New Under the Sun': Old Fears and New Media*, 8 INT'L J.L. & INFO. TECH. 338, 365–66 (2000).

142. E-Commerce Directive, *supra* note 134, Art. 15.

143. *See* Reuters' Report, *Europe's ISPs Overrun with Website Take-Down Orders* (Dec. 11, 2002), *available at* http://www.ispa.org.uk/html/media/coverage.html.

144. *See* Sutter, *supra* note 141, at 368–70.

145. The Internet Watch Foundation ("IWF") is one of the means of private ordering, which we discuss *infra*, text accompanying note 214. This is a clear example of the mixed European approach, of combining an indirect public ordering approach (rules pertaining to ISP liability) and private action (the IWF).

146. This is the immunity provided under 47 U.S.C. § 230 (2000). In one horrifying case, the Florida Supreme Court found that an ISP enjoyed immunity, despite its refusal to take down advertisements for the sale of videotapes and photographs posted in chat rooms depicting the rape of an 11-year-old child. The child's mother sued the ISP for negligently failing to act to remove the messages. The majority found the ISP to enjoy immunity, over the dissent of Justice

must act. In this manner, the cost imposed on it is minimized; it is cheaper to establish a system to receive notices and act upon their receipt than to build and maintain a monitoring system. The protected interest (privacy, reputation, etc.) is protected. The ISP need not act accordingly, and then it might risk a suit, in which it will be able to litigate the issue and raise "regular" defenses under the relevant cause of action.

Another form of public-ordering operates through the kind of policies at libraries and other areas where the public can access the Internet. In the UK, public libraries are run by local authorities (or library authorities) and have a duty "to provide a comprehensive and efficient library service for all persons desiring to make use thereof."[147] However, this duty does not extend to stocking pornography. Some materials stocked may be suitable for adults but deemed harmful to children. In this case, a restriction to prevent offense or any other harm being caused by stocked materials may not be out of step with the past practices of some libraries. In the past, libraries have been reported to have taken controversial books off the shelves but made them available upon request, for example with *The Satanic Verses*.[148] The problem discussed in this paper is distinct in that it is not practicable to request permission for access to every web site. The closest control is to provide filtering software or a system of ratings and to require permission before the software is to be disabled, which will be discussed below. Such measures would also raise the issues of privacy and the chilling effect considered by Etzioni.

D. Private-Ordering

A third regulatory regime avoids any public interference with the market, including the *pornography chain*. It leaves the regulation to the market. Obviously, this approach might not satisfy many. In the absence of regulation, some players will do nothing to prevent the distribution of harmful material, as well as illegal material, or children's access thereto. But quite a few will opt for this approach on

---

Lewis. *See* Doe v. Am. Online, Inc., 783 So. 2d 1010 (Fla. 2001) (applying *Zeran v. Am. Online, Inc.*, 129 F.3d 327 (4th Cir. 1997)).

147. Public Libraries and Museums Act, 1964, c.75, § 7(1). While this service is provided by local authorities, the Secretary of State for Culture Media and Sport superintends the system under Section 1. If a complaint is made that a library authority has failed to fulfill its duty, then the Secretary of State can hold an enquiry on the matter, after which an order declaring it in default and directing it to carry out its duties can be made.

148. Nicolette Jones, *For Your Eyes Only?*, TIMES (London), Jan. 7, 1999, at 36.

their own initiative, and the stronger players in the market are more likely to do so. The reason is simple: it is good for their business.[149] The ISPs, like any player in the market, prefer to take care of their own business and avoid external interference. ISPs know that in the absence of a serious effort on their behalf, the legislature or the courts are likely to fill the vacuum. In addition, the ISPs are interested (or should be interested) in providing a better service to their clients. If clients demand a "clean" environment, the wise ISP will do its best to supply it. A refusal of an ISP, for instance, to remove child pornography from its servers and block access to it might result in a public relations disaster.[150] Furthermore, as long as the legal climate is uncertain, ISPs fear they might be found jointly liable for the illegal acts of third parties.

Private-ordering can take many forms. Establishing a clear method of communication of users to the ISP for complaints of harmful or illegal material, *i.e.*, a "notice" system, followed by a "take-down" policy, is one example.[151] Adopting a clear code of conduct aimed at both the employees and the users, accompanied with sanctions, is another way. The "terms of use" can embody this code of conduct, and if formalities are met, they can be designed as a valid and enforceable contract. The contract can determine, for example, that a user who posts harmful material will be disconnected and his or her service terminated. The contract can further include a disclaimer that will immunize the ISP from breach of contract if it terminates an account in such circumstances. Another mechanism is one of labeling, or rating, which we will discuss shortly.

Private-ordering, or self-regulation, can also make use of technology. There are some technologies, and we are likely to see more in the future, that purport to "take care" of some of the concerns discussed here. Age-verification technologies, discussed earlier, are one example.[152] Filtering software is another interesting technology which we examine.

Self-regulation need not be left all to its own. Government can provide "background rules" either in the form of direct requirements (in which case it is no longer *self*-regulation), or a sanction for misrep-

---

149. Obviously, there might be other reasons as well, such as the moral views of the shareholders and executives of the ISPs.
150. *See, e.g.*, the sad case of *Doe v. Am. Online, Inc.*, 783 So. 2d at 1010.
151. *See supra* text accompanying note 140.
152. *See* Reno v. ACLU, 521 U.S. 844 (1997); *supra*, text accompanying note 19.

resentation,[153] in which case it is better to view this as an indirect public-ordering. But government can have an active role in supporting self-regulation, without direct or indirect interference. This is the European way thus far.

We begin by surveying the rating system and the filtering software, and then turn to survey the European measures undertaken thus far.

1. Ratings

The Platform for Internet Content Selection ("PICS") provides a technological method for rating the content of web sites. PICS is an industry standard that allows labels to be attached to web materials that can be read by a computer receiving the information. The technology differs from a filtering system which blocks sites that contain certain keywords.[154] Instead, PICS recognizes a predetermined label for the material. It has been used to develop ratings systems in which pages are given a label describing the content. Much depends on which system is used. Initial ratings were developed by the Recreational Software Advisory Council ("RSACi"), and its successor was launched in December 2000 by the Internet Content Rating Association ("ICRA").[155] This system is the most prominent and is built into both Internet Explorer and Netscape Navigator. The system places material into categories, including language, nudity, sex, and violence. The producers of the web pages then provide details of the content of their material in accordance with these categories.

The PICS standard was devised to create a means for material to be rated by content providers and by third parties for use by parents and teachers.[156] It allows the users to choose what sort of content they would like to see, and allows sites to be blocked according to those preferences. For example, if a viewer wishes to view violence but not nudity, then the software can be set in such a way. Those developing PICS recognized that a diversity of standards exists among Internet users and that access should be provided to a wide range of ratings

---

153. Compare to the U.S. Federal Trade Commission's power to prevent the use of "unfair methods of competition in or affecting commerce, and unfair or deceptive acts or practices in or affecting commerce." See 15 U.S.C. § 45(a)(1) (2000).
154. Harry Hochheiser, Computer Professionals for Social Responsibility, *Filtering FAQ*, http://www.cpsr.org/filters/faq.html#3.1 (Apr. 6, 2001).
155. See Internet Content Ratings Association, http://www.icra.org/.
156. World Wide Web Consortium, *PICS Statement of Principles*, http://www.w3.org/PICS/principles.html (last visited Nov. 3, 2003).

products.¹⁵⁷ In theory, PICS allows a method of preventing access to harmful material, but in a manner that is not imposed from a centralized source unrepresentative of any views as to what is harmful and what is not.

In practice, PICS raises some difficulties similar to filtering, which we will discuss shortly. PICS has been criticized, as one or two ratings systems dominate the market and create a uniform method of rating content that is built into Internet software.¹⁵⁸ Most users are likely to use whatever software comes with the browser, thereby undermining the commitment to diversity that is cited as one of PICS' strengths. Not every producer of material has the expertise or resources to provide a rating. However, some of the filtering software will automatically prevent access to sites that are unrated. Web content that goes unrated will thereby be harder to access, unless the user has the knowledge to remove the ratings software.¹⁵⁹ Consequently, large corporate sites that have the support and resources to self-rate may dominate the Internet. In addition, fears exist that rating will become the first step on a path to greater regulation—that when the software is found not to fulfill the high expectations of preventing access to harmful content, as seems inevitable, calls will be made for more direct censorship.¹⁶⁰ Furthermore, critics argue that it is impossible for all types of speech to neatly fall into the categories of rating.¹⁶¹ If a website about sexually transmitted diseases is labeled as sexual content, then it may be placed out of the reach of younger users. The last concern, but not the least, is who determines the standards. PICS envisions that software companies or website operators determine the rating. But these companies, benevolent as their motivations might be, are unelected and are not accountable. If PICS succeeds, the result is that the public has delegated its power to make moral judgments to technology designed by for-profit corporations.

2. Filtering Software

Filtering software purports to identify the content of on-line material and, accordingly, separate the "good" from the "bad." At the

---

157. World Wide Web Consortium, *Statement of the Intent and Use of PICS: Using PICS Well*, http://www.w3.org/TR/NOTE-PICS-Statement (June 1, 1998).
158. American Civil Liberties Union, *Fahrenheit 451.2: Is Cyberspace Burning?*, http://archive.aclu.org/issues/cyber/burning.html (Mar. 17, 2002).
159. *Id.*
160. *Id.*
161. *Id.*

heart of the software lie lists of URLs,[162] and IP addresses,[163] or an algorithm that reflects the choices and decisions of the code's designers. Today there are several commercial products available on the market, at an average annual cost of less than a $100. The software can be installed at various points, either on the user's computer or the ISP's servers. The technology aims at breaking the *pornography chain* just before it arrives at its destination: the minor user. The software, once installed, is supposed to block access to sites and other content it recognizes as harmful.

How is the algorithm composed? The various corporations engaging in this field adopt various methods.[164] Some employ parents who simply check websites and label them according to their content and according to the corporation's criteria; the label is then used to classify the website into "white lists" and "black lists." The algorithm reflects the lists. A second method is based on the software "reading" the web site, access to which is requested by the user. If the software identifies certain words (in the URL (the web address), the meta-tags, or the body of the website) such as "sex," it will block access. A third method is more sophisticated and is based on as many features as possible of pornographic web sites, such as the size of the letters, internal and external links, colors, text, etc. These are combined into a complex algorithm.[165] Of course, the various methods can be used in conjunction with each other.[166]

What are the exact criteria that guide each software-producer? The designers maintain this as a secret, or to be more precise, a trade secret.[167] This is not just a peculiar feature of intellectual property law.

---

162. A URL is a "Uniform Resource Locator," which is the "web address" of a web site, either numerically (the IP address) or a textual domain name. *See* Am. Library Ass'n, Inc. v. United States, 201 F. Supp. 2d 401, 417 (2002).

163. For an analysis of the limits of IP-based filtering software, see Benjamin Edelman, Berkman Center for Internet and Society, Harvard Law School, *Web Sites Sharing IP Addresses: Prevalence and Significance*, http://cyber.law.harvard.edu/people/edelman/ip-sharing/ (last updated Sept. 12, 2003).

164. For discussion of some of these methods, see *Am. Library Ass'n, Inc.*, 201 F. Supp. 2d at 427–36.

165. *See* iCognito, *Technology Overview*, http://www.puresight.com/technology/about.shtml (last visited Nov. 3, 2003).

166. *Am. Library Ass'n, Inc.*, 201 F. Supp. 2d at 427–36.

167. Trade secret law allows reverse engineering of the product, to reveal its "secret." However, reverse engineering of software inevitably requires its temporary copying. One US court found such a copying which was accompanied with a "bypass" code, to be copyright infringement. *See* Findings of Fact and Conclusions of Law at 11, Microsystems Software, Inc. v. Scandinavia Online AB (D. Mass. 2000) (No. 00-10485-EPH), *available at* http://www.epic.org/free_speech/censorware/cp_conclusions.html. Later on, the Librarian of Congress exempted this kind of reverse engineering from liability under the anti-circumvention

It is a matter of freedom of expression; it disables public supervision of the moral criteria chosen by unelected and unaccountable corporations. Indeed, the software producers can do so, and parents can buy the product. A parent is allowed to knowingly substitute his or her own moral reasoning and beliefs about education with the unknown choices of a corporation. The free speech problem arises when the software is installed in public institutions, such as public libraries. By installing filtering software, a public library in practice delegates its constitutional obligation not to interfere with freedom of speech, though it may itself be vulnerable to constitutional challenge.[168] This is an unfortunate event, especially in light of the many shortcomings of the software.

This problem is enhanced by the features of the industry at stake. There is a fierce competition among the producers of filtering software, and it seems that the competition drives them to block more rather than less. Another deficiency is that most of the programs are aimed at the American market, *i.e.*, at English content, and to the extent that moral judgments are made, they attempt to address the American taste.[169]

The current level of technology does not filter out 100 percent of pornographic content; it filters much more and, at the same time, much less. There is a long list of web sites which were erroneously blocked,[170] ranging from sites on breast cancer, sexual education, gays and lesbians, and planned parenthood, to sites of political organizations and candidates, a site which watches and criticizes the filtering software industry,[171] and even some sites containing legal documents

---

rules of the Digital Copyright Millennium Act, according to the authority given to him to do so under 17 U.S.C. § 1201(a)(1)(C) (2000). *See* Exemption to Prohibition on Circumvention of Copyright Protection Systems for Access Control Technologies, 65 Fed. Reg. 64556 (Oct. 27, 2000) (to be codified at 37 C.F.R. pt. 201).

168. Indeed, a US court found that a public library's decision to install filtering software runs afoul of the First Amendment. *See* Mainstream Loudoun v. Bd. of Trs. of the Loudoun County Library, 24 F. Supp. 2d 552, 570 (1998). An act that conditioned certain financial support to public libraries on installing filtering software has also been found unconstitutional, *Am. Library Ass'n v. United States*, 201 F. Supp. 2d 401 (E.D. Pa. 2002), but this decision was reversed by the Supreme Court, 123 S. Ct. 2297 (2003). *See* Children's Internet Protection Act ("CIPA") § 1712, Pub. L. No. 106-554, 114 Stat. 2763A-335, 2763A-340 (2000).

169. For a similar argument, see Carolyn Penfold, *The Online Services Amendment, Internet Content Filters, and User Empowerment*, 2000 NAT'L L. REV. (Austrl.), *at* http://pandora.nla.gov.au/parchive/2001/Z2001-Mar-13/web.nlr.com.au/nlr/HTML/Article/penfold2/penfold2.htm.

170. Some of these cases are mentioned in *Mainstream Loudoun*, 24 F. Supp. 2d at 556 n.2; *Am. Library Ass'n, Inc.*, 201 F. Supp. 2d at 446–47, and others are documented in a web site devoted to critically analyzing filtering software. *See* PeaceFire, http://www.peacefire.org/.

171. *See* Peacefire, *Blocking Software FAQ*, http://www.peacefire.org/info/blocking-software-faq.html (last visited Nov. 3, 2003).

and cases.[172] At the same time, independent surveys and studies found that the current software fails in filtering up to 20 percent of the pornographic sites.[173]

For the sake of the discussion, we are willing to assume that this is a transitional problem, and that technology will be developed that can block no more and no less than it is designed to. In many cases, this is a rather easy task which does not raise many constitutional difficulties. If an image of child pornography is blocked, neither an American nor a European court will object. But here lies the problem: in many cases it is unclear, and it cannot be made clear, in advance whether the content at stake is harmful to children or not. A court can decide so in retrospect, but what is the line between that which is harmful to a child and that which is not? Is an image of a naked person necessarily harmful? It might be a picture in a biology book, or a painting in a museum, or Michelangelo's statute of *David*. The answer lies in several factors that technology cannot "know" and cannot "understand": the context in which the content appears, the age of the user, the time and place, and the community's moral standards.[174] The conclusion is that the filtering software limits the "breathing space" which is so necessary for free speech. All of these problems are further enhanced by the inability to distinguish a child from a teenager from an adult, for whom the same content is not considered "harmful."

The problems caused by filtering can be illustrated by the experience in the UK. The UK government is committed to extending access to the Internet in schools and public libraries. No obligation exists in the UK for libraries to use filtering and rating software on public Internet stations. Instead, whether such guards are to be used is the decision of the local authority responsible for the library. The Library Association, in not endorsing filtering software, warns that it

---

172. Thus, for instance, a decision of the Israeli Supreme Court on gays' rights was blocked by a filtering program installed in the Hebrew University of Jerusalem. *See* Letter of Adv. Dori Spivak to Adv. Pappi Yakirevitch (July 19, 2000) (on file with the authors).

173. *See* Consumer Reports, *Digital Chaperones for Kids: Which Internet Filters Protect the Best? Which Get in the Way?*, March 2001, *available at* http://www.consumerreports.org/main/detail.jsp?CONTENT%3C%3Ecnt_id=18867& FOLDER%3C%3Efolder_id=18151&bmUID= 996766578117; Victoria Rideout et. al., Henry J. Kaiser Family Foundation, *See No Evil: How Internet Filters Affect the Search for Online Health Information* (Executive Summary), http://www.kff.org/content/2002/3294/Internet_Filtering_exec_summ.pdf (Dec. 2002).

174. The *Miller* test, applied in the US to define obscenity relies, *inter alia*, on "community standards." But once we go online, what is the relevant "community"? The Supreme Court struggled with this issue in *Ashcroft v. ACLU*, 535 U.S. 564 (2002), concluding that the statute's failure to define "community" does not in itself render it unconstitutional. *Id.* at 585–86.

is not always effective and may lead to a false sense of security that harmful material can no longer be accessed.[175] The Association further argues that "such software is inconsistent with the commitment or duty of a library or information service to provide all publicly available information in which its users claim legitimate interest."[176] Before allowing public access, libraries are encouraged to develop an "Acceptable Use Policy," which will determine who has access, what charges apply, whether it is filtered, and a code of conduct for users. This may include whether users will be required to have a user password or at least register their name at the time of use, either of which can raise privacy issues. A policy paper issued by the Networked Policy Task Group advises those managing libraries on the pros and cons of filtering, and while not coming to any specific recommendation, emphasizes the need to have a policy in place.[177] While the government has not legislated or made funding conditional on installing filtering software, information is provided to schools and on-line centers to help decide what measures are most suitable to limit harmful Internet content.[178] Such guidance accepts that some people may find a "culture of responsible use amongst their adult users is preferable to software filtering," and directs centers to consider the flexibility of a system for different-aged users before fitting filtering software.[179]

Much is therefore left to the local authorities to decide what approach to take. The different approaches taken in various areas have been shaped by a process of trial and error, rather than through courtroom battles as in the US. For example, Gloucester Council initially required written parental consent for children to gain unfiltered access to the Internet in a public library, as filtering software was thought to block out legitimate sites.[180] However, this was argued to be inadequate, as parents do not always know what sites their children will access once on-line.[181] Consequently, the policy was changed so that children were allowed access only to computers fitted with a

---

175. Library Association, *Guidance Notes on the Use of Filtering Software in Libraries*, http://www.la-hq.org.uk/directory/prof_issues/filter2.html (2000).
176. Library Association, *The Use of Filtering Software in Libraries*, http://www.la-hq.org.uk/directory/prof_issues/filter.html (1999).
177. Sara Ormes, *An Introduction to Filtering*, http://www.ukoln.ac.uk/public/earl/issuepapers/filtering.html (last visited Nov. 25, 2003).
178. *See, e.g.*, Superhighway Safety, http://www.saftey.ngfl.gov.uk/.
179. Superhighway Safety, *Internet Filtering Systems*, http://safety.ngfl.gov.uk/ukonline/pdf/d3.pdf (last visited Nov. 3, 2003).
180. GLOUCESTER ECHO, Apr. 30, 2002.
181. *Id.*

filter, with a number of unfiltered machines being left aside for adult use only.[182] Other libraries are more stringent and fit all computers with a filter, allowing teenagers and children under sixteen access to the machines only with parental permission.[183] On one occasion, a local authority in Glasgow shut down all access to the Internet temporarily after a school child gained access to pornography in a public library.[184] Prior to this, filters on all computers had been scrapped as they were found to be blocking legitimate sites.[185]

While the policies of these libraries may not have been met with same legal challenges as were libraries in the US, that is not to say that all attempts to protect children have been met with approval. Complaints have been made that blocking software prevents access to certain political and religious sites.[186] In one instance, a local authority was threatened with legal action after the filtering software blocked the far right British National Party web site, and later allowed the site to be accessed.[187] Such complaints are supported by concerns voiced in a recent study that that filtering software can block access to important information on health issues.[188]

The lesson is that technology can be an aid in shielding children from on-line pornography, but it has a substantial social cost: it means that, in essence, we desert the educational avenue; it means we delegate our moral judgments as moral agents and as a society at large to technology, a technology that is quite resistant to inspection; and it means we pay a price in terms of free speech. At the end of the day, technology cannot, and should not, substitute for human judgments.

3. The European Action Plan

The institutions of the European Union started looking into the matter of protecting children from harmful material available in the

---

182. Gloucester Council fitted eighty of its 200 library computers with filters following a complaint from a mother that witnessed teenagers accessing indecent images in a library. GLOUCESTER ECHO, Nov. 21, 2002.

183. For the Manchester Library Policy, see Manchester City Council, *Using the Internet*, http://www.manchester.gov.uk/libraries/ict/internet.htm (last updated Jan. 20, 2003).

184. Graeme Murray, *Schools in New Bid to Block Web Porn*, EVENING TIMES (Glasgow), Sept. 16, 2002, at 2 (referring to the East Ayrshire Council).

185. Gerry Braiden, *New Rules to Protect Scots Pupils from Net Perverts: Youngsters Will Not Be Given E-Mail Names*, EVENING TIMES (Glasgow), Mar. 21, 2001, at 5.

186. *See Censorship Concern*, ESSEX CHRON., Nov. 15, 2002, at 2.

187. *Around Wales: Monmouthshire*, WEST. MAIL, Oct. 29, 2001, at 6.

188. Rideout et. al., *supra* note 173.

digital environment in the mid-1990s.[189] From the very beginning, it was determined that the interest in protecting children was an issue of "overriding public interest."[190] In 1996 a comprehensive study, the *Green Paper*, was published, and set the agenda:

> The full potential of such developments will depend on society as a whole striking the right balance between freedom of speech and public interest considerations, between policies designed to foster the emergence of new services and the need to ensure that the opportunities they create are not abused by the few at the expense of the many.[191]

The goal was carefully crafted so not to cover illegal materials.[192] It was further observed that adults have a different interest than children: "[t]he aim is therefore limited to preventing minors from encountering, by accident or otherwise, material that might affect their physical and/or mental development."[193] Interestingly, the document also raises the issue of variance between minors of various ages.[194] These principles were expressed based on the assumption that the digital environment carries with it many advantages, and that the new technology requires a different treatment than the old media.[195] Accordingly, the *Green Paper* proposed that different solutions are adopted for these different problems.

The possible negative effect on freedom of expression was also recognized, as the above quoted passage illustrates.[196] The *Green Paper* instructed that any regulation should take it into account under the criteria set forth in the ECHR.

Private-ordering was the option advocated by the *Green Paper* (termed there self-regulation), and was then adopted by the Euro-

---

189. *See* European Union Communication on Illegal and Harmful Material, COM(96) 487 (proposing policy options for immediate actions); Telecommunications Council of Ministers' resolution concerning dissemination of illegal material over the Internet (Sep. 1996), *available at* http://europa.eu.int/ISPO/legal/en/internet/98-97en.html.

190. *See* Green Paper, *supra* note 22, at 1.

191. *Id.*

192. *Id.* at 6, ch. I, § 1. The illegal material was classified as a "general category of material that violates human dignity," and includes primarily "child pornography, extreme gratuitous violence and incitement to racial and other hatred, discrimination, and violence." To combat child pornography, for example, the EU established police hotlines for users to complain, special police units, a system of cooperation among the Member States, and several international operations took place.

193. *Id.*

194. *Id.* at 19, ch. II, § 2.2.2 ("... it is doubtful whether children of four have the same problems as adolescents of 15").

195. For the comparison of the new media to the old media, *see id.* at 7–11, ch. I, §§ 2–2.5.

196. *See id.* at 13, ch. II, §§ 1–1.1.

pean Commission and the European Parliament.[197] But this approach should not be mistaken for a liberal hands-off attitude. Rather, the underlying but clear direction was that self-regulation can be assisted by the State. The goal was set accordingly, to create a common framework for self-regulation,[198] and national frameworks within the Member States.[199] The idea was to foster a climate of confidence with the relevant industries.[200]

Self-regulation was to be achieved in cooperation with the industries. Some of the measures mentioned were drafting codes of conduct and identifying areas where there might be a need for common standards of labeling material.[201] In addition, it was suggested to raise awareness of users, especially of parents. Codes of conduct, as later EU documents elaborated, should inform users of any risks from the content, provide a warning page, visual signal, or a sound signal, have a descriptive labeling or classification of content, and apply a system to check the users' age, support parental control measures, and handle complaints.[202] In most EU countries there are now operative codes of conduct.[203]

These recommendations were followed up.[204] The most ambitious project undertaken was the multi-annual, 25-million-Euro *Safer Internet Action Plan*, which took place between 1999 and 2002.[205] The

---

197. *See* Council Recommendation, *supra* note 22, at 50–51 (detailing the various measures taken following the Green Paper).
198. *See* Green Paper, *supra* note 22, at 24–25, ch. III, § 3.1.
199. *See* Council Recommendation, *supra* note 22, at 50, Art. I (1).
200. *See id.*, recital 10, at 49.
201. *See* Green Paper, *supra* note 22, at 25, ch. III, § 3.2.
202. *See* Council Recommendation, *supra* note 22, at 52–55.
203. *See* Evaluation Report From the Commission to the Council and the European Parliament on the Application of Council Recommendation of 24 Sept. 1998 Concerning the Protection of Minors and Human Dignity, COM(01)106 final at 5–6, *available at* http://europa.eu.int/comm/avpolicy/regul/new_srv/ermin_en.pdf; *see also* Council Conclusions of 23 July 2001 on the Commission's Evaluation Report on the Application of the Recommendation Concerning the Protection of Minors and Human Dignity, 2001 O.J. (C 213) 10, 11, *available at* http://europa.eu.int/eur-lex/pri/en/oj/dat/2001/c_213/c_21320010731en00100011.pdf; Communication from the Commission to the Council, the European Parliament, the Economic and Social Committee and the Committee of the Regions: Intermediate Evaluation of the Implementation of the Multiannual Community Action Plan on Promoting Safer Use of the Internet by Combating Illegal and Harmful Content on Global Networks, COM(01)690 final at 2–7, *available at* http://europa.eu.int/eur-lex/en/com/cnc/2001/com2001_0690en01.pdf.
204. *See* Council Conclusions of 17 Dec. 1999 on the Protection of Minors in Light of the Development of Digital Audiovisual Services, Art. 9, 2000 O.J. (C 8) 8, 9, *available at* http://europa.eu.int/eur-lex/pri/en/oj/dat/2000/c_008/c_00820000112en00080009.pdf.
205. *See* Decision No. 276/1999/EC of the European Parliament and of the Council of 25 Jan. 1999 Adopting a Multiannual Community Action Plan on Promoting Safer Use of the Internet by Combating Illegal and Harmful Content on Global Networks, Art. 1, 1999 O.J. (L

European Union supported dozens of projects designed to create the common and national framework for self-regulation, focusing on three avenues: (1) *Hotlines*: creating a European network of hotlines to report illegal material. The projects resulted in a group of fourteen countries (including some non-European countries, such as the US and Australia) that have hotlines grouped together into a cooperative system, known as INHOPE.[206] (2) *Rating and Filtering*: several projects examined technological solutions. The intermediate assessment was that self-labeling and filtering schemes were not a practical solution for Europeans, at least in the year 2000, and that third-party filtering software products require major improvements.[207] One of the problems identified was the English language focus of most current filtering technology.[208] (3) *Awareness*: various projects to inform users of the risks and chances online. Furthermore, associations of ISPs were established throughout the EU, including in a pan-European association.[209] The Action Plan has been considered by the EU to be a success, and it is considering extending it for two more years.[210]

The UK government follows the European preference for self-regulation of the Internet. While the government has proposed the creation of a new criminal offense of "sexual grooming" on the Internet, for which individuals can be prosecuted before any sexual act has taken place,[211] it also launched the Internet Taskforce on Child Protection in 2001 to review Internet content rating systems, develop a "kitemarking" scheme for chat rooms, and promote "safe surfing"

---

33) 1, 3, *available at* http://europa.eu.int/eur-lex/pri/en/oj/dat/1999/l_033/l_03319990206en 00010011.pdf; *see also* Safer Internet, http://www.saferinternet.org.

206. *See* http://www.inhope.org.

207. *See* the European Union's announcement, Commission Issues Reports on Parental Control Technologies Aimed at Enhancing Safety of Internet, IP/00/621 (Brussels, 15 June 2000), *available at* http://europa.eu.int/ISPO/docs/services/docs/2000/June/ip_00_621_en.pdf.

208. Accordingly, one of the projects funded by the E.U. Action Plan aims at developing a multi-lingual filtering tool. *See* http://www.net-protect.org/en/scope.htm.

209. *See* European Internet Services Providers Association, http://www.euroispa.org.

210. *See* Follow-up to the Multiannual Community Action Plan on Promoting Safer Use of the Internet by Combating Illegal and Harmful Content on Global Networks, COM(02)152 final at 3, *available at* http://europa.eu.int/eur-lex/en/com/pdf/2002/en_502PC0152.pdf. The proposal is to extend the action plan also to new technologies that have been developed in the meantime, such as mobile and broadband content, peer-to-peer file sharing systems, chat rooms, instant messaging and more.

211. In one area, calls were even being made to prevent children from using Internet chat rooms in public libraries, following concerns that two murdered girls had possibly been "groomed" through a chat room. *See Chatroom Ban in Bid to Cut Abuse*, KENT & SUSSEX COURIER, Aug. 30, 2002, at 5. The use of chat rooms in this way shows a new problem caused by the Internet, as such text would not always infringe the Obscene Publications Act. *See* Akdeniz & Srossen, *supra* note 38, at 223.

education and awareness for parents and children.[212] The UK government also recently launched a new campaign warning of the dangers to children from using chat rooms, publishing a code of practice for operators of chat rooms developed jointly by the government, Internet bodies, and child protection agencies.[213]

The Internet Watch Foundation ("IWF") was established in 1996 with the support of the UK government, although it is not a government agency. The IWF operates a hotline to which Internet users can report illegal material, which is then passed on to the police, and which leads to ISPs being issued a notice to take down the illegal material.[214] The fear exists that if this means removing newsgroups that contain some illegal material,[215] this could also remove much legal content along with it. The organization promotes the use of voluntary ratings systems and filtering systems among parents, teachers, and others responsible for children. The IWF also seeks to educate users about the dangers on the Internet, especially for children, and ways of dealing with such problems. The division of functions in this way is argued to reflect the division between illegal and harmful material, promoting merely educational and voluntary measures in relation to the latter.[216]

---

212. *See* Improving Child Protection on the Internet: A Partnership for Action, http://www.homeoffice.gov.uk/docs/childprotnet.pdf (Mar. 28, 2001).

213. *See* http://www.thinkuknow.co.uk. The Code of Practice suggests chat room operators should include a virtual panic buttons and safety messages for child users. *See* Stuart Millar, *Chat Room Danger Prompts New Safety Code*, GUARDIAN, Jan. 6, 2003, *available at* http://www.guardian.co.uk/uk_news/story/0,3604,869385,00.html.

214. In 2001, the hotline processed 11,357 reports, "leading to notices to UK ISPs to take down 3332 web sites and newsgroup articles containing images of child abuse and 2949 reports to police for investigation." *See* Internet Watch Foundation, 2001 Annual Review (2002), http://www.iwf.org.uk/about/annual_report/ar2002/css/ar2002_2.htm [hereinafter IWF 2001 Annual Review]. The Hotline has been criticized in previous years on the basis that the figures
> tell us little as the actual amount of child pornography on the Internet is unknown. It is, therefore, difficult to judge how successful the UK hotline has been. Another downside is that the efforts of the organisation are concentrated on the newsgroups carried by the UK ISPs. This means that while illegal material is removed from the UK ISPs servers, the same material will continue to be available on the Internet carried by the foreign ISPs in their own servers.

Cyber-Rights and Cyber-Liberties (UK), *Who Watches the Watchmen Part II: Accountability & Effective Self-Regulation in the Information Age*, http://www.cyber-rights.org/watchmen-ii.htm (Sept. 1998) [hereinafter *Who Watches the Watchmen*].

215. The IWF recommends to ISPs not to host newsgroups that regularly contain child pornography or newsgroups that appear to advertise pedophile content or activity. *See* Internet Watch Foundation, *IWF Tightens Net on Child Abuse*, http://www.iwf.org.uk/news/press/detail_press.epl?INFO_ID=106 (Feb. 13, 2002); Internet Watch Foundation, *National Crime Squad Raids Confirm Newsgroup Strategy*, http://www.iwf.org.uk/news/press/detail_press.epl?INFO_ID=102 (Nov. 28, 2001).

216. Sutter, *supra* note 141, at 370–71.

While the IWF is an independent body, it works closely with the government and in the 2001–02 financial year, it received state funds.[217] It also liaises with other regulators, such as the Broadcasting Standards Commission and the British Board of Film Classification, and with industry, such as the Internet Service Providers Association and the London Internet Exchange. In addition to this, the organization has frequent exchanges with children's charities, and has attempted to work internationally with groups on similar projects. Concern was initially expressed about the lack of involvement from civil liberties groups,[218] though later the Board was reconstituted to gain more balance. While the IWF has been seeking a broader base of subscribers, most of its funding has come from ISPs,[219] and it is still very much linked to the industry.

The IWF has been criticized on the grounds that it operates as a regulatory body providing a public function involved in the development of government policy, but with the status of a private body.[220] This function will remain in the hands of the IWF even after the establishment of a new communications regulator.[221] This is indeed one of the constitutional difficulties with a legal regime of private-ordering—the private bodies are, depending on the extent of horizontal effect, beyond the reach of constitutional law.[222] Consequently, it has been argued that the body has avoided many of its public responsibilities, such as providing open information on why certain policies such as filtering are to be preferred, without proper public consultation or the normal channels of accountability.[223] In advising whether material is illegal, the IWF can be argued to be performing a quasi-

---

217. *See* IWF 2001 Annual Review, *supra* note 213.

218. Akdeniz & Strossen, *supra* note 38, at 224–25.

219. *See* IWF 2001 Annual Review, *supra* note 213. However, it has also received grants from the European Commission.

220. *Who Watches the Watchmen*, *supra* note 214. For similar criticisms, see Cyber-Rights and Cyber-Liberties (UK), *Memorandum for the Internet Content Summit 1999*, http://www.cyber-rights.org/reports/summit99.htm (Sept. 1999).

221. The initial remit of the new Office of Communications ("OFCOM"), established in the Office of Communications Act of 2002, does not include regulation of the Internet. Instead, OFCOM will support and promote the existing mechanisms of tackling illegal materials, in particular working with the IWF. *See* Department of Trade and Industry, *Communications White Paper*, ¶ 6.10, *available at* http://www.communicationswhitepaper.gov.uk/pdf/ch6.pdf (last visited Nov. 25, 2003).

222. Even if the IWF was deemed to be a public authority for the purposes of Section 6 of the Human Rights Act, a constitutional challenge would be harder to sustain given that the IWF's role is largely advisory rather than providing a direct restraint.

223. Cyber-Rights and Cyber-Liberties (UK), *Who Watches the Watchmen: Internet Content Rating Systems, and Privatised Censorship*, http://www.cyber-rights.org/watchmen.htm (Nov. 1997). However, steps have been taken to improve the situation.

judicial function that draws the boundary between illegal and harmful content. A further concern is that, by encouraging systems that go beyond the realm of the illegal material, and seeking to restrict harmful content through the promotion of filters and ratings, the body will step into the realms of censoring legitimate speech in the name of protecting children. Critics have argued that it would be preferable to have political action by a democratically elected and accountable government rather than "random censorship by law enforcement authorities or by self regulatory bodies."[224] A defense can be made that much of the IWF's work concerns child pornography, a subject about which there is an overall consensus and little in the way of legal ambiguity.[225] Furthermore the IWF enjoys the support of the industry in that it helps shield ISPs from potential liability by pointing out the illegal material that should be removed, without the ISPs having to provide resources to monitor and handle complaints.[226]

Conclusion

Digital technology crosses borders, and brings with it promises for empowerment and democratization, but in the short history of the Internet, it is also clear that pornography is global. In this Article, we looked at the European approach to the harms caused to children by the easy availability of such material online.

While the problem is not unique to Europe, thus far, European countries have opted for a very different approach than that opted for in the US. While the US attempted, and failed, the direct public-ordering approach—the 1996 CDA and its replacement, COPA—Europe chose a mixed approach of indirect public-ordering and private-ordering. Broadly speaking, this approach better fits Etzioni's analysis than the US approach.

There might be many political, sociological, and other explanations for this difference in approach. By way of conclusion, we wish to propose that the legal environment might be one such explanation: freedom of speech is recognized in Europe and protected, but its structure enables balancing it with other public interests. In light of this approach, the puzzle is even greater: European versions of the US CDA or COPA are far more likely to survive scrutiny in Europe

---

224. Akdeniz & Strossen, *supra* note 38, at 224.
225. *See* Edwards, *supra* note 14, at 296.
226. *Id.*

than in the US. Nevertheless, this path was not taken. We want to speculate that the reason for the different approaches can be found here: The strong protection that the First Amendment provides in the US does not leave much room for other interests, or balancing rights and interests, and the constitutional clash between freedom of speech and the interest in protecting children from harmful material is inevitable.

In Europe, the initial point is different. It is one of compromise and balancing. This renders the clash between the rival interests and rights less charged, at least on a legal-constitutional level. Once the rival interests are on a par, normatively speaking, there is a more relaxed atmosphere to devise a delicate balance. Private-ordering, which minimizes governmental interference, is such a solution.

However, this is not to say that the issue of protecting children on-line is less important in Europe than it is in the US. The newspapers are full of stories of the dangers to children that lurk on the Internet. In spite of the more relaxed legal environment, the European approach has taken note of the difficulties in US public-ordering and attempts to avoid such conflicts or embarrassments. Imposing direct public-ordering may threaten to create an issue that will polarize competing interest groups and lobbies, and thereby undermine the more balanced approach to expression rights that has been described. Furthermore, the difficulties in defining and enforcing direct public controls provide another deterrent. Regulating with the cooperation of the Internet industry is more likely to be practically administered and less divisive.

This form of regulation does not come without difficulties. It is unlikely to prevent children from accessing all material deemed harmful. Furthermore, it has the potential to act as censorship, but in a more subtle form. The actions of those deciding what types of material is harmful is less likely to fall foul of the schemes of constitutional protection. The less charged environment in Europe may also permit this type of approach. While fewer groups demand outright prohibition of harmful material, fewer groups also cry foul when such private organizations attempt to act as censors. The approach in Europe may be less divisive and high profile than the legislative attempts in the US, but it nevertheless carries costs.

Justice Louis Brandeis once noted, "[i]t is one of the happy incidents of the federal system that a single courageous State may, if its citizens choose, serve as a laboratory; and try novel social and economic experiments without risk to the rest of the country."[227] Nowadays, culture, economy, and politics are more global than ever before, the technology of the Internet is borderless, but principles of political morality are still universal, even if their concrete application is local. Europe and the United States can each serve as a laboratory to each other.

---

227. New State Ice Co. v. Liebmann, 285 U.S. 262, 311 (1932).

Part III
Structural Regulation: Media Concentration and Ownership

[7]

Rethinking European Union competence in the field of media ownership: the internal market, fundamental rights and European citizenship

Rachael Craufurd Smith*

> *Though the European Parliament has long pressed for a Directive to limit media ownership, European Commission attempts during the nineties to formulate such a measure stalled owing to Member State opposition and disagreement within the European Commission itself. Since then, media concentration has continued unabated and the case for European intervention remains strong, particularly given the role of the media in keeping European citizens informed. This article considers why, with many new media services and modes of distribution, it remains necessary to place limits on media ownership and examines the various bases for European Union action. It concludes that there is considerable potential for the European Union to support and supplement Member State initiatives in relation to media ownership and media pluralism more generally. Given the previous impasse and Member State sensitivities, consideration should be given to using a range of different regulatory techniques, in particular framework directives and the open method of co-ordination. A focused and integrated approach is clearly called for, but this may prove difficult at the European level because of the fragmented nature of the European Union institutions.*

Introduction

A brief survey of the media in the various European Union Member States might lead one to conclude that, in terms of access to information, Europeans have "never had it so good". Digital stations now supplement traditional analogue television and radio channels, cable and satellite services compete with terrestrial stations, individual access to the information cornucopia of the World Wide Web is rapidly expanding. These developments bring into question the continuing need for media ownership controls of the type found in many Member States. Though some states principally rely on competition law to control excessive media concentrations, many retain complex rules which, *inter alia*, prohibit certain organisations (political, religious etc.) from owning particular media interests; impose ownership limits in specific media sectors; or limit the accumulation of interests across two or more media sectors.[1] The main rationale put forward for these

---

* Jean Monnet Fellow, European University Institute; Senior Lecturer in Law, University of Edinburgh. I would like to thank Anja Wichmann for her help in the preparation of this article. This paper was completed in March 2004 but has been updated to include references to the European Parliament Resolution of April 22, 2004 *on the risks of violations, in the EU and especially in Italy, of freedom of expression and information*, P5_ TA-PROV(2004) 0373, detailed at n.48 below.

[1] For an overview see P. A. Bruck, D. Dörr, J. Favre, S. Gramstad, R. Monaco and Z. P. Culek, *Media Diversity in Europe*, (H/APMD (2003) 001), report prepared by the Council of Europe Advisory Panel on media diversity available at: *www.coe.int/t/e/human_rights/media/5_Documentary_Resources/*. The use of terminology in this area can be confusing, with "pluralism" or "external pluralism" frequently used when discussing media ownership and "diversity" or "internal pluralism" when considering content diversity. For

restrictions is a democratic one, namely, that for citizens to play a meaningful and active role in the political process they require access to varied information and opinions from a diversity of sources. The reflection by the media of diverse views and opinions is also considered to have cultural benefits: facilitating integration in multicultural societies and helping to preserve and promote cultural diversity.

But does fragmentation in media ownership, at least the fairly minimal level of fragmentation required in many Member States of the European Union, offer any guarantee that the media will in fact provide such diverse information? And can we not now rely on the market to provide the diversity we require? The phenomenon known as the "Hotelling effect" after the mathematician Howard Hotelling who described it, suggests that where a limited number of companies share a market between them they will tend to target the same middle ground in order to maximise sales, providing a relatively homogenous range of services.[2] According to Hotelling's theory as new firms enter the market and competition becomes increasingly intense a point is reached at which companies start to distance themselves from the mainstream by offering goods or services which appeal to niche customers, thereby avoiding damaging price competition. This indicates that, as the number of television and radio channels increases, the range of programmes on offer will begin to diversify.[3]

The application of this theory to the audio-visual sector is, however, complicated by the existence of different sources of funding—advertising, subscription, and licence fees —leading to a number of distinct markets operating, and interacting, side by side.[4] Nor does economic logic dictate that an increase in quantitative choice will be matched by an increase in substantive choice, it may instead lead to greater programme duplication.[5] Whether certain programmes are broadcast or not will depend on a range of factors, including viewer preferences, viewer profiles and affluence, method of channel finance, the level of market competition, as well as distribution and programme costs. Those seeking to offer "niche" services may find that there is insufficient consumer or advertiser demand to cover programme costs. Their ability to provide such services may also be affected by competition from the public sector, raising a nice question whether market

---

further analysis see D. Westphal, "Media Pluralism and European Regulation" [2002] E.B.L.R. 459, at p.476, and UK Department for Culture, Media, and Sport ("DCMS"), *Consultation on Media Ownership Rules* (2001), para.1.3, available at:*www.culture.gov.uk/broadcasting/QuickLinks/consultations/default.htm?properties=2001+closed+consultations%2C%2C*.

[2] H. Hotelling, "Stability in Competition" (1929) 39/153 *Economic Journal* 41.

[3] Jack Beebe suggests that under certain circumstances an increase in competition *may* lead to greater programme diversity: "Institutional Structure and Program Choice in Television Markets" (1977) 91/1 *Quarterly Journal of Economics* 15. A helpful overview of the economic literature on this topic is provided by B. C. Cunningham and P. J. Alexander, "A Theory of Broadcast Media Concentration and Commercial Advertising", FCC Staff Research Paper, Sept. 2002, available at: *www.fcc.gov/ownership/studies.html*.

[4] See the report prepared for the European Commission by Bird and Bird, *Market Definition in the Media Sector, Comparative Legal Analysis*, December 2002, available at: *http://europa.eu.int/comm/competition/publications/studies/legal_analysis.pdf*. Consumer willingness to watch second or third choice programmes rather than engage in other activities is also a complicating factor, on which see Beebe, n.3 above.

[5] Esther Gal-Or and Anthony Dukes suggest that in advertiser funded competitive media markets there are inbuilt incentives which lead to minimum differentiation in programming: "Minimum Differentiation in Commercial Media Markets" (2003) 12/3 *Journal of Economics and Management Strategy* 291. See also K. S. Brown, and R. J. Cavazos, "Empirical Aspects of Advertiser Preferences and Program Content of Network Television" FCC Media Bureau Staff Research Paper, December 2003, available at: *www.fcc.gov/mb/*.

654 Rethinking European Union competence

failure necessitates public intervention or whether public intervention is itself a major cause of market failure.[6]

Many commercial thematic television services cluster around popular programme genres, and even the more specialist channels, such as film or sports channels, still aim for a wide, though more targeted, audience.[7] There is, however, some evidence from America that particular pay television stations are deliberately "counter-programming" to the advertiser-funded networks, offering darker and more subtle programmes which are valued by audiences but not advertisers.[8] But even these stations seek to serve a large number of viewers and may find it difficult to move beyond the more popular programme formats.[9] Despite the potential increase in consumer satisfaction, there thus remains the risk that commercial services will fail to provide, or provide in sufficient quantity, the challenging, informative, and minority interest programmes which for social and democratic reasons we expect our audiovisual media to transmit.[10] This is one reason why Member States continue to subsidise the production of public interest radio and television programmes.[11]

But even if one does accept a link between the addition of new television or radio stations and the development of specialist services it does not follow that media ownership regulations are now redundant. This is because there is more than one possible rationale for controlling media ownership, among which the containment of private power is arguably as, if not more, compelling than the contribution such rules make on a day-to-day basis to content diversity.[12] In consequence, the case for state intervention is not inevitably tied to the degree of diversity existing at any given time in the print or audiovisual media. Nor does there appear to be any simple correlation between the number of owners in a given media market and content diversity. United States attempts during the 1980s, for example, to promote minority ownership in the radio and television sectors did not significantly enhance programme diversity, largely because of the "flattening and homogenising effect of the commercial market place".[13] Indeed, it is sometimes argued that diverse programming is *more* likely to result from concentration in, as opposed to fragmentation of, ownership. This is because large commercial companies with significant presence in the market can more easily bear the risk of innovation and cross-subsidise experimental or specialist channels from the profits derived from their popular services.[14]

---

[6] For further discussion see the report of the Independent Review Panel (chair: Gavin Davies), *The Future Funding of the BBC* (DCMS, London, 1999), particularly at p.136.

[7] Enhanced access to foreign stations may, however, prove valuable to ethnic or minority groups not well served by domestic channels.

[8] For discussion of HBO's programme strategy and the biases towards light and unchallenging programming created by advertising see Brown and Cavazos, n.5 above.

[9] *ibid.* pp.16–17.

[10] See Report of the Independent Review Panel, n.6 above, pp.206–208. Programmes which challenge widely shared beliefs, are unsettling, or question certain industry practices are not obviously attractive vehicles for advertising, whatever the level of market competition.

[11] Though see observation at text accompanying n.6 above.

[12] See DCMS (2001), n.1 above, at para.1.7.

[13] M. Price and J. Weinberg, "The Telecommunications Act of 1996 and US Media Ownership" (1996) *The Yearbook of Media and Entertainment Law* 99, p.105.

[14] This argument was put forward by the petitioners in *Sinclair Broadcasting Group Inc v FCC*, 284 F.3d 148 (D.C. Cir.2002), at III, on which see text accompanying n.89 below. See also, in relation to limited markets, P. O. Steiner, "Programme Patterns and Preferences, and the Workability of Competition in Radio Broadcasting" (1952) 66/2 *Quarterly Journal of Economics* 194, and Beebe, n.3 above.

Whether companies will do so without external "encouragement" is, of course, a different matter: research into the impact of ownership deregulation in the US radio market suggests that firms with significant market presence, rather than address the full range of niche markets, may instead "super-serve" a median audience of young adults.[15]

Media ownership controls remain necessary because the media industry is unlike any other, in that it has a unique capacity to promote or damage those goods, services, persons, or organisations that come within its sights. Whether particular events are reported or not, and how they are reported, affect how people perceive the world around them, it can determine whether they turn out to vote in national or regional elections, and their level of participation in the society around them.[16] The owner of a car factory will be concerned solely to promote the sale of his or her company's cars; the fact that that person has interests in other industries or supports a particular political party will not affect the management of the car company or the appearance of the cars. Although most media owners are also concerned to maximise profits, which, as indicated above, poses its own particular challenge for media diversity, they may also wish to use the power of the media to further other interests. In certain contexts this may be perfectly acceptable, even desirable, but it can also lead to important information being suppressed, to a lack of rigour in the face of official information, or a blurring of the line between news and opinion. The damaging consequences of bias or restricted coverage are likely to increase where media ownership is concentrated with only one or two potential countervailing voices. Media ownership controls should not, therefore, slip from the regulator's sights simply because we now live in a communications environment which offers multiple channels and alternative means of delivery.

Reviewing the current situation in Europe, the Parliamentary Assembly of the Council of Europe has concluded that media concentration is "a serious problem across the continent".[17] It expressed particular concern at the levels of concentration in certain Central and East European countries where a small number of companies control the printed press.[18] Nor does the internet offer a simple solution to these problems since it, too, is subject to the forces of consolidation. At a public hearing in January 2003, US Federal Communications Commissioner Michael Copps noted that most of the top 20 internet news sites were controlled by the same companies which provide television and newspaper news.[19] These developments undoubtedly pose complex regulatory questions for European states but do they also necessitate action on the part of the European Union and, if so, what form should this take? This article sets out to explore these issues by considering, first, why media ownership should not be left solely in the hands of the Member States and, secondly, what steps the European Union can take to either support or supplement domestic regulation in this area.

---

[15] DCMS, n.1 above, para.1.5.
[16] J. Waldfogel, "Consumer Substitution among Media", FCC Media Ownership Working Group Paper, September 2002, p.10, available at: *www.fcc.gov/ownership/studies.html*.
[17] Parliamentary Assembly of the Council of Europe, Recommendation 1589 (2003) *On Freedom of Expression in the Media in Europe*, available at: *http/assembly.coe.int/*.
[18] *ibid.* para.13.
[19] M. J. Copps, introductory remarks at Columbia Law School Forum on Media Ownership, January 16, 2003. "Is the Internet", he asked, "really giving us new voices, or just recycling existing ones?"

656 Rethinking European Union competence

Is media ownership a matter solely for the Member States?

One of the main objections to Community intervention in the field of media ownership is that media pluralism can be realised perfectly adequately by the Member States acting alone. Action at the European level, it is argued, offers no real advantages and could even lead to a reduction in controls in certain countries. National competence in this area consequently accords with the principle of subsidiarity established in Art.5 of the EC Treaty. The case for national competence certainly seems compelling when one considers the very different press and broadcasting traditions in the Member States and that markets for media services vary significantly from one country to another depending on factors such as language use, population size and relative affluence. These factors greatly complicated the Commission's attempt during the course of the 1990s to establish a single regulatory template for media ownership applicable in all member countries.[20] In consequence, the Commission introduced into its draft a "flexibility clause" which would have allowed states to exceed the set thresholds provided they adopted alternative compensatory measures.[21] Though properly recognising the need to take into account the disparity in market conditions across the EU, the flexibility provision threatened to undermine the project from within.

Member States, by contrast, can tailor measures to their own domestic markets and have a wide range of regulatory resources at their disposal. Moreover, the setting of ownership limits is not an exact science and involves drawing a number of discretionary lines.[22] In such circumstances the Community should be hesitant about substituting European for domestic regulations. Where a Member State fails adequately to control media concentrations redress may be sought either through the force of public pressure or before a national or international court depending on the constitutional structure and international undertakings of the state concerned. Most Member States have enshrined in their domestic constitutions the right/s to freedom of expression and information, and in some countries, for example Germany, there is an express right to media freedom.[23] All Member States are signatories to the European Convention on Human Rights, Art.10 of which guarantees freedom of expression, a freedom which explicitly embraces the freedom to receive and impart information.[24]

These arguments against direct Community intervention, though raising important issues, are not ultimately conclusive. First, there are a number of well documented political and industrial considerations that render Member States, or at least their governments, unreliable guardians of media pluralism.[25] When regulating a highly competitive and

---

[20] On which see G. Doyle, "From 'Pluralism' to 'Ownership': Europe's Emergent Policy on Media Concentrations Navigates the Doldrums" (1997) 3 *Journal of Information Law and Technology* (JILT), at: http://elj.warwick.ac.uk/jilt/commsreg/97_3doyl/; A. Harcourt, "EU Media Ownership Regulation: Conflict over the Definition of Alternatives" (1998) 36/3 *Journal of Common Market Studies* 369; and D. Ward, *The European Union Democratic Deficit and the Public Sphere: An Evaluation of EU Media Policy* (Amsterdam/Washington, IOS Press, 2002), pp.78–79.
[21] *ibid.*
[22] See text accompanying n.88 below.
[23] For general discussion see E. M. Barendt, *Broadcasting Law: A Comparative Study* (Oxford, Clarendon Press, 1995) and R. Craufurd Smith, *Broadcasting Law and Fundamental Rights* (Oxford, OUP, 1997).
[24] Though the right to information has been construed narrowly by the European Court of Human Rights, as a right only to receive that which others wish to communicate free from state intervention: *Leander v Sweden* (1987) 9 EHRR 433, para.[74].
[25] Ward, n.20 above, p.74.

increasingly international industry, states will be tempted to allow consolidation in order to build up domestic companies. Thus, the UK government when consulting on media ownership rules prior to the adoption of the Communications Act 2003 explicitly stated that "[w]hilst the need for a plurality of media sources remains clear, we are committed to a deregulatory approach to media markets. From a commercial point of view, further liberalisation would benefit existing and potential new investors, providing for further consolidation, greater efficiency, more scope for investment, and a more significant international presence".[26] The UK Communications Act 2003 opened the way to additional concentration in the terrestrial television broadcasting sector and removed certain restrictions on the cross ownership of newspapers and television stations.[27] Similar deregulatory policies are currently being pursued in other Member States, for example, in Italy,[28] and have caused considerable controversy in the United States.[29]

Equally, one cannot exclude the possibility that short-term party political considerations may have an impact on domestic media policy. It is not generally in a government's interest to antagonise powerful press or television interests, and there is the risk that regulatory concessions may be given in the hope that this will lead to improved media coverage, particularly at election time. Though the Community also ascribes substantial weight to industrial considerations and has sought to create a regulatory environment in which large media enterprises, capable of competing at an international level, can survive, it may be less prone to the sort of short-term political pressures indicated above. Given the rather limited and unsystematic reporting of European affairs in many newspapers and audiovisual services, individual Commissioners and Members of the European Parliament are unlikely to see their day-to-day survival tied in quite the same way to media coverage as domestic politicians.

Secondly, there are distinct limits to what domestic or international courts can do where a state introduces legislation that increases the level of media concentration or, more problematically, fails to take steps to control an ongoing process of consolidation. Litigation is a time consuming, costly process and addresses a problem only after it has occurred. Moreover, courts differ quite markedly in their understanding of what media pluralism, freedom of expression and information actually entail. In Italy, for example, the Constitutional Court has held that to allow the terrestrial television market to be dominated by one public and one private company would contravene the constitutionally derived principle of pluralism,[30] while the Court of Human Rights concluded in 2002 that the Austrian state terrestrial television monopoly did not infringe the Art.10 rights of a

---

[26] See DCMS (2001), n.1 above, para.1.8.

[27] Available at: *www.legislation.hmso.gov.uk/acts/acts2003/20030021.htm*.

[28] For discussion see, *inter alia*, R. Zaccaria, *La Legge Gasparri. Televisione Con... Dono: a Mediaset il 'Torrone' e alla Rai il 'Carbone'*, (Roma, Nuova Iniziativa Editoriale S.p.A., 2003) and J. Hooper, "Italy's President Refuses to Sign Pro-Berlusconi Media Bill" *Guardian*, December 16, 2003. The controversial Gasparri legislation was finally adopted in May 2004, Legge 3 maggio 2004, n.112, which can be obtained from: *http://xoomer.virgilio.it/ggdecesare/giurisprudenza_costituzionale.htm*, a useful website providing access to legislation, constitutional court cases, and articles relating to Italian media regulation and reform.

[29] For details see the FCC website at: *www.fcc.gov/ownership/* and for discussion of *Prometheus Radio Project v FCC*, Case No.03-3388, US Court of Appeals Third Circuit, see: *www.mediaaccess.org/mediaownershipcase.html*.

[30] This was stated most clearly by the Constitutional Court in sentenza n.826 of 1988, (1988) 33/1 *Giur. Cost.* 3893, in particular at para.19. For online access to the case and more recent rulings on this issue, including sentenza n.466 of November 20, 2002, see the website at n.28 above.

658   Rethinking European Union competence

would-be broadcaster because there were alternative methods of distribution available, namely cable.[31] This was so, even though only about half the number of terrestrial viewers was then connected to cable networks. In terms of remedies, although courts may be able to strike down legislation or call for the disactivation of particular broadcasting stations they are not legislators and do not have the capabilities to fashion a general scheme of ownership rules. Courts, at least in this context, tend to act as system breakers rather than system builders.[32]

In conclusion, though Member States are capable of establishing effective media ownership controls, there is a real risk that they will undervalue the importance of media pluralism, with governments prioritising international competitiveness or, at the extreme, their own political survival. Where these failures occur the existing remedies, particularly judicial, are likely to prove slow and of limited effect. The possibility of remedial action at the Community or Union level should consequently be taken seriously. This is so even though such action risks stretching the subsidiarity principle beyond the confines of Art.5 EC into a test of political will rather than substantive capabilities. Domestic media policies do, however, have an impact on citizens' (national and EU) access to information during national, regional, and European Parliament elections, and, given the importance of these elections for the system of European governance, it is possible to identify a European dimension on which Community intervention might be based.[33] The subsidiarity principle poses less of a problem for the Community where it frames its action in internal market, as opposed to democratic or cultural, terms. Where the ultimate objective is to facilitate the free movement of services, by reducing the complexities which diverse national ownership rules pose for business, then the case for action at the European level is much clearer.

It is thus apparent that the question of respective capabilities cannot be addressed in isolation from the question of objectives and the Treaty basis for Community or Union action. If we then proceed to examine the competence question in more detail we see that the choice is not simply between action at the European level or at the domestic level, but that there is considerable scope for co-ordinated action at both levels.

What competence does the European Union have in the field of media ownership?

European Union competence in the field of media ownership is generally considered problematic, given that neither the EC Treaty nor the Treaty on European Union ("TEU") provides expressly for action to guarantee media pluralism. The problem is not, however, one of legal incapacity, in that there are a number of bases on which such action might be founded, the problem stems rather from a lack of Member State support for European

---

[31] *Tele 1 Privatfernsehgesellschaft v Austria* (2002) 34 E.H.R.R. 181. The Court of Human Rights has tended to regard pluralism as a potential justification for state restrictions on individual speech, *Informationsverein Lentia v Austria* (1994) 17 E.H.R.R. 93, but more recent rulings suggest that Art.10 ECHR may also impose positive obligations on states to ensure that private individuals or bodies do not infringe the free speech rights of others, see *Özgür Gundem v Turkey* (2001) 31 E.H.R.R. 1082 and *Fuentes Bobo v Spain* (2001) 31 E.H.R.R. 1115, discussed by Bruck *et al*, n.1 above, Pt A.I, para.10.

[32] This is illustrated by the situation in Italy where, despite extremely clear indications by the Constitutional Court that the RAI/Mediaset domination of the terrestrial television sector under present technical capabilities is unconstitutional, political forces have continued to impede meaningful change. For further details see materials cited at n.28 above.

[33] See discussion at text accompanying n.61 below.

level intervention. There are also genuine concerns regarding the extent to which the available EC legislative procedures facilitate consideration of the democratic and cultural concerns which are here at play, and the respective roles of the Council and European Parliament in those procedures

Action by the institutions of the European Union or the European Community in relation to media ownership can take a number of different forms, considered in detail in the four subsections below, but outlined briefly here by way of introduction. First, Art.7 of the TEU enables the European Union to monitor the activities of the Member States that have an impact on media pluralism. Political pressure to take remedial action can then be exerted on those states where existing or potential infringements of human rights or democratic principles have been identified. Secondly, the European Community has competence to set standards with regard to media ownership, notably in order to complete the internal market. Community intervention need not, however, take the form of a harmonising directive and "softer" forms of intervention, such as recommendations or the provision of financial assistance for measures designed to promote media pluralism, are also possible. Thirdly, it is necessary to be aware of the potential impact of Community competition and merger control rules on media ownership. Fourthly, there is considerable scope for the Community to co-ordinate and commission research in order to enhance our understanding of the impact of ownership regulations within the EU and worldwide.

Initiating change from within: EU monitoring of domestic conformity with human rights and democratic values

Article 7(1) of the TEU, as amended by the Treaty of Nice, enables one-third of the Member States, the European Parliament, or Commission, to call on the Council to determine whether "there is a clear risk of a serious breach by a Member State of the principles mentioned in Art.6(1), and address appropriate recommendations to that State". Action by the Council requires a four-fifths majority and the assent of the European Parliament. The principles mentioned in Art.6(1) TEU are "liberty, democracy, respect for human rights and fundamental freedoms, and the rule of law", all of which are stated to be common to the Member States. On one view, this requires no more than that those institutions with the power of referral take seriously any potential infringements which happen to be brought to their notice. This approach would be essentially reactive. The Commission has indicated, however, that it sees the revised Art.7 TEU as opening the way for systematic monitoring of Member State conformity with fundamental rights and the identification of situations likely to lead to breaches in the future.[34]

Article 7(2) of the TEU authorises the Council, this time acting by unanimity and with the assent of the European Parliament, to determine that there has been a serious and persistent breach of the principles set out in Art.6(1) TEU. In this context only one-third of the Member States or the Commission, not the European Parliament, have competence to refer the issue to the Council. The pressure which can be exerted under Art.7 TEU on a particular state is of an essentially political "naming and shaming" nature, though in extreme cases of persistent breach all or some of a Member State's Treaty rights could be

---

[34] Commission *Communication on Art.7 of the Treaty on European Union. Respect for and Promotion of the Values on which the Union is Based*, COM(2003) 606 final, particularly p.9.

660 Rethinking European Union competence

suspended.[35] The role that Art.7 TEU can play in the control of media concentrations is illustrated by the petition made to the European Parliament in 2003 under Art.194 EC concerning the level of concentration in the Italian television sector.[36] The main objective of the petitioners was for the Italian situation to be referred to the Council for evaluation under Art.7(1) TEU, on the basis that the degree of ownership concentration put at risk a fundamental value of the European Union, namely "freedom and pluralism of the media".[37]

In 2002 the Commission, on the recommendation of the European Parliament, established an EU network of independent experts on fundamental rights. An important rationale for its creation was to assist with the monitoring process that the Commission considered Art.7 of the TEU to require. The network reported in March 2003 (for the year 2002, and the report is hereafter referred to as the "2002 Network Report") and explicitly noted the position of the Italian media under a section entitled "concentration of the media and pluralism of information".[38] Apart from the much documented question of media concentration in the terrestrial television sector, currently dominated by the state broadcaster RAI and the three commercial channels of the Mediaset company, the report noted the potential conflict of interest which arises where a politician, such the present Italian Prime Minister, maintains significant shareholdings in media companies.[39] The report also expressed concern at a number of legal actions previously brought against journalists and the decision by the public broadcaster RAI to cut from its schedules programmes by journalists previously critical of Mr Berlusconi. Italy was not, however, the only country to receive criticism for its treatment of the media: the 2002 Network Report also referred to the use of libel actions in Austria to chill political debate, to cases in the United Kingdom, Belgium, and Luxembourg concerning the protection of journalists' sources, and to previously problematic situations in both Greece and Portugal.

Although Art.7 TEU makes no express provision for Commission interaction with the Member States, the Commission observed in its 2003 communication on Art.7 TEU that it "could" contact the state concerned in order "to present the facts of which the Member State is accused and [to] allow that Member State to make its views known" but that it did not consider "such informal contacts . . . [to] be mandatory".[40] It is, however, difficult to imagine on what basis the Commission would decide not to present the facts to the Member State concerned, or to present the facts to one state and not another. It is interesting that in the context of Art.226 EC proceedings the Commission is required, once

---

[35] Art.7(3) TEU, extended to the sphere of the EC Treaty by Art.309 EC.
[36] Details of the petition and its sponsors can be obtained from:*http://save+democracy.net/petition/index.html*.
[37] *ibid.*, part one.
[38] EU Network of Independent Experts in Fundamental Rights, *Report on the Situation of Fundamental Rights in the European Union and its Member States in 2002*, available at: *www.europa.eu.int/comm/justice_home/cfr_cdf/doc/rapport_2002_en.pdf, p.110.*
[39] Mr Berlusconi and his family together own 96 per cent of the Fininvest Company which has controlling interests in the audiovisual company Mediaset and magazine/book publisher Mondadori. Mediaset runs, *inter alia*, three commercial terrestrial television channels-Canale 5, Italia 1, and Retequattro-which together attract 45 per cent of the television audience and 60 per cent of the advertising sales for commercial television. Details obtained from Katupa.net, a Caslon Analytics media resource, at: *www.ketupa.net/berlusconi.htm#introduction*, and the Reporters without Borders, *Annual Report 2003 for Italy* at: *www.rsf.fr/article.php3?id_article=6521*.
[40] Commission *Communication on Art.7 TEU*, n.34 above, para.2.2.

it considers that a Member State has failed to fulfil a Treaty obligation, to provide it with a reasoned opinion and an opportunity to submit observations. The reason why there is no such requirement here may be because those who framed Art.7 TEU saw ultimate responsibility for determining the existence of a clear risk of a serious breach, or a serious and persistent breach, of the Art.6(1) TEU principles, to rest with the Council, albeit with the assent of the Parliament for negative determinations. The Council is required to hear the Member State in question and, in the context of Art.7(1), there is specific provision for the commissioning of a report from an independent person on the situation in that state. Although the Court of Justice has no competence to review the substantive nature of the decision it does have limited power under Art.46(e) TEU to review the "purely procedural stipulations in Art.7".

The 2002 Network Report indicates the potential that Art.7(1) TEU opens up for regular and wide-ranging monitoring of Member State compliance with human rights and democratic values. The report identified not only structural problems in the media which affect citizens access to information but also more covert forms of censorship and the use of legal proceedings in individual cases to intimidate reporters. The process thus has a number of clear advantages: it involves independent assessors, enables systemic problems to be identified, and solutions to be tailored to the cultural and legal environment of a particular country. But what impact is the article likely to have on state behaviour in practice? This will almost certainly depend on the cost-benefit analysis carried out by states in relation to any given infringement. In many instances the serious nature of a reference to the Council and the political stigma attached to a finding that a state has infringed one of the fundamental principles of the Union should make constructive dialogue and remedial action attractive options. Where, however, such action would have particularly damaging repercussions for a government, or interests closely related to it, then a state may be prepared to brazen it out. This may well be the case in relation to domestic media ownership rules. States are prepared on occasion to ignore European Court of Justice rulings, even with the background threat of a heavy fine, and there is no reason to think they will act differently in relation to Council determinations under Art.7 TEU.

Two factors, in particular, may encourage a state to hold out in negotiations with the Commission or Council. First, human rights infringements are often the subject of intense debate with different interests pulling in different directions. This is one reason we use independent courts to resolve these disputes, but in the context of Art.7 TEU the Court of Justice has competence only, as the Commission notes with regret, in relation to procedural issues.[41] There is scope for considerable legal debate as to the meaning of, and relationship between, freedom of expression, freedom of information and media pluralism and, as previously noted, judicial approaches vary quite markedly.[42] Although Art.11 of the Charter of Fundamental Rights of the EU may help to cut through some of these uncertainties by stating that "the freedom and pluralism of the media shall be respected", use of the term "respect" could suggest merely a negative obligation not to impede rather than a positive obligation to actively promote.[43] The situation is also complicated by the

---

[41] *ibid.* para.1.2.
[42] See text accompanying nn.30 and 31 above.
[43] The term "respected" replaced the more robust "guaranteed" which appeared in earlier drafts of the Charter, on which see Westphal, n.1 above, p.486.

662 Rethinking European Union competence

fact that legislation concerning media ownership involves the drawing of discretionary, one is tempted to say somewhat arbitrary, lines.[44] Only in extreme cases of concentration, as illustrated by the Italian terrestrial television market, might one expect the Commission or Council to engage in a bruising exchange with a Member State, involving, *inter alia*, complex questions of market definition and media substitutability.

Secondly, Art.7 TEU may prove something of a double-edged sword for the Union with its very public system of criticism and potentially extreme, though rather open-ended, sanctions. Thus, although the procedure was not formally invoked against Austria in 2000, it has been suggested that the pariah status which Austria acquired consequent on the refusal of 14 Member States to conduct bilateral relations with it while the Freiheitliche Partei Östereichs was in the government, led to considerable anti-EU feeling within the country.[45] Moreover, the formal sanction provided for in Art.7(3) TEU, namely the suspension of rights, in particular voting rights, could undermine the legitimacy of initiatives taken without the participation of the state in question.[46] Though the whole point of the process is to underline the fact that a given state does not meet the membership requirements of the European Union club, and it is certainly desirable that electors in that country are aware of the fact, initiation of the Art.7 TEU procedure is not without risks for cohesion within the European Union.

It is probable, therefore, in the controversial area of media ownership, that the Commission, and ultimately Council, will take action only where the level of concentration appears particularly serious and anomalous.[47] It is notable that even the European Parliament has not responded to the requests made to it to refer the Italian situation on to the Council for consideration under Art.7(1) TEU, preferring instead to request the Commission to initiate more general initiatives in the field of media pluralism.[48] The resolve to act will certainly be stiffened where there has been authoritative condemnation from other institutions with credibility in the field of human rights, particularly national or international courts. Ongoing legal proceedings at the national or international level could, however, discourage action under Art.7 TEU in that neither the Commission nor Council are likely to want to prejudge the outcome of legal proceedings. There is also a risk that the "diplomatic" nature of the Art.7 TEU proceedings could lead to accommodation and political compromise among the Council and Member State concerned. One should not, however, discount the possibility that the very existence of Art.7 will be used by political factions as a strategic tool when hammering out media legislation or by journalists and media organisations seeking to resist political pressure or negotiate changes in the law on, say, the confidentiality of sources or the use of libel law in the context of political reporting.

---

[44] See text accompanying n.88 below

[45] For full documentation on this conflict see (2000) 12/3 *Revue universelle des droits de l'homme* 147.

[46] The Commission in its introduction to the communication on Art.7 TEU, n.34 above, was interestingly not prepared "to speculate" on penalties, preferring to approach Art.7 TEU in a "spirit of prevention".

[47] See Commission *Communication on Art.7 TEU*, n.34 above, para.1.4. Arguably the Italian situation must be coming very close to such a categorisation, on which see the Freedom House 2004 country survey on the Italian media at: *www.freedomhouse.org/research/pressurvey.htm*.

[48] European Parliament Resolution of April 22, 2004 *on the risks of violations, in the EU and especially in Italy, of freedom of expression and information*, P5_ TA-PROV(2004) 0373, available at the European Parliament website *www.europarl.eu.intl*. See also the earlier motion for a resolution of July 9, 2003, B5–0363/2003, by Sylviane Ainardi and 37 MEPs calling for an Art.7(1) reference to the Council, available at the same website.

Article 7 TEU on its own, therefore, is not an adequate basis for the development of EU policy in relation to media pluralism. Article 7 TEU can undoubtedly perform an extremely useful role in a wider regulatory framework: identifying potential abuses, exerting political pressure for change, and feeding information into the system.[49] But its existence should not be seen as a justification (or excuse) for refraining from considering other forms of intervention, of a more programmatic and pro-active nature, which the EC Treaty makes possible.[50] It may be noted that in the introduction to the 2002 Network Report the monitoring system is envisaged as having wider ramifications, leading to the accumulation of examples of good practice, the development of guidelines or benchmarks, comparison of various solutions to particular problems, and the exchange of new regulatory ideas. If Art.7 TEU does develop along these "open method of co-ordination" lines then it could exert a more systematic influence on domestic systems and play a role which goes beyond mere troubleshooting.

Standard setting by the European Community

Two main bases for Community standard setting in the field of media ownership are usually put forward: completion of the internal market and Art.308 EC. Perhaps the strongest justification for Community involvement is that media ownership rules restrict the freedom to provide services under Art.49 EC and that the complexity created by different national rules discourages companies from entering new markets.[51] These impediments may be removed under the combined force of Arts 49, 55, and 57(2) EC, or the rules approximated under Art.95 EC. To be legitimate, a genuine objective of any such measure must be to complete the internal market, but there remains considerable scope during the drafting of such legislation to take into account various policy objectives, among them media pluralism.[52] Moreover, all EC legislation must comply with fundamental rights and also, by virtue of Art.151(4) EC, take cultural aspects into account "in particular in order to respect and promote the diversity of its cultures".

Commission attempts to develop a directive on media ownership in the 1990s had just such an internal market basis and it is apparent that, though coming at the issue indirectly, it is possible for democratic and cultural as well as economic perspectives to be taken into account.[53] The other attraction, at least of Art.95, is that it enables the Parliament to be fully integrated in the legislative (co-decision) process and requires merely a qualified majority in the Council. The concerns sometimes expressed over an internal market basis are not, therefore, convincing. The major constraint which an internal market basis imposes is that it militates towards adoption of harmonised rather than minimum standards, the latter allowing states to adopt more exacting controls and thus to continue to fragment the market.

---

[49] There are thus synergies to be drawn with Community research initiatives discussed in the final sub-section below.

[50] See sub-section immediately below and the European Parliament Resolution at n.48 above.

[51] For further consideration see Ward, n.20 above, pp.76–77.

[52] See Case C–376/98 *Germany v European Parliament and Council of the European Union* [2000] E.C.R. I–8419, in particular para.[84].

[53] Though with inevitable tensions among the various Commission directorates-general, on which see Harcourt, n.20 above.

664 Rethinking European Union competence

There are mixed views about how successful the Commission's media ownership proposals would have proved to be in practice, and the move by certain states, such as the UK in its Communications Act 2003, to rely more heavily on case by case competition law controls, backed by a public interest test, is at odds with the Commission's "old-style" approach of fixing in advance media ownership thresholds.[54] Ultimately, as noted above, the Commission proposals foundered largely on Member State opposition and the complexity created by the variation in domestic markets. The facts which caused concern in the 1990s have not, however, gone away and in September 2003 the European Parliament called again on the Commission to prepare a Green Paper on media concentrations with a view to putting forward a draft directive on the subject.[55] Gillian Doyle has suggested that a more sophisticated measure could be drawn up, one which, though complex, takes account of the size of particular markets.[56] This would not seek to adopt uniform standards for the whole of the EU but would instead adopt a "tiered" approach, responsive to market conditions. Though there is little point in filling the legislative agenda with proposals which have little chance of success, it is arguable that if such a provision could be framed, the Parliament should be given the chance to see whether it can broker an agreement with the Council in the co-decision process.

The second potential basis for Community competence in this area is Art.308 EC. This enables the Community to take action in order to attain "one of its objectives" when the Treaty has not provided an express basis for action. It does not stipulate the adoption of any particular type of measure and thus leaves the Community with discretion to either establish binding standards in the form of a Directive or merely propose guidelines in the form of a Recommendation. Dietrich Westphal has suggested that because Art.95 EC, in connection with Art.55 EC, enables harmonisation of national ownership rules there is no basis to apply Art.308 EC. On the other hand, the objective of Art.95 EC is to remove obstacles to the internal market, and only indirectly can it be used to further media pluralism.

Media pluralism is not expressly mentioned in the Community's objectives set out in Art.2 EC, nor in the list of policies and measures set out in Art.3 EC. Similarly, the TEU offers little explicit assistance in this respect, in that the objectives stated in Art.2 TEU, though broad, are to be achieved "as provided for" in the TEU itself, which could be understood as action under Art.7 TEU, considered above. Article 6(4) TEU is, however, more open-ended, stating that "[t]he Union shall provide itself with the means necessary to attain its objectives and carry through its policies". The position of this provision at the end of Art.6 TEU also suggests that ensuring respect for the principles and rights set out in Art.6 is an objective of the TEU. Moreover, there are links between the TEU and EC Treaty in relation to Art.7 TEU, in that Art.309 EC requires that, where voting rights have been suspended under the TEU, a similar suspension is to take place with regard to voting rights under the EC Treaty, and authorises, but does not require, the suspension of other rights in parallel with action taken under Art.7(3)TEU. Together, these various provisions lend

---

[54] See the various articles cited at n.20 above. Among them, Ward, at p.78, suggests that the proposal finally hammered out by the Commission was a "fairly sophisticated and well thought through initiative".
[55] European Parliament, Resolution on Television Without Frontiers, P5_TA-PROV(2003)0381, para.41. The resolution also called for a commitment to diversity of media ownership to be introduced into any Directive reforming the 1989 Television Without Frontiers Directive (Directive 89/552 [1989] O.J. L298/23), or a future Directive on audiovisual content, para.39.
[56] Doyle, n.20 above.

some support for the conclusion that ensuring respect for the principles and rights in Art.6(1) TEU is also an objective of the EC Treaty.

Philip Alston and Joseph Weiler have suggested that Art.308 EC could be used, admittedly within carefully circumscribed limits, as the basis for the development of a Community human rights policy.[57] This view was endorsed by the independent experts on fundamental rights in their 2002 Network Report.[58] In particular, they make reference in the introduction to the Report to Regulation 976/1999, a measure designed to promote democracy and respect for human rights in third countries.[59] The Regulation was based on Art.235 EC, the precursor to Art.308 EC, and pursues, as its title and preamble make clear, the "general objective of developing and consolidating democracy and ... respecting human rights".[60] It would be strange if this "general objective" applied only in the context of external relations and not internal affairs. Though Art.308 EC is thus a potential basis for Community initiatives designed to guarantee media pluralism it is procedurally problematic, in that the Parliament is afforded only a right to be consulted and unanimity is required in the Council. For measures which are at all controversial, therefore, Art.308 is likely be of limited assistance.

Another potential Treaty basis for intervention is Art.22 which concerns Union citizenship. This Article provides an explicit basis for the adoption of provisions to "strengthen or to add to the rights" in Arts 17 to 21.[61] Under Art.19 EC European citizens enjoy rights to vote in European Parliament and municipal elections in Member States where they reside but do not have nationality. The members of the national or federal governments which they elect also have an influence on the policies adopted by the Council. In order for citizens to be able to participate meaningfully in these elections they require access to varied information from diverse sources.[62] Political parties undoubtedly have a responsibility to inform citizens about their policies and programmes, but many individuals rely heavily on domestic media to provide them with relevant information and opinions.[63] A parallel may here be drawn with the *Nationwide News* and *Capital Television* decisions of the Australian High Court, which, in the absence of an Australian bill or charter of rights, derived a right to freedom of expression from the democratic nature of the constitution.[64] Article 22 EC suffers, however, from similar limitations to Art.308 in that it requires

---

[57] P. Alston (ed.) with M. Bustelo and J. Heenan, *The EU and Human Rights* (Oxford, 1999), Ch.1, in particular pp.26–27.

[58] See n.38 above, pp.16–17.

[59] Council Regulation 976/1999, laying down the requirements for the implementation of Community operations, other than those of development co-operation, which, within the framework of Community co-operation policy, contribute to the general objective of developing and consolidating democracy and the rule of law and to that of respecting human rights and fundamental freedoms in third countries [1999] O.J. L120/8.

[60] *ibid.*, in particular see the first Recital to the Preamble.

[61] This provision thus precludes recourse to Art.308 in combination with the citizenship articles.

[62] The Italian petition to the European Parliament also drew this link between European citizenship and media pluralism, see n.36 above, at section 3.

[63] It has, *e.g.* been suggested that 82 per cent of Italians depend only on television for news, the highest percentage in the EU, figures cited at www.ketupa.net/berlusconi.htm, drawing on the work of P. Norris, *A Virtuous Circle: Political Communications in Post-Industrial Democracies* (Cambridge, CUP, 2000).

[64] *Nationwide News Pty Ltd v Wills* (1992) 177 C.L.R. 1 and *Australian Capital Television Pty Ltd v Commonwealth of Australia* (1992) 177 C.L.R. 106. For critical commentary suggesting a failure on the part of the majority in *Capital Television* to adequately address the implications of disparities in wealth for the democratic process see D. Tucker, "Representation-Reinforcing Review: Arguments about Political Advertising in Australia and the United States" and D. Z. Cass, "Through the Looking Glass: the High Court of

666 Rethinking European Union competence

Council unanimity and allows solely for consultation of the European Parliament. Just one dissenting voice in the Council would thus be enough to kill off any proposal.

A fourth, and rather less obvious, basis for action is Art.86 EC which relates to services of general economic interest. Article 86(3) EC enables the Commission to direct decisions or directives to the Member States to ensure that such services comply with the Treaty rules, but also requires the Commission to balance the impact of the Treaty rules on the operational capabilities of the public service and vice versa. In its *Green Paper on Services of General Interest* the Commission noted that "existing Community law instruments" only very indirectly aim to deliver "information to the citizen" and asked whether the Community should re-consider taking action in relation to media pluralism.[65] It is not, however, clear why the Commission has decided to raise this issue in the context of Art.86 EC, in that only organisations officially entrusted with public service obligations fall within its scope. The provision cannot, therefore, be a suitable basis for introducing media ownership controls across the commercial and public service media sectors. It is nevertheless possible that Art.86 EC, in co-ordination with Art.16 EC, could be used by the Commission to ensure that public service broadcasters are able to contribute meaningfully to the realisation of media diversity.[66] One possibility would be for the Commission to examine how political or commercial pressures affect the way in which public service broadcasters handle sensitive or controversial matters, and the various structural or regulatory steps which could be taken to reduce these pressures.[67]

Despite the existence of these various bases for Community action to control media concentrations there is little evidence of the political will needed for the establishment of European ownership thresholds. This leads one to consider two distinct questions: first, whether there are other regulatory strategies open to the Community which could prove more fruitful than attempts to agree a harmonising Directive and, secondly, whether the Commission should adopt a broader perspective and consider other measures which could contribute to media pluralism, other than controls on media ownership. On the first point, given that Member States are wary of losing autonomy in this controversial area, they may prove more receptive to a framework Directive that establishes specific objectives, yet leaves considerable discretion at the implementation stage. Under such a Directive Member States would be expected to draw up a programme of action, or explain why they consider no action to be necessary, and report on progress. To be effective, there would need to be a regular system of Community monitoring and a willingness on the part of the Commission to use the enforcement mechanism under Art.226 EC. Though similar in many respects to the Art.7 TEU procedure discussed above, this is potentially wider ranging in scope, could lead to the establishment of more specific objectives, and allows for ultimate recourse to the Court of Justice. The development of a European level media/

Australia and the Right to Political Speech" both in T. Campbell and W. Sadurski (eds.), *Freedom of Communication* (Aldershot, Dartmouth, 1994), pp.161 and 179 respectively.

[65] Commission, *Green Paper on Services of General Interest*, COM (2003) 270 final, paras 73–75.

[66] Art.16 EC provides that the Member States and Community, each within their respective powers and the scope of application of the EC Treaty, are to take care that services of general economic interest operate on the basis of conditions and principles which enable them to fulfil their missions.

[67] The Parliamentary Assembly of the Council of Europe has recently proposed measures to ensure that Member States undertake the appropriate legislative, political and practical measures in support of public service broadcasting: Recommendation 1641/2004 on Public Service Broadcasting, available at: *http://assembly.coe.int/Main.asp?link=http://assembly.coe.int/Documents/AdoptedText/TA04/EREC1641.htm*.

communications network involving national administrators, representatives from the media and interest groups, as well as consumers/citizens, could also prove effective in facilitating the comparison of state and industry practices, the identification of potential problems, and the gradual ratcheting up of standards.

In relation to the second question, it may be noted that "media pluralism" is a vague and rather open-ended term that could justify examination of a range of issues extending well beyond media ownership. For example, how can the media be protected from undue political or economic influence liable to affect the reliability of what is printed or broadcast? Should certain groups be guaranteed access to the media to enhance political or cultural diversity, or quotas established for certain types of programming to increase minority representation or the expression of diverse views? These issues have traditionally been seen as matters for the Member States to resolve, but it is arguable that, if the Community is to take media pluralism seriously, it should also consider whether action is also required at the European level.[68]

Two areas which appear to be particularly problematic are access to, and reporting by, the media during political elections, an issue which could extend to political advertising, and the potential conflict of interest which arises when political and media power is combined.[69] Though action in the first area could have an internal market rationale given the development of cross border broadcasting, and in both areas Art.22 and the citizenship provisions appear relevant, detailed Community standard setting is likely to be strongly resisted by the Member States. In relation to election reporting, a framework Directive designed to accommodate constitutional and political differences, could ultimately prove more effective in remedying deficiencies in particular states. The case for Community intervention in the second area is underlined by the findings of the 2002 Network Report, which noted the present Italian Prime Minister's extensive audiovisual and press interests.[70] A bill is currently under consideration in the Italian Parliament to address this issue which would prohibit members of the government from taking an active role in the management of commercial companies, including media organisations, but would not prevent them from owning (directly or indirectly) a controlling interest in such an organisation.[71] Under the proposal, a conflict of interest would arise where a member of the government uses his or her position for personal gain, contrary to the public interest. The various ways in which conflicts of interest can be addressed, for example, by media legislation, parliamentary disqualification and reporting regulations, or internal ethical

---

[68] The Community has already dipped its toe into the water with provisions on European and independent production quotas and a right of reply in the 1989 "Television Without Frontiers" Directive 89/552 [1989] O.J. L298/23, Arts 3(a), 4–9 and 23. For an extensive review of those areas where Community action might helpfully supplement Member State initiatives see the European Parliament Resolution detailed at n.48 above. The resolution, at para.79(p), calls in particular for further investigation into election reporting and media access.

[69] Both issues are, again, topical in Italy, see J. Hooper, "Berlusconi Kicks up a Storm over Plan to Curb TV Access" *Guardian*, December 20, 2003; U. Eco, "Le Regole del Potere nel Regime Mediatico" *La Repubblica*, January 9, 2004; Zaccaria, n.28 above, Ch.8; and Reporters Without Borders, n.39 above.

[70] EU Independent Experts in Fundamental Rights, n.38 above, p.110. Concerns have also been expressed in relation to Spanish media coverage, in particular, reporting in the immediate aftermath of the Madrid bombings, on which see the April 2004 European Parliament Resolution at n.48 above, para.53.

[71] Draft legislation (Frattini) No.1206-B, *norme in materia di risoluzione dei conflitti di interessi*, Art.2, detailed in Zaccaria, n.28 above. See also Reporters Without Borders, "A media Conflict of Interest: Anomaly in Italy" April 23, 2003, available at: *www.rsf.fr/article.php3?id_article=6392*. The article notes that family members and associates have been in put in charge of running Mr Berlusconi's businesses.

668 Rethinking European Union competence

codes adopted by political parties or institutions, would need to be considered to ensure that any Community proposal does not conflict with constitutional arrangements in the Member States.[72] Nevertheless, in this context one could envisage the formulation of quite specific prohibitions at Community level.

Finally, there is some scope for the Community to help finance new and innovative media projects. Article 151(5) EC enables the Community to adopt incentive measures for cultural purposes which could be used to support fledgling press or radio services, to help minority groups report on matters of interest to them, or to produce and distribute programmes which reflect their particular culture and perspectives.[73] The funding made available for programmes under Art.151 has, however, been limited. The Culture 2000 Programme, for example, received only €167 million for its first five years,[74] while the MEDIA training and plus programmes, designed to enhance the competitiveness of the European audiovisual industry, together received €400 million for a similar period.[75] Although both the MEDIA training and plus programmes recognise the need to take cultural issues into account in their operation, and offer not only training opportunities but also assistance in project development, distribution, and promotion, the focus on European competitiveness may render them less accessible to applicants seeking the sort of market supplementing initiatives considered here.[76] The various, and significant, funds linked to the Community's structural policy could possibly prove more useful in supporting initiatives designed to promote media diversity where these are linked to the development of the regions or deprived areas and offer employment opportunities.

The above discussion indicates that there are a number of bases on which the Community can work alongside Member States to promote media pluralism, but there is also a risk that Community law, whether competition law or the internal market freedoms, could impede genuine Member State initiatives in this area. The line between economic protectionism and the pursuit of general interest objectives is often hard to draw and requires further elucidation by the Community institutions.[77] The relevance of this issue for media pluralism is illustrated by the Treaty infringement proceedings brought by the Commission against Germany regarding the preference given by Rheinland-Pfalz to groups linked to the local community when granting radio licences.[78]

---

[72] In the United Kingdom, ownership by an M.P. of a controlling shareholding in a company with a broadcasting licence (effectively the present Italian case) is apparently precluded by the tortuous combination of Sch.2, Pt 1, para.1.(1) and Pt II, para.1(1)(d)(f)(i) of the Broadcasting Act 1990. This does, however, require one to accept that M.P.s are "officers" of Parliament and does not address potential problems arising from the combination of press interests and political office. These provisions more clearly cover officers of political parties.
[73] Art.151(5) EC explicitly excludes harmonisation measures, but allows for "incentive" measures and Council Recommendations.
[74] European Parliament and Council Decision 508/2000 (Culture 2000) [2000] O.J. L63/1.
[75] Council Decision 2000/821 (concerning the Media plus programme) [2001] O.J. L13/34 and Council Decision 163/2001 (concerning training) [2001] O.J. L26/1.
[76] See Recitals 14 and 18 of Council Decision 2000/821, n.75 above, and Recital 16 of Decision 163/2001, n.75 above.
[77] Discussed in relation to cultural policy by N. Nic Shuibhne, "Labels, Locals and the Free Movement of Goods" and R. Craufurd Smith, "Community Intervention in the Cultural Field: Continuity or Change?" in *Culture and European Union Law* (Oxford, OUP, 2004), forthcoming.
[78] Commission Press Release: "Broadcasting Services, Commission Refers Germany to Court over Allocation of Radio Spectrum" (July 24, 2003) DN: IP/03/1103, available at: *http://europa.eu.int/rapid/start/cgi/guesten.ksh*.

Competition law and media ownership

The main Community mechanism for controlling concentrations of media ownership has been the 1989 Merger Regulation, recently replaced by a new regulation that came into force on May 1, 2004.[79] The new Regulation concerns concentrations with a Community dimension,[80] and prohibits those concentrations that would significantly impede competition in the common market, or a substantial part of it.[81] Commission action is here focused on preserving effective competition and it has no competence, in line with the 1989 version, to pursue wider general interest objectives when determining whether or not to approve a merger. The reference in Art.2(1)(b) to "the interests of intermediate and ultimate consumers" could potentially be given a broad scope, but it is clear from Art.21(4) of the new Regulation (Art.21(3) of the old) that this cannot be pushed too far. Although the Community has sole competence to consider concentrations with a Community dimension, Art.21(4) provides that Member States can still review a proposed merger, approved by the Commission, to ensure conformity with "legitimate interests *other than those taken into consideration by this Regulation*".[82] These interests are expressly stated to include "plurality of the media".[83] This explains why public interest considerations have not to date been discussed by the Commission in any detail in the course of its merger appraisals.[84]

The Merger Regulation clearly plays an important role in keeping markets open to competition, but there are certain forms of concentration which may be acceptable from a competition law point of view, for example, where the merger involves companies each operating in different geographic markets, but which remain problematic from a pluralism point of view. Given the increasing number of media concentrations which fall within the scope of the Merger Regulation and the Community interest in fostering companies large enough to compete at the international level, it is arguable that a requirement to consider the impact of a proposed concentration on media pluralism should be built into the Regulation. This proposal does, however, raise a number of fundamental questions, not least the desirability of leaving to the competition directorate-general the task of carrying out such a politically contentious evaluation, one involving a distinct form of market

---

[79] Council Regulation 139/2004 [2004] O.J. L24/1, replacing, for agreed/announced mergers after its commencement date, Regulation 4064/89 [1989] O.J. L395/1. For discussion in the media context see I. Nitsche, *Broadcasting In the European Union: The Role of Public Interest in Competition Analysis* (The Hague, T.M.C. Asser Press, 2001), Ch.6, and Ward, n.20 above, Ch.6. Arts.81 and 82 of the EC Treaty also have considerable potential to affect media pluralism through, *e.g.* their impact on access to essential services or the purchasing of key programme rights; it is not possible, however, to deal adequately with these issues in the present article but, for further details, see Nitsche, above; European Parliament Resolution, n.48 above, paras 77 and 79(i)(j); and M. Arino Guttierez, "The Contribution of EU Competition Law to Media Pluralism" paper given at the workshop on Promoting Media Pluralism and Diversity in Europe, at the Robert Schuman Centre, European University Institute, Florence, May 21, 2004, available at: *www.iue.it/RSCAS/Events/*.

[80] Arts 1(2) and (3) of the 2004 Merger Regulation, n.79 above.

[81] Art.2(3) of the 2004 Merger Regulation, n.79 above.

[82] Art.21(4) of the 2004 Merger Regulation, n.79 above.

[83] *ibid.* Nitsche, n.79 above, p.127, notes that Member States have not made use of Art.21(3) of the 1989 Regulation. This contrasts with their willingness to request that a case be referred back to the domestic authorities under Art.9 of the Regulation, on which see C. Palzer and C. Hilger, "Media Supervision on the Threshold of the 21st Century-Structure and Powers of Regulatory Authorities in the Era of Convergence" (2001) 8 *IRIS plus*, available at the European Audiovisual Observatory website at: *www.obs.coe.int/oea_publ/iris/iris_plus/iplus8_2001.pdf.en, at p.8*

[84] Nitsche, n.79 above, p.127.

670 Rethinking European Union competence

analysis.[85] The creation of a specialist communications directorate-general, entrusted, *inter alia*, with advising the competition directorate-general on media pluralism issues, or even a European media concentration authority, could here help to defuse some of these concerns, though the latter proposal has in the past been rejected by the Commission.[86] Ultimately, Member States are unlikely to back such a proposal, given their limited influence in the field of Community competition law, and any moves in this direction would probably be seen as an illegitimate attempt to do indirectly what could not be agreed directly through the "appropriate" regulatory procedures.

Research and co-ordination at the European level

In order to develop credible policies designed to further media pluralism, whether at the national or European level, we need to know more than simply which companies own which media interests. We need to know how different ownership structures in the various media affect what is actually communicated and how different people use, and are in turn influenced by, the various media available to them. Under what circumstances, for example, when print and television companies merge will separate news gathering units be maintained for each service? If consolidation does occur what impact does this have on the range and quality of reporting and how do we evaluate this? One can ask similar questions with regard to media mergers at the regional and national levels. Thus, when a national media company, whether print or audiovisual, takes over a regional company what impact does this have on the range and coverage of local issues?

The European Union, in co-ordination with the Council of Europe, is already actively involved in supporting research into various aspects of media concentration and media diversity. One can point, for example, to the work of the Council of Europe advisory panel on media concentrations, pluralism and diversity, and the European Audiovisual Observatory which has a wide membership among the EU and Council of Europe Member States. Given the current level of market consolidation and the deregulatory policies being pursued by certain Member States a rather more focussed and integrated programme of research is, however, required. A parallel may here be drawn with the group of studies commissioned by the Community as part of its review of the Television Without Frontiers Directive, as well as with the set of working papers relating to media ownership commissioned by the FCC prior to its recent relaxation of US ownership regulations.[87]

Without such information those who seek to regulate the mass media are operating largely in the dark.[88] At a national level, the shaky empirical foundations for ownership controls and regulations designed to promote media diversity render them particularly vulnerable to being dismantled by legislative or executive bodies with other priorities. This

---

[85] For discussion of some of the problems inherent in such a proposal see A. Scheuer and P. Strothman, "Media Supervision on the Threshold of the 21st Century-What are the Requirements of Broadcasting, Telecommunications and Concentration Regulation?" (2002) 2 *IRIS plus*, available at the Audiovisual Observatory website at: www.obs.coe.int/oea_publ/iris/iris_plus/iplus2_2002.pdf.en, p.6.

[86] Palzer and Hilger, n. 83 above, p.5.

[87] At *http://europa.eu.int/comm/avpolicy/stat/studi_en.htm* and *http://www.fcc.gov/ownership/studies.html*.

[88] FCC Commissioner Michael Copps has been remarkably frank on this point, stating at a public hearing in January 2003 that the FCC did not "have the foggiest idea right now about the potential consequences" of its actions, see n.19 above.

uncertainty may also tempt interested parties to challenge the maintenance or dismantlement of such regulations by legal means, in particular on human rights grounds. In the United States, for example, judicial proceedings have revealed considerable weaknesses in a number of FCC ownership regulations. In the *Sinclair Broadcasting* case the petitioners challenged the maintenance by the FCC of a rule which allowed common ownership of local television stations only if there remained eight other independent television stations in the same area.[89] Sinclair argued that the Commission had "plucked the number eight out of thin air" and that its refusal to treat the press and other media as potential substitutes for local television was arbitrary.[90] Ultimately, the evidential record in *Sinclair Broadcasting* and the related *Fox Television* case, revealed major gaps in the Commission's knowledge and a lack of coherence across its ownership regulations.[91] Though US cases such as *Sinclair Broadcasting* may be considered products of the 1996 Telecommunications Act, in particular,[92] and American judicial review proceedings, in general, and consequently of limited interest in Europe, they illustrate quite graphically how ownership regulations can become extremely soft targets.

One should not lose sight of the fact that the object here is not to protect legislation or administrative decisions from legal challenge but to ensure that such measures are well founded in the first place. This is true not only of regulations adopted at the national but also European level. Attempts by the European Union to enact new measures or to require Member States to modify internal rules or practices under Art.7 TEU are only likely to be successful if they can be shown to be backed by convincing research and a sound understanding of the relevant markets.

Conclusion

The above discussion indicates that there is considerable potential for the Union to support and, where necessary, supplement Member State measures regarding media pluralism. Such a programme could address concerns arising in the commercial and public service sectors, and could deploy the odd financial carrot as well as the occasional regulatory stick. Above all, there is scope for the Union to tailor its intervention to be more or less prescriptive depending on the subject area and domestic sensitivities. In certain key areas, in relation, for example, to potential conflicts of interest or extreme forms of concentration, specific European requirements are arguably appropriate, while in others, objectives could be framed in more general terms to be implemented by the Member States in the light of domestic circumstances, but with fall-back powers vested in the Union. The development of a European level network of interested parties would assist such

---

[89] *Sinclair Broadcasting*, n.14 above.

[90] Ultimately, the Court "left for another day" the issue of the FCC's choice of the number eight, but held the FCC's divergent approach to market definition in the local radio and television orders to be arbitrary and capricious, *ibid.* at III.B.

[91] *Fox Television Stations, Inc v FCC*, 280 F.3d 1027, reh'g en banc, 293 F.3d 537 (D.C.Cir. 2002). The FCC's controversial decision to deregulate further in this area has recently been challenged by the Prometheus Radio Project, see n.29 above.

[92] This is because the 1996 Telecommunications Act requires the Commission to review its ownership rules biennially, and to repeal or modify any rule it considers no longer in the public interest. This requirement was found by the US Court of Appeals in *Fox Television*, n.91 above, at 1048, to create a presumption of repeal, so that the FCC has authority to retain a rule only where it can establish this to be "necessary in the public interest", a more exacting test than that normally applicable where the issue is "elusive" or "not easily defined", on which see *Sinclair Broadcasting* (Pt III), n.14 above.

672 Rethinking European Union competence

initiatives, enabling practices to be compared, guidelines or codes of conduct to be developed, and overall standards to be raised.

Establishing such a co-ordinated programme of action will not, however, be easy for the Union because of its institutional fragmentation.[93] This fragmentation is particularly noticeable in the Commission, with media-related issues divided across at least four directorates-general: the Internal Market, Information Society, Competition, and Education and Culture. Within the directorate-general for Justice and Home Affairs there is also a unit responsible, *inter alia*, for citizenship and fundamental rights. Increasingly, Member States are establishing integrated communications regulators able to address content and infrastructure issues across the different media and it is arguable that the development and coherence of Community policy in the media sector would benefit from a centralised Communications directorate-general.[94] Though mechanisms are in place to ensure dialogue and co-ordination across the various directorates-general, the present arrangement may lead to proposals on a given matter looking distinctly different depending on the directorate-general from which they originate. In particular, the "heavyweight" directorates-general in this area, namely Competition, Internal Market, and Information Society, are all primarily concerned with European competitiveness.

The Commission has emphasised on a number of occasions that "the protection of media pluralism is *primarily* a task for the Member States".[95] Unfortunately, it is a task that elicits varying degrees of enthusiasm from those to whom it is entrusted. Considering the present Italian situation one might be forgiven for thinking that this is rather like entrusting the friendly fox with the task of looking after the chickens. But we are not here merely discussing domestic problems: Union governance and domestic governance intersect in complex and quite subtle ways. For many European citizens domestic television, radio, and newspapers constitute their main source of information, however imperfect, on European affairs. Media pluralism cannot be categorised simply, or primarily, as a domestic or economic issue—it is an issue that goes to the very heart of the democratic legitimacy of the European Union itself.

---

[93] The April 2004 European Parliament Resolution, n.48 above, noted with regret, at par.2, the "fragmentation of the EU regulatory situation as regards the media", and called for the development of a more coherent Community programme on media pluralism.
[94] *e.g.* OFCOM in the United Kingdom.
[95] See, *e.g.* Commission, *Green Paper on Services of General Economic Interest*, n.65 above, para.74.

[8]

THE GOAL OF PLURALISM AND THE OWNERSHIP RULES FOR PRIVATE BROADCASTING IN GERMANY: RE-REGULATION OR DE-REGULATION?

Dr. Peter Humphreys*

Introduction

In Germany, as elsewhere in Western Europe, the past decade has witnessed a paradigmatic transformation of broadcasting from a traditional public-service monopoly with relatively few channels to a "dual," public/private, multi-channel system. In large measure this transformation has been driven by a major state-led infrastructural investment program, commenced during the 1980s, to promote the "new media" of cable and satellite. These new media rendered redundant the "scarcity of frequency" rationale for a continued public broadcasting monopoly and provided the opportunity to launch multiple private commercial channels. Not least in view of Germany's distinctive regulatory culture,[1] the process required a rigorous exercise in re-regulation. However, in important respects, the *formal* re-regulation masks a *substantive* de-regulation. Within the realm of broadcasting policy, economic goals have tended to prevail over cultural ones. This is particularly clear with regard to rules designed to limit the concentration of private media power. Regulatory policy has had a symbolic quality, legitimating what was from the start an already concentrated media industry structure and opening the legal doors to further concentration. The Federal Constitutional Court's guarantee of the vital role of

---

* Peter Humphreys is Reader in Government at Manchester University, where he has taught Comparative West European Politics since 1986. His principal research interest is comparative media policy. He has been a Fellow of the European Institute for the Media (Düsseldorf) since 1988. He is the author of Mass Media and Media Policy in Western Europe (1996); Media and Media Policy in Germany (1994); and the co-author of Broadcasting and New Media Policies in Western Europe (1988). The research upon which this Article is based was funded by the United Kingdom Economic and Social Research Council under its Media Economics and Media Culture program. The three year project (Grant No. L 12625109), which began in January 1996, is entitled "Regulating for Media Pluralism: Issues in Competition and Ownership." It examines policy-making relating to media concentration in Britain and Germany, and also at the European level. The author can be contacted at Peter.Humphreys@man.ac.uk

[1] Regulation has traditionally played a central role in German public policy. This is nowhere more true than in the field of broadcasting policy. For a general discussion of Germany's "regulatory culture," see Kenneth Dyson, *Theories of Regulation and the Case of Germany*, in The Politics of German Regulation 1-28 (1992).

the public broadcasters, strictly regulated to provide balanced and diverse programming (i.e., pluralism), is therefore of crucial significance. Strong public broadcasters are needed as a pluralistic counterbalance to the media power accumulated by a few large, less regulated private concerns.

I. THE *Status Quo Ante*: Public Service Monopoly

In the Federal Republic of Germany, it has always been axiomatic that broadcasting fulfils a special public-service role. Accordingly, "broadcasting freedom" cannot be assured by a *laissez faire* approach; it has to be *both* protected by negative restrictions (e.g., against political interference or dominance by strong social and economic actors) *and* positively promoted (e.g., through strong, publicly accountable and pluralistically representative, public-service broadcasters). German history has demonstrated both the potential for misuse of the medium and its sheer communicative power. Thus, in the view of no lesser authority than Germany's Federal Constitutional Court, broadcasting is not simply a "medium" but also a key "factor" for the functioning of democracy; it has the power to influence and shape opinion.[2] Moreover, the constraint of a relative "scarcity of frequencies" coupled with the high entry costs meant that broadcasting has historically been an activity open to only a limited number of operators. Until the 1980s, therefore, it was widely accepted that broadcasting should be exclusively the task of nine, non-profit-making, regional or occasionally inter-regional (i.e., *Land*-based or inter-*Land*), public corporations responsible for both radio and television.[3] Through their network, the Association of German Public Service Broadcasters (*Arbeitsgemeinschaft der öffentlich-rechtlichen Rundfunkanstalten der Bundesrepublik Deutschland, ARD*), these corporations have provided the First German TV channel (*Erstes Deutsches Fernsehen*), popularly known as "ARD"; they have also provided their own regional channels, the "Third Channel(s)." A second national television channel has been provided by a corporation organized collectively by the *Länder*, the Second German Television corporation (*Zweites Deutsches Fernsehen, ZDF*). The public broadcasters' remit has been the universal provision of a comprehensive, balanced and diverse range of programming that caters to German society's pluralism.

---

[2] *See* Silke Ruck, *Development of Broadcasting Law in the Federal Republic of Germany*, 7 EUR. J. COMM. 219, 223 (1992).

[3] Unification added two more corporations. Meanwhile, however, an Inter-*Land* Treaty of July 31, 1997, has provided for the merger of two of the southern German ones into *Südwestrundfunk*.

The public corporations have been funded by income from a license-fee levied on TV households, supplemented by revenue from strictly limited amounts of advertising air-time. Free from the commercial pressures that typically arise from competition for advertising revenue and viewer ratings in competitive, commercial broadcasting systems, German public broadcasters have delivered serious and socially responsible programming, informing and educating as well as entertaining the public.

The accountability mechanisms for German public broadcasting are rather unique. They originated in the Western Allies' policies, immediately after the Second World War, to ensure that broadcasting in (West) Germany should be decentralized and controlled pluralistically. The post-war German elites, too, accepted that broadcasting should be controlled in a way that safeguarded its independence from the state (the hallowed principle of *Staatsferne*, literally "distance from the state"). They also accepted the fact that the mass medium should on no account fall under the control of any powerful social interest or interests. Formally organized as corporations under public law (*Anstalten des öffentlichen Rechts*), Germany's public broadcasting institutions are classic examples of distinctly non-state, non-market media. They have been controlled by internal broadcasting councils (*Rundfunkräte*)—or, in the case of the ZDF, by a television council (*Fernsehrat*). These internal regulatory bodies have each contained representatives of the country's "socially significant groups" (*sozial relevante Gruppen*): i.e., cultural bodies, churches, employers' associations, trade unions, and so on, alongside directly political representatives. This kind of "internal control" (*Binnenkontrolle*) of the broadcasters is designed to guarantee the balanced and diverse character—the "internal pluralism" (*Binnenpluralismus*)—of their programming. The state in Germany only exercises a background regulatory role through the enactment and limited supervision[4] of broadcasting laws. Yet, this activity too is decentralized; broadcasting legislation and supervision is covered by the "cultural sovereignty" (*Kulturhoheit*) of the constituent federal states (*Länder*) of the Federal Republic.[5]

---

[4] *Rechtsaufsicht*, i.e., ensuring that all the broadcasters' affairs are conducted within their legal remit.

[5] For an account of the early development of the West German broadcasting system, see PETER HUMPHREYS, MEDIA AND MEDIA POLICY IN GERMANY ch. 3 (1994).

II. THE INTRODUCTION OF THE NEW MEDIA

The introduction of private commercial broadcasting from the mid-1980s onwards initially produced much controversy. On the one hand, a powerful coalition of forces favored commercial broadcasting. The advertising industry wanted an expansion of television advertising and competition between carriers of advertising (i.e., the abolition of the public broadcasters' monopoly of air-time that kept advertising charges high). Since the early 1970s, the electronics industry had lobbied energetically for policies that would promote the competitiveness of German industry in the looming Information age. Specifically, industry envisaged exciting new markets for cable systems, broadcasting satellites, satellite reception equipment, pay-TV decoders, new television sets, and so forth. The (then) state telecommunications operator—the *Bundespost*—wanted to develop the new technologies in order to assert its monopoly over the provision of infrastructure and key communications services.[6] Policy makers (national and *Länder*) were naturally eager to promote German industry in emerging international markets for new information and communications technologies. The CDU/CSU/FDP[7] coalition, in power since 1982 in Bonn, has clearly prioritized this goal over other concerns (e.g., competition policy, cultural concerns, etc.). Finally, German press groups, too, sought the opportunity to diversify into commercial broadcasting, not least in order to secure their share of media advertising revenue against future competition.

For their part, Germany's SPD were discomfited. They too wanted to reap the commercial advantages—for Germany, and for the *Länder* they governed— that were promised by the new media. Against this, they recognized that the mass introduction of the new media would mean the inevitable abolition of the public service broadcasting monopoly which they had always supported. In office in Bonn (1969-1982), the SPD had actually promoted the development of broadcasting satellites but at the same time imposed what the Christian Democrats had called a "cable blockade."[8] This rather contradictory policy reflected a deep-seated ambivalence

---

[6] *Id.* at 195-203.

[7] Chancellor Helmut Kohl's *Christlich-Demokratische Union* or Christian Democratic Union ("CDU"), together with its Bavarian sister party, the *Christlich-Soziale Union* or Christian Social Union ("CSU"), have been in a coalition with Germany's small Liberal party, the *Freie Demokratische Partei* or Free Democratic Party since the latter switched from its earlier "social-liberal" coalition (1969-1982) with the *Sozialdemokratische Partei Deutschlands* or "Social Democratic Party of Germany" ("SPD").

[8] It should be noted that the SPD did agree to a limited number of experimental cable television pilot projects.

about the new media on the part of the SPD. Sections of the SPD grass-roots (*Parteibasis*), the unions, and the Green party were extremely negative about the new media; even the churches worried about the possible negative effects of commercial broadcasting.[9]

When the CDU/CSU came to power in Bonn in 1982, the SPD's "cable blockade" was immediately overturned. Although broadcasting law and regulation were in the jurisdiction of the *Länder*, telecommunications policy was within federal jurisdiction. Therefore, the new CDU *Bundespost* minister was able to press full speed ahead with a program of massive state support for the introduction of cable and satellite (i.e. "telecoms") infrastructure. At the same time, CDU/CSU *Länder* began to enact their own laws to allow private commercial broadcasters to operate within—and from—their areas of jurisdiction. This, in turn, placed overwhelming pressure on the SPD *Länder* to do the same; otherwise, they would be effectively denying themselves the *Bundespost*'s cable investment, without being able to obstruct viewers in their *Länder* from receiving private satellite channels broadcast out of CDU/CSU *Länder*. Faced with these new realities, leading SPD politicians in the *Länder* began to prioritize the economic goals of broadcasting policy. The name-of-the-game for many *Länder* politicians—regardless of political color—became how to attract private media investment to their regions. From this point on, media policy in Germany became increasingly subsumed into what economists call locational policy (the term is *Standortpolitik*).[10]

III. THE CONSTITUTIONAL COURT'S KEY ROLE IN THE RE-REGULATION OF GERMAN BROADCASTING

In post-war Germany's "constitutional-legalistic" political culture, major controversies over the broadcasting system have been resolved by the Federal Constitutional Court in Karlsruhe (the federal republic's "Supreme Court"), which has therefore acted "much like a legislature" in establishing basic guidelines for the broadcasting order.[11] Since the beginning of the 1960s the case law of the Federal Constitutional Court has established the broad parameters of broadcasting regulation within which the legislators of the individual *Länder* have exercised their freedom to maneu-

---

[9] HUMPHREYS, *supra* note 5, at 204-11.
[10] In 1984 the SPD national party conference also recognized the sheer futility of continuing to oppose private commercial broadcasting, albeit after a fierce debate and much grass-roots opposition.
[11] WOLFGANG HOFFMANN-RIEM, REGULATING MEDIA: THE LICENSING AND SUPERVISION OF BROADCASTING IN SIX COUNTRIES 119 (1996).

ver.[12] Thus, in its famous "First TV Ruling" of 1961,[13] the Federal Constitutional Court had ruled against the introduction by the federal government of a commercial national television service. This confirmed both the *Länder* jurisdiction for broadcasting and, on the grounds of the scarcity of frequencies and high entry costs of broadcasting, the public-service broadcasting monopoly. However, two decades later the Federal Constitutional Court, with a 1981 ruling,[14] actually paved the legal path for the introduction of commercial broadcasting by recognizing that the new media of cable and satellite (and new possibilities for terrestrial broadcasting) rendered obsolete the "scarcity of frequencies" rationale for a public broadcasting monopoly. In other words, in 1981 the Constitutional Court gave the *Länder* the legal green light to legislate for private commercial broadcasting. At the same time, however, the Court made it very clear that broadcasting regulation should not be characterized by a "free-for-all" (*freies Spiel der Kräfte*). Private broadcasting, too, had public service obligations; above all, it had to supply pluralism in the expression and formation of opinion (*Meinungsvielfalt*).

In this vein, the Constitutional Court suggested that multi-channel broadcasting now provided the scope for a new "external pluralistic" model of broadcasting wherein individual program services might reflect an imbalance, so long as the traditional goals of balance and diversity were supplied by the totality of broadcasting output. The supervision of multi-channel broadcasting could accordingly be provided by an "external control" authority placed above the broadcasters, rather than an "internal control" body within them as had always been the case with public corporations (as described above). From 1984 onwards the individual federal states (*Länder*) enacted a wave of legislation for private commercial broadcasting. The new *Land* laws varied considerably in their details, but essentially they attempted to do two things: to open up opportunities for private media investment; and to safeguard time-

---

[12] *See* HUMPHREYS, *supra* note 5, at 338-442 (summarizing the Federal Constitutional Courts' key rulings on broadcasting through 1991); *see also* VINCENT PORTER & SUZANNAH HASSELBACH, PLURALISM, POLITICS AND THE MARKETPLACE: THE REGULATION OF GERMAN BROADCASTING (1991); Ruck, *supra* note 2, at 219-39. The Constitutional Court's rulings are reproduced in the journal *Media Perspektiven*.

[13] BVerfGE 12 (1961), 205, 1 Rundfunkentscheidung (Deutschland-Fernsehen). *See* HUMPHREYS, *supra* note 5, at 338.

[14] BVerfGE 57 (1981), 295, 3 Rundfunkentscheidung (FRAG/Saarlaendisches Rundfunkgesetz) (Third TV Ruling). Meanwhile, in 1971, a "Second TV Ruling," not referred to here, had confirmed the public-service monopoly. *See* HUMPHREYS, *supra* note 5, at 339-40.

honored standards and values, especially the constitutional-legal goal of pluralism, in broadcasting.

To this end, all of the individual *Länder* laws provided for the licensing and supervision of private broadcasters by a new tier of legally autonomous *Länder* regulatory authorities (*Landesmedienanstalten*). Following Unification there were fifteen of these authorities.[15] Generally, these regulatory authorities had pluralist supervisory boards composed of representatives of the "socially significant groups."[16] These provided for the pluralistic social control of private broadcasting in a manner that was analogous to the above-mentioned public-service broadcasting councils. Their key purpose was to ensure that private broadcasting was sufficiently pluralistic, especially with regard to diversity of opinion (*Meinungsvielfalt*). Following the Constitutional Court's 1981 ruling, this goal might now be achieved across a range of program services (i.e. "external pluralism"). Some (SPD) *Land* laws required "internal pluralism" from the private broadcasters, some (CDU/CSU) ones opted for "external pluralism," and the rest adopted mixed or transitional models.[17]

In 1986 the Court further clarified how pluralism in the emergent public/private "dual system" should be guaranteed. In its so-called "Fourth TV Ruling,"[18] the Court stressed that the public-service broadcasters future role was crucial: they should continue universally to provide a basic comprehensive service (*Grundversorgung*). The very constitutionality of private commercial broadcasting and its exemption from the same high programming requirements and from the same degree of close regulation (i.e. "internal control") depended on the public broadcasters' fulfillment of this key role.[19] The Court also stipulated that the private broadcasters should observe a basic standard of pluralism (*Grundstandard gleichgewichtiger Vielfalt*—literally "basic standard of balanced diversity"). This meant that all directions of opinion—including minority ones—should "have the possibility" of being expressed. Moreover, adequate measures should be taken by the leg-

---

[15] One regulatory authority for each of the united Germany's 16 *Länder*, except Berlin and Brandenburg, who shared one.

[16] In some cases, there were smaller executive boards. For a detailed, critical overview, see HOFFMANN-RIEM, *supra* note 11, at 125.

[17] HUMPHREYS, *supra* note 5, at 343; PORTER & HASSELBACH, *supra* note 12, at 57-59.

[18] BVerfGE 73 (1986), 118, 4 Rundfunkentscheidung (Landesrundfunkgesetz Niedersachsen).

[19] The Court has confirmed the bedrock nature of public service broadcasting in several subsequent rulings. Most notably, in 1987, it guaranteed their "continuity and development" (*Bestand und Entwicklung*), specifying that they should be able to expand into new technological and programming areas.

islators in the *Länder* and the media regulators to prevent the appearance of "dominant influence over the expression and formation of opinion" (*vorherrschende Meinungsmacht*). In other words, it was a constitutional-legal principle that media concentration should be counteracted.

IV. THE "MEDIA OWNERSHIP" MODEL OF CONCENTRATION CONTROL

The 1986 Federal Constitutional Court ruling at last provided the basis for the enactment by the *Länder* in 1987— after four years of hard negotiation in the inter-*Land* forum of the conference of *Länder* premiers (*Ministerpräsidentenkonferenz*)—of the "First Inter-*Land* Treaty on Broadcasting."[20] This, refined by a "Second Inter-*Land* Treaty on Broadcasting" of 1991,[21] included national framework rules for the new media and the regulation of the "dual" system along the lines recommended by the Constitutional Court. With regard to pluralism, the treaties first specified that the threshold for achieving "external pluralism" in the supply of new private television services was the availability of three national private commercial "generalist" channels provided by different companies. In the event of the availability of fewer than three such channels, then each private channel had to provide "internal pluralism."[22] Secondly, the treaties contained restrictions on the concentration of ownership of private commercial television channels. The "ownership model" of concentration control exhibited both "external pluralistic" and "internal pluralistic" elements. The former were fairly liberal, the latter more restrictive. External pluralism was to be assured by limiting the number of licenses held by a single nationwide private commercial broadcaster to a maximum of two channels each in radio and television, only one of which might be for a "generalist service" (*Vollprogramm*) or an "information service" (*Informations-Spartenprogramm*, i.e., a news and current affairs channel). Internal pluralism was to be guaranteed by strict limitations on individual shareholdings in generalist and information chan-

---

[20] *Staatsvertrag zur Neuordnung des Rundfunkwesens (Rundfunkstaatsvertrag)*, Mar. 12, 1987, *reprinted in* MEDIA PERSPEKTIVEN, DOKUMENTATION 81-102 (1987).

[21] *Staatsvertrag über den Rundfunk im vereinten Deutschland*, Aug. 31, 1991, *reprinted in* MEDIA PERSPEKTIVEN, DOKUMENTATION 105-72 (1991). This treaty catered to German Unification—which amounted to an effective extension of the West German model of broadcasting into the "Five New *Länder*." It also translated European legislation eliminating market barriers into German law. It also rendered more precise the ownership rules regarding private commercial television services.

[22] *See* Treaty of 1987, art. 8, I-III, VI; Treaty of 1991, par. 20 I-IV. This requirement was dropped from the 1996 Inter-*Land* Treaty, but by this time there were three national "generalist" channels operating.

nels. To be precise, any individual shareholder in such channels was limited to one single stakeholding below 50% and two more below 25%. Theoretically, these shareholding restrictions would ensure that broadcasting operations would be "pluralistic" associations of enterprises interested in broadcasting (*Anbietergemeinschaften*). Furthermore, the law stipulated that "comparable" means of influence (to direct shareholding) over programming, such as the possession of a dominant position in the supply of programs, should also be taken into consideration by the regulators. Yet cross-media ownership between press and nationwide broadcasting services was left conspicuously unrestricted.[23] Nonetheless, it was possible to argue that the inter-*Land* treaties, together with the wave of *Land* laws and the foundation of fifteen new regulatory authorities, constituted an extensive *formal* re-regulation of German broadcasting.

V. The Ineffectiveness of the "Media Ownership Model": Symbolic Regulation?

It is also possible to argue, however, that behind this formal re-regulation there was a *substantive* de-regulation. Thus Wolfgang Hoffmann-Riem, a leading authority on German broadcasting law and regulation, had suggested that the broadcasting re-regulation of the 1980s might have a mainly symbolic function, serving to provide "a politically and economically "well-ordered" entrance into a new age of broadcasting in accordance with the market model."[24] Hoffmann-Riem specifically referred to the work of the American political scientist, Murray Edelman, who had argued that business regulation was often largely symbolic in function: serving to produce political quiescence.[25] For example, Edelman had claimed that the U.S. Federal Trade Commission "ha[d] long been noted for its hit-and-miss attacks on many relatively small firms involved in deceptive advertising while it continue[d] to overlook much of the really significant activity it [wa]s ostensibly established to regulate: monopoly, interlocking directorates, and so on." Turning to

---

[23] *Staatsvertrag über den Rundfunk im vereinten Deutschland*, Aug. 31, 1991, *reproduced in* Media Perspektiven, Dokumentation 105, ¶ 21 I-V (1991). Cross-media ownership rules were contained in individual *Länd* laws, but the absence of an inter-*Land* standard encouraged *Standortpolitik*-led deregulation. Several *Land* laws saw cross-media ownership as something to be positively welcomed. *See* Hoffmann-Riem, supra note 11, at 136; *see also* Porter & Hasselbach, *supra* note 12, at 121-29.

[24] Wolfgang Hoffmann-Riem, *Law, Politics and the New Media: Trends in Broadcasting Regulation, in* The Politics of the Communications Revolution in Western Europe 125-46, 144. (Kenneth Dyson & Peter Humphreys eds. 1986).

[25] Murray Edelmann, The Symbolic Uses of Politics 22-43 (1964).

broadcasting regulation, Edelmann singled out decisions permitting greater concentration of control as one of the areas where the Federal Communications Commission gave "rhetoric to one side"—waxing "emphatic in emphasizing public service responsibility"—while it gave "the decision to the other."[26] The U.S. antitrust laws also seemed to promote the growth of great industrial organizations.[27]

It is certainly significant that anti-concentration policy is the principal area in which the new regulators of private sector broadcasting in Germany have fallen short of adequately fulfilling the Constitutional Court's injunction to protect pluralism in German broadcasting. While the regulators (the *Landesmedienanstalten*) have generally been active in pursuing other infringements of broadcasting law (such as advertising quotas and standards), they have failed to curb media concentration and thus failed to adequately address the more serious threat of "dominant influence over the expression and formation of opinion" (*vorherrschende Meinungsmacht*)" against which the Court explicitly warned. What, then, explains their failure?

It has already been mentioned that during the 1980s media policy became increasingly subsumed into locational policy (*Standortpolitik*). The introduction of private commercial broadcasting quickly gave rise to fierce inter-regional economic rivalries. At the same time, the federal structure of the German broadcasting regulation—with no fewer than fifteen regional regulatory authorities—rendered effective regulation very difficult. A national satellite broadcasting service could apply for and receive a license in a *Land* where the media ownership rules were suitably lax; its programs could then be transmitted for direct-to-home reception (by rooftop satellite receiving dishes), or re-transmitted via the country's rapidly expanding cable infrastructure, right across the federal republic (though terrestrial transmission still needed a license from each *Land* concerned).[28] Clearly, those *Länder* with liberal regulation were placed strategically to attract the new media investment. Many observers have commented on the way regional legislators, in translating the Constitutional Court's injunctions on media pluralism into *Land* media laws, and the *Landesmedienanstalten*, in

---

[26] *Id.* at 38-39.
[27] *Id.* at 40.
[28] This principle of recognition of transmissions lawfully broadcast elsewhere is spelled out in paragraph 35 of the Inter-*Land* Treaty. In fact, it not only applies to Germany. The principle applies across the European Union ("EU") following the 1989 EU Directive *Television Without Frontiers*. *See* Ruck, *supra* note 2, at 229.

translating the law (both *Land* media laws and inter-*Land* treaties) into regulatory practice, were under intense political pressure to prioritize the economic interests of their respective *Länder* which were vigorously competing to attract—or retain—media investment (i.e. *Standortpolitik*). A German political scientist Axel Zerdick neatly summarized the problem: "the objectives of the *Land* media laws, orientated less towards media policy than towards *Standortpolitik*, [had] led to a state of affairs in which too many *Landesmedienanstalten* controlled too few broadcasters who, at the same time, [were] being courted avidly by the respective *Land* governments."[29]

From a political-science perspective, the weakness of cross-media ownership regulation— amidst all the stricter formal restrictions—could be simply explained. As already noted, the large press companies had constituted one of the principal lobbies pushing for the introduction, and their own participation in, private broadcasting. Keen on stimulating regional investment, legislators and regulators in the *Länder* had readily obliged by actively encouraging diversification of the press into broadcasting. "This," Hoffmann-Riem has noted, "resulted in intensive, multimedia integration at the very outset of private broadcasting in Germany, which has increased ever since."[30] The major influence of press companies in German broadcasting was a "general trend" to which even those *Länder* that initially sought countermeasures "were forced to capitulate."[31] Moreover, by the time that the inter-*Land* treaties could be enacted for nationwide broadcast services, extensive cross-media involvement in commercial broadcasting had already occurred (e.g., the dominance of press ownership of the SAT 1 channel). In theory, of course, this did not prevent the formulation of cross-media rules to compel the *ex post-facto* de-concentration of the mass media. However, with respect to cross-media ownership, the rules for nationwide commercial broadcast services in Germany could be judged to have been tailored to market *faits accomplis* from the start.

Apart from sins of omission or commission, there were also serious problems of regulatory design and implementation. A major problem with the rules was that they were rendered difficult to implement effectively by the complex *Verflechtung* (interlocking webs of interests), which soon came to characterize the ownership

---

[29] Axel Zerdick, *Zwischen Frequenzen und Paragraphen: die Landesmedienanstalten als institionalisierter Kompromiß*, 129 BERTELSMANN BRIEFE 60-62 (1993).

[30] *See* HOFFMANN-RIEM, *supra* note 11, at 136; *see also* PORTER & HASSELBACH, *supra* note 12, at 124-26.

[31] *Id.* at 136.

structure of the commercial television sector as influential media investors acquired multiple holdings in media companies and co-operated in several broadcasting operations (*Anbietergemeinschaften*—see above).[32] Moreover, it soon became obvious that there were even less transparent kinds of linkage and control at work in the private television sector (e.g., finance, program supply). Further, control of broadcasting operations could be assured through intermediaries, front-men, and even family members.[33] To cap it all, the *Landesmedienanstalten* were not endowed with adequate powers to determine the legal reality behind such complex relations of control. Notably, they did not have the same rights of investigation of broadcasters' internal affairs that the Federal Cartel Office enjoyed for ensuring fair competition in German industry at large. Therefore, the regulators of private commercial broadcasting were largely dependent on the cooperation of the media companies they were supposed to regulate.[34] The greatest obstacle to regulatory effectiveness, though, was the "competitive deregulation of competition policy" (*Wettbewerb um die Wettbewerbspolitik*) that derived from the decentralized regulation of a highly concentrated national media industry. Evidently, the control of media concentration needed to be conducted at a higher level than that of the individual *Land* authorities in order to prevent "soft competition policy" from serving as an instrument of regional *Standortpolitik*.[35] Yet coordinated, collective action by the *Landesmedienanstalten* themselves was marred by difficulties and disputes—occasionally reaching the courts—between regulators who appeared to have come to view themselves as "advocates of 'their' companies, that is those headquartered within their jurisdictions."[36]

The symbolic and ineffectual nature of anti-concentration regulation became quickly manifest in an oligopoly of two giant national commercial "broadcasting families": notably, the Springer (press)/Kirch Group versus Bertelsmann/*Compagnie Luxembourgeoise de T'él'édiffusion* (CLT).[37] Both "broadcasting families"

---

[32] HOFFMANN-RIEM, *supra* note 11, at 137-38.

[33] HORST RÖPER & ULRICH PÄTZOLD, MEDIENKONZENTRATION IN DEUTSCHLAND: MEDIENVERFLECHTUNGEN UND BRANCHENVERNETZUNGEN 192 (1993).

[34] Dieter Dörr, *Kontrolle braucht Durchsetzunsbefugnisse*, in MEDIA PERSPEKTIVEN 42-47 (1995).

[35] Jürgen Heinrich, *Keine Entwarnung bei Medienkonzentration: ökonomische und publizistische Konzentration im deutschen Fernsehsektor 1993/94*, in MEDIA PERSPEKTIVEN 297, 298 (1994).

[36] HOFFMANN-RIEM, *supra* note 11, at 137.

[37] *See* Europäisches Medieninstitut, *Bericht über die Entwicklung der Meinungsvielfalt und der Konzentration im privatem Rundfunk gemäß* ¶ 21, *abs.* 6 Staatsvertrag über den Rundfunk im

1998] RULES FOR PRIVATE BROADCASTING IN GERMANY 539

(*Senderfamilien*) had significant interests in a number of program services; above all, they each controlled one of the two main "generalist" private channels (*Vollprogramme*). Bertelsmann/CLT controlled RTL and Kirch/Springer controlled SAT 1. Between them RTL and SAT 1 accounted for the lion's share of private sector viewing time and television advertising revenue in Germany. Bertelsmann is the world's third largest multimedia concern, with global interests in printing, book and magazine publishing, the record industry, and recently film and broadcasting production. Leo Kirch is Germany's leading dealer in program rights,[38] and also a major shareholder (35%) in Springer, which has a large share of the newspaper market. Leo Kirch's son, Thomas Kirch, appeared to have a controlling interest in Germany's third and only other major generalist private broadcasting channel, PRO 7.[39]

The PRO 7 case presented a classic illustration of the problems for effective regulation outlined above. The fifteen *Landesmedienanstalten* were divided over whether the channel should be counted as belonging to Leo Kirch's broadcasting "family." Thomas Kirch held a 47.5% share in PRO 7 and 3% more were held by PRO 7's managing director, formerly a top executive of Leo Kirch's. Moreover, it was widely believed that PRO 7 was dependent on the Kirch group for finance and program supply. Since Leo Kirch already had a major (direct and indirect) holding in the major "generalist" channel SAT 1, he should in theory—under the media ownership rules described above—have been prevented from having a significant interest (i.e., 25% or more) in another "generalist" channel.[40] However, the Schleswig-Holstein regulatory authority which had licensed PRO 7 declined to accept that PRO 7 was under the control of Leo Kirch. The *Landesmedienanstalten* repeatedly failed to collectively resolve the issue.[41]

---

vereinten, (*Deutschland, in* DIE LANDESMEDIENASTALTEN (HSRG), DIE SICHERUNG DER MEINUNGSVIELFALT; BERICHTE, GUTACHTEN UND VORSCHLÄGE ZUR FORTENTWICKLUNG DES RECHTS DER MEDIENKONZENTRATIONSKONTROLLE VOM HERBST 1994, at 127-220 (1995). The European Institute for the Media was commissioned to produce this report on concentration in the private broadcasting sector for the Association of the private broadcasting regulatory authorities (*Landesmedienanstalten*). It presents detailed information on the then ownership and control relations.

[38] He had started out as a supplier of programming to the public broadcasters.

[39] PRO 7 also controlled a subsidiary called Kabelkanal (now Kabel 1).

[40] Or in an information-orientated thematic service, e.g., a news channel.

[41] Schleswig-Holstein later tightened up its rules, introducing a prohibitive "family clause." Meanwhile, however, PRO 7 transformed itself into a publicly quoted company, partly to avoid the charge that it was part of the Kirch group, and the broadcaster moved to Berlin. One condition for the award of a Berlin license was that Thomas Kirch reduce his

The rival media "families" were loosely identified with the two main political parties.[42] Bertelsmann, which appears to have a special relationship with the SPD, is based in Social-Democratic North Rhine-Westphalia, as is the principal Bertelsmann/CLT television operation, RTL. The politically conservative Kirch group is based in Christian Social Union (CSU) governed Bavaria. These two *Länder*—Social Democratic NRW and Christian Social (i.e. conservative) Bavaria—were key protagonists in the *Standortpolitik*.[43] The high degree of media concentration in the Federal Republic, and the regulatory weakness of the *Landesmedienanstalten* led to serious discussion from 1994 onwards about how Germany's anti-concentration rules might be reformed. Predictably, "the question of ownership rules in private television has been . . . one of the most intensely discussed topics in the German media policy debate."[44] Numerous and loud were the calls from many quarters for more restrictive ownership rules and for more effective regulation. The unions, churches, regulatory bodies, media experts and academics, and many SPD party organizations called for action. The CDU/CSU, too, recognized the need to take action.[45]

In 1994 the *Länder* regulatory bodies (*Landesmedienanstalten*) themselves proposed a re-regulation of media ownership.[46] They suggested that the problem of the debasement of regulation by inter-*Land* competition for investment—the *Standortpolitik* problem—could be overcome by their closer cooperation (e.g., a special joint review body for license applications). They also suggested that media ownership should no longer be based on the size of an interest's stakeholding, but rather on the audience share controlled by individual media owners. An "audience share model" of regulation would, it was felt, most likely overcome the thorny problem of *Verflechtung* (complex interlocking webs of interest);[47] it

---

holding in PRO 7 below 25%. However, the new media ownership rules of 1996 have allowed him to increase his holding again, to 60%!

[42] *See, e.g.*, Hans Kleinsteuber & Bettina Peters, *Media Moguls, in Germany in* JEREMY TUNSTALL & MICHAEL PALMER, MEDIA MOGULS 184-205 (1991).

[43] The fact that they are especially large and populous *Länder* gave them extra weight in the inter-*Land* media policy debate.

[44] Christa-Maria Ridder, *Germany, in* MEDIA OWNERSHIP AND CONTROL IN THE AGE OF CONVERGENCE 65 (Int'l Inst. of Communication ed. 1996).

[45] The positions of the various parties and interests, including the relevant media industry interests, are presented in HERMANN KRESSE, PLURALISMUS, MARKT UND MEDIENKONZENTRATION: POSITIONEN (1995). The journal *Media Perspektiven* has also presented the positions of the main parties to the debate. *See also* Landesmedienastalten (Hsrg.), *supra* note 37, at 127-220.

[46] *See Lübeck Resolution of Sep. 17,1994, cited in* KRESSE, *supra* note 45, at 162-72. .

[47] *Verflechtung* would be considered redundant when media companies could control outright as many companies as they wanted up to the audience share limit.

would be conducive to improved transparency in the ownership and control relations in the sector. The German regulators suggested that the critical limit for undue ownership concentration should be set at the point when an interest controlled more than 25% of the television audience market. However, the German media industry immediately responded by claiming that a 25% limit of audience share would severely damage German competitiveness and encourage foreign takeovers of German broadcasting interests. The Kirch group called for a 35% audience limit.[48] Bertelsmann called for a limit of at least 30%.[49] These limits, of course, were liberal enough not to affect these concerns' current investments or even their further expansion into private broadcasting. Indeed, these large media concerns—which dominated the industry lobby— saw the "re-regulatory" exercise as an opportunity to carve out more freedom for their future expansion in the emerging age of global media operations, digital TV, and the much discussed "convergence" between telecoms, computing, and broadcasting.[50] The pro-liberalization lobby made much of the competitive threat posed by foreign media interests.[51] The current rules had, it was pointed out, already led to the German channel VOX being rescued by Rupert Murdoch's News International (he took a 49.9% stake in the channel) because current rules ruled out the obvious German media interests. As seen, the Luxembourg-based broadcasting multinational CLT was a major player in the German market.

There followed a fairly protracted process of political tugging and hauling. At first, the SPD called for stricter regulation along existing lines to protect pluralism in broadcasting, and was skeptical about the audience-share model. The CDU/CSU, by contrast, did not see a significant threat to broadcasting pluralism given the number of channels, both public and private, on offer to the German viewer. The CDU/CSU therefore lent its weight broadly to the industry position. Indeed, the CDU/CSU employed the concurrent issue about how much to raise the household broadcasting licence-fee, the public broadcasters' principal source of income, as a lever in the negotiations concerning re-regulating media ownership laws. The license-fee issue presented an opportunity to pres-

---

[48] Kirchgruppe, *Die Zukunft gestalten - Perspektiven einer vorwärtsgerichteten Medienpolitik*, in KRESSE, *supra* note 45, at 159.

[49] Bertelsmann AG, *Die medien-, kommunikations- und technologiepolitische Position des Hauses Bertelsmann*, in KRESSE, *supra* note 45, at 123.

[50] MARTIN STOCK ET AL., MEDIENMARKT UND MEINUNGSMACHT: ZUR NEUREGELUNG DER KONZENTRATIONSKONTROLLE IN DEUTSCHLAND UND GROBBRITANNIEN 2 (1997).

[51] Ridder *supra* note 44, at 67. *See e.g.*, Bertelsmann AG, *supra* note 49, at 118.

surize the SPD, the "champion" of the public-service broadcasters, into conceding to a liberal regulatory framework for media ownership.[52] However, the SPD position was ambiguous: influential elements within the SPD were very sensitive to the interests of Bertelsmann, a concern which—as mentioned—was friendly to the SPD and which was headquartered and invested heavily in the SPD heartland of North-Rhine Westphalia, Germany's most populous *Land*. In reality, not much separated the most influential policy makers; behind the political symbolism, both parties catered to considerations of regional economic policy (*Standortpolitik*).

Consequently, after a series of negotiations on policy details, which was steered by a small number of high-ranking *Land* politicians and legal experts in the state chancellories of the *Länder*, a new regulatory framework was agreed upon. It embraced the audience-share model and abolished the existing media ownership rules that had sought to prevent the excessive accumulation of channels and majority shareholdings in broadcasting companies. Henceforth "generalist" and other information-relevant channels could be accumulated and also owned in their entirety by individual investors, until their combined audience share exceeded the stipulated limit. The lengthy section of the "Third Inter-Land Treaty on Broadcasting"[53] devoted to the "Safeguarding of Diversity of Opinion" (*Sicherung der Meinungsvielfalt*, Paras. 25-34) provided a mixture of clear-cut numerical rules and less clear-cut, general stipulations which would allow for considerable regulatory flexibility in their implementation.[54]

The treaty established a precise threshold of 30% of the total national television audience, including that of the public-service broadcasters. Once a concern's channels had reached this threshold, that concern would be assumed to have acquired "dominant influence over the expression and formation of opinion" (*vorherrschende Meinungsmacht*), and its further expansion would be

---

[52] Since 1970, decisions on raising the level of the license-fee had been made by the premiers (*Ministerpräsidenten*) of the *Länder*. To help them, they had established their own advisory body called the Commission for Assessing the Financial Requirements of the broadcasting corporations (*Kommission zur Ermittlung des Finanzbedarfs der Rundfunkanstalten* - KEF). This latter body, however, was not really independent of the politicians and the process of setting the licence fee presented scope for political bargaining. Further scope for political leverage during the period under review arose from the fact that the *Länder* politicians were in the process of negotiating a reform of the license-fee procedure following the Federal Constitutional Court's "Eighth Broadcasting Ruling" of February 1994. *See infra* notes 73-76 and accompanying text.

[53] *See Dritter Staatsvertrag zur Änderung rundfunkrechtlicher Staatsverträge vom 26.8/ 11.9.1996* (*Dritter Rundfunkänderungs-staatsvertrag*), *reprinted in* MEDIA PERSPEKTIVEN, DOKUMENTATION (1996). The treaty came into force on January 1, 1997.

[54] STOCK ET AL., *supra* note 50, at 2-3.

1998] RULES FOR PRIVATE BROADCASTING IN GERMANY 543

blocked. In calculating the audience shares attributable to individual media interests, only holdings of 25% or more of the capital or voting shares in broadcasting operations were to be counted. Once an individual investor went beyond this so-called "insignificance threshold" (*Bagatellgrenze*), then the audience share of the broadcasting operation in question would be fully attributed to the individual investor. Below the "insignificance threshold," an individual investor's influence was deemed not to be significant and no audience share whatsoever would be attributed to that investor. Far less clear-cut was the provision that other, "comparable" sources of influence (e.g., program supply) would also be taken into account by the Commission for Determining Media Concentration (*Kommission zur Ermittlung der Konzentration im Medienbereich*—"KEK"), the expert body established to implement the new audience share rules. The KEK was empowered to investigate excessive cross-media ownership (through the *Landesmedienanstalten*) but "media-relevant related markets" (*medienrelevante verwandte Märkte*) would be taken into consideration only when a television concern approached the 30% audience share threshold. What constituted "approaching" 30% was left to the KEK to decide.

The KEK was a new inter-*Land* body composed of six experts in broadcasting and business law, to be appointed by the *Länder* premiers (and reappointable once). The KEK was the body ultimately responsible for safeguarding pluralism and counteracting media concentration in the national broadcasting market (i.e., not local or regional). The KEK examines license applications to the individual *Landesmedienanstalten* and has the last say about licensing decisions in so much as they might affect media concentration. Its decisions may only be overturned if at least three quarters of the *Landesmedienanstalten* agree. The KEK is also charged with the task of ensuring transparency of ownership and control in the television sector, in cooperation with the *Landesmedienanstalten* which are at last endowed with stronger information rights and powers of investigation *vis-a-vis* the media industry. As intended, the establishment of the KEK may well help overcome the problems arising from the decentralized regulation of national private broadcasting, in particular the influence of *Standortpolitik*. However, critics have pointed out, the mode of selection of the members of the KEK—by the prime ministers of the *Länder*—carries the danger that the KEK will not be sufficiently independent of the state (*staatsfern*), a key principle of German broadcasting regulation. Some have suggested, too, that the body is too "technocratic," and fails to represent the range of German society (e.g., in the manner of the *Rundfunkräte* of the public broadcasters).[55]

Another promising innovation is the provision that broadcasters with an audience share of 10% or more should make available air-time ("windows") to "independent third parties" *(unabhängige Dritte)*, i.e., small broadcasters who are independent of their "host" major broadcasters and who commission or produce their own window programs.[56] This rule therefore guarantees a measure of "internal pluralism." It is based on policy as already practised in SPD North-Rhine Westphalia (for broadcasters applying for terrestrial frequencies in this populous *Land*) and the concept was promoted enthusiastically by the SPD *Länder* in the negotiations over the new inter-*Land* treaty. The 10 per cent threshold for triggering window requirements had been described by the industry lobby as a "crass interference in the broadcasting freedom of private broadcasters."[57] The large broadcasters worried that such programs might be "ratings killers." Nonetheless, it was accepted by the CDU/CSU as part of the overall compromise package. On the one hand, the concept has seen the appearance of interesting "window" programs like *Spiegel-TV* and *Stern-TV*, and a range of regional "windows"; on the other hand, some have questioned the genuine independence of certain "third party broadcasters."[58] While doubts remain about the practical implementation of the rule, this innovative idea certainly offers one feasible way of promoting media pluralism within the oligopolistic broadcasting market that has developed in Germany.

However, the new media ownership limitations have left the oligopoly largely untouched; the rules even allowed scope for the further expansion of firms like KirchGruppe and Bertelsmann in the broadcasting sector. Moreover, the new rules contain only a single paragraph specifically referring to digital broadcasting, which is the technical delivery system of the future. This paragraph demands "equal, appropriate and non-discriminatory" conditions of access for all TV services to a digital platform (and to the

---

[55] Dieter Dörr, *Massnahmen zur Vielfaltssicherung gelungen*, in MEDIA PERSPEKTIVEN (1996).

[56] This should amount to a minimum of 260 minutes per week, of which a minimum of 75 minutes should be in prime viewing time. A window producer would have to be independent of the broadcaster and would be licensed by the relevant *Landesmedienanstalt*. If a window producer does not emerge consensually, the broadcaster is allowed to shortlist three candidates. The *Landesmedienanstalt* would then choose the one that promises to contribute the most to diversity.

[57] VPRT, *Zusammenfassung der VPRT Position zum materiellen Medienkonzentrationsrecht für den privaten Rundfunk*, BONN, May 23, 1996.

[58] Klaus Ott, *Zufall, na klar*, SÜDDEUTSCHE ZEITUNG, Feb. 4, 1998.

1998] RULES FOR PRIVATE BROADCASTING IN GERMANY 545

platform's electronic program guide). Significantly, the Kirch group had already launched a 17-channel digital TV service in the summer of 1996 and, already a dominant player in the German program rights market, secured a United States billion dollar deal with five Hollywood studios for the exclusive pay-TV rights to their output over the next few years. For its part, Bertelsmann has been able to proceed with the merger, already announced in the spring of 1996, between its television subsidiary Ufa and the giant Luxembourg-based broadcasting multinational CLT, thereby increasing its control over the RTL broadcasting "family." Despite years of controversy about media concentration and a good deal of *formal* re-regulation, the new ownership rules seemed designed to accommodate precisely this kind of merger. Moreover, the enduring weakness of cross-media ownership regulation combined with the fact that the "insignificance threshold" had been raised from 10% to 25% in the final stage of the political negotiations, meant that the mighty Springer press concern's[59] direct stakes in the major "generalist" channel SAT 1 (20%) and the country's principal sports TV channel DSF (24.9%) counted for nothing. (This is certainly different from the 16 percent of audience-share that a 10% "insignificance threshold" would have meant). Also uncounted was the fact that Springer had an indirect holding of a further 20% in SAT 1 through its 40% holding (but significantly not a majority holding) in another direct SAT 1 investor, the Aktuelles Presse Fernsehen company.[60]

The largely symbolic nature of the inter-*Land* treaty of 1996 could not be clearer. In the words of one respected commentator, it amounted to a "capitulation on the part of the policy makers to Germany's most powerful media concerns Kirch and Bertelsmann." Virtually the only concentration that the latest regulation ruled out now was *their* merger.[61] Not ruled out by the audience-share model, however, were joint ventures between them. Thus, in June 1997 Kirch and Bertelsmann announced plans to embark, together with Deutsche Telekom, on a digital pay-TV joint venture. "Bertelkirch," as some commentators now playfully referred to them, proposed to merge *Premiere*, the analogue pay-TV

---

[59] In which Kirch had a 35% interest.
[60] Horst Röper, *Mehr Spielraum für Konzentration und Cross ownership im Mediensektor*, MEDIA PERSPEKTIVEN 609-20 (1996).
[61] *See* Klaus Ott, *Der Triumph des Leo Kirch*, SÜDDEUTSCHE ZEITUNG, Jan. 4, 1997, at 4.

channel which they already shared with Kirch's recently-launched DF 1 digital pay-TV venture.[62]

VI. THE PUBLIC SERVICE BROADCASTERS: A PLURALISTIC COUNTERBALANCE

At first sight, Germany's "dual" broadcasting system appears to be very pluralistic. Most German homes now receive multi-channel television. By the beginning of 1997, two thirds of German homes (24.9 million) were passed by cable, and 44.6% of homes (16.7 million) were actually connected. Moreover, by mid-1997 around 28% of homes in West Germany (7.4 million) and more than 50% in less-densely cabled East Germany (3.3 million) had satellite television receivers.[63] Alongside the main "generalist" private commercial broadcasting channels—SAT 1, RTL and PRO 7—these homes can receive a range of smaller or thematic services: VOX, Deutsches Sport Fernsehen (DSF), Kabel 1, Premiere (an analogue pay-tv service), n-tv (a news channel), RTL 2, Super RTL, and two new music channels. Cable and satellite also carry a number of foreign channels as well as all the main German public channels. Furthermore, regional public-service channels are broadcast nationwide by satellite, and new public-service satellite channels have been introduced in joint ventures with other European public broadcasters (e.g., the Franco-German cultural channel ARTE). However, as seen, the ownership of the private television sector is highly concentrated.

Moreover, the three main private channels—SAT 1, RTL, and PRO 7—account for around 40% of total viewing, as much as the public-service channels (with around 20% of viewing being spread among other channels). Between them, SAT 1, RTL, and PRO 7 account for three quarters of the private sector's share of the overall audience market.[64] These three "generalist" channels also account for 83% of the private television sector's advertising revenue, and three quarters of the overall German television advertising market.[65]

Given this degree of concentration in the private sector, the role of the public-service sector in the dual broadcasting system

---

[62] The Federal Cartel Office, however, has expressed its concern about the joint venture on competition policy grounds. At the time of writing, the joint venture is subject to European Commission scrutiny.
[63] Detailed data on the German media are published every year in *Media Perspektiven. Basisdaten*, and *Frankfurt am Main*. The 1997 edition is the most recent edition as of the time of writing.
[64] 1996 figures, published in MEDIA PERSPEKTIVEN 72 (1997).
[65] *Id.* at 11, 20.

becomes all the more important as a pluralistic counterbalance. However, recent developments have by no means offered encouragement to the public-service broadcasters. Firstly, the commercial broadcasters have significantly eroded their audience share. Although audience trends appear to have stabilized at the levels noted above, there remains the fear that a further fall in viewing levels would be very damaging for the legitimacy of the television license-fee. Secondly, the public broadcasters have become much more dependent upon their license-fee income, as private commercial broadcasting has dramatically damaged their supplementary advertising revenues. The public broadcasters' advertising airtime has remained strictly limited[66] and they have no longer been able to charge monopoly prices for it. By contrast, the private broadcasters have been allowed very liberal advertising limits.[67] The latter have been able to undercut the public broadcasters' advertising rates and still see their advertising revenue increase dramatically. Thus, between 1989 and 1996, the total advertising revenue of the private television sector grew from DM 642.3 million to a staggering DM 6.2 billion.[68] At the same time, the public broadcasters have seen an absolute decline in their advertising revenue. In 1985 the ARD and ZDF drew no less than DM 1.4 billion in advertising revenue from television. By the end of 1996, the figure had collapsed to DM 648.4 million.[69] Thirdly, competition for program rights has inflated prices particularly for the strategic kind of programming that attracts mass audiences, notably sports events and recently released films. Commercial concerns, with very deep pockets, have been able to out-bid the public broadcasters for these program rights. Witness Leo Kirch's recent deals with Hollywood's major studios; Kirch has also acquired the rights for the 2002 and 2006 soccer World Cups! Finally, the public broadcasters have also faced political attacks. The commercial broadcasting lobby, and some allied politicians, have called for the eventual abolition of all advertising by the public-service broadcasters. Politicians as important as Bavaria's prime minister (*Ministerpräsident*) Edmund Stoiber (CSU) and Saxony's prime minister Kurt Biedenkopf (CDU) have even suggested that the ARD's First Chan-

---

[66] They are allowed 20 minutes of advertising per day, except on Sundays.
[67] They are allowed to devote up to 20% of their air-time, including Sundays, to advertising.
[68] American billions, i.e., 1,000,000,000.
[69] *See Id.*

nel might be closed down to cut costs and rationalize the public-service sector.[70]

VII. THE FEDERAL CONSTITUTIONAL COURT RULES AGAIN IN DEFENSE OF BROADCASTING PLURALISM

In 1994 Germany's public broadcasters received support from a familiar quarter: namely, the Federal Constitutional Court. As long ago as 1984 the Greens (*die Grünen*) had lodged a complaint with the Bavarian Administrative Court in Munich against the financing of the early cable television pilot-projects from a (very small) share of the broadcasting license-fee, the "cable penny."[71] Ever since, the "cable penny" has been used to finance the private broadcasting regulatory authorities (the *Landesmedienanstalten*). This case was passed all the way up to the Federal Constitutional Court which used it as an opportunity to take up a much broader position on the future funding of the public-service broadcasters by the license-fee.[72] In its "Eighth Broadcasting Ruling" of February 22, 1994,[73] the Court stated categorically that the public broadcasters' role was all the more important in view of the evident shortcomings of the commercial broadcasters. It noted their inadequate "breadth of content" and "thematic variety."[74] The Court reiterated the principle first expressed in its famous 1986 "Fourth TV Ruling" introducing the dual system—that the constitutional-legal *conditio sine qua non* for allowing private commercial broadcasting was the "continuity and future development" of a strong public broadcasting sector. Specifically, the Federal Constitutional Court ruled in favor of establishing a new procedure for setting the license-fee that would depoliticize the business. The Court ruled that the advisory Commission for Assessing the Financial Requirements of the Broadcasters (the KEF), which had been effectively an

---

[70] *See Thesen zur Strukturreform des öffentlich-rechtlichen Rundfunks, reprinted in* MEDIA PERSPEKTIVEN 104-08 (1995) (originally published by the state chancelleries of Bavaria and Saxony). In their view, the public broadcasters provided more "by far" than their constitutional-legal remit to provide a "basic comprehensive service" (*Grundversorgung*) required. The ARD, they argued, had evolved from being an association serving the *Länder* broadcasters collectively into a large, excessively staffed and resourced concern.

[71] The complainants argued that the "cable penny" was illegitimate on the grounds that 99% of license-fee payers would receive no service for the amount they paid. They further argued that the cable penny was a special tax that would primarily benefit private commercial interests.

[72] This was quite characteristic of the Federal Constitutional Court. In several of its TV rulings the Court had used very specific and narrow issues to make far-reaching pronouncements on the broadcasting system.

[73] BVerfGE 90 (1994), 60, 8 Rundfunkentscheidung (Rundfunkgebühren), *reprinted in* MEDIA PERSPEKTIVEN DOKUMENTATION (1994).

[74] *Id.* at 40-41.

1998] RULES FOR PRIVATE BROADCASTING IN GERMANY 549

instrument of the *Länder* politicians, should in the future be established legally as a body that is genuinely independent of the politicians and broadcasters. The broadcasters, though, should be involved in the process; their own calculation of what is required financially to allow them to fulfill their constitutional-legal remit would be the central focus of the committee's future deliberations. Subsequently, the "Third Inter-Land Treaty on Broadcasting" of 1996, along with providing the new media ownership rules, implemented this latest ruling in detail.[75]

The most recent ruling of the Federal Constitutional Court came in January 1997. Again, this ruling originated in a very specific case, dating back several years, that had been passed up through the various levels of the German legal system. Following the award by the Bavarian media regulators of a broadcasting license to the DSF sports channel in 1992, the Berlin regulatory authority complained to the Bavarian Administrative Court. The license had been awarded despite the reservations of several of the *Länder* regulatory authorities, based on the suspicion that the channel was under the effective control of Leo Kirch and that this amounted therefore to a breach of the then operative media ownership laws.[76] When the Bavarian Administrative Court granted an injunction suspending DSF's license pending further legal deliberation, the Bavarian regulatory authority appealed to the Bavarian Constitutional Court on the grounds that its "broadcasting freedom" was being infringed by the license suspension while the main case was still under review at the Bavarian (and later Federal) Administrative Court.[77] When the suspension was overturned, the Berlin regulatory authority appealed directly to the Federal Constitutional Court to have it re-instated. However, in its 1997 Ruling the Federal Constitutional Court ruled against the Berlin authority on the formal procedural grounds that it had not first directed this appeal for reinstatement of the injunction to the Federal Administrative Court where the main case was still *sub judice*. Characteristically, though, the Federal Constitutional Court used the opportunity to re-state some constitutional-legal first principles concerning media concentration which the Federal Administrative

---

[75] *Dritter Rundfunkänderungsstaatsvertrag*, Artikel 5.

[76] At the time, Kirch had a minority 24.5 per cent share in DSF. Springer, in which Kirch had a 35 per cent share, had a 24.9% share in DSF. Silvio Berlusconi's Rete Invest had a 33.5% share in the channel. Kirch was a business partner of Berlusconi's in Italian and Spanish pay-TV.

[77] According to Bavarian media law, the Bavarian regulatory authority was actually the "broadcaster" of private television services in Bavaria in a manner analogous to the former British Independent Broadcasting Authority.

Court should take into account in making its ruling on the central case. Thus, the Federal Constitutional Court restated the importance of preventing the formation of any dominant influence over the expression of opinion (*vorherrschende Meinungsmacht*). The Court emphasized the preventative aspect, since media concentration was especially difficult, if not impossible, to rectify in the media industry. Significantly, the Federal Constitutional Court denied that regulation was rendered redundant in the age of multichannel broadcasting. The ever closer integration of the different parts of the media industry, both horizontally and vertically, posed the threat that dominance in one part might be used to gain influence in other parts as well. Once again, the Court had made a principled statement about the danger of media concentration.

However, it was unclear how general principles would translate into detailed regulatory practice. The Federal Administrative Court did rule, in March of 1997, that the Bavarian media authority had insufficiently determined the real control relations within DSF, and therefore it did revoke DSF's license. However, by this time the "media ownership model" of concentration control in the broadcasting industry had been reformed to the "audience share model," which rendered this ruling completely irrelevant. As seen, the 1996 Inter-*Land* Treaty that had already taken effect on January 1, 1997 removed all previous ownership restrictions, provided that the audience share of a broadcasting company did not exceed 30% of total audience share. Therefore, the channel received a provisional license to continue to broadcast from the Bavarian regulatory authority for private broadcasting. In fact, DSF is now set to become jointly (50:50) owned by Kirch and Bertelsmann subsidiary Ufa-CLT within the auspices of their new digital TV joint venture. Whether this degree of concentration will be acceptable depends on the decision of the newly-established KEK, and ultimately the European Commission.

VIII. Conclusion and Outlook

In German media law it is axiomatic that broadcasters fulfil a key democratic role: they exercise a potentially powerful influence on the process of democratic opinion formation. It has always been deemed crucial, therefore, that public policy be geared towards guaranteeing a plurality of media sources and a diversity of opinions in media output. Indeed, in post-war Germany pluralism in broadcasting has been a fundamental injunction of no lesser authority than the Federal Constitutional Court. Accordingly, the introduction of the new media and the expansion of commercial broadcasting has occasioned a considerable amount of *formal* re-regulation including various safeguards for "diversity of opinion" (*Meinungsvielfalt*) and media ownership rules designed to forestall the development of "dominant influence over the expression and formation of opinion" (*vorherrschende Meinungsmacht*).

However, with regard to media ownership rules, this formal re-regulation has appeared to amount to little more than "symbolic politics" (as conceptualized by Murray Edelman). The principal determinant of media policy in the late 1980s and 1990s has been the policy makers' perceptions of what is in the economic interest of their jurisdictions (*Standortpolitik*). The private broadcasting sector has been characterized by a high degree of media concentration (by any measure, whether ownership, audience share, or advertising revenue). It is certainly the case that the latest media ownership rules contain some important innovations and corrections for the failings of the old ones. The "windows" for independent third parties (*unabhängige Dritte*, i.e., small window broadcasters independent of their "host" major broadcaster) provide a constructive means of safeguarding pluralism in an oligopolized market. The extension to the *Landesmedienanstalten* of information and investigation rights equivalent to those of the Federal Cartel Office is an unqualified—and long overdue—improvement. Even if the individual *Landesmedienanstalten* have lost much of their independent authority to the KEK, these new powers should make them, together with the KEK, more effective regulators. The establishment of the KEK itself should help counteract the problems arising from the decentralized regulation of a national industry, notably *Standortpolitik* (regional competition to attract or retain media investment). Nonetheless, the fact remains that the audience share model legitimizes the status quo—namely, the highly oligopolistic structure of the German private broadcasting industry—and even allows for the further expansion of the key players.

There remain dissenting voices. Academic media experts have carefully analyzed the weaknesses of the new rules.[78] Journalists in certain newspapers (e.g., the "alternative" *tageszeitung* and the liberal *Süddeutsche Zeitung*) have drawn attention to the inadequacy of the rules. Germany's media union IG Medien censured the new rules for "kow-towing" to the media industry and being unlikely to

---

[78] *See generally* STOCK ET AL., *supra* note 50 (providing a comprehensive critique of the new rules).

effectively counteract media concentration.[79] Similarly, Germany's Green party condemned all the mainstream parties (i.e., SPD, CDU/CSU, and the small liberal FDP) for their "conniving with" (*Kungelei*) Kirch and Bertelsmann; the SPD's "genuflection" (*Kniefall*) before the media concerns attracted their particular scorn.[80] There was a degree of disquiet among the SPD grassroots as well. However, these forces were—and still are— marginal to the regulatory policy process which was, as seen, effectively a negotiation between the SPD and the CDU/CSU *Länder* governments. Once the SPD pragmatists, motivated by regional *Standortpolitik*, gained the upper hand within the party, the SPD *Länder* accepted the CDU/CSU's "package deal" that effectively acknowledged the need for the future development of the public broadcasters in return for more liberal media ownership rules. Although there may exist a significant critical constituency in the wider media policy community, there remains little prospect of an effective political coalition for a future tightening up of the rules.

In the future, therefore, the role of the public-service broadcasters will be more vital than ever, as a counterbalance to the accumulated media power of Germany's large private broadcasting concerns. In Germany, this special role of the public broadcasters has repeatedly been underlined by the Federal Constitutional Court. However, the public broadcasters in Germany—as elsewhere in Western Europe—face a very severe challenge; the very legitimacy of the license fee may be jeopardized by possible future developments in the media marketplace (e.g., a fragmentation of TV audiences, a serious decline in their audience share). The big question for the coming decade is whether the German public broadcasters' constitutional-legal guarantee of their continuity and further development is enough to sustain them against market forces. They are functioning in a political arena where economic interests appear to be overriding social and cultural goals of media policy.

As for concentration in the private broadcasting sector, the legal process so far has proved to be too slow and unwieldy to cope with market *faits accomplis*. In the past, the Federal Constitutional Court has played almost a pro-active role (strictly speaking, of course, it always has to react to appeals brought to it by other actors) in broadcasting policy, defining the parameters of regulation

---

[79] IG Medien, Konzentrationsregelung ist Kotau vor Konzernen, *Presseinformation der IG Medien*, STUTTGART, Mar. 11, 1996.

[80] Bündnis 90/Die Grünen, *Kniefall vor den Medienkonzernen*, PRESSEDIENST, June 11, 1996.

at key stages in the development of the broadcasting system. For example, the Court ruled on matters such as whether the central government might control a television channel (the First Broadcasting Ruling), the constitutionality of private television (the Third Broadcasting Ruling) and the balance between private and public broadcasting (several rulings since 1986). As far as media concentration is concerned, in its Fourth Broadcasting Ruling (1986) the Court warned about the dangers and called upon the policy makers to prevent "dominant influence over the expression and formation of opinion" (*vorherrschende Meinungsmacht*). However, critics point out that the Court has failed to seize the one obvious opportunity since then to translate principle into practice: namely, the DSF case.[81] In defense of the Court, its ability to intervene in DSF was constrained by the legal process itself which actually commenced in the administrative courts and therefore had to follow the formal "ladder" of appeals; in any event, the Court was bound to point out that other legal means had not been exhausted. The real problem, though, is that by the final DSF verdict, both the media market's evolution and the re-regulation of media concentration control had rendered the verdict obsolete. Therefore, all that remains is yet another Constitutional Court ruling reiterating the *principle* of media pluralism and pointing out that this principle will not be rendered obsolete by new technological developments.

The European Union may be the ultimate obstacle to the German media concerns' ambitions. The European Commission has a track record of already having thwarted one digital pay-TV joint venture between Bertelsmann and the Kirch group together with Deutsche Telekom. In 1994, Brussels blocked the establishment of the Multimedia Service Gesellschaft ("MSG") on competition policy grounds. The joint venture, it was argued, would have very likely created, or reinforced, a dominant position in three separate markets: namely, Bertelsmann and Kirch in the market for pay-TV services (they already jointly-control Germany's only pay-TV channel); MSG in the market for the technical services associated with digital pay-TV (i.e., "gateway systems" for conditional access and subscriber management, electronic program guides, etc.); and Deutsche Telekom in the market for cable network services (DT had a quasi-monopoly of cable systems).[82] Currently, the Euro-

---

[81] Klaus Ott, *DSF wird weitersenden*, SÜDDEUTSCHE ZEITUNG, Mar. 21, 1997.
[82] Emma Tucker, *Brussels Closes Off A Multimedia Gateway*, FIN. TIMES, Nov. 10, 1994, at 3.

pean Commission is examining the Kirch group's and Bertelsmann's latest plans to embark on yet another joint venture to launch a digital TV platform, again in association with Deutsche Telekom, in what looks like an "MSG Mark Two." While EU competition commissioner Karel van Miert conducts his investigations into the latest proposed joint venture, the development of digital pay-TV in Germany remains temporarily halted. This state of affairs has drawn the German federal government into the affair, since German national economic interests are plainly at stake. The German media industry's complaint is that while "Europe" is blocking German companies, American companies are free to expand in digital TV, the delivery system of the future. Chancellor Helmut Kohl himself is reported to have lobbied the Commission in support of the Bertelsmann/Kirch joint venture and made clear to Commission President Jacques Santer his belief that Germany's media ownership rules suffice to protect media pluralism.[83]

In the multi-channel future, it has been suggested that the most appropriate kind of regulation for television will be general economic regulation; pluralism will be adequately safeguarded by competition policy. However, when top Bertelsmann executive Mark Wössner dared to moot such ideas in Germany, at the annual Munich media conference in 1996, he met with fierce opposition. In Germany, he was emphatically reminded, media regulation fell under the cultural sovereignty of the *Länder*. It could never be allowed to become a federal or EU competence. However, the theme has not died away.[84] It is clear that the economic stakes involved in media policy are high and, with the costs and risks of investment in digital TV, they are becoming higher all the time, not least for the German *Länder* North-Rhine Westphalia and Bavaria, respectively the host states of Germany's leading media concerns Bertelsmann and KirchGruppe. As this article has argued, the successive regulatory frameworks for limiting media concentration, produced by the *Länder*, have not been notably effective thus far. A factor contributing to this ineffectiveness has been the economic ambitions in the media field of the *Länder* (*Standortpolitik*). It is, of course, too early to judge the efficacy in safeguarding pluralism of the latest rules provided by the 1996 inter-*Land* treaty (which came

---

[83] *See* Klaus Ott, *Kohl für Kirch und Vaterland; Wie sich der Kanzler für die digitale Fernseh-Allianz einsetzt*, SÜDDEUTSCHE ZEITUNG, Jan. 6, 1998; *see also* Klaus Ott, *Freund Kohl in geheimer Mission*, Süddeutsche Zeitung, Dec. 17, 1997; Martin Walker, *Brussels Risks War with Kohl*, THE GUARDIAN, Feb. 2, 1998, at 12 .

[84] Klaus Ott, *Länder sind schlechte Medienwächter: die Kontrolle des Kommerz-Fernsehens wäre in Bonn und Brüssel besser aufgehoben*, SÜDDEUTSCHE ZEITUNG, Oct. 15, 1997.

1998] RULES FOR PRIVATE BROADCASTING IN GERMANY 555

into force on January 1, 1997), yet it is clear that they allow for the continuance of a concentrated industry structure. Therefore, it is entirely legitimate to suggest that competition policy—at the national and EU level—should play a greater role in broadcasting regulation in the future.[85] However, since broadcasting is also about democracy—and in Germany this understanding is particularly strong—regulation has to be about more than merely economic regulation.

---

[85] *Id.*

[9]
ARCHITECTURAL CENSORSHIP AND THE FCC

CHRISTOPHER S. YOO[*]

ABSTRACT

Most First Amendment analyses of U.S. media policy have focused predominantly on "behavioral" regulation, which either prohibits the transmission of disfavored content (such as indecent programming) or mandates the dissemination of preferred content (such as children's educational programming and political speech). In so doing, commentators have largely overlooked how program content is also affected by "structural" regulation, which focuses primarily on increasing the economic competitiveness of media industries. In this Article, Professor Christopher Yoo employs economic analysis to demonstrate how structural regulation can constitute a form of "architectural censorship" that has the unintended consequence of reducing the quantity, quality, and diversity of media content. The specific examples analyzed include (1) efforts to foster and preserve free television and radio, (2) rate regulation of cable television, (3) horizontal restrictions on the number of outlets one entity can own in a local market, and (4) regulations limiting vertical integration in television and radio. Unfortunately, current First Amendment doctrine effectively immunizes architectural censorship from meaningful constitutional scrutiny, and it appears unlikely that existing doctrine will change or that Congress or the Federal Communications Commission will step in to fill the void.

---

[*] Associate Professor of Law, Vanderbilt University. This Article benefited from questions by participants at the Conference on Federal Regulation and the Cultural Landscape, sponsored by the Curb Center for Art, Enterprise, and Public Policy at Vanderbilt University, as well as the 32nd Annual Telecommunications Policy Research Conference. I am also grateful to Stuart Benjamin, Owen Fiss, Jonathan Levy, Richard Nagareda, Robert Pepper, and Bob Rasmussen for their comments on earlier drafts, and Kate Albers and Paul Werner for their expert research assistance. I would like to offer special thanks to my friend, Ed Baker, for his willingness to engage in the lively and constructive intellectual exchange about my ideas appearing in this Article. I can offer no higher praise than to say that I have spent much of my career inspired by and responding to his work.

TABLE OF CONTENTS

INTRODUCTION 671
I.   ARCHITECTURAL CENSORSHIP OF MEDIA CONTENT 675
     A. THE PREFERENCE FOR FREE RADIO AND TELEVISION 676
         1. Impact on the Quantity of Television Produced 679
         2. Impact on the Quality and the Diversity of
            Programming 681
         3. Distortions Resulting from Allowing Advertisers to
            Serve as Intermediaries..683
     B. RATE REGULATION OF CABLE TELEVISION 685
     C. RESTRICTIONS ON HORIZONTAL CONCENTRATION.................688
         1. The Complex Relationship Between Market
            Concentration and Program Diversity 692
            a. Steiner Models 693
            b. Limitations of Steiner Models 696
         2. The Role of Efficiencies from Horizontal Integration.......699
     D. RESTRICTIONS ON VERTICAL INTEGRATION 701
         1. Structural Preconditions Implicit in Vertical Integration
            Theory...706
         2. Applying the Structural Preconditions to the Television
            Industry 707
II. ARCHITECTURAL CENSORSHIP'S IMMUNITY FROM
     MEANINGFUL FIRST AMENDMENT SCRUTINY 713
     A. THE NATURE OF ARCHITECTURAL CENSORSHIP.....................713
     B. *MINNEAPOLIS STAR* AND THE SHORT-LIVED PROSPECT OF
         STRICT SCRUTINY 715
     C. RATIONAL BASIS VS. INTERMEDIATE SCRUTINY.....................718
     D. APPLYING THE STANDARD OF REVIEW 723
III. POSSIBLE SOLUTIONS TO THE PROBLEM OF
     ARCHITECTURAL CENSORSHIP..725
     A. RELIANCE ON THE POLITICAL BRANCHES 725
     B. INTENSIFYING *O'BRIEN* SCRUTINY 727
CONCLUSION 731

INTRODUCTION

Recent events have suddenly turned the media ownership regulations promulgated by the Federal Communications Commission ("FCC") into a hot topic. In 2001 and 2002, a remarkable series of decisions by the U.S. Court of Appeals for the D.C. Circuit invalidated significant portions of the FCC's media ownership restrictions.[1] Moreover, the reasoning of the opinions, which at times chided the FCC for failing to honor its statutory obligation to "repeal or modify any regulation it determines to be no longer in the public interest,"[2] casts doubt on the validity of a number of the FCC's other media ownership provisions. With its regulatory scheme thrown into disarray, the FCC undertook its most comprehensive reexamination of media ownership regulations in decades, which resulted in a mammoth order that loosened many of the most prominent restrictions.[3]

The prospect of widespread media consolidation touched off a political firestorm.[4] Congress responded by enacting legislation partially scaling back the most salient of the FCC's regulatory changes.[5] Numerous

---

1. *See* Sinclair Broad. Group, Inc. v. FCC, 284 F.3d 148, 162–64 (D.C. Cir. 2002) (invalidating the FCC's rule restricting ownership of more than one television station in any local market); Fox Television Stations, Inc. v. FCC, 280 F.3d 1027 (D.C. Cir.) (invalidating the FCC's rules limiting the number of television stations one entity can own nationally and prohibiting joint ownership of a television station and local cable operator in the same city), *modified on reh'g*, 293 F.3d 537 (D.C. Cir. 2002); Time Warner Entm't Co. v. FCC, 240 F.3d 1126 (D.C. Cir. 2001) (invalidating the FCC's rule limiting the number of cable subscribers one entity can reach nationwide).

2. Telecommunications Act of 1996, Pub. L. No. 104-104, § 202(h), 110 Stat. 56, 111–12. The scope of this statutory mandate has generated substantial controversy. The D.C. Circuit initially interpreted section 202(h) as erecting "a presumption in favor of repealing or modifying the ownership rules." *Fox Television Stations*, 280 F.3d at 1048. *See also Sinclair Broad. Group*, 284 F.3d at 159 (citing with approval the quoted language from *Fox Television Stations*). Subsequent decisions have been somewhat more circumspect. *See* Prometheus Radio Project v. FCC, 373 F.3d 372, 394, 423 (3d Cir. 2004) (rejecting the idea that section 202(h) serves as a "one-way ratchet"); Cellco P'ship v. FCC, 357 F.3d 88, 97–98 (D.C. Cir. 2004) (concluding that *Fox Television Stations* and *Sinclair Broadcasting Group* left open whether section 202(h) created a presumption in favor of eliminating existing regulations).

3. 2002 Biennial Regulatory Review—Review of the Commission's Broadcast Ownership Rules and Other Rules Adopted Pursuant to Section 202 of the Telecommunications Act of 1996, Report and Order and Notice of Proposed Rulemaking, 18 F.C.C.R. 13,620 (2003) [hereinafter 2003 Biennial Review Order].

4. *See, e.g.*, Ben Scott, *The Politics and Policy of Media Ownership*, 53 AM. U. L. REV. 645 (2004) (reviewing the political controversies surrounding the FCC's media ownership decision).

5. Consolidated Appropriation Act of 2004, Pub. L. No. 108-199, § 629, 118 Stat. 3, 99–100 (scaling back the FCC's decision to liberalize the number of television stations one entity could own nationwide). For further discussion, see *infra* note 170 and accompanying text.

parties challenged the FCC's actions in court, some contending that the amendments were too sweeping and others arguing that they did not go far enough. The U.S. Court of Appeals for the Third Circuit stayed, and ultimately invalidated, the FCC's order.[6] These decisions gave the FCC precious little guidance regarding the types of changes that will be necessary in order for the media ownership regulations to survive judicial review. The resulting uncertainty threatens to undermine forthcoming mergers whose legality depend on the less restrictive limits that the FCC sought to impose. What will happen next is anyone's guess.

Although the bulk of the commentary on these events has focused on the relative merits of the FCC's actions and the court's decision to strike them down, I would like to analyze these events from a somewhat broader perspective. What I find most interesting are the specific grounds invoked by the courts to invalidate the media ownership rules. In most instances, the courts based their actions on principles of administrative law while largely rejecting challenges based on the First Amendment.[7] The failure of these challenges is consistent with the conventional wisdom concerning the constitutionality of ownership restrictions. It has long been recognized that measures directly regulating the behavior of media speakers—either by prohibiting the transmission of disfavored content, such as indecent programming,[8] or by mandating the dissemination of preferred content, such as children's educational programming and political speech[9]—raise serious First Amendment problems. Ownership restrictions and other forms of structural regulation are generally thought to pose fewer constitutional concerns.[10] Consequently, although the constitutionality of behavioral

---

6. *See Prometheus Radio Project*, 373 F.3d at 372 (remanding portions of the order), *petition for cert. filed*, 73 U.S.L.W. 3466 (U.S. Jan. 28, 2005) (Nos. 04-1020 & 04-1036), *and* 73 U.S.L.W. 3466 (U.S. Jan. 31, 2005) (Nos. 04-1033 & 04-1045); Prometheus Radio Project v. FCC, No. 03-3388, 2003 WL 22052896 (3d Cir. Sept. 3, 2003) (staying the FCC's media ownership order).
7. *See Prometheus Radio Project*, 373 F.3d at 401–02; *Sinclair Broad. Group*, 284 F.3d at 167–69; *Fox Television Stations*, 280 F.3d at 1045–47. For notable exceptions, see *infra* notes 78, 164, 282 and accompanying text.
8. *See* 18 U.S.C. § 1464 (2000).
9. *See* 47 U.S.C. §§ 303b(a)(2), 335(b)(1) (requiring broadcasters to offer children's educational programming); *id.* §§ 312(a)(7), 315 (requiring broadcasters to provide access to political candidates).
10. *See, e.g.*, Am. Family Ass'n v. FCC, 365 F.3d 1156, 1168–69 (D.C. Cir. 2004); Ruggiero v. FCC, 317 F.3d 239, 244 (D.C. Cir. 2003) (en banc); *Sinclair Broad. Group*, 284 F.3d at 167–68; *Fox Television Stations*, 280 F.3d at 1046; Leflore Broad. Co. v. FCC, 636 F.2d 454, 458 n.26 (D.C. Cir. 1980); David L. Bazelon, *The First Amendment and the "New Media"—New Directions in Regulating Telecommunications*, 31 FED. COMM. L.J. 201, 212 (1979); Timothy G. Gauger, Comment, *The Constitutionality of the FCC's Use of Race and Sex in the Granting of Broadcast Licenses*, 83 NW. U. L. REV. 665, 673 (1989).

regulation has been the subject of extensive academic commentary, the constitutionality of structural regulation has received considerably less attention.[11] A complete analysis of the impact of structural regulation on program content has yet to appear in the literature.

This Article seeks to move beyond those previous analyses by offering a more comprehensive discussion of the ways that structural regulation affects media content. Part I explores a series of examples in which structural regulation has had a dramatic influence on the content of speech. The specific examples include: (1) efforts to foster free television over pay television, (2) rate regulation of cable television, (3) restrictions on the number of media outlets one entity can own in any media market, and (4) regulations limiting vertical integration in television and radio. Each of these examples was enacted to further three interests that the Supreme Court has determined to be unrelated to the content of expression: the preservation of free, local broadcasting; the promotion of competition; and the need to foster a diversity of sources and viewpoints.[12]

Each case demonstrates how structural regulation can have unintended effects on media content. Not only do these structural regulations tend to reduce the overall quantity and quality of media programming, they also

---

11. One analysis focused on the relatively narrow issue of whether particular structural regulations were enacted out of conscious effort to promote a diversity of viewpoints. *See* Jonathan W. Emord, *The First Amendment Invalidity of FCC Ownership Regulations*, 38 CATH. U. L. REV. 401 (1989). Other scholars have offered general discussions of how media concentration supposedly threatens the democratic values that they see underlying the constitutional commitment to free speech. *See* C. Edwin Baker, *Media Concentration: Giving up on Democracy*, 54 FLA. L. REV. 839 (2002) [hereinafter Baker, *Media Concentration*]; Yochai Benkler, *From Consumers to Users: Shifting the Deeper Structures of Regulation Toward Sustainable Commons and User Access*, 52 FED. COMM. L.J. 561 (2000); Ronald J. Krotoszynski, Jr. & A. Richard M. Blaiklock, *Enhancing the Spectrum: Media Power, Democracy, and the Marketplace of Ideas*, 2000 U. ILL. L. REV. 813; Lawrence Lessig, The Censorships of Television (Mar. 8, 1999) (unpublished manuscript, *available at* http://cyber.law.harvard.edu/works/lessig/tv.pdf). For my criticism of efforts to reconceptualize free speech in civic republican terms, see Christopher S. Yoo, *The Rise and Demise of the Technology-Specific Approach to the First Amendment*, 91 GEO. L.J. 245, 306–46 (2003). More importantly for our purposes, these analyses have not engaged in any extended analysis of the precise relationship between media concentration and media content. Other scholars have analyzed the First Amendment implications of a single type of structural regulation without offering a more general analysis of the relationship between structural regulation and content. *See* C. Edwin Baker, *Merging Phone and Cable*, 17 HASTINGS COMM & ENT. L.J. 97 (1994) (discussing the constitutionality of a provision barring crossownership of local telephone and cable operations); Stuart Minor Benjamin, *The Logic of Scarcity: Idle Spectrum as a First Amendment Violation*, 52 DUKE L.J. 1 (2002) (discussing the free speech implications of federal spectrum policy). The most complete discussion of the issue is C. EDWIN BAKER, MEDIA, MARKETS, AND DEMOCRACY 20–62 (2002) (discussing the impact of advertising support and local concentration on content). Even Baker's analysis stops short of exploring the full range of complexities of how structure and content interact.

12. *See* Turner Broad. Sys., Inc. v. FCC, 512 U.S. 622, 662 (1994) [hereinafter "*Turner I*"].

affect the diversity of media content. Put another way, structural regulation can represent a form of "architectural censorship"[13] that can have a tangential, but substantial, adverse impact on speech.[14] In so doing, my analysis reveals that previous First Amendment discussions of structural regulation have simultaneously been too broad and too narrow. They have been too broad in their tendency to simply posit that media concentration necessarily represents a threat to free speech without engaging in any searching analysis of the precise nature of the relationship between concentration and content. This analysis suggests that the relationship between media concentration and the quantity, quality, and diversity of media content is more complex than is generally realized. At the same time, prior analyses have been too narrow in restricting their focus to media concentration. My analysis identifies other structural features that pose even more serious dangers of architectural censorship than do the concerns about industry concentration patterns that have dominated the existing scholarship.

Part II examines how the identified instances of architectural censorship would fare when measured against current First Amendment doctrine. Given the potentially adverse impact that structural regulation can have on the content of speech, one would hope that the First Amendment would provide a basis for identifying and redressing architectural

---

13. Professor Baker correctly notes that the term "censorship" is most applicable to situations in which the government deliberately attempts to affect the content of speech. *See* C. Edwin Baker, *Media Structure, Ownership Policy, and the First Amendment*, 78 S. CAL. L. REV. 733, 754–755 & n.71 (2005). I concede that I use the term in part to add a touch of rhetorical flourish to my argument. That said, the censorship label may be more apt than initially appears. Many early attempts to regulate media structure were intended to influence media content. *See, e.g.*, Amendment of Sections 3.35, 3.240 and 3.636 of the Rules and Regulations Relating to Multiple Ownership of AM, FM and Television Broadcasting Stations, Report and Order, 18 F.C.C. 288, 291–93 (1953) (noting that national ownership rules were designed in part to "maximize diversification of program and service viewpoints") [hereinafter 1953 Multiple Ownership Order]; FED. COMMUNICATIONS COMM'N, REPORT ON CHAIN BROADCASTING 65 (1941) (imposing the Chain Broadcasting Rules in part because network control hampers local stations' ability to "broadcast[] such outstanding local events as community concerts, civic meetings, local sports events, and other programs of local consumer and social interest"). *See also Turner I*, 512 U.S. at 676–77 (O'Connor, J., concurring in part and dissenting in part) (arguing that seemingly structural measures designed to protect free broadcasting were really motivated by a desire to promote more local, educational, and public affairs-related content); Christopher H. Sterling, *Television and Radio Broadcasting*, *in* WHO OWNS THE MEDIA?: COMPETITION AND CONCENTRATION IN THE MASS MEDIA INDUSTRY 299, 310 (Benjamin M. Compaine ed., 2d ed. 1982) (noting that "the FCC has followed an unwritten but fairly clear policy of seeking to modify the ownership of broadcasting facilities as a means of effecting changes in content").

14. In some respects, my analysis bears some similarity to Lawrence Lessig's claim that Internet protocols represent architectural elements that can censor in much the same manner as the government. *See* LAWRENCE LESSIG, CODE AND OTHER LAWS OF CYBERSPACE 6 (1999); Lessig, *supra* note 11. The analytical tools that I employ and the claims that I advance are quite different from Professor Lessig's.

censorship when it arises. Unfortunately, such hopes are misplaced. Recent judicial decisions indicate that the most stringent standard of review that might be applied to structural regulation is the intermediate scrutiny announced in *United States v. O'Brien*.[15] *O'Brien* doctrine has been widely criticized as being too deferential.[16] As a result, current Supreme Court precedent effectively insulates instances of architectural censorship from meaningful constitutional scrutiny.

Part III briefly explores possible solutions to the de facto constitutional immunity enjoyed by architectural censorship. Although courts could leave resolution of these constitutional issues to the political branches, doing so would represent an abdication of the proper role of courts and would charge Congress and the executive with responsibilities that they are loath to bear. The only other alternative is to revise *O'Brien* doctrine to take the individual's interest in speech and the availability of alternative means of communication seriously. The failure of the Supreme Court's recent efforts to put teeth in *O'Brien* scrutiny, however, makes it unlikely that architectural censorship will be subject to meaningful First Amendment review in the foreseeable future.

I. ARCHITECTURAL CENSORSHIP OF MEDIA CONTENT

This Part employs economic analysis[17] to examine four ways in which the current regime of structural regulation can give rise to architectural

---

15. United States v. O'Brien, 391 U.S. 367 (1968).
16. *See infra* Part II.D.
17. Professor Baker spends a significant portion of his commentary criticizing my work for taking an economic or "commodity-based" approach to media policy rather than framing the issues in terms of democracy. *See* Baker, *supra* note 13, at 742–747. I have offered two basic criticisms of attempts to base media regulation on democratic principles elsewhere and will only sketch my conclusions here. First, by valuing speech for its contributions to democracy, these theories adopt a consequentialist approach that is at odds with the autonomy-centered vision that has long dominated free speech theory. Second, the existing democracy-centered theories are too insufficiently theorized to yield a workable system of media regulation. These theories recognize that their Jeffersonian vision of small speakers might have to yield to other considerations (including economics), yet fail to provide a coherent framework for determining how to balance these countervailing considerations. In this respect, it is telling that such luminaries as Lillian Bevier, Vincent Blasi, Robert Bork, Cass Sunstein, Owen Fiss, and Harry Kalven have each advanced theories of media regulation that began from similar, democracy-based premises, and yet have implemented their theories by drawing radically different conclusions. Yoo, *supra* note 11, at 306–46. *See also* Christopher S. Yoo, *Copyright and Democracy: A Cautionary Note*, 53 VAND. L. REV. 1933, 1953–62 (2000). In this respect, the debate between economic and democratic visions of media policy parallels a similar debate in antitrust. In that case, the populist approach to antitrust failed in no small part because of its inability to offer a basis for resolving the trade-offs between competing considerations. *See* Christopher S. Yoo, Beyond Network Neutrality pt. II.E.2 (unpublished manuscript, on file with author). Professor Baker appears to recognize the

censorship. Although most of the features of the current regime were not always created out of a desire to affect media content, they nonetheless have precisely that effect.

A. THE PREFERENCE FOR FREE RADIO AND TELEVISION

The desire to promote free (advertising-supported[18]) radio and television over pay versions of the same media has long represented one of the central tenets of U.S. media policy.[19] Policymakers have exhibited hostility toward radio services that were provided on a fee basis since the earliest days of radio regulation. This hostility is reflected in the FCC's longstanding hostility toward subscription-based radio technologies,[20] discernible most recently in its resistance to satellite radio—known technically as Digital Audio Radio Services ("DARS")—such as XM and Sirius.[21]

---

problem, and indeed his work on "complex democracy" is among the most promising in the field. *See* Baker, *supra* note 11, at 143–47. Even his laudable efforts fall short of articulating a basis sufficient to make difficult trade-offs inherent in any system of media regulation.

18. As I have noted elsewhere, the term "free" is something of a misnomer. "Free" radio and television is only possible because the broadcast industry receives hundreds of billions of dollars worth of spectrum for free. Indeed, the industry members are the only significant commercial users of spectrum that do not have to pay for their frequencies. The commitment of these resources inevitably increases the cost of all other spectrum-based technologies. As a result, the public bears the costs of "free" radio and television by paying higher fees for cellular telephony and other spectrum-based technologies. *See* Christopher S. Yoo, *Rethinking the Commitment to Free, Local Television*, 52 EMORY L.J. 1579, 1712–14 (2003).

19. The discussion that follows draws on the more complete analysis appearing in Yoo, *supra* note 18, at 1668–82. For other related analyses that draw somewhat different policy conclusions, see BAKER, *supra* note 11, at 24–40; Jora R. Minasian, *Television Pricing and the Theory of Public Goods*, 7 J.L. & ECON. 71 (1964); Michael Spence & Bruce Owen, *Television Programming, Monopolistic Competition, and Welfare*, 91 Q.J. ECON. 103, 118–19 (1977).

20. *See* KMLA Broad. Corp. v. Twentieth Century Cigarette Vendors Corp., 264 F. Supp. 35, 41 (C.D. Cal. 1967). *See generally* Howard A. Shelanski, *The Bending Line Between Conventional "Broadcast" and Wireless "Carriage"*, 97 COLUM. L. REV. 1048, 1052–57 (1997) (detailing the hostility toward subscription radio services historically exhibited by the FCC and its predecessor agency, the Federal Radio Commission). One of the few early exceptions was the transmission of background music pioneered by the Muzak Corp., which the FCC allowed to be provided on a subscription basis. *See* Muzak Corp., 8 F.C.C. 581, 582 (1941). Even then, such subscription services are generally heavily restricted. *See KMLA Broad. Corp.*, 264 F. Supp. at 37–38 (describing how the FCC required radio stations to provide background music services solely via subcarrier frequencies and mandated that those services not interfere with the main-channel transmissions that are available for free to the entire listening public).

21. *See* Thomas W. Hazlett, *All Broadcast Regulation Politics Are Local: A Response to Christopher Yoo's Model of Broadcast Regulation*, 53 EMORY L.J. 233, 248–52 (2004) (detailing the manner in which FCC regulations have hampered DARS).

The hostility toward subscription media services is also manifest in U.S. television policy.[22] When the development of scrambling technology made subscription television feasible, the FCC acted fairly quickly to stifle the industry's growth.[23] The bias against pay television services was even more evident in the FCC's policies toward cable television, most particularly in the relentless campaign to require local cable operators to provide free carriage to all full-power broadcast stations operating in their service area (commonly known as "must-carry").[24] The desire to foster free

---

22. *See* Yoo, *supra* note 18, at 1669–75.
23. The FCC declined to authorize subscription television as a general service and instead, the FCC merely authorized it on an experimental basis. Amendment of Part 3 of the Commission's Rules and Regulations (Radio Broadcast Services) to Provide for Subscription Television Service, Third Report, 26 F.C.C. 265 (1959). When the FCC eventually authorized more widespread deployment in 1968, it saddled the technology with a host of restrictions. Amendment of Part 73 of the Commission's Rules and Regulations (Radio Broadcast Services) to Provide for Subscription Television Service, Fourth Report and Order, 15 F.C.C.2d 466 (1968), *aff'd sub nom.* Nat'l Ass'n of Theatre Owners v. FCC, 420 F.2d 194 (D.C. Cir. 1969). It was only after the D.C. Circuit's adverse decision in *Home Box Office, Inc. v. FCC*, 567 F.2d 9, 28–51 (D.C. Cir. 1977) (per curiam), that the FCC relented and lifted the restrictions on subscription television. *See* Repeal of Programming Restrictions on Subscription Television, Report and Order, 43 Fed. Reg. 15,322 (F.C.C. Apr. 7, 1978); Amendment of Part 73 of the Commission's Rules and Regulations in Regard to Section 73.642(a)(3) and Other Aspects of the Subscription Television Service, Third Report and Order, 90 F.C.C.2d 341 (1982).
24. The FCC foreshadowed the imposition of must-carry in the very first decision in which it asserted jurisdiction over cable systems. Carter Mountain Transmission Corp., 32 F.C.C. 459, 465 ¶ 17 (1962), *aff'd*, 321 F.2d 359 (D.C. Cir. 1963). The FCC later imposed must-carry on cable systems receiving programming through microwave transmission, *see* Amendment of Subpart L, Part 11, to Adopt Rules and Regulations to Govern the Grant of Authorizations in the Business Radio Service for Microwave Stations to Relay Television Signals to Community Antenna Systems, First Report and Order, 38 F.C.C. 683, 705 ¶ 57, 716–17 ¶¶ 85–90 (1965) [hereinafter CATV First Report and Order], and extended must-carry to systems that retransmitted over-the-air television broadcasts, Amendment of Subpart L, Part 91, to Adopt Rules and Regulations to Govern the Grant of Authorizations in the Business Radio Service for Microwave Stations to Relay Television Signals to Community Antenna Systems, Second Report and Order, 2 F.C.C.2d 725, 746 ¶¶ 48–49, 752–53 ¶ 66 (1966) [hereinafter CATV Second Report and Order] (extending the same rules to all cable systems). *See also* Amendment of Part 74, Subpart K, of the Commission's Rules and Regulations Relative to Community Antenna Television Systems, Cable Television Report and Order, 36 F.C.C.2d 143, 170–71 ¶ 74, 173–76 ¶¶ 78–87 (1972) (reaffirming must-carry); Implementation of the Provisions of the Cable Communications Policy Act of 1984, 50 Fed. Reg. 18,637 (F.C.C. May 2, 1985) (final rule) (same).

The FCC justified must-carry in large part by a desire to prevent those who are unable or unwilling to pay for television service from being deprived of it. *See* Cable Television Syndicated Program Exclusivity Rules, Report and Order, 79 F.C.C.2d 663, 744 ¶ 185 (1980), *aff'd sub nom.* Malrite T.V. of N.Y. v. FCC, 652 F.2d 1140 (2d Cir. 1981); CATV Second Report and Order, *supra*, at 788–89 ¶ 155; CATV First Report and Order, *supra*, at 699 ¶ 44, 700 ¶ 48(1).

Eventually, congressional intervention was necessary before must-carry could withstand judicial review. *See* Cable Television Consumer Protection and Competition Act of 1992, Pub. L. No. 102-385, §§ 4–5, 106 Stat. 1460, 1471–81 (codified as amended at 47 U.S.C. §§ 534, 535 (2000)). The must-carry statute would eventually be sustained by the Supreme Court as a valid means to further the government's interest in "preserving the benefits of free, over-the-air local broadcast television." Turner Broad. Sys., Inc. v. FCC, 520 U.S. 180, 189 (1997) [hereinafter "*Turner II*"].

television is apparent in the steps taken to regulate direct broadcast satellite ("DBS") systems, such as DirecTV and the Dish Network,[25] and underlies the FCC's decision to deploy digital television through broadcasters rather than through cable and satellite providers.[26]

Historically, efforts by Congress and the FCC to promote free radio and television have not been driven by content-based motivations. Instead, they are the result of a desire to preserve access for households that cannot afford subscription services.[27] Although these goals are quite laudable, application of economic analysis reveals that fostering advertising-supported radio and television has had a hidden, deleterious effect on the quantity, quality, and diversity of programming provided.

---

25. *See* H.R. CONF. REP. NO. 106-464, at 101 (1999) (noting that the Satellite Home Viewer Improvement Act was "intended to preserve free television for those not served by satellite or cable systems"); S. REP. NO. 106-51, at 1 (1999) (recognizing that the purpose of the legislation was "protecting the availability of free, local over-the-air television"); *id.* at 13 (finding that "maintaining free over-the-air-television is a preeminent public interest" and identifying "protecting the viability of free, local, over-the-air television" as one of the statute's purposes); H.R. REP. NO. 100-887(II), at 26 (1988), *reprinted in* 1988 U.S.C.C.A.N. 5638, 5655 (expressing the concern that, if unregulated, satellite television would "undermine the base of free local television service upon which the American people continue to rely").

26. The FCC has repeatedly justified the importance of deploying digital television through broadcasting rather than other television services on the grounds that broadcasting, unlike subscription services, represents a "free" service that is available to almost all U.S. households. *See* Advanced Television Systems and Their Impact upon the Existing Television Broadcast Service, Fifth Report and Order, 12 F.C.C.R. 12,809, 12,811–12 ¶ 5, 12,820 ¶¶ 27–29 (1997); Advanced Television Systems and Their Impact upon the Existing Television Broadcast Service, Fifth Further Notice of Proposed Rule Making, 11 F.C.C.R. 6235, 6249 ¶ 36 (1996); Advanced Television Systems and Their Impact upon Existing Television Broadcast Service, Second Report and Order and Further Notice of Proposed Rulemaking, 7 F.C.C.R. 3340, 3342 ¶ 4 (1992); Advanced Television Systems and Their Impact on the Existing Television Broadcast Service, Tentative Decision and Further Notice of Inquiry, 3 F.C.C.R. 6520, 6525 ¶¶ 38–39 (1988). *See also* Advanced Television Systems and Their Impact upon the Existing Television Broadcast Service, Fourth Report and Order, 11 F.C.C.R. 17,771, 17,787–88 ¶ 33 (1996) (noting that the goals of digital television deployment include preserving a free, universal broadcasting service); Advanced Television Systems and Their Impact upon the Existing Television Broadcast Service, Memorandum Opinion and Order, Fourth Further Notice of Proposed Rulemaking and Third Notice of Inquiry, 10 F.C.C.R. 10,540, 10,541 ¶ 6, 10,543 ¶ 22 (1995) (same). Concerns about preserving free television have also animated the FCC's proceedings regarding the extension of the must-carry rules to digital programming. *See* Carriage of Digital Television Broadcast Signals, First Report and Order and Further Notice of Proposed Rulemaking, 16 F.C.C.R. 2598, 2600 ¶ 3, 2648 ¶ 113 (2001); Carriage of the Transmissions of Digital Television Broadcast Stations, Notice of Proposed Rulemaking, 13 F.C.C.R. 15,092, 15,114–15 ¶ 43 (1998).

27. *See, e.g.*, Cable Television Consumer Protection and Competition Act of 1992, Pub. L. No. 102-385, § 2(a)(12), 106 Stat. 1460, 1461 (finding a "substantial government interest in promoting the continued availability of such free television programming, especially for viewers who are unable to afford other means of receiving programming"); *Turner II*, 520 U.S. at 190 (relying on the need to preserve free television to uphold must-carry).

1. Impact on the Quantity of Television Produced

Reliance on advertising support is likely to lead to a systematic underfinancing of media programming. When broadcasters derive revenue solely from advertising, one would expect the total revenue to be determined by viewers' and listeners' responsiveness to the advertising contained within programs. In other words, advertisers will increase their spending so long as the revenue generated by exposing audiences to an additional commercial exceeds the cost of purchasing an additional commercial.

Although it is possible that audiences' responsiveness to advertising might yield the same net revenue as direct payments for the underlying programs, there is no theoretical reason to expect that these levels would be the same.[28] In fact, the available empirical evidence indicates that advertisers place a significantly lower value on programming than viewers and listeners. One oft-cited study conducted in the 1970s estimated that viewers were willing to pay seven times more for television programming than were advertisers.[29] A pair of recent event studies confirmed those results by showing that television programs financed by pay-per-view generate significantly greater revenue than programs financed by advertising support.[30]

---

28. *See* Minasian, *supra* note 19, at 74–75. *See also* Spence & Owen, *supra* note 19, at 104–05.

29. *See* ROGER G. NOLL, MERTON J. PECK & JOHN J. MCGOWAN, ECONOMIC ASPECTS OF TELEVISION REGULATION 23 (1973). *See also* Harvey J. Levin, *Program Duplication, Diversity, and Effective Viewer Choices: Some Empirical Findings*, 61 AM. ECON. REV. 81, 82, 88 (1971) (concluding that entry by pay television supported more informational programming and other special interest programming than would advertising-supported television); Spence & Owen, *supra* note 19, at 118–19 (drawing on the Noll-Peck-McGowan data to show that reliance on advertising support was suppressing the emergence of a fourth television network). Although other economists have quibbled with the precise size of this disparity, they do not dispute the fundamental conclusion that consumers are willing to pay far more for television than advertisers. *See* Stanley M. Besen & Bridger M. Mitchell, *Noll, Peck, and McGowan's* Economic Aspects of Television Regulation, 5 BELL J. ECON. & MGMT. SCI. 301, 308–11 (1974) (book review); Bryan Ellickson, *Hedonic Theory and the Demand for Cable Television*, 69 AM. ECON. REV. 183, 188–89 (1979).

30. *See* Claus Thustrup Hansen & Søren Kyhl, *Pay-Per-View Broadcasting of Outstanding Events: Consequences of a Ban*, 19 INT'L J. INDUS. ORG. 589, 590, 601, 604 (2001); Steinar Holden, *Network or Pay-Per-View?: A Welfare Analysis*, 43 ECON. LETTERS 59, 62–64 (1993). It is interesting to note that these two studies drew different normative implications from the same empirical findings. The difference results from the fact that the Hansen and Kyhl study employed the generally accepted welfare metric of total surplus, while the Holden study focused solely on consumer surplus. This disagreement over the proper welfare metric should not obscure the conclusion drawn by both studies that a shift to pay television would cause the total revenue captured by the programmer to increase and would make possible programming that would not exist if advertising support were the sole source of revenue.

These studies indicate that advertising support drastically understates the intensity of consumers' preferences for television and radio programming and that reliance on advertising support causes revenue to drop far below efficient levels. Put another way, favoring advertising support over direct payments systematically starves programming of resources. Programmers would be able to generate substantially greater revenues (and thus devote greater resources to production) if they were allowed to charge directly for programs.[31] Preventing them from doing so has the effect of reducing the total amount of television and radio programming produced.

The policy commitment to foster advertising-supported television has also had the indirect effect of hindering the development of multichannel television technologies.[32] This preference was implicit in the regulations requiring cable and satellite television providers to carry all full-power local broadcast stations.[33] The bias against multichannel technologies was made explicit during proceedings to determine how to deploy digital television.[34] The unfortunate result of this bias against new, multichannel technologies is a restriction on the amount of channel capacity available in any local market.[35] As we shall see, limitations on channel capacity play a

---

31. I do not mean to suggest that advertising support should be banned, but rather that television and radio providers should be allowed to rely on subscription fees or advertising as they see fit. I would not expect the market to rely exclusively on either form of financing. On the contrary, the most likely result would be a mix of networks, some relying solely on advertising, some relying solely on direct viewer payments, and some relying on a combination of the two, resembling the current market for newspapers in many cities. *See* Yoo, *supra* note 18, at 1682.

32. *See id.* at 1703.

33. *See supra* notes 23–26 and accompanying text.

34. The initial regulations encouraged digital broadcasters to transmit a single stream of high definition television rather than multiple streams of standard definition television. Advanced Television Systems and Their Impact on the Existing Television Broadcast Service, First Report and Order, 5 F.C.C.R. 5627, 5627 ¶ 1, 5629 ¶ 12 (1990).

35. This effect is the most dramatic with respect to the local carriage obligation imposed on satellite television services, such as DBS. By its nature, DBS provides service on a national scale. As a result, DBS providers who wish to offer programming from the major broadcast networks (namely, ABC, CBS, NBC, and Fox) in Nashville must necessarily transmit that programming to the entire country even though no one outside of Nashville would be legally allowed to receive those signals. Although the DBS providers are in the process of deploying "spot beam" technology that should allow them to restrict the geographic coverage of particular channels, such technologies are not likely to be operational for several years. Requiring DBS providers to carry all local stations in any market in which they would like to provide local service forces them to dedicate large amounts of their limited channel capacity to transmitting redundant signals that only a small portion of the country can legally receive. This has the inevitable effect of reducing the number of channels that viewers in any particular city can receive.

critical role in causing media markets to underproduce programming that appeals to relatively small audience segments.[36]

2. Impact on the Quality and the Diversity of Programming

Reliance on advertising also reduces program quality and diversity. Limiting programs' ability to generate revenue necessarily increases the minimum audience size needed for a program to break even. This in turn has the inevitable effect of skewing the market against programming that appeals only to a relatively small segment of the audience.[37]

Conventional markets provide a straightforward mechanism for encouraging the production of low-volume products that enrich the product mix, as evidenced by the survival of high-priced boutiques in a world increasingly dominated by mass-market discounters. So long as consumers who prefer those low-volume products are willing and able to pay more for them, the total revenue generated will be sufficient to cover costs, even if those costs are substantially higher. Stated more formally, low-volume products can exist so long as consumers can use prices to signal the intensity of their preferences.

Advertising support effectively forecloses viewers and listeners from using prices to signal the intensity of their preferences. Simply put, advertising support provides viewers and listeners with only a single degree of freedom with which to signal the intensity of their preferences. They can either choose to view the programming offered by the network, in which case the network derives revenue equivalent to that type of viewer's responsiveness to advertising, or they can choose not to watch, in which case the network receives nothing. This limits viewers to an all-or-nothing signal of their preferences.[38] It makes revenue largely a function of audience size,[39] thereby preventing small audiences from obtaining the programming they want no matter how much they are willing to pay for it.

The recent financial and critical success of HBO provides an eloquent demonstration of these dynamics. Viewers' ability to signal intensity of

---

36. *See infra* notes 101, 108, 124–126 and accompanying text.
37. *See* Spence & Owen, *supra* note 19, at 112.
38. This distortion is analogous to the problem endemic in many election schemes, in which voters simply vote "yes" or "no" for a particular candidate without being able to signal the intensity of their preferences. *See* Richard L. Hasen, *Vote Buying*, 88 CAL. L. REV. 1323, 1332 (2000); Saul Levmore, *Voting with Intensity*, 53 STAN. L. REV. 111 (2000).
39. *See, e.g., Turner II*, 520 U.S. 180, 208–09 (1997) (collecting empirical research confirming the "direct correlation [between] size in audience and station [advertising] revenues" (alterations in original and internal quotation marks omitted)).

preference through direct payments allows HBO to generate more than half the revenue of CBS even though its prime time audience is almost fifteen times smaller.[40] In other words, HBO is able to generate roughly eight times more revenue per viewer than CBS. This makes it far easier for HBO to produce programs that appeal to relatively small audiences. In addition, to the extent that program quality is correlated with the amount spent producing each program, a shift to subscription services also causes program quality to improve. Indeed, HBO's dominance of recent Emmy Awards provides a powerful demonstration of this effect.[41]

Reliance on advertising support has the inevitable effect of excluding programming that appeals only to small audiences, regardless of both the strength of viewers' and listeners' preferences and their willingness to pay. Reliance on advertising support thus tends to reduce the diversity of the programming mix by preventing the survival of economically viable programs that appeal only to small audiences.[42] Indeed, recent empirical studies focusing on black and Hispanic audiences, who represent precisely the type of small audiences with nonmainstream preferences that advertising support tends to disfavor, indicate they are in fact underserved in precisely the manner that this theory predicts.[43]

Conversely, allowing direct payments for preferred programming would make it far easier for programming strongly desired by a small portion of the audience to appear. To use a somewhat fanciful example, suppose that there is a small group of ten thousand opera lovers who each would be willing to pay up to $1000 to view the entire season of the New York Metropolitan Opera on television.[44] If these opera lovers were able to make direct payments to the television network, they would be able to offer

---

40. *See* John M. Higgins, *Still Strutting after All These Years: Although NBC Remains No. 1, CBS Is Close Behind*, BROADCASTING & CABLE, Dec. 13, 2004, at 20, 21, 24 (reporting that CBS had 2004 revenue of $4.45 billion with an average audience of 13.3 million, while HBO had 2004 revenue of $2.4 billion with an average audience of 893,000).

41. *See* Mike Duffy, *Sunday Belongs to HBO: Cable Network Is the Emmy Powerhouse to Beat*, DET. FREE PRESS, Sept. 21, 2003, at 1E; Bernard Weinraub, *HBO Is Big Winner at Emmy Awards*, N.Y. TIMES, Sept. 20, 2004, at A22;

42. *See* Spence & Owen, *supra* note 19, at 113; SIMON P. ANDERSON & STEPHEN COATE, MARKET PROVISION OF PUBLIC GOODS: THE CASE OF BROADCASTING 23–28 (Nat'l Bureau of Econ. Research, Working Paper No. 7513, 2000), *available at* http://www.nber.org/papers/w7513.

43. Peter Siegelman & Joel Waldfogel, *Race and Radio: Preference Externalities, Minority Ownership, and the Provision of Programming to Minorities*, 10 ADVERTISING & DIFFERENTIATED PRODUCTS 73, 80–83 (Michael R. Baye & Jon P. Nelson eds., 2001); Joel Waldfogel, *Preference Externalities: An Empirical Study of Who Benefits Whom in Differentiated-Product Markets*, 34 RAND J. ECON. 557 (2003).

44. For those with different tastes, the example applies equally well to a small group of loyal fans of a team located in a different city.

a total of $10 million to a station willing to provide such programming, in which case the programming might well appear. If the network offering this programming were forced to rely solely on advertising support, the amount of revenue that such a program would capture would be limited by the amount of advertised products that this relatively small group of opera lovers would be willing to buy. In this case, the revenue generated by advertising support would likely be only a fraction of that generated by direct payments.[45]

3. Distortions Resulting from Allowing Advertisers to Serve as Intermediaries

Reliance on advertising support introduces additional market distortions by allowing advertisers to serves as intermediaries in the relationship between audiences and program producers. Although reliance on advertising support tends to make the impact of any particular audience member more uniform than would be possible under a system of direct payments, the fact that individuals with certain demographic characteristics are likely to be more responsive to advertising inevitably makes some audience members more valuable to advertisers than others.[46]

This, in turn, can skew the markets away from an audience's true preferences. For example, reliance on advertising support encourages television and radio programmers to be consumerist in focus and tends to make them excessively sensitive to the preferences of those demographic groups that are the most responsive to advertising.[47] Consequently, it tends

---

45. Professor Baker chides me for ignoring externalities. *See* Baker, *supra* note 13, at 749. This criticism is in tension with one of the central economic lessons of the past half-century, which is that, so long as transaction costs are low, markets are far more effective at dealing with externalities than previously thought. *See* R.H. Coase, *The Problem of Social Cost*, 3 J.L. & ECON. 1 (1960). For my analysis of the implications of externalities and transaction costs on media ownership policy, see Christopher S. Yoo, *Vertical Integration and Media Regulation in the New Economy*, 19 YALE J. ON REG. 171, 193–200, 213–17, 232–37 (2002). To the extent that the relevant externalities are positive externalities enjoyed by audiences, however, the collective action problems created by the large number of people involved may cause markets to fail. *See* Christopher S. Yoo, Rethinking the Coasean Critique of Broadcast Regulation (2005) (unpublished manuscript, on file with author). Even if transaction costs prevent markets from fully internalizing the extant externalities, the classic solution would be subsidies (or perhaps liability rules) rather than ownership restrictions. *See* A.C. PIGOU, THE ECONOMICS OF WELFARE 192–94 (4th ed. 1932) (subsidies); Guido Calabresi & A. Douglas Melamed, *Property Rules, Liability Rules, and Inalienability: One View of the Cathedral*, 85 HARV. L. REV. 1089 (1972) (liability rules).
46. *See* Franklin M. Fisher, John J. McGowan & David S. Evans, *The Audience-Revenue Relationship for Local Television Stations*, 11 BELL J. ECON. 694 (1980).
47. *See* BAKER, *supra* note 11, at 26; CASS R. SUNSTEIN, DEMOCRACY AND THE PROBLEM OF FREE SPEECH 71 (1993).

to bias the market against programming preferred by those who are the least responsive to advertising. For example, one would expect that reliance on advertising support would tend to lead to a systematic underproduction of children's educational television, since purchasing decisions are typically made by supervising parents whose responsiveness to the commercials contained in children's programming is constrained by the fact that they frequently do not see the commercials at all.[48] Allowing parents to make direct payments for programming would provide a much more straightforward means for signaling their preferences. It is almost certainly no accident that most of the best children's educational programming on commercial television appears on cable.[49]

Reliance on advertising support also allows the biases of particular advertisers to influence the program mix. Anecdotal evidence suggests that some advertisers have discouraged networks from offering programming that addresses controversial issues or that casts their products in a poor light.[50] This reliance also leaves programmers vulnerable to the political biases of advertisers and special interest groups. Consider, for example, the recent controversy surrounding the miniseries *The Reagans*, originally scheduled to air on CBS. When dissatisfaction with the portrayal of the former President and First Lady threatened to erupt into a consumer boycott of any products advertised during the miniseries, Viacom shifted the program from CBS to Showtime, a premium movie channel that does not depend on advertising support.[51]

This episode bears a striking resemblance to the reaction to a pair of programs on abortion aired during the 1970s. When NBC tried to broadcast its movie version of *Roe v. Wade*, it faced such a backlash from advertisers that it eventually opted to show the movie without commercials, which in turn caused it to incur significant economic losses on the project.[52] This is in sharp contrast to the relative ease with which HBO was able to air a documentary on the same subject. The fundamental difference is that HBO's survival does not depend on its ability to assuage sponsors. As one

---

48. *See* Policies and Rules Concerning Children's Television Programming, Report and Order, 11 F.C.C.R. 10,660, 10,675 ¶¶ 32–33 (1996).

49. *See* Yoo, *supra* note 11, at 327–28.

50. *See* BAKER, *supra* note 11, at 24–30; SUNSTEIN, *supra* note 47, at 63–65; Steven Shiffrin, *The Politics of the Mass Media and the Free Speech Principle*, 69 IND. L.J. 689, 696–713 (1994).

51. *See* Meg James, Greg Braxton & Bob Baker, *The Vetoing of "Reagans": How Protests and Bad Timing Led CBS to Cancel a Movie About the Former First Couple*, L.A. TIMES, Nov. 10, 2003, at E1; Emily Nelson & Joe Flint, *CBS Pulls "Reagans" amid Opposition from Conservatives*, WALL ST. J., Nov. 5, 2003, at A3.

52. *See* SUNSTEIN, *supra* note 47, at 65; Shiffrin, *supra* note 50, at 698.

HBO executive explained, "We're not any braver than the networks. It's just that our economic basis is different."[53]

It is thus clear that the FCC's historical commitment to promoting a radio and television industry supported by advertising represents a form of architectural censorship that has had the unintended consequence of reducing the overall quantity, quality, and diversity of radio and television programming. Although a number of other scholars recognizing the problems associated with advertising support have proposed second-order corrective measures,[54] I would prefer the more straightforward solution of eliminating the hostility toward fee-based services. Advertising-supported media would appear to be a singularly inefficient mechanism for ensuring that all U.S. households have access to media regardless of their socioeconomic status. The evidence suggests that a targeted subsidy system, in which households falling below the poverty line are given discounted service, would be far more effective than the current system of untargeted subsidies.[55]

Recent pronouncements by Congress, the Supreme Court, and the FCC, however, make it quite likely that the commitment to preserving free television will remain one of the central aims of U.S. media policy for the foreseeable future.[56] As long as that is the case, this policy will continue to have unintended and adverse impacts on the content of speech.

B. RATE REGULATION OF CABLE TELEVISION

Another common feature of U.S. media policy has been the imposition of rate regulation on the cable television industry.[57] These efforts were clearly designed to protect consumers against excessive prices charged by

---

53. Jan Hoffman, *TV Shouts "Baby" (and Barely Whispers "Abortion")*, N.Y. TIMES, May 31, 1992, at H1, *quoted in* Shiffrin, *supra* note 50, at 698.
54. *See* BAKER, *supra* note 11, at 98–99, 114–21; SUNSTEIN, *supra* note 47, at 84–88.
55. *See* Yoo, *supra* note 18, at 1675–76; Yoo, *supra* note 11, at 354–55.
56. *See* Cable Television Consumer Protection and Competition Act of 1992, Pub. L. No. 102-385, § 2(a)(12), 106 Stat. 1460, 1461 (codified as amended at 47 U.S.C. §§ 534–535 (2000)); *Turner I*, 512 U.S. 622, 663 (1994) (recognizing that "nearly 40% of American households still rely on broadcast stations as their exclusive source of television programming" and holding that "protecting noncable households from loss of regular television broadcasting service" is an important federal interest (internal quotation marks omitted)); 2003 Biennial Review Order, *supra* note 3, at 13,674–75 ¶ 148 (identifying "the preservation of free, universally available local broadcast television in a digital world" as an important goal).
57. For a useful overview of the early history of cable rate regulation, see *Time Warner Entertainment Co. v. FCC*, 56 F.3d 151, 178–80 (D.C. Cir. 1995) (opinion of Randolph, J.).

local cable monopolists.[58] Because of its economic focus and its unrelatedness to program content, rate regulation represents a classic example of structural regulation. As a result, conventional wisdom presumes that rate regulation has little to no impact on the content of speech.[59]

The on-again/off-again history of cable rate regulation[60] provides an ideal opportunity for using event studies to assess its effectiveness empirically. Somewhat surprisingly, these studies indicate that rate regulation has largely been a failure. Despite the fact that rate regulation was designed to protect consumers against excessive prices charged by cable operators who did not face effective competition, the evidence suggests that rate regulation failed to yield any real welfare benefits for consumers.[61]

The key to understanding why rate regulation proved to be such a disappointment is to acknowledge the regime's inherent limitations. Rate regulation has always worked best when applied to commodity services, in which the quality and type of service provided does not vary. The would-be monopolist has only one dimension—price—with which it can extract surplus from consumers. When that is the case, limiting the prices that monopolists charge may well prove effective in limiting the exercise of market power.

A different situation obtains when the regulated service is not a commodity.[62] Where products vary in terms of quality, price represents

---

58. *See* 47 U.S.C. § 543(b)(1) (2000); *Time Warner*, 56 F.3d at 184–85.
59. *See Time Warner*, 56 F.3d at 183.
60. Rate regulation was widely imposed by cities until 1984, at which point it was effectively abolished by Congress. *See* Cable Communications Policy Act of 1984, Pub. L. No. 98-549, § 2, sec. 623(b), 98 Stat. 2779, 2788 (codified as amended in scattered sections of 47 U.S.C.) (allowing rate regulation unless cable operators faced "effective competition"); Implementation of the Provisions of the Cable Communications Policy Act of 1984, 50 Fed. Reg. 18,637, 18,648–50 ¶¶ 91–100 (F.C.C. May 2, 1985) (final rule) (defining "effective competition" in a way that exempted 96% of all cable systems). Congress reinstated cities' authority to regulate cable rates in 1992. *See* Cable Television Consumer Protection and Competition Act of 1992, Pub. L. No. 102-385, § 3(a), 106 Stat. 1460, 1464. It abruptly changed course once again four years later by passing another statute largely deregulating cable rates. *See* Telecommunications Act of 1996, Pub. L. No. 104-104, § 301(b), 110 Stat. 56, 114–15.
61. *See generally* THOMAS W. HAZLETT & MATTHEW L. SPITZER, PUBLIC POLICY TOWARD CABLE TELEVISION (1997); Gregory S. Crawford, *The Impact of the 1992 Cable Act on Household Demand and Welfare*, 31 RAND J. ECON. 422 (2000).
62. *See generally* David Besanko, Shabtai Donnenfeld & Lawrence J. White, *Monopoly and Quality Distortion: Effects and Remedies*, 102 Q.J. ECON. 743 (1987); David Besanko, Shabtai Donnenfeld & Lawrence J. White, *The Multiproduct Firm, Quality Choice, and Regulation*, 36 J. INDUS. ECON. 411 (1988); Kenneth S. Corts, *Regulation of a Multi-Product Monopolist: Effects on Pricing and Bundling*, 43 J. INDUS. ECON. 377 (1995).

only one of several dimensions along which producers can appeal to customers. Unless the regulator imposes comprehensive controls over quality as well as price, the regulated entity may evade any price restrictions simply by degrading the quality of its product offerings.

Indeed, the empirical evidence strongly suggests that this is precisely what has occurred in the cable industry. Although rate regulation caused nominal cable prices to drop, once other characteristics—such as the total number and quality of channels offered—are taken into account, the empirical evidence indicates that rate regulation caused quality-adjusted rates to increase and that deregulation caused quality-adjusted rates to fall.[63] This implies that consumers would have preferred larger, higher quality bundles of channels than they received under rate regulation, even if acquiring them meant paying higher prices. Placing a cap on cable rates simply limited cable operators' ability to move their product packages closer to consumers' ideal preferences.[64]

It thus appears that rate regulation did little to prevent local cable operators from exercising whatever monopoly power they possessed. Instead, rate regulation had the unintended consequence of degrading the quality of existing cable offerings and foreclosing the emergence of higher quality channel packages despite viewers' willingness to pay for them.[65]

---

63. *See* HAZLETT & SPITZER, *supra* note 61, at 2, 69–177, 208; Crawford, *supra* note 61, at 444–45.

64. Cable operators wishing to add high-end programming did have another option. They could have purchased it on an à la carte/premium channel basis. Forcing cable operators, however, to offer such channels on a stand-alone basis can have a dramatic impact. It prevents the operator from obtaining the benefits of bundling, which in turn makes it possible for the cable operator to offer a wider range of programming. *See* Yoo, *supra* note 18, at 1706–12.

65. Baker argues that distributive concerns, particularly the need to preserve access to the media by the poor, might justify cable rate regulation. *See* Baker, *supra* note 13, at 748. As is typical of economic analyses, my argument is not focused on distribution. That said, Baker's position is somewhat inconsistent with the basic structure of media policy. As noted earlier, preserving access to television by all citizens represents one of the central commitments of U.S. policy with respect to *broadcasting*. *See supra* notes 17–22, 25 and accompanying text. Broadcast stations remain the only commercial users who are not required to pay for their spectrum. The justification for what some Senators have condemned as an unsupportable act of corporate welfare is the need to preserve access to television. Yoo, *supra* note 18, at 1673–74, 1700. Conservative estimates place the value of this spectrum giveaway at $450 billion. *See* Hazlett, *supra* note 21, at 252 & n.44. The need to preserve access to broadcast television also represented one of the central justifications for must-carry. *See supra* notes 23, 26 and accompanying text. Having already set aside *broadcasting* as the medium for ensuring universal access to television and after having committed so many resources to ensure that it is available to everyone, regulating *cable* to accomplish the same end would seem excessive. Even if that were the goal, it is quite likely that direct subsidies would represent a far more effective means for promoting indigent access to cable television. *See* Yoo, *supra* note 18, at 1675–76; Yoo, *supra* note 11, at 354–55.

C. RESTRICTIONS ON HORIZONTAL CONCENTRATION

The FCC has long restricted the number of media outlets that one entity can own in any local media market. Some rules focus on *intra*media crossownership. They originated as a preference in licensing hearings against holding licenses to two AM radio stations operating in the same city.[66] The FCC formalized this licensing preference in a "duopoly rule" promulgated in 1940, which explicitly prohibited anyone from holding licenses for two television stations or two FM radio stations that served substantially the same area.[67] The duopoly rule was extended to AM radio in 1943.[68]

Other restrictions focus on *inter*media crossownership. Like the intramedia crossownership restrictions, intermedia crossownership restrictions began in licensing hearings as a preference in favor of diversification of media ownership.[69] In the 1970s, the FCC formalized these preferences into a series of explicit intermedia crossownership restrictions. The principal intermedia crossownership restrictions included (1) the "one-to-a-market" rule, which prohibited combined ownership of a radio and television station in the same local market;[70] (2) the

---

66. *See* Genesee Radio Corp., 5 F.C.C. 183, 186–87 (1938).
67. *See* 6 FCC ANN. REP. 68 (1940).
68. *See* Multiple Ownership of Standard Broadcast Stations, 8 Fed. Reg. 16,065 (F.C.C. Nov. 27, 1943). The FCC tightened the duopoly rule in 1969, abolishing the more permissive standard that only prohibited joint ownership if the stations served "substantially the same area" in favor of a more stringent restriction forbidding joint ownership of radio stations that had any overlap in their primary service contours, no matter how small. The rule was even more restrictive for television, which prohibited joint ownership of stations whenever there was any overlap in their secondary service contours. *See* Amendment of Sections 73.35, 73.240 and 73.636 of the Commission's Rules Relating to Multiple Ownership of Standard, FM and Television Broadcast Stations, Memorandum Opinion and Order, 45 F.C.C. 1728 (1964).
69. *See, e.g.*, Port Huron Broad. Co., 5 F.C.C. 177, 182 (1938); Newspaper Ownership of Radio Stations, 9 Fed. Reg. 702 (F.C.C. Jan. 18, 1944) (notice of dismissal of proceeding). *See generally* Policy Statement on Comparative Broadcast Hearings, 1 F.C.C.2d 393, 394–95 (1965) (identifying "[d]iversification of control of the media of mass communications" as a "factor of primary significance" in comparative licensing proceedings) (italics omitted). The FCC's early application of this criterion was far from consistent. *See* HENRY J. FRIENDLY, THE FEDERAL ADMINISTRATIVE AGENCIES: THE NEED FOR BETTER DEFINITION OF STANDARDS 64–69 (1962); Bernard Schwartz, *Comparative Television and the Chancellor's Foot*, 47 GEO. L.J. 655, 673–78, 685–94 (1959).
70. *See* Amendment of Sections 73.35, 73.240 and 73.636 of the Commission Rules Relating to Multiple Ownership of Standard, FM and Television Broadcast Stations, First Report and Order, 22 F.C.C.2d 306, 308 ¶ 8 (1970) [hereinafter 1970 Multiple Ownership Order]. Shortly thereafter, the FCC liberalized the one-to-a-market rule to permit AM-FM combinations in the same market and to allow existing radio licensees to acquire UHF stations in the same market. *See* Amendment of Sections 73.35, 73.240 and 73.636 of the Commission's Rules Relating to Multiple Ownership of Standard, FM and Television Broadcast Stations, Memorandum Opinion and Order, 28 F.C.C.2d 662 (1971).

newspaper/broadcast crossownership rule, which banned common ownership of a newspaper and broadcast station when the broadcast station's service contour completely encompassed the newspaper's city of publication;[71] (3) the cable/broadcast crossownership rule, which effectively prohibited the owner of a local cable system from also owning a local broadcast station;[72] and (4) the cable/local telephone company crossownership rule, which prohibited local telephone companies from providing video programming to subscribers in their respective local service area.[73]

The FCC has long justified its restrictions on horizontal concentration with two rationales: the need to protect competition[74] and the need to promote a diversity of programming and viewpoints.[75] The first is completely economic in focus and unrelated to the content of speech. The second implicates First Amendment concerns more directly because "ownership carries with it the power to select, to edit, and to choose the methods, manner and emphasis of presentation."[76]

---

71. *See* Amendment of Sections 73.34, 73.240, and 73.636 of the Commission's Rules Relating to Multiple Ownership of Standard, FM and Television Broadcast Stations, Second Report and Order, 50 F.C.C.2d 1046 (1975) [hereinafter 1975 Multiple Ownership Order].

72. As a formal matter, this rule only prohibits a cable television system from carrying the signal of any broadcast television station if it owns a broadcast station in the same local market. Amendment of Part 74, Subpart K, of the Commission's Rules and Regulations Relative to Community Antenna Television Systems, Second Report and Order, 23 F.C.C.2d 816, 820–21 689 ¶ 12–14 (1970) [hereinafter Community Antenna Order]. When combined with the cable operators' must-carry obligations, this rule effectively prohibits cable/broadcast crossownership. *See* Fox Television Stations, Inc. v. FCC, 280 F.3d 1027, 1035 (D.C. Cir.), *modified on reh'g*, 293 F.3d 537 (D.C. Cir. 2002).

73. Applications of Telephone Companies for Section 214 Certificates for Channel Facilities Furnished to Affiliated Community Antenna Television Systems, Final Report and Order, 21 F.C.C.2d 307, 323–25 ¶¶ 43–49 (1970), *aff'd sub nom.* Gen. Tel. Co. of the Southwest v. United States, 449 F.2d 846 (5th Cir. 1971). This requirement was codified by the 1984 Cable Act. *See* Cable Communications Policy Act of 1984, Pub. L. No. 98-549, § 2, sec. 613(b)(1), 98 Stat. 2779, 2785 (originally codified at 47 U.S.C. § 533(b)).

74. *See, e.g.*, Amendment of Section 73.3555 of the Commission's Rules, the Broadcast Multiple Ownership Rules, First Report and Order, 4 F.C.C.R. 1723, 1724 ¶ 8, 1727 ¶¶ 32–33 (1989) [hereinafter 1989 Multiple Ownership Order]; 1975 Multiple Ownership Order, *supra* note 71, at 1074 ¶ 99; 1970 Multiple Ownership Order, *supra* note 70, at 307 ¶ 3 ; Amendment of Sections 73.35, 73.240, and 73.636 of the Commission's Rules Relating to Multiple Ownership of Standard, FM and Television Broadcast Stations, Report and Order, 45 F.C.C. 1476, 1476–77 ¶¶ 2–3 (1964) [hereinafter 1964 Multiple Ownership Order]; Genesee Radio Corp., 5 F.C.C. 183, 186–87 (1938).

75. *See, e.g., Genesee*, 5 F.C.C. at 186–87; 1989 Multiple Ownership Order, *supra* note 74, at 1723–24 ¶ 7, 1727 ¶ 31; 1975 Multiple Ownership Order, *supra* note 71, at 1074 ¶ 99; 1970 Multiple Ownership Order, *supra* note 70, at 307 ¶ 3; 1964 Multiple Ownership Order, *supra* note 74, at 1476–77 ¶¶ 2–3.

76. 1975 Multiple Ownership Order, *supra* note 71, at 1050 ¶ 14.

Several forces led the FCC to relax a number of these rules in the ensuing years. The first was a series of deregulatory initiatives launched during the administrations of Ronald Reagan and George H.W. Bush.[77] Furthermore, a series of lower federal court decisions handed down during the mid-1990s voided the cable/local telephone company crossownership rule on First Amendment grounds.[78] The issue had already been briefed and argued before the Supreme Court when it was rendered moot by a provision of the Telecommunications Act of 1996 that eliminated the rule.[79]

The 1996 Act also contained a number of provisions raising the thresholds needed to trigger various horizontal ownership restrictions.[80] In

---

77. In 1989, the FCC relaxed the duopoly rule. Under the old rule, common ownership of two broadcast stations in the same service was prohibited if there was any overlap in the two stations' primary service contours. Under the new rule, same-service common ownership would be prohibited only if the two stations' principal city contours overlapped. 1989 Multiple Ownership Order, *supra* note 74, at 1723. The next day, the FCC relaxed the one-to-a-market rule to allow for a presumptive waiver for failed stations and for crossownership in the top twenty-five markets so long as thirty independent voices remain in the market. Amendment of Section 73.3555 of the Commission's Rules, the Broadcast Multiple Ownership Rules, Second Report and Order, 4 F.C.C.R. 1741 (1989) [hereinafter 1989 Second Multiple Ownership Order]. In 1992, the FCC repealed the network/cable crossownership rule. Amendment of Part 76, Subpart J, Section 76.501 of the Commission's Rules and Regulations to Eliminate the Prohibition on Common Ownership of Cable Television Systems and National Television Networks, Report and Order, 7 F.C.C.R. 6156, 6162–63 ¶ 10 (1992) [hereinafter Order to Eliminate the Prohibition on Common Ownership]. The FCC also relaxed the duopoly rule with respect to radio, allowing a single entity to own two AM and two FM stations in the same market so long as the market contained fifteen or more commercial stations and so long as the radio combinations did not exceed a designated audience share. In smaller markets, the 1992 amendments permitted a single entity to own three radio stations, no more than two of which could be in the same service. Revision of Radio Rules and Policies, Report and Order, 7 F.C.C.R. 2755, 2757–61 ¶¶ 4–12 (1992) [hereinafter Radio Rules and Policies]. The FCC also initiated proceedings to revisit the rules with respect to television. *See* Broadcast Services, 60 Fed. Reg. 6,490 (F.C.C. Feb. 2, 1995) (further notice of proposed rulemaking). Television Broadcast Services; Video Marketplace, 57 Fed. Reg. 28,163 (F.C.C. June 24, 1992) (notice of proposed rulemaking); Broadcast and Cable Services, Effect of Changes in the Video Marketplace, 56 Fed. Reg. 40,847 (F.C.C. Aug. 16, 1991) (notice of inquiry).

78. *See* US West, Inc. v. United States, 48 F.3d 1092 (9th Cir. 1995), *vacated and remanded*, 516 U.S. 1155 (1996); Chesapeake & Potomac Tel. Co. v. United States, 42 F.3d 181 (4th Cir. 1994), *vacated*, 516 U.S. 415 (1996); S. New England Tel. Co. v. United States, 886 F. Supp. 211 (D. Conn. 1995); BellSouth Corp. v. United States, 868 F. Supp. 1335 (N.D. Ala. 1994); Ameritech Corp. v. United States, 867 F. Supp. 721 (N.D. Ill. 1994); NYNEX Corp. v. United States, Civ. 93-323-P-C, 1994 WL 779761 (D. Me. Dec. 8, 1994). *See generally* Glen O. Robinson, *The New Video Competition: Dances with Regulators*, 97 COLUM. L. REV. 1016, 1018–24 (1997) (reviewing these cases).

79. *See* Telecommunications Act of 1996, Pub. L. No. 104-104, § 302(b)(1), 110 Stat. 56, 124 (repealing 47 U.S.C. § 533(b) (1994)).

80. Specifically, the Act substantially relaxed the one-to-a-market rule with respect to radio. *Id.* § 202(b), 110 Stat. at 110. It also directed the FCC to conduct a proceeding to determine whether to retain, modify, or eliminate the duopoly rule with respect to television. *Id.* § 202(c)(2), 110 Stat. at 111. It expanded a presumptive waiver to the radio/television crossownership rule for the top twenty-five markets discussed above, *supra* note 70, to cover the top fifty markets. § 202(d), 110 Stat. at 111. The

addition, Congress directed the FCC to create a biennial review process in which it would revisit all of its ownership rules every two years to "determine whether any of such rules are necessary in the public interest as a result of competition" and to "repeal or modify any regulation it determines to be no longer in the public interest."[81] The FCC amended a number of its rules during its initial biennial review, but left many others in place.[82] This was followed by a pair of decisions issued by the D.C. Circuit in 2002 striking down the FCC's refusal to revisit the cable/broadcast crossownership[83] and the revised duopoly rules.[84]

Judicial invalidation of these ownership restrictions prompted the FCC to undertake a massive reassessment of the regulations as part of its 2002 biennial review proceeding. Rejecting calls for repeal of most of its ownership rules, the FCC instead replaced the hodgepodge of local ownership rules with a new, integrated approach based on a "diversity index" designed to take into account all media when assessing the overall competitiveness of the local market.[85] The FCC supplemented the traditional concerns of competition and diversity of viewpoints[86] with one additional policy consideration that was often asserted in connection with television and radio policy, but had not previously been invoked with

---

Act also repealed the statutory provision prohibiting cable/telephone company crossownership. *Id.* § 302(b)(1), 110 Stat. at 124. It also repealed the provision codifying the cable/broadcast crossownership rule. *Id.* § 202(i), 110 Stat. at 112. Repealing the statutory ban on cable/broadcast crossownership left in place the parallel regulatory requirement imposed by the FCC. *See* Fox Television Stations, Inc. v. FCC, 280 F.3d 1027, 1035 (D.C. Cir.), *modified on reh'g*, 293 F.3d 537 (D.C. Cir. 2002).

81. Telecommunications Act of 1996, § 202(h), 110 Stat. at 111–12.

82. Review of the Commission's Regulations Governing Television Broadcasting, 64 Fed. Reg. 50,651 (F.C.C. Sept. 17, 1999) (final rule). Of particular note is the manner in which the FCC relaxed the one-to-a-market rule and the duopoly rule for television. In each case, the FCC incorporated an "independent voices" test into the rule that allowed a greater degree of crossownership if a sufficient number of independent ownership groups remained after the merger. There was one key difference between the two independent voices tests devised by the FCC. With respect to the duopoly rules, the FCC took an expansive view of what constituted an independent voice, including other media such as radio stations, daily newspapers, and local cable systems. *Id.* at 50,659–60. The Commission took a much narrower approach when determining what constituted an independent voice for purposes of the one-to-a-market rule, limiting its scope only to other television stations. *Id.* at 50,655 ¶ 30.

83. *See Fox Television Stations*, 280 F.3d at 1049–52. On remand, the FCC declined the opportunity to attempt to generate an alternative justification for the rule and instead simply repealed it. 2003 Biennial Review Order, *supra* note 3, at 13,620. The D.C. Circuit also invalidated the FCC's national television station ownership rule. *Fox Television Stations*, 280 F.3d at 1040–47. As that rule is primarily vertical, rather than horizontal, in focus, it is discussed *infra* notes 171–172 and accompanying text.

84. *See* Sinclair Broad. Group, Inc. v. FCC, 284 F.3d 148, 162–65 (D.C. Cir. 2002).

85. 2003 Biennial Review Order, *supra* note 3, at 13,775–807 ¶¶ 391–481.

86. *Id.* at 13,627–43 ¶¶ 18–72.

respect to horizontal ownership restrictions: localism.[87] Interestingly, in each instance, the FCC concluded that relaxation of the horizontal ownership restrictions would have no adverse impact on the responsiveness of media outlets to the needs and interests of their local communities.[88] Unlike vertical integration, which can give national networks the power to dictate local programming decisions, horizontal integration has no effect on localism, since the locus of programming decisions remains within the community.[89] In many cases, the record suggested that permitting greater horizontal concentration would actually promote localism by allowing media outlets to realize the efficiencies associated with crossownership.[90]

Shortly after the issuance of the biennial review order, the Third Circuit issued a stay preventing it from going into effect pending judicial review.[91] The court, somewhat remarkably, held that it could ignore the traditional requirement that the party seeking the stay demonstrate a likelihood of success on the merits—generally regarded as one of the standard requirements for the grant of a stay—if the issues were sufficiently complex and the hardships sufficiently severe.[92] The Third Circuit subsequently remanded the changes to the horizontal ownership restrictions that would have been effected by the biennial review order.[93]

1. The Complex Relationship Between Market Concentration and Program Diversity

There is general agreement that horizontal concentration affects program diversity, although theorists differ as to the precise nature of the relationship. On the one hand are commentators who are largely critical of increases in media concentration and warn that the likely result will be a reduction in the quantity and diversity of media content.[94] On the other

---

87. *Id.* at 13,643–45 ¶¶ 73–79. *See also id.* at 13,738 ¶ 304 (noting that the FCC had not previously emphasized localism as a justification for restricting the number of radio stations one entity could own in any one locality).
88. *See id.* at 13,737–38 ¶¶ 302–304, 13,753–54 ¶ 342, 13,772–73 ¶ 383.
89. *Id.* at 13,738 ¶ 304.
90. *See id.* at 13,678–85 ¶¶ 155–169, 13,753–60 ¶¶ 342–354, 13,772–73 ¶¶ 382–385; *infra* Part I.C.2.
91. Prometheus Radio Project v. FCC, No. 03-3388, 2003 WL 22052896 (3d Cir. Sept. 3, 2003).
92. *Id.* at *1 (citing Wash. Metro. Area Transit Comm'n v. Holiday Tours, Inc., 559 F.2d 841, 843 (D.C. Cir. 1977)).
93. Prometheus Radio Project v. FCC, 373 F.3d 372 (3d Cir. 2004) (affirming the FCC's power to regulate ownership restrictions, but remanding several portions of the order as not sufficiently supported by the record).
94. *See, e.g.*, BEN H. BAGDIKIAN, THE MEDIA MONOPOLY (6th ed. 2000); EDWARD S. HERMAN & NOAM CHOMSKY, MANUFACTURING CONSENT: THE POLITICAL ECONOMY OF THE MASS MEDIA 3–

hand are scholars who adopt the less intuitive position that increases in market concentration can promote program quality and diversity.[95] This section outlines a more complex approach that captures the nuances of both positions. As with most economic issues, the truth lies somewhere in between.

a. Steiner Models

Reconciliation of these two divergent inferences requires an understanding of how it is possible for media monopolies to produce greater program diversity than competitive markets. The argument has its roots in the model of local radio markets proposed by Peter Steiner,[96] which has subsequently been adapted to the television industry[97] and which has gained substantial attention from courts,[98] commentators,[99] and the FCC.[100]

---

14 (1988); Baker, *Media Concentration, supra* note 11; Krotoszynski & Blaiklock, *supra* note 11, at 832–34, 859–80.

95. For my initial review of this literature, see Yoo, *supra* note 17, at 1935–48. For other surveys, see BRUCE M. OWEN & STEVEN S. WILDMAN, VIDEO ECONOMICS 64–100, 141–44 (1992); Matthew L. Spitzer, *Justifying Minority Preferences in Broadcasting*, 64 S. CAL. L. REV. 293, 304–17 (1991).

96. Peter O. Steiner, *Program Patterns and Preferences, and the Workability of Competition in Radio Broadcasting*, 66 Q.J. ECON. 194 (1952).

97. *See* Jack H. Beebe, *Institutional Structure and Program Choices in Television Markets*, 91 Q.J. ECON. 15 (1977); Jerome Rothenberg, *Consumer Sovereignty and the Economics of TV Programming*, 4 STUD. PUB. COMM. 45, 47–48 (1962); P. Wiles, *Pilkington and the Theory of Value*, 73 ECON. J. 183 (1963).

98. *See* Schurz Communications, Inc. v. FCC, 982 F.2d 1043, 1054–55 (7th Cir. 1992).

99. *See, e.g.*, Benjamin, *supra* note 11, at 97 n.278; Yochai Benkler, *Siren Songs and Amish Children: Autonomy, Information, and Law*, 76 N.Y.U. L. REV. 23, 94–95 (2001); Daniel L. Brenner, *Government Regulation of Radio Program Format Changes*, 127 U. PA. L. REV. 56, 63–69 (1978); Jim Chen, *The Last Picture Show (On the Twilight of Federal Mass Communications Regulation)*, 80 MINN. L. REV. 1415, 1448, 1491 (1996); Krotoszynski & Blaiklock, *supra* note 11, at 868 & n.366; Spitzer, *supra* note 95, at 305–12.

100. *See* 2002 Biennial Regulatory Review—Review of the Commission's Broadcast Ownership Rules and Other Rules Adopted Pursuant to Section 202 of the Telecommunications Act of 1996, Notice of Proposed Rule Making, 17 F.C.C.R. 18,503, 18,530 ¶ 82 & n.159 (2002); Revision of Radio Rules and Policies, Second Memorandum Opinion and Order, 9 F.C.C.R. 7183, 7186 ¶ 21 (1994). *See also* 2003 Biennial Review Order, *supra* note 3, at 13,740–42 ¶¶ 310–15 (discussing Steiner); Review of the Commission's Regulations Governing Television Broadcasting, Further Notice of Proposed Rule Making, 10 F.C.C.R. 3524, 3550–51 ¶¶ 62–63 (1995) (same); Revision of Radio Rules and Policies, Notice of Proposed Rule Making, 6 F.C.C.R. 3275 (1991) (same); Reexamination of the Commission's Cross-Interest Policy, Policy Statement, 4 F.C.C.R. 2208, 2212 ¶ 30 (1989) (same); Consideration of the Operation of, and Possible Changes in, the Prime Time Access Rule, § 73.658(k) of the Commission's Rules, Second Report and Order, 50 F.C.C.2d 829, 894 (1975) (Robinson, Comm'r, dissenting) (same). *But see* 2003 Biennial Review Order, *supra* note 3, at 13,742 ¶ 314 (declining to embrace Steiner's model).

The counterintuitive nature of Steiner's argument can best be understood through a simple numerical example. Steiner assumed that the preferences of an audience in a particular local market could be divided into four discrete program formats of the following sizes:

FIGURE 1. Steiner's model of program diversity

	Program format			
	Type 1	Type 2	Type 3	Type 4
Audience size	210	75	50	31

The first station to enter the market would naturally offer programming targeted at the largest segment, Type 1. The second entrant would face a choice of either offering programming targeted toward the second largest segment, Type 2, in which case it would capture an audience of 75, or duplicating the same type of programming offered by the first entrant, in which case it would split the Type 1 audience with the first entrant and capture an audience of 105. So long as half of the largest segment exceeds the size of the second largest segment, the second entrant will duplicate existing programming format.

The problem, from a welfare standpoint, is that the entire volume captured by the second entrant consists of audience members who were already being served by the first (an effect sometimes called "demand diversion"). Because the first entrant was already serving these listeners, entry by the second station creates no welfare benefits. If the second entrant had instead offered Type 2 programming, its audience would have consisted entirely of incremental listeners who were not previously being served by the incumbent (an effect sometimes called "demand creation"). Thus, to the extent that the audience captured by a new entrant results from demand creation, entry is welfare enhancing. To the extent that the new entrant's audience results solely from demand diversion, it creates no consumer benefits and instead simply wastes resources.

Steiner recognized that competitive entrants would target their programming without taking into account whether the audience it captured was the product of demand creation or demand diversion. As a result, they may offer redundant programming notwithstanding the fact that doing so creates no welfare benefits. In addition, to the extent that channel capacity is limited, duplication of existing formats tends to crowd out other program types.[101] This logic suggests that a third entrant would offer programming targeted toward Type 2,[102] while a fourth entrant would again duplicate Type 1 programming.[103] Type 3 programming would not appear until the arrival of the sixth station, and Type 4 until the arrival of the tenth.[104]

The tendency toward duplication of program types disappears, however, if the entrants are jointly owned.[105] Unlike a competitive entrant, a monopolist would consider whether the revenue captured by an additional station resulted from demand creation or from demand diversion. In fact, a monopolist controlling all stations would focus solely on generating new audiences and would eschew any strategy that simply cannibalized listeners from its other stations.

Stated in the terms of the numerical hypothetical described above, if the initial two entrants were jointly owned, the owner would not use both stations to target Type 1, since the audience captured by the second station would come entirely at the expense of the first. Instead, the owner would direct each successive station at a different market segment. Thus, Steiner was able to show that, under his assumptions, monopoly control of a local radio market can satisfy more viewers and yield greater program diversity than can competition.[106]

Steiner's analysis also has implications for program quality. In the case of competitive entry, multiple entrants divide the revenue generated by any particular program type. In the case of monopolistic control of a local radio market, each audience segment is served by precisely one station. Therefore, each station under the monopoly solution will capture more revenue than under competition. To the extent that quality correlates with program cost, monopoly provision should cause program quality to increase.[107]

---

101. *See* Beebe, *supra* note 97, at 23, 30–31; Rothenberg, *supra* note 97, at 48.
102. The third entrant would find that the size of the second largest segment (75) exceeds the audience it would capture if it divided the largest segment with the two other entrants (70).
103. The fourth entrant would find that one-third of the Type 1 audience (70) would still be larger than half of the Type 2 audience (37.5) or the entirety of the Type 3 audience (50).
104. *See* Steiner, *supra* note 96, at 200.
105. *See id.* at 206–07; Wiles, *supra* note 97, at 188.
106. Steiner, *supra* note 96, at 206–07.
107. *See* OWEN & WILDMAN, *supra* note 95, at 144–48. The extent to which increased revenue will result in increased expenditures on programming depends on the elasticity of demand. Bruce Owen and Steven Wildman note the theoretical possibility that competitive entry might stimulate the production of higher-quality programming. *See id.* at 85. In either scenario, governmental restrictions on horizontal concentration would have a direct impact on program quality.

b. Limitations of Steiner Models

Steiner models suffer from a number of limitations, some well recognized, other less so. Theorists building on Steiner's work have pointed out that the correlation between monopoly and program diversity that he found depends on a host of assumptions: the particular skewness found in the distribution of demand, the willingness of audiences to view second-choice programming, the magnitude and variability of program cost, and the availability of excess channel capacity.[108] These limitations have been analyzed elsewhere[109] and those arguments will not be repeated here.

Other fundamental limitations to Steiner's analysis have largely gone unnoticed. For example, his approach necessarily presupposes that programming can be segregated into one of several discrete formats.[110] Experience has shown that radio and television programming defies easy categorization. Consider the popular "oldies" radio format that, in a fairly short period, multiplied from one format to several, as different stations targeted listeners of different ages. The FCC has recognized that radio and television formats are far too dynamic and varied to be classified in such a simple, categorical manner, and the Supreme Court has given its blessing to this conclusion.[111]

Equally problematic is Steiner's assumption that entry by an additional station into an occupied format simply duplicates existing programming.[112] In effect, he assumes that, within any particular format, programming is completely fungible. The most casual perusal of the radio market falsifies this assumption—the popularity of radio stations offering the same format category varies widely.[113] Stations that appear to be offering the same type of programming typically provide very different levels of utility to listeners. This suggests that program types might be better understood not as falling into discrete categories, but rather as occupying a position along a spectrum of program characteristics. Revising the model of program selection in this manner would call into question the

---

108. *See* Beebe, *supra* note 97, at 23–31; Rothenberg, *supra* note 97, at 49–50.
109. *See* Yoo, *supra* note 17, at 1938–42.
110. For similar efforts, see Edward Greenberg & Harold J. Barnett, *TV Program Diversity—New Evidence and Old Theories*, 61 AM. ECON. REV. 89, 90 (1971); Levin, *supra* note 29, at 84–87.
111. *See* Development of Policy re: Changes in the Entertainment Formats of Broadcast Stations, Memorandum Opinion and Order, 60 F.C.C.2d 858, 861–63 ¶¶ 11–15 (1976) [hereinafter Format Policy Statement], *aff'd sub nom.* FCC v. WNCN Listeners Guild, 450 U.S. 582 (1981).
112. *See* Steiner, *supra* note 96, at 199. *See also id.* at 206 (relaxing this assumption).
113. *See* Format Policy Statement, *supra* note 111, at 863–64 ¶ 18.

assumption that duplication of an existing program type by a new entrant necessarily yields no welfare benefit, since it remains possible that a new entrant might attract new listeners or provide greater satisfaction to members of the audience who were already listening.[114]

Finally, Steiner's approach measured welfare through a voting model that simply counted the number of viewers in any audience.[115] The inability of such voting-oriented models to take intensity of preferences into account limits their ability to assess economic welfare properly.[116] In addition, omitting any aspect of price competition eliminates the possibility that welfare gains created by increased program diversity might be offset by welfare losses incurred through the exercise of oligopoly power in a concentrated market. Although Steiner's voting model might have made sense at a time when radio broadcasters could not typically charge for their programs,[117] it makes less sense in a world in which fee-based radio and television services are a reality.

These weaknesses of the Steiner model indicate that local media markets might be better analyzed under the more general model of spatial competition pioneered by Harold Hotelling. This model assumes that producers compete by occupying a position along a continuous product spectrum, rather than by placing themselves into one of a discrete number of product categories.[118] The legal literature[119] and the FCC have largely overlooked these models.[120]

I hope to offer a more complete application of spatial competition models to the FCC's media ownership regulations in my future work. For

---

114. *See* Steiner, *supra* note 96, at 204.
115. *See id.* at 196–97.
116. *See id.* at 197.
117. *See id.* at 198.
118. *See* Harold Hotelling, *Stability in Competition*, 39 ECON. J. 41 (1929). For a general introduction to spatial competition models, see Christopher S. Yoo, *Copyright and Product Differentiation*, 79 N.Y.U. L. REV. 212, 241–46 (2004). For applications to television programming, see Eli M. Noam, *A Public and Private-Choice Model of Broadcasting*, 55 PUB. CHOICE 163 (1987); Alessandro Vaglio, *A Model of the Audience for TV Broadcasting: Implications for Advertising Competition and Regulation*, 42 RIVISTA INTERNAZIONALE DI SCIENZE ECONOMICHE E COMMERCIALI 33 (1995); David Waterman, *Diversity and Quality of Information Products in a Monopolistically Competitive Industry*, 4 INFO. ECON. & POL'Y 291 (1991).
119. The only discussion of any significance appearing in the law review literature is Spitzer, *supra* note 95, at 314–16.
120. The only FCC reference to this literature of which I am aware is the bare citation of a paper by Richard Schmalensee that employed a spatial competition model. *See* Review of the Commission's Regulations Governing Television Broadcasting, Further Notice of Proposed Rule Making, 10 F.C.C.R. 3524, 3551 n.81 (1995).

now, it suffices to note that spatial models suggest that the relationship between horizontal concentration and welfare may be more complex than Steiner's model suggests. First, Hotelling-style spatial competition acknowledges that entry by a similar product can yield welfare benefits, both by capturing incremental demand and by allowing some audience members who were already viewing to consume programming that offers a better fit with their ideal preferences. These models also reflect how joint ownership can cause welfare to increase by inducing firms to pay attention to whether their revenue is the product of demand creation or demand diversion.[121] Finally, the more sophisticated spatial models take into account the fact that any economic benefits resulting from a monopolist's refusal to duplicate existing programming must be offset by the welfare losses associated with the reduction in price competition.[122]

Spatial models thus provide reason to be somewhat skeptical of Steiner's simplistic conclusion that market concentration necessarily promotes greater program variety as well as the supposition advanced by many commentators that media concentration invariably reduces the diversity of media content.[123] Although monopolists' unwillingness to cannibalize audiences from their own stations may tend to promote product diversity, their willingness to withdraw stations from the market and their tendency to charge supercompetitive prices works in the opposite direction.

Which of these two countervailing effects dominates is an empirical question that cannot be determined a priori. Formal models have shown that either too much or too little program diversity may exist in equilibrium and that monopoly may or may not produce greater program diversity or generate greater economic benefits.[124] Attempts to resolve this question empirically have yielded mixed results. While one leading study concluded that increases in horizontal concentration in local radio markets tended to

---

121. *See* JOHN BEATH & YANNIS KATSOULACOS, THE ECONOMIC THEORY OF PRODUCT DIFFERENTIATION 57 (1991); Severin Borenstein, *Price Discrimination in Free-Entry Markets*, 16 RAND J. ECON. 380, 388–89 (1985); Roger W. Koenker & Martin K. Perry, *Product Differentiation, Monopolistic Competition, and Public Policy*, 12 BELL J. ECON. 217, 226–27 (1981); N. Gregory Mankiw & Michael D. Whinston, *Free Entry and Social Inefficiency*, 17 RAND J. ECON. 48, 49, 52, 54–55 (1986).

122. *See* JEFFREY CHURCH & ROGER WARE, INDUSTRIAL ORGANIZATION 395–404 (2000); B. Curtis Eaton & Myrna Holtz Wooders, *Sophisticated Entry in a Model of Spatial Competition*, 16 RAND J. ECON. 282 (1985); Steven C. Salop, *Monopolistic Competition with Outside Goods*, 10 BELL J. ECON. 141, 143–45 (1979).

123. *See supra* note 94 and accompanying text.

124. *See* ANDERSON & COATE, *supra* note 42, at 19–23.

increase program diversity,[125] other studies have confirmed the tendency toward duplication and underscored the critical role played by channel capacity.[126] Yet another study of the television industry focusing on product differentiation concluded that program variety approached optimal levels,[127] while another study of the radio industry found excess entry.[128] Still other studies have drawn somewhat different conclusions.[129]

Fortunately, for the purposes of this Article, the precise relationship between market concentration and program diversity need not be resolved. It is sufficient to show that a relationship does exist, even if the direction and magnitude of the effect remain somewhat uncertain.[130] This relationship reveals that the degree of horizontal concentration permitted under current media ownership regulations will have a direct impact on media content.

2. The Role of Efficiencies from Horizontal Integration

Horizontal integration also affects program diversity by allowing media groups to realize cost efficiencies. Horizontal integration enables entities that own multiple stations to economize on costs, which in turn can support increases in the quantity, quality, and diversity of programming

---

125. *See* Steven T. Berry & Joel Waldfogel, *Do Mergers Increase Product Variety? Evidence from Radio Broadcasting*, 116 Q.J. ECON. 1009 (2001).

126. *See* August E. Grant, *The Promise Fulfilled? An Empirical Analysis of Program Diversity on Television*, 7 J. MEDIA ECON. 51, 62 (1994); Robert P. Rogers & John R. Woodbury, *Market Structure, Program Diversity, and Radio Audience Size*, 14 CONTEMP. ECON. POL'Y 81 (1996).

127. *See* Ronald L. Goettler & Ron Shachar, *Spatial Competition in the Network Television Industry*, 32 RAND J. ECON. 624 (2001).

128. *See* Steven T. Berry & Joel Waldfogel, *Free Entry and Social Inefficiency in Radio Broadcasting*, 30 RAND J. ECON. 397 (1999). This study acknowledged, however, that the radio industry is somewhat unusual in that it serves two different groups of customers—advertisers and listeners—only one of which (advertisers) is able to make direct payments for programming. What appears to be excess entry when measured solely in terms of benefits to advertisers may in fact be efficient when measured in terms of both advertisers and listeners. *Id.* at 412–14.

129. *See* 2003 Biennial Review Order, *supra* note 3, at 13,740–42 ¶¶ 310–315 (reviewing the literature).

130. As Professor Baker points out, my conclusion that horizontal concentration has an ambiguous impact on media content is not completely consistent with my overarching claim that the forms of architectural censorship I have identified reduce the quantity, quality, and diversity of media programming. *See* Baker, *supra* note 13, at 739–740. I concede that my attempt to reduce the central thesis of this Article into a pithy catchphrase represents something of an overstatement in this limited respect. That said, I do believe that my summation does accurately reflect the negative impact that the other forms of architectural censorship I have identified have on media content. Furthermore, the fact remains that horizontal restrictions are having a direct effect on the content of media speech regardless of the direction of the effect. The fact that governmental actions are altering program content should raise First Amendment concerns regardless of the precise nature of the effect.

offered by allowing the media industry to invest a larger proportion of its revenue in program production. The FCC has repeatedly recognized that local crossownership provides precisely these benefits by allowing the station owner to combine administrative, programming, sales, marketing, promotion, and production costs.[131] Indeed, some data suggest that crossownership can reduce the cost of these functions by 30% to 35%.[132]

In addition, crossownership can help newspapers realize more efficient use of their efforts to collect local news. Like all forms of television and radio programming, local news bears many of the classic indicia of a pure public good. In particular, consumption of local news is nonrivalrous, in that consumption of it by one person does not reduce the supply available for others. In economic terms, this is usually modeled by assuming that once a media entity has incurred the fixed costs associated with gathering the news, the marginal cost of sharing with others is zero. Thus, once the costs of collecting the local news have been incurred, economic success depends on disseminating that information to as many paying customers as possible.[133] Thus, as a theoretical matter, the greater return on investment made possible by crossownership may enable media outlets to provide more diverse programming.[134] Empirical studies have largely borne this out.[135]

---

131. *See* Review of the Commission's Regulations Governing Television Broadcasting, Report and Order, 14 F.C.C.R. 12,903, 12,920–22 ¶¶ 34–36 (1999); Radio Rules and Policies, *supra* note 77, at 2760–61 ¶ 11, 2774 ¶ 37; 1989 Second Multiple Ownership Order, *supra* note 77, at 1746–47 ¶¶ 39–51; 1989 Multiple Ownership Order, *supra* note 74, at 1727 ¶ 36.

132. Amendment of Section 73.3555 of the Commission's Rules, the Broadcast Multiple Ownership Rules, Notice of Proposed Rulemaking, 2 F.C.C.R. 1138, 1140–41 ¶ 20 (1987).

133. *See* Yoo, *supra* note 18, at 1657–59.

134. *See* 2003 Biennial Review Order, *supra* note 3, at 13,678 ¶¶ 155–156, 13,753–61 ¶¶ 342–358, 13,772–73 ¶¶ 382–385; 1989 Second Multiple Ownership Order, *supra* note 77, at 1747 ¶ 44.

135. *See* John C. Busterna, *Television Station Ownership Effects on Programming and Idea Diversity: Baseline Data*, 1 J. MEDIA ECON. 63 (1988); Robert B. Ekelund, Jr., George S. Ford & Thomas Koutsky, *Market Power in Radio Markets: An Empirical Analysis of Local and National Concentration*, 43 J.L. & ECON. 157, 180 (2000); David Pritchard, *A Tale of Three Cities: "Diverse and Antagonistic" Information in Situations of Local Newspaper/Broadcast Cross-Ownership*, 54 FED. COMM. L.J. 31 (2001). These studies largely corroborated earlier research finding that media ownership had little to no impact on the diversity of program content. *See* 1975 Multiple Ownership Order, *supra* note 71, at 1073 ¶ 97 (noting that empirical studies of the impact of ownership on content were "inconclusive"); WALTER S. BAER, HENRY GELLER, JOSEPH A. GRUNDFEST & KAREN B. POSSNER, CONCENTRATION OF MASS MEDIA OWNERSHIP: ASSESSING THE STATE OF CURRENT KNOWLEDGE 121–40 (1974) (surveying the empirical literature and concluding that crossownership plays a "minor role, if any" in influencing media content); STANLEY M. BESEN & LELAND L. JOHNSON, REGULATION OF MEDIA OWNERSHIP BY THE FEDERAL COMMUNICATIONS COMMISSION: AN ASSESSMENT 28–31, 52, 57–59 (1984) (reviewing the empirical literature and concluding that crossownership has no clear impact on program diversity); Benjamin M. Compaine, *The Impact of Ownership on Content: Does It Matter?*, 13 CARDOZO ARTS & ENT. L.J. 755, 770 (1995) ("Multiple studies have concurred that

It is thus clear that the degree of horizontal integration permitted can have a fairly dramatic impact on the quantity, quality, and diversity of speech. Horizontal ownership restrictions represent a little-recognized, but important, form of architectural censorship.

D. RESTRICTIONS ON VERTICAL INTEGRATION

The FCC has long been concerned that vertical integration in the radio and television industry would harm competition.[136] The focus has been on whether vertical integration or vertical contractual agreements can allow a firm to use a dominant position in one market (called the primary market) to harm competition in another market (called the secondary market). These concerns animated the FCC's first major regulatory initiative, commonly known as the Chain Broadcasting Rules. The Rules were driven by the belief that the then-existing triopoly of radio networks was hindering the emergence of competition from new networks[137] and was inhibiting local control of the programming carried by any particular station.[138] As a result, the Chain Broadcasting Rules strictly limited radio networks' ability to own broadcast stations[139] and restricted the networks' ability to use affiliation agreements to limit the autonomy of local stations.[140] The Supreme Court sustained the Rules in the seminal decision on broadcast regulation, *NBC v. United States*.[141] The FCC subsequently extended the Chain Broadcasting Rules to television in 1946.[142] In time, the FCC would

---

programming differences related to group ownership are mixed and, even at that, are quite small." (footnote omitted)).

136. For a more detailed review of the history and theory of the FCC's regulation of vertical integration in the television industry, see Yoo, *supra* note 45, at 181–248. The primary focus of this discussion is downstream vertical integration by television and radio networks. It bears mentioning that at times the FCC has also regulated upstream vertical integration by networks into program supply. For critiques of the now-notorious and defunct "prime time access rule" ("PTAR") and the financial interest and syndication rules ("finsyn"), see THOMAS G. KRATTENMAKER & LUCAS A. POWE, JR., REGULATING BROADCAST PROGRAMMING 72–74, 99–100 (1994); Chen, *supra* note 99, at 1454–58.

137. *See* FED. COMMUNICATIONS COMM'N, *supra* note 13, at 51, 59, 66.

138. *Id.* at 64–65.

139. Specifically, the FCC prohibited networks from owning more than one station in any market and from owning any stations in markets in which competition was substantially restrained. *See id.* at 92, *repealed in part by* Review of the Commission's Regulations Governing Television Broadcasting, Report and Order, 10 F.C.C.R. 4538, 4540 ¶ 10 (1995) [hereinafter 1995 Chain Broadcasting Order]. This rule was overshadowed by the national television station ownership limits discussed below.

140. FED. COMMUNICATIONS COMM'N, *supra* note 13, at 51–66.

141. NBC v. United States, 319 U.S. 190 (1943).

142. Amendment to Part 3 of the Commission's Rules, 11 Fed. Reg. 33 (F.C.C. Jan. 1, 1946).

repeal them with respect to radio[143] and roll back some of the restrictions with respect to television as well.[144] Certain television-related provisions still remain in effect.[145]

Congress has also taken steps to limit vertical integration in the cable industry.[146] The "channel occupancy" provision authorized the FCC to limit the channel capacity that cable operators could devote to their vertically affiliated networks.[147] Congress also enacted a series of access requirements designed to protect against the dangers of vertical integration. For example, the leased access provision requires all cable systems with more than thirty-five channels to set aside part of their channel capacity for use by unaffiliated programmers.[148] The program access provisions prevent vertically integrated programmers from discriminating against unaffiliated operators[149] or from entering into exclusive dealing contracts.[150] Most importantly, Congress enacted the must-carry provisions, requiring cable operators to provide free carriage to all full-power television stations broadcasting in their service area.[151] Although enacted in part to preserve

---

143. Review of Commission Rules and Regulatory Policies Concerning Network Broadcasting by Standard (AM) and FM Broadcast Stations, Report, Statement of Policy, and Order, 63 F.C.C.2d 674 (1977).

144. *See* 1995 Chain Broadcasting Order, *supra* note 139; Review of Rules and Policies Concerning Network Broadcasting by Television Stations: Elimination or Modification of Section 73.658(c) of the Commission's Rules, Report and Order, 4 F.C.C.R. 2755 (1989).

145. *See* 47 C.F.R. § 73.658 (2004). A proposal to repeal these remaining restrictions has been pending without action since 1995. *See* Review of the Commission's Regulations Governing Programming Practices of Broadcast Television Networks and Affiliates, Notice of Proposed Rule Making, 10 F.C.C.R. 11,951 (1995).

146. Even before Congress acted, the FCC placed some limits on vertical integration in the cable industry when it promulgated regulations prohibiting national television networks from holding ownership stakes in cable operators. *See* Community Antenna Order, *supra* note 72, at 821 ¶ 15. The FCC relaxed this restriction in 1992. *See* Order to Eliminate the Prohibition on Common Ownership, *supra* note 77. Congress abolished it altogether in 1996. *See* Telecommunications Act of 1996, Pub. L. No. 104-104, § 202(f)(1), 106 Stat. 56, 111.

147. 47 U.S.C. § 533(f)(1)(B) (2000). The FCC set this limit at 40% of the operators' channel capacity. Implementation of Sections 11 and 13 of the Cable Television Consumer Protection and Competition Act of 1992, Horizontal and Vertical Ownership Limits, Second Report and Order, 8 F.C.C.R. 8565, 8592–96 ¶¶ 64–70 (1993). The channel occupancy limit applied only to the first seventy-five channels of any cable operator's capacity. Channel capacity in excess of seventy-five channels was not subject to the limit. *Id.* at 8601–02 ¶ 84. The D.C. Circuit overturned the 40% limit. *See* Time Warner Entm't Co. v. FCC, 240 F.3d 1126, 1137–39 (D.C. Cir. 2001).

148. The amount of channel capacity that must be set aside varies from 10% to 15%, depending on the size of the cable operator. 47 U.S.C. § 532(b)(1). Enactment of this statute overturned a previous Supreme Court decision holding that the FCC lacked the authority to mandate leased access. *See* FCC v. Midwest Video Corp., 440 U.S. 689 (1979).

149. 47 U.S.C. § 548(c)(2)(B).

150. *Id.* § 548 (c)(2).

151. *Id.* §§ 534, 535.

horizontal competition in local advertising markets,[152] must-carry was also intended to guard against vertical integration.[153]

In addition, the FCC has historically limited the number of television and radio stations that any one entity can own nationwide.[154] It justified these restrictions with the need to foster competition,[155] the need to promote a diversity of sources,[156] and the desire to encourage local initiative.[157] Congress eventually eliminated the national station ownership limits for radio and amended the national television station ownership limit to permit ownership of any number of television stations reaching less than 35% of the national audience.[158] Congress also passed legislation authorizing the FCC to establish a limit on the number of cable subscribers

---

152. *See Turner II*, 520 U.S. 180, 200–01 (1997).

153. *See id.* at 198–99 (justifying must-carry in part on testimony indicating that vertical integration gives "cable operators . . . an incentive to drop local broadcasters and to favor affiliated programmers"). *See also* Cable Television Protection and Competition Act of 1992, Pub. L. No. 102-385, § 2(a)(5), 106 Stat. 1460, 1460–61 (codified as amended at 47 U.S.C. §§ 534–535) (finding that vertical integration in the cable industry has given "cable operators . . . the incentive and ability to favor their affiliated programmers" and "could make it more difficult for noncable-affiliated programmers to secure carriage on cable systems"); S. REP. NO. 102-92, at 25 (1992), *reprinted in* 1992 U.S.C.C.A.N. 1133, 1158 (noting that "vertical integration gives cable operators the incentive and ability to favor their affiliated programming services" and might lead a cable operator to refuse to carry unaffiliated programmers).

154. The FCC initially set the national cap for television at three stations. *See* Rules and Regulations Governing Experimental Television Broadcast Stations, § 4.226, 6 Fed. Reg. 2283, 2284–85 (F.C.C. May 6, 1941). The national cap for radio was set at five stations. *See* Multiple Ownership, 9 Fed. Reg. 5442 (F.C.C. May 23, 1944). By 1954, a series of subsequent amendments eventually turned both the national radio and television station ownership limits into what became known as a "Rule of Seven." *See* 1953 Multiple Ownership Order, *supra* note 13, at 291 (limiting any one owner to five television stations and seven radio stations nationwide); Amendment of Multiple Ownership Rules, Report and Order, 43 F.C.C. 2797 (1954) (increasing the national limit for television from five to seven stations so long as two stations were UHF). The Rule of Seven was sustained against a judicial challenge by the Supreme Court. *See* United States v. Storer Broad. Co., 351 U.S. 192 (1956). The limit was later liberalized into a "Rule of Twelve." *See* Amendment of Section 73.3555 [formerly Sections 73.35, 73.240 & 73.636] of the Commission's Rules Relating to Multiple Ownership of AM, FM & Television Broadcast Stations, Report and Order, 100 F.C.C.2d 17 (1984) [hereinafter 1984 Multiple Ownership Order] (authorizing group ownership of up to twelve stations), *on reconsideration*, 100 F.C.C.2d 74 (1985) (adding the additional requirement that the twelve-station group reach no more than 25% of the national audience).

155. *See Storer Broad.*, 351 U.S. at 203 (concluding that the FCC's public interest mandate requires it to "assure fair opportunity for open competition in the use of broadcasting facilities"); 1984 Multiple Ownership Order, *supra* note 154, at 38–46 ¶¶ 64–86, 50–51 ¶¶ 97–99.

156. *See Storer Broad.*, 351 U.S. at 203; 6 FCC ANN. REP. 68 (1941); 1984 Multiple Ownership Order, *supra* note 154, at 24–38 ¶¶ 24–63.

157. *See* 6 FCC ANN. REP. 68 (1941).

158. Telecommunications Act of 1996, Pub. L. No. 104-104, § 202(c), 110 Stat. 56, 110 (codified as amended at 47 C.F.R. § 73.3555(b) (2004)).

that any one company can reach nationwide.[159] The FCC eventually set that limit at 30%.[160]

The first round of judicial challenges to these provisions proved unsuccessful.[161] More recent decisions have exhibited the courts' greater willingness to invalidate vertical ownership restrictions. In *Time Warner Entertainment Co. v. FCC*,[162] the D.C. Circuit invalidated the 30% cable subscriber limit set by the FCC based on a failure to implement the provision in the manner prescribed by Congress.[163] The court also struck down the FCC's channel occupancy limit on First Amendment grounds.[164] Furthermore, in *Fox Television Stations, Inc. v. FCC*,[165] the D.C. Circuit overturned the FCC's decision not to eliminate the national television station ownership cap during its first biennial review.[166] The court held that refusal to repeal the rule violated both the Administrative Procedure Act and the FCC's obligation under the Telecommunications Act of 1996 to "repeal or modify any regulation it determines to be no longer in the public

---

159. *See* 47 U.S.C. § 533(f)(1)(A) (2000).

160. Implementation of Sections 11 and 13 of the Cable Television Consumer Protection and Competition Act of 1992, Horizontal and Vertical Ownership Limits, Second Report and Order, 8 F.C.C.R. 8565, 8576–79 ¶¶ 24–29 (1993) (setting this limit at 30% of all nationwide subscribers). Cable systems were allowed to reach up to 35% of nationwide cable homes provided that such additional cable systems were minority-controlled. *Id.* at 8578–79 ¶ 28. After seeking additional comment, the FCC subsequently reaffirmed these limits. *See* Implementation of Section 11(c) of the Cable Television Consumer Protection and Competition Act of 1992; Horizontal Ownership Limits, Memorandum Opinion and Order on Reconsideration and Further Notice of Proposed Rulemaking, 13 F.C.C.R. 14,462, 14,467–83 ¶¶ 9–51 (1998); Implementation of Section 11(c) of the Cable Television Consumer Protection and Competition Act of 1992; Horizontal Ownership Limits, Third Report and Order, 14 F.C.C.R. 19,098, 19,113–27 ¶¶ 36–70 (1999).

161. *See Turner II*, 520 U.S. 180 (1997) (sustaining must-carry against a facial challenge); Time Warner Entm't Co. v. United States, 211 F.3d 1313 (D.C. Cir. 2000) (sustaining the subscriber limit and the channel occupancy provision against a facial challenge); Time Warner Entm't Co. v. FCC, 93 F.3d 957, 967–71, 977–79 (D.C. Cir. 1996) (sustaining the leased access and vertically integrated programmer provisions against a facial challenge). Interestingly, the district court did initially sustain a facial challenge to the subscriber limit provision, only to see its decision overturned on appeal. *See* Daniels Cablevision, Inc. v. United States, 835 F. Supp 1, 10 (D.D.C. 1993), *rev'd sub nom.* Time Warner Entm't Co. v. FCC, 211 F.3d at 1316–20.

162. Time Warner Entm't Co. v. FCC, 240 F.3d 1126 (D.C. Cir. 2001).

163. *Id.* at 1133–36.

164. *Id.* at 1137–39. Interestingly, the distinction seems to turn on the fact that *Sinclair Broadcasting Group* and *Fox Television Stations* involved broadcasting and thus were only held to rational basis scrutiny, whereas *Time Warner* involved regulation of the cable industry and thus was held to intermediate scrutiny. On the problematic nature of this distinction, see *infra* Part II.C.

165. Fox Television Stations, Inc. v. FCC, 280 F.3d 1027 (D.C. Cir.), *modified on reh'g*, 293 F.3d 537 (D.C. Cir. 2002).

166. *See* 1998 Biennial Regulatory Review—Review of the Commission's Broadcast Ownership Rules and Other Rules Adopted Pursuant to Section 202 of the Telecommunications Act of 1996, Biennial Review Report, 15 F.C.C.R. 11,058, 11,072–75 ¶¶ 25–30 (2000).

interest."[167] The FCC responded by revising the national television station ownership rule to permit companies to own any number of stations so long as the station group could reach no more than 45% of the nation's television households.[168] Again, the FCC analyzed the issues in terms of the policy goals of competition, diversity, and localism, placing primary reliance on localism considerations.[169] The ensuing controversy over the decision led Congress to enact legislation setting the national television station ownership cap at 39% and exempting the restriction from mandatory periodic review by the FCC.[170]

The national television station ownership and cable subscriber limits are often misconstrued as being horizontal in focus.[171] Properly evaluated, horizontal restrictions bar mergers among direct competitors who would otherwise be serving the same customers. In the case of U.S. media regulation, excess horizontal concentration is prevented by the rules prohibiting crossownership of media outlets in the same city described in the preceding subsection. The national television station ownership and cable subscriber limits are more properly regarded as prohibiting joint ownership of television stations or cable systems in different cities. Even though these jointly owned properties occupy the same product market, their geographic markets are distinct, and thus they do not compete with one another. In other words, allowing a television station operating in New York City to merge with one operating in Los Angeles does not involve a merger between direct competitors and does not have any impact on options available to any viewer.

Although group ownership of broadcast stations does not enhance horizontal market power with respect to viewers, it may enhance vertical market power by increasing the group's bargaining leverage with respect to networks and other program suppliers. As a result, the national television

---

167. *Fox Television Stations*, 280 F.3d at 1040–49.
168. 2003 Biennial Review Order, *supra* note 3, at 13,842–45 ¶¶ 578–84.
169. *Id.* at 13,818–42 ¶¶ 508–578.
170. Consolidated Appropriation Act of 2004, Pub. L. No. 108-199, § 629, 118 Stat. 3, 99–100. Setting the national television ownership cap at 39% had the practical advantage of making it unnecessary for Fox and Viacom to divest the television stations they had acquired in excess of the previous 35% cap pursuant to temporary waivers granted by the FCC. Making it possible for Fox and Viacom to retain these stations removed much of the political impetus for further liberalization of the national ownership cap.
171. *See* S. REP. NO. 102-92, at 32–33 (1991), *reprinted in* 1992 U.S.C.C.A.N. 1133, 1165–66 (referring to the growth of multiple system operators as "horizontal integration" and "horizontal concentration"); *Turner II*, 520 U.S. 180, 197 (1997) (referring to the growth of multiple system operators as "[h]orizontal concentration"); Time Warner Entm't Co. v. FCC, 240 F.3d 1126, 1128 (D.C. Cir. 2001) (referring to the subscriber limit provision as a "horizontal" restriction).

and radio ownership restrictions are more properly regarded as protecting against vertical market power rather than horizontal market power.[172]

Concerns about vertical integration are also evident in the furor that has surrounded many recent mega-mergers in the television industry, including Disney's acquisition of ABC, Viacom's merger with CBS, Time Warner's acquisition of Turner Broadcasting, and America Online's subsequent acquisition of Time Warner. Each merger was accompanied by a spate of commentary warning of dire consequences should the mergers be permitted.[173]

1. Structural Preconditions Implicit in Vertical Integration Theory

The nature of the economic threat posed by vertical integration has long been one of the most hotly contested issues in competition policy.[174] Although proponents of the leading schools of antitrust law and economics have often disagreed sharply over the extent to which vertical integration can harm competition, they do share common ground on some basic points.[175] Both sides in the debate agree that certain structural preconditions must be satisfied before vertical integration can pose a threat to competition. All of the vertical integration models explicitly or implicitly acknowledge that the primary market must be concentrated before vertical integration can harm competition. If this precondition is not met, the allegedly anticompetitive firm has no dominant position to use as leverage. Furthermore, the secondary market must be protected by barriers to entry if attempts to reduce competition in the secondary market are to have any hope of success. In addition, even if these structural preconditions are met, both approaches acknowledge the possibility that efficiencies may exist that nonetheless make vertical integration economically desirable.

In fact, these structural preconditions have become so much a part of the conventional wisdom that they are incorporated into guidelines

---

172. Yoo, *supra* note 45, at 219, 222.

173. *See, e.g.*, John H. Barton, *The International Video Industry: Principles for Vertical Agreements and Integration*, 22 CARDOZO ARTS & ENT. L.J. 67 (2004); Symposium, *Viacom-CBS Merger*, 52 FED. COMM. L.J. 499 (2000); Patrick M. Cox, Note, *What Goes Up Must Come Down: Grounding the Dizzying Height of Vertical Mergers in the Entertainment Industry*, 25 HOFSTRA L. REV. 261 (1996).

174. *See, e.g.*, Andy C.M. Chen & Keith N. Hylton, *Procompetitive Theories of Vertical Control*, 50 HASTINGS L.J. 573, 575 (1999) ("Few subjects in American antitrust law have undergone as many changes and generated as much debate among economists and lawyers as the regulation of vertical arrangements.").

175. The discussion that follows draws on the more complete analysis appearing in Yoo, *supra* note 45, at 187–205.

employed by the Justice Department and the Federal Trade Commission to evaluate the impact of vertical mergers on competition.[176] These guidelines explicitly acknowledge that vertical mergers are unlikely to harm competition unless the primary market is concentrated.[177] The measure of market concentration employed by the guidelines is the Hirschman-Herfindahl Index ("HHI"), which is calculated by squaring the market share of each competitor and then summing the resulting numbers. For example, a market of four firms with market shares of 30%, 30%, 20%, and 20% would have an HHI of 2600.[178] The result is a continuum that situates the concentration of a market on a scale from 0 (in the case of complete market deconcentration) to 10,000 (in the case of monopoly). When the post-merger HHI of the primary market is below 1800, vertical integration is considered unproblematic.[179] This standard is somewhat more lenient than that applied to horizontal mergers, which are more likely to create competitive problems.[180] The D.C. Circuit recognized the importance of these structural preconditions in striking down the FCC's attempt to implement the channel occupancy provision enacted by Congress.[181]

2. Applying the Structural Preconditions to the Television Industry

Determining whether a particular market is concentrated depends on proper market definition, which in turn requires the identification of the relevant product and geographic markets. The relevant market is best understood if the television industry is viewed as the multilevel chain of

---

176. U.S. DEP'T OF JUSTICE, NON-HORIZONTAL MERGER GUIDELINES §§ 4.212, 4.213, 4.221, 4.24 (promulgated in 1984 and reaffirmed in 1992 and 1997), *available at* http://www.usdoj.gov/atr/public/guidelines/2614.pdf.

177. *Id.* § 4.213.

178. $30^2 + 30^2 + 20^2 + 20^2 = 900 + 900 + 400 + 400 = 2600$.

179. U.S. DEP'T OF JUSTICE, *supra* note 176, § 4.213. *See also id.* § 4.131 (using the 1800 HHI threshold for determining when vertical integration can harm potential competition); *id.* § 4.221 (using the 1800 HHI threshold for determining when vertical integration can facilitate collusion).

180. *Id.* § 4.0. Unlike vertical mergers, which are thought to raise competitive problems only if the post-merger HHI exceeds 1800, horizontal mergers are open to challenge even when post-merger HHI is as low as 1000. Specifically, the Horizontal Merger Guidelines classify markets in which the post-merger HHI is between 1000 and 1800 as "moderately concentrated." U.S. DEP'T OF JUSTICE & FED. TRADE COMM'N, HORIZONTAL MERGER GUIDELINES § 1.51(b), *available at* http://www.usdoj.gov/atr/public/guidelines/horiz_book/ 15.html (Apr. 8, 1997). Horizontal mergers in markets that fall within this range "potentially raise significant competitive concerns" and may be subject to challenge if they increase HHI by more than 100 points. *Id.* The Horizontal Merger Guidelines treat markets in which the post-merger HHI exceeds 1800 as "highly concentrated." *Id.* § 1.51(c). In these markets, mergers that raise post-merger HHI by more than 50 points "potentially raise significant competitive concerns" and may be challenged. Mergers that raise post-merger HHI 100 points are "presumed . . . to create or enhance market power or facilitate its exercise" and are likely to be challenged. *Id.*

181. Time Warner Entm't Co. v. FCC, 240 F.3d 1126, 1138–39 (D.C. Cir. 2001).

distribution depicted in Figure 2.[182] The uppermost level is occupied by the networks and movie studios that create television programs. The intermediate level is occupied by local television stations and local cable operators, who acquire programming from program suppliers and deliver them locally. The bottommost level is occupied by end users, who obtain television service from local television stations and cable operators.

Many mistakenly assume that the relevant market is the one in which households obtain television programming from broadcast stations and cable operators (denoted in the figure by the letter B). B is a local market because, until recently, households could only obtain television from an outlet located within their local community. In addition, because the number of entities from which households could obtain television programming has historically been rather limited, if this were the relevant market, it would appear to be sufficiently concentrated to make vertical integration a real anticompetitive threat.[183]

---

182. This is a somewhat simplified version of the description of the industry advanced in Yoo, *supra* note 45, at 182–83, 220–21. The more complex analysis presented in that paper disaggregated the first stage depicted in Figure 2 into two different stages rather than lumping program producers and television networks into the same category. Because that distinction is not as central to the argument presented here, the basic framework can be simplified in this manner without any loss of analytic power.

183. It is unclear whether this is still true. The arrival of DBS as a significant multichannel video programming distributor ("MVPD") has made the market for local delivery of television signals much more competitive. The FCC's most recent data indicate that as of June 2004, DBS had captured over 25% of the MVPD market. *See* Annual Assessment of the Status of Competition in the Market for the Delivery of Video Programming, Eleventh Annual Report, FCC 05-13, slip op. at 38–39 ¶ 54, 115 tbl.B-1 (F.C.C. Feb. 4, 2005) [hereinafter Eleventh Annual Report on Video Competition], *available at* http://hraunfoss.fcc.gov/edocs_public/attachmatch/FCC-05-13A1.pdf. This exceeds the 15% threshold established by Congress for determining when a cable operator faces effective competition from other MVPDs. *See* 47 U.S.C. § 543(*l*)(1)(B)(ii) (2000).

FIGURE 2. Vertical chain of production in the television industry

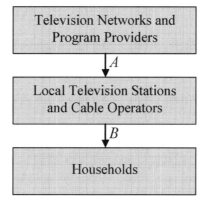

The problem with this analysis is that limits on the number of viewers that one station or cable operator group can reach nationwide have no impact whatsoever on the degree of market concentration in any local market. This fundamental insight can be seen most clearly by conducting the following thought experiment. Suppose the FCC banned vertical integration in the television industry altogether and required every television station owner and cable operator to divest any ownership interests in any network or program supplier. Would doing so decrease the ability of television stations and cable operators to exercise market power in market B? Clearly, the answer is no. Market power in B exists by virtue of the relatively small number of options any particular household has for obtaining television service. Preventing television stations and cable operators from holding ownership interests in networks would not increase or decrease the number of those options one iota. Forcing owners of television stations and cable operators to sell their proprietary interests in television programming would have no impact on market power in market B.

Vertical disintegration could potentially have an impact on market A, the upstream market in which local television stations and cable operators meet networks and program suppliers. The economics of producing television programming (particularly the fact that it requires the incurrence of substantial up-front costs) leaves program producers vulnerable to strategic behavior by local television stations and cable operators. Restrictions on the number of television stations and cable operators one entity can own nationwide has the inevitable effect of reducing program producers' ability to use vertical integration or vertical contractual arrangements to internalize these risks.[184] In addition, the national television station ownership and cable subscriber limits also affect the relative bargaining power of the players in market A by ensuring that the networks and other program suppliers negotiate with station and cable operator groups that represent smaller proportions of the national audience.

The proper focus, then, is on market A, in which television stations and cable operators bargain with networks and program suppliers. On reflection, it becomes clear that the geographic scope of this market is national, not local. Even in the extreme case, where the local cable operator possesses monopoly power over viewers in a particular city, that operator is unlikely to be able to exert any significant market power against a television network that can reach a sufficient number of other viewers located elsewhere in the nation. A program producer cares less about whether it is able to reach viewers in any particular city and more about how much of the national market it is able to access. In other words, it is the network's national reach, not its local reach, that matters. The network would, of course, prefer to reach all viewers nationwide. That it may be unable to reach certain customers is of no greater concern than it would be to manufacturers of particular brands of cars, shoes, or other conventional goods who are unable to gain access to the entire country. Their inability to reach customers in any particular geographic area does not threaten competition so long as they are able to obtain access to a sufficient number of customers located elsewhere. The proper question is not whether local television stations and cable operators wield market power in the local market for television viewers in any particular city, but rather whether groups of television stations and local cable systems possess sufficient market power to harm competition in the nationwide market for obtaining television content.

When viewed in this manner, it becomes relatively clear that the relevant primary market (as in the national market for household delivery of television programming) is unconcentrated. Consider, for example, the current national television station ownership rule, which prohibits television station groups that can reach more than 39% of the U.S. television audience.[185] It would be a mistake to assume that this limit would permit a television station group to control 39% of the market. This is because no broadcast network is able to capture more than 15% of the

---

184. *See* Yoo, *supra* note 45, at 192–200, 213–17, 232–37.
185. *See supra* note 170 and accompanying text.

potential audience that it reaches.[186] Thus, even if a group were able to reach 39% of the U.S. market, it would only be able to capture less than one sixth of those viewers. Setting the national audience cap at 39% effectively guarantees that no group of television stations will control more than 6% of the national audience. In that case, there are at least sixteen independent players bidding in the national market for television programming, more than enough to ensure that the market remains competitive.[187] Indeed, these numbers suggest that there would have been little danger setting the national audience cap at the 45% level that was overturned by Congress.

Similar reasoning applies to the national cable subscriber limits. As of June 2002, no multichannel video program distributor ("MVPD") controlled more than 15% of the national market, and the HHI of the total market was 884.[188] By June 2004, Comcast's acquisition of AT&T's cable properties caused HHI to rise to 1097.[189] Even this higher number falls well below the enforcement threshold under the vertical merger guidelines.[190] The level of concentration in the market for MVPDs is thus too diffuse to give any MVPD market power sufficient to give rise to anticompetitive concerns.[191]

---

186. For example, the highest ranked network during the November 2004 sweeps period (CBS), was only able to capture 12% of adult viewers. *See* Jim Finkle, *How Fall Played Out; CBS Triumphs, ABC Improves, NBC Falters*, BROADCASTING & CABLE, Dec. 6, 2004, at 14.

187. Even if the entire industry were composed of station groups of the largest size, the HHI would be less than 700, well below the levels thought to raise competitive concerns.

188. *See* Annual Assessment of the Status of Competition in the Market for the Delivery of Video Programming, Ninth Annual Report, 17 F.C.C.R. 26,901, 26,913 tbl.B-3 (2002).

189. *See* Eleventh Annual Report on Video Competition, *supra* note 183, at 77 ¶ 144, 118 tbl.B-3, 119 tbl.B-4

190. In fact, the level of concentration in the market for MVPDs approaches the level of nonenforcement under the more stringent guidelines governing horizontal mergers. Indeed, as the FCC has noted, economic theory and empirical studies suggest that a market need not have more than five participants of roughly equal size. *See* 2003 Biennial Review Order, *supra* note 3, at 13,731 ¶ 289 & n.609 (citing economic commentary). This suggests that HHIs as high as 2000 might well be unproblematic. That said, the Third Circuit rejected the FCC's finding that five equal-sized competitors would be sufficient to protect competition as arbitrary and capricious. *See* Prometheus Radio Project v. FCC, 373 F.3d 372, 432–34 (3d Cir. 2004).

191. Professor Baker invokes Edward Chamberlin's classic analysis of monopolistic competition as support for his belief that media entities typically earn high operating profits. *See* Baker, *supra* note 13, at 737 & n.15. *See also id.* at 750–751. I offer a more complete analysis of the implications of Chamberlinian monopolistic competition for media policy in Yoo, *supra* note 18, at 1602–28, 1633–36. For the time being, it is sufficient to point out that Chamberlin himself did not believe that firms engaged in monopolistic competition would earn sustainable economic profit. Although firms might earn some profit in the short-run, entry by other firms selling similar products would eventually dissipate any supercompetitive returns. *See* EDWARD HASTINGS CHAMBERLIN, THE THEORY OF MONOPOLISTIC COMPETITION: A RE-ORIENTATION OF THE THEORY OF VALUE 83–85 (8th ed. 1969).

Furthermore, an empirical analysis of the relevant secondary market—comprised of television networks and program providers—reveals that it is sufficiently unprotected by barriers to entry to obviate any anticompetitive concerns. The FCC reports that the total number of television networks has steadily increased, swelling from seventy networks in 1990[192] to a total of 388 networks in 2004, with another seventy-eight networks in the planning stages.[193] In addition, the percentage of vertically integrated networks has declined more or less steadily over the past decade.[194]

It is also likely that vertical integration in the radio and television industry will yield sufficient efficiencies to justify condoning it. The FCC has acknowledged that permitting broader network station ownership could yield substantial managerial, technical, and operational efficiencies.[195] Furthermore, because the creation of television programming typically requires the incurrence of substantial sunk costs, program producers are often vulnerable to hold-up, free riding, and other forms of strategic behavior.[196] The classic solution to such problems is through vertical integration or through some form of vertical contractual restraint.[197] Empirical studies confirm that, on balance, vertical integration in the cable industry tends to be welfare enhancing.[198]

Chamberlin's zero-profit result is in turn subject to several caveats. *See* Yoo, *supra* note 18, at 1607–09. Subsequent research has shown that the general validity of Chamberlin's zero-profit result depends on the magnitude of the fixed costs relative to the overall market. *See* Yoo, *supra* note 118, at 240. Baker's observation would have had greater applicability during earlier eras, when television and radio markets were protected by entry barriers. The emergence of alternative transmission technologies and new networks has largely dissipated the danger of supercompetitive returns. *See* Yoo, *supra* note 18, at 1633–36. This suggests that media policy would be better served by focusing on lowering barriers to entry, which in turn would require an abandonment of the commitment to fostering free radio and television as well as lowering the barriers to vertical integration.

192. Annual Assessment of the Status of Competition in the Market for the Delivery of Video Programming, Tenth Annual Report, 19 F.C.C.R. 1606, 1691 tbl.8 (2004) [hereinafter Tenth Annual Report on Video Competition].

193. Eleventh Annual Report on Video Competition, *supra* note 183, at 78 ¶ 145, 81 ¶ 152.

194. Tenth Annual Report on Video Competition, *supra* note 192, at 1690–91 ¶ 142 & tbl.8 (noting that the percentage of vertically integrated networks declined steadily from 50% in 1994 to 30% in 2002 before rising slightly to 33% in 2003); Eleventh Annual Report on Video Competition, *supra* note 183, at 78 ¶ 145 (noting that in 2004, the percentage of vertically integrated networks once again declined to 23%).

195. *See* 1995 Chain Broadcasting Order, *supra* note 139, at 4540 ¶ 11.

196. *See* Yoo, *supra* note 45, at 213–17, 232–37.

197. *See, e.g.*, Benjamin Klein, Robert G. Crawford & Armen A. Alchian, *Vertical Integration, Appropriable Rents, and the Competitive Contracting Process*, 21 J.L. & ECON. 297 (1978); Lester G. Telser, *Why Should Manufacturers Want Fair Trade?*, 3 J.L. & ECON. 86 (1960).

198. *See* Tasneem Chipty, *Vertical Integration, Market Foreclosure, and Consumer Welfare in the Cable Television Industry*, 91 AM. ECON. REV. 428, 430, 448–50 (2001).

It thus appears that the structure of the television industry makes it unlikely that vertical integration will harm competition, as demonstrated eloquently by the failure of the Disney-ABC, Viacom-CBS, Time Warner-Turner Broadcasting, and AOL-Time Warner mergers to generate significant anticompetitive harms. Instead, the existing regulations limiting vertical integration only serve to prevent industry participants from realizing the available efficiencies, which in turn reduces total quantity, quality, and diversity of speech. As a result, the regulatory restraints on vertical integration appear to represent still another form of architectural censorship.

II. ARCHITECTURAL CENSORSHIP'S IMMUNITY FROM MEANINGFUL FIRST AMENDMENT SCRUTINY

Many of the extant structural regulations thus constitute forms of architectural censorship that can have a dramatic impact on the quantity, quality, and diversity of radio and television programming. As a result, one would expect that the incidental impact structural regulation can have on speech would be subject to scrutiny under the First Amendment. This Part analyzes the level of scrutiny to which structural regulation should be subject under current First Amendment doctrine. Unfortunately, my analysis suggests that the identified types of architectural censorship will be effectively insulated from meaningful judicial review.

A. THE NATURE OF ARCHITECTURAL CENSORSHIP

The impact that structural regulation can have on media content should raise First Amendment concerns. Consider first the reductions in the total quantity of television and radio programming. Regulations that impede all forms of speech without regard to content still impair the free flow of expression. That a regulation may have affected all viewpoints equally does not change the fact that the reduction in opportunities for expression effects a First Amendment harm, whether viewed from the perspective of individual liberty or the proper functioning of the democratic process.[199] Scholars have also cautioned that media-specific regulations allow special interest groups to redirect the regulatory process toward rent

---

[199]. *See* Benjamin, *supra* note 11, at 32–35; Martin H. Redish, *The Content Distinction in First Amendment Analysis*, 34 STAN. L. REV. 113, 128–31 (1981); Frederick Schauer, *Cuban Cigars, Cuban Books, and the Problem of Incidental Restrictions on Communications*, 26 WM. & MARY L. REV. 779, 782–83 (1985).

seeking at the expense of the general public.[200] Moreover, the government may not merely be an innocent bystander in the process of rent seeking by politically powerful groups; it may actually be following a policy of "rent extraction," in which it deliberately restricts or threatens to restrict speech to create a pool of rents that can then be redistributed through the regulatory process.[201]

In addition, liberty-oriented theorists would find interference with individual speakers' editorial discretion to be a First Amendment harm, even in the absence of evidence that particular content was favored or disfavored. Access requirements are particularly problematic in this regard.[202] Tellingly, the Supreme Court has found preserving editorial discretion to be an important First Amendment value even with respect to broadcasting, the medium of communications that receives the lowest level of constitutional protection.[203] The Court has also repeatedly recognized that cable operators' selection of the content they transmit represents an exercise of their free speech rights.[204] Acknowledging that the interest in editorial discretion may be offset by other considerations[205] does not change the fact that interference with a speaker's liberty interest implicates important First Amendment values.

---

200. *See* Neil Weinstock Netanel, *Locating Copyright within the First Amendment Skein*, 54 STAN. L. REV. 1, 61–67 (2001).

201. *See* Benjamin, *supra* note 11, at 35–36 (suggesting that regulators reduce the total amount of spectrum-based speech in order to generate monopoly rents). For a general discussion on the process of rent extraction, see FRED S. MCCHESNEY, MONEY FOR NOTHING: POLITICIANS, RENT EXTRACTION, AND POLITICAL EXTORTION (1997).

202. *See* Pac. Gas & Elec. Co. v. Pub. Util. Comm'n, 475 U.S. 1, 11 (1986) (plurality opinion); Miami Herald Publ'g Co. v. Tornillo, 418 U.S. 241, 258 (1974). *Cf.* Martin H. Redish & Kirk J. Kaludis, *The Right of Expressive Access in First Amendment Theory: Redistributive Values and the Democratic Dilemma*, 93 NW. U. L. REV. 1083, 1114–17 (1999) (describing the cognitive and dignitary harms associated with imposing affirmative content obligations on the media).

203. *See* Ark. Educ. Television Comm'n v. Forbes, 523 U.S. 666, 673–75 (1998); FCC v. League of Women Voters, 468 U.S. 364, 379–80 (1984); CBS, Inc. v. Democratic Nat'l Comm., 412 U.S. 94, 105–11, 118–21, 124–25 (1973).

204. As the Court noted in the *Turner I* decision, "[a]t the heart of the First Amendment lies the principle that each person should decide for himself or herself the ideas and beliefs deserving of expression, consideration, and adherence. Our political system and cultural life rest upon this ideal." *Turner I*, 512 U.S. 622, 641 (1994). *See also* Leathers v. Medlock, 499 U.S. 439, 444 (1991); City of Los Angeles v. Preferred Communications, Inc., 476 U.S. 488, 494 (1986).

205. The Court has acknowledged that the interest in preserving broadcasters' editorial discretion must be balanced against the benefits to the public of being exposed to views that would otherwise be barred from the airwaves. *See Ark. Educ. Television*, 523 U.S. at 673–74; *League of Women Voters*, 468 U.S. at 377–78; *CBS*, 412 U.S. at 101–02. With respect to cable, the Court has held that bottleneck control of cable operators justifies permitting some restriction of their editorial discretion. *Turner I*, 512 U.S. at 656–57.

Lastly, as I have detailed above, many of the FCC's structural restrictions have the unintended consequence of skewing media content toward certain demographic groups and stifling the emergence of more diverse programming. There can be little doubt that such content-specific effects raise serious constitutional concerns.[206]

B. *MINNEAPOLIS STAR* AND THE SHORT-LIVED PROSPECT OF STRICT SCRUTINY

Even though the structural regulations described above affect the quantity and mix of media content in ways that implicate the First Amendment, it is not completely clear what standard the courts would apply when evaluating the constitutionality of these regulations. It is now well established that regulations restricting speech on the basis of its content are subject to strict scrutiny.[207] At the same time, the Supreme Court has squarely established that it does not regard the rationales that underlie structural regulation as content-based.[208]

---

206. Professor Baker suggests that because the market can have as much of an adverse impact on media content as structural regulations, market distortions should raise similar concerns under my approach to the First Amendment. *See* Baker, *supra* note 13, at 755–759. This ignores the state action doctrine, which represents one of the central underpinnings of classic liberal theory. *See* Yoo, *supra* note 11, at 331–34 (describing the difficulties in reconciling democratic theories of media policy with the state action doctrine). The role that state action plays in defining the relationship between the individual and the state explains why adverse speech effects resulting from governmental actions might be problematic, whereas similar effects resulting from private ordering would not.

207. *See* City of Los Angeles v. Alameda Books, Inc., 535 U.S. 425, 434 (2002); United States v. Playboy Entm't Group, Inc., 529 U.S. 803, 813 (2000); *Turner I*, 512 U.S. at 641–43; Simon & Schuster, Inc. v. Members of N.Y. State Crime Victims Bd., 502 U.S. 105, 115, 118 (1991); Ark. Writers' Project, Inc. v. Ragland, 481 U.S. 221, 230–31 (1987).

208. *Turner I* squarely concluded that each of the three policy goals underlying structural regulation—(1) the preservation of free, local television; (2) the promotion of a diversity of information sources, and (3) the promotion of competition—were unrelated to the content of message conveyed. *Turner I*, 512 U.S. at 662. *See also* FCC v. Nat'l Citizens Comm. for Broad., 436 U.S. 775, 798–801 (1978) (holding that the promotion of diverse views is content-neutral).

This conclusion is far from unassailable. As noted earlier, the goal of promoting diversity is intimately intertwined with who has the power to select, edit, and present speech. *See supra* note 76 and accompanying text. Similarly, the preference for localism clearly signifies the government's conclusion that a particular type of speech is especially valuable. Indeed, Justice O'Connor's dissent in *Turner I* vigorously disputed the conclusion that promoting diversity and localism was content-neutral. *See Turner I*, 512 U.S. at 677–78 (O'Connor, J., dissenting). Courts that have recognized the problematic nature of the conclusion that regulations designed to promote viewpoint diversity and localism are content-neutral have felt constrained to follow *Turner I*'s resolution of the issue. *See* Horton v. City of Houston, 179 F.3d 188, 192–94 (5th Cir. 1999). *See also* Am. Family Ass'n, Inc. v. FCC, 365 F.3d 1156, 1169–70 (D.C. Cir.) (holding that promotion of a diversity of views and localism to be content-neutral), *cert. denied*, 125 S. Ct. 634 (2004); Fox Television Stations, Inc. v. FCC, 280 F.3d 1027, 1041–42, 1046 (D.C. Cir.) (identifying the promotion of competition, diversity, and localism as the interests underlying the national television station ownership limits and holding them to be content-

One line of decisions, associated with *Minneapolis Star & Tribune Co. v. Minnesota Commissioner of Revenue*,[209] appeared to entertain the possibility of subjecting structural restrictions to strict scrutiny even in the absence of facial content discrimination or content-based motive. In *Minneapolis Star*, the Court expanded on a precedent invalidating a state tax that applied only to newspapers[210] and applied strict scrutiny to strike down a generally applicable tax whose burden fell disproportionately on a small group of newspapers. In so doing, the Court framed the issues in a manner almost ideally suited for redressing the problems of architectural censorship. Strict scrutiny was not limited to instances in which the government acted out of an illicit motive.[211] Instead, the Court recognized that "even regulations aimed at proper governmental concerns can restrict unduly the exercise of rights protected by the First Amendment."[212] As a result, any restriction "that singles out the press, or that targets individual publications within the press, places a heavy burden on the State to justify its action."[213] This language suggests that the doctrine is not designed solely to ferret out regulations that are mere facades for suppressing speech of a particular content or by particular speakers. Rather, *Minneapolis Star* could arguably be construed as applying to economic regulation that, though innocently enacted, has the unintended byproduct of adversely affecting the content of speech.

The Supreme Court reinforced this line of jurisprudence in *Arkansas Writers' Project, Inc. v. Ragland*,[214] in which it struck down a sales tax that exempted newspapers and religious, professional, trade, and sports journals. The Court held that the reasoning of *Minneapolis Star* applied a fortiori to a tax that differentiated on its face among different types of magazines on the basis of their content.[215] Because the differential taxation of magazines represented sufficient grounds for striking down the sales tax, the Court declined to address whether the distinction drawn between newspapers and magazines also violated the First Amendment.[216]

---

neutral), *modified on reh'g*, 293 F.3d 537 (D.C. Cir. 2002); CBS Broad., Inc. v. Echostar Communications Corp., 265 F.3d 1193, 1210 (11th Cir. 2001) (holding the promotion of localism to be content-neutral).

209. Minneapolis Star & Tribune Co. v. Minn. Comm'r of Revenue, 460 U.S. 575 (1983).
210. *See* Grosjean v. Am. Press Co., 297 U.S. 233 (1936).
211. *Minneapolis Star*, 460 U.S. at 592 ("Illicit legislative intent is not the *sine qua non* of a violation of the First Amendment.").
212. *Id.*
213. *Id.* at 592–93.
214. Ark. Writers' Project, Inc. v. Ragland, 481 U.S. 221 (1987).
215. *Id.* at 229–30.
216. *Id.* at 232–33.

The Court would soon foreclose any prospect that *Minneapolis Star* and its progeny would serve as a check on architectural censorship. In *Leathers v. Medlock*,[217] the Court upheld a sales tax that applied to cable television but exempted satellite television providers as well as certain newspapers and magazines. The Court regarded the tax as a law of general applicability that did not single out the press for differential treatment.[218] The tax was not structured in a way that raised suspicions that it was intended to fall solely on a small group of media speakers.[219] Even though the exemption for satellite television providers effectively created differential treatment for media that were functionally similar, the fact that the tax affected approximately one hundred cable suppliers obviated any suggestion that it penalized any particular speaker or the expression of any particular idea.[220]

The Court reaffirmed the idea that the *Minneapolis Star* line of precedents only applies when a statute of general application affects a small number of speakers in its first *Turner Broadcasting* decision (*"Turner I"*).[221] Rejecting the argument that must-carry should be subject to strict scrutiny, the Court distinguished the *Minneapolis Star* line of cases by pointing out that the restriction in question applied to large numbers of cable systems. As a result, it "d[id] not pose the same dangers of suppression and manipulation that were posed by the more narrowly targeted regulations in *Minneapolis Star*" and its progeny.[222]

The limitations imposed by *Leathers* and *Turner I* drastically limit the *Minneapolis Star* line of cases' potential for redressing the problem of architectural censorship.[223] So long as the restriction in question applies to a sufficiently large number of entities, it does not matter that it favors one form of communication over another. The type of structural regulations that represent the focus of this Article will almost invariably apply to a sufficiently large number of entities to take them outside of this scope.

---

217. Leathers v. Medlock, 499 U.S. 439 (1991).
218. *Id.* at 447.
219. *Id.* at 448.
220. *Id.* at 449.
221. *Turner I*, 512 U.S. 622 (1994).
222. *Id.* at 661. The Court alternatively noted that differential treatment may also be "'justified by some special characteristic of' the particular medium being regulated." *Id.* at 660–61 (quoting Minneapolis Star & Tribune Co. v. Minn. Comm'r of Revenue, 460 U.S. 575, 585 (1983)). The Court concluded that the bottleneck monopoly power exercised by cable operators represented just such a special characteristic. *Id.* at 661.
223. Indeed, the D.C. Circuit has suggested that *Minneapolis Star* and *Arkansas Writers' Project* only apply to tax cases. *See* BellSouth Corp. v. FCC, 144 F.3d 58, 68 n.11 (D.C. Cir. 1998); Walsh v. Brady, 927 F.2d 1229, 1236 (D.C. Cir. 1991). *See also* Benjamin, *supra* note 11, at 29–30.

Indeed, *Leathers* and *Turner I* fundamentally altered the spirit of the *Minneapolis Star* line of cases, in effect suggesting that differential impacts caused by laws of general applicability only raise constitutional concerns when they betray some indicia of a clandestine desire to suppress expression. As such, *Minneapolis Star* no longer offers much promise of addressing architectural censorship that arises from the unintended consequences of economically motivated regulation.

C. RATIONAL BASIS VS. INTERMEDIATE SCRUTINY

Since the Supreme Court's foreclosure of any real possibility of subjecting structural regulation to strict scrutiny, courts have struggled to determine whether the proper standard should be one of rational basis or intermediate scrutiny. The problem was presented quite nicely by the D.C. Circuit in *News America Publishing, Inc. v. FCC*.[224] *News America* is the result of a rider buried in a massive, 471-page continuing resolution appropriating funds for the entire federal government for fiscal year 1988.[225] The rider forbade the FCC from using any funds to extend any temporary waivers to the current newspaper/television crossownership rule. As the court noted, the statute was "general in form but not in reality."[226] At the time, only one such temporary waiver had been issued: the one held by Rupert Murdoch that allowed him to own both WXNE-TV and the *Boston Herald*.

Because this generally applicable statute had the effect of burdening a single speaker, it appeared to represent precisely the type of provision that would be subject to strict scrutiny under *Minneapolis Star*. The court instead evaluated the constitutionality of the rider under the lower level of First Amendment scrutiny applied by the Supreme Court to the newspaper/broadcast crossownership rule in *FCC v. National Citizens Committee for Broadcasting ("NCCB")*.[227] There, the Court upheld the newspaper/broadcast crossownership rule as a "reasonable means of promoting the public interest in diversified mass communications."[228]

---

224. News Am. Publ'g, Inc. v. FCC, 844 F.2d 800 (D.C. Cir. 1988).
225. This is in contrast to the usual practice, in which the federal budget is enacted through a series of thirteen appropriations acts. The continuing resolution was also unusual in that the text of the legislation was printed only in a 1194-page conference report. *See id.* at 801–02.
226. *Id.* at 802.
227. *Id.* at 810–11 (citing FCC v. Nat'l Citizens Comm. for Broad., 436 U.S. 775 (1978)). *See also* NBC v. United States, 319 U.S. 190 (1943) (applying a lower level of First Amendment scrutiny to broadcasting to sustain the Chain Broadcasting Rules).
228. *Nat'l Citizens Comm. for Broad.*, 436 U.S. at 802.

Other courts addressing constitutional challenges to structural regulation of the broadcast industry have felt obligated to follow *NCCB*.[229]

Several aspects of this decision are quite problematic. *NCCB* was based on the longstanding rationale that the physical scarcity of the electromagnetic spectrum justifies conferring a lesser degree of First Amendment protection on broadcasting than on other media.[230] Over the years, however, a stream of commentary has undermined the vitality of the scarcity doctrine by demonstrating its analytical incoherence.[231] In addition, technological developments allowing for more intensive use of the spectrum and the advent of cable television have lessened the extent to which the spectrum serves as a bottleneck for transmitting media speech. The Supreme Court seems to have backed away from the doctrine as well. Not only has the Court declined invitations to extend it to other forms of communication,[232] its recent decisions raise serious questions as to its continuing vitality even with respect to broadcasting.[233]

---

229. *See* Prometheus Radio Project v. FCC, 373 F.3d 372, 401–02 (3d Cir. 2004); Sinclair Broad. Group, Inc. v. FCC, 284 F.3d 148, 167–68 (D.C. Cir. 2002); Fox Television Stations, Inc. v. FCC, 280 F.3d 1027, 1045–46 (D.C. Cir.), *modified on reh'g*, 293 F.3d 537 (D.C. Cir. 2002). *See also Sinclair Broad. Group*, 284 F.3d at 172 (Sentelle, J., dissenting); 2003 Biennial Review Order, *supra* note 3, at 13,625–27 ¶¶ 13–16. As the D.C. Circuit noted in another case involving the newspaper/broadcast crossownership rule, "We are stuck with the scarcity doctrine until the day that the Supreme Court tells us that the *Red Lion* no longer rules the broadcast jungle." Tribune Co. v. FCC, 133 F.3d 61, 69 (D.C. Cir. 1998).

230. *Nat'l Citizens Comm. for Broad.*, 436 U.S. at 799. The scarcity doctrine has its roots in the seminal decision on broadcast regulation, *NBC v. United States*, 319 U.S. 190, 226–27 (1943), and has been reaffirmed many times since then. *See* Metro Broad., Inc. v. FCC, 497 U.S. 547, 566–67 (1990), *overruled on other grounds by* Adarand Constructors, Inc. v. Peña, 515 U.S. 200 (1995); FCC v. League of Women Voters, 468 U.S. 364, 374–77 (1984); CBS, Inc. v. FCC, 453 U.S. 367, 394–96 (1981); CBS, Inc. v. Democratic Nat'l Comm., 412 U.S. 94, 101–02 (1973); Red Lion Broad. Co. v. FCC, 395 U.S. 367, 388–89 (1969).

231. The academic criticism of the constitutionality of the scarcity doctrine is voluminous. *See generally* Yoo, *supra* note 11, at 266–92 (reviewing and extending the leading critiques of scarcity). Tellingly, even proponents of broadcast regulation no longer attempt to defend the scarcity doctrine. *See, e.g.*, LEE C. BOLLINGER, IMAGES OF A FREE PRESS 87–90 (1991); SUNSTEIN, *supra* note 47, at 110; Ronald J. Krotoszynski, Jr., *Into the Woods: Broadcasters, Bureaucrats, and Children's Television Programming*, 45 DUKE L.J. 1193, 1247 (1996); Charles W. Logan, Jr., *Getting Beyond Scarcity: A New Paradigm for Assessing the Constitutionality of Broadcast Regulation*, 85 CAL. L. REV. 1687, 1701–05 (1997); Jonathan Weinberg, *Broadcasting and Speech*, 81 CAL. L. REV. 1101, 1106 (1993).

232. The Supreme Court has rejected attempts to extend the broadcast regime to the mail, telephony, and the Internet. *See* Reno v. ACLU, 521 U.S. 844, 868 (1997) (Internet); Sable Communications of Cal., Inc. v. FCC, 492 U.S. 115, 124 (1989) (telephony); Pac. Gas & Elec. Co. v. Pub. Utils. Comm'n, 475 U.S. 1, 10 n.6 (1986) (plurality opinion) (mail); Bolger v. Youngs Drug Prods. Corp., 463 U.S. 60, 74 (1983) (mail); Consol. Edison Co. v. Pub. Service Comm'n, 447 U.S. 530, 542–43 (1980) (mail). *But see* Time Warner Entm't Co. v. FCC, 93 F.3d 957, 974–77 (D.C. Cir. 1996) (extending the broadcast rationale to DBS). For a time, the Court appeared to entertain the possibility of extending the broadcast justification to cable television. *Compare Turner I*, 512 U.S. 622, 637–39

In addition, this broad reading of *NCCB* is hard to square with *Turner I*, which rejected extending the Court's broadcast precedents to cable television. Notwithstanding the Court's acknowledgement of other courts' and commentators' criticisms of the scarcity doctrine's analytical coherence,[234] the Court declined to revisit the applicability of the scarcity doctrine to broadcasting in a case that did not properly present the issue.[235] Because cable does not depend on the broadcast spectrum, it does not suffer from the "inherent limitations" and "danger of physical interference" that supposedly confront broadcasting.[236] At the same time, the Court held that "laws that single out the press, or certain elements thereof, for special treatment" must be subject to some measure of heightened scrutiny.[237] As a result, the Court followed a line of D.C. Circuit cases[238] and concluded that the proper standard was the level of scrutiny applicable to content-neutral restrictions that impose an incidental burden on speech as announced in *United States v. O'Brien*.[239] Unlike the *NCCB* standard, which is stated in

---

(1994) (rejecting the application of the broadcast regime to uphold must-carry), *with* Denver Area Educ. Telecomms. Consortium, Inc. v. FCC, 518 U.S. 727, 737–48, 755 (1996) (plurality opinion) (suggesting that *Turner I* did not foreclose applying the broadcast regime to uphold behavioral regulation of cable television). This possibility was subsequently foreclosed by the Court's decision in *United States v. Playboy Entertainment Group, Inc.*, 529 U.S. 803, 811–14 (2000) (5-4 decision).

    233.   The Supreme Court's recent decisions have avoided reliance on the traditional justifications and have instead turned to other doctrines to justify holding the regulation under review to a lower level of First Amendment scrutiny. *See* Greater New Orleans Broad. Ass'n v. United States, 527 U.S. 173 (1999) (commercial speech); Ark. Educ. Television Comm'n v. Forbes, 523 U.S. 666 (1998) (public forum doctrine); United States v. Edge Broad. Co., 509 U.S. 418 (1993) (commercial speech).

    234.   *Turner I*, 512 U.S. at 638 & n.5.

    235.   *Id.* at 638 (citing FCC v. League of Women Voters, 468 U.S. 364, 376, n.11 (1984)).

    236.   *Id.* at 638–39.

    237.   *Id.* at 640–41.

    238.   *See* Century Communications Corp. v. FCC, 835 F.2d 292, 298–304 (D.C. Cir. 1987) (invalidating revised must-carry regulations), *clarified by* 837 F.2d 517 (D.C. Cir. 1988); Quincy Cable TV, Inc. v. FCC, 768 F.2d 1434, 1454–62 (D.C. Cir. 1985) (invalidating initial must-carry regulations); Home Box Office, Inc. v. FCC, 567 F.2d 9, 14, 48–50 (D.C. Cir. 1977) (invalidating regulations restricting pay television). The must-carry decisions did not formally decide that *O'Brien* provided the appropriate basis for evaluating the constitutionality of must-carry. Because the courts concluded that the restrictions under review failed the more lenient level of scrutiny announced in *O'Brien*, they found it unnecessary to resolve whether the regulations should be subjected to a more stringent standard of review, such as strict scrutiny. *See Century Communications*, 835 F.2d at 298; *Quincy Cable*, 768 F.2d at 1448, 1450–54.

    239.   *Turner I*, 512 U.S. at 661–62 (citing United States v. O'Brien, 391 U.S. 367 (1968)). *Turner I* thus represented the culmination of a fairly remarkable transformation of *O'Brien* doctrine. Originally applicable only to general regulations that had a tangential impact on speech, following *Turner I*, *O'Brien* doctrine is now applicable to direct regulations of speech so long as they are content-neutral. For an insightful discussion of pre-*Turner I* cases applying *O'Brien* to direct restrictions of speech, see Keith Werhan, *The O'Brien*ing *of Free Speech Methodology*, 19 ARIZ. ST. L.J. 635, 649–58 (1988).

terms reminiscent of rational basis,[240] the *O'Brien* standard employs language that suggests an intermediate level of scrutiny.[241]

The language in *Turner I* holding that all regulations targeting a certain element of the press were necessarily subject to some form of heightened scrutiny seems to apply with equal force to structural regulation imposed on broadcasting. Courts have struggled to reconcile these two precedents. Some courts have attempted to rely on a technology-based distinction, applying the *NCCB* standard to the structural regulation of broadcasting,[242] while applying the *Turner I* standard to structural regulation of the cable industry.[243] They point out that, although *Turner I* acknowledged the analytical deficiencies with the scarcity doctrine, it explicitly declined to question its continuing validity with respect to broadcasting.[244]

Attempts to draw technology-based distinctions suffer from severe analytical problems. Most obviously, they do not provide a basis for determining the appropriate standard of review to be applied to crossownership of broadcast and nonbroadcast media. Indeed, the FCC's 2003 Biennial Review Order recognizes the conundrum posed by crossownership restrictions. Although the FCC maintains that

---

240. *See supra* note 228 and accompanying text.

241. Specifically, *O'Brien* requires that the restriction in question "further[] an important or substantial governmental interest; if the governmental interest is unrelated to the suppression of free expression; and if the incidental restriction on alleged First Amendment freedoms is no greater than is essential to the furtherance of that interest." *O'Brien*, 391 U.S. at 377. The first prong focuses on the constitutional authority to impose the regulation rather than the First Amendment. The third prong is the equivalent of a threshold inquiry into whether the restriction is content-based. The remaining two prongs, which require a "substantial government interest" that is "no greater than is essential to the furtherance of that interest," are analogous to classic intermediate scrutiny. *See* Michael C. Dorf, *Incidental Burdens on Fundamental Rights*, 109 HARV. L. REV. 1175, 1202 (1996); Srikanth Srinivasan, *Incidental Restrictions of Speech and the First Amendment: A Motive-Based Rationalization of the Supreme Court's Jurisprudence*, 12 CONST. COMMENT. 401, 404 (1995).

242. *See* Sinclair Broad. Group, Inc. v. FCC, 284 F.3d 148, 167–69 (D.C. Cir. 2002); Fox Television Stations, Inc. v. FCC, 280 F.3d 1027, 1046–47 (D.C. Cir.), *modified on reh'g*, 293 F.3d 537 (D.C. Cir. 2002).

243. *See, e.g.*, Time Warner Entm't Co. v. FCC, 240 F.3d 1126, 1129–30 (D.C. Cir. 2001); Time Warner Entm't Co. v. United States, 211 F.3d 1313, 1316–19 (D.C. Cir. 2000); Time Warner Entm't Co. v. FCC, 93 F.3d 957, 966–67, 967–73, 978–79 (D.C. Cir. 1996); Time Warner Entm't Co. v. FCC, 56 F.3d 151, 181–86 (D.C. Cir. 1995); US West, Inc. v. United States, 48 F.3d 1092, 1100–06 (9th Cir. 1995) (striking down cable/telephone company crossownership), *vacated and remanded*, 516 U.S. 1155 (1996).

244. *See Sinclair Broad. Group*, 284 F.3d at 161–62; *Fox Television Stations*, 280 F.3d at 1046. *Cf.* Prometheus Radio Project v. FCC, 373 F.3d 372, 402 (3d Cir. 2004).

crossownership restrictions will be subject only to rational basis scrutiny, it acknowledges that because they

> [w]ill limit the speech opportunities not only for broadcasters, but also for other entities that may seek to own and operate broadcast outlets (including those with the fullest First Amendment protection—newspapers), we should draw the rule as narrowly as possible in order to serve our public interest goals while imposing the least possible burden on the freedom of expression.[245]

At the same time, the FCC acknowledged the possible relevance of the cable precedents by ensuring that the crossownership rules are "narrowly tailored."[246]

The distinction between broadcast and nonbroadcast media is likely to be clouded still further by the growing functional similarity between different television technologies.[247] For example, television broadcasters are now in a position to use the enhanced efficiency made possible by digital transmission to begin to provide multichannel service.[248] In addition, the emergence of DBS systems, such as DirecTV and the Dish Network, has rendered spectrum-based and wireline television technologies largely interchangeable. As a result, it would seem quite strange to subject functionally identical technologies to drastically different First Amendment standards. Indeed, courts have reacted with some confusion as to the proper standard of review to be applied to DBS regulations. While some courts have applied the more lenient broadcast standard to DBS, other courts have

---

245. 2003 Biennial Review Order, *supra* note 3, at 13,793 ¶ 441.

246. *Id.* at 13,798 ¶ 455 & n.988 (citing *Time Warner*, 240 F.3d at 1135). Courts have largely been able to avoid addressing the merits of this issue. On a few occasions, it arose in the context of the newspaper/broadcast crossownership restrictions and thus was squarely controlled by *NCCB. See* Tribune Co. v. FCC, 133 F.3d 61, 69 (D.C. Cir. 1998). When the issue arose in the context of the cable/broadcast crossownership rule, the D.C. Circuit was able to avoid the issue by disposing of it on statutory grounds. *See Fox Television Stations*, 280 F.3d at 1049. The only court that attempted to reconcile the ambiguity created by these competing standards of review held that the heightened scrutiny mandated by *Turner I* applied only when a regulation singles out a subclass of broadcasters and did not apply to regulations imposing obligations on broadcasters as a whole. *See Sinclair Broad. Group*, 284 F.3d at 168. This resolution is inconsistent with *Turner I*, which concluded that heightened scrutiny is applicable to any laws that "single out the press, or certain elements thereof, for special treatment." *Turner I*, 512 U.S. 622, 640 (1994). Indeed, *Turner I*'s limitation of *Minneapolis Star* to cases in which regulations single out small numbers of media speakers suggests that the distinction identified in *Sinclair Broadcasting* is better suited to identifying situations subject to strict scrutiny than to determining whether to apply rational basis or intermediate scrutiny. *Id.* at 659–61.

247. *See* Yoo, *supra* note 45, at 227–29.

248. *See id.* at 213, 227.

subjected structural regulation of DBS to the higher level of scrutiny mandated by *Turner I*.[249]

Even courts that agree that *NCCB* provides the appropriate First Amendment standard have expressed confusion over the proper way to apply that standard of review. Some courts have construed *NCCB* as holding that structural regulation of the broadcast industry is subject only to rational basis scrutiny.[250] Other courts have construed *NCCB* as requiring them to apply intermediate scrutiny.[251] Still others have applied a standard of review that falls somewhere in between.[252] Thus, even if one were to settle on the particular constitutional standard to be applied, considerable confusion would remain as to precisely what that standard requires.

D. APPLYING THE STANDARD OF REVIEW

Ultimately, it may not matter precisely how this dispute is resolved. This is because even the most stringent of these tests—intermediate scrutiny under *O'Brien*—has long been criticized as too deferential. As noted earlier, the heart of the *O'Brien* standard requires that the restriction

---

249. *Compare* Time Warner Entm't Co. v. FCC, 93 F.3d 957, 975–77 (D.C. Cir. 1996) (applying the more lenient broadcast standard to sustain a statute requiring a DBS provider to set aside channel capacity for "noncommercial programming of an educational or informational nature" (citation omitted)), *with* Satellite Broad. & Communications Ass'n v. FCC, 275 F.3d 337, 352–66 (4th Cir. 2001) (applying the intermediate scrutiny of *Turner I* to sustain a statute requiring satellite broadcasters to carry local stations). These precedents cannot be squared with either the Supreme Court's broadcast or cable precedents. Under the broadcast precedents, one would have expected structural regulation of DBS to be subject to rational basis scrutiny under *NCCB* and behavioral regulation to be subject to intermediate scrutiny under *League of Women Voters*. *See* FCC v. Nat'l Citizens Comm. for Broad., 436 U.S. 775 (1978); FCC v. League of Women Voters, 468 U.S. 364, 380 (1984). Because *Satellite Broadcasting & Communications Ass'n* applied intermediate scrutiny to structural regulation, these cases do not place DBS within the broadcast paradigm. Under the Supreme Court's cable precedents, one would have expected structural regulation of DBS to be subject to intermediate scrutiny under *Turner I* and behavioral regulation to be subject to strict scrutiny under *United States v. Playboy Entertainment Group, Inc.*, 529 U.S. 803, 813–15 (2000). Because *Time Warner* applied something less than strict scrutiny to behavioral regulation, *see Time Warner*, 93 F.3d at 975, these cases fall outside the cable paradigm as well.

250. *See, e.g.*, Prometheus Radio Project v. FCC, 373 F.3d 372, 401–02 (3d Cir. 2004); *Sinclair Broad. Group*, 284 F.3d at 167–68. *Cf.* Fox Television Stations, Inc. v. FCC, 280 F.3d 1027, 1046–47 (D.C. Cir.) (limiting heightened scrutiny to content-based restrictions on broadcast speech while holding that structural regulations were subject only to "deferential review"), *modified on reh'g*, 293 F.3d 537 (D.C. Cir. 2002).

251. *See, e.g.*, News Am. Publ'g, Inc. v. FCC, 844 F.2d 800, 812 (D.C. Cir. 1988). *See also* Benjamin, *supra* note 11, at 54–64 (arguing that broadcast regulation is subject to intermediate scrutiny).

252. *See* Ruggiero v. FCC, 317 F.3d 239, 243–45 (D.C. Cir. 2003) (en banc) (holding that more than minimal rationality is required when a structural regulation has the effect of completely prohibiting an individual from using a particular communications medium).

"further[] an important or substantial governmental interest" and that "the incidental restriction on alleged First Amendment freedoms [be] no greater than is essential to the furtherance of that interest."[253] The requirement that the regulation further a "substantial" governmental interest has been construed to require only that the interest be nontrivial without requiring it to be particularly significant.[254] Any lack of substantiality can also be obviated by raising the level of generality until the requirement is met.[255]

O'Brien's tailoring requirement has proven to be equally permissive. Although initially stated in somewhat restrictive terms, the Court has subsequently reinterpreted it to be satisfied whenever the underlying government interest "would be achieved less effectively absent the regulation."[256] This reconstruction of the tailoring requirement represents a comparison of the various means available to the government, rather than an inquiry into whether the strength of the government interest justifies the intrusion on individual liberty. As a result, *O'Brien* doctrine devolves into a regulatory inquiry focusing solely on the extent to which the means chosen promote the government's goals.[257]

The result is a level of scrutiny that has been repeatedly criticized as tantamount to the presumption of nonprotection associated with rational basis review,[258] reaching only "laws that engage in the gratuitous inhibition of expression."[259] Unless *O'Brien* scrutiny is given more bite,[260] it is of little practical consequence whether any particular instance of structural regulation is formally subject to rational basis scrutiny, intermediate scrutiny, or something in between.[261]

---

253. *See* United States v. O'Brien, 391 U.S. 367, 397 (1968).
254. *See* Dean Alfange, Jr., *Free Speech and Symbolic Conduct: The Draft-Card Burning Case*, 1968 SUP. CT. REV. 1, 23.
255. *See* John Hart Ely, *Flag Desecration: A Case Study in the Roles of Categorization and Balancing in First Amendment Analysis*, 88 HARV. L. REV. 1482, 1486 n.17 (1975); Geoffrey R. Stone, *Content-Neutral Restrictions*, 54 U. CHI. L. REV. 46, 51 (1987).
256. *Turner I*, 512 U.S. 622, 662 (1994) (internal quotation marks omitted) (quoting Ward v. Rock Against Racism, 491 U.S. 781, 799 (1989) (quoting United States v. Albertini, 472 U.S. 675, 689 (1985))).
257. Werhan, *supra* note 239, at 672. Werhan further notes, "There is no speech side to the Court's balance. The Justices assess only the operational efficiency of the government's regulatory agenda, avoiding any consideration of whether that program is 'commensurably more important' than the [F]irst [A]mendment values advanced by the expression at issue." *Id.* at 641–42 (footnote omitted).
258. *See, e.g.*, Schauer, *supra* note 199, at 787–88; Stone, *supra* note 255, at 50–52.
259. Ely, *supra* note 255, at 1485–86.
260. *See infra* Part III.B (exploring this possibility).
261. Professor Baker argues that evaluating the constitutionality of structural regulation on the basis of the rationales proffered by the FCC risks overlooking the true rationales underlying structural regulations and their most important effects. *See* Baker, *supra* note 13, at 759–760. The Supreme

III. POSSIBLE SOLUTIONS TO THE PROBLEM OF ARCHITECTURAL CENSORSHIP

Thus, First Amendment doctrine does not appear to provide for meaningful judicial review of architectural censorship. This Part explores two possible solutions to this problem. First, it entertains the possibility of leaving matters unchanged and relying on Congress and the FCC to protect against architectural censorship. Second, it explores the possibility of revising *O'Brien* doctrine to allow for more meaningful judicial review.

A. Reliance on the Political Branches

One alternative is to leave the responsibility for protecting against the dangers of architectural censorship squarely in the hands of the political branches. A long and distinguished heritage offers support for such a proposal. Indeed, the authority of each coordinate branch to interpret the Constitution has been endorsed by such historical luminaries as James Madison, Thomas Jefferson, Andrew Jackson, Daniel Webster, Stephen Douglas, Abraham Lincoln, and Felix Frankfurter,[262] as well as by a veritable "all-star list of constitutional law scholars."[263] Such a claim might seem somewhat jarring to those steeped in the ringing declaration of *Marbury v. Madison*[264] that "[i]t is empathically the province and duty of the judicial department to say what the law is."[265] To say, however, that the courts have authority to construe the Constitution is not to say that they have the *exclusive* authority to do so. Indeed, *Marbury* is based on the

---

Court's adoption of the "substantial evidence" test in *Turner I*, 512 U.S. 622, 666 (1994), would seem to justify focusing solely on the rationales and factual inferences that were before Congress and the FCC, and would limit reviewing courts to assessing the goals and the means asserted by the government.

262. *See* LOUIS FISHER & NEAL DEVINS, POLITICAL DYNAMICS OF CONSTITUTIONAL LAW 1–26 (1992); GERALD GUNTHER, CONSTITUTIONAL LAW 21–28 (11th ed. 1985); WALTER F. MURPHY, JAMES E. FLEMING & WILLIAM F. HARRIS, II, AMERICAN CONSTITUTIONAL INTERPRETATION 195–247 (1986); Gary Apfel, *Whose Constitution Is It Anyway? The Authority of the Judiciary's Interpretation of the Constitution*, 46 RUTGERS L. REV. 771, 777–82 (1994); Michael Stokes Paulsen, *The Merryman Power and the Dilemma of Autonomous Executive Branch Interpretation*, 15 CARDOZO L. REV. 81, 84–97 (1993).

263. Thomas W. Merrill, *Judicial Opinions as Binding Law and as Explanations for Judgments*, 15 CARDOZO L. REV. 43, 49 n.26 (1993) (noting that Alexander Bickel, Edward Corwin, Philip Kurland, Gerald Gunther, Henry Monaghan, and Herbert Wechsler had each endorsed the authority of all three coordinate branches to interpret the Constitution). *See generally* Steven G. Calabresi & Christopher S. Yoo, *The Unitary Executive During the First Half-Century*, 47 CASE W. RES. L. REV. 1451, 1463–72 (1997) (providing an overview of the debate on coordinate construction).

264. Marbury v. Madison, 5 U.S. (1 Cranch) 137 (1803).

265. *Id.* at 177.

premise that "[t]hose who apply the rule to particular cases, must of necessity expound and interpret that rule."[266] Legislators and executive branch officials routinely apply the Constitution to particular factual contexts. Thus, in firmly establishing the judiciary's right to interpret the Constitution, *Marbury* implicitly recognized the other branches' authority to do so as well.[267]

The fact that Congress and the executive branch are competent to interpret and enforce the Constitution does not necessarily justify leaving important issues of constitutional interpretation exclusively in their hands. The judiciary bears an obligation to exercise its independent constitutional judgment even when other branches are in a position to offer their own assessment of the constitutionality of a particular governmental action.[268] From this perspective, allowing instances of architectural censorship to evade meaningful judicial scrutiny would represent a disturbing abdication of responsibility.

Whether the political branches will prove particularly effective in protecting against the dangers of architectural censorship is also questionable. Members of Congress are typically loath to consider constitutional issues. As Abner Mikva, who as a judge and former member of the House of Representatives was uniquely well situated to comment on the relationship between the judiciary and the legislature on matters of constitutional interpretation, once observed, "The fastest way to empty out the chamber [of Congress] is to get up and say, 'I'd like to talk about the constitutionality of this bill.' Members of Congress believe that's what courts are for."[269] Agencies are often equally reluctant to address constitutional issues,[270] as has been the case for the FCC.[271]

---

266. *Id.*
267. *See, e.g.*, LAURENCE H. TRIBE, AMERICAN CONSTITUTIONAL LAW § 3-2, at 25 (2d ed. 1988); Paul Brest, *Congress as Constitutional Decisionmaker and Its Power to Counter Judicial Doctrine*, 21 GA. L. REV. 57, 63 (1986); Merrill, *supra* note 263, at 51; William W. Van Alstyne, *A Critical Guide to Marbury v. Madison*, 1969 DUKE L.J. 1, 37 (1969).
268. *See Turner I*, 512 U.S. 622, 666 (1994); Sable Communications of Cal., Inc. v. FCC, 492 U.S. 115, 129 (1989); Landmark Communications, Inc. v. Virginia, 435 U.S. 829, 843–44 (1978).
269. Linda Greenhouse, *What's a Lawmaker to Do About the Constitution?*, N.Y. TIMES, June 3, 1988, at B6. *See also* Abner J. Mikva, *How Well Does Congress Support and Defend the Constitution?*, 61 N.C. L. REV. 587 (1983) (arguing that Congress should do more to discover constitutional shortcomings in legislation).
270. *See, e.g.*, Johnson v. Robison, 415 U.S. 361, 368 (1974) (noting that "adjudication of the constitutionality of congressional enactments has generally been thought beyond the jurisdiction of administrative agencies" (internal quotation marks and alterations omitted)); Henry P. Monaghan, *First Amendment "Due Process"*, 83 HARV. L. REV. 518, 523 (1970) (describing how agencies can suffer from "institutional 'tunnel vision'" that makes them more likely to frame questions of speech in terms of the regulatory issues with which they have been charged than in terms of the First Amendment).

B. INTENSIFYING *O'BRIEN* SCRUTINY

The other alternative is to refine First Amendment doctrine to give the courts a larger role in reviewing instances of architectural censorship. A plurality of the Supreme Court in *Turner I* experimented with this option when it incorporated into *O'Brien* scrutiny the requirement that the "recited harms [be] real, not merely nonconjectural, and that the regulation... alleviate these harms in a direct and material way."[272] To determine whether legislative findings satisfied this requirement, the Court balanced two opposing considerations. On the one hand was the fact that the legislative branch is better suited institutionally to make predictive judgments and is not required to produce the kind of record generally required of administrative agencies.[273] On the other hand was the recognition that blanket deference to legislative findings would constitute abdication of the judiciary's role in protecting the Constitution. To balance these two considerations, the Court borrowed the administrative law principle requiring the government to have "drawn reasonable inferences based on substantial evidence."[274]

The overall thrust of this development led many commentators to speculate whether the addition of this requirement would turn *O'Brien* scrutiny into a more meaningful form of judicial review.[275] Historically, courts have been quite reluctant to second guess the evidentiary findings made by Congress and the FCC. As the Court has noted, its "opinions have repeatedly emphasized that the Commission's judgment regarding how the public interest is best served is entitled to substantial judicial deference."[276] On other occasions, the Court has been slightly more circumspect, declining to "defer" to the other branches, but nonetheless "afford[ing]

---

271. For example, the FCC initially declined to repeal the Fairness Doctrine notwithstanding serious doubts as to its constitutionality. Inquiry into Section 73.1910 of the Commission's Rules and Regulations Concerning the General Fairness Doctrine Obligations of Broadcast Licensees, 102 F.C.C.2d 145, 147 ¶ 6, 148–57 ¶¶ 8–21, 246–47- ¶¶ 175–176 (1985), *vacated sub nom.* Radio-Television News Dirs. Ass'n v. FCC, 831 F.2d 1148 (D.C. Cir. 1987). The FCC's refusal to address the issue drew a sharp rebuke from the D.C. Circuit, which chided, "we are aware of no precedent that permits a federal agency to ignore a constitutional challenge to the application of its own policy merely because the resolution would be politically awkward." Meredith Corp. v. FCC, 809 F.2d 863, 874 (D.C. Cir. 1987).
272. *Turner I*, 512 U.S. at 664 (citing Edenfield v. Fane, 507 U.S. 761, 770–71 (1993)).
273. *Id.* at 665–66.
274. *Id.* at 666.
275. *See* Dorf, *supra* note 241, at 1201 n.101; Robert Post, *Recuperating First Amendment Doctrine*, 47 STAN. L. REV. 1249, 1263 n.67 (1995).
276. FCC v. WNCN Listeners Guild, 450 U.S. 582, 596 (1981) (citing FCC v. Nat'l Citizens Comm. for Broad., 436 U.S. 775 (1978)). *See also* FCC v. WOKO, Inc., 329 U.S. 223, 229 (1946).

great weight to the decisions of Congress and the experience of the Commission" and "pay[ing] careful attention to how the other branches of Government have addressed the same problem" when confronted with "a complex problem with many hard questions and few easy answers."[277]

There are some indications that the Court may now be willing to engage in more searching scrutiny of the factual predicate underlying statutory enactments. For example, the Court has shown its willingness to scrutinize the sufficiency of the evidentiary record in other contexts, including the Commerce Clause,[278] warrantless searches under the Fourth Amendment,[279] and most notably Congress's exercise of its authority under Section 5 of the Fourteenth Amendment.[280] The Court's willingness to rely on the absence of a real, nonconjectural harm to strike down restrictions of commercial speech also makes this argument quite plausible.[281] Indeed, shortly thereafter, various courts invoked this consideration to invalidate a number of restrictions on the cable industry.[282] Some courts have seen in *Turner I* the emergence of a stricter standard, one that will govern all content-neutral regulations that discriminate amongst the media.[283]

Subsequent developments have substantially reduced the likelihood that the factual review announced by the *Turner I* plurality will

---

277. CBS, Inc. v. Democratic Nat'l Comm., 412 U.S. 94, 102–03 (1973). *Accord* Metro Broad., Inc. v. FCC, 497 U.S. 547, 569 (1990) (quoting and following the above-quoted language from *CBS*), *overruled on other grounds*, Adarand Constructors, Inc. v. Peña, 515 U.S. 200 (1995).
278. *See* United States v. Morrison, 529 U.S. 598, 614–18 (2000); United States v. Lopez, 514 U.S. 549, 557 n.2 (1995).
279. *See* Chandler v. Miller, 520 U.S. 305, 318–22 (1997).
280. *See* Bd. of Trustees v. Garrett, 531 U.S. 356, 368–72 (2001); Kimel v. Fla. Bd. of Regents, 528 U.S. 62, 88–90 (2000); City of Boerne v. Flores, 521 U.S. 507, 530–32 (1997). *But see* Nev. Dep't of Human Res. v. Hibbs, 538 U.S. 721, 729–40 (2003) (sustaining the sufficiency of the legislative record underlying the Family and Medical Leave Act).
281. *See* Rubin v. Coors Brewing Co., 514 U.S. 476, 489–91 (1995); Ibanez v. Fla. Dep't of Bus. & Prof'l Regulation, 512 U.S. 136, 144–49 (1994); Edenfield v. Fane, 507 U.S. 761, 771–73 (1993); Zauderer v. Office of Disciplinary Counsel of the Sup. Ct., 471 U.S. 626, 648–49 (1985). *See also* United States v. Nat'l Treasury Employees Union, 513 U.S. 454, 476–77 (1995).
282. *See* Time Warner Entm't Co. v. FCC, 240 F.3d 1126 (D.C. Cir. 2001) (invalidating cable broadcast crossownership rule); Horton v. City of Houston, 179 F.3d 188 (5th Cir. 1999) (holding that summary judgment was granted in error in evaluating the fee charged on non-locally produced cable programs); US West, Inc. v. United States, 48 F.3d 1092 (9th Cir. 1994) (invalidating the cable/telephone company crossownership ban), *vacated and remanded*, 516 U.S. 1155 (1996); Chesapeake & Potomac Tel. Co. v. United States, 42 F.3d 181 (4th Cir. 1994) (same), *vacated*, 516 U.S. 415 (1996); Preferred Communications, Inc. v. City of Los Angeles, 13 F.3d 1327 (9th Cir. 1994) (invalidating the issuance of an exclusive cable franchise). *See also* Comcast Cablevision of Broward County, Inc. v. Broward County, 124 F. Supp. 2d 685 (S.D. Fla. 2000) (stating in dicta that ordinance requiring open access to cable modem systems would have failed intermediate scrutiny).
283. *See* Netanel, *supra* note 200, at 55–58.

significantly check architectural censorship. When the Court restated these principles in its second *Turner* decision (*"Turner II"*), it employed a far different tone. Noticeably missing from the opinion was any reference to judicial exercise of "independent judgment" or inquiry into whether the harm was "nonconjectural." Instead, the language and the structure of the opinion emphasized deference.[284] Later decisions have raised further doubts as to whether *Turner I*'s imposition of a substantial evidence requirement will actually lead to more searching judicial review. In *Nixon v. Shrink Missouri Government PAC*,[285] the Court noted that "[t]he quantum of empirical evidence needed to satisfy heightened judicial scrutiny of legislative judgments will vary up or down with the novelty and plausibility of the justification raised."[286] In particular, the Court acknowledged the possibility, first noted by the plurality opinion in *City of Renton v. Playtime Theatres, Inc.*,[287] that the government could rely on a factual record developed in another context or jurisdiction so long as the evidence on which the record is based is reasonably believed to be relevant.[288] A plurality of the Court reaffirmed this position in *City of Erie v. Pap's A.M.*,[289] concluding that the City of Erie could rely on the evidentiary foundations laid out in *Renton*[290] and *Young v. American Mini Theatres, Inc.*[291]

---

284. *Turner II*, 520 U.S. 180, 199 (1997). *See also* Glen O. Robinson, *The Electronic First Amendment: An Essay for the New Age*, 47 DUKE L.J. 899, 935, 937–38 (1998) (noting that "[w]ithout a doubt, the Court's decision in *Turner II* undercut what many thought to be the effect of *Turner I*" and lamenting the opportunity to engage in meaningful scrutiny of the relationship between ends and means). For an excellent analysis of the differences between *Turner I* and *Turner II*, see Stuart Minor Benjamin, *Proactive Legislation and the First Amendment*, 99 MICH. L. REV. 281, 301–03 (2000).

285. Nixon v. Shrink Mo. Gov't PAC, 528 U.S. 377, 391 (2000).

286. *Id.* It bears noting that this language is clearly dicta. After noting this possibility, the Court explicitly acknowledged the existence of a sufficient factual basis. *Id.* at 393–94.

287. City of Renton v. Playtime Theatres, Inc., 475 U.S. 41, 51–52 (1986).

288. *Shrink Mo. Gov't*, 528 U.S. at 393 n.6.

289. City of Erie v. Pap's A.M., 529 U.S. 277, 296–97 (2000) (plurality opinion).

290. *City of Renton*, 475 U.S. at 50–51 (relying on the factual record recited in a decision of the Supreme Court of Washington upholding a restriction on nude dancing in Seattle).

291. Young v. Am. Mini Theatres, Inc., 427 U.S. 50, 71 & n.34 (1976) (plurality opinion). Even more disturbing is the suggestion that the City of Erie's invocation of the Supreme Court's decision in *Barnes v. Glen Theatre, Inc.*, 501 U.S. 560 (1991), would have been sufficient alone to sustain the restriction under review. *See Pap's A.M.*, 529 U.S. at 297. As Justice Souter noted in dissent, the plurality opinion in *Barnes* did not purport to rely on any factual evidence indicating the existence of a problem. *Id.* at 315 (Souter, J., concurring in part and dissenting in part). Permitting a mere citation of *Barnes* to satisfy *Turner I*'s substantial evidence requirement would effectively gut the substantial evidence standard and would condone a form of constitutional bootstrapping that would be quite unprincipled.

In any event, even if the substantial evidence requirement advanced by the *Turner I* plurality survives as a basis for more searching scrutiny under the *O'Brien* standard, it is unlikely to redress the type of architectural censorship discussed in this Article. Commentators have long regarded *O'Brien* doctrine as uncovering restrictions driven by an improper government motive.[292] Indeed, the search for illicit purpose best supports putting the government to its proof in the manner dictated by *Turner I*.[293]

As such, adding this element is unlikely to bear on architectural censorship, which is generally the unintended byproduct of truly innocent governmental actions.[294] Even under this invigorated form, *O'Brien* scrutiny would do little to balance the importance of the governmental interest asserted vis-à-vis the individual's interest to engage in speech. Nor would it lead courts to inquire whether alternative avenues of communication exist or whether the same goals could be accomplished in a less intrusive manner. Architectural censorship would be better addressed through a test focusing on a regulation's effects on speech. Such tests, however, are generally disfavored, largely due to concerns that employing an effects test would open an unacceptably large swath of governmental action to constitutional scrutiny.[295]

---

292. *See* Paul Brest, *The Conscientious Legislator's Guide to Constitutional Interpretation*, 27 STAN. L. REV. 585, 590 (1975) (suggesting that the real teaching of *O'Brien*, despite the Court's contrary language, was that "some motives are unconstitutional"); Elena Kagan, *Private Speech, Public Purpose: The Role of Governmental Motive in First Amendment Doctrine*, 63 U. CHI. L. REV. 413, 438–42, 491–505 (1996) (arguing that *O'Brien* is primarily designed to expose regulations animated by improper governmental motives); Jed Rubenfeld, *The First Amendment's Purpose*, 53 STAN. L. REV. 767, 775–76 (2001) (concluding that "the *O'Brien* test itself is centrally concerned with legislative purpose, despite the Court's protests to the contrary"); Srinivasan, *supra* note 241, at 420 (synthesizing the Court's jurisprudence on incidental restrictions on speech as focusing on "a concern with speech-suppressive administrative motivation"). Interestingly, this reading of *O'Brien* is inconsistent with *O'Brien* itself, which disavowed that it was designed to identify illicit legislative motive. *See* United States v. O'Brien, 391 U.S. 367, 382–83 (1968).

293. *See* Netanel, *supra* note 200, at 61–62 (arguing that the invigorated intermediate scrutiny of *Turner I* is designed to root out improper governmental motive).

294. *See* Redish, *supra* note 199, at 130–31 (concluding that content-neutral regulations enacted without illicit motives can nonetheless skew speech markets in impermissible ways); Stone, *supra* note 255, at 106–07 (observing that properly motivated regulations may still have an adverse incidental impact on speech); Susan H. Williams, *Content Discrimination and the First Amendment*, 139 U. PA. L. REV. 615, 658 (1991) (noting that "even regulations serving a noncommunicative purpose can have a discriminatory effect on the speech market available to would-be listeners").

295. *See* Richard H. Fallon, Jr., *The Supreme Court, 1996 Term—Foreword: Implementing the Constitution*, 111 HARV. L. REV. 54, 84–86 (1997); Dorf, *supra* note 241, at 1178; Schauer, *supra* note 199, at 784, 790. *Cf.* Kagan, *supra* note 292, at 413–14 (criticizing the effects tests).

CONCLUSION

The analysis advanced in this Article demonstrates that the current debate has taken a far too simplistic approach to the impact that media ownership rules can have on television and radio program content. The analysis set forth reveals that the relationship between structural regulation and media content is much more complex than is generally recognized. Even worse, the current regulatory regime has all too often unintentionally degraded the quantity, quality, and diversity of programming available. In other words, structural regulation can represent a form of architectural censorship that can reduce the quantity, quality, and diversity of media programming. Unfortunately, current First Amendment doctrine effectively immunizes architectural censorship from meaningful constitutional scrutiny. As a result, either Congress or the FCC must bear the primary responsibility for safeguarding free speech values against these dangers, or the courts must revise *O'Brien* doctrine to permit more searching review capable of protecting the important speech interests at stake. Neither outcome appears likely at this point.

[10]
COMMENTARY

MEDIA STRUCTURE, OWNERSHIP POLICY, AND THE FIRST AMENDMENT

C. EDWIN BAKER[*]

Ever since Mark Fowler's 1982 article laid down the gauntlet to those who favor structural media regulation,[1] legal academia has produced a host of free market acolytes advancing his views. These young academics increasingly dominate media law teaching and the FCC. Professor Christopher Yoo is one of this group's best (as well as a personal friend). This short Comment on his article, *Architectural Censorship and the FCC*,[2] is written not because I consider it uniquely objectionable, but rather because its fundamental errors and characteristic distortions are representative of this influential group of scholars. This Comment will start with observations about Yoo's policy and economic analyses and then conclude with a critique of his desired constitutional regime.

---

[*] Professor of Law, University of Pennsylvania. I wish to thank Professor Christopher Yoo for encouraging me to publish my comments and the Curb Center for Art, Enterprise and Public Policy for inviting me to comment on Professor Yoo's article at its Conference on Federal Regulation and the Cultural Landscape, held at Vanderbilt University (March 2004). I also benefited from helpful comments of Margaret Jane Radin, Fritz Kubler, and Charlotte Gross.

[1] Mark S. Fowler & Daniel L. Brenner, *A Marketplace Approach to Broadcast Regulation*, 60 TEX. L. REV. 207 (1982). Actually, regulatory efforts had already peaked in the early 1970s. By the late 1970s, deregulatory efforts had begun, and some significant legal scholars—including Lucas Powe and Douglas Ginsburg—were already criticizing FCC structural broadcast regulation. Still, Fowler and Brenner's article, possibly because Fowler was FCC Chairman, represents the open declaration of a new era.

[2] Christopher Yoo, *Architectural Censorship and the FCC*, 78 S. CAL. L. REV. 669 (2005).

I. CASE STUDIES OF ARCHITECTURAL CENSORSHIP: OWNERSHIP CONCENTRATION

Yoo offers four examples of architectural censorship that he finds objectionable and favors having legally challenged and invalidated on First Amendment grounds.[3] Each architecture or legal framework has, he says, the (usually unintended) consequence of "degrad[ing] the quantity, quality, and diversity of programming available."[4] My critique is threefold: his discussion represents (1) inadequate normative or policy analyses; (2) simplistic, if not simply incorrect, economic analyses; and (3) misguided constitutional wishes. Of course, not all of the article is wrongheaded. Yoo's description of existing constitutional doctrine, for example, is careful and quite insightful, as far it goes. The task here, though, is to identify real problems, not real merits, in the article. Putting the constitutional points aside until later, and out of a need to be brief, the focus here will be on the normative framework and economic analysis in his discussions of ownership regulation—both horizontal concentration and what he describes as vertical integration. Because ownership issues are presently the most alive politically, the most important, and the most legally unresolved, and because they are the issues to which Yoo devotes the most attention, they are the chief focus of this Comment.

A. REASONS TO RESTRICT OWNERSHIP CONCENTRATION

Any normative evaluation of a legal regulation depends on an understanding of the goals it purports to further or the values it purports to embody. Thus, before turning to Yoo's analysis, it is important to roughly outline the most significant policy reasons for regulatory limits on media ownership concentration.[5]

The first and single most important reason to resist concentration of media ownership derives from a certain vision of democracy. Of course, normative theories of democracy are controversial.[6] Still, the major visions overlap in important ways. For many people (and most theories), true

---

3. Yoo identifies each as illustrating architectural censorship. He comments that "one would hope that the First Amendment would provide a basis for identifying and redressing architectural censorship when it arises." Yoo, *supra* note 2, at 675.

4. *Id.* at 731.

5. This subsection is adapted from testimony given before the Senate Committee on Commerce, Science & Transportation, Full Committee Hearing on Media Ownership (Sept. 28, 2004), *available at* http://commerce.senate.gov/hearings/testimony.cfm?id=1321&wit_id=3847. *See also* C. Edwin Baker, *Media Concentration: Giving Up on Democracy*, 54 FLA. L. REV. 839, 902–19 (2002).

6. *See infra* note 96.

democracy implies as wide as practical a dispersal of power within public discourse.[7] Dispersal of ownership also may promote the availability and consumption of diverse content—but no theorist of whom I am aware believes that this will always be true.[8] But democratic values mean that it makes a huge difference whether any lack of a particular type of diversity is imposed by a few powerful actors or reflects the independent judgments of many different people, for example, owners, with the ultimate power to determine content. The key goal, the key value, served by ownership dispersal is that it directly embodies a fairer, more democratic allocation of communicative power. This *distributive* value, combined with the points raised in the next paragraph, were probably what prompted nearly two million people to write, petition, or e-mail the FCC in opposition to reducing restrictions on concentration.[9] Without more, and regardless of empirical investigations or controversial economic analyses, this value judgment provides a proper reason to oppose any media merger or to favor any policy designed to increase the number of separate owners of media entities. Of course, in some circumstances, countervailing considerations might properly provide a basis to limit the sway of an anti-concentration or anti-merger principle. The Supreme Court approved the propriety of essentially this value judgment when it held that strict limits on media crossownership were appropriate to prevent an "undue concentration of economic power" in the communications realm.[10]

Second, the widest practical dispersal of media ownership provides two safeguards of inestimable democratic significance. Concentrated ownership in any local, state, or national community creates the possibility of an individual decisionmaker exercising enormous, undemocratic, largely unchecked, and potentially irresponsible power. Although this power may seldom or never be exercised, no democracy should risk the danger. Like the Constitutional separation of powers provisions, the fourth estate role of the press is designed in part to reduce the risk of abuses of power. In this sense, the widest possible dispersal of media power serves a structural role

---

   7. As argued below, Yoo is often misled by an almost exclusively commodity-based conception of value. He did occasionally offer, however, a more active or speaker-based view of free speech, for example, when he invoked "the *individual's* interest to engage in speech." Yoo, *supra* note 2, at 730 (emphasis added). Surely, however, contrary to Yoo, most people would find this interest as a reason to limit media concentration, not a reason to treat limits as presumptively objectionable because they restrict the communicative power of corporate conglomerates.
   8. *See infra* note 20.
   9. Prometheus Radio Project v. FCC, 373 F.3d 372, 386 (3d Cir. 2004) (noting two million individual communications to the FCC).
   10. FCC v. Nat'l Citizens Comm. for Broad., 436 U.S. 775, 780 (1978).

in the democratic process independent of any commodity that it produces or distributes on a day-to-day basis. Structurally, dispersal of ownership prevents a potential "Berlusconi" effect. Whether American owners of media conglomerates will use their power to unequally and partisanly affect the public sphere cannot be predicted abstractly. Whether or not this is true of William R. Hearst's yellow journalism or Rupert Murdoch's commercial or political agenda,[11] the undemocratic distribution and use of communicative power presents real dangers. German democracy did not benefit from Alfred Hugenberg's ability to use Germany's first media conglomerate to substantially contribute to Hitler's rise to power.[12] Note, however, that this serious risk is not the sort that economic or typical social science empirical content analyses are well equipped to identify.

Ownership dispersal can also provide safety by increasing the number of ultimate decisionmakers who have the power to commit journalistic resources to exposing government or corporate corruption or identifying other societal problems. This larger number of independent decisionmakers is likely to increase the chances that at least one will identify contexts in which to make this socially valuable commitment of resources. Similarly, this larger number will normally make it more difficult for any outsider to persuade or bribe the salient media into silence. Roughly thirty-five years ago, the FCC relied on this point when emphasizing that its concern is with diversity of *sources*:

> A proper objective is the maximum diversity of ownership . . . . We are of the view that 60 different licensees are more desirable than 50, and even that 51 are more desirable than 50 . . . . It might be the 51st licensee that would become the communication channel for a solution to a severe local social crisis.[13]

The third argument is largely an economic one, combined with a sociological prediction. Economic theory predicts that media markets will radically fail to provide people with the media content they want.[14] One reason for this relates to externalities, both positive and negative. For

---

11. *See* James Fallows, *The Age of Murdoch*, ATLANTIC MONTHLY, Sept. 2003, at 81, *available at* http://www.theatlantic.com/doc/prem/200309/fallows.

12. *See, e.g.*, Daniel C. Hallin & Paolo Mancini, COMPARING MEDIA SYSTEMS: THREE MODELS OF MEDIA AND POLITICS 155 (2004).

13. Amendment of Sections 73.35, 73.240 and 73.636 of the Commission Rules Relating to Multiple Ownership of Standard, FM and Television Broadcast Stations, First Report and Order, 22 F.C.C.2d 306, 311 ¶ 21 (1970) [hereinafter 1970 Multiple Ownership Order]. Given this view, the only empirical issue affecting whether a merger is desirable is whether, after the merger, there are fewer independent media owners. Except under unusual circumstances, the answer should be clear.

14. *See generally* C. EDWIN BAKER, MEDIA, MARKETS, AND DEMOCRACY pt. I (2002).

example, if many nonreaders of a newspaper benefit from its high quality investigative journalism that deters or exposes corruption, those benefits to nonreaders do not generate revenue for the paper. The paper therefore has too little profit-based incentive to produce the good journalism that produces net value for society (that is, value net of its costs). People get less and the market-oriented firm produces less good journalism than people want because transaction costs and free rider effects prevent payments for results that, as individuals, people value. (Corresponding points apply to negative externalities, which include media consumers' antisocial or misguided behavior, ranging from violence to stupid voting, that affect people other than the immediate media consumer.) Additionally, monopolistic competition in the media realm results in successful media entities typically having particularly high operating profits.[15] (High operating profits are generally recognized to exist,[16] although the accuracy of my account of why this is true is not especially important here.)

---

15. The classic text on this type of competition and industry structure is EDWARD HASTINGS CHAMBERLIN, THE THEORY OF MONOPOLISTIC COMPETITION: A RE-ORIENTATION OF THE THEORY OF VALUE (8th ed. 1969). Yoo has also applied Chamberlin's theory of monopolistic competition to the media realm. In doing so, he recognized conditions—high fixed costs (compare high first copy costs), a sufficiently low number of (major?) competing products, strong product differentiation, and "asymmetric preferences" (that is, consumers differing among themselves about the comparative desirability of different products)—which allow firms to obtain long-term (as well as short term) supracompetitive profits. *Compare* Yoo, *supra* note 2, at 711 n.191, *with* Christopher S. Yoo, *Rethinking the Commitment to Free, Local Television*, 52 EMORY L.J. 1579, 1607–08, 1627 (2003) [hereinafter *Rethinking the Commitment*]. In both Yoo's and my work (which in this regard is substantially similar), monopolistic competition is only one element of the analysis, with the nonrivalrous use element of public good theory also being central; this combination supports my claim that supracompetitive profits will be endemic to monopolistic competition that exists in the media realm. *Compare Rethinking the Commitment*, *supra*, *with* C. EDWIN BAKER, ADVERTISING AND A DEMOCRATIC PRESS 3, 20–24, 73–76 (1994), *and* BAKER, *supra* note 14, at 20–40. There are two primary differences between our accounts: (1) my analysis of how this competition in the media realm, particularly as exacerbated by advertising, mega-products, and price discrimination, can sometimes cause a reduction in total surplus (that is, how competition can exacerbate inefficiency by causing the failure of many media products that produce large consumer surpluses); and (2) an explanation for why precisely democratically important media products are likely to be especially subject to this negative effect. Thus, though our accounts are substantially similar (except for Yoo's much more informed and extensive reliance on existing economic literature), our greatest differences lie in my empirical predictions (which seem supported, *see infra* note 16) and my greater engagement in a democratic-oriented normative analysis.

16. According to an FCC study, in 2000 average profits as a percentage of revenues of network affiliate TV stations was over 30.2%, and for the comparatively few independent stations it was 42.3%, as compared to 6.8% for the 500 largest industrial corporations. JONATHAN LEVY, MARCELINO FORD-LIVENE & ANNE LEVINE, BROADCAST TELEVISION: SURVIVOR IN A SEA OF COMPETITION 34 tbl.16 (Office of Plans & Policy, FCC, 37 Working Paper Series, 2002), *available at* http://hraunfoss.fcc.gov/edocs_public/attachmatch/DOC-226838A22.doc. Given heavy debt in the media industry paid in the form of interest, cash flow may be the more relevant criterion and it is much higher, often adding ten to twenty or more percentage points to the profitability percentage. *See id.* at

Given these two facts about typical media markets, the structural or architectural goals ought to include getting ownership into the hands of people *most likely* to devote a large portion of the media entities' potentially high operating profits to providing better journalism—producing positive externalities—rather than to maximizing the bottom line. My sociological prediction is that both high- and mid-level executives of most large, publicly traded media companies often measure their success and are rewarded largely based on the profits their enterprises produce.[17] In contrast, heads of smaller, more local, especially individual or family-based entities, often self-identify with the quality of their firms' journalistic efforts and service to community. They value the acknowledgment of this service among people in their community as well as by their fellow journalistic professionals.

Direct *structural* pressures caused by corporate mergers can exacerbate the resulting firms' undesirable focus on profit-maximization. The buyer most willing and best able to capitalize the purchased entity's potential profits is able to make the highest buyout bid. But when successful, this price locks the purchaser into needing to maximize operating profits to cover the cost of the indebtedness created by the bid. In contrast, the original or long term owner is freer to use, often she will have been "inefficiently"[18] using, that potential income to provide better quality

---

33. Similar strong gross profitability exists in the radio industry—and Clear Channel, the country's largest radio station operator with about 1200 stations, reportedly obtained an annual gross profit to revenue rate of between 26% and 29% between 2001 and 2003. GEORGE WILLIAMS & SCOTT ROBERTS, FORMAT AND FINANCE: RADIO INDUSTRY REVIEW 2002, at 4 (FCC, Media Bureau Staff Research Paper, 2002), *available at* http://hraunfoss.fcc.gov/edocs_public/attachmatch/DOC-226838A20.doc; Clear Channel Communications, Inc., LEXIS CoreData U.S. Institutional Database (on file with the *Southern California Law Review*). The average profit margins in 1998 of seventeen publicly traded newspapers was 18.6, while the reported operating cash flow or EBITA (earnings before interest, taxes, depreciation, and amortization) of the thirteen companies reporting this figure was 27%. GILBERT CRANBERG, RANDALL BEZANSON & JOHN SOLOSKI, TAKING STOCK: JOURNALISM AND THE PUBLICLY TRADED NEWSPAPER COMPANY 34, 37 (2001).

17. *See* CRANBERG ET AL., *supra* note 16, at 8. Research into newspapers has generally found (unsurprisingly) that increasing the number of journalists leads to a higher quality paper. Relevant to the current point, a study of mid-sized newspapers found that, statistically and holding other factors constant, public ownership (that is, subjecting the newspapers to shareholder demand) leads to higher profits and a staff of fewer journalists. In fact, a higher profit margin and public ownership independently correlate with fewer journalists and, presumably through this effect, with a lower quality paper. *See* Stephen Lacy & Alan Blanchard, *The Impact of Public Ownership, Profits, and Competition on Number of Newsroom Employees and Starting Salaries in Mid-Sized Daily Newspapers*, 80 JOURNALISM & MASS COMM. Q. 949 (2003).

18. The quotes reflect my claim that this practice *is not* inefficient from a social perspective but seems that way from the perspective of anyone who tends to equate efficiency with profit-maximizing use of resources—especially from the perspective of the buy-out firm for whom this creates a profit-

products—by hiring more journalists, providing more hard news, and doing more investigations.

The fourth point begins with the recognition that the ideal legislative or regulatory media policy is not only debatable and reflects experience but also changes with changed circumstances. Entrenched groups always have momentum and political advantage over undeveloped alternatives. The media context, because of the media's unique role in relation to politically salient public opinion, intensifies this problem. This problem can be further exacerbated if a few of these entities become too powerful. Then the likelihood diminishes that subsequent debates and decisions, especially in Congress, but also at the FCC, will reflect lawmakers' or FCC professionals and commissioners' true, informed, and thoughtful evaluation of the public interest. Rather, concentration increases the likelihood that the economic interests of these huge media conglomerates will largely control the policy debates and legal outcomes.

B. OWNERSHIP RESTRICTIONS AS (UNDESIRABLE) CENSORSHIP

With this backdrop of objections to media concentration, consider Yoo's argument. He suggests that the FCC's rationales for ownership restrictions, and the only justifications he considers, have been premised on "the need to protect competition and the need to promote a diversity of programming and viewpoints."[19] Yoo observes that commentators have come out both ways on whether ownership limits actually serve diversity. Originally,[20] he (incorrectly) asserted that one group (which includes this commentator) adopts the categorical position that increases in media

---

making opportunity. To put it provocatively, the question is whether to see efficiency from the perspective of capital or of people—efficient at producing profit or welfare respectively.

19. Yoo, *supra* note 2, at 689 (footnote omitted).

20. Although Yoo now agrees that this mischaracterizes the alternative positions and has accordingly changed his final text, I have retained my original reference primarily because it illustrates more starkly the perspective that continues to plague Yoo's analysis. Both his original and revised versions want to characterize two conflicting positions about the effect of concentration on "the quantity and diversity" of media content and then present his more nuanced third position. The curiosity is that he presents no reason to think that the first group, those critical of media concentration, could not completely agree with both the less intuitive observation of the second group and with Yoo's claim of complexity. In fact, at least some of them do (and the others are not concerned enough about what Yoo seems to view as the central issue to comment on it). *Compare id.* at 692, *with* Ronald J. Krotoszynski, Jr. & A. Richard M. Blaiklock, *Enhancing the Spectrum: Media Power, Democracy, and the Marketplace of Ideas,* 2000 U. ILL. L. REV. 813, 831–32, *and* BAKER, *supra* note 14, at 34. In other words, Yoo has not managed to describe contending positions. Yoo's difficulty reflects his continued assumption that scholars' policy-based disagreements will involve concerns with and predictions about commodities. *See infra* note 28. In contrast, all the commentators critical of concentration treat the primary issue as concentration's relevance for democracy.

concentration will reduce "the quantity and diversity of media content"[21] but he (correctly) argues that the issue is more complex.[22] As to horizontal concentration, however, Yoo seems content to claim that concentration will have an impact on program diversity, "even if the direction . . . of the effect remain[s] somewhat uncertain."[23] This conclusion seems somewhat a letdown from his initial claim: that the "structural regulations . . . [that he examines not only] reduce the overall quantity and quality of media programming, they also affect the diversity of media content."[24] Still, this assertion of complexity does not seem to be his final word. He ends the section by first suggesting that efficiencies made available by merger, which lead to a "greater return on investment[,] . . . may enable media outlets to provide more diverse programming."[25] Then, implicitly converting "may enable" into "will cause," he is able to conclude that "[h]orizontal ownership restrictions represent . . . architectural censorship"[26]—by which he apparently means they reduce total "quantity, quality, and diversity of speech."[27]

Though surely right that structure affects content—in fact, I know no one who thinks differently[28]—that point does not alleviate the fundamental problem with Yoo's analysis: diversity and quality of content are not the only, or even the primary, considerations justifying regulation. True, if under particular circumstances effects on quantity, quality, and diversity—

---

21. Yoo, *supra* note 2, at 692. *See also id.* at 697 n.118. *But see infra* note 28. I would categorically maintain one point about content diversity and ownership distributions: given that different structures will produce different contents, whichever produces the most diversity is not a simple empirical observational or counting matter, but rather depends in part on inevitably contested criteria of relevant content distinctions—criteria generated by necessarily evaluative judgments concerning how significant various differences are.
22. Yoo, *supra* note 2, at 692–699.
23. *Id.* at 699.
24. *Id.* at 674.
25. *Id.* at 700.
26. *Id.* at 701.
27. *Id.*
28. In an earlier draft Yoo incorrectly asserted that various others and I take the categorical position that horizontal concentration always reduces quantity and diversity of media content. *See supra* note 20. I suspect this initial misreading reflects his commodity-focused view of the value of media. From that perspective, any categorical objection to media concentration must be based on a categorical conclusion that concentration negatively affects the commodity—presumably either the quantity, quality, or diversity of content made available to the consumers. As I explain more fully in the text above and in the next subsection, the commodity value, while real, is simply not the reason for the categorical objection. Moreover, few theorists make their objection to concentration categorically. The authors he cites, for example, are unlikely to have such categorical objections to a large media entity or to substantial concentration if (1) the concentrated *private* media is effectively regulated as a common carrier, or (2) a degree of concentration occurs within an appropriately structured public broadcaster subject to appropriate public service obligations.

which Yoo recognizes might be positive or negative—are too severely negative, that would give a strong reason for an exception to an otherwise firm anti-concentration principle. From the beginning, the FCC recognized (as did Congress when it adopted the Newspaper Preservation Act) that media combinations are sometimes beneficial. Given adequate contextual justifications, for example, that local broadcast service depends on joint ownership, the FCC regularly provides for waivers of ownership restrictions. The Court has always praised the FCC for so providing.[29] But Yoo ignores the real issues. None of the reasons noted above for opposing media concentration depend on any empirical assumption about typical effects on quantity, quality, or diversity of content. That is, Yoo ignores the entire set of reasons described above for favoring restrictions. This represents a failure to engage in even a minimally adequate normative or policy analysis of the issue.[30]

---

29. *See, e.g.*, FCC v. Nat'l Citizens Comm. for Broad., 436 U.S. 775, 802 n.20 (1978).

30. Yoo raises two objections to my focus on democracy in evaluating media policy and, implicitly, in defense of his commodity focus. Yoo, *supra* note 2, at 675 n.17. First, he suggests that this democracy focus is inconsistent with "the autonomy-centered vision that has long dominated free speech theory." *Id.* Although I would be happy if Yoo were right about speech theory, he is entirely wrong about Press Clause theory. No Supreme Court decision makes the almost incoherent suggestion that autonomy theory applies to structural regulation of the media, and virtually every theorist or Justice who has addressed the constitutional role of the press has interpreted it instrumentally in relation to serving a free and democratic society. *See* BAKER, *supra* note 14 (discussing democratic interpretations of the Press Clause); C. Edwin Baker, *First Amendment Limits on Copyright*, 55 VANDERBILT L. REV 891 (2002) (discussing the difference between the autonomy rationale of the Speech Clause and the democratic rationale of the Press Clause); C. Edwin Baker, *Turner Broadcasting: Content-Based Regulation of Persons and Presses*, 1994 SUP. CT. REV. 57 (same) [hereinafter *Turner Broadcasting*]. If Yoo is interested in media policy, he ought to be concerned with the useful contributions that media makes to human life—and though those contributions do include production and distribution of commodities that audiences value, any analysis that stops there is demonstrably impoverished.

Yoo's second point appears to be that since implications of democratic theory for the media are controversial, media policy should ignore the differential contributions that the choice of media policies makes to the media's democratic role. Yoo, *supra* note 2, at 675 n.17. Would uncertain and controversial contributions of different medical policies to human health and well-being properly lead an analyst to ignore these impacts, and instead consider only how they will affect the profitability of medical providers? The error seems the same as that which plagued many social scientific evaluations of newspaper concentration. Studies would find that chain ownership of newspapers is beneficial because it leads to better front page graphics or harmful because it leads to smaller staff size and a smaller newshole, without considering more important social and democratic impacts because they are both controversial and (often) not quantifiable. *See* C. EDWIN BAKER, OWNERSHIP OF NEWSPAPERS: THE VIEW FROM POSITIVIST SOCIAL SCIENCE (Joan Shorenstein Ctr., John F. Kennedy Sch. of Gov't, Harvard Univ., Research Paper R-12, 1994). Of course, Yoo is right that the contributions of media are normatively controversial. In the middle of the nineteenth century, would an abolitionist paper make a more valuable contribution than a more slickly written, pro-slavery, local booster paper? People would disagree and the First Amendment properly bars censorship of either. But given its inevitable architectural effects, should postal policy make it easier or harder for the abolitionist paper to circulate in the South? A slightly, but only slightly, less value-laden normative judgment is whether a newspaper

C. REDUCTIONIST COMMODIFICATION

Why would someone as smart and careful as Yoo go so wrong? One possibility is that Congress, the FCC, and the courts only offered the limited sort of reasons that he considered. Even if true, this provides no real excuse. Often people know that they value a result without being well able to articulate the reason. Of course, such intuitive judgments should be subject to serious examination—they might reflect prejudices or possibly unconscious, but false, empirical or indefensible normative assumptions. Still, a serious scholarly inquiry should, certainly before rejecting a popular policy as unjustified censorship, consider whether any reasons for the policy are normatively persuasive—and, maybe, consider whether these reasons are consistent with others that society or lawmakers normally consider important. If convinced by the four concerns with ownership concentration that are provided above—or convinced of at least the permissibility of viewing these concerns as important—the proper inquiry would evaluate ownership rules from this perspective, including their costs (in various dimensions[31]). Failure to engage in that effort is my central normative criticism of Yoo.

In any event, blindness to these concerns cannot be explained by a historical failure of the government to assert them. Rather, the failure reflects a tin ear caused by a complete commodity focus, a problem that, although not inherent to economic logic, does seem common among many modern market economists both at the FCC and in the legal academy. Once attuned to political process concerns and noncommodified values described

---

engaged in careful and expensive reporting about abuses of private and public power generates more positive externalities or greater democratic contributions (which Yoo would ignore) than does a newspaper that reports mostly on buying opportunities in the city's wonderful new shopping mall. The fact that the normative valences of a policy are controversial is not a reason to ignore them in favor of objectively measurable, but possibly much less significant, efficiencies. Rather, it is a reason to recognize, first, that the best policy often will be politically contested and, second, if scholarly inquiry is to provide useful and relevant policy guidance, it must engage in the inevitably inconclusive investigation into the most important issues, issues for which abstract economic analyses often simply do not provide determinative guidance.

31. A key mistake of the FCC's reasoning in its recent ownership decision is that it treats as a cost an interference with the media corporation's purported First Amendment right to speak to as large an audience as it can generate ownership capacity to reach. *See* 2002 Biennial Regulatory Review— Review of the Commission's Broadcast Ownership Rules and Other Rules Adopted Pursuant to Section 202 of the Telecommunications Act of 1996, Report and Order and Notice of Proposed Rulemaking, 18 F.C.C.R. 13,620 (2003) [hereinafter 2003 Biennial Review Order]. This confuses corporate entities with flesh and blood people. The Court has consistently treated the media's rights as valued only instrumentally in relation to how it serves the democratic interests of audiences. *See, e.g.*, Associated Press v. United States, 326 U.S. 1, 20 (1945).

in Part I.A, Yoo's characterizations can be seen to be false. To support the proposition that the two rationales for ownership restrictions are the "the need to protect competition and the need to promote a diversity of programming and viewpoints,"[32] Yoo cites, for example, a 1970 FCC decision that emphasized not diversity of *programming* content but "the importance of diversifying *control*."[33] This concern led the FCC to conclude that "[a] proper objective is the maximum *diversity of ownership* that technology permits."[34] More recently, Congress stated that its purpose in the Cable Communications Policy Act of 1984 was to "assure that cable communications provide ... the widest possible diversity of information *sources* and services to the public."[35] If real value can lie only in diversity and quality of content, which is a plausible view within a solely commodity framework, this language could be easily mistaken as expressing a purpose to achieve diversity of programming and viewpoints combined with an empirical assumption that diversifying ownership would be instrumental to achieving the result. The Court has stated that this empirical hypothesis is presumptively acceptable—and sometimes it will be correct, but also, as Yoo explains, sometimes it will not. In contrast, from the perspective of distribution of power within the public sphere, and certainly from the perspective of a broader sharing of communicative power, the statements would instead be understood to mean what they say—that the goal is as many different sources as possible. This second conclusion would hold whether or not the diversity of sources produces any particular content diversity. It is the democratic as opposed to commodity vision well stated by the essayist and writer, E.B. White, as quoted by Justice Felix Frankfurter:

> The controlling fact in the free flow of thought is not diversity of opinion, it is diversity of the *sources* of opinion—that is, diversity of ownership .... There are probably a lot more words written and spoken in America today than ever before, and on more subjects; but if it is true ... that these words and ideas are flowing though fewer channels, then our first freedom has been diminished, not enlarged.[36]

A similar tin ear for noncommodified values is evident in Yoo's understanding of the Supreme Court's view that diversification of mass

---

32. Yoo, *supra* note 2, at 689 (footnote omitted).
33. *See* 1970 Multiple Ownership Order, *supra* note 13, at 311 ¶ 21 (emphasis added).
34. *Id.* (emphasis added).
35. Cable Communications Policy Act of 1984, Pub. L. No. 98-549, § 2, 98 Stat. 2779, 2780 (codified at 47 U.S.C. § 521(4) (2000)) (emphasis added). *See also* 47 U.S.C. § 532(a).
36. Pennekamp v. Florida, 328 U.S. 331, 354 n.2 (Frankfurter, J., concurring) (quoting E.B. White, *Book Review: The First Freedom*, NEW YORKER, Mar. 16, 1946, at 97).

media ownership serves the public by "preventing undue concentration of economic power."[37] Clearly, the Supreme Court was not duplicating the early trustbusters' concern with a large total amount of private economic power. Though legitimate, that concern would hardly justify focusing on the mass media, where no company that is primarily a media content provider ranks among the largest fifty companies. Wal-Mart, the oil companies, and car companies are the largest, but under existing law and policy, they are not treated as necessarily too large.[38] Rather, the Court's reference was surely to concentrated economic power in the special realm of mass media. After saying that, however, there is still the question of what precisely makes some power "undue."

For Yoo, economic power in the media realm is "undue" if it allows an enterprise to operate noncompetitively. His characterization of the first of the FCC's two rationales for restricting ownership as "the need to protect competition"[39] illustrates this understanding. Still, the value of competition can be given various interpretations. Yoo asserts that this concern with competition "is completely economic in focus and unrelated to the content of speech."[40] His interpretation appears to duplicate antitrust policy's limited concerns—to restrict enterprise power to raise prices to noncompetitive levels (or otherwise exercise monopoly power to increase profits). This interpretation explains his view that limits on national ownership (ownership of multiple media entities that do not compete geographically) are only explicable in terms of the mostly unreal dangers of vertical integration. It also explains his view that after analysis, these dangers do not justify existing FCC restrictions.[41] Finally, this interpretation explains his invocation of the Hirschman-Herfindahl Index as a means to identify competitive problems. So long as an analyst adopts a

---

37. FCC v. Nat'l Citizens Comm. for Broad., 436 U.S. 775, 780 (1978) (citation omitted).

38. In *Fortune*'s latest list of largest companies by revenue, Wal-Mart ranks first with revenues of $263 billion, followed by three oil companies and four car companies, respectively. The largest company prominent in the mass media field is General Electric ("GE") at ninth (at about half the revenues of Wal-Mart), but mass media is a relatively minor part of GE's business. Verizon at twenty-eighth and Sony at thirtieth are primarily communications service or appliance companies. From there, media companies are absent until Time Warner at eighty-third. But again, a significant portion of its revenue comes through its communication service, AOL, rather than being solely a mass media content provider—although the two are intertwined. Paola Hjelt, *The Fortune Global 500*, FORTUNE, July 26, 2004, at 159, 163.

39. Yoo, *supra* note 2, at 689.

40. *Id.*

41. *Id.* at 701, 713.

single-minded focus on supplying commodities valued by consumers, Yoo's interpretation of undue economic power makes sense.[42]

From broader perspectives, there are at least two other significant possibilities for understanding undue concentration of economic power. The first offers a different understanding of the value of competition and the second more fundamentally avoids reducing the concern with power to a concern with competition in provision of commodities. Although hardly the received paradigm, the proper issue in antitrust policy generally[43] and even more so in the media realm[44] is not limited to power over price (and monopoly profits), but instead involves a broader power over the consumer's (content or product) choice. Although still commodity oriented, this emphasis has the advantage of recognizing that a firm can have inappropriate power over consumer choice (power over content) even when it has no power to increase profits to monopoly levels (power over price). Consider two newspapers, Paper 1 and Paper 2, competing inside a single city without the ability to reap monopoly profits from either readers or advertisers. Still, without losing customers, changing production costs, or any decline (or increase) in profitability, either paper may be able to choose between alternative editorial orientations. For example, Paper 1 may be able to attract an equally large (although presumably discrete) audience by distinguishing itself from Paper 2 from either the left or the right. This amounts to power over content, a power that may be relatively pervasive in the media field. Wherever it exists, a concern with power involved in concentration is appropriate. An entity with this power can restrict or orient consumer choice.

Measurement of consumer choice, like identification of diversity, is not a merely technical, economic, or empirical matter, but one that requires (usually normative) judgments identifying the precise choices or diversities that matter. Yochai Benkler expressed this point quite simply when he wrote, "If I am interested in learning about the political situation in

---

42. I should emphasize, however, that my disagreement with Yoo does not reduce to a mere disagreement about interpretation of a phrase in a Supreme Court opinion. The real issue is which interpretation relates best to the actual issues at stake. Our disagreement is solidly normative. To the extent there is room for interpretative choice relating to legal language, when the interpretation relates to giving the language legal force, the interpreter's obligation is to understand the language in its most normatively justifiable sense.

43. *See generally* Neil W. Averitt & Robert H. Lande, *Consumer Sovereignty: A Unified Theory of Antitrust and Consumer Protection Law*, 65 ANTITRUST L.J. 713 (1997).

44. *See generally id.* at 750–53; Baker, *supra* note 5; Robert Pitofsky, *The Political Content of Antitrust*, 127 U. PA. L. REV. 1051 (1979); Maurice E. Stucke & Allen P. Grunes, *Antitrust and the Marketplace of Ideas*, 69 ANTITRUST L.J. 249 (2001).

Macedonia, a news report from Macedonia or Albania is relevant, even if sloppy, while a Disney cartoon is not, even if highly professionally rendered."[45] Lack of market power over price has no bearing on whether competitors are providing for a person's desires for different goods. Only a normative theory can suggest how to weigh whether choice in music formats, political perspectives, or between music and political content, for example, is more significant when evaluating whether existing competition adequately provides for diversity. Showing that there is a reason to expect a Steiner monopolist to provide one does not necessarily provide a reason to expect that it will provide the other. That is, there is no reason to expect that a monopolist will provide the form of diversity that normative theory or even consumers consider more important.[46]

At various points, Congress clearly has concluded that power over content was more significant than power over price. This judgment, for example, explains its repeated concern with national media concentration. Likewise, this judgment is implicit in the Newspaper Preservation Act,[47] conditionally allowing merger of competing newspapers' business operations. So long as the joint newspaper operating agreement keeps power over content dispersed, the Act gives the papers greater opportunities to extract money from customers, presumably both readers and advertisers.

Nevertheless, this consumer choice interpretation of the value of competition is still removed from the reading that anyone except a person attuned only to commodified values would give to the Supreme Court's concern with concentration of economic power. The earlier quote of E.B. White had it right.[48] Agreed, the commodity-oriented concern with audiences' choices of content is worthy and an improvement on the narrow focus on power over price. More fundamental, however, are concerns with

---

45. Yochai Benkler, *Coase's Penguin, or, Linux and The Nature of the Firm*, 112 YALE L.J. 369, 383 (2002).
46. Interestingly, while the Steiner explanation of monopolistically created diversity is often cited in this country as a reason to favor some degree of media concentration, recognition of the same point (though not articulated using the same economic language) was also seen in Europe to justify monopoly. This was true, however, only under circumstances where the resulting diversity would not represent merely market forces or ownership whim but was subject to public (legal) policy mandates to provide for the country's varying political and cultural divisions. Essentially, the view was that the government—indirectly, by creating a new independent, non-profit institution—but not the private monopolist would respond to the need for the right kinds of diversity. Thus, the Steiner insight was thought to justify either a strong or monopoly public broadcasting system in Europe. *See, e.g.*, PETER J. HUMPHREYS, MASS MEDIA AND MEDIA POLICY IN WESTERN EUROPE 119 (1996).
47. Newspaper Preservation Act, Pub. L. No. 91-353, 84 Stat. 466 (1970).
48. *See supra* note 36.

a speaker's power within the cultural and political public sphere. These concerns were highlighted in my initial list of reasons for opposing media concentration. None of Yoo's empirical or economic discussion shows why these concerns do not justify all existing, or even greater, restraints on concentration. The presumably "unintended consequences" of following his commodity-centric recommendations could be to undermine basic democratic values.

D. BAD ECONOMICS

The critique so far has been with Yoo's inadequate and economistic normative orientation. Proper economic analysis can provide empirical explanations and predictive insights to which it is hard to object in principle.[49] A poorly executed economic analysis, however, can be seriously misleading. For example, to be informative, any policy-relevant economic analysis must at least take into account all the preferences or values that people have in relation to the matter under examination. Certainly, in the media or political realms, many of these values will not relate to possession of commodities. On this basis, this subsection notes problems with Yoo's economic analysis. Although there is much more to say about various specifics of Yoo's presentation, this subsection briefly illustrates only four generic concerns: whether he gave sufficient or correct attention to (1) distribution of power or influence within the communications realm; (2) externalities, especially as they involve noncommodified values; and, in light of the first two points, (3) consequences of most media firms' high operating profits, which exist probably due to the nature of monopolistic competition; and (4) identification of true efficiencies.

Enterprise-based and welfare-based economics highlight different goals. For an enterprise's economist, a merger's impact on competitive opportunities and possible cost saving are both important. The competitive opportunities could involve a greater ability to produce or sell products or greater market pricing power. The telling point, however, is that this perspective does not make directly relevant any of the four structural economic issues noted in the above paragraph—the merger's distributional effects, the existence of positive or negative externalities, the use of

---

49. *But see* Margaret Jane Radin, CONTESTED COMMODITIES 6–15, 115–22 (1996) (noting problems associated with commodification as a world view). Economic reasoning is, however, limited. The most important and most difficult tasks for law and legal scholarship are to understand, interpret, and reason about values and normative visions—and for this, economics is largely irrelevant.

monopoly profits, or whether a cost saving is truly an efficiency improvement. Of course, the enterprise's economist might be sensitive to some of these for instrumental, but sometimes socially perverse, reasons. The economist might check for newly created opportunities to externalize costs cheaply or identify someone from whom to collect (internalize) some of the enterprise's otherwise positive externalities. Neither of these, however, and certainly not the first, should be treated as welfare enhancing or efficient even though beneficial to the firm. Similarly, although a firm or its economists might loosely refer to cost savings as efficiency gains, technically this can be incorrect. The overlap between the firm's cost saving actions, some made possible by mergers, and efficiency or welfare gains vary, most obviously depending on whether and how the cost saving action affects externalities. If the action creates negative externalities or eliminates positive externalities—which can be viewed as economic language for my four general objections to media concentration—cost savings may well represent efficiency losses. The critique here is that Yoo's analysis greatly resembles the limited concerns that a firm may have but that it fails as sound economic theory or as the economic analysis needed for public policy purposes. His economics privileges the firm's profit maximization over the goal of maximizing social welfare. Thus, I want to briefly remark on the four inquires noted above.

1. Distribution

In a brief discussion, Yoo argued that the "conventional wisdom [wrongly] presumes that [cable] rate regulation has little to no impact on the content of speech."[50] His primary claim was that although rate regulation "caused nominal cable prices to drop," it "caused quality-adjusted rates to increase," and because "consumers would have preferred larger, higher quality bundles," it "failed to yield any real welfare benefits for consumers."[51] On the contrary, conventional wisdom was not so stupid. Virtually everyone designing cable rate regulation recognized that it would inevitably have some content effects, given the inevitability that cable operators would try to respond by providing lower quality or lower quantity offerings at the mandated reduced rates. The recognized design difficulty was to regulate in a manner that first protected consumers at the basic tier from excessive charges while minimizing the effects on content. On the evidence offered, anyone sensitive to distributive issues should find Yoo's

---

50. Yoo, *supra* note 2, at 686.
51. *Id.* at 686–687.

report about consumer preferences to be much too sweeping. *Even if* consumers as a group mostly prefer higher rates combined with a better product, there remains the question of whether this is true for all consumers, especially the poor. Would some groups prefer lower prices even if it meant some degradation of quality? Yoo provides no information on this question. Rate regulation could even require a basic-service tier priced below cost, achieving a clear distributive gain for those who limited their purchase to this package, as a condition for selling higher-priced tiers from which providers could reap monopoly profits and finance losses in the basic tier. To be effective in distributional aims, this pricing of basic-tier service would be combined with a requirement to offer service throughout a specified geographic service area. Of course, questions—both empirical and normative—remain. But any economic analysis that ignores these possibilities, or that treats consumers as an undifferentiated group, shows itself to be blind to distributive values.

2. Externalities

Externalities, positive or negative, received virtually no discussion in Yoo's article. This absence had two serious consequences. First, it led his article to ignore the most significant justifications for ownership regulation—the four concerns discussed above. Each of the values of ownership dispersion are, like democracy itself, matters about which many people are passionate but are not purchasable in the context of buying a media product. No welfare or efficiency analysis can provide policy relevant information if it ignores these highly valued externalities. Second, this blindness contributed to an inability to determine which cost savings are efficiency enhancing.[52]

A third consequence of ignoring these externalities is that it may contribute to many economists' inability to understand the political protests against government actions permitting greater concentration. This failure often leads to an elite assumption that the protests show only people's inability to understand unintended effects or represent people's deluded belief that ownership limitations assure a better selection of commodities— the only result that many free market disciples can imagine people valuing. On the contrary, the primary way people are able to express the importance of these externalities (or noncommodified values—the terminology is hardly crucial) is to act politically. As was suggested earlier, these values may provide the most insightful explanation of the mass public objection to

---

52. *See infra* Part I.D.4.

reducing restrictions on media ownership. Here, politics works better than the market in showing people's true preferences—their willingness to pay.

3. Use of Monopoly Profits

At various points Yoo noted that media mergers could produce cost savings, but he paid scant attention to the abnormally high level of the typical media firm's operating profits. Any economist engaged in a policy-oriented discussion of media structure should provide an explanation for this phenomenon of high operating profits.[53] More important than providing this explanation, however, is a discussion of predictable effects of different structures on how a media firm's potential profits are used. Here, there are two plausible policy goals. A distributive goal would be to lower prices to the advantage of consumers. Alternatively, given media entities' potential to produce high externalities, these profits could usefully be spent on high externality-producing activities like investigative journalism.

The closest Yoo comes to these issues is in his discussion of cable rate regulation. There, his main claim is negative: attempts to prevent monopoly profits did not and, presumably, could not succeed.[54] He also periodically makes what may seem a naive inference: a merged enterprise would use any cost saving to provide the consumer with more or better content rather than improve its bottom line. For example, in discussing the Steiner paradigm, Yoo rightly observes that the monopoly owner would "capture more revenue than under competition"[55] and then curiously asserts that "[t]o the extent that quality correlates with program cost, monopoly provision should *cause* program quality to increase."[56] No reason is ever given for predicting this particular use of the new revenue. Even outside of monopoly, media owners typically have the capacity to spend more than they often do or are competitively required to spend on content. Doing so would often be desirable if more was spent on content that has positive externalities (or that avoids negative externalities). Both sociological and structural factors justify the prediction that dispersed rather than

---

53. *See supra* note 15.
54. *But see supra* Part I.D.1.
55. Of course, the point that the monopolist could "capture more revenue" has long been accepted in the academic literature and, I suspect, is understood by media firms seeking to merge. Yoo actually said, however, that "each station" would capture more revenue, which does not follow from Steiner's model. Yoo, *supra* note 2, at 695. Rather, both the combination, whose more diverse commodity offerings picks up new viewers, and the station appealing to the dominant taste capture more; the individual station(s) serving marginal tastes presumably do not.
56. *Id.* at 695 (emphasis added). *See also id.* at 699–701.

concentrated ownership would empower decisionmakers more likely to do so.[57]

4. True Efficiencies

Business enterprises and some economists instinctively interpret any cost saving, for example, a cost saving available due to a merger, as an efficiency. This is simply wrong from the perspective of total social welfare and the typical economic meaning of efficiency. If a merged entity can provide and sell news with fewer journalists, but if each journalist engaged in news gathering produces positive externalities, lay-offs would produce cost-saving for the firm but also reduce the number of journalists producing positive externalities. The real possibility, indeed the likely consequence, is that the lay-offs would reduce costs and increase profits but be inefficient.

The FCC, and Yoo by reference, relied on empirical evidence that purportedly showed that local media crossownership generally does not have undesirable effects on news quality.[58] Think a moment about what situation one would expect to find in examining crossownership of a newspaper and broadcaster. To maximally realize possible synergies, a newspaper would predictably own the broadcaster that specialized most in news, a station that provided better than average, probably the best, news coverage in the locale. Not surprisingly, empirical studies bear this out. But what does this show? That crossownership benefits the public? Not at all.

---

57. This prediction receives strong support from the Project on Excellence in Journalism ("PEJ") report. *See* PROJECT FOR EXCELLENCE IN JOURNALISM, DOES OWNERSHIP MATTER IN LOCAL TELEVISION NEWS: A FIVE-YEAR STUDY OF OWNERSHIP AND QUALITY (2003), *at* http://www.journalism.org/resources/research/reports/ownership/Ownership2.pdf

58. *See, e.g.*, 2003 Biennial Review Order, *supra* note 31, at 13,754–55 ¶¶ 343–345, 13,761 ¶ 358 (discussing THOMAS C. SPAVINS, LORETTA DENISON, SCOTT ROBERTS & JANE FRENETTE, THE MEASUREMENT OF LOCAL TELEVISION NEWS AND PUBLIC AFFAIRS PROGRAMS (FCC, Media Ownership Working Group Study No. 7, 2002), *at* http://hraunfoss.fcc.gov/edocs_public/attachmatch/DOC-226838A12.pdf; and PROJECT FOR EXCELLENCE IN JOURNALISM, *supra* note 57). Although the PEJ study examined only six cases of crossowned stations, which provides a reason not to find its results in respect to crossownership to be statistically significant, most ownership categories that it considered contained at least fifty sample stations, making the FCC decision to ignore the study's primary results questionable. Thus, the FCC briefly mentioned but made no use of the PEJ's much better-supported finding that small ownership groups (three or fewer stations) as compared to larger ownership groups, and network affiliated stations as compared to network owned and operated stations, produce "higher quality newscasts." *Compare* 2003 Biennial Review Order, *supra* note 31, at 13,840–42 ¶¶ 572–577, *with* PROJECT FOR EXCELLENCE IN JOURNALISM, *supra* note 57, at 1. This contradicts Yoo's conclusion that national ownership caps amount to censorship that reduces the quality of speech. Yoo, *supra* note 2, at 713.

A thorough evaluation requires consideration of two other questions. First, did the ownership "cause" better broadcast news or did it merely profit from synergies while owning the best local television news station, or did it do both? The best empirical evidence would probably examine whether the crossowned station provides better quality news than the finest news broadcaster in an equivalent market. If not, the crossownership may have merely merged control of the two dominant providers of local news, dangerously concentrating communicative power without increasing in any way the quality of local news. Unfortunately, the studies cited by the FCC do not provide evidence relevant to or otherwise evaluate this issue. Second, what about the effect of the merger on the newspaper? Newspapers are arguably the most significant source of information placed within the local public sphere.[59] If so, the newspaper's journalistic performance is arguably the most relevant issue in relation to crossownership, and this is ignored by the studies. Elsewhere, however, some more ethnographic empirical investigations suggest that the greater workload placed on journalists who are required to perform for both mediums is having a negative effect.[60] That is, the cost savings of crossownership may be real, but whether these should be seen as efficiencies, especially efficiencies that promote provision of a community's information or journalism needs, seems doubtful.

5. Vertical Concentration

Yoo identifies a broad category of regulations of vertical integrations that includes not only limitations on cable operators' freedom to program all their channels with their own or their affiliated companies' content,[61]

---

59. Sources that people rely on for news are not nearly as significant as the source of correct and relevant information that people actually assimilate. As a source of accurate and assimilated news, newspapers may be much more important than commercial television. *See, e.g.*, Steven Kull, Clay Ramsay & Evan Lewis, *Misperceptions, the Media, and the Iraq War*, 118 POL. SCI. Q. 569 (2003). Moreover, a newspaper can be a media source of information even for a person who does not read it— if, for example, news knowledge comes from conversations and the conversation partners report information learned from the paper.

60. *See* Eric Klinenberg, *Convergence: News Production in a Digital Age*, ANNALS AM. ACAD. POL. & SOC. SCI. (forthcoming 2005).

61. Yoo notes that the D.C. Circuit struck down this "architecture" on First Amendment grounds. Time Warner Entm't Co. v. FCC, 240 F.3d 1126, 1137–39 (D.C. Cir. 2001). Given his apparent approval of the decision, *see, e.g.*, Yoo, *supra* note 2, at 704, it is not surprising that he did not discuss its truly remarkable nature. The Court repeatedly invoked the First Amendment "test" from *Turner Broadcasting System, Inc. v. FCC*, 512 U.S. 622 (1994) [hereinafter "*Turner I*"], but completely ignored the decision's actual holding and substantive dicta. The majority found must-carry rules possibly valid (actual validity had to await *Turner Broadcasting System, Inc. v. FCC*, 520 U.S. 180 (1997) ("*Turner II*")). Essentially, mandated carriage of stations other than those of the cable operator

but also rules capping national ownership of broadcast stations or of cable systems.[62] In his discussion, he argues that a careful economic analysis shows that the "integrations" examined do not harm competition.

Put aside here are more specific challenges to Yoo's economic analysis.[63] Another point is more telling. Yoo places all these rules into the category of vertical integration rather than horizontal because "[p]roperly evaluated, horizontal restrictions bar mergers among direct competitors who would otherwise be serving the same customers."[64] This claim is crucial. He purportedly can then demonstrate that, as a form of vertical integration, the restricted mergers are unlikely to create sufficient economic power to restrict market competition and, therefore, the "existing regulations . . . serve to prevent industry participants from realizing the available efficiencies, which in turn reduces total quantity, quality, and diversity of speech . . . [thus] represent[ing] still another form of architectural censorship."[65] Viewing these concerns as vertical integration however, already implies a commodity focus. If the policy concern is power over price, or even over content conveniently available to an individual consumer, Yoo is obviously right that sellers of content in

---

was the issue in both *Turner I* and *Time Warner*. Thus, the *Turner I* majority should have had little trouble finding that the same interests in diversity and power in the marketplace of ideas justify Time Warner ceding similar control over a portion of its channels to independent content providers. Even more interesting is Justice O'Connor, whose dissent offered cable operators greater First Amendment protection. After noting the "danger in having a single cable operator decide what millions of subscribers can or cannot watch," she suggested that "Congress might . . . conceivably obligate cable operators to act as common carriers for some of their channels, with those channels being open to all through some sort of lottery system or time-sharing arrangement." *Turner I*, 512 U.S. at 684 (O'Connor, J., dissenting). This is the precise scheme that the D.C. Circuit struck down. Thus, the D.C. Circuit relied most heavily on a Supreme Court decision where both the majority and dissent clearly supported the opposite of the Circuit's result.

62. In this category, Yoo also includes chain-broadcasting rules that attempted to limit networks' power over affiliates and must-carry rules that require cable carriage of local broadcasters.

63. Yoo, like the D.C. Circuit in *Time Warner*, 240 F.3d at 1126, finds that vertical integration creates problems only when there is concentrated power in the primary market (for example, the market for cable channels or programming). Yoo, *supra* note 2, at 707–711. Apparently on this basis, the Circuit rejected, as would Yoo, the FCC decision to limit a single company to owning no more than 30% of the country's cable systems—a requirement that guarantees the country will have at least four cable owners. The court could only see a competitive reason to require a 60% ownership limit, which would guarantee just two cable purchasers of cable programming nationwide. To emphasize his point, Yoo notes with approval the Justice Department's use of the Hirschman-Herfindahl Index ("HHI") that does not, at least as a rule of thumb, find a danger of vertical integration if the HHI is below 1800. *Id.* at 707. The Circuit's approved standard, however, which allows a 60% and 40% two-firm market, could produce a HHI of 5200. Even under the FCC repudiated rule, a four firm industry of 30%, 30%, 30%, and 10% produces an HHI of 2800.

64. Yoo, *supra* note 2, at 705.

65. *Id.* at 713.

Sacramento and Des Moines are not direct competitors.[66] If the concern, however, is power or influence over public opinion in a national public sphere, these two stations, and every other station in the country, compete directly to influence the opinions that become nationally dominant. That is, Yoo's construction of "direct competitors" already embodies a normatively inadequate commodity focus. This focus dramatically reduces and reshapes the values that once led the FCC to create stringent national caps on broadcast ownership and that today generate popular opposition to the expansion of national (or international) media conglomerates.

II. CENSORSHIP AND THE CONSTITUTION

Medievalists consider, with affection, Chaucer to be the C-word. In contrast, for First Amendment devotees, though with considerably less affection, the C-word is censorship. Yoo describes as "architectural censorship" those "unintended effects" of structural regulation that have an important "adverse impact on speech."[67] This characterization provides crucial support for the only doctrinal move that would realistically satisfy his "hope that the First Amendment would provide a basis for identifying and redressing architectural censorship when it arises."[68] A constitutional objection to "unintended byproduct[s]" must lie in their "effects."[69] Given the constitutional odiousness of censorship, finding censorship in effects provides the link that could motivate constitutional invalidation.

This Section first objects to Yoo's rhetorical move, and secondly argues that, if followed, his move suggests substantive conclusions against which I suspect even Yoo himself would rebel. Finally, this Section suggests that proper First Amendment theory, as well as existing black letter law, rejects Yoo's hoped-for doctrinal approach.[70]

Possibly because of its extraordinary negative connotations, the term "censorship" is most at home when referring to *conscious attempts to suppress* particular expressive content. Other uses are normally derivative, aiming to add force to some criticism.[71] Of course, First Amendment

---

66. *Id.* at 705.
67. *Id.* at 673–674.
68. *Id.* at 675.
69. *Id.* at 730.
70. Yoo recognizes that existing doctrine rejects his approach. *Id.* at 730–731.
71. I was guilty of seeking this rhetorical gain, though without trying to turn the claim into a constitutional argument, when I asserted that advertisers today are possibly the greatest censors in America. *See* C. EDWIN BAKER, ADVERTISING AND A DEMOCRATIC PRESS 3 (1994). Sometimes advertisers specifically seek to suppress content that negatively portrays their product, company, or

objections to government actions are possible without characterizing the problem as censorship. Consider, for example, challenges to content neutral regulations—so called time, place, and manner regulations—or objections to judicially ordered disclosures of a reporter's confidential sources. Censorship does not even provide the only basis for objection to content-oriented laws. For example, a possible inquiry is whether requiring drug manufacturers to include warning labels is constitutionally objectionable in the way that compelling a school child to salute the flag is,[72] although the only interesting insight comes from seeing why the two are not normatively or constitutionally analogous. Still, censorship seems to be neither the well-articulated objection to the compelled flag salute nor the rejected objection to warning labels.

Certainly, applying the term "censorship" to unintended consequences requires a considerable stretch. Two taxes, one media specific, the other general, could have the same adverse impact on media provision of expression without being equally identified as censorship. Characterizing Huey Long's 2% "tax on lying" as censorship would not mean that this characterization fits today's general sales taxes.[73] Not only is the term usually reserved for cases where the lawmaker is conscious of the effects, but also implicitly for where these adverse effects are desired. Any intelligent policymaker recognizes that almost any media architecture, as compared to the universe of possible structures, will affirmatively promote some and negatively affect other expressive opportunities.[74] For these reasons, Yoo's use of the term censorship to indicate rhetorical disapprobation for various structural or architectural rules is problematic.

Assume, for discussion purposes, the propriety of Yoo's usage, and that censorship can refer to architecture that has serious adverse consequences for speech even where these consequences are not affirmatively desired or intended. What must be recognized is that the

---

political agenda—conduct that might narrowly be viewed as censorious. I and others, however, have also described as censorship the media's provision of content aimed at attracting an advertiser-desired audience with full knowledge of, but no particular desire for, the serious adverse consequences for the creation and availability of media content desired by other people. *See* William B. Blankenburg, *Newspaper Ownership and Control of Circulation to Increase Profits*, 59 JOURNALISM Q. 390 (1982).

72. *Compare* Va. State Bd. of Pharmacy v. Va. Citizens Consumer Council, 425 U.S. 748, 771 n.24 (1976), *with* W. Va. State Bd. of Educ. v. Barnette, 319 U.S. 624 (1943).

73. *See* Grosjean v. Am. Press, 297 U.S. 233, 250 (1936) (invalidating Louisiana's gross receipts tax on newspapers and magazines).

74. *Cf.* Duncan Kennedy & Frank Michelman, *Are Property and Contract Efficient?*, 8 HOFSTRA L. REV. 711 (1980) (showing that neither state of nature, collective ownership, nor a private property regime can be shown to be Pareto superior to any of the others, but that each has characteristically different Pareto optimal solutions).

market is a form of architecture. Not only is it architecture, it can have severe adverse effects on the quality and diversity of mass media communications. In Yoo's terminology, the market is censorship.[75] Precisely this observation has been central to a Western European view that a predominantly public broadcast architecture is needed in order to prevent this serious censorship.[76]

Certainly, market architecture's potential for adverse consequences for diversity, quality, and some groups' speech opportunities has been central to many disputes about media policy. During the late 1920s and early 1930s, corporate interests pushing a market-based, commercial broadcasting system and labor, religious, educational, and public interest groups favoring a partially or completely noncommercial system were fully aware that "architectural" choices have substantial effects on content. Their disagreement was primarily over evaluation—although they may have also differed somewhat in empirical predictions.[77] Congress realized that either licensing spectrum to noncommercial interests, licensing largely to commercial users, creating private property rights in spectrum, or some other regulatory scheme such as managed common carriage would promote valuable speech opportunities in the face of the tragic chaos of the commons. It also realized that adoption of any scheme would have important adverse consequences for some speech content and speakers in comparison to how they would fare under an alternative regime. The Supreme Court, however, seemed little troubled by Congress's authority to choose a preferred architecture.[78] It also observed that, once having made a choice, Congress could, but was not required to, impose further architectural or affirmative behavioral requirements on favored spectrum users on behalf of the speech interests of those disadvantaged by the chosen architecture.[79] The Court properly found that the choice of which speech to architecturally advantage and which to disadvantage is normally a matter of democratic normative choice about the type of communications order, that is, the type of content, to favor.[80]

---

75. Implicitly, but famously, Justice Hugo Black made this point in Associated Press v. United States, 326 U.S. 1, 20 (1945).
76. *See* HUMPHREYS, *supra* note 46, at 119.
77. *See generally* ROBERT W. MCCHESNEY, TELECOMMUNICATIONS, MASS MEDIA, AND DEMOCRACY: THE BATTLE FOR CONTROL OF U.S. BROADCASTING, 1928–1935 (1993) (describing the political battles around this choice).
78. Red Lion Broad. v. FCC, 395 U.S. 367 (1969).
79. *Id. See also* CBS v. Democratic Nat'l Comm., 412 U.S. 94 (1973).
80. *Turner Broadcasting, supra* note 30.

In contrast, as he makes clear in his preference for an activist judicial review of "architectural censorship," Yoo presumably must recommend that any reliance on market architecture be subject to strict constitutional review.[81] Interestingly, the German Constitutional Court accepts Yoo's activist ambitions, although it accepts neither his normative nor economic views. It has held that the German constitution mandates provision of the services of public broadcasting; moreover, the German Court found that a mixed system that includes private broadcasting is constitutionally permissible only if private broadcasters are subject to considerable regulation, including regulation aimed at promoting diversity.[82]

Yoo's approach could also lead to examination of other architectural features. In *Miami Herald Publishing Co. v. Tornillo*,[83] the Court glorified the editor's freedom to decide on content.[84] From the perspective of Yoo's article, it may be questionable whether an architecture that allows an owner to fire, or otherwise control, an editor for editorial choices is

---

81. Yoo offers the state action doctrine as the reason why market structures that censor speech, in his sense of censor, are not problematic, while structures chosen by government that have similar censorious effects "might be [constitutionally] problematic." Yoo, *supra* note 2, at 715 n.207. His article's primarily enterprise is to use economic analysis to identify media structures that have adverse consequences and then to propose changes in constitutional doctrine to invalidate objectionable structures. Thus, it would seem evenhanded and consistent to seek changes in "state action doctrine," just as he proposes changes in "effects doctrine"—unless his real concern was not media performance but protection of corporate power. Clearly the more evenhanded move would be logically easy—and actually, probably more desirable and less dangerous than giving courts the role of monitoring effects. As for logic, the government creates the ownership, contractual, property, corporate, and other rights that determine the form of any market structure. If constitutional law should invalidate various architectures chosen by government (as opposed to, for example, focusing on interferences with the media's content choices), the consistent position should be to evaluate all government created and maintained market structures. Nothing in the abstract logic of state action doctrine prevents this. In fact, as noted below, the Court in an unusual context invalidated on First Amendment grounds the government's attempt to leave content choices to a market entity. Individual Justices—and other countries—have also found reliance on media market structures to create constitutional problems in the media context. *See infra* text accompanying notes 82–89. Of course, my main point is not to advocate this activist result—although, if consistent, Yoo should—but to argue that the Court had it right when it treated media structural choices as not raising serious constitutional problems as long as those choices can be reasonably interpreted as attempting to promote a democratic communications order.

82. *See* Cable Penny Case, BVerfGE 90, 60 (1994); North Rhine-Westphalia Broad. Case, BVerfGE 83, 238 (1991); Third Broad. Case, BVerfGE 57, 295 (1981); HUMPHREYS, *supra* note 46, at 137–38. These cases are translated into English in 2 DECISIONS OF THE BUNDESVERFASSUNGSGERICHT, FEDERAL CONSTITUTIONAL COURT, FEDERAL REPUBLIC OF GERMANY, PARTS I & II: FREEDOM OF SPEECH 199–219, 493–534, 587–619 (1998).

83. Miami Herald Publ'g Co. v. Tornillo, 418 U.S. 241 (1974).

84. *Id.* at 258 (holding that the "statute [at issue] failed to clear the barriers of the First Amendment because of its intrusion into the functions of editors . . . . [and that] [t]he choice of material to go into a newspaper . . . constitute[s] the exercise of editorial control and judgment").

unconstitutional because it "intrud[es] into the functions of editors."[85] Again, some western democracies provide editors some legal protection—often described as providing "internal freedom"—precisely from improper interference by publishers, that is, owners.[86]

Likewise, either market or ideological incentives can lead broadcasters to deny access to those who would pay to present views on controversial public matters, as opposed to advertising their products. Two dissenting Justices generally very sensitive to First Amendment freedoms, Justices Brennan and Marshall, concluded that the government violated the First Amendment by failing to mandate that commercial broadcasters open up their advertising time to paid discussions of controversial issues. These Justices concluded that free market architecture here was unconstitutional, in part because of its severe adverse effects on the diversity of, and opportunities for, expression.[87] Although better explained as a finding that government had engaged in an indirect but content-based attempt to rid cable of indecency, the Court similarly found a peculiar use of free market architecture to be unconstitutional in *Denver Area Educational Telecommunications Consortium, Inc. v. FCC*.[88] The Court invalidated a congressional attempt to recognize the "market-based" authority of cable systems to determine whether certain public, educational, or governmental ("PEG")[89] channels may broadcast indecent material.

Despite objections to the market's important adverse consequences, in this country free market architecture is presumptively not unconstitutional. Still, this result may obtain not because Americans are unable to see that the market is merely one among many possible architectures. Rather, it may and should reflect judicial acceptance of broad legislative choice over communication systems' architecture. Most congressional structural choices that restrict an unregulated media market are also routinely upheld. Before *Turner Broadcasting System, Inc. v. FCC* ("*Turner I*"),[90] the "reasonableness" standard articulated in *FCC v. National Citizens*

---

85. *Id.* at 258.
86. *See* DANIEL C. HALLIN & PAOLO MANCINI, COMPARING MEDIA SYSTEMS: THREE MODELS OF MEDIA AND POLITICS 40 n.4, 175 (2004); HUMPHREYS, *supra* note 46, at 108–09.
87. CBS v. Democratic Nat'l Comm., 412 U.S. 94 (1973) (Brennan, J., and Marshall, J., dissenting).
88. Denver Area Educ. Telecomms. Consortium, Inc. v. FCC, 518 U.S. 727 (1996).
89. *Id.* at 732.
90. *Turner I*, 512 U.S. 622 (1994).

Committee for Broadcasting[91] (*"NCCB"*) arguably represented the courts' treatment of structural regulation in all communications media, not just in the case's "official" domain of broadcasting.[92] It allowed, without specific empirical support, virtually any structural regulation that could be understood as reasonable and not an attempt at speech suppression, even though the regulation might seriously and obviously disadvantage some speech or speakers.

In contrast, in order to restrict what he purports to be censorship in his four case studies, Yoo argues for more stringent review. He believes that *Turner I* could have helped considerably if interpreted to require strong empirical evidence of the necessity for and benefits of regulation.[93] But given the apparent failure of this approach to eventuate, Yoo's activist conclusion is that only an "effects" analysis is likely to achieve the desired results[94]—although he clearly recognizes that the Court is unlikely to adopt such an approach.[95]

This Comment abstains from presenting a general critique of the necessary instrumentalism of any serious constitutional scrutiny of structural regulations (an instrumentalism not inherent in constitutional objections to facial "censorship"[96]). Instead the Comment notes two difficulties with Yoo's proposed activism, both unintentionally illustrated

---

91. FCC v. Nat'l Citizens Comm. for Broad., 436 U.S. 775 (1978) (holding that regulations are valid if "based on permissible public-interest goals . . . so long as [they] . . . are not an unreasonable means for seeking to achieve these goals").

92. *See generally Turner Broadcasting, supra* note 30. As Yoo observes, *NCCB* self-consciously limited not just broadcasters but ownership opportunities of entities with "'the fullest First Amendment protection—newspapers.'" Yoo, *supra* note 2, at 722 (quoting 2003 Biennial Review Order, *supra* note 31, at 13,793 ¶ 441). Once the Court in *Turner I* made clear that it viewed *Miami Herald Publishing Co. v. Tornillo*, 418 U.S. 241 (1974), as representing an objection to penalizing content choices and not as an objection to a structural rule limiting editorial control, *see Turner I*, 512 U.S. at 653–56, and once *Denver Area Education Telecommunications Consortium, Inc. v. FCC*, 518 U.S. at 727, is seen as invalidating a complex design intended to promote content-based suppression of indecency, only *Los Angeles v. Preferred Communications, Inc.*, 476 U.S. 488 (1986), remains as an example of the Supreme Court identifying a structural (architectural) regulation it might invalidate. Even there, invalidation would occur, if at all, only because the government improperly authorized a monopoly without any justification in physical or economic needs.

93. Yoo, *supra* note 2, at 727–728.

94. *Id.* at 730.

95. *Id.* at 731.

96. *See* Simon & Schuster v. N.Y. State Crime Victims Bd., 502 U.S. 105, 124–28 (1991) (Kennedy, J., concurring) (holding that once a general content-based restriction on protected speech is found, the inquiry should end rather than continue to the issue of justification). *See also* Hustler Magazine v. Falwell, 485 U.S. 46 (1988); Brandenburg v. Ohio, 395 U.S. 444 (1969); N.Y. Times v. Sullivan, 376 U.S. 254 (1964). *See generally* C. Edwin Baker, *Harm, Liberty, and Free Speech*, 70 S. CAL. L. REV. 979 (1997).

by his own analyses, and then observes a final and more basic problem with Yoo's suggestion. First, his hoped-for review, like many instrumentalist scrutinies, slides too easily into improperly finding constitutional violations by first misidentifying a law's legitimate purpose and then showing the law unnecessary, maybe even counterproductive, in achieving that end. This occurred in Yoo's analysis of both horizontal and assertedly vertical ownership restrictions (and arguably is common to many proposed constitutional critiques of non-media structural regulations that limit corporate moneymaking activities). Second, potentially informative economic analyses often mislead because the analyst unjustifiably ignores a law or policy's most important positive effects (especially noncommodified effects). Here too, Yoo's article provides an illustration. By negative lesson, the article should serve as counsel against authorizing rigorous constitutional review of structural or architectural choices.

The final problem is more fundamental. Even without disagreement about the likely overt, immediate consequences, people differ over the significance and even valence of these consequences and, hence, over the value of the structural and distributional rules that produce them. Nowhere is this more true than in the communications sphere. Different respectable democratic theories suggest different, only somewhat overlapping, visions of an ideal communications order. Different structural regulations—including different degrees of reliance on unregulated markets—advance different conceptions of press freedom that are themselves related to different conceptions of democracy.[97] Empirical evidence assessable by courts will not resolve which visions of democracy, and hence which communications orders, are best. The democratic theory that I find most persuasive—which I call "complex democracy"[98] and Jürgen Habermas, in rejecting both republican and liberal theories as adequate in themselves, calls "discourse theory of democracy"[99]—suggests an even greater difficulty for constitutional review. Given real but somewhat conflicting requirements of ideal democratic public discourse, which of many opposing media policies will better serve society's communication needs cannot be determined on the basis of abstract—read "constitutional"—principles. Rather, the best answer will reflect which of several valuable but often competing communicative discourse forms currently needs

---

97. BAKER, *supra* note 14, at 125–53, 193–213.
98. *Id.* at 143–47.
99. JÜRGEN HABERMAS, BETWEEN FACTS AND NORMS: CONTRIBUTIONS TO A DISCOURSE THEORY OF LAW AND DEMOCRACY 166–67, 283–86, 296–302 (1996); Jürgen Habermas, *Three Normative Models of Democracy*, 1 CONSTELLATIONS 1, 4 (1994).

nurturing most. If this claim is right, identifying proper goals of media policy not only will be normatively controversial, it will also be very finely contextually variable. Does society need more common media[100] or more media pluralism? All major democratic theories largely agree that traditional censorship is objectionable. In contrast, the value of varying architectures (and their intended and unintended consequences) is properly subject to contextual debate, a debate that is short-circuited by Yoo's broader use of the censorship label. Commentators, of course, act appropriately in identifying unintended consequences—although they should be modest about any claim that these consequences were really unpredicted. Their effort, however, should aim at informing political debate, not supporting judicial invalidation.

* * *

At the country's beginning, Congress provided a huge subsidy for newspaper delivery. Many newspapers were mailed under rules that provided for free postage, while the general rate for sending a newspaper 500 miles was 1.5¢ as compared to 25¢ for a letter. The result was that letter writers, primarily businessmen dealing with commercial matters, were disadvantaged (censored?) while subsidizing the delivery of newspapers. Nevertheless, support for postal subsidies "transcended national-local, Federalist-Republican cleavages . . . . [since] [a]ll who wanted the fragile union to survive saw the urgency of ensuring a widespread flow of public information."[101] Still, debate about the subsidies' precise content was sharp. All sides saw that the architecture held differential advantages for various contents and different political, cultural, and sectional agendas. Some significant postage charge and distance-sensitive postage would help protect rural papers from being overwhelmed by competition from city papers; a single low rate or, even better, free postage would advantage urban papers and, purportedly, lead to outlying districts being better informed. The initial result in 1792 can be seen as a compromise—free postage for exchange papers (that, though traditional, possibly most benefited the local country papers) plus two distance zones combined with very low rates for newspapers in both zones, as compared to nine zones and higher rates for letters. In fact, these architectural disputes continued. The adamant opposition of Jacksonian forces, for example, defeated free postage for papers in the Senate by only

---

100. *Cf.* CASS SUNSTEIN, REPUBLIC.COM (2001).
101. RICHARD B. KIELBOWICZ, NEWS IN THE MAIL: THE PRESS, POST OFFICE, AND PUBLIC INFORMATION, 1700–1860s 38 (1989). This paragraph is based on *id.* at 31–38, 58–64, 82–86. The material and its implications are summarized in *Turner Broadcasting, supra* note 30, at 98–99, 105–08.

one vote. They clearly feared that free postage would lead to urban papers utterly destroying their local papers, maybe even local culture—"annihilat[ing] at least one-half of [their] village newspapers."[102] By contrast, in 1845 both sides could see benefit in establishing free local delivery for papers as compared to the continued (although still greatly subsidized) charges for delivery of distant papers.

So what is to be learned from this postal history? That Yoo is right in seeing that architectural consequences are real. But who could have thought otherwise? Certainly not those who choose policies that sometimes rely on and sometimes diverge from the market. Those consequences are the point. What else do we learn? Clearly, without these and many other architectural or structural interventions we would not have arrived at the media system that has often served the public and the country well.[103] Without congressional structural interventions, even with the negative censorious effects they had on some publications, the newspaper industry may have never played its crucial political role at the country's beginning. I doubt that the Constitution provides the resources to determine which form of postal architecture is best. I am confident that, whether or not ideal, the country was better off, and possibly its survival achieved, because of political structural choices not being ruled impermissible by the courts. Despite the possibility of severe negative consequences for disfavored competing publications, censorship is the wrong term to apply to, and constitutional law the wrong mechanism to judge, these choices that offered crucial support to the project of creating a nation.

---

102. KIELBOWICZ, *supra* note 101, at 61 (citation omitted) (quoting Senator Isaac Hill, a former publisher of a "zealous Jackson organ").

103. *See generally* PAUL STARR, THE CREATION OF THE MEDIA: POLITICAL ORIGINS OF MODERN COMMUNICATIONS (2004).

[11]

Control over Technical Bottlenecks – A Case for Media Ownership Law?[1]

Professor Thomas Gibbons
School of Law, University of Manchester

During the debates about the implications of media convergence during the last decade, the bottleneck problem has been consistently and widely recognised as a significant issue for regulatory attention. Convergence has forced a closer analysis of the various components of the value chain for producing media content and it has become apparent that anybody who controls entry to, or movement along, the chain will exert considerable power. Understandably, the exercise of such power has been regarded as analogous to control over media ownership. It has seemed to follow, therefore, that, as sector-specific ownership rules become increasingly inappropriate for many media markets, their place may be taken by measures directed at bottleneck problems. The recent implementation of the EC's Access Directive appears to reflect that approach to regulating control in the new media, with its combination of competition law and conditional access rules. However, in this paper, I will argue that the Directive is not a sufficient means for regulating concentrations of power in new and converged media. The kinds of bottleneck issues that it regulates have been allowed to distract attention from the policy requirements that underlie media ownership regulation. Although the form of regulation may be becoming outmoded, those policy requirements are still relevant to the new media and it will be necessary, therefore, to find new regulatory devices for implementing them. It may be that the Directive provides a model for doing that, but it is inadequate in its present form.

Media Pluralism

At the outset, it is worth reminding ourselves that the rationale for regulating media ownership and control is the protection of media pluralism.[2] Although media pluralism has different shades of meaning, it embraces three basic concerns. First, there should be a diversity of content, namely, the availability of a range of views, opinions and subject matter. Secondly, there should be a diversity of sources, namely, a variety of programme or information producers, editors or owners. Thirdly, there should be a diversity of outlets for delivering material directly to the audience.

Amongst the differing conceptions of media pluralism, it is diversity of content that constitutes the strongest version. However, neither diversity of source nor diversity of outlet can be guaranteed to produce diversity of content.[3] The most direct way of regulating for diversity of content is found in

---

1) An earlier version of this paper was presented to the Observatory/IViR Workshop on "Vertical Limits - New Challenges for Media Regulation", Amsterdam, 27th September 2003. I am grateful for the helpful comments made by the participants and especially by Natali Helberger. Any errors or omissions are my own responsibility.
2) For a fuller discussion of these issues, see: T. Gibbons, "Concentrations of Ownership and Control in a Converging Media Industry" in C. Marsden & S. Verhulst (eds.), Convergence in European Digital TV Regulation (1999), Blackstone Press, pp. 155-173. This paper develops a number of points made in that chapter. See also: T. Gibbons, Regulating the Media (1998, 2nd ed.), Sweet & Maxwell.
3) For discussion of media pluralism, see: R. Craufurd Smith, Broadcasting Law and Fundamental Rights (1997) Oxford: Oxford University Press, chap. 7; T. Congdon et al., The Cross Media Revolution: Ownership and Control (1995) London: John Libbey; R. Collins & C. Murroni, New Media, New Policies (1996) London: Polity Press, chap. 3.

public service broadcasting obligations that require a range of material to be provided, generally with news and controversial issues being presented impartially. But this approach may need to be supplemented with a diversity of other viewpoints from other sources, since public service broadcasting may well not represent all perspectives.

Regulation of media ownership has therefore attempted to provide a diversity of sources of content, thereby intending to limit any power exercised by the domination of only a few voices, by imposing structural control over the number of firms in a sector. However, the relationship between ownership control and media pluralism is not straightforward[4] and the presence of many players in a market does not *guarantee* a diversity of output. All that ownership control can do is to provide a set of conditions that are most favourable towards pluralism and thereby seek to minimise the adverse effects of media concentration.

Until relatively recently, the regulatory approach to achieving that outcome was to focus on concentration at the point of content (programming) outlets. There were a limited number of such outlets, partly due to licensing restrictions for allocating broadcasting spectrum and partly due to prioritising public service broadcasting policies. In addition, the outlets were characterised by a high degree of vertical integration whereby the ownership effect was experienced at all levels in the value chain, from production to delivery to the consumer. Since the originators of content were so closely linked to the mode of distribution, regulation was, therefore, sector-specific.

In a converged media industry, the segmenting of media markets and the creation of a producer-consumer nexus are both made possible by the use of common, non-sector-specific delivery platforms (the outlets). In theory, it does not matter who owns the platform or outlet, provided that the end-users are able to obtain a diversity of content. The platforms, as modes of outlet, will not be capable of maintaining a premium over their rivals because each will be capable providing the same content, so that the economic focus will shift and increasing emphasis will be laid on the production of content and the audience's access to it. In the light of such changes, it is generally recognised that it will no longer make sense, to regulate against concentrations of control over content by reference to programming sectors because, structurally, the different sectors will not be able to be kept separate. A possible response to this development is to treat conditional access to outlets, together with the creation of content or acquisition of rights, as the most important factors for regulation. In this context, the possibility that firms will seek to integrate vertically, with the same organisation controlling all aspects of content production, packaging and distribution to the end-user, may be considered a particular danger to media pluralism.[5] It may seem to follow, therefore, that the remedy in the new media will be to ensure that competition law operates effectively and that conditional access regulation is adequate to deal with the bottleneck problems that arise. The prior question, however, is whether those bottleneck problems are the real difficulty.

The Use of Competition Law

Under European Community law, Article 81 of the European Treaty prohibits, as incompatible with the common market, all agreements between undertakings that may affect trade between Member States and have as their object or effect the prevention, restriction or distortion of trade within the common market. Article 81 is aimed at collusion between undertakings that restricts competition, and agreements that infringe the prohibition are void. However, provision is made for notice of such agreements to be given to the European Commission, which may grant exemptions, individually or by group, provided that the relevant agreements contribute to the improvement of production or distribution of goods, or promote technical or economic progress. Article 82 prohibits any abuse by one or more undertakings of a dominant position within the common market, or in a substantial part of it, as incompatible with the common market in so far as it may affect trade between Member States. Article 82 is more concerned with the extent of market power enjoyed by a firm and the freedom that that gives from the constraints of market pressures. The prohibition is not directed against the existence or acquisition of market power as such, but against its use by a firm to affect adversely those with whom it deals.

---

4) See T. Gibbons, "Concentrations of Ownership and Control in a Converging Media Industry", n.1, above.
5) Such concerns were recently expressed by the Council of Europe: Report Prepared by the AP-MD, Media Diversity in Europe (2002) H/APMD(2003)001.

Both provisions have been used to test the competitiveness of strategic alliances or vertical arrangements between suppliers and distributors in the media industry.[6] A recent example is that of Telenor/Canal+/Canal Digital,[7] which involved an agreement about distribution of pay-TV premium content channels in the Nordic region through Canal Digital's satellite platform. It is interesting in the context of bottlenecks because it demonstrates how the provisions could operate in relation to a set of agreements that were designed to preserve the effect of undoing a previous vertical structural link. Telenor was a subsidiary of the Norwegian telecommunications operator and provided satellite and cable transmissions and distribution. It was jointly running Canal Digital with Groupe Canal+, the film and TV division of the Vivendi Universal Group. Groupe Canal+ also produced pay-TV channels and distributed bouquets of pay-TV channels through its subsidiary, Canal+ Nordic. When the parent company, Groupe Canal+, decided to transfer its shareholding in Canal Digital to Telenor, a series of agreements were reached concerning the distribution of Canal+ Nordic's pay-TV channels via Canal Digital and concerning the supply of pay-per-view (PPV) and near-video-on-demand (NVOD) channels by Canal+ Nordic to Canal Digital. There were provisions allowing pay-TV channels exclusivity to Canal+ Nordic for ten years and committing Telenor not to launch a similar competing channel for ten years. In addition, Canal+ Nordic was given an exclusive supply agreement in relation to supply of PPV and NVOD in the Nordic region for five years. In the light of the European Commission's objections that these arrangements would breach Article 81, the parties revised the arrangements so as to reduce the lengths of the agreements (to four years in relation to pay-TV exclusivity, and to three years in relation to the channel non-compete for PPV and NVOD), and to define the scope of the agreements more narrowly. The case demonstrates that vertical integration in itself is not so important as detecting anti-competitive arrangements. But it is also clear that the Commission's intervention was intended to enhance competition in the markets for supply of programming and any effect on viewer would be indirect, flowing from the general assumption that competition will improve consumer welfare.

In the United Kingdom context, a similar kind of analysis was recently applied by the competition regulator, the Office of Fair Trading (the OFT), to BSkyB's operations in the pay-TV market.[8] BSkyB is vertically integrated in this market by virtue of its channel production operations (possessing exclusive sports rights, especially for Premier League football, and having exclusive contracts with Hollywood studios to show films on its premier film channels), together with its distribution operations (bundling the channels and distributing them on the only satellite platform in the UK). Certain of BSkyB's competitors had complained that it was abusing its dominant position in the wholesale supply of its premium content in three ways. First, it was alleged that BSkyB exercised a "margin squeeze" in relation to its premium channels. It could do this by wholesaling the channels to other distributors at a price that allowed an insufficient margin for them to make a profit, even if they were as efficient as the vertically integrated BSkyB operation. The implication would be that, in such a situation, the vertically integration would be subsidising the distribution of premium channels, thereby distorting the market. The OFT concluded that an analysis of the market and BSkyB's behaviour could only lead to a borderline finding. Although the distribution operation had turned in a small loss, BSkyB's roll-out costs were very high. So, on balance, it had not abused its dominant position. The second complaint was that the practice of "mixed bundling" was an abuse of dominant position. This might happen if a competitor was foreclosed from supplying premium channels, because it was not worth doing since BSkyB had offered the channel at a discount to subscribers of its other premium channels. In this case, the OFT could not find sufficient evidence of the practice. Significantly, one of the reasons for this was that BSkyB had exclusive contractual rights to premium content, the inference being that rival suppliers did not have anything to supply in any event! The third allegation was that BSkyB had abused its dominant position by offering discounts to distributors in favour of BSkyB's own channels. Again, there was no evidence that this was happening.

While these and other similar cases show that competition principles can be applied to the media, it is not clear that they show how competition law can promote media pluralism. Removing barriers to entry and preventing foreclosure certainly provide the opportunity for diversity of viewpoint to surface. However, the main benefit is to the players in the markets for supplying content. The impact

---

6) Earlier examples include: *Re Astra* [1993] OJ L 20/23. [on appeal, Case T-22/93, British Telecommunications v European Commission]; *Centre Belge d'études de marché – Télémarketing (CBEM)* v *CLT and Information publicité Benelux* [1985] ECR 3261; Auditel IV/32031 [1993] OJ L 306/50. For subsequent, ancillary proceedings, see [1995] ECR II-0239; Re German TV Films: The Community v Degeto Film GmbH [1990] 4 CMLR 841; Magill TV Guide/ITP, BBC and RTE [1989] 4 CMLR 745; upheld by the CFI in RTE, BBC and RTE v European Commission [1991] 4 CMLR 586, 669 & 745; Re European Broadcasting Union [1987] 1 CMLR 390; Re the Application of the European Broadcasting Union [1991] 4 CMLR 228. See generally, T. Gibbons, "Concentrations of Ownership and Control in a Converging Media Industry", n.1, above.
7) COMP/C2/38.287 (2003/C 149/10)
8) Office of Fair Trading, *BSkyB:* The Outcome of the OFT's Competition Act Investigation (2002) OFT 623.

on the recipients of content is marginal and of incidental interest. As far as the wholesalers are concerned, one viewer is interchangeable with another; their objective is to reach an aggregate audience that delivers an acceptable turnover.

A competition measure that appears more directly relevant to concentrations of media control is merger regulation. The European Community's Merger Regulation[9] applies to all concentrations with a Community dimension, that is where the combined aggregate worldwide turnover of all the undertakings concerned is more than ECU 5,000 million and where the aggregate Community-wide turnover of at least two of the undertakings is more than ECU 250 million (unless each of the undertakings concerned achieves more than two-thirds of its aggregate Community-wide turnover in one Member State, in which case national jurisdiction applies). Proposed mergers must be notified to the Commission and a concentration that creates or strengthens a dominant position, as a result of which effective competition would be significantly impeded in the common market or in a substantial part of it, must be declared incompatible.

A number of high-profile merger cases in the mid-1990s demonstrated that, notwithstanding the infancy of various new media markets and the high start-up costs of developing new technologically intensive enterprises, the Commission was still prepared to intervene to prevent ventures that created dominant positions.[10] But, consistently, in the absence of such dominance, mergers were not opposed.[11] More recent cases provide examples of similar reasoning, but they also tend to demonstrate its irrelevance to media pluralism. In Microsoft/Liberty Media/Telewest,[12] Microsoft (a software company with telecommunications and multimedia links) and Liberty Media (a subsidiary of AT&T) proposed to acquire joint control of Telewest (a UK broadband company that provides digital television, telephony and high speed Internet access). The European Commission was concerned that the acquisition would have an adverse effect on competition in the market for digital set-top boxes because it could strengthen Telewest's existing dominant position as an exclusive supplier of cable services within its franchise area. In the event, notice of the merger was withdrawn.

In Kirch/BskyB/Kirch Pay-TV,[13] Kirch and BSkyB proposed to take joint control over Kirch Pay-TV. BSkyB's activities were outlined above, and Kirch was an audio-visual media firm with interests in sports, film and television rights, television production, pay-TV, and technical services for digital broadcasting and encryption. The European Commission was concerned that Kirch Pay-TV would strengthen its dominant position in the German and Austrian market because it would have access to BSkyB's financial resources and marketing and distribution know-how; this would raise the already high barriers to entry. The Commission was also concerned about the impact of the joint venture on the burgeoning market for digital interactive television services. Here, Kirch Pay-TV would be likely to enter the market ahead of its rivals and therefore consolidate its own set-top box as the standard decoder for such interactive services. In response to these objections, the parties offered an undertaking to allow other digital interactive television operators to manufacture their own technical platform and to have access to Kirch Pay-TV's conditional access system. They also undertook to increase interoperability by allowing other operators to run their services more easily on Kirch Pay-TV's platform. The Commission accepted these as lowering barriers to entry and therefore did not oppose the merger.

There are many more cases from the media sector[14] but a good instance of the considerations involved is the Newscorp/Telepiù decision in April 2003.[15] Newscorp intended to acquire control of Telepiù from the Vivendi group and merge it with its own pay-TV platform, Stream. Although the merger would create a near-monopoly in the Italian pay-TV market, it seemed clear that Stream would

---

9) Council Regulation (EEC) No 4064/89 of 21 December 1989 on the control of concentrations between undertakings (1989 OJ L 395, 1990 OJ L 257) with amendments introduced by Council Regulation (EC) No 1310/97 of 30 June 1997 (1997 OJ L 180, p.1). The basic approach will not be altered by the package of reforms that was adopted by the European Commission in December 2002 and is anticipated to be introduced during 2004.
10) See generally, T. Gibbons, "Concentrations of Ownership and Control in a Converging Media Industry", n.1, above. Major cases were: MSG Media Services [1994] IV/M.469, OJ L 364/1; the Nordic Satellite Distribution case: [1995] IV/M.490, OJ L 053/20; RTL-Veronica-Endemol [1995] IV/M.553, OJ L 134/32; [1996] OJ L 294/14.
11) Kirch-Richemont- Telepiu [1994] IV/M.410, OJ C 225/3; Bertelsmann-CLT [1996] IV/M.779, OJ C 364; Bertelsmann-News International-Vox [1994] IV/M.489, OJ C 274/9.
12) JV.27, 22 March 2000.
13) JV.37, 21 March 2000.
14) Vivendi/Canal+/Seagram, M.2050, 13 October 2000; Kabel Nordrhein Westfalen, JV.46, 20 June 2000; Kabel Baden-Württemberg, JV.50, 2 August 2000.
15) This is reported in V. Baccaro, "The Commission closes probe into pay-TV industry in Italy approving Newscorp/Telepiù merger deal" (2003) Competition Policy Newsletter 8-11. Another example is Sogecable/Canalsatelite Digital/Via Digital, M2845, 16 August 2002.

otherwise be closed down by Newscorp. But Newscorp would have become the gatekeeper of the only satellite platform and its conditional access system would be adopted; both situations would create barriers to entry to the market for pay-TV services. In addition, because Stream and Telepiù would together hold exclusive rights to premium films and major Italian football coverage, their potential rivals would be foreclosed from obtaining so called "driver" content for pay-TV. To deal with these obstacles, the parties offered undertakings to provide third party access to the premium content, the technical platform and the conditional access system, until 2011. The Commission accepted these, giving responsibility for their supervision to the Italian competition regulator.

This case illustrates that, typically, consumers' interests in general, let alone their interest in media pluralism, do not feature strongly but an interesting exception is the case involving the proposed Vizzavi Internet portal venture.[16] Here, the idea was to provide a multi-access portal to provide a set of web-based services across a range of platforms such as fixed and mobile telephones, PCs and television sets. In the light of the Commission's concern that the development of alternative Internet portals might be impeded, the parties gave undertakings to allow the end-users the ability to change the default portal if they so wished. The Commission considered this to be significant for allowing users to choose their content provider independently of their access provider. Nevertheless, in a media pluralism context, it is clear that such a choice would not necessarily achieve diversity of sources or of outlets.

More generally, the ability of competition law to protect media pluralism remains open to doubt. To summarise points that I have made elsewhere,[17] competition law does not recognise a firm's dominance, in itself, to be a problem, so it allows large firms to exist provided the market is contestable. However, the impact of such firms on diversity of content is not taken into account. In addition, any contribution by competition law to media pluralism is assumed to result from a supposed relationship between diversity of source and diversity of content for the audience.[18] Most importantly, competition law does not analyse market effects from the perspective of the end-user's experiences, whereas the main objective underlying pluralism is that individuals are provided with a range of media content at the point of use. For that reason, it is a matter of concern that individuals may obtain knowledge from a range of outlets, whether newspapers, radio or television, that are controlled in their locality by one supplier, whether or not that supplier is dominant in the broader market.

Conditional Access and Access-related Regulation

In the 1999 Communications Review,[19] it was accepted that the nurturing of emerging digital television markets required special measures to supplement competition regulation. Furthermore, the interests of the end-users were given specific recognition. However, Recital 10 of the Access Directive[20] is over-ambitious in implying that provisions of the Directive are a response to the recognition that "Competition rules alone may not be sufficient to ensure cultural diversity and media pluralism in the area of digital television". Certainly an aim of conditional access regulation is to "make sure that a wide variety of programming and services is available" and it may well be considered "necessary to ensure accessibility for end-users to specified digital broadcasting services". But the focus of the Access Directive, and indeed the other Communications Directives, is on communications networks, associated facilities and services rather than content, and it appears to be simply assumed that making material available is sufficient to deliver media pluralism.

The Access Directive in general deals with access relationships between providers of networks, associated facilities and services in wholesale markets (Article 1, paragraph 2, states that access in this context does not include access by end-users). It seeks to secure access and interconnection within such markets by providing for specific, *ex ante*, obligations to be imposed on undertakings with significant market power. Two provisions of the Directive, however, authorise the imposition of conditions regardless of significant market power. Under Article 5, national regulatory authorities are given discretion to impose obligations to ensure end-to-end connectivity, including interconnection

---

16) Vodafone/Vivendi/Canal+, JV 48, 20 July 2000; Sogecable/Via Digital, COMP/C2/37.652, 8 May 2003 and M2845, 16 August 2002.
17) In: T. Gibbons, "Concentrations of Ownership and Control in a Converging Media Industry", n.1, above.
18) The so-called Hotelling effect points in the opposite direction: see H. Hotelling, "Stability in Competition" (1929) 39 Economic Journal 41-52. See also, M. Cantor & J. Cantor, Prime Time Television: Content and Control (1992) London: Sage.
19) COM (1999) 657 final. See also the Communication from the Commission: ... Orientations for the New Regulatory Framework, COM (2000) 239 final.
20) 2002/19/EC of the European Parliament and of the Council of 7th March 2002.

of their networks, on undertakings that control access to end-users. The regulators are also given discretion to require operators to provide access to application program interfaces (APIs) and electronic programme guides (EPGs) on fair, reasonable and non-discriminatory terms, where that is necessary to secure access for end-users to digital radio and television broadcasting services. Under Article 6, a set of conditions dealing with technical conditional access systems must be applied to all operators. Essentially, there must be facilities for transcontrol of such systems and operators must offer access to them on fair, reasonable and non-discriminatory terms and maintain separate financial accounts of their conditional access activities.

It must be conceded that the provisions of Article 5 are likely to be helpful in working towards achieving media pluralism. In particular, the end-to-end network interconnection obligation, in Paragraph 1, has the potential to alleviate the problem of localised bottlenecks whereby end-users are unable or unwilling to subscribe to more than one distributor's outlet. Reflecting the thinking in the Vizzavi Internet portal case, mentioned above, it has the effect of allowing users to choose their content provider independently of their access provider. But if the content provided is insufficiently diverse, that will not advance media pluralism. However, Article 5's purpose is actually different (albeit none the worse for being so), namely to prevent the loss of consumer welfare that results from viewers being bound, economically, to one distributor's outlet and thereby experiencing constraints on the free flow of information.

Similarly, the provisions of Article 5, Paragraph 2 are not aimed at media pluralism, although that might be incidentally benefited. Access to APIs on fair, reasonable and non-discriminatory terms is intended to assist the development of digital interactive television services. Many of those are likely to deal with commercial transactions rather than information flows. While there is no doubt that an API could provide the bottleneck that concentrated a flow of information, removal of the bottleneck will not necessarily provide the diversity of material that the broader social policy requires. Similarly, whilst access to EPGs will enable a wider selection of material to become available to end-users, it is the "presentational" aspect of such facilities – implicitly outside the Directive under Article 6, Paragraph 4 – that are likely to have a greater impact on control of content.

Article 6 is also limited in its potential for enhancing media pluralism. Again, the point is that the regulation of conditional access systems may be necessary but it is not sufficient for this purpose. The application of the Directive in the UK will serve to illustrate the point.[21] The regulator (OFTEL, acting on behalf of OFCOM, under s.408 of the Communications Act 2003) has emphasised that conditional access conditions must be imposed on a particular person (company) under the Directive. Furthermore, they apply to operators, "whose access services broadcasters depend on to reach any group of potential viewers or listeners ...". Applied in the UK context, most broadcasters on the digital satellite platform reach their viewers by being included in a BSkyB pay-TV package, so they do not need conditional access services. Similarly, broadcasters on cable systems also reach the viewers via the packaging arrangements made by the cable operators; they do not need any technical conditional access system to do it. Similarly, free-to-air digital television (Freeview) does not use conditional access to reach its viewers. In principle, therefore, none of these digital outlets appear liable to conditional access conditions being imposed. However, in the case of digital satellite, some broadcasters reach their viewers without being part of a BSkyB package but, to do so, they must use the conditional access service provided on digital satellite, a system provided by Sky Subscriber Services Ltd (SSSL, a BSkyB controlled firm). Since the broadcasters "depend" on this service, SSSL falls within the scope of the Directive, and OFTEL have accordingly imposed conditions on it.[22]

Here, OFTEL are not interpreting the Access Directive narrowly, because Article 6 is precisely concerned with problems arising from dependence on proprietary set-top boxes. But, equally, it is clear that the Article is limited in its potential to enhance media pluralism. It cannot exert any effect on the contractual bundling arrangements made by the programme packagers, because they are not obliged to offer access to such bundles on fair, reasonable and non-discriminatory terms. That kind of access is presumed to be capable of being regulated by competition law, without such an *ex ante* condition being applied.

---

21) Communications Act 2003, ss.45, 73, 75 and 76.
22) OFTEL, The Regulation of Conditional Access: Setting of Regulatory Conditions (2003) 24th July.

Is Vertical Integration the Problem for Media Pluralism?[23]

The main impact of competition law and conditional access regulation is on the market for the supply of programming content. But this does not guarantee that end-users actually experience a pluralistic choice of material. Indeed it is easy to become sidetracked by the need to deal with vertical integration in the new media, both generally and through its more technical manifestations, without appreciating that that has not been the major policy objective in media ownership regulation. In the early broadcasting media, vertical integration was endemic: the typical company would produce the programmes and then deliver them across the transmission network that it owned. Concerns about competition led to modifications of the basic model, for example, requirements for independent production quotas, on the one hand, and divestment of transmission networks on the other. But those measures did not alter the fundamental locus of control, which was the broadcaster-as-editor. For that reason, in the United Kingdom for example, the attention of ownership regulation was directed to the licensee who was responsible for the content that ultimately reached the viewer or listener. Ownership regulation (as opposed to competition law) never moved further up the value chain to scrutinise the bottlenecks that did exist in the way that content packages were assembled. Various deals relating to production, use of library material and copyright, were not transparent. Nor, from a media pluralism perspective, did they need to be, because it was the net outcome that determined the range of content available.

This suggests that the drift towards vertical integration in the new media does not raise new issues for media pluralism. The more traditional (composite) firms had similar characteristics to the merged and joint ventures of the new media. If there was a case for regulating the former, there is a case for regulating the latter. Similarly, the well-aired older debates, about structural ownership regulation versus general competition law, will apply equally to the new vertically integrated entities. However, this does not mean that the advent of the new media has made no difference to configuring media regulation. Rather, the nature of new media practice, with the segmentation of different parts of the value chain, has the effect of highlighting the critical points of control over media content. Ownership regulation was crude by comparison, because there was no need to analyse the reason why diversity at the point of output was compromised; it was sufficient to know that, somehow, the result of the organisation's activities was to concentrate the diet of material available.

The transparency of the value chain in the new media may actually enhance some opportunities for regulating for media pluralism. One, possibly counter-intuitively, beneficial effect is that some access to the media will become commercialised and take advantage of bottleneck regulation. In a single vertically integrated media firm, any claim for access to the media (gaining access to the programming schedule) has to be framed in terms of political theory, centred on freedom of speech. Such claims are difficult to substantiate. Partly, this is because it may not be obvious why one person's speech should take priority in using scarce media resources compared to another's; partly, it is because the media firm's owners will assert their own free speech right to publish what they wish; again, partly, it is because those owners will claim a property right to use their resources as they wish. Generally, therefore, editorial selection tends to take priority over individual access claims. However, to the extent that separate acts of speech are reflected in independent commercial entities, competition law and conditional access rules can assist in enabling a greater diversity of expression to reach the audience. But the benefits must not be overstated, since the limitations are very obvious. If there are no anti-competitive effects or the abuse of a dominant position,[24] no regulator will force a company to do a deal with another that wishes to gain access to a schedule. Ironically, it may be that the more effective bottlenecks provide the best opportunities for mandating access!

Solutions?

Generally, what the transparency of the new media reveals is that the most critical bottleneck for the purposes of promoting media pluralism is the bundling of a programming package. It is this activity, analogous to the editorial decision, which determines what content reaches the audience. In the terms used by ownership regulation, in order to promote pluralism, we need to locate the entity that "controls" the decision making that is significant for diversity of content and, ultimately, for diversity of outlet. Clearly, regulating ownership of the platform for delivery of material will be less significant. So what alternative techniques may be adopted?

---

23) For an argument that, in any event, vertical integration does not impose a problem in economic terms, and therefore should not be regulated, see: C.S. Yoo, "Vertical integration and media regulation in the new economy" (2002) Yale Journal of Regulation 171-300.
24) Including circumstances where there is an "essential" facility providing a bottleneck.

One approach is to regulate content directly to secure a measure of pluralism. Although this may not be appropriate for all kinds of programming, it remains one of the objectives of public service broadcasting. Even then, some diversity in public service programming is needed to stimulate creativity: content regulation alone cannot generate the creativity and originality that a diversity of sources can facilitate. Whatever the diversity of public service broadcasting, however, "must-carry" obligations are essential to preserve its prominence in the new media.[25]

Another approach to promoting pluralism is to focus on the delivery to the audience and ensure that a choice of material is available. My criticism, above, of the potential for access measures to promote pluralism does not mean that such choice should not be encouraged. Access measures effectively entail the imposition of some kind of common carrier obligations on operators controlling access to end-users. This should mean that the end-users can sample the whole range of material that those operators control. However, diversity of content depends on what has been produced by others and access provisions cannot determine that.

The most focused approach is to target programme production and supply, in order to guarantee diversity of output. One way of doing that could be to concentrate on behavioural measures under competition law and another way could be to impose some structural constraints. Taking competition measures first, although it has been suggested, above, that they are unlikely to protect pluralism in themselves, they may be able to contribute broadly to a freer flow of content. In the United Kingdom, for example, the broadcasting regulator has acted to facilitate greater consumer choice in the pay-TV market, by imposing restrictions on premium channel bundling by distributors. The imposition of minimum carriage requirements by programme suppliers is prohibited, the aim being to minimise anti-competitive effects of "buy-through" arrangements that required premium channels to be purchased only in conjunction with (typically large) packages of other channels. Buy-through itself has not been prohibited, because small, basic packages are permitted, but premium channels must be made available on an à la carte basis from those basic packages.[26] Of course, exactly which channels are made available will depend on commercial arrangements made upstream.

If general competition principles are insufficient, it may be that some modifications to the regulatory criteria could help to promote pluralism. One such modification might be to use a presumptive audience share threshold, of the kind found in media ownership rules, to trigger a market investigation.[27] However, in the new media, with so many combinations of programme rights and distribution arrangements, that may prove too complex. A different modification might be to incorporate a public interest test into the standard competition regime. For example, in the United Kingdom, the Communications Act 2003 has amended the main merger regime under the Enterprise Act 2002 to allow special tests to be applied in relation to media mergers. For mergers involving newspapers, the need for accurate presentation of news and free expression of opinion, together with a sufficient plurality of views in each newspaper market, must be taken into account alongside questions about substantial lessening of competition. For broadcasting mergers, the need to secure a plurality of control over the media serving each audience, the need to secure a wide range of quality broadcasting and the need for broadcasting controllers to be genuinely committed to programme standards, must be taken into account alongside questions about substantial lessening of competition. In addition, where one of the parties to a merger is a newspaper or broadcaster that controls 25% or more of the relevant national market or a substantial part of it, the same criteria may be applied independently of any competition test.[28] This modified regime has the potential to provide significant safeguards for media pluralism. However, in the United Kingdom, it applies only to proposed mergers, which are identifiable events, and where significant thresholds exist. Even if the criteria were to be applied to a general competition regime, and that may not be easy in practice, their application would depend on *ex post* behavioural analysis by the competition regulator, and the concern is that media pluralism could be compromised before the regulator was able and willing to act.

Another way of dealing with the issue, therefore, may be to adopt some form of structural constraint on programme supply. This would be analogous to media ownership regulation but applied in the

---

25) In the United Kingdom, must-carry provisions are implemented by s.64 of the Communications Act 2003. Those are complemented by "must-offer" requirements imposed on the public service broadcasters under s.272.
26) Independent Television Commission, Guidance on ITC's Bundling Remedies (1998) 26th June. The ITC's measure was upheld on judicial review: see T. Gibbons, R v Independent Television Commission ex parte Flextech plc (1999) 4 Communications Law 70-71.
27) See T. Gibbons, "Concentrations of Ownership and Control in a Converging Media Industry", n.1, above.
28) Communications Act 2003, ss.373-389, in particular amending and supplementing the Enterprise Act 2002, ss.58, 58A, 59, 59A, 61 and 61A.

context of new digital media, where sector-specific measures are no longer appropriate.²⁹ But, consistent with the aims of media ownership regulation, the main focus would have to be on the editorial decision-maker, in effect, the programme bundler (including multiplexer). The objective would be to ensure that each member of the audience had access to programming from a minimum number of independent bundlers.

Here, there may be a role for a degree of vertical disintegration to be imposed by structural regulation. For example, the basic effect of Article 6 of the Access Directive is to separate the technical platform from content services, and that will enable a diversity of content to be made available through any conditional access service. Similarly, in the United Kingdom, digital terrestrial television was established in a vertically disintegrated form when a new economic entity, the multiplex, was created to package and deliver programming.³⁰ But the rationale was less the need for diversity than the desire to foster greater competition for services. Again, the independent productions quota in the "Television without Frontiers" Directive was raised from 10% to 25% in the United Kingdom,³¹ not so much to enhance pluralism as such but to improve the market setting for independent producers. In all these cases, a potentially more pluralistic environment can be formed through measures of vertical disintegration but it is unlikely to be a sufficient response.

If a structural method is considered desirable, it has to focus on the audience's experience of pluralism and seek to prevent major media players from controlling an unacceptable share of the audience's exposure to their content. Exactly what counts as an acceptable share is a political decision and cannot be subject to an economic formula. However, there seems to be widespread European consensus that there should be at least three players in each market, if pluralism is to be protected. I have suggested already that the method used in the Access Directive cannot achieve this. However, there may be some scope for adopting the basic technique and applying it, not to wholesale markets, but to content received by end-users. Article 5, in particular contains concepts that could be adapted in this way: the necessity to ensure end-to-end connectivity and accessibility.

What I envisage is that, where end-users in any local market encounter fewer than three separate sources of programming material, the distributor should be required to secure sufficient interconnection facilities to make a greater diversity of sources available. Such interconnection and access would, of course, have to be provided on fair, reasonable and non-discriminatory terms.³² It might also be supplemented by regulation of EPGs content, to ensure that sufficient prominence is given, not only to public service broadcasters, but also to a diversity of alternative sources.³³

Whether such a structural approach is necessary, however, depends on the nature of the media markets. It may not be so in the United Kingdom at present, because diversity of content is an important driver for commercial operators, so that duplication of content provision is unlikely to be resisted. On the contrary, "must-offer" provisions were written into the recent new legislation to enable such operators to secure public service broadcast content. However, in the absence of a general willingness to supply each other's content (on commercial terms), it is not obvious that consumers would find it easy to obtain a diversity of material because, until technical advances can provide true convergence, it is unlikely to be easy for them to migrate across the three main platforms of digital terrestrial television, cable and satellite. My conclusion is that, whilst it may well be sufficient to rely on a combination of competition regulation and the *ex ante* conditions authorised by the Access Directive, for the present, the need for what amounts to a surrogate form of media ownership must be anticipated for the future. The kind of interconnection requirement that I have suggested would not guarantee pluralism of content but it would impose diversity of source (bundling) and protect the end-user from concentrations of power over content distribution. It seems plain that the way that the markets are developing at present could pose dangers of unacceptable concentration before long. Arguably, unless constraints are put in place now, it may be too late to reverse the trend.

---

29) Partly as a response to anticipated convergence across communications and media, the United Kingdom's Communications Act 2003 abolishes most of the previous restrictions on ownership, including those relating to foreign ownership and accumulations of interest. Two of the more significant of the few that are retained include a cross-ownership rule preventing newspaper proprietors who control 20% of national or local markets from controlling Channel 3, and a rule requiring local radio to have at least two commercial operators in each market in addition to the BBC. See generally, Schedule 14 to the Act.
30) Under the Broadcasting Act 1996, Part I.
31) Council Directive (89/552/EEC) of 3 October 1989 and Directive 97/36/EC of the European Parliament and of the Council; Broadcasting Act 1990, ss.16(2)(h) and 186.
32) It is appreciated that these terms have yet to be given clear meaning.
33) For the UK's general approach, see: Independent Television Commission, ITC Code of Conduct on Electronic Programme Guides (1997).

Part IV
Issues in Regulating New Media

[12]

The Regulation of Interactive Television in the United States and the European Union

Hernan Galperin* and François Bar**

I.	INTRODUCTION	61
II.	THE THREE GENERATIONS OF BROADCASTING	65
III.	WHAT IS ITV?	68
IV.	POLICY CONCERNS RAISED BY ITV	70
	A. Transmission System	70
	B. Return Path	73
	C. Home Terminal	75
V.	THE CASES: THE AOL/TIME WARNER MERGER AND BRITISH INTERACTIVE BROADCASTING	78
	A. The AOL/Time Warner Merger	78
	B. The British Interactive Broadcasting Case	80
VI.	CONCLUSION	83

I. INTRODUCTION

The broadcasting industry is rapidly entering the era of digitization, distributed intelligence, and interactivity. Despite lingering standardization issues, digital transmission is replacing analog transmission in the three

---

* Ph.D., Stanford University, 2000; Assistant Professor at the Annenberg School for Communication, University of Southern California. An earlier version of this article was presented at the 29th Telecommunication Policy Research Conference, Alexandria, Virginia, Oct. 27-29, 2001.
** Ph.D., University of California—Berkeley, 1990; Senior Research Fellow, Berkeley Roundtable on the International Economy (BRIE), University of California—Berkeley; Assistant Professor of Communication at Stanford University.

major delivery platforms (terrestrial, cable, and Direct Broadcast Satellite ["DBS"]). Programmable user terminals built upon personal computer hardware and software technology are replacing "dumb" analog television sets. More importantly, after several failed attempts, interactive television ("ITV") services are finally poised for large-scale deployment. This transition opens many exciting opportunities for businesses and users, ranging from television-based electronic commerce (known as "t-commerce") to interactive educational programming. These new applications will evolve as broadcasters, software vendors, equipment makers, and users experiment with novel ways to enhance and perhaps transform the television experience altogether.

These changes, however, have also raised several questions about who will shape the architecture of these emerging broadcasting networks, and hence determine business models, communication patterns, and the dynamics of technological innovation for the next generation of television. Will programmers or network operators alone decide which interactive services will be made available to users? Will electronic marketplaces develop as open transactional spaces or "walled gardens"?[1] Will users be able to connect new terminal equipment to the network and experiment with new network uses such as peer-to-peer applications? As they address these questions, policymakers face several pressing concerns. Who will create incentives for firms to invest in this infant marketplace and at the same time protect competition in services and applications, foster decentralized innovation, and secure users' access to a wide range of information and transaction services? Would ex ante regulation squelch the success of a sector that, after many failed attempts, now appears ready for prime time? What regulatory principles and tools should be used to confront the questions raised by ITV?

Far from hypothetical, these questions have already surfaced in several high-profile cases, in particular the merger of America Online ("AOL") and Time Warner, which combined the world's largest Internet Service Provider ("ISP") and early entrant in the ITV market with the United States's second-largest cable operator and major worldwide programmer. In reviewing the merger, the Federal Trade Commission ("FTC") and the Federal Communications Commission ("FCC") found that the combination of distribution facilities, service operations, and content

---

1. "Walled garden" refers to a network architecture which prevents users from accessing content or services provided by parties unaffiliated with the network operator. *See* Nondiscrimination in the Distribution of Interactive TV Services Over Cable, Comments of Consumers Union, Consumer Federation of America and The Center for Media Education, CS Dkt. No. 01-7, at 5-6 (Mar. 19, 2001), *at* http://www.cme.org/access/press/Itvfin.pdf.

held by AOL/Time Warner raised competition concerns in three markets: broadband Internet access service, broadband Internet transport service, and ITV. While regulators imposed several merger conditions relating to broadband Internet access and transport services, those relating to ITV were, in comparison, rather minor.

In this paper we analyze the development of ITV in the United States and Western Europe and the policy debates that have accompanied it. We argue that despite the nascent character of the market, there are important regulatory issues at stake that will determine the future architecture of this new information distribution platform. In most local markets, cable operators function as monopolies. There is evidence that even in markets where competition exists, it does not significantly affect cable operators or the rates they charge.[2]

Absent rules that provide for non-discriminatory access to network components and a degree of standardization for terminal equipment, these platform operators will have strong incentives to leverage their ownership of delivery infrastructure into market power over ITV services and content. While in the short term, integration between platform operator, service provider, and terminal vendor is likely to facilitate the introduction of services, the lasting result could be a collection of fragmented "walled gardens" offering only the content and applications approved by the infrastructure incumbent. If ITV develops under such a model, the exciting opportunities for broad-based innovation and widespread access to multiple information, entertainment, and educational services in the next generation of television may never materialize.

We recognize that given the incipient nature of the market, particularly in the United States, it would be premature for regulators to attempt to implement detailed industry-wide rules for ITV platforms and services. There is simply too much uncertainty about which services users will want and at what price, how the technology will evolve, and what business models will emerge. It could be argued that platform owners will have incentives to open their networks in order to stimulate the proliferation of content and services upon which they could levy a distribution fee.[3] The dynamics of market competition would then stimulate a migration from proprietary technologies and "walled garden" business

---

2. *See generally* Implementation of Section 3 of the Cable TV Consumer Prot. and Competition Act of 1992, *Statistical Report*, 12 F.C.C.R. 22756, 10 Comm. Reg. (P & F) 977 (1997), *available at* http://hraunfoss.fcc.gov/edocs_public/attachmatch/FCC-02-07A1.pdf (last visited Oct. 15, 2002).

3. *See generally* James B. Speta, *Handicapping the Race for the Last Mile?: A Critique of Open Access Rules for Broadband Platforms*, 17 YALE J. ON REG. 39 (2000).

models to open standards and interconnected networks, thus making regulatory safeguards less necessary. Experience to date with the development of ITV, however, is not encouraging in this regard. We contend that it is not too early to establish general rules and first principles against which market developments can be monitored. ITV provides another instance where digital convergence calls for adaptation of existing broadcasting and telecommunications policies to balance industry development with the economic and social benefits associated with open network access. The debate over broadband cable Internet offered a first approach to the problem and some important lessons.[4] While technologies may vary from case to case, ultimate policy goals should not.

The case of ITV offers an opportunity to investigate how desirable policy goals—among them competition, broad-based innovation, and widespread access to information "from diverse and antagonistic sources"[5]—should be implemented in the post-convergence environment. In this Article we first review the evolution of the broadcasting industry through three successive models: the traditional "Fordist"[6] television model, the current multichannel television model, and the emerging ITV model. Second, we characterize the basic components of ITV and explore the concerns raised by the evolution of multichannel video programming distributors ("MVPDs") into ITV platform operators. Our conclusion is that dominant MVPDs are likely to have the ability and the incentive to leverage control over the transmission infrastructure into the ITV applications environment, engineering market outcomes in favor of affiliated programmers, electronic retailers, and ITV service providers. We note that, in contrast to the case of broadband cable Internet, the policy concerns go beyond infrastructure access control and include in particular the use of proprietary terminal equipment technology. Third, we review how regulators in the United States and the European Union ("EU") have so far responded to these concerns by contrasting two prominent cases: the AOL/Time Warner merger and British Interactive Broadcasting joint venture. We conclude that the wait-and-see approach taken by American regulators risks tolerating the deployment of a network architecture that

---

4. *See generally* François Bar et al., *Access and Innovation Policy for the Third-Generation Internet*, 24 TELECOMM. POL'Y 489 (2000) (arguing that the success of the Internet in the United States fundamentally rests on FCC policies aimed at maintaining network openness by making key network components available to all on cost-oriented terms).

5. Associated Press v. United States, 326 U.S. 1, 20 (1945).

6. The term "Fordism" is used to describe the particular system of production and consumption that emerged during the earlier part of the twentieth century. *See* MICHAEL J. PIORE & CHARLES F. SABEL, THE SECOND INDUSTRIAL DIVIDE (1984).

could restrict competition in ITV services, hamper innovation, and leave second-class digital economy citizens with access to a limited array of entertainment, transaction, and educational services. We also note that the imposition of limited open-access requirements in the United Kingdom ("UK") market has hardly hampered investments in ITV. Finally, we outline a general framework for regulatory thinking about open network access that reflects the convergence of communications industry sectors and the need to integrate seemingly conflicting policy goals.

II. THE THREE GENERATIONS OF BROADCASTING

The broadcasting industry has developed through three technological generations—each characterized by different types of services, business models, control strategies, and regulatory environments—as shown in Table 1.[7] It is interesting to note that each new generation has not thoroughly replaced the pre-existing industry structure, but rather added a layer of complexity to it. From the start of commercial broadcasting in the post-war period to about the mid-1970s, television consisted essentially of one-way terrestrial broadcasting of a limited number of channels that each aggregated and sold large audiences to advertisers. Their operators were protected by rules that restricted competition both within the industry and from new entrants. The regulatory model was based on the idea that broadcasters (both public and private) are trustees of a public resource (the radio spectrum) and thus under obligation to serve the public interest as defined by the government. While government protection from competition ensured the profitability of most broadcasting operations, fulfillment of public interest obligations was, at best, questionable.[8]

During the 1970s, a series of technological and regulatory developments created the conditions for the rapid growth of cable, and later DBS These new platforms essentially offered more of the same service: one-way delivery of branded packages of television programming. A new business model emerged, however, based on the collection of payments directly from subscribers, spawning the growth of specialized channels with a limited audience base.[9] The regulatory model was fashioned as a mix of traditional broadcasting and utility regulation. Cable operators were for the most part granted monopolistic franchises by local authorities in return

---

7. *See infra* p. 67. *See* Eli M. Noam, *Towards the Third Revolution of Television* (Dec. 1, 1995) (symposium presentation, *available at* http://www.columbia.edu/dlc/wp/citi/citinoam18.html).

8. ROBERT BRITT HORWITZ, THE IRONY OF REGULATORY REFORM: THE DEREGULATION OF AMERICAN TELECOMMUNICATIONS (1989).

9. BRUCE M. OWEN, THE INTERNET CHALLENGE TO TELEVISION 3 (1999).

for payments and limited access obligations (the so-called PEG, or Public, Education and Government channels, and leased access channels). The federal government later imposed restrictions on cable operators' editorial control by limiting the number of channels that can be occupied by affiliated video programmers.[10] Notwithstanding these obligations, cable essentially developed as a closed network with tight integration between network layers (transmission infrastructure, service provision, and terminal equipment).

The regulatory model for the second generation of broadcasting thus evolved in remarkable contrast with that of the telecommunications network, particularly after the FCC, starting in the late 1960s, progressively forced open the monopoly phone network by encouraging open attachment of terminal devices,[11] network interconnection,[12] and third-party access to unbundled network elements.[13] As a result of the implementation of these rules, over the last three decades the telecommunications industry has experienced a period of unprecedented innovation based on experimentation by network users and third-party service providers. In comparison, innovation and the introduction of new services in the cable industry has been limited, reflecting only the resources and the narrow economic incentives of those in control of the transmission infrastructure.[14]

---

10. Cable Television Consumer Protection and Competition Act of 1992, Pub. L. No. 102-385, § 11, 106 Stat. 1460 (1992).

11. *See generally* Use of the Carterfone Device in Message Toll Tel. Serv., *Decision*, 13 F.C.C.2d 420, 13 Rad. Reg.2d (P & F) 597 (1968) (establishing that the Carterfone terminal equipment created no harm to the AT&T network).

12. *See generally* Establishment of Policies and Procedures for Consideration of Application to Provide Specialized Common Carrier Servs. in the Domestic Public Point-to-Point Microwave Radio Serv. and Proposed Amendments to Parts 21, 43, and 61 of the Comm'n's Rules, *First Report and Order*, 29 F.C.C.2d 870, 22 Rad. Reg.2d (P & F) 1501 (1971) (ruling that AT&T would have to make its local telephone exchanges available to new entrants under reasonable terms).

13. *See generally* Regulatory Policies Concerning Resale and Shared Use of Common Carrier Servs. and Facilities, *Report and Order*, 60 F.C.C.2d 261, 38 Rad. Reg.2d (P & F) 141 (1976) (requiring AT&T to make the Bell System network and services available to its competitors on an unbundled basis).

14. In fact, there have been notable failures in the introduction of new services by cable operators, such as Qube, an interactive cable service launched by Warner Amex Cable and available in the early 1980s, and more recently Full Service Network, announced by Time Warner in the early 1990s. *How Blind Alleys Led Old Media to New*, N.Y. TIMES, Jan. 16, 2000, at C1; *see also* Press Release, VR1 Entertainment, VR1 Signs Agreement with Time Warner Cable's Full Service Network (Feb. 26, 1996) *available at* http://www.vr1.com/press_releases/full_pr/1996/022696_fsn.html.

Table 1. The Three Generations of Broadcasting

	1st Generation: Fordist Television	2nd Generation: Multichannel Television	3rd Generation: ITV
Service	One-way broadcasting of few video channels	One-way broadcasting of multiple video channels	Two-way delivery of multiple video channels and other services
Business Model	Mass advertising and/or license fees	Mass advertising, license fees, and subscriptions	Targeted advertising, subscriptions, and transaction fees
Control Strategies	Property rights over spectrum license	Integration of distribution and content assets	Access control and proprietary standards
Regulatory Model	Public trustee (incumbent protection)	Mix of public trustee and limited utility regulation	* Yet to be defined

After much delay, the revolution in digital processing and transmission of information is finally ushering the broadcasting industry into a new era. As a long-time industry analyst described it, "[a]fter a half-century of glacial creep, television technology has begun to change at the same dizzying pace as the wares of Silicon Valley."[15] But as MVPDs evolve from distributors of video programming into operators of a network that supports a variety of information services, regulators are confronted with a fundamental policy question: Under what regulatory model should the next generation of television services develop? Should we use the model of the second generation of broadcasting—even though cable and satellite operators may effectively act as providers of telecommunications infrastructure rather than as content aggregators and distributors—or that of open network access that has guided much telecommunications policy over the last decades? This question was an important thread of the debate

---

15. OWEN, *supra* note 9, at 3.

leading to the Telecommunications Act of 1996,[16] and continues to run through current policy discussions. So far, however, television and telecommunications continue to be regulated under separate regimes.

III. WHAT IS ITV?

Due to the infancy of the market, any description of what constitutes ITV is necessarily a working definition. In the context of this article, we follow the definition proposed by the British broadcasting regulator, Independent Television Commission ("ITC"): ITV services are "pull" services initiated by the subscriber to a MVPD that are not necessarily related to any specific video programming.[17] This definition allows for an understanding of ITV that goes beyond a simple extension of current television. In fact, we differentiate between two types of ITV services: program-related and dedicated. The first type of service constitutes relatively straightforward extensions of current television, while the second type of service offers modes of interaction that are fundamentally new.

Program-related services refer to ITV services that are directly related to one or more video programming streams. They enhance and extend the broadcaster's core business. For example, these services allow users to obtain additional data related to the content (either programming or advertising) to select from a menu of video feeds, to play or bet along with a show or sports event, to interact with other viewers of the same program, or to initiate transactions of goods or services featured in the video programming. In this case, ITV enhancements (such as Advanced Television Enhancement Forum ["ATVEF"] "triggers")[18] are overlaid onto the Moving Pictures Expert Group ("MPEG") video programming stream. These enhancements, when selected, direct viewers to content stored either in the set-top box or on a remote server. In the latter case, the enhanced content is delivered either through the same video pipeline or through a separate transmission line (e.g., an Internet connection). Examples of services already available or in the deployment stage are: the delivery of on-demand financial information and stock quotes, along with a business news channel; enhanced television commercials that allow viewers to request more information about the product; enhanced educational

---

16. *See generally* PETER HUBER, LAW AND DISORDER IN CYBERSPACE (1997).

17. *See* Independent Television Commission, *Interactive Television: An ITC Public Consultation* (2000) (on file with Journal).

18. ATVEF is a cross-industry group formed by programmers, broadcasters, ITV service providers, hardware makers, and software developers intended to create standard protocols for the delivery of ITV enhancements. Advanced Television Enhancement Forum *at* http://www.atvef.com/ (last visited Sept. 30, 2002).

programming; and services that allow users to play or bet along with quiz shows, reality shows, and live sports events.

In the case of program-related services, the programmer or advertiser will typically contract with an ITV service provider for the creation of programming enhancements, storage of interactive content, and management of return channel data. Nevertheless, the agreement of the network operator is still needed to deliver the downstream program enhancements, to allow compatibility between ITV applications and operator-provided customer equipment (unless a stand-alone box is used, which is unlikely for reasons discussed below), and possibly to provide the high-speed return path needed for certain applications. The ability of programmers and ITV service providers to experiment with and to deploy services is, therefore, de facto dependent on access to both the transmission infrastructure and the home terminal functions. As we argue in the next Section, unless regulatory safeguards guarantee such access on nondiscriminatory terms, the network operators will, as in the past, control the terms of innovation for the next generation of broadcasting services.

By contrast, dedicated services are independent from any specific programming stream. Typically, these will be entertainment, information, and transaction services provided by electronic retailers on the basis of contracts with the MVPD, which essentially acts as a platform operator, offering third parties a "window" for t-commerce. Examples of these services already available or in the deployment stage are electronic programming guides ("EPGs"), video-on-demand, e-mail, games, gambling, and electronic banking. While some of these electronic retailers may already have Internet-based services, these typically need to be re-authored for the different systems used by television network operators (though, as discussed below, there are several standardization efforts under way). In contrast with the Internet-based services, however, these television platforms are strictly "walled-garden" environments: The network operator selects a limited number of electronic merchants that are made available to subscribers and typically charges an up-front fee for access control (e.g., authentication) and for billing services as well as a commission on sales.[19]

While the market for ITV is still maturing, the pace of development has accelerated dramatically in recent years. Growth has been fueled by decreasing equipment costs (of both network hardware and home

---

[19]. For example, BSkyB reportedly takes an eight percent commission on sales of one of its services. Emma Duncan, *At Least Television Works*, THE ECONOMIST, Oct. 7, 2000, at 27, 30.

terminals)[20] and related infrastructure investments that facilitate the provision of ITV services, in particular the slow but steady migration to digital transmission in terrestrial, cable, and satellite television.[21] Some have argued that such rapid growth makes a case against regulatory action. On the contrary, we argue that it is precisely in this formative stage of network development that rules are critical, much like policy intervention favoring open network access in telecommunications networks starting in the late 1960s that allowed the Internet revolution to unfold several years later. Important architectural features of the ITV network will be established and solidified in this early period. It is precisely during this formative period of the industry that policymakers have an opportunity to favor innovation and competition in the next generation of television services.

IV. POLICY CONCERNS RAISED BY ITV

Opportunities for dominant network operators to foreclose competition in the adjacent market for ITV services exist with three network components: the transmission system, the return path, and the home terminal (typically a digital set-top box). In this Section we examine these opportunities and discuss the incentives for discriminatory behavior by vertically integrated network operators. We argue that as a result of technical, economic, and regulatory factors, a dominant platform for the delivery of ITV services is likely to emerge in every geographic market (in the U.S. case, the local cable franchises). Unaffiliated ITV service providers and third-party programmers may consequently face discriminatory access to network components as the dominant platform operator would have incentives to favor its affiliated ITV service, thus reducing innovation and discouraging entry in this infant market.

A. Transmission System

In the case of program-related services, the most apparent opportunity that exists for network operators to discriminate in favor of affiliated ITV service providers and programmers consists of "stripping" the ITV enhancements from the video signal of an unaffiliated programmer, thus blocking access to the enhanced features offered by competitors. Time

---

20. For example, the cost of video servers, the core component of video-on-demand systems, has dropped ninety percent over the last ten years. The cost of digital set-top boxes has also dropped dramatically in recent years following the decline in prices for computer components. *See* Ken Kerschbaumer, *Interactive Television: Fulfilling the Promise*, BROAD. & CABLE, July 10, 2000, at 22, 23.

21. *See* Hernan Galperin, *Can the US Transition to Digital TV be Fixed?: Some Lessons from Two European Union Cases*, 26 TELECOMM. POL'Y 3 (2002).

Warner Cable, for example, has repeatedly blocked subscriber access to Guide Plus+, a free EPG offered by ITV provider Gemstar that is carried over the Vertical Blanking Interval ("VBI").[22] By stripping out the data inserted by Gemstar in the VBI of local television broadcast stations, Time Warner was favoring a competing EPG offered by its own cable subsidiaries.[23]

It is important to note that in the case of program-related services, the issue is not of programmers' access rights to cable distribution per se. Even when the network operator has agreed (or is forced by statute, as in the case of local television stations) to carry an unaffiliated programmer, it has the ability to favor its own related programmer (e.g., AOL/Time Warner's Cartoon Network as compared to Disney's Disney Channel) by stripping the interactive features of a rival's video signal (e.g., the ATVEF "triggers"). Alternatively, the platform operator can slow down the rate of transmission of the downstream interactive data, thus interfering with the synchronization between the interactive service and the programming to which it is related. The ultimate effect is similar: to make an unaffiliated video signal less compelling as an information/entertainment experience.

In the case of dedicated ITV services, the bundling of transmission and ITV service presents questions similar to those discussed in the context of the debate over broadband cable Internet. Nonetheless, in this case the concerns are exacerbated by the fact that, unlike ISPs, ITV service providers are faced from the start with the closed network architecture of the second generation of broadcasting, rather than the end-to-end architecture of the first-generation Internet. Hence, if a single transmission network emerges as the only viable alternative to compete in the provision of ITV services (an assumption we explore *infra*), the network operator does not need to re-engineer the network in order to favor its affiliates because entry will be, from the outset, by invitation only. As the ITC explains:

> The distinctiveness of interactive television services as compared with the [I]nternet is manifested in the "walled garden" concept, where a limited number of sites or parts of sites are selected by the interactive licensee.... In this environment an interactive licensee has the

---

22. The VBI is the interval between television frames in analog broadcasting, which allows for a limited capacity of data transmission. OWEN, *supra* note 9, at 103-05.

23. Gemstar Int'l Group, Ltd. and Gemstar Dev. Corp. Pet. for Special Relief, *Memorandum Opinion and Order*, 16 F.C.C.R. 21531, para. 28, 25 Comm. Reg. (P & F) 333 (2001). Due to regulatory scrutiny of the AOL/Time Warner merger, Time Warner Cable has reportedly ceased such practice. Ex Parte Filing of The Walt Disney Company to the FCC, CS Dkt. No. 00-30, at 28 (July 25, 2000), *available at* http://gullfoss2.fcc.gov/prod/ecfs/retrieve.cgi?native_or_pdf=pdf&id_document=6511458333.

potential to exercise a degree of pre-selection and control of content through their contractual relationships with the providers of the walled garden content. This factor... suggests that a somewhat different treatment is needed from that applied to the [I]nternet.[24]

Critics of ex ante regulatory action on ITV nonetheless contend that network operators are unlikely to have incentives to discriminate against unaffiliated ITV providers or programmers, and that any rules imposed will have costly effects on investments and service efficiency:

[I]t is implausible that any local ITV platform could hope to raise entry barriers by denying access to rival ITV service providers. Because it could not raise entry barriers, it would have incentives to deny access if and only if such a denial were efficient: either because the denied service provider would not efficiently fit the platform or because vertical integration of ITV platforms and services is more efficient. Any interference with such decisions would make ITV markets inefficient, with higher costs or lower quality for consumers.[25]

In our opinion, the argument that network operators will lack incentives to discriminate, and will therefore offer access to as many programmers and service providers as would "fit the platform" is weak for a number of reasons. First, while it is clear that a network operator will want to maximize available content in order to attract new subscribers or to sell more products to existing subscribers, it is not clear that it will have incentives to grant users access to competing ITV service providers, particularly given the existence of close substitutes in the programming market.[26] Further, where there are capacity constraints that prevent carriage of all possible content, operators naturally will privilege affiliated content. In addition, the very nature of many interactive services creates further incentives to discriminate. These new services routinely generate transactions that bring revenues in addition to basic subscription—for example, payments to play online games, product or services purchases, or commissions. Network operators certainly will want to capture some share of these additional revenues, and therefore will have strong incentives to favor affiliated providers of such interactive services.

Furthermore, the above argument is based on a static notion of market efficiency and consumer welfare that overlooks two fundamental goals in

---

24. Independent Television Commission, *supra* note 17, at 7.

25. Nondiscrimination in the Distribution of Interactive Television Services Over Cable, *Comments of National Cable Television Association*, CS Dkt. No. 01-7, Attachment A at 35 (March 19, 2001), *available at* http://gullfoss2.fcc.gov/prod/ecfs/retrieve.cgi?native_or_pdf=pdf&id_document=6512562663.

26. *See* Nondiscrimination in the Distribution of Interactive Television Services Over Cable, *Declaration of J. Gregory Sidak & Hal J. Singer*, CS Dkt. No. 01-7 (June 8, 2001), *at* http://gullfoss2.fcc.gov/prod/ecfs/retrieve.cgi?native_or_pdf=pdf&id_document=6512569104.

communication policymaking: that of fostering dynamic innovation in broadcasting services and that of promoting widespread access to information "from diverse and antagonistic sources."[27] The very existence of a gatekeeper between ITV services and end users will suffice to discourage entry by application developers and programmers. The ultimate result would be an efficient (in static terms) but highly constrained environment for the conduct of commerce and speech. Finally, the argument simply contradicts the actual evidence of discriminatory behavior by cable operators against unaffiliated ITV service providers, which as discussed *infra* has been amply documented in the AOL/Time Warner merger review.

B. Return Path

ITV is based on the existence of a return path that provides upstream communication between the home terminal and the service provider. This return path can potentially take many forms: It may be a standard dial-up Internet connection (used for example by WebTV), a proprietary version of a dial-up connection (used by AOLTV), an "out-of-band" reverse data channel (as used by most cable operators), or even a wireless two-way radio connection (used by Gemstar's GuidePlus+).[28] For most dedicated ITV services, the speed and synchronization of the return path with the video signal do not pose significant market-entry barriers. For program-related services and dedicated services that do not tolerate latency or require full-screen video streaming, the availability of a high-speed, high-capacity return path that works in close coordination with the related video feed is essential to create a compelling ITV experience. In most cases, this is best achieved through a broadband Internet connection.

As a result, cable is likely to become the dominant platform for ITV services, at least in the United States, where the cable plant is already installed and rapidly being upgraded to provide two-way digital services (we discuss *infra* the European case where DBS seems to have a first-mover advantage over cable). As the FTC explains:

> Cable has distinct advantages over alternative ITV transport and connection methods. The television signal is already transmitted over cable, which makes synchronizing viewer interaction with the programming easier. Neither satellite nor DSL connections can

---

27. Associated Press v. United States, 326 U.S. 1, 20 (1945).
28. *See generally* Nondiscrimination in the Distribution of Interactive TV Servs. Over Cable, *Notice of Inquiry*, 16 F.C.C.R. 1321, para. 21 (2001) [hereinafter *Nondiscrimination in the Distribution of ITV Servs. Notice of Inquiry*] (describing the various methods of distribution of video access signals).

integrate the cable video programming and the interactive functionality as smoothly as cable.[29]

Cable networks also provide extensive transmission capacity in both directions (downstream and upstream), a critical factor for the new generation of broadcasting services. Furthermore, operators have already made substantial investments in upgrading facilities to offer digital television packages and broadband cable Internet, upon which ITV services could be piggybacked.[30] As the FCC concludes, "[o]ur understanding of the current state of technology suggests that the cable platform is likely to be the best suited for delivering ITV services, particularly high speed services, for at least the near term."[31]

The lack of a credible competitor to discipline cable operators opens several avenues for discriminatory behavior in favor of affiliated programmers and ITV service providers. Cable operators can simply refuse to provide a return path to third parties. In fact, during the AOL/Time Warner merger review the FCC received several complaints from unaffiliated programmers about Time Warner Cable's refusal to provide guarantees in terms of nondiscriminatory use of the return path.[32] The network operator may also degrade the quality of the return path (in terms of speed or reliability) offered to third parties. In addition, it could seek charges for t-commerce transactions originated through its platform. This would be similar to an ISP seeking compensation from electronic retailers such as Amazon.com for every item sold to its subscribers. Rather than simply enabling transactions under the end-to-end principle, the transport operator would erect a tollgate between buyers and sellers.[33] Lastly, valuable customer data can be obtained from the return path even when the platform operator is not a party of the commercial transaction taking place. This has raised concerns not only from third-party programmers and ITV

---

29. Compl., America Online, Inc. & Time Warner Inc., FTC Dkt. No. C-3989 at 4 (2000) [hereinafter FTC Complaint].

30. The National Cable & Telecommunications Association estimates that by the end of 2001, seventy percent of households were passed by cable broadband. CABLE & TELECOMM. INDUS., Mid-Year Overview 2, chart 2 (2002), *at* http://www.ncta.com/pdf_files/Mid'02Overview.pdf.

31. *Nondiscrimination in the Distribution of ITV Servs. Notice of Inquiry, supra* note 28, para. 21.

32. *See* Application of America Online, Inc. & Time Warner Inc. for Transfers of Control, *Ex Parte Submission of The Walt Disney Company*, CS Dkt. No. 00-30 (filed Oct. 25, 2000), *available at* http://gullfoss2.fcc.gov/prod/ecfs/retrieve.cgi?native_or_pdf=pdf&id_document=6511960660.

33. As Time Warner Cable executive Kevin Leddy stated, "If a programmer wants to offer its advertisers the ability to have two-way communication with viewers, the cable operator has to be part of that." Saul Hansell, *AOL-Time Warner Rivals Preparing for Interactive TV Fight*, N.Y. TIMES, Sept. 11, 2000, at C1.

service providers, but also from consumer groups alarmed about viewers having little control over how the return-path data will be compiled and used.[34]

C. Home Terminal

The third necessary component of an ITV system is the home terminal or digital set-top box. As the number and complexity of ITV services increases, so will the processing and storage capacity of the home terminal in order to perform the different tasks. In essence, a digital set-top box is similar to a stripped-down personal computer. There are at least two components within the digital set-top box that, absent regulatory safeguards or open industry standards, present opportunities for discriminatory behavior by dominant platform operators. The first is the Application Program Interface ("API"), which is the software layer between the operating system and the different applications running on the terminal. Unlike the more mature personal computer industry, there is no de facto industry standard for set-top box APIs. If such a standard were to develop in the future, and if its technical specifications were available to application developers on nondiscriminatory terms, the competitive concerns associated with the API would be mitigated. There are a number of industry consortia working to create an open platform for ITV. Among them are OpenCable's OpenCable Applications Platform ("OCAP"),[35] the Digital Video Broadcasting ("DVB") group's Multimedia Home Platform ("MHP"),[36] and even a Linux-based platform sponsored by the TV Linux Alliance.[37] For the foreseeable future, however, proprietary (i.e., non-interoperable) APIs will be deployed by network operators, forcing developers to rewrite ITV applications for several different environments.

In order to enter the market, an ITV service provider (assuming it has secured both downstream and upstream carriage) faces two options: it can either contract with the dominant platform operator to gain access to the

---

34. *See, e.g.*, CENTER FOR DIGITAL DEMOCRACY, TV THAT WATCHES YOU: THE PRYING EYES OF INTERACTIVE TELEVISION (June 2001), *at* http://www.democraticmedia.org/privacyreport.pdf (last visited Oct. 1, 2002).

35. OpenCable is an initiative of CableLabs, a Research and Development consortium formed by U.S. cable operators. *See generally* http://www.cablelabs.com (last visited Oct. 7, 2002).

36. The DVB group is an European consortium formed by equipment manufacturers, broadcasters, content producers, software developers, and representatives of national regulatory bodies. *See generally* http://www.dvb.org (last visited Oct. 7, 2002).

37. The TV Linux Alliance is a U.S.-based consortium of technology suppliers to cable, satellite and telecommunications network operators. *See generally* http://www.tvlinuxalliance.org (last visited Oct. 7, 2002).

installed base of terminals, or it can deploy a stand-alone box and bypass the proprietary terminal components altogether. The second option, while theoretically possible, is nonetheless uneconomical for most potential entrants. It is highly unlikely that users will be willing to buy a new box for every new ITV application. Who would be willing to buy a separate personal computer for every new application? The failure of stand-alone boxes marketed by companies like TiVo (which allowed digital video recording) and WebTV (despite heavy marketing spending by its parent Microsoft) has shown that consumers prefer a single box that integrates traditional video programming with new services.[38] Furthermore, the evidence from the introduction of DBS, wireless telephony, and digital television shows that heavy terminal subsidies are necessary. Thus, as a European competition official said, "[T]he scale of investment required means that the new entrants' most realistic option is to provide a . . . service using the set top boxes which already exist."[39]

Access to the API specifications and related facilities (authoring tools, authorization keys, memory control, etc.) is therefore critical for potential entrants in the ITV services market. This creates several opportunities for strategic behavior by dominant network operators such as refusing to provide authoring tools, discriminatory access pricing, discriminatory allocation of set-top boxes facilities (e.g., set-top box memory for caching), and bundling of API access with other services (e.g., conditional access or subscription management). That fair competition in ITV services requires either standardization or nondiscriminatory licensing of API specifications has been long recognized by European regulators. Accordingly, the new EU communications regulatory package contains several measures to promote API standardization in member-states, and allows national regulatory authorities to take steps to ensure nondiscriminatory access to API specifications and related facilities.[40]

The second component that raises policy concerns is the EPG, a navigation tool that allows users to browse and select television channels

---

38. *See generally* CONSUMER ELECTRONICS ASSOCIATION, DIGITAL AMERICA 2002, THE U.S. CONSUMER ELECTRONICS INDUSTRY TODAY, *at* http://www.ce.org/publications/books_ references/digital_america/default.asp. In fact, Personal Video Recorders ("PVRs") by TiVo and others are now being embedded into cable and satellite receivers. *See generally* Jennifer 8. Lee, *In the US, Interactive TV Still Awaits an Audience*, N.Y. TIMES, Dec. 31, 2001, at C1.

39. Linsey Mc Callum, *EC Competition Law and Digital Pay TV*, 1 COMPETITION POL'Y NEWSL. 4, 11 (1999).

40. *See, e.g.*, Directive 2002/19/EC of the European Parliament and of the Council of 7 March 2002 on Access to, and Interconnection of, Electronic Communications Networks and Associated Facilities, 2002 O.J. (L 108) 7, *available at* http://europa.eu.int/eur-lex/ pri/en/oj/dat/2002/l_108/l_10820020424en00070020.pdf.

and services. With the manifold increase in the number of channels and applications made possible by the transition to digital television, the EPG is expected to become to the broadcasting industry what Web portals have become to the Internet: powerful tools to direct traffic and obtain advertising revenues. From a regulatory standpoint, the main concern is that dominant platform operators do not use the EPG to leverage their power onto the market for content and ITV services. As European regulators explain:

> Issues of ensuring listing of third-party services or programming, and the quality of such listings, will be of critical importance. Exclusive arrangements tying particular EPGs to particular service bundles may become a problem requiring regulatory intervention to ensure third-party access on fair, transparent and non-discriminatory terms.[41]

U.S. regulators increasingly have grown concerned about issues of first-screen and presentation bias in EPGs, although regulatory action so far has been limited. For example, a few nondiscriminatory provisions were adopted in the Telecommunications Act of 1996[42] in the case of EPG services offered by Open Video Systems operators,[43] as well as in the Satellite Home Viewer Improvement Act of 1999[44] in the case of EPGs offered by DBS operators. Yet, these rules do not extend to cable systems. In 1999, the merger of the two major EPG providers (Gemstar and TV Guide) was the subject of an antitrust investigation by the U.S. Department of Justice, which ultimately declined to challenge the deal.[45] In Europe, by contrast, regulators have taken a more active role in regulating EPG services, either to protect third-party programmers and service providers or to favor publicly funded broadcasters. In the UK, for example, the Office of Telecommunications ("OFTEL") has interpreted EPGs as covered by the nondiscriminatory rules for telecommunications access services,[46] while the ITC has adopted a "code of conduct" for EPG providers that, among other things, mandates that the visual interface grants public service channels "due prominence."[47] As discussed *infra* in the British Interactive

---

41. Green Paper on the Convergence of the Telecommunications, Media and Information Technology Sectors, and the Implications for Regulation, Commission of Eur. Cmtys., COM(97)623EC 24-25 (Mar. 12, 1997) (citations omitted).
42. Pub. L. No. 104-104, 110 Stat. 56 (codified at scattered sections of 47 U.S.C.).
43. 47 U.S.C. § 573(b) (2000).
44. 47 U.S.C. § 338 (2000).
45. Christopher Grimes, *Gemstar Closes TV Guide Deal*, FIN. TIMES, July 13, 2000, at 34, *available at* LEXIS, News Library, Financial Times.
46. *See* OFFICE OF TELECOMM., DIGITAL TV AND INTERACTIVE SERVICES: ENSURING ACCESS ON FAIR, REASONABLE, AND NONDISCRIMINATORY TERMS (1998), *at* http://www.oftel.gov.uk/publications/1995_98/broadcasting/dig398.htm. (Mar. 1998).
47. *See generally* INDEPENDENT TV COMMISSION, CODE OF CONDUCT ON ELECTRONIC

Broadcasting ("BiB") case, European competition authorities also have acted against exclusivity arrangements between EPG providers and dominant network operators.

V. THE CASES: THE AOL/TIME WARNER MERGER AND BRITISH INTERACTIVE BROADCASTING

The debate about the proper tools and scope of regulatory action vis-à-vis ITV services already has surfaced in a number of cases. In this Section we analyze two of the most prominent ones: the AOL/Time Warner merger and the BiB case. BiB was a joint venture for the launch of ITV services in the UK created by BSkyB, British Telecommunications ("BT"), Midland Bank (part of the HSBC banking group), and Matsushita, the Japanese consumer electronics giant.[48] We contrast the approach taken by American and European regulators with the issues raised by these cases and analyze the implications of the regulatory obligations imposed in each case.

A. The AOL/Time Warner Merger

The January 2000 announcement of the merger between AOL and Time Warner triggered close scrutiny by federal regulators. The investigation conducted by the FTC concluded that the combination of AOL's Internet properties with Time Warner's cable holdings and content assets had anticompetitive effects in three distinct markets: broadband Internet access service, broadband Internet transport service, and ITV services.[49] While most of the debate about the competitive effect of the merger focused on the first two issues, the FTC findings brought attention to the architecture of next-generation broadcasting networks. The main concerns that were raised related to AOLTV, AOL's ITV product.[50] The existing generation of the AOLTV service consisted of a stand-alone set-top box that connected to a cable or DBS receiver and blends this video programming with interactive content transmitted via a narrow-band dial-up modem.[51] While regulators raised few concerns about this service, AOL's plan to upgrade it by embedding AOLTV within Time Warner cable boxes and utilizing the broadband Internet platform of the cable

---

PROGRAMME GUIDES (1997), at http://www.itc.org.uk/itc_publications/codes_guidance/electronic_programme_guide/epg_code.asp (last visited Sept. 30, 2002).

48. David Teather, *Closing time: BskyB is set to close Open, the interactive TV firm, in order to boost its own business. But has the market lost its taste for TV shopping?*, THE GUARDIAN, May 7, 2001, at 30.

49. America Online, Inc. and Time Warner Inc., *Complaint*, FTC Dkt. No. C-3989, at 3-4 (2000).

50. *Id.* at 4.

51. *Id.*

operator troubled competition authorities. As the FTC explained:

> AOL recently launched AOL TV [sic], a first generation ITV service, and is well positioned to become the leading ITV provider. Local cable companies will play the key role in enabling the delivery of ITV services. After the merger, AOL/Time Warner will have incentives to prevent or deter rival ITV providers from competing with AOL's ITV service. Thus, the merger could enable AOL to exercise unilateral market power in the market for ITV services in Time Warner cable areas, which also affects the ability of ITV providers to compete nationally.[52]

Despite the strong wording of these findings, the FTC ultimately imposed rather weak remedies related to ITV. The consent decree that authorized the merger simply prohibits AOL/Time Warner from interfering with its subscribers' ability to use the interactive signals or "triggers" provided by programmers that it has agreed (or is forced by statute) to carry.[53] In essence, the FTC order only addressed one of the possible anticompetitive strategies discussed *infra*, that of network operators "stripping" the signals of unaffiliated programmers from its interactive content. Other discriminatory practices related to downstream transmission, upstream transmission (the return path), and the home terminal were left unaddressed.

The FCC investigation concurred with the findings of the FTC review:

> AOL Time Warner would have the potential ability to use its combined control of cable system facilities, video programming and the AOLTV service to discriminate against unaffiliated video programming networks in the provision of ITV services. We also find that AOL Time Warner may have incentives to engage in such discriminatory behavior.[54]

The FCC analysis is broader in scope and acknowledges that the anticompetitive strategies available to AOL/Time Warner go beyond the "stripping" of interactive content of unaffiliated programmers. It also notes that the Memorandum of Understanding, by which the merger parties committed to provide customers with a choice of ISPs, does not obligate the company to provide access for ITV uses. Nonetheless, the FCC declined to impose additional conditions on the parties pending further

---

52. America Online, Inc. and Time Warner Inc., *Analysis of Proposed Consent Order to Aid Public Comment*, FTC Dkt. No. C-3989, at 2 (2000).

53. America Online, Inc. and Time Warner Inc., *Decision and Order*, FTC Dkt. No. C-3989, at 11 (2000).

54. Applications for Consent to the Transfer of Control of Licenses and Section 214 Authorizations by Time Warner Inc. and America Online, Inc., Transferors, to AOL Time Warner Inc., Transferee, *Memorandum Opinion and Order*, 16 F.C.C.R. 6547, para. 217, 23 Comm. Reg. (P & F) 157 (2001).

examination of market developments and the potential incentives for discriminatory behavior by AOL/Time Warner. In the Commission's analysis, the FTC's prohibition on "stripping," coupled with the conditions relating to the availability of multiple ISPs, suffice to protect competition, at least during the initial stages of the ITV market. In the words of then-Commissioner (and now Chairman) Michael Powell, "[A]lthough it is surely possible to hypothesize public interest harms flowing from a cable operator's control of assets like those at issue in this merger, the [ITV] market is too immature to conclude with any confidence whether such harms are sufficiently probable to warrant direct government intervention."[55]

B. The British Interactive Broadcasting Case

BiB operates one of the largest and most advanced ITV services worldwide. It is available to the more than six million subscribers of BSkyB's digital television satellite service, offering a variety of dedicated services such as e-mail, electronic banking, games, and gambling, as well as program-related services tied to channels offered by BSkyB.[56] It provides ITV services in the UK by means of satellite broadcasting (leased from BSkyB, with BT responsible for the uplink) in combination with a narrow-band return path through a standard telephone line. The terminal equipment required to use BiB services is embedded in the BSkyB digital television set-top box, which BiB partly subsidizes (this includes a proprietary API developed by OpenTV and BSkyB's EPG).[57] Revenues come from end-users, from retailers and from ITV service providers that BiB carries on its platform.[58]

European competition authorities raised two main concerns about BiB. First, that the company would use its control of the set-top box software components to foreclose competition in ITV services, denying third parties access to the boxes being deployed. Second, that BiB would enhance the already-dominant position of BSkyB and BT in the markets for

---

55. Press Statement, FCC Commissioner Michael Powell, Approval of AOL-Time Warner Merger (Jan. 11, 2001), *available at* http://www.fcc.gov/Speeches/Powell/Statements/2001/stmkp0101.html.

56. BRITISH SKY BROADCASTING GROUP PLC, ANNUAL REPORT AND ACCOUNTS (2002), *available at* http://media.corporate-ir.net/media_files/lse/bsy.uk/reports/BSKYB_ar_2002.pdf.

57. Notice published under Article 19(3) of Council Regulation No. 17 Concerning an Application for Negative Clearance or an Individual Decision to Grant an Exemption Pursuant to Article 85(3) of the EC Treaty (Case No IV/36.539—BiB), 1998 O.J. (C 322) 6 (Oct. 21, 1998).

58. Duncan, *supra* note 19, at 27, 30.

pay television and telecommunications local loop respectively. In October 1998, the European Commission approved the joint venture subject to a number of conditions.[59] In contrast with the AOL/Time Warner case, the main regulatory concern was to ensure that "third parties, whether operators of digital television or digital interactive TV services, have fair, reasonable, and nondiscriminatory access to all proprietary components of the digital set top box which BiB will subsidise."[60] The difference in focus is due to the fact that while cable operators effectively control the transmission infrastructure, satellite television operators lease capacity from (often unaffiliated) satellite carriers.[61] Market power, therefore, stems not from control over transmission infrastructure but rather from first-mover advantages and switching costs associated with proprietary home terminals.[62]

One of the conditions imposed concerned the recovery of the set-top box subsidy. The Commission forced BiB to establish a separate company to manage the subsidy payments in order to ensure that the recovery is evenly distributed among service operators and broadcasters, whether affiliated with BiB and its partners or not. It also demanded that the subsidy was not linked to a subscription to BSkyB's pay-television service.[63]

Another condition related to the terms of access to the home terminal components. BiB agreed to provide, upon request, the API specifications and other proprietary technical information to third parties. The Commission also forced BiB to end its exclusivity agreement with BSkyB, whereby BiB would be the only available ITV service on BSkyB's EPG. In addition, the Commission also imposed several obligations on the joint venture partners. BSkyB agreed to offer access services to programmers and ITV service providers (including BiB) on fair, reasonable, and nondiscriminatory terms regulated by OFTEL. It also agreed to supply, upon request, a "clean feed" (i.e., stripped of interactive applications) of its film and sports channels to other MVPDs (e.g., cable operators) in order to

---

59. Mc Callum, *supra* note 39, at 6.

60. *Id.* at 13.

61. In the case of BSkyB, it leases satellite capacity from Société Européenne des Satellites. OFFICE OF FAIR TRADING, DIRECTOR GENERAL'S REVIEW OF BSKYB IN THE WHOLESALE PAY TV MARKET (1996) *at* http://www.oft.gov.uk/NR/rdonlyres/eupiymw 5hitxeuawrxfukg44iowwyubhp33q4nimzdezdp33wgfi2kg2vigxp6r5ma55pmtejiigl7a3fottls ocjwb/oft179.pdf.

62. *See* Martin Cave, *Regulating Digital TV in a Convergent World*, 21 TELECOMM. POL'Y 575 (1997).

63. Notice published under Article 19(3) of Council Regulation No. 17 Concerning an Application for Negative Clearance or an Individual Decision to Grant an Exemption Pursuant to Article 85(3) of the EC Treaty (Case No IV/36.539—BiB), 1998 O.J. (C 322) 6 (Oct. 21, 1998).

prevent bundling strategies that would favor BiB. Finally, BT agreed to divest from its existing cable interests.[64]

The conditions imposed by the EC on the BiB venture are consistent with the established doctrine among community competition authorities that ex post rules are insufficient to remedy the problem of access to telecommunications facilities, and thus need to be supplemented by ex ante, sector-specific obligations.[65] This doctrine has been implemented through a series of Council Directives under the so-called Open Network Provision ("ONP") framework, which imposes on telecommunications operators having significant market power certain nondiscriminatory obligations that go beyond those that would normally apply under general competition law.[66] It is interesting to note that a few weeks before the BiB decision, the Commission adopted the Access Notice that explicitly stated that the ONP framework extends not only to telecommunications facilities, but also to "access issues in digital communications sectors generally."[67] This doctrine was crystallized in the recently passed Access Directive, which specifies the instruments for extending interconnection obligations to providers of ITV facilities such as the API and the EPG.[68] Through these efforts, EU policymakers are progressively bringing about a policy convergence that mirrors technological convergence, aiming for technology-independent rules that govern communication activities according to general principles.

---

64. *Id.*

65. *See* HERBERT UNGERER, ACCESS ISSUES UNDER EU REGULATION AND ANTITRUST LAW: THE CASE OF TELECOMMUNICATION AND INTERNET MARKETS (Harvard Univ. Weatherhead Ctr. for Int'l Affairs, Working Paper No. 00-05, 2000).

66. *See generally* Council Directive of 28 June 1990, Establishment of the Internal Market for Telecommunications Services Through the Implementation of Open Network Provision, 1990 O.J. (L 192) 1 (discussing the general framework provided by the ONP Framework Directive); *see also* Council Directive 92/44/EEC of 5 June 1992 Application of Open Network Provision to Leased Lines, 1992 O.J. (L 165) 2; Council Recommendation 92/382 of 5 June 1992, Harmonized Provision of a Minimum Set of Packet-Switched Data Services (PSDS) in Accordance with Open Network Provision (ONP) Principles, 1992 O.J. (L 200) 1; Directive 95/62/EC of the European Parliament and of the Council of 13 Dec. 1995, Application of Open Network Provision (ONP) to Voice Telephony, 1995 O.J. (L 321) 6.

67. Application of the Competition Rules to Access Agreements in the Telecommunications Sector—Framework, Relevant Markets and Principles, Notice, 1998 O.J. (C 265) 2, 3.

68. European Council Directive of 7 March 2002, Access to, and Interconnection of, Electronic Communications Networks and Associated Facilities (Access Directive), 2002 O.J. (L 108) 7.

VI. CONCLUSION

In the aftermath of the AOL/Time Warner merger, the debate about open cable access has faded considerably. The more general problem of nondiscriminatory access to the basic layers of communications infrastructure, whether cable lines,[69] the local loop,[70] the emerging wireless data networks,[71] or the digital television user terminal,[72] is arguably the crucial issue for industry and regulators in the post-convergence era. In this Article we examined how this problem unfolded in the migration to the third generation of broadcasting services, that of ITV. We argued that absent regulatory safeguards that provide for nondiscriminatory access to several network components (including digital set-top box components), dominant platform operators are likely to leverage ownership of delivery infrastructure into market power over ITV services and content, foreclosing competition and discouraging third parties and users from experimenting with yet-unimagined ways to use television.

In the case of ITV, the question of open access is not about extending existing regulatory principles to the new generation of technologies. Rather, it is about seizing the opportunities offered by these new technologies to better serve our policy goals. Broadcasting regulation has traditionally taken distribution scarcities and closed network architecture as a fact of life dictated by the available technology, thus relying on ownership caps, content obligations, must-carry rules, and other instruments of structural regulation to attain its goals. It is now widely acknowledged that this approach has not only largely failed on its own merits, but also is inadequate for a post-convergence world. The third generation of television calls for shifting the focus of regulatory action from government "tinkering with the configuration of a mass media market"[73] to rules that ensure nondiscriminatory access to the capacity to

---

69. *See generally* Mark Lemley & Lawrence Lessig, *The End of End-to-End: Preserving the Architecture of the Internet in the Broadband Era*, 48 UCLA L. REV. 925 (2001) (arguing that cable modem services threaten the open architecture of the Internet, and thus its growth and innovation).

70. *See generally* UNGERER, *supra* note 65 (outlining the ONP doctrine used to introduce competition in the European telecommunications sector).

71. *See generally* Eli M. Noam, The Next Frontier for Openness: Wireless Communications (Sept. 25, 2001) (paper for the 2001 Telecommunications Policy Research Conference, Alexandria, Va.), *at* http://arxiv.org/ftp/cs/papers/0109/0109102.pdf.

72. *See generally* Galperin, *supra* note 21 (discussing how problems of access to the terminal equipment threaten competition in digital television services).

73. Yochai Benkler, *From Consumers to Users: Shifting the Deeper Structures of Regulation Toward Sustainable Commons and User Access*, 52 FED. COMM. L.J. 561, 562 (2000).

experiment with and provide information, entertainment, and transaction services over broadcasting networks.

American regulators prefer to wait and observe the unfolding of the new media. They have been content so far with rather toothless safeguards to prevent discriminatory behavior by incumbent network operators in the ITV market. Furthermore, these rules are dispersed across statutes addressing different platforms, thus distorting market competition. European authorities, by contrast, are in the process of fashioning a comprehensive framework that addresses problems of access and interconnection across electronic communications networks. This framework does not impose specific remedies but rather lays out general principles to tackle problems as they arise. By addressing access in a piecemeal, ad hoc fashion, U.S. policymakers may undermine the very basis of the unprecedented innovation in communications technology of the last decade, and, in the case of television, forgo the possibility to revamp a failed regulatory regime. As media converge, it is becoming clear that access will be a generic issue, relevant to wireless, cable, broadcast, telephone, and the Internet alike. Whether one agrees with specific EU rules or not, there is a compelling logic in the European endeavor to revert to basic principles and to apply them uniformly across communication media.

[13]

The "Right to Information" and Digital Broadcasting: About Monsters, Invisible Men and the Future of European Broadcasting Regulation

NATALI HELBERGER

INSTITUTE FOR INFORMATION LAW, UNIVERSITY OF AMSTERDAM*

LT Access to information; Commercial broadcasting; Digital television; EC law; Public service broadcasting; Television

Introduction

"What kind of a monster would do this to your child—would come into your home and put a padlock on his TV fun? What kind of a monster would force you to feed your TV set bucketfuls of dollars—or suffer the humiliation of being labeled a 'cheapskate' in the eyes of your children? There is such a monster. It's a greedy thing called Pay TV."[1]

This was the slogan of the "Californian Crusade for Free TV" in the 1960s.[2] The Californian Crusade for Free TV declared war on Subscription Television, Inc, a company whose goal it was to launch pay-TV services in Los Angeles and San Francisco in the summer of 1964. The Californian Crusade for Free TV found broad support among the public, movie theatre owners, publishers, electronics manufacturers, stockbrokers and, last but not least, the Dodgers and Giants baseball organisation. As the argument went, the public had the privilege of receiving broadcasting free of charge.[3] The campaign finally led to the end of Subscription Television, Inc.

Is it "monstrous" to control access to broadcasting content and demand money for access? Does the electronic control of access to information, a crucial element of a range of new business models in digital broadcasting, conflict with what is often referred to as the public's "right of access to information"? And, if so, does existing European broadcasting law adequately address the conflict?

These are timeless questions that later played a prominent role in the revision of the Television Without Frontiers ("TWF") Directive[4] and the modernisation of provisions that protect the "right to information". In the course of the revision, the European Commission issued in 2003 two discussion papers under the heading "Access to events of major importance for society".[5]

---

* Comments are welcome to helberger@ivir.nl. Parts of this article are an adaptation of chapters from Helberger, Controlling Access to Content—Regulating Conditional Access to Digital Broadcasting (Kluwer International, Den Haag, 2005). The book discusses in greater depth some of the subjects that are raised in this article. The author would like to express her gratitude to P.B. Hugenholtz, E. Dommering, N.A.N.M van Eijk, A. Nieuwenhuis, W. Hins, T. Gibbons and W. Schulz for their comments on earlier drafts. All mistakes or omissions are entirely my own.

1. "Darn that pay-TV!", advertisement in *Los Angeles Times*, October 12, 1964, sec.3, p.5.
2. For a thorough overview on the Californian Crusade for Free TV, see Gunzerath.

3. This was the official argument. Strategic and economic considerations of potential competitors of Subscription Television, Inc will have played no lesser role.
4. Directive 89/552 of October 3, 1989 on the co-ordination of certain provisions laid down by law, regulation or administrative action in Member States concerning the pursuit of television broadcasting activities, Brussels, October 17, 1989 [1989] O.J. L298/23 ("Television Without Frontiers Directive") and Directive 97/36 of the European Parliament and of the Council of June 30, 1997 amending Council Directive 89/552 on the co-ordination of certain provisions laid down by law, regulation or administrative action in Member States concerning the pursuit of television broadcasting activities, Brussels, July 30, 1997 [1997] O.J. L202/60 ("Directive 97/36 Amending the Television Without Frontiers Directive"). In the following, references to the Television Without Frontiers ("TWF") Directive refer to the directive in the form as amended by Directive 97/36 Amending the Television Without Frontiers Directive.
5. European Commission, discussion paper, "Events of major importance for society", available at http://europa.eu.int/comm/avpolicy/regul/review-twf2003/twf2003-theme1_en.pdf (last visited November 11, 2005) and "Access to short extracts of events subject to exclusive rights", available at http://europa.eu.int/comm/

The Commission invited comments on Art.3a of the TWF Directive, a provision that allows Member States to draw up a so-called list of important events that may not be shown exclusively on pay-TV. It was also suggested modelling a European right to short reporting on Art.9 of the European Convention on Transfrontier Television.[6] This part of the revision process was driven by the wish to modernise the European rules on private exclusionary legal or electronic control over viewers' access to broadcast television coverage of "important events" or news reports, and to bring European broadcasting law in line with Art.10 of the European Convention on Human Rights ("ECHR")—Freedom of Expression.[7]

The results from the consultation were summarised in the Commission's communication on the future of the European audiovisual regulatory policy. Here, the Commission concluded that "the issue of right to access to newsworthy events needs further attention".[8] As a consequence, in December 2004, focus group three, a group of experts from academia, industry, regulatory authorities and other stakeholders, was invited to debate on possible revisions of the European TWF Directive to realise what the European Commission then called the "right to information".[9] Based on the work of focus group three, the European Commission launched a second discussion paper with the same title in the summer of 2005.[10] Again, comments were invited, also in preparation for a major audiovisual conference, "Between culture and commerce," that was held in September 2005 in Liverpool and where focus group three presented its final conclusions.[11]

The goal of this article is to discuss and critically evaluate the conclusions of the focus group and the different solutions that were suggested to protect the "right of the public to information". It will do so from the perspective of the viewers, whose right of access to information the measures are meant to protect. Viewers of broadcasting content, or their representatives, were rather invisible during the work of the focus group[12] and the consultations.[13] Moreover, the suggested approach for a modernisation of the TWF Directive was characterised by a prevailing perception of the anonymous viewer as passive receiver of broadcasting. The article will explain that the arrival of electronic access control substantially changes the way viewers access broadcasting content. It will show why the traditional perception of the viewer no longer matches the realities in modern broadcasting markets and why measures that incorporate the said user perception are not very effective in guaranteeing broad access to broadcasting content. This article concludes with concrete suggestions for a complementary course of action that guarantees that access for the individual members of the audience to broadcasting content is not obstructed by private exclusionary control of access.

Whose "right to information"?

The notion of the "right to information" is often used by lawmakers, academics and stakeholders alike to refer to the fundamental role that access to information, culture and knowledge plays in our cultural, democratic and social life. The notion of a "right to information" is often understood as an entitlement of the individual citizen to have access to all kinds of information. The way the European Commission and the Council of Europe talk about the "right to information" contributes to the shaping of this idea. What is often overlooked, however, is that the "right to information" as it is referred to in existing broadcasting law has little to do with a right of the individual citizen to access broadcasting information, notably information of major importance for democracy, culture, society, or the creation of knowledge. For the time being, at least in the concept of broadcasting regulation, such a right has yet to be created. This is at least the conclusion if, as both the European Commission and the Council of Europe do, one grounds the "right of access to information" in Art.10 of the ECHR—Freedom of Expression.[14] Article 10 of the ECHR is a provision that has fundamentally influenced in one form or another national broadcasting regulation and here in particular the provisions concerning the availability and accessibility of broadcasting content.

At this point, proponents of a right of access to information will intervene and point to the landmark decisions of the ECHR in *Guerra* and the *Sunday Times* case. It was in the *Sunday Times* case that the ECHR coined the term of the right of the public to receive certain information.[15] And, in *Guerra*, the ECHR left no doubt that

---

avpolicy/regul/review-twf2003/twf2003-theme6_en.pdf (last visited November 11, 2005).
6. Council of Europe, European Convention on Transfrontier Television, Strasbourg, May 5, 1989, text amended according to the provisions of the Protocol (ETS No.171) which entered into force on March 1, 2002 ("European Convention on Transfrontier Television—ECTT").
7. European Commission, "The right to information and the right to short extracts", October 2004, available at http://europa.eu.int/comm/avpolicy/regul/Focus%20groups/fg3_extracts_en.pdf (last visited November 11, 2005).
8. Communication from the Commission to the Council, the European Parliament, the European Economic and Social Committee and the Committee of the Regions on the future of European regulatory audiovisual policy,COM (2003) 784 final., December 15, 2003, available at http://europa.eu.int/eur-lex/en/com/cnc/2003/com2003_0784en01.pdf (last visited November 11, 2005).
9. European Commission, fn.7 above.
10. European Commission, Issues paper for the audiovisual conference in Liverpool, "Right to information and right to short reporting", July 2005, available at http://europa.eu.int/comm/avpolicy/revision-twf2005/ispa_shortreport_en.pdf (last visited November 11, 2005).
11. Rapport final du groupe de travail 2 [should be 3: the author], "Droits à l'information et aux courts extraits", available at http://europa.eu.int/comm/avpolicy/revision-twf2005/docs/liverpool-wg2-fr.pdf (last visited November 11, 2005) ("Rapport final du groupe de travail 2").

12. See list of the members of the focus groups, available at http://europa.eu.int/comm/avpolicy/regul/Focus%20groups/list_participants_fg_new2.pdf (last visited November 12, 2005).
13. See the list of written Contributions to the Public Consultation for the review of the "Television without Frontiers" Directive,http://europa.eu.int/comm/avpolicy/regul/review-twf2003/contribution.htm (last visited November 11, 2005).
14. Barendt, at p.49. See also De Meij/Hins/Nieuwenhuis/Schuijt, at pp.299–303, 306; Mackaay, at p.171. Extensively, Helberger 2005, at pp.74 et seq.
15. European Court of Human Rights, *Sunday Times*, Strasbourg, April 26, 1979, Series A, No.30 ("*Sunday Times*"), at

"[i]n cases concerning restrictions on freedom of the press it [the court] has on a number of occasions recognized that the public has a right to receive information".[16]

To jump from these findings to the conclusion that the court acknowledged a right of the individual viewer to access certain information that is of particular interest would, however, go a step too far. Instead, the "right of the public to be properly informed" must be seen within the context of the traditional role of the media to make information accessible for the audience:

"whilst the mass media must not overstep the bounds imposed on the interests of the proper administration of justice, it is incumbent on them to impart information and ideas concerning matters that come before the courts just as in other areas of public interest. Not only do the media have the task of imparting such information and ideas: the public also has a right to receive them."

And, in *Guerra*, the court continues:

"[the public has a right to receive information] as a corollary of the specific function of journalists, which is to impart information and ideas on matters of public interest."[17]

In other words, the right of access to information is the right of an undefined public to receive information and ideas on matters of public interest that the media impart, not the right of individual viewers to have access to media content. The mass media functions as an intermediary and carrier of the public interest to be informed. In this conception, the role of the individual citizen is a passive one, namely to receive information and ideas that the media choose to impart.

Meanwhile, the mass media's perception of its own function is changing. Programmes are often no longer broadcast freely into the air for whoever wishes to receive them. Broadcasting programmes are aggregated via central marketing platforms that are choosy when it comes to deciding to whom they impart broadcasting content and under which conditions. Thanks to electronic access control technologies, also referred to as conditional access, the media can suddenly choose not to impart information to certain viewers, for example, viewers who refuse or cannot afford to pay a subscription fee, who reside in a different Member State or who have acquired a set-top box that is able to support the programme of one but not another pay-TV operator. As will be shown below, there are good reasons to argue that economic and technological changes call for a fresh look at the "right to information", the media's mission to inform the public properly and the role the viewer plays in this process. This fresh look is missing in the final recommendations of focus group three, which ignores, when dealing with the impact of electronic access control, the changes electronic access control is making to the traditional distribution pattern of broadcasting services.

How conditional access changes the way viewers access broadcasting

Conditional access is a technical solution for controlling access to electronic services,[18] and as such a vital element of the business model of any pay-TV service. Conditional access has been widely welcomed as a driving factor behind the prospering of the "information economy". More specifically, conditional access refers to a combination of hardware (set-top box) and software devices (encryption, Application Program Interface) that, combined, enable access to a service that is transmitted electronically to be blocked, and subject access to an automated authorisation process. Conditional access builds an "architecture of identification"[19] in which compliance with set conditions is an integral part of the authorisation process.

From the viewers' perspective, electronic access control fundamentally changes the conditions under which electronic content is delivered to the viewer. The electronic control of access to content allows for new, more sophisticated pricing and marketing strategies than the concepts of financing through fees or advertising allow for. Using electronic access control, service providers can send targeted services to single members of the audience based on geographical location, market segment or personal preferences.[20] More importantly, services are no longer "set off in the air" but marketed individually to subscribers. As long as broadcasting signals were uncontrollable, there was no tangible matter that could be sold to viewers. In free-TV, the contractual relationship is between the broadcaster and the advertiser; viewers "pay" for broadcasting content in the form of a public broadcasting fee and their attention. This changed when electronic access control was introduced to the distribution process. Conditional access systems implement structures of individualised control over the viewers. Viewers become consumers, and the individual relationship between citizen-consumer and service provider is governed in the first place by contract law instead of broadcasting law.

This also means, however, that operators of electronic access control can exercise considerable influence over market conditions, competition and individual access to content. Technical or contractual conditions or the level of transparency in the subscription arrangement can ultimately influence viewers' access to information.

Access to information becomes subject to technical conditions

In pay-TV, the economic power of a technical platform or elements thereof can be influenced by the popularity of a certain embedded standard. This has to do with the close economic links that are

---

[65]; European Court of Human Rights, *Lingens*, Strasbourg, July 8, 1986, Series A No.103 ("*Lingens*"), at [41]. *Lingens* concerned the case of an Austrian journalist who complained about his conviction because of defamation after he had written an article about SS crimes during the Second World War.
16. European Court of Human Rights, *Guerra v Italy*, Strasbourg, February 19, 1998, No.116/1996/735/932 ("*Guerra*"), at [53].
17. ibid.

18. For a more detailed explanation, see Helberger 2005, fn.14 above, at pp.4 et seq.
19. Lessig, at p.34.
20. O'Driscoll, at p.14.

often found between the technical and the service platforms, the dynamics of the market in general, and the influence of indirect network effects and first mover advantages in particular.[21] This is why pay-TV operators will often require viewers to rent or purchase a set-top box that supports only the conditional access technology of a particular provider. The consequence is that (1) subscribers who own set-top boxes that support the standard of a provider will often not be able to access services from other providers; and (2) owners of other set-top boxes might not be able to access the broadcasting content that is offered by the first provider. This is the argument behind the so-called decoder towers, namely the assumption that viewers are less likely to subscribe to a second pay-TV platform if both platforms require the purchase of different, incompatible set-top boxes. Some argue that the importance of this argument will vanish if set-top boxes are offered at lower prices or are subsidised by the pay-TV platform operator. On the other hand, as the example of the Apple iPod shows, exclusive control over a technical standard is an important and effective means of binding subscribers and content producers to a particular service platform and of preventing other service providers from gaining access to the consumer.

Closely related is the aspect of audience fragmentation. Arguably, digitisation will favour the development of more specialised niche channels and hence increase the fragmentation of the viewer base. The use of electronic access control can further contribute to this process by dividing the audience into different zones of incompatible conditional access standards. Audience fragmentation can also take place along national borders. Today's access-controlled services such as pay-TV are often restricted to a national territory and the required smart cards are only sold to residents.[22] This is often due to the licensing practice of content rights, which are often issued on a territorial or language basis. Other reasons are divergent broadcasting laws (for example, in youth protection), the character of a service as national service for citizens of that state (such as public fee-financed broadcasting), or, again, the use of different conditional access standards by the different pay-TV operators in the respective countries. The effect can be to reinstall territorial borders in transborder media such as satellite distribution. A Danish citizen living in France, for example, may be prevented from accessing the encrypted Danish public service broadcasts of DR1 and DR2 and hence from accessing information from his or her home country and cultural heritage.

Contractual conditions

Another strategy used to monopolise the viewer and to which relatively little attention has been paid in the pay-TV discussion is that of contractual viewer lock-ins. In this context, the duration of the subscription contract is important as is the ease with which viewers can terminate the agreement.[23] Binding viewers to long-term subscription contracts or making it difficult to terminate the contract is a form of bundling in time. For pay-TV, this is usually 12 to 24 months. This time frame may have a negative effect on the viewers' mobility and willingness to switch to other sources of broadcasting content before the end of their initial contract.[24] Subscription contracts frequently contain very far-reaching provisions about their automatic extension that are not always easy to detect.[25] Contractual conditions that "sanction" a contract termination can also have a discouraging effect. Examples include an obligation to return a set-top box at the end of the contract or losing an email address.[26] Here, terminating the contract has the additional consequence of effectively barring viewers from receiving any digital, access-controlled or other information services before they have invested in new equipment and/or services. Such contractual conditions may be legitimate, reasonable and common in other sectors (for example, mobile phone subscriptions), but they can be a major problem in sectors such as the broadcasting sector where broad access to content and content from different sources is the ruling public policy objective.

Sub-forms of contractual lock-ins are programme bundling strategies that oblige viewers to subscribe to a whole package of services even if they only wish to access one particular channel or that make the provision of certain information services (for example, premium channels) conditional on subscription to others (such as basic channels).[27] Again, this can discourage viewers from subscribing to additional services that resemble services they already have in their package.

Access to comparable service information

An important factor for viewer choice is information about what is on offer under which conditions and for which prices.[28] The significance of access to comparable information about content services is even greater in an environment of access-controlled information. In the unencrypted world, viewers can search and choose freely, for example, by flicking through broadcasting channels. When viewers come across channels or programmes that are subject to electronic access control, it will be difficult for them to determine whether they contain relevant content because they are encrypted or otherwise protected against access. This

---

21. Shapiro, at pp.3 et seq.; Evans, at p.32.
22. See, for example, the subscriber conditions at www.sky.com/ordersky/home and www.canalplus.nl (last visited November 12, 2005).
23. Aghion/Bolton, at pp.389 et seq. (making a distinction between nominal length and effective length of a contract). See also Farrell/Shapiro, at p.125; Klemperer, at p.376
24. Differentiating Aghion and Bolton, fn.23 above, at p.399.
25. See, for example, BBC World Service, Terms and Conditions, No.2 (Terms): "The Agreement shall be automatically extended for further periods of twelve months, subject to payment of the Subscription by the Subscriber, unless terminated by either party giving to the other party not less than fifteen days written notice to expire on the last day of the then current term." Also Canalplus, "Algemene voorwaarden van Canal+ N.V. voor de doorgifte en ontvangst van televisieprogramma's via de kabel en voor de doorgifte en ontvangst van digitale aardse televisiesignalen via de infrastructuur van Digitenne" ("Terms and conditions Canal+ Nederland"), No.A4, available at www.canalplus.nl (last visited November 12, 2005). Note in the small print that the contract must be terminated by registered letter.
26. See Canal+, Terms and Conditions Canal+ Nederland, fn.25 above, No.C18. See also Shapiro, fn.21 above, at p.11.
27. On the effects for consumer switching costs in case of pay-TV bundling see also Galbiati/Nicita/Nizi, at p.23; and Harbord/Szymanski.
28. Fritsch/Wein/Ewert, at p.294.

is even truer in the case of electronic access control to multi-channel service platforms such as pay-TV. Here, viewers find themselves in front of closed doors knowing that the marketplace lies somewhere behind them. The opposite, however, is also true: how will viewers who subscribe to one service platform know about the services available outside the "walled garden"? This highlights the importance of electronic programme guides ("EPG"), which are used by most pay-TV platforms. EPGs give their controllers enormous potential to manipulate the way viewers access and receive content, particularly where no independent alternatives that would allow viewers to compare and access different services are available.[29]

To summarise, individual control over access to content services introduces new features to the broadcasting world as we know it. The classic broadcasting model involves the undirected one-way transmission of electronic content toward a multitude of (anonymous) recipients. Once sent, electronic content can be received by anyone who has the necessary technical equipment and is within reach of the respective transmission medium, be it the footprint of a satellite or the local cable or telephone network. The possibility of exclusionary control over access to broadcasting content (with the help of conditional access technology) is changing the general distribution structure for what was commonly known as "broadcasting". Pay-TV services address individual viewers separately to authorise access, influence their viewing behaviour, send them bills and provide them with specialised advertising or a specific service they requested. The new business models can influence viewers' access to broadcasting content in many ways.

The recommendations of the focus group for a modernisation of the European rules on access to audiovisual content

Focus group three embraced two approaches to the problem of exclusionary control over access to broadcasting content and the "public's right to access to information".[30] First, there was a strong majority for maintaining the list-of-important-events concept in its unchanged form. Ideas suggesting that the European Commission make the list concept mandatory for all Member States, demand more involvement in the process of drafting the lists, or harmonise the interpretation of certain conditions ("substantial part of the public") were rejected. Secondly, it was concluded, though not unanimously, that the list-of-important-events concept was not enough to protect the public's access to information and that a right to short reporting should be introduced in addition. Still unclear are the conditions of, for example, whether to include press agencies, and the limitations to such a right. These two suggestions were brought forward during the consultations and will be taken into account in the final phase of the process of modernising the European rules on audiovisual content. In the following, both suggestions are discussed critically from the perspective of their value to realising actual access to broadcasting content for Europe's citizens and to addressing the problems that were described in the previous section. It will be shown that instead of modernising the European framework they are focused on maintaining the status quo of an analogue past.

List of important events
Concept
The recommendations of focus group three rely strongly on the existing lists of important events in Art.3a of the TWF Directive. The basic concept behind Art.3a of the TWF Directive is to limit the exclusive exploitation of transmission rights for the sake of a general public interest in the wide accessibility of certain content. Recital 18 to the revised TWF Directive stipulates that Member States should be able to take measures to protect the public interest, and here more specifically the "right to information", and to ensure wide access by the public to television coverage of national or non-national events of major importance for society. The list-of-important-events concept has already been described as a "new category of universal service". Pay-TV operators are not entitled to the exclusionary exploitation of such contents. It is unclear what the scope of the limitation on the exclusive exploitation of such events is, what its duration is and under which conditions a partial or deferred coverage is acceptable. Bearing in mind that the rights for deferred or partial coverage are sold separately, this question is important for the future licensing policy of such rights.

Article 3a recognises the right of Member States to draw up so-called "lists of important events". The lists identify events of particular public importance that should be shown—in their entirety or partially—on free-TV. Pay-TV operators are not banned from showing designated events providing the public has the possibility of following such events on free-TV as well. Note that Art.3a of the TWF Directive applies only to transmission rights for organised events and not to other content, for example, content that is subject to intellectual property right law such as films and documentaries.

Assessment
It is not the intention of this article to expose a number of more general concerns regarding the list concept. The biggest challenge probably consists of defining the types of information that are of major importance to the public, the criteria used to define them, and, no less importantly, who will be allowed to define them.[31] Instead, this article will focus on the viewers' perspective and the extent to which the list concept makes content accessible to them.

The list concept deals with content that must be publicly accessible, meaning accessible for all viewers irrespective of whether they subscribe to a service or not. For the time being, such content is restricted to major sports events and some cultural events. Access to major sports events may, indeed, be of particular value to some members of the public. The truth is, however, that

---
29. Even where alternative EPGs are available, they may be not supported by the technical platform that consumers subscribe to.
30. Rapport final du groupe de travail 2, fn.11 above, aao.

31. See in more depth on the list concept, Helberger 2005, fn.14 above, at pp.96 et seq.; Hins, at pp.318 et seq.; Helberger 2002, at pp.292 et seq.

we cannot say which events citizens consider of major importance because citizens are not involved in the list-making process. The list concept reflects and protects a perception of the state of the content viewers should be able to see on free-TV. The list concept also fails to provide a solution for all the other content that is subject to electronic access control, for example, sports events that are not listed but that can still be of major interest to the public or individual members of the public. Finally, it does little to restore the balance between controllers of access to content and those individual viewers who are seeking access to information (see above). With the exception of the listed events, pay-TV providers are entirely free to foreclose electronic access to all kinds of content and content services and make access to such content subject to their own conditions and requirements.

The European Commission acknowledges that Art.3a of the TWF Directive might not suffice to take care of national public information policy concerns when dealing with content that is subjected to electronic access control.[32] Accordingly, the European Commission concludes that "the issue of rights to access to newsworthy events needs further attention"[33] and that it would invite focus group three to discuss the right to short reporting as a possible solution.

The right to short reporting
Concept
For the time being, European Union broadcasting law does not have a right to short reporting. Modelling such a right on Art.9 of the ECTT was suggested:

"Each Party shall examine and, where necessary, take legal measures such as introducing the right to short reporting on *events of high interest* for the public to avoid the right of the public to information being undermined due to the exercise by a broadcaster within its jurisdiction of exclusive rights for the transmission or retransmission, within the meaning of Article 3, of such an event" (emphasis added).[34]

---

32. Explanation by the author: in the above quote, the European Commission addresses the question of the need at the European level for an additional right to short reporting, and whether Art.5(3)c of the European Copyright Directive satisfies the public information policy concerns of Member States when dealing with electronic access control; Council Directive 2001/29 of the European Parliament and of the Council of May 22, 2001 on the harmonisation of certain aspects of copyright and related rights in the information society, Brussels, June 22, 2001 [2001] O.J. L167/10 ("Copyright Directive").
33. European Commission, Communication from the Commission to the Council, the European Parliament, the European Economic and Social Committee and the Committee of the Regions on the future of European regulatory audiovisual policy, fn.8 above, aao, at p.16.
34. See also Council of Europe, Recommendation No.R (91)5, of the Committee of Ministers to Member States on the right to short reporting on major events, where exclusive rights for their television broadcast have been acquired in a transfrontier context, Strasbourg, April 11, 1991 ("Recommendation No.R(91)5 on the Right to Short Reporting"). See also Draft Recommendation on the Right to Short Reporting on major events where exclusive rights have been acquired, updating Recommendation No.R(91)5 of the Committee of Ministers to Member States on the right to short reporting on major events where exclusive rights for their television broadcast have been

The right to short reporting found its way into the Convention at the end of the 1980s. It was drafted to protect the public's "right to information" long before the issue of pay-TV attracted wider attention in Europe. Again, reference is made to Art.10 of the ECHR and a "right of access of the public to information".[35] Pluralism is a second important aspect behind the right to short reporting, namely to encourage competition between several broadcasters.[36]

The right to short reporting does not guarantee coverage of the full event in free-to-air television but only the possibility of reporting about the event by those broadcasters that have not acquired the exclusive rights for full coverage. On the other hand, unlike the list-of-important-events-concept, "the public's right of access to information" is not restricted to cultural or sports events of "major importance for society". It also applies to political and social events of "only" high public interest such as less important sports and cultural events, and social or political newsworthy events such as a report about an accident, a natural disaster or an armed conflict.

Conceptually, the right to short reporting in Art.9 of the ECTT probably comes the closest to an access right; only it is reserved for broadcasters. This is certainly true where it provides a right of access to an event's site. As far as the right to record a signal for the purpose of using it in a short report is concerned, it resembles an exception to exclusive rights in the public interest similar to Art.5(3)c of the Copyright Directive.[37] Unlike Art.3a of the TWF Directive, the right to short reporting does not impose any restrictions on the exclusivity of the transmission—the event can still be broadcast on an exclusive basis. But, similarly to an exception in copyright law, it does oblige the entity that carries out the exclusive transmission to allow certain uses, namely the making of short reports. In other words, the right to short reporting makes the exercise of exclusive transmission rights subject to certain limitations.

Assessment
The right to short reporting is more flexible than the list-of-important-events concept in that it does not work with predefined definitions of "high public interest". It is up to the broadcasters, and in the last instance the courts, to decide if an event is of "high public interest" or not. This increases the range of events that can be subject to a right to short reporting and are not restricted to a predefined list. One may wonder whether the courts are the best place to pass judgment on the rather political question of whether an event is in the public interest or not. Still, the right to short reporting provides more ad hoc flexibility and room for a more audience-oriented decision of the events that are in the

---

acquired in a transfrontier context, Strasbourg, April 16, 2003, MM-Public(2003)003 ("Draft Recommendation updating Recommendation No.R(91)5 on the Right to Short Reporting").
35. Council of Europe, Recommendation No.R(91)5 on the Right to Short Reporting, fn.34 above, Explanatory Memorandum, paras 4 and 5.
36. ECTT, fn.6 above, Explanatory Memorandum, at para.174; Council of Europe, Recommendation No. R(91)5 on the Right to Short Reporting, *ibid.*, at paras 1–2.
37. For an extensive comparison see Helberger 2005, fn.14 above, at pp.109 et seq.

public interest. An implementation of the right to short reporting in the revised Directive should be considered seriously. Having said that, focus group three did not consider a very practical, major problem with the right to short reporting and access of the public to information, at least where it is invoked against pay-TV operators. The right to record a signal for the purpose of making a short report will not be effective if the primary broadcaster is not obliged to provide the signal in unencrypted form. The secondary broadcaster would first have to gain access to the encrypted signal or the event itself before being able to make a short report. Article 9 of the ECTT, however, does not include a corresponding obligation for the primary broadcaster. In practice, this could lead to lengthy negotiations and the risk that the interest of the public to be informed about a particular event becomes obsolete with the passage of time.

It must be noted that the audience plays a passive role in the right to short reporting; again, it is the mission of the media to keep the audience informed and to decide which events it needs to known. The right to short reporting focuses on keeping the public informed more than on ensuring the public's full access to the content in question. This is easily explained by the fact that the right to short reporting protects informational interests only. The public does not need to watch the full event to be properly informed. What the right to short reporting does not guarantee to viewers is that content which is of importance to them personally is accessible at fair conditions and prices.

General critique

The goal of the list of important events and the right to short reporting is to ensure the universal accessibility of content that is subject to exclusive rights in free-TV. Article 3a of the TWF Directive imposes limits on the exclusive exploitation of transmission rights for certain events. The right to short reporting grants broadcasters a right to use content or access events that are subject to exclusive rights. The right to short reporting primarily protects the interest of citizens to be properly informed. Article 3a of the Television Directive also serves other public interest objectives, namely to promote access to content of importance for society in order to foster social cohesion, competition between free-TV and pay-TV, and, at least this is the interpretation of the author, public broadcasting.

One can doubt whether both tools, the list of important events and the right to short reporting, are adequate and effective in achieving their stated goals: to protect the public's access to event and news reports of public importance. It exceeds the scope of this article to expose the different lacunae in both concepts in their present form.[38] The author's main point of criticism in this article is that both initiatives focus entirely on maintaining a concept of broadcasting that belongs to the analogue past. The existing solutions are incomplete as they do not really take into account the varied and individual interests or needs of modern viewers of broadcasting content, notably viewers of access-controlled broadcasting.

It was explained that one of the most significant changes brought by digitisation in combination with electronic access control to the traditional broadcasting world is that viewers enter into an individualised, commercial, pay-directly-for-broadcasting-content relationship. The production of information goods and services is not for "free" and often requires substantial investment. It is neither economically rational[39] nor realistic to assume that viewers will satisfy their need for access to information in the future exclusively via free-TV. This will be even more so once the process of digitisation has been completed and broadcasters and network providers are confronted with the need to recoup their investments in the digitisation process. From the viewers' perspective, they will increasingly depend on access to an access-controlled platform before they can access particular content. The distribution of broadcasting content is shifting from a previously public sphere to a more personal sphere where the conditions for access to a service are directly negotiated between the service provider and the requester. It is the contractual relationship between the viewer and the platform operator that will determine which content can be viewed under which conditions. The relationship between them is no longer governed by broadcasting law alone but also by the terms and conditions imposed by the platform operator. As explained above, the technical, contractual or other conditions of the subscription package and the lack of a sufficiently transparent information environment can disadvantage subscribers and non-subscribers. They can pose serious obstacles not only to viewers' access to a commercial service, but also viewers' access to information. Neither the list-of-important-events concept nor the right to short reporting takes these effects of electronic access control into account.

To conclude, both the right to short reporting and the list of important events are ill-prepared for the resulting changes to the distribution structure, meaning the change from broadcasting to individual access, from viewer to consumer, and from public mission to profit-driven services. Where access to broadcasting content becomes a matter of private contracts and control over access to content, traditional concepts do not help much. Ensuring that some events or excerpts thereof remain available on free television may be a means of preserving the importance and competitiveness of free-TV in general and public broadcasting in particular. But the more pay-TV prospers and is perceived by viewers as a third possible form of financing broadcasting, the

---

38. See instead Helberger 2005, fn.14 above, at pp.96 et seq. and pp.107 et seq.

39. Making access to information subject to price negotiations is not new in the media world. Consumers are used to paying for their newspapers. The same is true for films shown in cinemas or the purchase of CDs and DVDs. Here too, citizens do not usually access the information stored on a CD or DVD without having to pay first. Even the reception of public and commercial broadcasting is far from being "for free". To receive public and commercial free-to-air broadcasting, consumers not only have to purchase a television or computer, they must also subscribe to, for example, a cable or satellite network and "pay" for some programmes in the form of public broadcasting fees. For commercial, advertisement-financed programmes they also pay in non-monetary but money-worth "assets", such as time and attention.

more urgent is the need to take into account the effect of electronic access control on viewers' access to digital content.

Reform proposal

Having concluded that the list-of-important-events concept and the right to short reporting are not sufficient safeguards of the viewers' access to information in a modern digital environment, the next question is how to reform the situation.

Individual right of access for viewers?

One option would be to plead for a right of access to information for individual viewers. There are, however, several reasons that would speak against such a right. It would go beyond the scope of this article to discuss the different reasons in more depth.[40] In short, an individual right of access to information could conflict with the equally valuable protection-worthy rights of others, notably the service providers' protection under Art.10 of the ECHR to impart information if and how they wish. It could conflict with the service providers' economic freedoms such as the freedom of contract and property. In other words, there is a need to carefully weigh differing interests before jumping to the conclusion that one private party should have a right of access to information over another private party. This process of weighing is, arguably, best done in parliament. This does not, however, prevent states from reaching the conclusion that there is a need to create a "right of access to information" over other private parties in the form of statutory rules.[41] If the states do reach this conclusion, they must carefully balance the interests of all parties concerned according to, among others, Art.10(2) of the ECHR. Another question is whether such a right is desirable. Do we want to create a media landscape in which one private party could make an enforceable claim against another private party to provide it with certain information it holds? What would this mean for the protection of property, the private sphere and personal autonomy? Even if providing such a right were restricted to a right against the media, where would the line between a private person and the media have to be drawn in a time in which transmission technologies can turn every viewer into a "broadcaster" or "press service"? In the case of pay-TV, a right of access to information might not even be very useful because it is not the intention of the provider of access-controlled broadcasting to refuse access but to sell access; albeit under his own conditions.

"Broadcasting-viewer protection law"

In other words, from the viewer perspective, it is not so much the question whether they will have access to certain kinds of information but under which conditions they will be able to access the information and whether the conditions are acceptable and affordable. This means that the key to finding a solution that takes the changing and increasingly interactive and individualised distribution patterns into account lies in the commercial relationship between content controller and viewer. The revision process of the TWF Directive confirmed again that the present approach in broadcasting law of dealing with electronic access control is still a top-down approach in which the state, state agencies or, as in the case of the right to short reporting, broadcasters determine which information is to remain accessible to viewers. This article suggests following a bottom-up approach, meaning that broadcasting regulators should, in addition, address the contractual relationship between service providers and viewers and its impact on viewers' access to information.

Bottom-up approach

Arguably, one way of promoting individual access to content without interfering disproportionately with the programming autonomy of the content provider (as would be the case with an individual access right), is to create the conditions for fair and affordable access to broadcasting content. Viewers' access to informational or cultural content should not be inhibited by unfair access conditions in subscriber contracts, contractual or technical lock-ins or lock-outs, or a lack of adequate service information. A future goal for the regulation of pay-TV should be to ensure that pay-TV platforms are publicly accessible, meaning that the terms and conditions of access are such that no members of the public are arbitrarily excluded for technical, financial or transparency reasons. In the information society, each citizen should be able to benefit from new services that become available by means of advanced communications:

> "The information society is not only affecting the way people interact but it is also requiring the traditional organizational structures to be more flexible, more participatory and more decentralised."[42]

Likewise, subscribers to one particular platform should not be unreasonably impeded from benefiting from access to various services and pluralism between different platforms because of technical, contractual or informational lock-ins. The fairness and openness of the individual commercial relationship between service provider and viewer is key to preventing electronic access control from being used to the detriment of competition, viewers and public information policy.

The commercial relationship between service providers (broadcaster or platform operators) and viewers is a relationship that has so far been ignored in European broadcasting law. Some aspects of subscription contracts may fall under e-commerce law. E-commerce law, however, is not designed to protect further-reaching public policy objectives attached to the product or service such as access to knowledge and the particular value that access to information and culture has for democracy and society. Looking for possible models for a bottom-up approach to broadcasting regulation leads to a field of law that is not often discussed in the context of broadcasting regulation: telecommunications law, and here in particular the Universal Service Directive.

---

40. See instead Helberger 2005, fn.14 above, at pp.74 et seq.
41. See also Mackaay, fn.14 above, at p.172, referring to information about criminals or hazardous products.

42. G7 Summit, Conclusion of G7 Summit Information Society Conference, Doc/95/2/, Brussels, February 26, 1995.

Example: Universal Service Directive

The Universal Service Directive (1) seeks to balance the commercial relationship between viewer and service provider to stimulate a functioning marketplace; and (2) acknowledges the particular value of accessibility and availability of electronic communication to society *and* individual consumers. First, the Universal Service Directive requires a minimum level of availability and affordability of basic electronic telecommunications services (the so-called universal service obligations). Secondly, it guarantees a set of consumer rights and consumer protection rules for the sector for users and consumers of electronic telecommunications services. One underlying idea of the Universal Service Directive is that functioning competition and the broad availability of a range of different telecommunications services for consumers is not only a matter of access for service providers to telecommunications networks and facilities, but also a matter of access for consumers to services. As in European broadcasting law, there is a need in telecommunications law to strike the right balance between relying as much as possible on market mechanisms and competition to achieve a high level of choice and quality, and ensuring regulatory intervention to uphold a minimum number of consumers' rights throughout the European Union[43] and protection-worthy interests in the broad availability and accessibility of services, including services providing informational content. Enterprises with significant market power that charge excessive or predatory prices to consumers, apply unreasonable bundling strategies or show undue preferences to certain consumers can inhibit individual consumers' access and, by so doing, the realisation of general public interests.[44] This is why the Universal Service Directive aims to

> "ensure the availability throughout the Community of good quality publicly available services through effective competition and choice and to deal with circumstances in which the needs of end-users are not satisfactorily met by the market" (Art.1(1) of the Universal Service Directive).

The following brief overview may provide an idea of how telecommunications law has already developed solutions for problems that broadcasting regulators are just starting to realise.

Fairness of contractual conditions

As far as consumer contracts are concerned, the Universal Service Directive provides a legal framework that gives National Regulatory Authorities (NRAs) instruments to ensure that service providers offer services to consumers at adequate, non-discriminatory conditions and fair prices, and refrain from unjustified bundling strategies.[45] The Universal Service Directive acknowledges that for

43. Proposal for a Directive of the European Parliament and of the Council on universal service and users' rights relating to electronic telecommunications networks and services, December 19, 2000 [2000] O.J. C365/238, Explanatory Memorandum, s.III.
44. Council Directive 2002/22 of the European Parliament and of the Council of March 7, 2002 on universal service and users' rights relating to electronic communications networks and services, Brussels, April 24, 2002 [2002] O.J. L108/51 ("Universal Service Directive"), Recital 30.
45. ibid., Art.17(2).

"reasons of efficiency and social reasons, end-user tariffs should reflect demand conditions as well as cost conditions, provided that this does not result in distortions of competition".[46]

The Universal Service Directive also stipulates that service providers must provide consumers with a minimum level of legal certainty in subscriber contracts concerning contractual terms and conditions, service quality, contract and service termination conditions, compensation measures and dispute resolution. Contracts must include information on prices, tariffs, and terms and conditions in order to increase the consumers' ability to optimise their choices and thus fully benefit from competition.[47] These provisions provide tools to tackle contractual lock-ins and unfair provisions in consumer contracts.

Adequacy of technical conditions

Regarding the problem of technical lock-ins, the Universal Service Directive offers a solution to the interoperability of consumer equipment. According to Art.24 of the Universal Service Directive, Member States must ensure that any analogue television set has at least one open interface socket to connect additional devices and ensure interoperability. The provision, however, only refers to digital television equipment, not to set-top boxes.

Questions concerning decoder interoperability are dealt with in the Access Directive and the Framework Directive. However, neither of the directives obliges operators in the pay-TV sector to provide for adequate interoperability solutions. In addition, under Arts 5(1)b and 6 of the Access Directive, NRAs do not have as many possibilities of imposing interoperability obligations on pay-TV providers as they do for all other telecommunications service and facilities providers according to Arts 8 to 13 of the Access Directive. It is true that the discussion about the adequacy and desirability of mandated interoperability can be very controversial. Nevertheless, the author concludes that there are valid and important arguments in favour of mandated interoperability solutions in areas in which service mobility and fair access opportunities would otherwise be at risk. It is difficult to see why the (un)willingness of major industry players towards standardisation in pay-TV would change in the future and why necessary initiatives in this field should be further postponed. Without imposing a particular standard, initiatives could focus on making consumer equipment interoperable through open interfaces and open standards for software and middleware. The approach of Art.24 of the Universal Service Directive to mandate a common interface could serve as a model.

Comparable service information

In terms of the transparency issue, the importance of comparable service information has already been acknowledged in the Communications Framework. The Universal Service Directive, and in particular the provisions on directory services and transparency obligations in Arts 21 and 22, demonstrate that transparency, as a precondition for functioning competition and the realisation of public information policy objectives, is taken very seriously

46. ibid., Recital 26.
47. ibid., Recital 30, Art.20(1), (2).

in Europe. Access to telephony directory services, a kind of search agent, is even subject to a universal service obligation. The examples of TV guides for the broadcasting sector or search engines and browsers for the internet demonstrate that there can be a market for independent information agents and that competition between such services is generally possible. Broadcasting regulators could seek to stimulate such competition. Article 21(2) of the Universal Service Directive, a provision that mandates providing consumers with comprehensive, comparable and user-friendly service information, could serve as a model.

Convergence

It is worth noting that, at present, none of the potentially relevant provisions in the Universal Service Directive (in particular Arts 17, 20, 21, 22, and 32), with the exception of Art.24 (interoperability of television equipment), applies to the broadcasting sector.[48] European telecommunications law is based on the idea that the sector-specific regulation of media services is divided into transport, or technical aspects, and content-related aspects. Both aspects fall under very distinct regulatory frameworks.[49] This is another indication that, as far as broadcasting services are concerned, regulators handle a fundamentally different idea of the consumer/service-provider relationship than they do for all other communication services. In the case of digital broadcasting, viewers are still not considered active market participants or "consumers"—which they are in pay-TV—but passive receivers. Pay-TV services are distributed to viewers, in a similar way as telephony and other telecommunications services, on an individualised basis. Consequently, in pay-TV too, technical or contractual conditions in subscriber packages or the lack of transparency can be a means to impede or even foreclose access.

With ongoing convergence, the differentiation between broadcasting and non-broadcasting services is no longer justified *as far as the modalities of the way services are marketed to viewers are concerned*. It is difficult to see why viewers of digital broadcasting services should receive less legal protection than consumers of other electronic services.

Conclusion

Electronic access control brings an end to the idea of the anonymous viewer and the undirected mass distribution of broadcasting services to whoever decides to watch. Today, business models for the distribution of broadcasting services are based on commercial relationships between service providers and individual viewers. Contractual and technical conditions overrule the free access culture that was once the essence of many traditional broadcasting services. It can be expected that digitisation will further accelerate this trend. For viewers, this has

48. *ibid.*, Recital 45; Framework Directive, Art.2(c).
49. European Commission, Towards a new framework for electronic communications infrastructure and associated services, The 1999 Communications Review, Brussels, November 10, 1999, COM (1999) 539 final ("1999 Communications Review"), pp.vi–vii.

far-reaching consequences in terms of their access to and use of electronic content.

In order to guarantee that the outcome of this trend respects the "public's right to information", it is necessary to concentrate more on the contractual and technical arrangements between service providers and the individual(ised) members of the public: the viewers of broadcasting content. Decisions about the accessibility and availability of content are made at the level of the viewer/provider relationship. This is also the level at which individual viewers express which kind of content they find particularly important. At this stage, this article cannot do much more than hint at the need for a new perspective in broadcasting law—the viewer's perspective—and for rules that protect the position of viewers in their dealings with broadcasting service providers. Further research should be carried out to explore the question of how specific "broadcasting-viewer protection rules" should be to guarantee the broad accessibility and availability of broadcasting services on fair, non-discriminatory conditions, and at an affordable price. Secondly, general consumer protection law applies to all kinds of services and is probably not designed to reflect the idea of information as a service of particular social and democratic relevance, continuous and reliable access to which should be available at affordable prices, with good quality, and on user-friendly terms. Further research is also needed to explore the potential of consumer protection law to realise public information policy goals, such as pluralism and the realisation of freedom of expression and democratic principles. The example of the Universal Service Directive could be a good starting point and serve as a model. Unlike general consumer protection law, the Universal Service Directive leaves room to combine consumer protection with the realisation of general competition and public information policy objectives in order to promote what the directive calls

> "the twin objectives of promoting effective competition whilst pursuing public interest needs, such as maintaining the affordability of publicly available services for some consumers".[50]

The goal of this reform proposal is not to replace but to complement existing concepts, notably the list-of-important-events concept and the right to short reporting. In particular the suggestion of implementing the latter into European broadcasting law should be taken seriously. Electronic access control should hamper neither competition between free-TV and pay-TV nor the functioning of the media. The same is true of the task of competing media to inform and to criticise. An important objective of the list of important events and the right to short reporting is to protect the competition between free and pay-TV. Both concepts, however, need further improvement to be effective.

Pay-TV is not a monster, and the arrival of electronic access control is not necessarily a threat to a flourishing media landscape; it could also be seen as an opportunity for new and more responsive content services. The precondition is, however, that individual viewers are no longer treated as if they were invisible to the broadcasting regulator. For an open and diverse broadcasting environment, it is vital that viewers

50. Universal Service Directive, fn.44 above, Recital 26.

can access access-controlled services on fair, affordable and non-discriminatory conditions, that they have a real choice between different competing platforms, including platforms from other countries, and that they have reliable information about the services available to them.

References

P. Aghion and P. Bolton, "Contracts as a Barrier to Entry" (1987) 77 *American Economic Review* 388

E. Barendt, *Broadcasting Law* (Clarendon Press, Oxford, 1993)

D. Evans, "The Antitrust Economics of Multi-Sided Platform Markets" (2003) 20(2) *Yale Journal on Regulation* 325

J. Farell and C. Shapiro, "Optimal Contracts with Lock-in" (1989) 79 *American Economic Review* 51

M. Fritsch, T. Wein and H. J. Ewers, *Marktversagen und Wirtschaftspolitik* (3rd edn, Vahlen, München, 1999)

R. Galbiati, A. Nicita and G. Nizi, "Regulation, Competition, and Institutional Design in Media Markets: The Evolution of pay-TV in UK, Australia and Italy" [2004] *American Law & Economics Association Annual Meetings*, No.76

D. Gunzerath, "Darn that PAYTV!: STV's Challenge to American Television's Dominent Economic Model" [2000] *Journal of Broadcasting & Electronic Media* 655

D. Harbord and S. Szymanski, *Restricted View. The Rights and Wrongs of FA Premier League Broadcasting* (Study, Consumers' Association, London, 2005)

N. Helberger 2002, "Brot und Spiele—Die Umsetzung der Listenregelung des Artikel 3a der Fernsehrichtlinie" [2002] *Archiv für Presserecht* 292

N. Helberger 2005, *Controlling Access to Content—Regulating Conditional Access in Digital Broadcasting* (Kluwer International, Den Haag, 2005)

A. W. Hins, "Uitzendrechten voor Belangrijke Evenementen en de EG-Televisierichtlijn" [1998] *Mediaforum* 318

P. Klemperer, "Markets with Consumer Switching Costs" (1987) 102 *Quarterly Journal of Economics* 375

L. Lessig, *Code and Other Laws of Cyberspace* (Basic Books, New York, 1999)

E. Mackaay, "The Public's Right to Information", in *Information Law Towards the 21st Century* (W. F. Korthals Altes, E. J. Dommering, P. B. Hugenholtz and J. J. C. Kabel ed., Kluwer Law International, Information Law Series, Deventer, 1992)

J. M. de Meij, A. W. Hins, A. J. Nieuwenhuis and G. A. I. Schuijt, *Uitingsvrijheid. De Vrije Informatiestroom in Grondwettelijk perspectief* (3rd edn, Otto Cramwinckel, Amsterdam, 2000)

G. O'Driscoll, *The Essential Guide to Digital Set-top Boxes and Interactive TV* (Prentice Hall PTR, Upper Saddle River, New York, 2000)

C. Shapiro, "Exclusivity in Network Industries" [1999] *George Mason Independent Law Review* 3

[14]

Access to content by new media platforms: a review of the competition law problems

Damien Geradin*

LT Audiovisual industry; Competition law; EC law; Media; Sporting events

> *Access to premium content, and in particular football games and Hollywood movies, is of crucial importance for operators active in the delivery of audio-visual content. Getting access to premium content is, however, a rather complex matter. The combination of scarcity of such content and the presence of exclusivity clauses in premium content rights contracts has translated into a spiralling of the costs involved in buying content. In most Member States, premium content has been monopolised by dominant pay-TV operators. As a result, new entrants, such as new media platforms, have no access to such content. This prevents them from acquiring market share. The main argument followed throughout this paper is that, while recent Commission decisions contain remedies, which will help new media platforms to gain access to premium content, such remedies are insufficient to create a level playing field in the market for the acquisition of such content. Numerous anti-competitive practices continue to plague this market and further competition law intervention is thus required.*

Introduction

The paper seeks to provide a discussion of the competition law issues raised by access to premium content (essentially blockbusters and football rights) by content delivery operators with a special emphasis on new media platforms. A significant amount of literature has been published on the application of competition rules to premium content rights agreements,[1] but the specific obstacles encountered by new media platforms have been relatively unexplored. This paper seeks to fill this gap in the literature.

The European Commission (hereafter, the Commission) has recognised in its decisions that premium content is an "essential input" for operators active in the delivery of audio-visual content.[2] There is indeed no substitution possible with other, less attractive, forms of content. In fact, premium content such as major football events represents "stand-

---

* Member of the Brussels bar. Professor of Law and Director of the Institute for European Legal Studies, University of Liège and Professor and Director of the Global Competition Law Centre, College of Europe, Bruges (d.geradin@ulg.ac.be). I would like to thank Nicolas Petit and David Henry for excellent research assistance. This paper has benefited from the financial support provided by the PAI P4/04 granted by the Belgian State, Prime Minister's Office, Science Policy Programming.
[1] See, *e.g.* Darren McAuley, "Exclusively for All and Collectively for None: Refereeing Broadcasting Rights between the Premier League, European Commission and BSkyB", [2004] E.C.L.R. 370; Stefan Szymanski and David Harbord, "Football Trials" [2004] E.C.L.R. 117.
[2] See Commission decision of October 13, 2000, *Vivendi/Canal+/Seagram*, Case No.IV/M.2050, [2000] O.J. C311/3 at para.19.

alone" driver content for pay-TV operators.³ Absent access to such content it is very difficult for content delivery operators to gain or retain market shares. Access to premium content is thus a matter of life or death for such operators.

Yet, getting access to premium content is not an easy matter. First, premium content is scarce as there are only a few blockbusters and a limited number of premium sport events every year. Moreover, premium content rights contracts usually involve some form of exclusivity pursuant to which dominant pay-TV operators often manage to monopolise such rights for several years at the expense of weaker competitors. The combination of scarcity and exclusivity has translated into a spiralling of the costs involved in buying premium content.⁴ For instance, while in 1992, broadcasters paid €434 million for the TV rights of the English Premier League, in 2000, they paid €2.6 billion for only three seasons.⁵

The lack of access to premium content represents a significant handicap for new entrants, such as new media platforms. If these platforms want to gain market share, they need to show programmes, which are able to compete with the content shown by dominant pay-TV operators.⁶ Access foreclosure to premium content would thus not only prevent new entries from taking place in the highly concentrated pay-TV market, but would also affect technological developments and consumer choice as the latter would be prevented from watching their favourite programmes on the platform of their choice. Thus, in a number of policy speeches, Commission officials have insisted on the importance that new media platforms gain access to premium content.⁷

The main argument followed throughout the paper is that, while recent Commission decisions contain remedies, which will help new media platforms to gain access to premium content, such remedies are insufficient to create a level playing field in the market for the acquisition of such content. Numerous anti-competitive practices continue to plague this market and further competition law intervention is thus required.

This paper is divided into seven sections. Following this introduction, the second section reviews the reasons why access to premium content is a major bottleneck affecting the content delivery market and in particular new entrants on that market. The third section outlines the legal framework in which the debate over access to premium content takes place. The fourth section addresses the complex issue of market definitions in the media industry. The fifth section reviews the competition law issues that have been addressed by the Commission in its decisional practice. Reference to relevant Court of Justice case law is also made. The sixth section analyses the extent to which the remedies adopted by the

---

³ See Commission decision 2004/311/EC of April 2, 2003, *Newscorp/Telepiù*, Case COMP/M.2876, [2004] O.J. L110/73 at para.66.

⁴ Other factors may also have contributed to the growing cost of acquiring premium content, such as for instance the salary increases of the football players in the 1990s.

⁵ See Torben Toft, "TV Rights of Sports Events", Brussels, January 15, 2003.

⁶ See Commission decision of July 23, 2003 relating to a proceeding pursuant to Art.81 of the EC Treaty and Art.53 of the EEA Agreement (COMP/C.2–37.398), [2003] O.J. L291/25 at para.83:"[. . .] content rights will be necessary for the development of the new services, in the same way as content rights are necessary for TV broadcasting services, where football content is being used to entice consumers to take up pay-TV subscriptions and to attract advertisers to TV channels."

⁷ See, *e.g.* Philip Lowe, "Media Concentration and Convergence: Competition in Communications", speech delivered in Oxford on January 13, 2004, available at *http/europa.eu.int/comm/competition/speeches/text/sp2004_002_en.pdf*.

Commission sufficiently address the difficulties encountered by new media platforms in acquiring premium content. The seventh section discusses the problem for new platform operators in acquiring TV channels. Finally, the eighth section contains a brief conclusion.

Premium content as a major bottleneck

As we have seen above, access to premium content is a major bottleneck affecting the content delivery industry. This is perhaps the most important concern for operators and this represents a very large proportion of their costs. For instance, football represents 30–65 per cent of the broadcasters' total rights expenditure.[8] We have seen in the preceding section that some factors, such as scarcity of content or exclusivity, may be a part of the problem, but other factors are also relevant. They are briefly summarised in the bullet points, which follow:

- Because of scarcity combined with exclusivity, buying rights may involve astronomical amounts of money and thus create important financial risks for new operators. This may in turn lead to barriers to entry, especially considering the commercial policy of sellers, such as the requirement of minimum guarantees;
- Rights holders try to extract maximum value from their rights, by a variety of commercial practices, such as, for instance, selling movies several times (the so-called windows system);
- Rights holders also want to protect the value of their rights by preventing content to be shown through certain delivery means (*e.g.* movie rights or football rights sellers tend to be reluctant to sell their rights to new media platforms because they believe this may diminish their value)[9];
- There is a tendency for content producers and content delivery companies to enter into long-term exclusive contracts. There may be good reasons for exclusivity, but at the same time it creates significant risks of foreclosure.[10]
- Buyers of rights will often negotiate holdback and pre-emption rights (*e.g.* for second window movies),[11] which may prevent new entrants from entering the market as "fringe" competitors. These protection rights may also have the effect of withdrawing content (*e.g.* second window movies) from the market.
- The rights purchased by content delivery operators may cover one platform (*e.g.* DTH),[12] but also several other platforms (*e.g.* UMTS[13] and internet). In the latter

---

[8] Herbert Ungerer, "Commercialising sport: Understanding the TV Rights debate", speech delivered in Barcelona October 2, 2003, available at *http/europa.eu.int/comm/competition/speeches/text/sp2003_024_en.pdf*.

[9] See Alexander Schaub, "Sports and Competition: Broadcasting rights for sports events", speech delivered in Madrid, February 26, 2002: "We have for example seen a reluctance of sports associations in granting Internet and UMTS-rights because broadcasters fear that the Internet will undermine the value of their TV rights.", p.6, available at *http/europa.eu.int/comm/competition/speeches/text/sp2002_008_en.pdf*.

[10] See Lowe, above n.7, at p.8.

[11] For a discussion of the "windows" system, see infra text accompanying nn.86–88.

[12] DTH, which stands for Direct-to-Home system, refers to the transmission over satellite of programming directly into small receiving antennas located at viewers' homes. DTH is an alternative to the cable system and is used in some Member States to provide pay-TV services. See *www.indiainfoline.com/nevi/cabl.html*.

[13] UMTS, which stands for Universal Mobile Telecommunications System, is the European standard for third generation mobile telephony. Data speeds will range from 114 to 2000 kbps, allowing a wide range of high bandwidth applications as well as voice telephony. See *www.flexibility.co.uk/helpful/glossary.htm*.

case, these alternative platforms are unable to deal with the content provider directly. Instead, they will have to negotiate with a competitor on the delivery segment (*i.e.* the company who bought the rights for the different platforms), which may decide not to sell them the rights to prevent entry.
- In the sports industry, rights are often sold jointly (*e.g.* the Premier League or the UEFA). In some cases, rights may also be purchased jointly (*e.g.* the Eurovision system);
- The media market has been subject to both vertical and horizontal mergers in recent years. There may good reasons for such mergers, but they will often involve significant risks for competition. Hence, such mergers have been only authorised provided that the parties agreed to accept significant commitments;
- On the delivery segment, new entrants face a chicken and egg problem: to gain market shares they need premium content, but to gain access to content they need significant market shares.

In sum, although some describe the market for the acquisition of rights as a "bidder market" (because of exclusivity, competition is rather "for" the market, rather than "in" the market), this seems to be hardly the case in Europe. Indeed, the market(s) for the acquisition of premium content is/are suffering from serious failures with the result that only dominant pay-TV operators have access to such content. Some form of public intervention through, for instance, the application of EC competition rules is therefore needed.

The legal framework

A distinction should be made here between, the sector-specific regulatory framework and the competition law framework.

As far as the media sector is concerned, the sector-specific regulatory framework comprises two main components. First, there is the new regulatory framework for electronic communications which, with some limited exceptions, regulates the non-content related aspects of electronic communications.[14] This framework contains the rules that should be applied to electronic communications operators holding significant market power.[15] The second component of the sector-specific framework is the TV without

---

[14] In substance, the new regulatory framework on electronic communications is composed of five directives: one framework directive and four specific directives respectively dealing with authorisations, universal service, access and interconnection, and data protection and privacy in the telecommunications sector. See Directive 2002/21 on a common regulatory framework for electronic communications networks and services, [2002] O.J. L108/33; Directive 2002/20 on the authorisation of electronic communications networks and services, [2002] O.J. L108/21; Directive 2002/22 on universal service and users' rights to electronic communications networks and services, [2002] O.J. L108/51; Directive 2002/19 on access to, and interconnection of, electronic communications networks and associated facilities, [2002] O.J. L108/7; Directive 2002/58 concerning the processing of personal data and the protection of privacy in the electronic communications sector, [2002] O.J. L201/37.

[15] For a good discussion of this framework, see Alexandre de Streel, "Remedies in the Electronic Communications Sector" in D. Geradin (ed.), *Remedies in Network Industries* (Intersentia, 2004).

72 New Media Platforms

borders directive, which regulates content-related issues.[16] There is thus a clear work sharing between these two components of the sector-specific regulatory framework.

This regulatory framework contains two relevant provisions discussing access to content. First, Art.3(a) of the TV without borders directive permits Member States to take measures to ensure wide access by the public to "free-to-air" television coverage of sports events that are regarded in the Member State as being of "major importance" for society. This covers both events of national importance (such as, for instance, the "Giro" in Italy or the "Vuelta" in Spain), but also events of international importance such as the Olympic Games or the football World Cup. Secondly, Art.31(1) of the Universal Service Directive permits Member States to impose proportionate and transparent "must carry" obligations on cable television network operators. They may be imposed when a significant number of end users use cable networks to receive radio and television broadcasts. They may also be imposed on terrestrial and satellite networks.

The main aspects of the competition law framework are Arts 81 and 82 EC, which respectively prohibits restrictive agreements between competitors and abusive conduct by dominant operators, as well as the Merger Control Regulation which prevents mergers that may "significantly impede effective competition".[17] As will be seen below, recent decisions such as the exemption of the UEFA joint selling agreement and the clearance of the *Newscorp/Telepiu* merger (in both cases after significant commitments made by the parties), bear testimony to the major influence of competition rules in the shaping of the legal environment of the media rights markets.

While there are often interactions between competition law and sector-specific regulation,[18] the rest of the paper will essentially focus on the application of competition rules to the problems raised by access to content. Indeed, I believe that it is on the basis of EC competition rules that effective remedies could be designed to address the market failures which are still plaguing the market(s) for acquisition of premium content.

Market definitions

Competition law is about markets and market definition is often the first necessary step in competition law analysis. Defining product markets is not an easy task in the media industry as the technological environment is evolving very quickly.[19] It is also an industry involving a range of operators, such as, for instance, content suppliers, channel suppliers, free-to-air TV operators, pay-TV operators, as well as a range of different products, such

---

[16] See Council Directive of October 3, 1989 on the coordination of certain provisions laid down by law, regulation or administrative action in Member States concerning the pursuit of television broadcasting activities (89/552), [1989] O.J. L298/23 as amended by Directive 97/36 of the European Parliament and of the Council of June 30, 1997, [1997] O.J. L202/60.

[17] See Art.2(3) of the Council Reg.139/2004 of January 20, 2004 on the control of concentrations between undertakings, [2004] O.J. L24/1.

[18] See Damien Geradin and Greg J. Sidak, "European and American Approaches to Antitrust Remedies and the Institutional Design of Regulation in Telecommunications", forthcoming in the *Handbook of Telecommunications Economics* (Vol.2), at p.18.

[19] For a discussion of market definition in the broadcasting sector, see Laurent Garzaniti, *Telecommunications, Broadcasting and the Internet—EU Competition Law and Regulation* (2nd ed., Sweet and Maxwell, 2004) at pp.452–491.

as video-on-demand, near-video-on-demand, pay-per-view, etc. Market structures thus tend to be complex.

From a general standpoint, it is useful to draw a distinction between upstream and downstream markets. While the upstream level is concerned with the production/ acquisition of content, the downstream level relates to the delivery of content to the consumers. There are, however, clear interactions between the two levels as, for instance, commercial practices taking place at the upstream level will affect the competitive structure of the downstream level.

At the upstream level, the Commission has traditionally segmented the purchasing activity for content rights into separate markets according to the nature of the content. Segmentation on the basis of content can be illustrated by the following examples:

- In *Canal+/RTL/GJCD/JV*, the Commission found that although sport broadcasting rights may constitute a distinct field from other television programming, that market ought to be further subdivided into separate product markets and that, at least within the EEA, football broadcasting rights may not be regarded as substitutes for other sports broadcasting rights.[20] The Commission therefore concluded that there was a separate market for the acquisition and resale of football broadcasting rights to events that are played regularly throughout every year.
- In *Vivendi/Canal+/Seagram*, the Commission distinguished between a market for broadcasting rights for feature films and a market for broadcasting rights for made-for-TV programmes.[21]
- In *Newscorp/Telepiu*, the Commission defined the markets affected by the transaction as the acquisition of: exclusive rights to premium films; exclusive rights to football events that take place every year where national teams participate (mainly national league, national cup, UEFA cup and UEFA Champions League); exclusive rights to other sport events; and acquisition of TV channels.[22]

The above markets were identified in the context of specific cases at a given point in time. Nothing would thus prevent the Commission from further sub-dividing these markets in the future. For instance, in *Newscorp/Telepiu*, the Commission invoked the possibility that the acquisition of exclusive broadcasting rights for popular sport events (*e.g.* important tennis tournaments, boxing matches, golf and motor bike races, etc.) could be considered as separate product markets according to single sports.[23] However, the Commission did not take a formal position on that point. Similarly, as will be seen below, a further market segmentation could be envisaged as, for instance, it could be argued that during a limited period of time, each blockbuster could be defined as a single market. As the Commission has never taken a formal decision on this issue, it still remains open.

At the downstream level, the Commission draws a distinction between different modes of delivery of audio-visual content to consumers. For instance, in *TPS*, the Commission

---

[20] See Commission Press Release of November 13, 2001, "Commission clears sports rights venture between Canal+, RTL and Groupe Jean-Claude Darmon", IP/01/1579.
[21] See Commission decision, above n.2, at para.[17].
[22] See Commission decision, above n.3, at para.[55].
[23] *ibid.*, at paras 71–72.

74 New Media Platforms

decided that the market for pay-TV was separate from that of free-to-air TV.[24] In *Newscorp/Telepiu*, the Commission went one step further as it distinguished between the pay-TV market, the cable market, the free-to-air TV market, and the new media platforms.[25] But here again, as at the upstream level, the above market definitions have been adopted in specific circumstances and these definitions might well evolve in the future. Moreover, further downstream markets have also been identified by the Commission, such as, for instance, in *UEFA* where the Commission refers to the "downstream markets on which broadcasters compete for advertising revenue depending on audience rates and pay-TV subscribers".[26]

Because of cultural and linguistic differences, the Commission generally defines markets as being national in scope or a wider area that is linguistically homogeneous.[27]

The competition law issues addressed by the Commission and the European Courts

Over the last few years, the Commission has adopted several important decisions over access to premium content. Interestingly, these decisions have been reactive rather than pro-active, *i.e.* they result from notifications under former Reg.17/62 (*i.e.* notification for exemption under Art.81(3) EC) or under the Merger Control Regulation. However, the Commission is now taking a pro-active attitude with its decision to launch a sector inquiry into the sale of sports rights to internet and 3G mobile operators.[28] In the discussion that follows, I draw a distinction between the Commission decisions regarding selling and buying and those regarding merger transactions. As far as the buying and selling agreements are concerned, three types of practices have been examined by the Commission: (i) joint selling; (ii) joint buying; and (iii) long term exclusivity contracts. For the sake of convenience, joint selling and long term-exclusivity contracts, which are often combined, will be discussed under a single heading.

Selling and buying practices

Joint selling and joint buying schemes are relatively frequent in media rights markets. From a general standpoint, joint selling agreements involve cooperation in the selling of products and services between undertakings operating at the same level of the supply chain. Joint buying agreements cover a wide range of different forms of coordination of purchase policy between undertakings.

Joint selling

The issue of joint selling of rights has been recently discussed by the Commission in its *UEFA* decision, which relates to the joint selling arrangement regarding the sale of

---

[24] See Commission decision of April 30, 2002, COMP/JV 57-TPS, at para.14.
[25] See Commission Decision, above n.3, at para.19.
[26] See Commission Decision, below n.6, at para.80.
[27] *ibid.*, at paras 88–89.
[28] See Commission press release of January 30, 2004, "Commission launches sector inquiry into the sale of sports rights to Internet and 3G mobile operators", IP 04/134.

commercial rights of the UEFA Champions League, a pan-European football club competition.[29]

The Commission considered that this joint selling arrangement restricted competition among the football clubs as it had the effect of coordinating the pricing policy and all other trading conditions on behalf of all individual football clubs producing the UEFA Champions League content (restriction of competition at the upstream level).[30] However, this agreement could nevertheless be exempted as it provided the consumer with the benefit of league focused media products from this pan-European football club competition that is sold via a single point of sale and which could not otherwise be produced or distributed equally efficiently.[31]

The Commission thus exempted this joint selling agreement provided that the Parties substantially modified the notified agreement. One of the particular difficulties of the agreement was that UEFA sold the free-TV and pay-TV rights on an exclusive basis in a single bundle to a single TV broadcaster per territory for several years in a row.[32] This created substantial risks of foreclosure since it made it possible for a single large broadcaster per territory to acquire all TV rights of the UEFA Champions League to the exclusion of all other broadcasters (restriction of competition at the downstream level).[33]

Following the intervention of the Commission, the re-notified agreement contains the following features:

- The media rights contracts will not be concluded for a period longer than three years[34];
- The award of rights contracts will follow an "invitation to tender" giving all qualified broadcasters an equal opportunity to bid for the rights in the full knowledge of the key terms and conditions[35];
- The UEFA will unbundle the media rights by splitting them into several rights packages that will be offered in separate packages to different parties[36];
- The UEFA will also allow the football clubs to sell on a non-exclusive basis in parallel with UEFA certain media rights relating to action in which they are participating[37];
- Both UEFA (in respect of all matches) and the football clubs (in respect of matches in which they participate) will have a right to provide video content on the internet one and a half hours after the match finishes, that is to say, as from midnight on the night of the match [...]. Both UEFA and the football clubs may choose to provide their services themselves or via internet service providers. The content will be based on the raw feed produced for television[38];

---

[29] See UEFA decision, above n.6.
[30] ibid., at para.1.
[31] ibid.,
[32] ibid., at para.19.
[33] ibid.,
[34] ibid., at para.25.
[35] ibid., at para.27.
[36] ibid., at paras 32–39.
[37] ibid., at para.34.
[38] ibid., at paras 40 and 42.

76 New Media Platforms

- Both UEFA (in respect of all matches) and the clubs (in respect of matches in which they participate) will have a right to provide audio/video content via UMTS services available maximum five minutes after the action has taken place (technical transformation delay). This content will be based on the raw feed produced for television [...]. UEFA intends to build a 3G/UMTS wireless product that will be based on an extensive video database to be developed by UEFA. UEFA will offer the rights on an exclusive or non-exclusive basis to operator(s) with an UMTS licence, initially and exceptionally for a period of four years and subsequently for periods of three years.[39]

The *UEFA* decision is important because it raises a number of important issues and provides helpful insights into how the Commission will address joint selling agreements combined with long-term exclusivity in the future. Already, similar principles as those found in the UEFA decision can be identified in the Art.19(3) notice (under Reg.17/62) released by the Commission in the *Premier League* case and in the Art.27(4) notice (under Reg.1/2003) in the *German Bundesliga* case.[40] Table 1 compares the main aspects found in these notices with the Commission decision in *UEFA*.

Table 1: Summary of UEFA, Premier League and German Bundesliga

	UEFA	Premier League	Bundesliga
Obligation to sell media rights through a public tender	Yes (para.27)	Yes (para.6)	Yes (para.6)
Unbundling of media rights in several packages	Yes (paras 32–39)	Yes (paras 18–24)	Yes (paras 7–8)
Authorisation for clubs to sell the rights of their own games by themselves	Yes (para.34)	Yes (para.19)	Yes (para.9)
Right to provide video content on the internet	Yes, with embargo until midnight on the day of the game (para.40)	Yes, with embargo until midnight on the day of the game (para.26)	Yes, but with a maximum showing of 30 minutes of content (to disappear in 2006) (para.10)
Rights to provide video content on UMTS	Yes, with a 5 minutes embargo (technical transformation delay) (para.44)	Yes, with no embargo except if needed for technical reasons (para.28)	Yes, without restriction (para.10)

---

[39] *ibid.*, at paras 44 and 45.
[40] See Notice published pursuant to Art.19(3) of Council Reg.17 concerning case COMP/C.2/38.173 and 38.453, joint selling of the media rights of the FA Premier League on an exclusive basis, [2004] O.J. C115/3. Notice published pursuant to Art.27(4) of Council Reg.1/2003 in Case COMP/C.2/37.214, joint selling of the media rights to the German Bundesliga [2004] O.J. C229/13.

Joint purchasing

As far as joint purchasing agreements are concerned, an interesting case relates to the investigation launched by the Commission following the notification by Sogecable and Via Digital of an agreement in which they pooled forces to acquire and exploit the broadcasting rights to Spanish First League football matches for 11 seasons ending in 2009 through the audiovisual sport joint venture.[41] The Commission took the view that the agreement amounted to an unacceptable monopolisation of the rights by the two main TV platforms for a very long period of time and warned that it would impose fines unless the agreement was terminated or significantly modified.[42] Following this, Telefonica and Sogecable announced that they would give entrants in the Spanish cable and digital terrestrial television markets access to the football rights and accepted that such competitors would be free to set their own pay-per-view prices. The Commission, however, closed the case following the decision of Via Digital and Sogecable to merge, a merger which was accepted by the Commission subject to undertakings by the Parties.

Another important decision of the Commission relates to several agreements concerning the joint acquisition of sports television rights, the sharing of the jointly acquired sport television rights, the exchange of the signal for sporting events, the sub-licensing scheme and the sub-licensing rules notified by the European Broadcasting Union ("EBU"), a professional non-profit association of radio and television organisations. This decision, adopted in May 2000, exempted these agreements under Art.81(3).[43] An important part of this decision concerned the sub-licensing scheme and the sub-licensing rules, which constituted the scheme for third party access to the Eurovision system. This access scheme was seen as necessary by the Commission to counterbalance the restrictions of competition created by the joint acquisition scheme put into place by the EBU. This decision was, however, annulled by the Court of First Instance ("CFI"), in October 2002 on the ground that contrary to what the Commission had concluded in its decision, the sub-licensing scheme did not guarantee competitors of EBU sufficient access to rights to transmit sporting events held by the latter organisation.[44]

Mergers and alliances

In recent years, there has also been a significant consolidation movement in the media industry. While some mergers essentially had a vertical dimension, others mainly had a horizontal dimension, while some others again had both horizontal and vertical components.[45] There are generally valid business justifications for mergers in the media sector, but such mergers may also trigger anti-competitive effects. For instance, while vertical media mergers between content providers and delivery companies may allow parties to realise economies of scope and offer new products and services to consumers, they may

---

[41] See Commission press release of April 12, 2000, "Commission ready to lift immunity from fines to Telefónica Media and Sogecable in Spanish football rights case", IP/00/372.
[42] *ibid.*,
[43] See Commission decision of May 10, 2000, Case IV/32.150, Eurovision, [2000] O.J. L151/18.
[44] See CFI, July 11, 1996, *Metropole télévision SA v Commission*, Joined Cases T-528/93, T-542/93, T-543/93 and T-546/93, [1996] E.C.R. II-649.
[45] For a review of such mergers, see Garzaniti, above n.17, at 452–491.

78 New Media Platforms

also create the risk of discriminatory access to content.[46] Similarly, while horizontal mergers may allow parties to realise economies of scale, they may also strengthen market power at the upstream and/or downstream level(s), thus triggering the risks of foreclosure effects.[47] Because of these concerns, such mergers were generally cleared provided that the Parties agreed to substantial commitments.

For instance, in *Newscorp/Telepiu*, the Commission expressed concern that the merger between the two major Italian pay-TV operators would create a quasi-monopolistic situation for the acquisition of exclusive rights for films.[48] This monopsonistic situation, in the absence of corrective measures, would foreclose access to content for third parties and was likely to restrict availability of content to consumers, thus reducing their possibilities of choice.[49] This problem would be further aggravated by the fact that the Parties would be able to further reduce the accessibility of content by exercising holdback and pre-emption rights as regards second window movies, as was envisaged in Telepiu's existing contract with most majors.[50] This would effectively prevent potential new entrants (such as new media platforms) from attempting to enter the market as "fringe" competitors. Moreover, these protection rights exercisable by the Parties would effectively withdraw second window rights from the market, thus harming consumers' welfare and their freedom to choose at what price and at what time to "consume" pay-TV products.

As far as the acquisition of content is concerned, the commitments included in the *Newscorp/Telepiu* case are important, especially as they provide access to content by new media platforms. With respect to these platforms, the commitments offered by Newscorp provided that:

> "*On-going exclusive contracts*
> [. . .]
> (b) *Newscorp* shall waive exclusive rights with respect to TV platforms other than DTH (terrestrial, cable, UMTS, internet, etc.). Furthermore, the parties shall waive any protection rights as regards means of transmission other than DTH.
> (c) *Newscorp* shall waive exclusive rights for pay-per view, video on demand and near video on demand on all platforms.
>
> *Future exclusive contracts*
> (d) *Newscorp* shall not subscribe contracts exceeding the duration of two years with football clubs and of three years with film Studios. The exclusivity attached to these contracts would only concern DTH transmission and would not apply to other means of transmission (for example, terrestrial, cable, UMTS and internet). Furthermore, the

---

[46] A similar issue was at stake recently, in the context of the Sony/BMG merger where the Commission raised concerns that Bertelsmann, who has a leading position in television and radio broadcasting, in Europe through its subsidiaries could give preferential access to Sony BMG music, thereby foreclosing competing record companies from equal access to the TV/radio markets in some countries. See Commission press release of February 12, 2004, "Commission opens in-depth investigation into Sony/Bertelsmann recorded music venture", IP/04/2000.
[47] See *NewsCorp/Telepiu*, above n.3.
[48] *ibid.*, para.150.
[49] The Commission also applied the same analysis with respect to the implications of the mergers on the market for the acquisition of exclusive rights for football events in which national teams participate. See para.162 of the decision.
[50] *ibid.*, at para.152.

parties shall waive any protection rights as regards means of transmission other than DTH. As regards football rights and world-wide sports events, the contractual counterparts shall be granted a unilateral right to terminate contracts on a yearly basis.
(e) *Newscorp* shall not acquire protection rights for DTH and will waive exclusive rights for pay-per view, video on demand and near video on demand on all platforms.
(f) *Newscorp* shall not acquire, through future contracts or re-negotiations of the terms of the existing contracts, any protection or black-out right with respect to DTH.

Relations with competitors/third parties: wholesale offer and access to the platform and technical services.
(g) *Newscorp* shall offer third parties, on a unbundled and non-exclusive basis, the right to distribute on platforms other than DTH any premium contents if and for as long as the combined platform offers such premium contents to its retail customers. Such offer will be made on the basis of the retail minus principle ... "[51]

These commitments represent a significant step made by the Commission to address the difficulties met by new media platforms active in the Italian market to acquire premium content. As will be seen in the next part, however, these commitments address only part of the difficulties. Further interventions on the basis of competition rules will thus have to be made to create a level playing field between existing and new platforms on the market for the acquisition of premium content.

How do the current remedies fare with the new media platforms?

The previous section laid out the main competition law problems regarding the acquisition of premium content, which were addressed by the Commission in some recent decisions. This section analyses the extent to which the remedies contained in these decisions sufficiently address the difficulties experienced by new media platforms in acquiring premium content, which, as will be seen below, in great part result from the restrictions of competition contained in the contracts concluded between content holders and dominant content delivery operators. As many of the issues discussed below have not yet been addressed by the Commission and the European Courts, the forthcoming developments are exploratory in nature. Some of the arguments developed may thus appear particularly "creative". Time will test their validity.

Although the markets for the acquisition of premium movies and the market for the acquisition of premium sport content hold distinctive features,[52] many of the difficulties encountered by new media platforms when it comes to acquiring premium content are similar for movies and sport. They are thus jointly addressed in the paragraphs which follow.

---

[51] *ibid.*, at para.225.
[52] For instance, sport events are highly perishable goods as, while the value of the rights for live broadcasting is very high, the value of the rights for deferred broadcasting will be considerably less. Movie rights keep their value for a considerably longer period of time.

80 New Media Platforms

Common issues

High minimum guarantees

One significant problem for new media platforms seeking to acquire premium content results from the high minimum guarantees that need to be paid by such platforms to premium movie or sport rights holders. These minimum guarantees represent a heavy burden for new entrants as, unlike fees to be paid to right holders on the basis of a model of revenue sharing, they have to be paid upfront by these new entrants independently of the number of subscribers they will eventually manage to capture. In practice, they operate as a barrier to entry as the payment of such rights involves substantial risks for these companies.

So far, the Commission has not addressed such minimum guarantees, although they could amount to competition law infringements. First, there are reasons to believe that such guarantees could violate Art.82 EC. The application of this provision requires that two conditions be present: (i) one or several companies are in a dominant position and (ii) an abuse is committed. The existence of a dominant position essentially depends on the definition of the relevant market. While a broad definition of the market will reduce the likelihood that one given company is in a dominant position, a narrow market definition will increase the likelihood of a finding of dominance. For instance, in the movie industry, one could argue that every blockbuster represents, during a limited period of time, a market in itself as other successful movies might not be valid substitutes. Although this would have to be backed up by serious market analysis, some movies, such as "Lord of the Rings" or "Gladiator" represent at a given point in time and for a limited period of time a "must have" for content delivery operators. During that period, the holder of the rights for such a movie might be dominant. This argument would, however, lead to extremely narrow market definitions and competition authorities may resist this approach.[53] Moreover, the fact that Hollywood movies are, with limited exceptions, generally sold in bulk (through output deals) would tend to suggest that defining blockbusters as a single market product even for a limited period of time might be met with some resistance.

But even if one were opting for a broader definition of the market, which would for instance amount to the market for the selling/buying of Hollywood movies, one could probably hold the majors collectively dominant. A situation of collective dominance arises when several companies (none of which is dominant alone) jointly enjoy market power on a given market.[54] This is, for instance, the case in tight oligopolistic markets where tacit coordination among the firms on the market can take place, leading to collective price increases or to parallel behaviour. For a situation of collective dominance to be found, it

---

[53] This issue can, however, be subject to discussion. Indeed, competition authorities generally seek to define relevant product markets as narrowly as possible. A field where narrow market definitions have, for instance, been found is intellectual property rights. Inventions protected by intellectual property rights have often been considered as a relevant product market with the consequence that the holder of the right is unavoidably dominant. See, for instance, ECJ, April 6, 1995, C-241/91 and C-242/91, *Radio Telefis Eireann and Independent Television Publications Ltd v Commission* [1995] E.C.R. 743.

[54] See, generally, on collective dominance and oligopolistic markets, see Richard Whish, *Competition Law*, (4th ed., Butterworths, 2001) pp.473–482. See also, Sigrid Stroux, "Is EC Oligopoly Control Outgrowing Its Infancy?" (2000) 23 *World Competition*, 3; Juan Briones and Atilano Jorge Padilla, "The Complex Landscape of Oligopolies under EU Competition Policy—Is Collective Dominance Ripe for Guidelines?", (2001) 24 *World Competition*, 307.

is necessary to prove the existence of "economic links" between the undertakings on the market.[55] As the ECJ held in *Compagnie Maritime Belge*, the concept of economic links is not limited "to the existence of agreements or of other links in law but can also be based on other connecting factors and would depend on an economic assessment and, in particular, on an assessment of the structure of the market".[56] The test for the required intensity of the links to ensure the stability of a collusive equilibrium was clarified by the CFI in *Airtours* where it was held that for collusion to occur, there must be (i) a certain amount of transparency, (ii) a punishment (or retaliatory) mechanism and (iii) the absence of reaction by consumers and potential competitors that could undermine the collusive policy (*i.e.* assessment of the barriers to entry).[57]

On the basis of this case law, it could be convincingly argued that the market for the selling/buying of Hollywood movies is characterised by a situation of collective dominance. In addition to a number of features which are prone to give rise to coordinated effects (limited number of firms, symmetry of market shares, high barriers to entry—*i.e.* huge financial investments for producing movies—similar degree of vertical integration among the firms, limited buyer power, etc.) it could be argued that the first and the third conditions laid down in *Airtours* are fulfilled. As far as the first condition is concerned, the market for the selling/buying of Hollywood movies is highly transparent because the majors often announce publicly and long in advance the new movies, their budgets, etc. As far as the third condition is concerned, it is unlikely that a collective price increase by the majors would lead to the entry of new competitors given the substantial barriers to entry on this market. Similarly, it is doubtful whether consumers would shift to new suppliers or whether a sufficient decrease in demand would dissuade firms from increasing their price.

As far as the presence of an abuse is concerned, the minimum guarantees could amount to an unfair trading condition contrary to Art.82(a) EC. Indeed, it is hard to believe that high minimum guarantees could be imposed by movie rights holders in a competitive market. Moreover, the Commission has already condemned such types of practices as abusive. For instance, in *BRT v. SABAM*, a case concerning a performing rights society, it was considered that restrictions imposed on the authors who were members of the society were unfair in so far as they were not necessary to allow the performing rights society to properly conduct its business.[58] Similarly, in *Tetra Pak II*, a clause obliging the payment of a rent, at the beginning of the contract, of almost the same amount as the value of the machine was considered unfair.[59] *Mutatis mutandis*, high minimum guarantees would in fact force new platform operators to spend extremely important sums on content, even before they were given the time to gain market shares, thus forcing them to take very significant financial risks.[60]

---

[55] See CFI, 10 March 1993, Società Italiano Vetro v Commission, Case T-68/89, ECR[1992] II-1403
[56] See ECJ, March 16, 2000, Case C-395, *Compagnie Maritime Belge Transports SA v Commission* [2000] E.C.R. I-1365, at para.[45].
[57] CFI, June 6, 2002, Case T-342/99, *Airtours Plc v Commission* [2002] E.C.R. II-2585. On this case, see Massimo Motta, "EC Merger Policy and the Airtours Case", [2000] E.C.L.R. 199; Ali Nikpay and Fred Houwen, "Tour de Force or Little Local Turbulence? A Heretical View on the Airtours Judgment", [2003] E.C.L.R. 193.
[58] ECJ, January 30, 1974, *Belgische Radio en Televisie v SV SABAM and NV Fonior* [1974] E.C.R. 51.
[59] CFI, October 6, 1994, Case T-83/91, *Tetra Pak International SA v Commission* [1994] E.C.R. II-755.
[60] A similar analysis can be made for sport rights.

82 New Media Platforms

Hence, it would be preferable that the compensation of premium content right holders be based on a revenue sharing model, which would allow new platforms to enter the market and gain market shares. To be compatible with EC competition law, the fees charged by the rights holder(s) under the revenue sharing mechanism should not be excessive and should be established in a non-discriminatory manner, while excessive prices would amount to another violation of Art.82(a).[61] Price discrimination among platforms, absent an objective justification, would amount to an abuse under Art.82(c) EC.[62]

Exclusivity agreement with the dominant operator in the pay-TV national market

Exclusivity is an issue which has been addressed by the Commission in the *UEFA* and *Newscorp/Telepiu* decisions (in the context of Art.81 in the former case and the Merger Control Regulation in the latter). The remedies imposed in these decisions such as, for instance, the reduction of the length of the contracts, the unbundling of the rights into smaller packages (*UEFA*), as well as the right for film studios and football clubs to unilaterally terminate contracts with the Parties with no applicable penalties, the Parties' waiver of exclusive rights *vis-à-vis* TV platforms other than DTH and for pay-per-view, video-on-demand,[63] and near-video-on-demand[64] for all platforms (*Newscorp/Telepiu*), clearly go a long way towards addressing exclusivity problems in the media industry. Of course, the terms of these decisions are only binding on the Parties, although they certainly represent important precedents, which will guide the Commission in further inquiries.

Yet, these decisions will not suddenly solve all problems linked with exclusivity and there will still be cases in the future where new media platform operators will be foreclosed access to premium content by exclusive agreements. Where such agreements are concluded by dominant pay-TV operators, it is subject to question whether they are compatible with Art.82 EC. By concluding exclusive agreements with premium rights holders, the dominant operator would in fact foreclose weaker competitors to gain access to premium content, thereby preventing them from entering the pay-TV market (or excluding them from that market) and thus strengthening its dominant position on this market. Of course, the purchasing of exclusive rights will generally involve the payment of very high sums to right holders by the dominant operator (thus making such an exclusionary strategy very costly), but, as in classic predatory strategies, the initial loss this may entail would be subsequently compensated by the monopolisation of the downstream market.

The dominant operator could try to objectively justify its purchase of exclusive rights by claiming, for instance, that such rights are needed to protect sunk investments. But the

---

[61] See ECJ, February 14, 1978, *United Brands v Commission* [1978] E.C.R. 207.

[62] See Commission decision 94/210/EC of March 29, 1994, Case IV/33.941, HOV SVZ/MCN, [1994] O.J. L104/34. See also, CFI, September 30, 2003, *Michelin v Commission*, Case T-203/01, not yet published.

[63] Video on demand ("VOD") is a technology to view and order video content as and when required in the home, generally on a pay-per-view basis. Users typically have some form of set-top box that provides the service to their television set. Users navigate menus with a remote control to select and order content, with the cost billed to an account. See http://encyclopedia.thefreedictionary.com/digital%20cable.

[64] NVOD systems, or near video on demand systems are systems in which users wanting to watch a film are batched up for the next start time. This is a reasonable model for films which are in high demand, as the video server can simply distribute the film at short intervals, preferably using multicast techniques. NVOD provides users with a video on demand service, but imposes a short latency delay before the film starts. See www.brainyencyclopedia.com/encyclopedia/v/vi/video_on_demand.html.

bulk of the physical investments carried out by pay-TV operators were initially made in the 1990s, and are thus largely amortised. Moreover, the need to protect investments would be a paradoxical argument in an industry where the principal costs of downstream operators relate to the buying of exclusive premium content rights. Exclusivity would thus be needed to help dominant operators to amortise the huge investments made in the purchasing of rights, the exclusivity of which would be largely responsible for their astronomical costs.[65] Thus, exclusivity triggers significant allocative efficiency losses without these losses being compensated by dynamic efficiency gains (as there is no reason to believe exclusivity is necessary to induce investments and innovation in the sector, quite the contrary as its primary victims are new media platforms).

It could also be argued that, on the supply side, it is necessary for premium rights holders to engage in exclusive agreements in order to protect the value of their rights. For instance, there seems to be a fear among premium rights holders that the showing of such content on new media platforms could destroy the value of their rights. But this concern seems to be unsubstantiated as such platforms are generally closed systems. Renting a movie through an internet-based video on demand service is not conceptually different than renting a movie from a video store. In both cases, the content holder will share revenues with the distributor and the movie will be watched once by no one else than the renter(s). In fact, the risk that the value of the rights might be undermined seems to be much higher with video-cassettes and DVDs as these can be easily replicated.

More generally, from a public policy standpoint, exclusive rights for premium content can hardly find any justification. The pervasiveness of such rights has transformed the competition paradigm in the media industry into a regime of competition "for" the market. Competition "for" the market is a model that can be recommended in markets holding natural monopoly features, such as water systems, but clearly should have no place in markets where competition "in" the market is possible.[66] But even if a model of competition "for" the market was to be recommended, it would not work in practice as, because of its imperfections, only dominant pay-TV operators seem to be in a position of acquiring premium content rights.

By contrast, the absence (or at least a strong reduction of the scope) of exclusivity would hold many benefits from a public policy standpoint. First, in the absence of exclusivity, the content delivery business would essentially consist in purchasing and selling bundles which satisfy the consumers.[67] This would force operators to work towards the satisfaction of consumer preferences as a key factor for gaining market shares would be the ability to provide the right bundle to the right customer. Other variables to the competitive process would be prices, quality and convenience of the service. Moreover, by allowing several platforms to compete for viewers/subscribers, this approach would comply with the principle of "technological neutrality" and enhance consumer choice as the latter could select the platform on which they wish to watch a given content. Thus, in contrast with a tight exclusivity model, an open model would trigger both allocative (in terms of price

---

[65] See Antonio Nicita and Giovanni Ramello, "Content Exclusivity and Competition Policy in media Markets: The Case of Pay-TV in Europe", forthcoming in *International Journal of the Economics of Business*.

[66] See Damien Geradin and Michel Kerf, *Controlling Market Power in Telecommunications: Antitrust vs Sector-specific Regulation* (Oxford University Press, 2003), at Ch.2.

[67] See Nicita and Romello, above n.61.

84 New Media Platforms

cuts, quality increase, etc.) and dynamic (in terms of investment into new networks, etc.) efficiency gains.

An interesting question at this stage is the extent to which a requirement of sub-licensing by the content delivery operator holding the rights could reduce the foreclosure effects of exclusive agreements. A sub-licensing requirement might certainly contribute to reduce such effects. But the value of sub-licensing depends in great part on the terms on the basis of which sub-licences are granted.[68] Sub-licences sold at excessive prices may be of little value to competitors. In addition, to allow such a scheme to be effective and in compliance with EC competition rules, the terms of the sub-licences should be known in advance and be transparent and non discriminatory.

An interesting form of sub-licensing can be found in the commitments made by the Parties in the *Newscorp/Telepiu* merger, which *inter alia* provide that:

> "(g) *Newscorp* shall offer third parties, on an unbundled and non-exclusive basis, the right to distribute on platforms other than DTH any premium contents if and for as long as the combined platform offers such premium contents to its retail customers. Such offer will be made on the basis of the retail-minus principle."[69]

In its discussion of the usefulness of this remedy the Commission explains that "[t]he underlying idea is that such wholesale offer will lower barriers to entry in the pay-TV market by allowing non-DTH pay-TV operators to access premium contents which would otherwise be too costly for them to purchase directly or which are locked away by means of long-duration exclusivity agreements entered into by the incumbent players with the content providers".[70] Responding to comments made by third parties, the Commission further explains that "[a]s regards the contractual availability of the necessary rights in order to provide a wholesale offer, Newscorp has submitted an undertaking including a 'best endeavours clause' concerning the acquisition of the necessary non-DTH rights for the wholesale offer to work".

This wholesale offer regime deserves several comments. First, the possibility given to non-DTH operators to buy premium content is a useful way to introduce some competition in the downstream content delivery market. But it provides no long-term solution to the problem of exclusivity as it does little to help new media platforms to buy rights directly from the premium content right holders. One could, of course, argue that there is no longer any necessity for these platforms to buy content directly from rights holders as they can buy the same rights on a wholesale basis from the pay-TV operators. But this is not entirely true. First, this wholesale regime essentially transforms non-DTH platforms into resellers, which is at best a form of fringe competition. Secondly, the retail minus pricing mechanisms provided for in the decision means that the margins that can be made by non-DTH operators are extremely small. In practice, this system continues to allow the content

---

[68] See John Temple Lang, "Media, Multimedia and European Community Antitrust Law" (1998) 21 Fordham Int. L.J. 1296.
[69] A good or service sold by firm A to firm B at retail-minus amounts to the retail price set by firm A minus the economies the retail costs fairly incurred by the retail activity of that operator. Thus, if the retail price of a service is €1 and the retail costs (marketing, billing, etc.) are 20 cents, the retail minus price is 80 cents.
[70] See para.246 of the decision, above n.3.

delivery operators to charge its monopoly profits to its competitors.[71] Thus, while this system might allow some competition on the fringe, this competition will not translate into cheaper prices. In some cases, the margin will be so tight that no entry will be possible. Such a situation could lead to a "margin squeeze", a practice that can amount to an abuse of a dominant position.[72] Finally, and perhaps more fundamentally, the wholesale model places new entrants at the mercy of the pay-TV incumbent. As the terms of the undertakings submitted by the merged entity are loosely drafted,[73] there is a danger that the pay-TV incumbent will drag its feet when asked to provide unbundled media rights to new entrants. Though the decision provides for an arbitration mechanism, as well as the possible intervention of the Italian Communications Regulator in case of disputes over the application of the commitments, these disputes will, however, always be extremely difficult to solve as they will usually involve issues such as the definition of premium content, cost-allocation strategies, etc. They will also inevitably introduce delays and costs, which will mainly affect new entrants.

Thus, while the introduction of a wholesale offer regime is a desirable step which should allow new entrants to gain access to premium content and gain market shares, it will never replace the ability of these operators to buy premium content rights directly from right holders.[74]

The presence of hold back clauses in contracts between rights holders and the exclusive buyers of such rights

Pursuant to holdback clauses, the acquirer of the content rights is prevented from selling such content to other media platforms. This issue was addressed in the *Newscorp/Telepiu* decision as the Parties committed to waive any protection rights as regards means of transmission other than DTH.

But more generally, the legality of such rights could be put into question. For the same reason of exclusivity agreements, holdback clauses could be considered as abuses of dominant position on the part of dominant pay-TV operators as they prevent alternative platforms from gaining access to premium content. Such operators have indeed much to gain by having such clauses inserted in the contracts rights they negotiate with content providers. Such clauses could thus contribute to a strategy of exclusion of new media platforms from the content delivery market.

In addition to the potential application of Art.82 in the context of a single dominance situation, it is subject to question whether the concept of vertical collective dominance as applied by the CFI in *Irish Sugar* could be applied with respect to holdback clauses.[75] In

---

[71] For a discussion of the pros and cons of retail minus schemes, which in practice amount to the Efficient Component Pricing Rule ("ECPR") methodology, see Geradin and Kerf, above n.66, at pp.39–41.

[72] For an example of margin squeeze in the telecommunications sector, see the *Deutsche Telekom* decision adopted by the Commission on May 21, 2003, COMP/C-1/37.451, 37.578, 37.579, [2003] O.J. L263/9. A discussion of margin squeeze in the media sector can be found in the OFT decision in the *BSkyB* case, see decision of the Office of Fair Trading under s.47 relating to decision CA90/20/2002: Alleged Infringement of the Ch.II Prohibition by BSkyB, July 29, 2003, para.16 *et seq.*

[73] See para.249 of the decision, above n.3.

[74] Torben Toft, "Football, Joint Selling of Media Rights" (2003) 3 Competition Policy Newsletter 47 at p.48. While sub-licensing arrangements can in some circumstances help to remedy competition problems, it is preferable to have direct contractual relationships between the original rights owners rather than contractual relationships among competitors.

[75] See CFI, October 7, 1999, Case T-228/97, *Irish Sugar Plc v Commission* [1999] E.C.R. II-2969.

86 New Media Platforms

the latter case, the CFI came to the conclusion that collusive practices between undertakings placed a different level of production aiming at protecting their national market from imports or at preventing the entry of new competing brands could be brought under the concept of joint dominance. As seen above, a condition for this is to prove the existence of "economic links". In *Irish Sugar*, a number of structural links (holding of shares by the producer in the distribution company) as well as behavioural commitments (exclusive supply commitment, etc.) have been evidenced by the Commission. A transposition of the *Irish Sugar* line of reasoning could be attempted by considering that holdback clauses are a collusive vertical strategy implemented between rights holders and content delivery operators in order to foreclose new entrants' entry on the market. The concept of "economic link" would flow from the tight contractual framework between the parties and a number of correlated interests.

Multi-year duration and automatic renewal of contracts

The Commission has always been attentive to ensure that the duration of the exclusivity in supply contracts is no longer than necessary. The length of the exclusivity that is tolerated essentially depends on the circumstances.[76] For instance, in the energy sector where considerable investments are made by parties, the Commission will generally tolerate relatively long periods of exclusivity (*e.g.* 15 years).[77] By contrast, in sectors where no major investments are made, the maximum length of exclusivity tolerated by the Commission might be considerably shorter. This seems to be the case in the media sector. For instance, in *UEFA*, the Commission required the parties to reduce the length of media rights contracts and the parties eventually proposed the principle that such contracts be concluded for a period not exceeding the UEFA Champions League seasons.[78] In *Newscorp/Telepiu*, the parties committed that "Newscorp shall not subscribe contracts exceeding the duration of two years with football clubs and of three years with film studios".[79]

Clauses providing for automatic renewal of premium content contract rights often amount to a disguised method to extend the duration of exclusivity. Indeed, it is only when one of the two parties decides not to renew the contract that third parties could be given a chance to bid for content rights. As both content providers and dominant content delivery operators generally favour exclusivity, this situation may never take place in practice.[80] As

---

[76] See ECJ, October 6, 1982, Coditel SA, Case 262/81, *Compagnie générale pour la diffusion de la télévision, v Ciné-Vog Films SA* [1982] E.C.R. 3381 at para.[19] ("It must therefore be stated that it is for national courts, where appropriate, to make such inquiries and in particular to establish whether or not the exercise of the exclusive right to exhibit a cinematographic film creates barriers which are artificial and unjustifiable in terms of the needs of the cinematographic industry or the possibility of charging fees which exceed a fair return on investment, or an exclusivity the duration of which is disproportionate to those requirements, and whether or not from a general point of view, such exercise within a given geographic area is such as to prevent, restrict or distort competition within the common market.").
[77] See Commission decision 91/329 of April 30, 1991, IV/33.473, *Scottish Nuclear* [1991] O.J. L178/31. See also notice pursuant to Art.19(3) of Council Reg.17/62, Case No.IV/E-3/35.485, *REN/Turbogás* [1996] O.J. C118/7; notice pursuant to Art.19(3) of Council Reg.17/62 Case No.IV/E-3/35.698, *ISAB Energy* [1996] O.J. C138/3. See, finally, 27th Annual Report on Competition Policy at p.127.
[78] See Commission decision, above n.6, at para.[25].
[79] *ibid.*, at para.225.
[80] See Ungerer, above n.8.

a result, the Commission has generally considered such clauses to be anti-competitive.[81]

In some cases, media rights agreements also contain a preferential renewal clause pursuant to which at the expiration of the agreements, the incumbent buyer is given an opportunity to match the highest bid received from any third party. Such clauses, which are often referred to as "English clauses", enhance transparency in the market and thus increase the likelihood that a competitor making an aggressive offer, will not gain the market, but simply force the incumbent buyer to match its offer.[82] As a result, English clauses will usually fare no better than automatic renewal clauses in the eyes of the Commission.[83] For instance, in *Sport 7*, the Commission considered that the granting by the Dutch football association of an exclusive licence to a new broadcaster, Sport 7, for the duration of seven years, was caught by Art.81(1) and could not be exempted as it eliminated competition for the rights for too long a period.[84] In addition, the re-negotiation process foreseen at the end of the contract gave *Sport 7* an advantage because it had the right to match the bid of its competitors.

Issues specific to the movie industry

In addition to these common issues, some additional issues are specific to the market for the acquisition of movie rights and the market for the acquisition of sport rights.

Exploitation of windows

As far as movie rights are concerned, a specific difficulty for the new media platforms relate to the systems of exploitation of windows used by the majors,[85] *i.e.* time periods during which a movie can be exploited for one purpose only.[86] Pursuant to that system, the first exploitation is generally theatrical release in cinemas, followed by home video, pay-per-view/video-on-demand, pay-TV, and free-to-air TV. The new media platforms' concern with this windows system is that it places video-on-demand, which is one of the core services they sell to their clients, at a competitive disadvantage compared with home video, which benefits from an "earlier" window while this is an analogous service.

One way to attack this system would be to say that it amounts to an abuse of dominance on the part of the movie producers. Application of Art.82 requires the identification of one or several dominant companies. As pointed out above, single or collective dominance can probably be identified in the market for the supply of blockbusters. As far as the abuse is concerned, it would take the form of discrimination between contract delivery operators contrary to Art.82(c) EC. This provision provides that: "applying dissimilar conditions to

---

[81] See Schaub, above n.9, at p.7.
[82] See Jonathan Faull and Ali Nikpay, *The EC Law of Competition* (Oxford University Press, 1999), at p.165.
[83] See, *e.g.* the *IRI/AC Nielsen* case, 26th Annual Report on Competition Policy at para.63.
[84] See the comments on the case *Sport 7/KNVB* in the 27th Annual Report on Competition Policy at p.122.
[85] Unlike movie rights, sports rights are not sold in accordance with exploitation mechanisms. TV rights to sports events are generally sold exclusively to a single broadcaster, account being taken, however, of the limits that have now been placed in the UEFA decision.
[86] See Faull and Nikpay, above n.82, at p.776.

88   New Media Platforms

equivalent transactions with other trading parties, thereby placing them at a competitive disadvantage" can be held to constitute an abuse of a dominant position.

The first condition of application is to show that the transactions are equivalent. In the present case, it could convincingly be argued that, from the point of view of the rights sellers, selling such rights for exploitation through video stores or through video-on-demand, represent equivalent transactions. The second condition of application of Art.82(c) requires that the dominant firm's trading parties be placed at a competitive disadvantage as a result of the discrimination. This is not hard to establish in the present case. Being only authorised to show a blockbuster two months after its release in video stores is a clear handicap for new media platforms as it considerably reduces the proceeds they can earn through their video-on-demand services. It also deprives them of a commercial asset (the ability to show blockbusters shortly after their release in cinemas), which could help them build a subscriber base.

As a general rule, it is a legitimate defence for a dominant operator accused of discriminatory behaviour to demonstrate that this behaviour can be objectively justified. One dominant supplier accused of price discrimination could, for instance, show that it was forced to discriminate to meet a competitor's offer.[87] But in this case, it is not clear why, for instance, the exploitation of movie rights through video stores and video-on-demand should be realised through different windows. As pointed out above, these two services are conceptually similar. In fact, video-on-demand is just a more technologically advanced way for customers to buy a movie for a single viewing.

Similarly, from a public policy standpoint, allowing that the exploitation of movie rights through video stores and video-on-demand would benefit consumers as it would allow them to watch a blockbuster at a given point in time on the platform of their choice. Giving video-on-demand a better window would also allow new media platforms to gain subscribers, hence allowing them to become stronger competitors on the content delivery market, as well as inducing them to invest in new technologies.

Most Favoured Nation clause

Contracts between Hollywood studios and content buyers usually contain "Most Favoured Nation" clauses (hereafter, "MFN clauses") pursuant to which every time a content buyer gives a special advantage to a studio (*i.e.* a higher revenue sharing offer to the benefit of this seller or an advantageous selling condition), this advantage should be extended to the other studios. To induce content buyers' compliance, these MFN clauses also provide for control mechanisms allowing studios to verify whether advantages are extended to all of them.

MFN clauses entail serious restrictions of competition as they have the effect of harmonising prices and other selling conditions. As far as prices are concerned, MFN clauses lead to a mechanism of price leadership whereby one firm's pricing movements are followed by its rivals. In view of the anti-competitive effects created by MFN clauses, the Commission decided a couple of years ago to initiate an investigation into such

---

[87] See Faull and Nikpay, above n.82, at p.176.

clauses. So far, the Commission has not taken a decision and it is thus not yet possible to know whether and how the Commission will address these restrictions of competition.

Perhaps the most obvious approach to attack MFN clauses would be to consider them as a price cartel. Obviously, studios share an interest in imposing MFN clauses in content rights contracts and this practice seems to be so pervasive that these operators know what they are doing. The problem for a plaintiff is that Art.81(1) requires that proof of an anticompetitive agreement between competitors be established. Absent material proof of such an agreement, a concerted practice will have to be inferred from circumstantial evidence. But the competition authorities will not be necessarily easy to convince that such evidence meets the evidentiary requirements of Art.81(1).[88] An alternative for the application of Art.81(1) would be to tackle the MFN clause as an anticompetitive vertical agreement. However, because its effect on competition is likely to be limited if examined on a stand alone basis, it is probable that such clause will not have a sufficient anti-competitive impact to forbid it pursuant to Art.81(1). Nonetheless, a solution for bringing MFN clauses within the remit of Art.81(1) could be to rely upon the doctrine of "cumulative effect" whereby the effect of an agreement may be reinforced if there are similar agreements between the same firms or between third parties.[89] Vertical practices which may, as such, be considered of minor importance, can nonetheless infringe Art.81(1) by virtue of the cumulative effect of parallel networks of similar agreements. Pursuant to this approach, it could be considered that the generalisation of MFN clauses in contracts between studios and content providers leads to a phenomenon of price harmonisation that restricts competition.

Another approach to tackling such clauses is to argue that they amount to an abuse of a dominant position committed by collectively dominant firms (*i.e.* the major Hollywood studios). The advantage of this approach is that it would lower the evidentiary requirement imposed for the application of Art.81. Indeed, as was said before, the movie industry demonstrates a number of oligopolistic features. Interdependent parallel behaviour by majors could thus be caught under the concept of collective dominance. In particular, the anticompetitive effect of the similar trading conditions imposed by the majors on their customers could be seen as a form of tacit coordination constitutive of an abuse pursuant to Art.82 EC. It is not sure, however, whether an action of this kind would be admissible before a competition authority. Indeed, the Commission has made a cautious use of the concept of collective dominance under Art.82 EC. Rather, it has mainly enforced the doctrine *ex ante*, in the context of merger proceedings.[90]

Issues specific to the sports industry

The sale of all rights to a single content delivery operator

One serious problem faced for new media platforms is that sports rights, and in particular football rights, have usually been sold to a single broadcaster (generally pay-TV

---

[88] In particular, the heavy burden of proof adopted in the *Woodpulp II* case. Joined Cases C-89/85, C-104/85, C-114/85, C-166/85, C-117/85 and C-125/85 to C-129/85, *A Ahlström Osakeyhtiö v Commission* [1993] E.C.R. I-1317.

[89] See, for instance, ECJ, December 12, 1967, Case 23/67, *Brasserie de Haecht*, [1967] E.C.R. 407; ECJ, February 28, 1991, Case C-234/89, *Stergios Delimitis v Henninger Bräu AG* [1991] E.C.R. I-935.

[90] Even in merger control, the Commission seems to be now more cautious about relying on a finding of collective dominance to prohibit a merger. For instance, in the BMG/Sony merger, the Commission has

90 New Media Platforms

operators) through joint selling arrangements. Thus, it has been impossible for these platforms to acquire football rights directly from the rights owner.

This problem has, however, been addressed by the Commission in its *UEFA* decision. The re-notified UEFA agreement implies an unbundling of the media rights by dividing them in different rights packages that will be sold to different parties. For instance, we have seen that, as reported by the Commission, the re-notified agreement provides that "[b]oth UEFA (. . . .) and the football clubs (. . . .) will have the right to provide video content on the internet one and a half hours after the match finishes, that is to say, as from midnight on the night of the match"[91] and that "[b]oth UEFA and football clubs may choose to provide their services themselves or via internet service providers".[92] Read literally, however, the second sentence does not offer any guarantee to new media platforms (*e.g.* internet portals selling content to its customers) that they will have access to the Champions League games.[93] Indeed, this sentence gives the UEFA and the football clubs a choice between providing video content on the internet by themselves or to sell the rights to internet service providers. Similarly, the *Premier League* and *Bundesliga* notices referred to above do not seem to provide any guarantee that new media will have access to the internet rights.[94]

The re-notified UEFA agreement seems to offer better guarantees to UMTS operators to acquire football rights. The re-notified agreement provides that "[b]oth UEFA (. . . .) and the clubs (. . . .) will have a right to provide audio/video content via UMTS services available 5 minutes after the action has taken place (technical transformation delay)"[95] and that "UEFA will offer the rights on an exclusive or non-exclusive basis to operator(s) with an UMTS licence, initially and exceptionally for a period of three years".[96] A similar approach is found in the *Premier League* notice.[97] The reason for this difference between internet and UMTS rights certainly comes from the fact that, unlike for internet rights, football leagues and clubs are not in a position to exploit UMTS rights by themselves. They thus need to sell these rights to operators holding an UMTS license.

Embargo time

Another problem that is faced by the new media platforms relates to the embargo that is imposed on them before they can show football games. As seen in the preceding section, the re-notified UEFA agreement provides that Champion League games could only be shown on the internet one and a half hours after the game finishes, *i.e.* from midnight on

---

recently backed off from the claim it initially made in its statement of objections that the merger raise major competition problems because BMG and Sony were part of an oligopolistic market. See Global Antitrust Weekly 290, June 12–18 2004, p.1.
[91] para.40.
[92] para.42.
[93] For a similar approach, see para.26 of the Art.19(3) Notice, which the Commission has adopted in the *Premier League* case ("Both the FALP (in respect of all matches) and the clubs (in respect of matches in which they participate) will have a right to provide video content on the Internet as of midnight on the night of the match").
[94] para.26 of the *Premier League* notice, above n.40 and para.10 of the *Bundesliga* notice, *ibid.*,
[95] *UEFA* decision, above n.6, at para.[44].
[96] para.45.
[97] para.28 of the notice, above n.40.

the day of the game).⁹⁸ Interestingly, however, it also provides that "[l]ive streaming will not be made possible because of the technical development of the internet at this stage, which does not permit the maintenance of a satisfactorily high quality. This will of course change over time, making it necessary to revisit the embargo in the foreseeable future".⁹⁹ Thus, as soon as the technology for live streaming is available (in fact, it seems to be already available), the re-notified agreement should be amended. Here again, a similar approach is taken in the *Premier League* notice.¹ By contrast, the *Bundesliga* notice does not refer to any embargo time.²

As we have seen, the re-notified UEFA agreement does not provide for a revision of the 5 minutes embargo that is imposed on the showing of football games on UMTS. The *Premier League* notice does not provide for an embargo, but simply states that the Football Association Premier League Limited ("FALP") will sell "an effectively live mobile clips package to all games (any delay being limited to the technical need to repurpose the clips)".³ As to the *Bundelisga* notice, it does not refer to any embargo/delay at all.⁴

Other potential competition law problems: Access to TV channels

So far, this paper has been concerned with the acquisition of premium movies and sports content. Yes, other types of contents may also be attractive, such as, for instance, TV channels. TV channels are increasingly seen as a commodity with firms producing TV contents (games, talk shows, news, series, etc.), and/or acquiring and assembling contents into channels, and/or delivering channels to the viewers.⁵ There is often some degree of vertical integration between these activities, thus raising the risk of discriminatory access to channels. Some TV channels are extremely popular and represent a "must have" for new media platforms. Among valuable channels figure not only commercial channels (Fox News, Disney Channel, etc.), but also in some cases public free-to-air TV channels, which are susceptible to attract a large TV audience (*e.g.* RAI, BBC, etc.).

As far as cable operators are concerned (some of which can be considered as new media platforms when they provide interactive services, multimedia products, etc.), getting access to TV channels may in some Member States (*e.g.* Belgium or France) be ensured through the "must carry" obligation, which is found in Art.31(1) of the Universal Service directive.⁶ But must carry obligations do not have to be imposed by the Member States. This is only an option that is left to their discretion.⁷

---

⁹⁸ This embargo had been explicitly criticised by the Internet service providers, which submitted observations on the *UEFA* re-notified agreement. See para.98 ("Internet service providers would like to have live rights. They consider that the embargo is too long for deferred exploitation and that Internet and TV are two distinct markets. They regret that deferred rights are reserved for UEFA and the football clubs and that Internet service providers are excluded from competing for the rights").

⁹⁹ para.40.

¹ See para.26 of the notice, above n.40.

² See para.10 of the notice, above n.40.

³ See para.28 of the notice, above n.40.

⁴ See para.10 of the notice, above n.40.

⁵ See Nicita and Ramello, above n.65, at p.4.

⁶ See above text accompanying n.14.

⁷ To the extent that some free-to-air channels are largely funded by the tax payers' money, it is subject to question whether such channels should be subject to a "must offer" obligation so as to allow consumers/tax payers to watch these channels on the platform of their choice.

92 New Media Platforms

Thus, in absence of regulatory requirements, how could a new media platform gain access to TV channels? Although to the best of the author's knowledge this issue has never been addressed by a competition authority, it is interesting for the sake of the argument to explore whether EC competition law could be helpful to these new media platforms. Let's take the hypothetical example of a new media platform that would seek to obtain the rights to broadcast the channels of the two main TV operators. These channels would indeed be extremely popular and, thus, an essential tool to attract viewers to the platform. The new media platform would rebroadcast these channels unedited. In particular, they would not remove the adverts from the programmes. Assuming that these operators were to refuse to sell the rights or only accept to sell these rights at an excessive price, what would be the legal avenues open to these new platforms?

As far as the refusal to sell TV channels rights is concerned, two strategies could be explored.

First, to the extent that this problem relates to access to valuable inputs, it is subject to question whether the case law of the ECJ on refusal to supply could be relied upon here.[8] In order for this case law to apply, several conditions must be met. First, it is essential to demonstrate that the provider(s) of the input is/are in a dominant position on the market for that input (the upstream market). In our hypothetical example, the upstream market would consist of the market for the designing and assembling of national TV free-to-air channels and two operators would be on that market. As far as establishing dominance is concerned, in the *Magill* case, the Commission's decision (subsequently upheld by the CFI and the ECJ) found that three Irish TV companies enjoyed a dominant position on their respective TV listings magazines (the input that was requested the plaintiff in that case).[9] *Mutatis Mutandis*, a similar approach, whereby each of the TV operators would be found dominant on the market for their own TV channels, could be envisaged here. Alternatively, as suggested by Professor Whish in relation to *Magill*,[10] a collective dominance approach could also be followed provided that the conditions imposed by the ECJ in *Compagnie Maritime Belge* are met.[11]

One would need to show that a refusal to supply the requested input (in this case TV channels) would be abusive. In order to identify an abuse, the strict conditions imposed by the ECJ in *Bronner* would have to be met.[12] According to the ECJ, Art.82 of the Treaty may only apply to refusal to supply cases where three conditions are fulfilled: (i) the refusal of access to a facility must be likely to prevent any competition at all on the applicant's market, *i.e.* the downstream market, in this case the market for the delivery of TV channels; (ii) the access must be indispensable or essential for carrying out the

---

[8] See ECJ, March 6, 1974, Case 6–7/73, *Commercial Solvents v Commission*, [1974] E.C.R. 223. See also, John Temple Lang, "The Principle of the Essential Facilities Doctrine in European Community Competition Law—The Position Since Bronner", (2000) 1 *Journal of Network Industries*, 375; Derek Ridyard, "Essential Facilities and the Obligation to Supply Competitors and Access to Essential Facilities", [1996] E.C.L.R. 438.

[9] See Commission decision 89/205 of December 21, 1988, IV/31.851, *Magill TV Guide/ITP, BBC and RTE* [1989] O.J. L 78/43.

[10] See Whish, above n.54, at p.474.

[11] See ECJ, above n.56.

[12] See ECJ, November 26, 1998, *Oscar Bronner v Mediaprint*, Case C-7/97 [1998] E.C.R. I-7791 at para.[41].

applicant's business; and (iii) the access must be denied without any objective justification.[13] These conditions are not necessarily easy to meet and whether or not they are met in a given case essentially depends on the specific circumstances of that case.

The other avenue would be to argue that the TV channels rights holders' reluctance to sell these rights to new media platforms is part of a strategy of predation. As in our hypothetical example, the new media platforms would rebroadcast the channel programmes with the adverts they contain, it would indeed be in the TV channels rights owners' best interests to have such channels rebroadcast on additional platforms. As the number of viewers would increase, so would the value of the advertising slots. Thus, by refusing to sell their rights to a new media platform, the TV channels rights holders would in fact sacrifice short-term profits for the long-term profits that would be reaped by excluding the rival platforms, a strategy that is assimilated to predation.[14]

Another problem with which new media platforms could be faced is that, while TV channel suppliers are willing to sell their rights, they may do so at an excessive price. In fact, excessive prices could be another way of deterring entry or excluding rival platforms. This would be part of the same strategy of predation analysed above. But even if these excessive prices were not part of a strategy of exclusion, they could still violate Art.82 as a form of exploitative abuse. However, proving such an abuse may not be so easy. In *United Brands*, the ECJ provided that a price is deemed to be excessive when "it has no reasonable relation to the economic value of the product supplied".[15] However, experience teaches that competition authorities as well as courts are generally reluctant to find a price excessive, if only because they do not want to be transformed into price regulators.[16] Moreover, determining the reasonableness of a price is not easy in practice (because of asymmetry of information problems) and it is only in the most blatant cases of abusive pricing, such as the sudden increase of a price by several hundred percent, that they will be willing to intervene.

Conclusions

Access to premium content is of critical importance for new media platforms as such content is necessary to attract viewers and gain market shares. So far, it has been difficult for these platforms to gain access to premium movies, sports, or even TV channels. Access foreclosure to premium content not only prevents new entry from taking place in the highly concentrated pay-TV market, but it also affects technological developments and consumer choice as the latter would be prevented from watching their favourite programmes on the platform of their choice.

This difficulty for new media platforms of gaining access to premium content is in great part due to the existence of long-term exclusive contracts between movies and sports

---

[13] In the case where the input is protected by intellectual property rights, the ECJ seems to impose an additional condition, which is that the refusal prevents the emergence of a new product for which there is a potential consumer demand. See ECJ, April 29, 2004, Case C-418/01, *IMS Health GmbH & Co OHG v NDC Health GmbH & Co KG*, not yet published.

[14] See Philip Areeda and Donald F. Turner, "Predatory Pricing and Related Practices under Section 2 of the Sherman Act" (1975) 88 Harv. L.R. 697.

[15] See ECJ, February 14, 1978, Case 27/76, *United Brands v Commission* [1978] E.C.R. 207 at para. [250].

[16] See Faull and Nikpay, above n.82, at para.3.298.

94   New Media Platforms

rights holders and dominant pay-TV operators. In some recent decisions, such as *UEFA* and *Newscorp/Telepiu*, the Commission took measures to reduce the length and the scope of the exclusivity in order to prevent the monopolisation of premium content rights by some pay-TV operators. These decisions also contain specific measures to facilitate the acquisition of premium content rights by new media platforms.

The decisions do not, however, appear sufficient to create a level playing field on the markets for the acquisition of premium content rights. Other commercial practices, such as high minimum guarantees, holdback clauses, exploitation of rights through a system of windows, and MFN clauses create barriers to new entrants, such as the new media platforms. As some of these practices represent severe restrictions on competition, it is suggested that the Commission and the national competition authorities should initiate investigations to put them to an end. The sector enquiry launched by the Commission into the sale of sports rights to internet and 3G mobile operators is a good positive first step.

[15]

TELEVISION AS SOMETHING SPECIAL? CONTENT CONTROL TECHNOLOGIES AND FREE-TO-AIR TV

ANDREW T KENYON[*] AND ROBIN WRIGHT[†]

[*Many areas of digital communication, including digital television, raise concerns about unauthorised reuse of content. Proposals exist in the United States and Europe for applying content control technologies to free-to-air digital television to limit the reuse of broadcast content. These proposals have implications for regulatory options, and for the social and cultural position of television in countries such as Australia. Each proposal also demonstrates the importance of current issues in copyright reform for questions of media law and policy. By examining the history and current status of the broadcast flag in the United States and the Content Protection and Copy Management standard being developed in Europe, this article suggests that Australian regulators are likely to face similar calls for action on digital broadcast content and explains some of the possible regulatory choices regarding the transmission and the reception of digital free-to-air content. As with the United States' and European plans, the choices made in relation to television may have wider implications for digital networked communications and the evolution of a diverse media environment.]*

CONTENTS

I   Introduction: Television's Digital Future 339
  A   Interaction between Copyright Law and Broadcasting Regulation 340
  B   Levels and Locations of Control over Reusing DTV Content 341
  C   Impact of Reuse Restrictions on Free-To-Air Television in Australia 341
II The US and the Broadcast Flag 343
  A   Background 343
  B   FCC Rulemaking and Technology Approval 346
  C   Review by the United States Court of Appeals 349
  D   Current Status of the Broadcast Flag 351
III Europe and CPCM 351
IV DTV and Content Control 355
  A   The Limits of Technological Control 355
  B   Control and Copyright 356
  C   Regulatory Powers 363
V   Conclusion 367

---

[*] LLB (Hons) (Melb), LLM (Dist) (Lond), PhD (Melb); Associate Professor, Faculty of Law, The University of Melbourne; Director, Centre for Media and Communications Law, The University of Melbourne; Editor, *Media & Arts Law Review*.
[†] LLB (Hons) (La Trobe), BMus (Melb), MA (Monash); Research Fellow, Centre for Media and Communications Law, The University of Melbourne.
  This research has been supported by the Australian Research Council under the Discovery — Projects scheme (DP0559783 (Kenyon)). Thanks to David Lindsay for his comments on early stages of the research.

I INTRODUCTION: TELEVISION'S DIGITAL FUTURE

Australian free-to-air television is becoming digital,[1] even if the transition has been slower than expected when conversion plans were announced in 1998.[2] Free-to-air digital television ('DTV') commenced in 2001 and reached all metropolitan and rural areas over the following three years.[3] However, digital uptake has been slight with only 15 per cent of households receiving free-to-air DTV at 31 December 2005 and 12 per cent subscribing to DTV.[4] The transition to digital poses many challenges for Australian industry and policymakers, with numerous broadcasting and copyright reviews relevant to DTV conducted since the late 1990s.[5] As well as the substantial work of the Productivity Commission in 2000,[6] more recent reviews have examined aspects of the DTV regulatory framework related to simulcasting, licences and quotas,[7] the treatment of technological protection measures ('TPMs') under copyright law,[8] and copyright exceptions including fair dealing.[9] Many of the reviews are ongoing.[10] Television broadcasting is a complex policy area, with multiple changes in technologies, industries and audiences relevant to the Australian environment. One area which has not received a great deal of attention in the policy process surrounding DTV is the concern of copyright owners about protecting digital audiovisual content from unauthorised reuse.

---

[1] Different terms are used for television delivery formats in various countries, such as 'terrestrial', 'broadcast', 'cable', 'satellite', 'pay', 'subscription' and 'multichannel'. In this article, the terms 'free-to-air' and 'broadcast television' are generally used to describe the longstanding Australian form for broadcast television that is received by viewers without direct charge. Cable and satellite-delivered subscription multichannel services are generally described in this article as 'subscription television'.

[2] See, eg, the second reading speech for the Television Broadcasting Services (Digital Conversion) Bill 1998 (Cth): Commonwealth, *Parliamentary Debates*, House of Representatives, 8 April 1998, 2830 (Warwick Smith, Minister for Family Services). Smith stated that the transition would be complete by 1 January 2004: at 2831. This transition to digital is focused on television transmission and reception; television production has already moved to digital technology, see, eg, Jock Given, 'A Digital Agenda' (2002) 35 *Southern Review* 21, 22–3.

[3] The reforms were contained in the *Television Broadcasting Services (Digital Conversion) Act 1998* (Cth) and the *Broadcasting Services Amendment (Digital Television and Datacasting) Act 2000* (Cth). See generally Lesley Hitchens, 'Digital Television Broadcasting — An Australian Approach' (2001) 12 *Entertainment Law Review* 112. For a brief overview, see Des Butler and Sharon Rodrick, *Australian Media Law* (2nd ed, 2004) 541–7.

[4] See Department of Communications, Information Technology and the Arts, Australia ('DCITA'), 'Meeting the Digital Challenge: Reforming Australia's Media in the Digital Age' (Discussion Paper, 2006) 15; Tom Loncar, Peter Fairbrother and Julie Dalziel, *Digital Media in Australian Homes* (2005) 20. Note, however, that satellite subscription television in Australia is digital: see Given, 'A Digital Agenda', above n 2, 22.

[5] Many of them are mandated in legislation governing DTV: see *Broadcasting Services Act 1992* (Cth) sch 4 cls 60–60B.

[6] Productivity Commission, *Broadcasting: Inquiry Report*, Report No 11 (2000).

[7] DCITA, *Reports on Reviews of the Digital Television Regulatory Framework* (2006).

[8] House of Representatives Standing Committee on Legal and Constitutional Affairs, Parliament of Australia, *Review of Technological Protection Measures Exceptions* (2006).

[9] Attorney-General's Department, Australia, *Fair Use and Other Copyright Exceptions: An Examination of Fair Use, Fair Dealing and Other Exceptions in the Digital Age — Issues Paper* (2005) ('*Fair Use Issues Paper*').

[10] See, eg, DCITA, 'Meeting the Digital Challenge', above n 4.

The emergence of broadband internet and associated equipment such as digital video recorders ('DVRs')[11] means that television material broadcast digitally 'in the clear' — without any form of technological protection — has the potential to be copied and redistributed online without any loss of quality. Such redistribution may well breach the copyright held by creators, investors or broadcasters. However, the experience of content owners with peer-to-peer music distribution suggests that protection by copyright law alone is unlikely to prevent unauthorised reuse of digital broadcast material. In addition, the tradition of open and free reception of television content may leave viewers with few qualms about reusing content that has already been made available somewhere in the world on broadcast television. File sharing of DTV content is now far from novel.[12] Since the 1990s, content owners have investigated options for the technological protection of free-to-air DTV content. In recent years, processes based in the United States and Europe have sought to develop specific technological and regulatory protection mechanisms for such content. These initial steps in developing, deploying and regulating content control technologies raise significant issues for broadcasters, associated creative industries and viewing publics. US and European plans for the technological protection of DTV content identify issues for the Australian legislative and regulatory context, three of which appear particularly relevant. These are: the interaction between copyright law and broadcasting regulation; the appropriate levels and locations of control over the reuse of DTV content; and the impact of reuse restrictions on the place of free-to-air television in Australia.

A *Interaction between Copyright Law and Broadcasting Regulation*

The first of the three issues involves the relationship between two areas of law and policy. Under contemporary network conditions, broadcasting and copyright regulation are of growing *joint* importance for digital communications.[13] As the proposals for content protection illustrate, choices that have been made in copyright policy may re-emerge in decisions about media regulation and vice versa. For example, the treatment of TPMs and their circumvention under copyright law — major issues in Australia and internationally[14] — also become significant for DTV policy once control technologies are proposed for content. Closer relations between broadcasting regulation and copyright law raise issues about policy formation, such as the range of participants involved and the

---

[11] These devices are also commonly known as personal video recorders ('PVRs').
[12] See, eg, David Smith and Alice O'Keeffe, 'TV: So How Will You Watch It?', *Focus, The Observer* (London), 12 March 2006, 24.
[13] See, eg, Jonathan Weinberg, 'Digital TV, Copy Control, and Public Policy' (2002) 20 *Cardozo Arts and Entertainment Law Journal* 277. On the repositioning of television and film policy as part of a 'whole of government' approach to service industries (within which copyright has an important economy-wide role), see Tom O'Regan and Ben Goldsmith, 'Making Cultural Policy: Meeting Cultural Objectives in a Digital Environment' (2006) 7 *Television and New Media* 68.
[14] See, eg, House of Representatives Standing Committee on Legal and Constitutional Affairs, above n 8; Kimberlee Weatherall, 'On Technology Locks and the Proper Scope of Digital Copyright Laws — *Sony* in the High Court' (2004) 26 *Sydney Law Review* 613.

opportunities for public involvement in the process.[15] They also highlight issues about the extent of regulatory power within the sector: to date, limited regulatory power has determined the fate of the US proposal for broadcast content protection. And they reveal the breadth of regulatory impact when choices are made about DTV. In particular, decisions aimed at DTV can have implications for networked communications more generally and for the open architecture of the internet.[16]

B Levels and Locations of Control over Reusing DTV Content

The second issue concerns control. The content control technologies proposed for broadcast television in the US and Europe suggest that choices about reusing content may move away from viewers towards copyright owners. In terms of authorised content use at least, the ability to reuse content may be left to market mechanisms, with copyright owners able to offer content on more or less restrictive terms. (Of course, if those terms are too unacceptable to viewers, they may look to unauthorised file sharing networks.) Such a shift from users to copyright owners accords with literature on the interaction of digital rights management ('DRM') and copyright exceptions, which recognises that digital technologies can allow restrictions to be placed on uses that, in the analogue environment, were not practical to control.[17] But the context of free-to-air television differs in an important way from many other instances of technological content control. It is not based on contracts where viewers purchase the ability to view particular content and, in doing so, limit their rights to reuse material. Rather, free-to-air television offers content widely in order to accumulate viewers and then sells their attention — or the model of it created through ratings measurements — to advertisers.[18] Examining the US and European proposals for DTV content control suggests that media policy may offer its own reasons to limit changes in control from viewers towards content owners — reasons that are distinct from those raised within existing debates about the interaction of TPMs, copyright exceptions and contracts.[19]

C Impact of Reuse Restrictions on Free-To-Air Television in Australia

Shifts in control also relate to a third issue: digital broadcast content control illustrates how in the digital environment free-to-air television may change from

---

[15] Such closer relations between broadcasting and copyright policy may also challenge the political weight of existing media entities, which is something that appears to have long been important in Australian media policy: see, eg, Mark Westfield, *The Gatekeepers: The Global Media Battle To Control Australia's Pay TV* (2000); Trevor Barr, *Newmedia.com: The Changing Face of Australia's Media and Communications* (2000) 206–7.

[16] See, eg, Susan P Crawford, 'Shortness of Vision: Regulatory Ambition in the Digital Age' (2005) 74 *Fordham Law Review* 695.

[17] See, eg, Dan L Burk, 'Legal and Technical Standards in Digital Rights Management Technology' (2005) 74 *Fordham Law Review* 537.

[18] See, eg, John Sinclair, 'Into the Post-Broadcast Era' in John Sinclair (ed), *Contemporary World Television* (2004) 42.

[19] On the latter, see, eg, Copyright Law Review Committee, Attorney-General's Department, Australia, *Copyright and Contract* (2002).

being 'something special' in terms of its cultural, economic and political position within other forms of media and within media policy. As Jock Given has suggested, broadcast media has a particular place in the history of countries like Australia:

> the simplicity of music, speech and images supplied by radio and TV broadcasters, and the almost completely intuitive operation of the receiving devices, got them to nearly 100 per cent of households throughout the industrialised world. This outcome proved much more difficult for other media such as print, which required literacy, and the telephone, which required direct payment and a physical connection to the home. Broadcasting was something special.[20]

Broadcasting was free — at least without direct charge to the viewer in Australia — and almost universal in its use. And free-to-air television remains the most widely used mass media form in Australia,[21] certainly for news and information.[22] While there is no 'inherent social magic' in the free availability of broadcast content,[23] free-to-air television has held a significant place in cultural, economic and political life in countries such as Australia for many years. From the mid-20th century, television was conceptualised as a nationally-based mass medium ascribed with important roles in identity formation, political life and public debate.[24] In recent decades, mass engagement with television has changed and it appears set to change much further in television's digital future. In part, this is because the technologies underlying DTV also support alternative delivery formats for audiovisual content. This suggests how Australian viewers might bypass broadcasters and access their favourite programs directly — whether via authorised or unauthorised avenues. Technically such a shift is possible, but histories about the adoption of many technologies suggest that the cultural, political and institutional weight of free-to-air television will sustain it — in some form and for some time — although its content and financial models will be far from unchanged.[25]

One important factor in these developments will be the scope and style of content protection measures. This article examines proposals for applying content control technologies to free-to-air DTV in the US and Europe, and examines implications for the Australian regulatory environment. Part II examines the progress and current status of the 'broadcast flag' proposal in the US, including the Federal Communications Commission ('FCC') rule-making process and the review by the US Court of Appeals for the District of Columbia of the FCC's authority to make rules in this area. It also outlines the types of

---

[20] Jock Given, *Turning Off the Television: Broadcasting's Uncertain Future* (2003) 40.
[21] Productivity Commission, above n 6, 61–74.
[22] David Denemark, 'Mass Media and Media Power in Australia' in Shaun Wilson et al (eds), *Australian Social Attitudes: The First Report* (2005) 220, 220–5.
[23] Given, *Turning Off the Television*, above n 20, 258.
[24] Sinclair, above n 18, 42–5.
[25] For a useful review of work on the reception of media technologies, see Leah A Lievrouw and Sonia Livingstone (eds), *Handbook of New Media: Social Shaping and Social Consequences of ICTs* (2006).

content control technologies which were approved before the FCC rules were struck down, and considers the concerns of public interest advocates about the impact of the broadcast flag rules on matters such as fair use under copyright law and the open architecture of the internet. Part III looks at a proposal for a Content Protection and Copy Management ('CPCM') standard, which is being developed in Europe by the Digital Video Broadcast ('DVB') consortium and incorporates more complex usage controls than the broadcast flag. The types of reuse of broadcast material which could potentially be controlled, and the differences between CPCM and the broadcast flag, are explored.[26] Part IV, while noting the limited effectiveness of content control technologies in general, focuses on the different regulatory choices that could be available under models provided by the US or European proposals. The possibilities for regulation are examined in relation to issues in both copyright and broadcasting law to illustrate a more general point: the examination of digital communications and content control can encompass multiple legal domains. This Part also analyses whether appropriate powers exist in Australia to implement options for regulating DTV reception equipment and transmission that are suggested by the international proposals. The analysis suggests how careful regulation of content control technologies might support the development of DTV in ways that maintain at least some of the cultural, economic and political roles of Australian television in the broader digital media environment.

II THE US AND THE BROADCAST FLAG

A Background

Unauthorised redistribution of digital broadcast content can be limited in various ways. For example, the signal can be encrypted at its source in an attempt to limit all unauthorised access to material, or watermarked by embedding data in signals to identify copyright owners and related information and to allow sophisticated tracing of copies. Subscription forms of DTV also offer the possibility of specific contractual terms under which subscribers consent to their reuse of content being limited and their activities being monitored — such as through networked set-top box reception equipment, which may be owned by subscription companies, and itself offers extensive possibilities to protect content technologically.[27] Free-to-air broadcast television lacks a

---

[26] Both the US and European developments could also be considered in light of the World Intellectual Property Organization's ('WIPO') proposed treaty on broadcasting: Standing Committee on Copyright and Related Rights, WIPO, *Revised Draft Basic Proposal for the WIPO Treaty on the Protection of Broadcasting Organizations*, 15th session, WIPO Doc SCCR/15/2 (2006), which is available at <http://www.wipo.int/meetings/en/doc_details.jsp?doc_id=64712>. Some commentators suggest that the treaty will underlie the adoption of the content control technologies internationally: see, eg, Crawford, 'Shortness of Vision', above n 16, 714. On the proposed treaty, see generally Kate Gilchrist, 'Internet Transmissions and the WIPO Broadcasters' Treaty' (2004) 7(8) *Internet Law Bulletin* 108. However, for present purposes, examining the US and European proposals is enough to highlight the central regulatory issues likely to develop in Australia.

[27] See, eg, Matt Carlson, 'Tapping into TiVo: Digital Video Recorders and the Transition from Schedules to Surveillance in Television' (2006) 8 *New Media and Society* 97.

contractual relationship with its viewers, meaning attention has focused on technological possibilities for controlling the reuse of content. In the US, one such approach in particular — referred to as the 'broadcast flag' — developed alongside wider processes of converting television to digital.

The broadcast flag aims to control 'indiscriminate redistribution' of DTV content.[28] The flag is a series of bits which is embedded in an Advanced Television Systems Committee ('ATSC') digital broadcast signal.[29] Under the scheme, which combined technical standards and regulation, DTV signals containing the broadcast flag would be recognised by reception equipment and the redistribution of that content in digital form would be restricted by approved technologies within that equipment. Regulations would mandate that all reception equipment respond to the flag. It is worth noting that this model for controlling free-to-air DTV content requires public regulation to ensure that reception equipment complies with the particular control technology.[30]

In 2002, content owners suggested they might withhold high value content, such as recent movies in high definition format, from US digital free-to-air television unless they were satisfied with the available content protection.[31] Such concerns appeared likely to hamper US government plans for DTV and threaten the viability of free-to-air DTV.[32] In this context, a report was issued in June 2002 by a US industry body known as the Broadcast Protection Discussion Subgroup ('BPDG').[33] The BPDG was formed to 'evaluat[e] technical solutions for preventing unauthorized redistribution ... of unencrypted digital terrestrial broadcast television'.[34] It was established within the Copy Protection Technical Working Group, which itself is a voluntary body of representatives from consumer electronics, information technology, motion picture, and cable and broadcast industries.[35] The *BPDG Report* proposed implementing an ATSC flag scheme and regulating the manufacture of all devices that could receive digital

---

[28] *Report and Order and Further Notice of Proposed Rulemaking Re Digital Broadcast Content Protection*, 18 FCCR 23550, [6] (4 November 2003), codified at 47 CFR §§ 73, 76 (2005) ('*Broadcast Flag Order*').

[29] ATSC is an international, not-for-profit organisation which develops voluntary standards for advanced television systems. Its standard has been adopted for US DTV, unlike the DVB standard which is used in Europe, Australia and parts of Asia: see below Part III.

[30] Broadly similar proposals were adopted earlier for cable television in the so-called 'plug and play' Order: *Second Report and Order and Second Further Notice of the Proposed Rulemaking Re Implementation of Section 304 of the* Telecommunications Act of 1996, *Commercial Availability of Navigation Devices, Compatibility between Cable Systems and Consumer Electronics Equipment*, 18 FCCR 20885 (10 September 2003) ('*Plug and Play Order*'). See further Susan P Crawford, 'The Biology of the Broadcast Flag' (2003) 25 *Hastings Communications and Entertainment Law Journal* 603, 616–18.

[31] See, eg, Pamela McClintock, *Viacom's Ultimatum — CBS Parent: No Piracy Protection, No Hi-Def* (16 December 2002) Variety.com <http://www.variety.com/article/VR1117877531?categoryid=1237&cs=1>.

[32] *Broadcast Flag Order*, 18 FCCR 23550, [4] (4 November 2003).

[33] Robert Perry, Michael Ripley and Andrew Setos, *Final Report of the Co-Chairs of the Broadcast Protection Discussion Subgroup to the Copy Protection Technical Working Group* (2002) <http://www.cptwg.org/Assets/TEXT FILES/BPDG/BPDG Report.DOC> ('*BPDG Report*').

[34] Ibid 1.

[35] Ibid.

broadcast signals to ensure they would recognise the existence of the flag in a broadcast and act on it by preventing unauthorised redistribution of that content.

The *BPDG Report* contained two proposed components: the broadcast flag standard and rules for reception equipment.[36] The flag would be a small amount of digital data that could be added to an ATSC digital signal. In 2003, this component of the report was adopted by the ATSC and formed part of ATSC Standard A/65B.[37] The second component comprised rules for reception equipment, known as 'compliance and robustness requirements'.[38] These would require digital broadcast reception equipment to recognise the broadcast flag and restrict redistribution of digital content in a manner that met the compliance and robustness requirements. Existing digital broadcast reception equipment — legacy devices — would be able to receive and handle DTV signals without limitation.

Despite the publication of the *BPDG Report*, there was considerable disagreement among participants about the process and the proposed solution. Some of their expressed concerns included issues of policy formation and copyright law. For example, the companies Philips, Thomson and Zenith objected 'fundamentally' to the BPDG process as 'the only meaningful negotiations were occurring behind closed doors' and were dominated by the studios and the 5C companies.[39] Other parties expressed concern that BPDG had entered too far into public policy, despite being a private technology discussion group with 'no official or unofficial governmental standing',[40] and in particular that the broadcast flag could be seen as restricting 'the fair use rights of consumers'.[41]

In any event, the *BPDG Report* was put forward to the FCC as the appropriate solution for protecting DTV broadcast content. Proponents of the report drew on the same arguments used by content owners in the 1990s when lobbying for the *Digital Millennium Copyright Act of 1998*[42] — that copyright owners would not make works available digitally without some ability to prevent widespread piracy — and the process appears to have been driven substantially by the major

---

[36] Ibid 2–10.
[37] See ATSC, *Program and System Information Protocol for Terrestrial Broadcast and Cable (Revision C) with Amendment No 1*, ATSC Standard A/65C (2006) <http://www.atsc.org/standards/a_65c_with_amend_1.pdf>.
[38] Perry, Ripley and Setos, *BPDG Report*, above n 33, 14.
[39] Philips, Thomson and Zenith, *Comments Submitted by Philips, Thomson and Zenith on the Report of the Broadcast Protection Discussion Subgroup to the Copy Protection Technical Working Group* (2002) [C] <http://www.cptwg.org/Assets/text files/BPDG/Tab_P-04.doc>. These comments were supported by Vereniging Open Source Nederland, Sharp Laboratories of America, DigitalConsumer.org, Electronic Frontier Foundation and Microsoft: see Email from Vereniging Open Source Nederland to BPDG, 30 May 2002; Email from Sharp Laboratories of America to BPDG, 30 May 2002; Email from DigitalConsumer.org to BDPG, 30 May 2002; Email from Electronic Frontier Foundation to BPDG, 31 May 2002. These emails are available at <http://www.cptwg.org/Assets/TEXT FILES/BPDG/Tab P-05.doc>. The 5C companies comprised Hitachi, Intel, Matsushita Electric Industrial, Sony and Toshiba.
[40] Perry, Ripley and Setos, *BPDG Report*, above n 33, 4 fn 4.
[41] Letter from Gary Shapiro (President and CEO, Consumer Electronics Association) to Michael Ripley, 16 May 2002 <http://www.cptwg.org/Assets/TEXT FILES/BPDG/Tab P-06.doc>.
[42] Pub L No 105-304, 112 Stat 2860.

movie studios. In response to the *BPDG Report*, the FCC issued a Notice of Proposed Rulemaking[43] in August 2002 to explore 'whether the FCC can and should mandate the use of a copy protection mechanism for digital broadcast television' and the consumer impact of such regulation.[44]

B *FCC Rulemaking and Technology Approval*

The FCC's Notice of Proposed Rulemaking elicited thousands of responses,[45] and in November 2003 the FCC issued a Report and Order and Further Notice of Proposed Rulemaking.[46] The *Broadcast Flag Order* required all products capable of receiving DTV signals broadcast over the air to comply with the broadcast flag by 1 July 2005.[47] In its *Broadcast Flag Order*, the FCC examined various content protection alternatives. Encryption at the transmission source was considered more effective than flag-based systems, but rejected because of its potential expense, uncertain implementation time line, and rendering obsolete of legacy televisions.[48] The FCC also examined watermarking technologies, which were considered insufficiently mature for implementation.[49] This led the FCC to conclude that 'an ATSC flag-based system is the best option for providing a reasonable level of redistribution protection at a minimal cost to consumers and industry'.[50] Under the system, DTV reception devices would need to recognise and give effect to the broadcast flag in a manner consistent with compliance and robustness rules which the FCC was still to formulate.[51] From July 2005, consumer electronic manufacturers would need to make devices that 'direct flag-marked content to digital outputs associated with approved content protection and recording technologies'.[52] However, the FCC's rule would not prevent flagged content being directed to analogue outputs on reception devices.

Two points are worth noting here. First, the *Broadcast Flag Order* did not determine the reuse standards that should be adopted within approved content protection technologies. However, the scope of allowable restrictions emerged subsequently in the FCC's interim process for approving various technologies, discussed below. Second, although the Order was directed at free-to-air DTV, its scope was much wider. The broadcast flag model is addressed to 'reception' and 'demodulator' devices, but technological convergence means the flag regulation

---

[43] *Notice of Proposed Rulemaking Re Digital Broadcast Copy Protection*, FCC No 02-231 (9 August 2002).
[44] FCC, 'FCC Explores Digital Broadcast Copy Protection: Goal Is To Facilitate Transition to Digital Television' (Press Release, 8 August 2002).
[45] *American Library Association v Federal Communications Commission*, 406 F 3d 689, 691 (Edwards J) (DC Cir, 2005) ('*American Library Association*').
[46] *Broadcast Flag Order*, 18 FCCR 23550 (4 November 2003).
[47] FCC, 'FCC Adopts Anti-Piracy Protection for Digital TV: Broadcast Flag Prevents Mass Internet Distribution; Consumer Copying Not Affected; No New Equipment Needed' (Press Release, 4 November 2003).
[48] *Broadcast Flag Order*, 18 FCCR 23550, [22]–[23] (4 November 2003).
[49] *Broadcast Flag Order*, 18 FCCR 23550, [26] (4 November 2003).
[50] *Broadcast Flag Order*, 18 FCCR 23550, [11] (4 November 2003).
[51] *Broadcast Flag Order*, 18 FCCR 23550, [40] (4 November 2003).
[52] *Broadcast Flag Order*, 18 FCCR 23550, [50] (4 November 2003).

would affect a broad range of digital equipment, including digital televisions, DVRs, DVD recorders, digital VHS recorders, various mobile devices including video-capable mobile phones, and computers with tuner cards. The flag concerns digital content protection across all media, which in contemporary contexts means, especially, the internet.[53] This breadth of effect underlies the criticism that the *Broadcast Flag Order* received from copyright users groups.[54] The flag was seen as a substantial threat to the historically open architecture model of the internet.[55]

The FCC established an interim procedure for approving content protection technologies under the *Broadcast Flag Order*.[56] The procedure clarified that technology complying with the broadcast flag could control content in a wide range of ways. The FCC considered 13 existing technologies that focused on output protection, recording methods and wider DRM techniques. Submissions were made by interested parties, generally commercial entities interested in licensing existing control technologies for use in DTV reception equipment and non-commercial entities with public interest concerns linked to the accessibility of copyright material. The FCC approved all of the 13 technologies considered, which varied greatly in the content control they allowed. For example, approved output protection technologies included: Digital Transmission Content Protection, which employs encryption to ensure content cannot be sent to noncompliant devices;[57] High Bandwidth Digital Content Protection, which does not permit any content to be copied;[58] and TiVoGuard Digital Output Protection Technology, which allows content to be transferred among a limited number of devices.[59] The approved recording methods also included technologies that encrypt content in a manner which is maintained when the content is copied, and those that use a proprietary recording and playback process which only allows copies to play on compliant devices.[60]

The FCC emphasised that its goal was a system of control 'that will prevent the mass indiscriminate redistribution of digital broadcast television content'.[61] At the same time, the FCC said it sought to avoid limits on domestic redistribution that would be legal under the fair use doctrine in US copyright law, and did not intend to limit internet redistribution in contexts in which content

---

[53] Crawford, 'Shortness of Vision', above n 16, 709–15. The FCC was aware of the general point: *Broadcast Flag Order*, 18 FCCR 23550, [40] (4 November 2003).

[54] See, eg, Center for Democracy and Technology, *Implications of the Broadcast Flag: A Public Interest Primer (Version 2.0)* (2003) <http://www.cdt.org/copyright/20031216broadcastflag.pdf>.

[55] See, eg, Crawford, 'Shortness of Vision', above n 16.

[56] *Broadcast Flag Order*, 18 FCCR 23550, [50]–[57] (4 November 2003). While the procedure for approval was interim, the approval was intended to be ongoing; it would only be reviewed if other circumstances changed.

[57] *Order Re Digital Output Protection Technology and Recording Method Certifications*, FCC No 04-193, [5]–[13] (4 August 2004) ('*Digital Output Protection Order*').

[58] *Digital Output Protection Order*, FCC No 04-193, [14]–[18] (4 August 2004).

[59] *Digital Output Protection Order*, FCC No 04-193, [19]–[23] (4 August 2004).

[60] *Digital Output Protection Order*, FCC No 04-193, [24]–[46] (4 August 2004).

[61] *Digital Output Protection Order*, FCC No 04-193, [61] (4 August 2004). See also above n 28 and accompanying text.

could be adequately protected. However, some commentators doubted these aims were achieved:

> the *Broadcast Flag Order* said that the flag was to be used only to prevent redistribution of digital broadcasts, not mere copying. The Order explained the importance of this limitation in terms of preserving valuable uses of broadcast programming: 'Consumers will continue to have the ability to make copies of broadcast content, including news and public interest programming.' The Order did not make it entirely clear how this purported limitation on the degree of permissible TPM constraint would be enforced. And, as it turned out, the FCC later approved ... some technologies that limited copying, explaining that the technologies 'were developed prior to adoption of the *Broadcast Flag Order*' and therefore 'carry with them certain legacy attributes that, while less than ideal from a broadcast flag perspective, may have been appropriate or necessary at the time and in the context that they were developed'.[62]

One of the areas of particular debate in the approval process for broadcast flag compliant technologies was described by the FCC as 'localization'.[63] It involved restricting content redistribution 'to a tightly defined physical space in and around the home'.[64] Localization was sought by the Motion Picture Association of America ('MPAA'), at least until further technical, legal and privacy-related matters could be addressed. This issue was discussed particularly in relation to one of the technologies, 'TiVoGuard'. TiVoGuard allows content to move between a group of TiVo DVRs, all of which are registered to one TiVo customer account.[65] In combination with other TiVo services, it allows content to be transferred from a TiVo DVR via the internet to a personal computer which has TiVo media player software. TiVoGuard was the only approved technology that allowed some internet transmission of content marked with the broadcast flag.[66] However, as Susan P Crawford has explained, the ability to transmit was quite narrow:

> TiVoGuard, the lone holdout against the MPAA's forceful demands to the FCC ... itself permitted transmissions only to a single computer with a 'dongle' (a

---

[62] Molly Shaffer Van Houweling, 'Communications' Copyright Policy' (2005) 4 *Journal on Telecommunications and High Technology Law* 97, 106–7 (citations omitted).
[63] *Digital Output Protection Order*, FCC No 04-193, [70] (4 August 2004).
[64] *Digital Output Protection Order*, FCC No 04-193, [70] (4 August 2004).
[65] TiVo is a proprietary DVR and associated subscription service available in the US and the UK. The TiVo software and an associated electronic program guide and genre database provide a high level of user functionality which allows television viewers to search for programs that match their interests and record them to an internal hard drive for time-shifted viewing. Programs can also be transferred to other devices, such as personal computers, or burned onto digital video discs. For a useful overview of the development of the TiVo DVR, see generally Carlson, above n 27.
[66] Three other technologies abandoned plans to allow some online transmission of flagged content during the interim process of the FCC. See Crawford, 'Shortness of Vision', above n 16, 711 fn 69: 'All save TiVo agreed to insert "time to live" (TTL) and "round trip time" (RTT) limitations in the packets generated by the protection technology. These limitations mean that packets can travel no more than three hops (in no more than seven milliseconds) before expiring'.

small device that plugs into a computer port that prevents illicit copies of software from being made) attached or within a constrained personal network.[67]

TiVo did not want to adopt proximity controls within TiVoGuard for the purpose of localization — relying instead on its own software and hardware restrictions — and it was not required to adopt them by the FCC. In approving the TiVo model, the FCC emphasised that the aim of the broadcast flag regulations was to prevent the *indiscriminate* redistribution of DTV content:

> Our goal was not to prevent 'unauthorized' redistribution as advanced by MPAA. Rather, we explicitly provided that the scope of the *Broadcast Flag Order* 'does not reach existing copyright law.' ... With respect to TiVoGuard, we note in particular that under the terms of TiVo's subscriber agreement, copyrighted content may only be used for personal, non-commercial purposes. The limit of 10 devices uniquely associated with a single secure viewing group additionally prevents content from being indiscriminately redistributed in a 'daisy chain' fashion. ... It is our hope that ... TiVoGuard ... will not only provide a reasonable level of redistribution control for digital broadcast content, but will also facilitate new and innovative consumer uses, such as remote access to content.[68]

The approval of TiVoGuard illustrates that, while approved broadcast flag technologies could be very restrictive of content reuse, the FCC did not limit all types of redistribution. However, neither did the FCC mandate viewer access to particular content by ruling that some content could not be 'flagged'. Decisions about the possible reuse of content by viewers would be left to broadcasters and copyright owners (in terms of applying or not applying the flag and in supplying content) and to the influences of the market (both in terms of attracting viewers and selling reception devices that allow greater or lesser content reuse). The possibility that factual information, public domain material, news and commentary could be restricted by the broadcast flag was noted as potentially problematic by FCC Commissioners Michael Copps and Jonathan Adelstein,[69] and raised many concerns with critics.

C Review by the United States Court of Appeals

There was significant opposition to the broadcast flag from representatives of copyright users, consumer advocates and open source communities. As well as general concerns about free speech — which have particular prominence under the First Amendment — critics raised the treatment of factual information, which would not receive copyright protection under US law, unlike at least some factual compilations in Australia which could be protected by copyright.[70] Some commentators suggested that 'the broadcast flag regulations could have a profound effect on the ability of consumers to watch, record, or use DTV and on

---

[67] Ibid 711 (citations omitted).
[68] *Digital Output Protection Order*, FCC No 04-193, [72] (citations omitted) (4 August 2004).
[69] *Broadcast Flag Order*, 18 FCCR 23550, [66]–[72] (4 November 2003).
[70] See, eg, *Desktop Marketing Systems Pty Ltd v Telstra Corporation Ltd* (2002) 119 FCR 491; cf *Feist Publications Inc v Rural Telephone Service Co Inc*, 499 US 340 (1991).

the design of devices that play, transmit, or store digital content, including computers'.[71] And the Electronic Frontier Foundation argued that the broadcast flag 'would give Hollywood unwarranted control over the development of digital television ... and related technologies to the detriment of creators and consumers'.[72]

In October 2004, nine non-profit organisations representing consumer, research, educational and library interests — including the American Library Association and the Consumer Federation of America — asked the US Court of Appeals for the District of Columbia to review the FCC's *Broadcast Flag Order*.[73] The petitioners argued the FCC had exceeded its statutory powers in requiring DTV receivers and other reception devices to include broadcast flag technology. They argued the FCC was seeking to protect copyright owners in a manner that had been rejected under earlier legislative enactments, which amounted to 'usurping the prerogative of Congress to create and define the scope of copyright'.[74] They also argued the FCC acted without evidence that the flag was needed or that it would solve the problem at which it was directed.[75]

The decision in *American Library Association* in May 2005 focused on the regulator's powers, particularly the limits of its delegated authority. The FCC claimed authority for the Rulemaking from the *Communications Act of 1934*,[76] under which the FCC was created and empowered to make rules '[f]or the purpose of regulating interstate and foreign commerce in communication by wire and radio'.[77] This includes jurisdiction over 'all instrumentalities, facilities [and] apparatus' associated with the overall circuit of messages sent and received.[78] The FCC claimed the provisions provided it with 'ancillary authority' to make the broadcast flag regulations. The previous leading case on the ancillary authority of the FCC, *United States v Southwestern Cable Co*,[79] had upheld certain regulations for cable television made before the FCC had any express regulatory authority for cable.[80] Similarly, with respect to the broadcast flag the FCC claimed that it had ancillary authority to regulate consumer DTV reception equipment.

The Court of Appeals ruled against the FCC. It found the 'insurmountable hurdle' facing the regulator was that its general jurisdiction does not include power to regulate 'consumer electronics products ... when those devices are not engaged in the process of radio or wire transmission'.[81] The Court of Appeals continued:

---

[71] Center for Democracy and Technology, above n 54, 3.
[72] Electronic Frontier Foundation, 'Electronic Frontier Foundation Rejects Broadcast Flag' (Press Release, 9 December 2002).
[73] *American Library Association*, 406 F 3d 689, 691 (Edwards J) (DC Cir, 2005).
[74] Ibid 698 (Edwards J).
[75] Ibid.
[76] 47 USC §§ 151, 154(i)–(j), 303, 403, 521 (1934) ('*Communications Act*').
[77] 47 USC § 151 (1934).
[78] 47 USC §§ 151, 153(50), 153(51) (1934).
[79] 392 US 157 (1968).
[80] See *Broadcast Flag Order*, 18 FCCR 23550, [29] fn 70 (4 November 2003).
[81] *American Library Association*, 406 F 3d 689, 700 (Edwards J) (DC Cir, 2005).

Because the *Flag Order* does not require demodulator products to give effect to the broadcast flag until *after* the DTV broadcast has been completed, the regulations adopted in the *Flag Order* do not fall within the scope of the Commission's general jurisdictional grant.[82]

Thus, manufacturers did not need to meet the July 2005 deadline for flag-compliant products. It is worth emphasising that the decision turned on the scope of FCC power; it did not address the interaction of the broadcast flag and copyright law, or the scope for other technological approaches to DTV content — issues which were raised before the Court of Appeals and in the earlier FCC rule-making process.

D Current Status of the Broadcast Flag

American Library Association removed the requirement for manufacturers to incorporate content control technologies in DTV products. Not surprisingly, copyright owners remained concerned that DTV broadcasts lacking appropriate protection would be redistributed without authorisation. If US legislators decided there was value in implementing the scheme, they could grant authority to the FCC to reinstate the broadcast flag regulations. Although legislative prediction is particularly difficult in the US context, a congressional response to the *American Library Association* decision appears plausible. As early as September 2005, a bipartisan group of 20 politicians wrote to the chair of the House of Representatives Subcommittee on Telecommunications and the Internet seeking federal law to support the broadcast flag model.[83] Interest has continued during 2006,[84] and legislation may be passed giving the FCC power to regulate DTV reception equipment. In any event, the growth in alternative distribution channels — authorised or not — suggests that demand for digital content controls will persist in the US. Such demand is present elsewhere, including in Europe.

III EUROPE AND CPCM

In Europe, the DVB consortium that was instrumental in creating DTV standards is developing specifications for another content control technology. The consortium includes broadcasters, manufacturers, network operators, software developers, regulators and others responsible for setting technical standards.[85] Australia has adopted DVB standards for DTV transmission along

---

[82] Ibid (emphasis in original). For further analysis of the ancillary jurisdiction of the FCC, see Crawford, 'Shortness of Vision', above n 16, 728–36.
[83] Declan McCullagh, *Politicians Want To Raise Broadcast Flag* (30 September 2005) CNET News.com <http://news.com.com/Politicians+want+to+raise+broadcast+flag/2100-1028_3-5886722.html>.
[84] See, eg, Brooks Boliek, *Stevens Wants 'Broadcast Flag' Vote by March* (25 January 2006) The Hollywood Reporter <http://www.hollywoodreporter.com/thr/television/brief_display.jsp?vnu_content_id=1001884261>; Paul Sweeting, 'Copy-Protection Bills Gain Senate Support: But Some Say It Will Limit Technology' (2006) 26(5) *Video Business* 8; Anne Broache, *Senators Endorse Broadcast Flag Plan* (2006) CNET News.com <http://news.com.com/Senators+endorse+broadcast+flag+plan/2100-1028_3-6088711.html>.
[85] See Digital Video Broadcasting Project <http://www.dvb.org>.

with the majority of European and Asian countries.[86] The planned DVB content protection standard is known as the CPCM system. As with the broadcast flag, the CPCM development process appears to have been strongly influenced by US film studios.[87] However, unlike the broadcast flag, CPCM has the potential to be applied across all forms of DTV content transmitted under DVB standards for terrestrial broadcast, cable and mobile television. The adoption of DVB in Australia makes CPCM particularly relevant. There may be pressure for its implementation in Australia if the CPCM standard is finalised and widely adopted in other countries using DVB transmission standards.[88]

The broadcast flag functions as a simple marker. Detection of the flag brings certain usage controls into play, controls which are set within the reception technology and which have been approved by the FCC. In contrast, CPCM appears to go further towards defining allowable uses within metadata embedded in the broadcast transmission itself. And the DVB subgroup responsible for developing CPCM has identified the importance of restricting unauthorised redistribution while still allowing certain uses including moving content within a user's domestic network environment 'to conform with traditional user experience and expectations based on the portability of pre-recorded content'.[89]

Planned specifications for the first three elements of the CPCM standard were released in November 2005 in the form of the document, *Digital Video Broadcast (DVB); Content Protection & Copy Management*, for information purposes.[90] This *Bluebook* contains proposed usage rules which define how certain operations on content would be controlled by the CPCM system. The usage rules are to be implemented through 'usage state information' ('USI') within the DTV signal. USI would cover five areas of control: copy and movement control; consumption control; propagation control; output control; and ancillary control.[91] The proposed CPCM specifications would allow highly

---

[86] Australia has applied standards known as Digital Video Broadcast-Terrestrial ('DVB-T'). European countries have either launched a DVB-T service, adopted the standard, are undertaking trials of the standard, or have industry recommendations to adopt it. Japan is a major Asian jurisdiction that has adopted different technology, known as Integrated Services Digital Broadcasting-Terrestrial ('ISDB-T'). For the global adoption of DVB, see DVB, *DVB Worldwide: Where Have DVB Standards Been Adopted?* <http://www.dvb.org/about_dvb/dvb_worldwide>.

[87] The DVB content protection technical subgroup is led by Chris Hibbert from Walt Disney TV International.

[88] One of the concerns that is likely to arise more prominently as the CPCM proposal develops, which would also have relevance in Australia, is that the mandate of a *single* content control technology may have adverse implications for innovation in communications technologies or for competition more widely. For example, technology innovation arguments featured in US debates about the broadcast flag (even though the flag-based system controls technological standards less closely than a system like CPCM): see, eg, Matt Jackson, 'Protecting Digital Television: Controlling Copyright or Consumers?' (2006) 11 *Media & Arts Law Review* (forthcoming). It is also notable that recent innovations in digital communications and their use — such as peer-to-peer networks — appear to have arisen with little initial concern about whether the law allows or prohibits the 'disruptive' innovation.

[89] Chris Hibbert, 'Copy Protection: Work in the DVB' (2003) 5 *DVB-Scene* 14, 15 <http://www.dvb.org/documents/newsletters/DVB-SCENE-05.pdf>.

[90] DVB, *Digital Video Broadcast (DVB); Content Protection & Copy Management*, DVB Doc No A094 (2005) <http://www.dvb.org/technology/dvb-cpcm/a094.DVB-CPCM.pdf> ('*Bluebook*').

[91] Ibid 31.

granulated control information to be embedded in digital broadcast streams by copyright owners, broadcasters or other distributors. This is a key difference to the broadcast flag, and it raises the possibility of similarly granular and flexible policy choices by regulators beyond pre-approved and fixed technologies embedded into receiving equipment as with the broadcast flag. However, like the broadcast flag, CPCM requires regulation of reception equipment to ensure that it recognises and implements the content control.

The *Bluebook* provides some detail about various types of content usage control that are envisaged. For instance, copy and movement control would include the ability to allow multiple copies, only a single copy, or no copies to be made.[92] Under consumption control, the proposed restrictions include providing a 'signalled time window' in which the content could be consumed,[93] and placing a limit on the number of concurrent content uses.[94] In relation to propagation control, the CPCM system could establish a number of 'propagation realms' within which content could be exchanged, but from which it could not be removed. The propagation realms include an 'authorised domain' which is defined as a 'set of DVB CPCM compliant devices, which are owned, rented or otherwise controlled by members of a single household'[95] as well as other local and geographically-constrained CPCM compliant devices. Another proposed realm is called a 'local environment', which would allow content to be restricted to exchange over a local area network within a domestic home-sized physical space, regardless of any device authorisation. Potentially, this could extend to a wider range of authorised proximate devices referred to as a 'localised authorised domain', which could include mobile devices.[96] Each of these propagation realms is intended to reduce the possibility of unauthorised redistribution of protected content beyond a particular group or territory of authorised receiving devices. In relation to output control, the USI could determine how and where digital and analogue outputs can be directed. It would include the ability to enable or disable export to 'untrusted spaces' or in specified analogue formats. In addition, ancillary control would provide the ability to assert the direction 'do not CPCM scramble', which allows owners delivering material to direct that broadcasters do not apply CPCM controls to specified content.[97]

---

[92] The 'no copies' option would still allow a secure temporary copy to be made 'solely for the purpose of pausing of play-back, or trick-play': ibid.
[93] This could be an absolute time window, a specified period after acquisition, or the first time the content is consumed: ibid 32.
[94] Time-based usage would operate in devices that play content as well as devices that store content, while the concurrent usage feature could also limit the number of devices to which an item could be supplied: ibid.
[95] Ibid 29.
[96] There could also be a 'geographical area' realm that would be 'intended as a special case for allowing Remote Access to CPCM Content' (which otherwise would be restricted to the Localised Authorised Domain) through authorising access by specific remote devices: ibid 36. Being able to authorise some form of geographical control would allow copyright owners or broadcasters to 'ensure that original broadcast footprints are not violated' and to 'enforce regional black-outs of certain broadcast Content Items': ibid 33.
[97] Ibid 37.

How the CPCM model might interact with existing copyright exceptions and statutory licences is yet to receive detailed discussion. However, as noted above, the model aims to conform to viewers' existing understandings about content use. In the European context, this raises issues about statutory schemes for private copying.[98] With digital communications, Europe appears to be moving, in general, from levy-based schemes toward using DRM to license specific uses.[99] However, the debate is not complete — either overall or with regard to DTV — and the questions can be expected to receive attention as CPCM or similar technologies move towards completion. The finely granulated control made possible by the CPCM system may provide additional regulatory possibilities for the protection of exceptions to copyright.

The FCC declined to prohibit the application of the broadcast flag to specific 'types' of content — for example, to allow wider access to news or public domain content, in order to avoid the 'practical and legal difficulties' in determining which content should be protected.[100] By comparison, a system like CPCM offers the possibility that regulators could protect specific 'uses' of all, or certain types of, content through limiting the settings which could be applied at the point of transmission. This may allow greater flexibility for regulators to fashion a regime which focuses on protecting both the public interest in information and the interests of copyright owners, which is explored below.[101]

There is another point to note about CPCM. As with the broadcast flag, it appears that CPCM will not affect legacy reception equipment. This could result in a substantial amount of non-controlled reception for some time after any introduction of CPCM.

---

[98] See, eg, Katerina Gaita and Andrew F Christie, 'Principle or Compromise? Understanding the Original Thinking behind Statutory Licence and Levy Schemes for Private Copying' (2004) 8 *Intellectual Property Quarterly* 422; Andrew F Christie, 'Private Copy Licence and Levy Schemes: Resolving the Paradox of Civilian and Common Law Approaches' in David Vaver and Lionel Bently (eds), *Intellectual Property in the New Millennium: Essays in Honour of William R Cornish* (2004) 248.

[99] See, eg, P Bernt Hugenholtz, Lucie Guibault and Sjoerd van Geffen, Institute for Information Law, University of Amsterdam, *The Future of Levies in a Digital Environment: Final Report* (2003) <http://www.ivir.nl/publications/other/DRM&levies-report.pdf>; High Level Group on Digital Rights Management, European Commission, *Final Report* (2004) <http://europa.eu.int/information_society/eeurope/2005/all_about/digital_rights_man/high_level_group/index_en.htm>.

[100] *Broadcast Flag Order*, 18 FCCR 23550, [38] (4 November 2003).

[101] While the FCC did not act in relation to particular types of *content*, its interim approval process generally protected certain types of *reuse*, in particular, time-shifting. Only one of the control technologies it approved for use in broadcast flag-compliant reception equipment completely prohibited copying (beyond the reproduction involved in initially displaying the content). The FCC noted this was a pre-existing control technology, and emphasised its intention was that the broadcast flag should not prevent viewers from copying digital broadcast content: see generally *Digital Output Protection Order*, FCC No 04-193 (4 August 2004). Placing regulation at the reception end of the equation allows for competition between different control technologies and for users to chose the reception equipment (with a particular control technology) which provides them with their preferred level of reuse.

IV DTV AND CONTENT CONTROL

Notable concerns about the content protection proposals include the ineffectiveness of technological controls in networked environments and the potential for content protection schemes to restrict reuses that copyright law would allow.[102] In addition, the application of content protection to television broadcasts raises questions about the power of authorities to regulate this area.

A *The Limits of Technological Control*

It is commonly understood that no TPM is completely effective.[103] The proposed reuse controls, aimed at domestic DTV viewers, may have little impact on those with the technological abilities to avoid them. More significantly, peer-to-peer distribution over the internet supports the wide availability of content once it has been 'released' from technological protection.[104] While any regulatory action is likely to have only partial success, the actual effectiveness of technological control systems in limiting redistribution deserves examination as the models evolve. Authorised and unauthorised markets for DTV content exist and the evolution of both markets may be hard fought. In that regard, it is notable that Australians reportedly have the highest per capita downloading rate for television programs,[105] an activity which, until recently, was in breach of copyright.[106] Case law about audio file sharing — such as recent litigation over Grokster and Kazaa in the US and Australia[107] — is likely to be influential for user-driven forms of unauthorised DTV content distribution.

Beyond this general point about technological control, concerns have been raised about particular limitations to the effectiveness of content control technologies in the DTV environment resulting from analogue reconversion,

---

[102] Concerns have also been expressed about the possible effects on innovation within the consumer electronics industry; the potential anti-competitive impacts on networked communications; and the potential impact on privacy. On this last point, see, eg, Debra Kaplan, 'Broadcast Flags and the War against Digital Television Piracy: A Solution or Dilemma for the Digital Era?' (2005) 57 *Federal Communications Law Journal* 325, 335–6.

[103] See *Broadcast Flag Order*, 18 FCCR 23550, [19] (4 November 2003): 'We are equally mindful of the fact that it is difficult if not impossible to construct a content protection scheme that is impervious to attack or circumvention'.

[104] See, eg, Paul Biddle et al, 'The Darknet and the Future of Content Distribution' (Paper presented at the 2002 ACM Workshop on Digital Rights Management, Washington, DC, 18 November 2002) <http://crypto.stanford.edu/DRM2002/darknet5.doc>.

[105] Mark Pesce, *Piracy is Good? New Models for the Distribution of Television Programming* (2005) 7 <http://www.aftrs.edu.au/download.cfm?DownloadFile=B0A6D409-2A54-23A3-69F5E21BEA2270EA>.

[106] On early legal movie downloading services in Australia, see, eg, Andrew Colley, 'PCs To Become Online Picture Palaces', *The Australian* (Sydney), 23 February 2006, 27.

[107] *Metro-Goldwyn-Mayer Studios Inc v Grokster Ltd*, 545 US 125 (2005); *Universal Music Australia Pty Ltd v Sharman License Holdings Ltd* (2005) 220 ALR 1. See also Jane C Ginsburg and Sam Ricketson, 'Inducers and Authorisers: A Comparison of the US Supreme Court's *Grokster* Decision and the Australian Federal Court's *KaZaa* Ruling' (2006) 11 *Media & Arts Law Review* 1; Graeme W Austin, 'Importing *Kazaa* — Exporting *Grokster*' (2006) 22 *Santa Clara Computer and High Technology Journal* 577. The Kazaa litigation was settled in July 2006: see, eg, APP with Louisa Hearn, *Kazaa Capitulates, Settles Piracy Case* (28 July 2006) The Age <http://www.theage.com.au/news/digital-music/kazaa-capitulates-settles-piracy-case/2006/07/27/1153816326515.html>.

international markets and legacy devices. The problem of analogue reconversion — sometimes called the analogue hole — arises when digital signals can be output in an analogue form and then reconverted to digital. The second digital version will have lost some quality in the process but it will no longer contain any of the reuse control information that was attached to the original digital signal. As the FCC noted, the broadcast flag 'could be easily circumvented, potentially through the use of digital to analog converters'.[108] The reconversion problem has been an ongoing concern to the content industry and in December 2005 the proposed Digital Transition Content Security Act of 2005, HR 4569, 109th Cong was presented to the US Congress to prohibit the manufacturing or selling of electronic devices that convert analogue video signals to digital. The proposal was referred to the House Committee on the Judiciary.[109] Unlike the broadcast flag model, it appears that the CPCM system could allow analogue redistribution to be limited. The manner of that limitation — and its interaction with legacy devices — is not yet clear, and the extent of such control is an important issue in the continued development of CPCM. Limits to technological control are also posed by international communications — unless regulation with similar effects is in place in other jurisdictions — and by legacy devices.[110] Notwithstanding these limitations, the FCC's *Broadcast Flag Order* considered that the broadcast flag scheme would provide adequate protection for content in the digital environment by 'creating a "speed bump" mechanism to prevent indiscriminate redistribution of broadcast content and ensure the continued availability of high value content to broadcast outlets'.[111]

B *Control and Copyright*

One way of approaching the broadcast flag is to view it as simply another site of ongoing US battles over copyright and free speech — in particular, over reconciling the rights granted to copyright owners and the public policy benefits perceived to flow from wide access to information.[112] Proposals for DTV content controls have generated strong criticism about their possible effect on US copyright doctrines such as fair use and the public domain:

---

[108] *Broadcast Flag Order*, 18 FCCR 23550, [17] (4 November 2003).
[109] See Library of Congress, *Bill Summary and Status File: HR 4569* (2006) <http://thomas.loc.gov/cgi-bin/bdquery/z?d109:h.r.04569:>; Jennifer LeClaire, *Congress May Require Embedded Copyright-Protection Tech in DVDs* (20 December 2005) E-Commerce Times <http://www.ecommercetimes.com/story/D6gdfc3mXOVN60/Congress-May-Require-Embedded-Copyright-Protection-Tech-in-DVDs.xhtml>.
[110] Lisa M Ezra, 'The Failure of the Broadcast Flag: Copyright Protection To Make Hollywood Happy' (2005) 27 *Hastings Communications and Entertainment Law Journal* 383, 391–4; Garrett Levin, 'Buggy Whips and Broadcast Flags: The Need for a New Politics of Expression' [2005] *Duke Law and Technology Review* 24, [18] <http://www.law.duke.edu/journals/dltr/articles/PDF/2005DLTR0024.pdf>.
[111] *Broadcast Flag Order*, 18 FCCR 23550, [19] (4 November 2003).
[112] See, eg, Neil Weinstock Netanel, 'Copyright and the First Amendment: What *Eldred* Misses — and Portends' in Jonathan Griffiths and Uma Suthersanen (eds), *Copyright and Free Speech: Comparative and International Analyses* (2005) 127.

concern is expressed most prominently regarding news or public internet-based content, or works that have already entered the public domain. Despite suggestions raised by consumer rights groups, the FCC has so far declined to adopt language to prevent content providers from using the broadcast flag on such programs, largely because of the 'practical and legal difficulties of determining which types of broadcast content merit protection from indiscriminate redistribution and which do not'.[113]

Many commentators consider that the proposed broadcast flag scheme could undermine fair use and restrict legitimate access to broadcast television material. The determination of a fair use defence in the US is made by the courts on the basis of the four factors outlined in § 107 of the *Copyright Act of 1976*.[114] This fact-sensitive approach makes technological determinations difficult. In the case of the broadcast flag, which would simply be set 'on' or 'off' when material is transmitted, there may be a danger that '[i]f there is not a secure technological way to prevent a possible fair use from turning into widespread online distribution, the use will be prevented by the approved technologies.'[115]

During the FCC broadcast flag rule-making process, Commissioner Copps submitted a partially dissenting statement due to concerns that the flag might restrict access to important content:

> I dissent in part ... because the Commission does not preclude the use of the flag for news or for content that is already in the public domain. This means that even broadcasts of government meetings could be locked behind the flag. Broadcasters are given the right to use the public's airwaves in return for serving their communities. The widest possible dissemination of news and information serves the best interests of the community.[116]

These US debates have been echoed within Australia in contexts such as the planned revision of anti-circumvention provisions under the *Australia–United*

---

[113] Angie A Welborn, Congressional Research Service, *Copyright Protection of Digital Television: The 'Broadcast Flag'*, CRS No RS22106 (11 May 2005) <http://www.cdt.org/righttoknow/crsreports/RS22106_20050511.pdf>.

[114] 17 USC § 107 (2000). The four factors concern: the purpose and character of the use (including whether it is commercial or for non-profit educational purposes); the nature of the work; the amount and substantiality of what is taken; and the effect on the potential market for the work: see, eg, Robert Burrell and Allison Coleman, *Copyright Exceptions: The Digital Impact* (2005) ch 9; and for a comparative analysis of § 107, see Gerald Dworkin, 'Copyright, the Public Interest, and Freedom of Speech: A UK Copyright Lawyer's Perspective' in Jonathan Griffiths and Uma Suthersanen (eds), *Copyright and Free Speech: Comparative and International Analyses* (2005) 153.

[115] Levin, above n 110, [22].

[116] Michael Copps, 'Statement of Commissioner Michael J Copps Approving in Part, Dissenting in Part — Re: Digital Content Broadcast Protection' (Press Release, 4 November 2003) <http://hraunfoss.fcc.gov/edocs_public/attachmatch/DOC-240759A4.pdf>. See also Robert T Numbers II, 'To Promote Profit in Science and the Useful Arts: The Broadcast Flag, FCC Jurisdiction, and Copyright Implications' (2004) 80 *Notre Dame Law Review* 439, 458, who comments that
> everything from local government meetings to images of important national events will be restricted ... Images such as the fall of the Berlin Wall, the protests in Tiananmen Square, and the September 11 attacks might not be available for public use, except in those ways which the content provider and those designing flag-compliant technology allow them to be used ...

States Free Trade Agreement,[117] and the ongoing consideration of reforming Australian copyright exceptions.[118] Australian copyright law may soon be reformed to allow more domestic reuses, for example, allowing domestic time-shifting of broadcast content.[119] While the final shape of any reforms will be relevant to how international plans for DTV content protection might best be adapted for Australia, the style of control offered by the European model offers interesting regulatory possibilities.

As noted above, CPCM aims to give broadcasters and copyright owners finely calibrated controls across five areas of reuse. As a result, CPCM may appear better suited than the broadcast flag to *allowing* content to be distributed in a form that enables reuses such as time-shifting, while still preventing unrestricted redistribution. This would require mandating the type of technological control to be used *and* regulating how the technology can be applied to transmission. The broadcaster, as well as the equipment manufacturer, would need to be subject to regulation. Although there is a separate question as to whether such an approach to reuse should be *required* for any or all content in light of public policy objectives, in adopting the CPCM model regulators would not necessarily face the decision of simply allowing or prohibiting all restrictions for certain categories of content. The decision would not just be whether some broadcast content should be prevented from having any technological protection, with protection left purely to copyright law. Rather, a more finely honed decision would be open about the types of uses which should not be restricted in certain instances. A simple 'on' or 'off' decision was what faced the FCC under the broadcast flag model with a range of pre-approved reuse controls then applying to any received content when the flag was set as 'on'. It seems that with CPCM, regulators could seek to prevent certain USI from being applied to particular areas of control — such as copy and movement control, consumption control or propagation control — and this could be done in relation to all or certain types of content. Regulators could decide, for example, that free-to-air broadcast content could be protected in any manner under CPCM so long as private time-shifting was allowed. Or allowing format-shifting within a viewer's authorised domain could be mandated. These options could be required only for particular types of free-to-air content such as news. The choices would be numerous. While they may not appear to be the usual types of choices for copyright policy, they are not

---

[117] Opened for signature 18 May 2004, [2005] ATS 1 (entered into force 1 January 2005) ('*AUSFTA*'); see also *US Free Trade Agreement Implementation Act 2004* (Cth). See further House of Representatives Standing Committee on Legal and Constitutional Affairs, above n 8.

[118] See, eg, Attorney-General's Department, *Fair Use Issues Paper*, above n 9.

[119] The broad policy of allowing some time-shifting for audiovisual content (and allowing some format-shifting and time-shifting for audio content) was announced in May 2006: Philip Ruddock, Attorney-General, 'Major Copyright Reforms Strike Balance' (Press Release, 14 May 2006). No draft Bill is yet available and nor is one expected to be introduced before 2007. On the general issue of allowing these types of use through free exceptions or remunerated licences, each of which might be better suited to different acts of private copying, see, eg, David Lindsay, 'Fair Use and Other Copyright Exceptions: Overview of Issues' (2005) 23 *Copyright Reporter: Journal of the Copyright Society of Australia* 4; Kimberlee Weatherall, 'A Comment on the Copyright Exceptions Review and Private Copying' (Working Paper No 14, Intellectual Property Research Institute of Australia, The University of Melbourne, 2005).

unusual under traditions of broadcasting regulation in Australia (and many comparable countries) where it has been common to apply particular requirements for different types of content.[120]

With regard to these possible regulatory choices it is worth noting that, while art 11 of the *WIPO Copyright Treaty* [121] imposes an obligation to ensure 'adequate legal protection and effective legal remedies' against the *circumvention* of TPMs, its terms do not require that owners be able to *apply* any form of content control technology that they wish to any form of content.[122] Debate may arise over whether this point should be inferred from these types of anti-circumvention provisions, debate which could develop the general issues canvassed in relation to contracting out of copyright exceptions.[123] However, it would seem difficult to mount such an argument in relation to *licensed* broadcast media, such as free-to-air television, where it is well-accepted that such services cannot be restricted by encryption or similar controls. Australian commercial broadcasting services, for instance, must be provided without encryption in a manner that can be received on commonly available equipment and without the need for special receivers.[124]

There is also evidence that, in the US, the FCC will limit the application of at least some content control technologies in order to facilitate viewer access to content.[125] For example, in the *Plug and Play Order*[126] — which regulated the manufacture of cable set-top boxes — the FCC imposed 'encoding rules' to prevent cable companies placing copy restrictions on retransmissions of free-to-air broadcasts.[127] As Molly Shaffer Van Houweling notes, the FCC is regulating control technologies 'in a way that limits their reach in order to preserve certain consumer uses that [they] might otherwise prohibit.'[128] This general policy approach was also taken in relation to the broadcast flag by

---

[120] For example, consider context codes and standards for television program content in Australia: see Butler and Rodrick, above n 3, 522–34.
[121] Opened for signature 20 December 1996, 36 ILM 65 (entered into force 29 February 2002).
[122] For a detailed background to art 11 of the *WIPO Copyright Treaty*, see especially Sam Ricketson and Jane C Ginsburg, *International Copyright and Neighbouring Rights: The Berne Convention and Beyond* (2nd ed, 2006) 966–82. See also Jörg Reinbothe and Silke von Lewinski, *The WIPO Treaties 1996: The* WIPO Copyright Treaty *and the* WIPO Performances and Phonograms Treaty (2002) 135–47; Mihály Ficsor, *The Law of Copyright and the Internet: The 1996 WIPO Treaties, Their Interpretation and Implementation* (2003).
[123] See, eg, Copyright Law Review Committee, above n 19.
[124] *Broadcasting Services Act 1992* (Cth) s 14. See also Butler and Rodrick, above n 3, 492, which draws on the Explanatory Memorandum, Broadcasting Services Bill 1992 (Cth).
[125] Shaffer Van Houweling, above n 62, 106–10.
[126] 18 FCCR 20885 (10 September 2003). In relation to this Order, it is worth noting that a large percentage of US viewers receive free-to-air television via cable delivery services. '[A]lmost 86 per cent of TV households subscribe to [cable or satellite television]': *Re Annual Assessment of the Status of Competition in the Market for the Delivery of Video Programming*, FCC No 06-11, [8] (3 March 2006). The large US cable market underlies one way in which content control technologies for free-to-air DTV in Australia arise in a different context than in the US. In addition, before the broadcast flag debates, the FCC regulated to ensure the compatibility of cable television systems with varied DTV receivers, and it limited the application of content controls technologies to free-to-air content delivered via cable in order to preserve user expectations, such as time-shifting.
[127] *Plug and Play Order*, 18 FCCR 20885, [11] (10 September 2003).
[128] Shaffer Van Houweling, above n 62, 109.

including '[t]he extent to which the digital output protection technology or recording method accommodates consumers' use and enjoyment of unencrypted digital terrestrial broadcast content' [129] as one of the criteria used when evaluating TPMs for use in broadcast flag compliant reception devices.[130] Despite approving one existing technology which restricted all copying (beyond that involved in the initial real-time display of content) under its interim procedure,[131] the FCC maintained the general ambition that 'technologies can protect content while facilitating consumer uses and practices'.[132]

Demand for an approach which restricts how content control technologies are applied to free-to-air DTV might increase if Australian copyright law is reformed to allow domestic reuse of copyright material.[133] In this regard, it is interesting to note Australian television interests' views on legalising certain instances of private copying. They appear in submissions made during 2005 to the Australian Government's inquiry into whether the *Copyright Act 1968* (Cth) should include 'a general exception associated with principles of "fair use" or specific exceptions to facilitate the public's access to copyright material in the digital environment'.[134] The government's general response to that review — which was not to introduce a general fair use provision, but to allow some private time-shifting of audiovisual content — was announced in May 2006.[135] In their submissions, many television-related entities displayed concerns to limit unauthorised reuse of content and protect existing revenue streams. For example, the joint submission of the Nine and Seven Networks considered that exceptions allowing copying for private use would unacceptably harm their commercial activities. Similarly, the Screen Producers Association of Australia opposed allowing time-shifting, format-shifting or introducing a statutory licence for private copying.[136] However, some broadcasting organisations were prepared to countenance a limited free exception allowing time-shifting for personal use, although concerns remained about possible damage to secondary markets. For

---

[129] *Digital Output Protection Order*, FCC No 04-193, [3] (4 August 2004).
[130] *Digital Output Protection Order*, FCC No 04-193, [3] (4 August 2004).
[131] See above nn 56–60 and accompanying text.
[132] *Broadcast Flag Order*, 18 FCCR 23550, [55] (4 November 2003).
[133] See above n 119 and accompanying text.
[134] Attorney-General's Department, *Fair Use Issues Paper*, above n 9, 4. In total, 162 submissions were made responding to the issues paper. These submissions are available at Attorney-General's Department, Australia, *Fair Use and Other Copyright Exceptions: An Examination of Fair Use, Fair Dealing and Other Exceptions in the Digital Age* (2005) <http://www.ag.gov.au/agd/WWW/agdHome.nsf/Page/Publications_2005_Copyright_-_Review_of_Fair_Use_exeption>.
[135] See above n 119. The announcement also included the introduction of 'flexible dealing' exceptions for parody and satire, and for non-commercial uses within the education and museum sectors, both of which would be subject to the 'three-step test' applicable under international treaties: see, eg, Sam Ricketson, *The Three-Step Test, Deemed Quantities, Libraries and Closed Exceptions* (2002).
[136] Submission to the Attorney-General's Department, Australia, in response to the *Fair Use Issues Paper*, 1 July 2005, Submission No 112 (Screen Producers Association of Australia). However, supporters for a statutory licensing scheme included the audiovisual collecting society, Screenrights, and the Australian Copyright Council. See Submission to the Attorney-General's Department, Australia, in response to the *Fair Use Issues Paper*, 8 July 2005, Submission No 142 (Screenrights); Submission to the Attorney-General's Department, Australia, in response to the *Fair Use Issues Paper*, June 2005, Submission No 61 (Australian Copyright Council).

example, Network Ten Pty Ltd suggested an exception allowing free-to-air broadcasts to be recorded only 'for the purpose of *private and domestic use of the maker of the copy* in order to allow a program to be viewed after the scheduled broadcast time *by the person who copied the broadcast.*'[137] The Australian Subscription Television Radio Association also supported reform to allow some time-shifting while suggesting the copyright owner should still retain the right to use TPMs to prevent such reuse.[138] The national broadcasters were more open to the introduction of provisions for time-shifting and format-shifting, noting in addition their dual role as both creators and users of copyright. For example, the Australian Broadcasting Corporation submitted:

> Fair dealing plays a key role in the ABC meeting its Charter obligations. However, the pace of technological change has far exceeded the ability of copyright law to maintain a 'balance' between the interests of both users and creators of copyright. With the focus now more prominently directed towards copyright piracy than creativity, the ABC has been restricted in its ability to access and deliver information, content and innovation to meet both its Charter obligations and public expectations.[139]

These examples illustrate how the possibilities for control (and regulation) offered by DTV content control technologies relate to issues within copyright law. In this article, the possible impact on copyright exceptions has been noted as a key issue. Consideration could equally be given to statutory licences or anti-circumvention provisions.[140] That is, an examination of DTV and content control needs to engage with questions from across at least three areas of copyright. The first involves copyright exceptions and statutory licences, both with regard to their legal scope and the degree to which copyright owners seek to police them. The second concerns technological controls: what style of control is applied to what content, and is that decision determined, at least in part, legally or economically? And the third raises exceptions to TPMs: in particular, who is able to circumvent a TPM and for what purposes? Depending on the type of copyright exceptions or licences within a jurisdiction, particular reuse might be better served by regulating the application of technological controls, the circumvention of controls, or both. For instance, given recent Australian litigation in *Network Ten Pty Ltd v TCN Channel Nine Pty Ltd* about reusing

---

[137] Submission to the Attorney-General's Department, Australia, in response to the *Fair Use Issues Paper*, July 2005, Submission No 161, 4 (emphasis in original) (Network Ten Pty Ltd).

[138] Submission to the Attorney-General's Department, Australia, in response to the *Fair Use Issues Paper*, July 2005, Submission No 107 (Australian Subscription Television and Radio Association).

[139] Submission to the Attorney-General's Department, Australia, in response to the *Fair Use Issues Paper*, July 2005, Submission No 152, 3 (Australian Broadcasting Corporation). See also Submission to the Attorney-General's Department, Australia, in response to the *Fair Use Issues Paper*, July 2005, Submission No 125 (Special Broadcasting Service Corporation).

[140] A technology like CPCM seems intended to come within the anti-circumvention provisions of Australian copyright law, although past Australian experience suggests such aims may not always be achieved: *Stevens v Kabushiki Kaisha Sony Computer Entertainment* (2005) 221 ALR 448.

television content under the copyright exception for fair dealing,[141] it is not surprising that several submissions to the recent Australian review on exceptions to copyright infringement proposed revising fair dealing to allow for parody or similar commentary.[142] These submissions appear to have been accepted, in general terms, by the Attorney-General and a new exception to parody and satire is planned for introduction.[143] Such an exception is a classic instance of reuse which cannot be accommodated by DRM technologies because all content is in principle open for parody. If such a reform were introduced, then allowing circumvention of TPMs may be more significant than limiting the application of technological controls for the practical availability of a parody exception — not that debates about how to provide for TPMs exceptions are likely to be any less fraught than on issues like the broadcast flag, especially with the reforms expected in 2006 to Australian law on TPMs exceptions under the *AUSFTA*.[144] But all three areas are important for analysis, with the variety of content, controls and reuses that can exist.

While general debates within digital copyright — drawn from across these three areas — are certainly significant for DTV content control, further arguments can be derived from traditions of *broadcast* policy and analysis. For example, we have outlined how the finely-tuned reuse controls available through content control technologies such as CPCM might allow the protection of certain content to be limited by regulators. Technically, it appears possible for regulators to require that such technologies do not prevent private time-shifting of free-to-air content. Even so, choices made in several other instances tell against regulators seeking to mandate how a technology like CPCM is applied: US experience to date with the broadcast flag,[145] emerging models of authorised broadband distribution,[146] and the use of DVRs by Australian subscription

---

[141] (2004) 218 CLR 273 (*'The Panel Case'*). See also *TCN Channel Nine Pty Ltd v Network Ten Pty Ltd [No 2]* (2005) 145 FCR 35. For discussion of these cases, see, eg, Melissa de Zwart, 'Seriously Entertaining: *The Panel* and the Future of Fair Dealing' (2003) 8 *Media & Arts Law Review* 1; Michael Handler and David Rolph, '"A Real Pea Souper": *The Panel Case* and the Development of the Fair Dealing Defences to Copyright Infringement in Australia' (2003) 27 *Melbourne University Law Review* 381.

[142] See Submission to the Attorney-General's Department, Australia, in response to the *Fair Use Issues Paper*, July 2005, Submission No 152, 14–17 (Australian Broadcasting Corporation); Submission to the Attorney-General's Department, Australia, in response to the *Fair Use Issues Paper*, July 2005, Submission No 107, 6–7 (Australian Subscription Television and Radio Association); Submission to the Attorney-General's Department, Australia, in response to the *Fair Use Issues Paper*, July 2005, Submission No 161, 3 (Network Ten Pty Ltd); Submission to the Attorney-General's Department, Australia, in response to the *Fair Use Issues Paper*, July 2005, Submission No 125, 5–6 (Special Broadcasting Service Corporation).

[143] See above n 119 and accompanying text.

[144] Opened for signature 18 May 2004, [2005] ATS 1 (entered into force 1 January 2005). See also House of Representatives Standing Committee on Legal and Constitutional Affairs, above n 8.

[145] Although the FCC has acted in relation to cable redistribution of free-to-air content: see above nn 125–8 and accompanying text.

[146] See, eg, Craig Birkmaier, 'The Real Digital TV Transition Begins' (2006) 48(1) *Broadcast Engineering* 14; Roulla Yiacoumi, 'Thinking Outside the Box', *The Sydney Morning Herald* (Sydney), 9 March 2006, 12; John Lehmann, 'Internet "Threat to Pay-TV"', *The Australian* (Sydney), 10 March 2006, 19.

television,[147] all suggest that regulators might not intervene. Instead, copyright owners might retain the ability to decide the terms on which their content is offered, subject to market pressures from content users to allow at least some form of reuse. It is important to appreciate that if a content protection system like CPCM was mandated for use *without* any regulation over how it could be applied to content, content owners would have greater technological abilities to limit reuse than under the broadcast flag scheme — that is, a CPCM-style scheme without some form of regulation over its application to content could be the most restrictive of content reuse. While the above instances might tell against regulatory action on this issue, an additional point which arises from broadcast policy tradition suggests that public interest claims might retain some purchase in supporting the regulation of how control technologies are applied to broadcast content. As noted at the outset of this article, free-to-air television has occupied a special place within the Australian media. To the degree to which that position continues, or the degree to which it is sought to be maintained, a different approach from that taken with the broadcast flag in the US may appeal to regulators, if content control technology is adopted for free-to-air DTV in Australia. Certain regulation of how control technologies are applied to free-to-air content — for instance, to mandate viewer ability to time-shift — need not prevent broadcasters limiting other valuable reuses. And allowing copyright owners to limit other reuses of content could support the development of alternative payment-based distribution platforms by existing media and communications players as well as other organisations.

C Regulatory Powers

Examining the US and European proposals illustrates how the appropriate site for regulation could vary depending on the technology's operation and the regulatory outcomes sought. In addition, these proposals illustrate how regulation at different sites would govern different actors. Regulation at the point of transmission would apply to the actions of those directly involved with content, such as broadcasters and content owners. Regulation applied at the reception end would govern the actions of reception equipment manufacturers. It is worth emphasising that both the US and European systems require the regulation of reception devices, but they differ on the regulatory options related to transmission. In the *Broadcast Flag Order*, the FCC decided not to regulate at the point of transmission because it believed that '[b]roadcasters and content owners have strong incentives to implement the ATSC flag'[148] and that they should have the latitude to decide if they did not wish to insert the flag into any broadcast. This was despite calls from consumer advocates to prohibit the application of the flag to news and public interest programming.[149] However, the

---

[147] The Australian subscription television provider, Foxtel, has a DVR called 'iQ' which allows some content to be recorded for later viewing with reuse restricted by contract and technology: Foxtel iQ <http://www.foxteliq.com.au>.
[148] *Broadcast Flag Order*, 18 FCCR 23550, [37] (4 November 2003).
[149] *Broadcast Flag Order*, 18 FCCR 23550, [38] (4 November 2003).

more finely granulated control offered by a technology such as CPCM offers additional possibilities for regulation at transmission to protect public use of certain types of material. While the FCC did not wish to involve itself in discussions about which types of broadcast content merited protection and which did not,[150] it was not presented with a technology which allowed more sophisticated levels of control at the point of transmission — for example, to allow universal time-shifting while still ensuring content owners could protect selected material from retransmission. Once the decisions made at transmission involve more than just 'on' or 'off', there is the potential for regulation to provide greater protection for public interest exceptions to copyright.

Regulatory options regarding the application of content control technologies to free-to-air television broadcasts depend on the scope of the regulator's powers. Two issues that arose for the FCC in relation to its power to regulate are relevant here: regulating reception equipment and limiting the application of reuse controls at the point of transmission.

First, as US experience with the broadcast flag illustrates, regulators may lack powers to control reception equipment. In the Australian context, to adopt a system such as CPCM may well require legislative reform, either to directly set standards for reception equipment or to augment the powers of the Australian Communications and Media Authority ('ACMA') to enable the regulator to set relevant standards. While the questions are complex and may depend on the final form of any control technology adopted to work with DVB transmissions, the need for legislative reform seems likely if there is pressure for content control mechanisms to be applied to free-to-air DTV transmissions in Australia. Such an approach would go beyond the generally non-mandatory standards for DTV transmission and reception equipment established by Standards Australia.[151]

Sections 7–12 of the *Australian Communications and Media Authority Act 2005* (Cth) ('*ACMA Act*') give ACMA various functions and the powers required to exercise them.[152] Section 10 of the Act sets out functions related to broadcasting, some internet content and datacasting. Along with the general function of regulating services in accordance with the *Broadcasting Services Act 1992* (Cth), *ACMA Act* s 10 lists functions such as administering the licensing process, researching community attitudes about content, developing program standards, and assisting providers to develop appropriate codes of practice. None of these powers directly relate to regulating reception equipment or content reuse. In addition, the *Broadcasting Services Act 1992* (Cth) itself sets out a number of roles for ACMA. Schedule 4 governs the introduction of DTV broadcasting in Australia and requires ACMA to formulate schemes for

---

[150] *Broadcast Flag Order*, 18 FCCR 23550, [38] (4 November 2003).
[151] Standards Australia is the longstanding peak non-governmental standards body of Australia: see Standards Australia <http://www.standards.org.au>.
[152] ACMA came into existence in 2005 through the merger of the Australian Broadcasting Authority and the Australian Communications Authority and carries on their regulatory functions under broadcasting, telecommunications and spectrum management legislation: see generally Australian Communications and Media Authority <http://www.acma.gov.au>.

converting commercial and national broadcasters to digital.[153] Under cl 36B, regulations may provide that domestic reception equipment must be accessible by each commercial and national broadcaster and datacaster.[154] This does not appear to extend to the power to require reception equipment to give effect to particular content control standards such as CPCM. Regulations can also determine format standards and technical standards for broadcasting DTV.[155] The technical standards are to focus on matters of transmission and ACMA's conversion plans: cl 39(1) provides that regulations can determine technical standards related to the transmission of DTV broadcasting to be followed by ACMA in creating or varying the conversion schemes for commercial and national television.[156]

However, technical differences between the ATSC broadcasting flag in the US and DVB's proposed CPCM system may affect the interpretation of the relevant regulatory powers. While the FCC could not legislate to cover reception devices where the content control technology operated *after* a flagged broadcast signal was received, it may be possible to argue that CPCM controls operate *as* the broadcast stream arrives in reception devices and therefore that the recognition and implementation of CPCM by reception equipment is a technical standard applying to DTV transmission. But such an argument does not appear to accord with the general approach to DTV in the *Broadcasting Services Act 1992* (Cth).[157] In any event, the US experience illustrates how the required regulatory power, even if not currently available in Australia, could be sought through legislative reform.

With regard to the second issue faced by the FCC — namely, limiting the application of reuse controls at the point of transmission — it should be noted that the introduction of technological restrictions on the reuse of free-to-air broadcast content relates to questions about the status of television in Australian society. If television is to remain 'something special' — although in quite different ways than before the multiplication of channels, audience fragmentation and the development of alternative delivery mechanisms — there could be

---

[153] *Broadcasting Services Act 1992* (Cth) sch 4 cls 6, 19.
[154] *Broadcasting Services Act 1992* (Cth) sch 4 cl 36C requires national broadcasters to comply with any such regulations.
[155] *Broadcasting Services Act 1992* (Cth) sch 4 cls 37–37D, 39.
[156] Another provision worth noting is cl 39(2), a limiting provision which states that technical standards related to 'conditional access systems' *must* be directed towards achieving the policy objective that systems will be open to all providers of 'eligible datacasting services' so far as that is practical. A similar provision exists for 'application program interfaces' which are defined as having 'the meaning generally accepted within the broadcasting industry'. The aim is to make interfaces open to all providers of 'eligible datacasting services': see *Broadcasting Services Act 1992* (Cth) sch 4 cl 39(2AA), (5). The mandatory wording of cl 39(2) suggests that technical standards could not relate to other objectives about conditional access.
[157] As occurred in the US, such arguments may seek to rely on the incidental powers of ACMA. For example, under *ACMA Act* s 10(1)(s), ACMA has the function of doing 'anything incidental to or conducive to the performance of any of [the other functions set out in that section].' ACMA also has delineated powers under *Telecommunications Act 1997* (Cth) s 376 to regulate technical standards relating to 'specified customer equipment' to protect the integrity and interoperability of the network. While these may have some relevance to regulating converged digital devices like video capable mobile telephones, the Act focuses on matters such as the allocation of carrier licences and interconnection across the telecommunications network.

reasons to regulate how control technologies are applied to various forms of DTV content.[158] Such an argument depends on deciding what type of digital access and reuse would serve similar social purposes to the near universal availability of analogue television. The significant social role of television in providing news and information remains one important factor in that assessment.[159] Given the general regulatory approach of the *Broadcasting Services Act 1992* (Cth), that type of argument for limiting the application of control technologies also depends on deciding that leaving their application to the market would not achieve desired results in relation to particular types of content or reuse. Existing subscription television practices in Australia suggest there is commercial value in allowing some reuse of content, for example, time-shifting.[160] Similar market pressures may influence how control technologies are applied to free-to-air DTV content. In addition, alternative distribution mechanisms can be expected to change the experience of television for audiences. While multi-channel subscription television has allowed niche programming, downloading content for viewing at a place and time of one's own choosing suggests a very different relationship with television. As Terry Flew has noted, being able to download content like popular movies and television programs from the internet is 'an instance of qualitatively "new" media, not so much because it changes the form, but because it changes the means of distribution and storage, and the associated business models, of those media.'[161]

If regulation of the way in which control technologies are applied to free-to-air content is pursued, it could apply to all free-to-air content or distinguish between different types of content.[162] Thus, within the general question of the power of ACMA to regulate DTV transmission lies an issue about its power to regulate differentially particular types of content and forms of reuse. ACMA, however, has limited powers in relation to content under the *Broadcasting Services Act 1992* (Cth). The powers include, for example, setting program standards for children's television and Australian content on commercial television.[163] The regulator also has a supervisory role in registering industry developed codes of practice about matters such as program classification, accuracy and fairness in news, and permitted amounts of advertising on commercial services.[164] In

---

[158] However, a larger and significant question remains about whether it is possible to support the continued delivery of the social value that has been seen to inhere in mass audience free-to-air television in societies like Australia as audiovisual content spreads across more varied platforms.

[159] See, eg, Butler and Rodrick, above n 3, 487; *Australian Broadcasting Tribunal v Bond* (1990) 170 CLR 321. See also Denemark, above n 22.

[160] See, eg, above n 147.

[161] Terry Flew, *New Media: An Introduction* (2nd ed, 2005) 2.

[162] Differentiation could also be expected between broadcasting services given that the powers of ACMA vary for different types of service. The *Broadcasting Services Act 1992* (Cth) differentiates services by their perceived degrees of influence on community views and free-to-air commercial television is thus subject to greater regulation: *Broadcasting Services Act 1992* (Cth) s 4(1). See Butler and Rodrick, above n 3, 490–8 for an overview of the eight categories of broadcasting service under the legislation.

[163] *Broadcasting Services Act 1992* (Cth) s 122. In addition, broadcasting services are subject to requirements related to medical advertising, tobacco advertising, the broadcasting of material of national interest etc: *Broadcasting Services Act 1992* (Cth) sch 2.

[164] *Broadcasting Services Act 1992* (Cth) ss 123–4.

addition, ACMA can set program standards where industry-developed codes of practice are failing to provide appropriate safeguards.[165] None of these powers appear to relate to controlling the reuse of content. That control has been a matter for copyright law alone, at least since the advent of domestic video recording made reuse of content accessible to many viewers. Thus, if content control technology for free-to-air television broadcasts was to be introduced, the question of reforming the powers of ACMA in relation to content would be likely to arise. Any decision to exercise regulatory power at either the point of transmission or reception could have a major influence on the development of DTV and the wider media environment in Australia. And if, for example, powers to control reception equipment were provided but regulation governing the transmission of content was not addressed, equivalent issues about the accessibility of particular content for certain types of permitted reuse could be expected to arise in the other regulatory domains of copyright exceptions and licences, and the circumvention of TPMs.

V CONCLUSION

The broadcast flag in the US was an industry-generated proposal, aimed at controlling indiscriminate reuse of digital broadcast content. It involved both technological and regulatory elements — which is to be expected in the free-to-air digital environment. It went through a public FCC process which offered relatively wide opportunities for input from the content industries and from groups representing users. Although the FCC's *Broadcast Flag Order* was successfully challenged in court, the decision leaves open a future legislative endorsement of the flag by the US Congress and this is being pursued by its supporters. The general shape of the US debate about limiting indiscriminate redistribution of digital content while protecting some content reuse was mirrored in the earlier FCC *Plug and Play Order* regarding cable television, and appears likely to be repeated in consideration of a digital audio flag.[166] While the European CPCM proposal is less developed, in public documents at least, if it or a similar system is adopted as a DVB standard and mandated for use by other countries with DVB transmissions, it is likely to be influential in Australia.

The broadcast flag and CPCM proposals illustrate ways in which copyright and media policy increasingly interact, which underscores the value of inclusive methods of developing and implementing law and regulation in this area. The proposals also suggest how technological controls may increase the ability of copyright owners to decide what reuse is possible — perhaps by default if careful policy consideration is not given to the interaction of technological controls with copyright exceptions and licences, and with anti-circumvention legislation. And the ways in which these factors affect future media policy in

---

[165] *Broadcasting Services Act 1992* (Cth) s 125(1). ACMA must also act if an industry code was not developed for one of the specified areas: *Broadcasting Services Act 1992* (Cth) ss 123(2), 125(2).
[166] See *Plug and Play Order*, 18 FCCR 20885 (10 September 2003); Tony Sanders, *Audio Flag Waived, at Least until March* (4 February 2006) Billboard.biz <http://www.billboard.biz/bb/biz/magazine/upfront/article_display.jsp?vnu_content_id=1001919574>.

Australia may well be influenced by the extent to which television is seen to retain special cultural, economic and political roles that warrant legal or regulatory intervention. As John Sinclair has noted:

> Instead of a medium for social communication, able to constitute its viewers as a national society, calling on them to participate, however vicariously, in the national culture and public sphere which broadcast television built, the new means of delivery and the variety of channels on offer mean that television viewers have less reason to think of themselves as an audience, let alone as citizens of a nation, but rather, as customers of a service.[167]

Content protection schemes for DTV are another element in these long-term changes in the social position of free-to-air broadcast television. Television is changing from an almost universally available media, based on mass advertising-generated revenue, into a market-differentiated one, where content is available via a variety of delivery mechanisms. It is important to recognise that carefully framed content control regulation could support DTV evolution in ways that might retain some of the social roles long offered by free-to-air television in Australia. However, if content protection schemes are not applied, or not applied carefully, free-to-air broadcasting may soon play a negligible role for many viewers. At the very least, difficulties in developing and implementing content control technologies may reinforce the tendency for free-to-air broadcast television to move to particular types of content focused on specially-created and often live events.[168] As Given commented in 2004 on the contemporary success of reality television programming:

> The explosion of Event Television over the last two years or so seems to me a potent counter-trend to the widely and accurately predicted fragmentation of a multi-channel and online universe. SMS, program websites, and other developments, strangely, seem to be boosting the social scale of those shared events, driving users back to the network screen when and where advertisers are waiting for them.[169]

It is not surprising that the degrees of control offered by technologies like CPCM are seen as valuable to copyright and broadcasting interests. But it is important to recognise what might be less obvious: such possibilities for control could also be of great interest to regulators and DTV viewers. As Danny Butt and Axel Bruns argue with regard to DRM and music, content control technologies should not be seen only for how they can limit access and use of content; the technologies also provide innovative ways for allowing particular types of uses.[170] In the context

---

[167] See, eg, Sinclair, above n 18, 43.
[168] See, eg, Pesce, above n 105, 7.
[169] Gerard Goggin and Geert Lovink, 'Histories, Trends, Futures: A Round Table on the Australian Internet' in Gerard Goggin (ed), *Virtual Nation: The Internet in Australia* (2004) 274, 283, citing Jock Given.
[170] Danny Butt and Axel Bruns, 'Digital Rights Management and Music in Australia' (2005) 10 *Media & Arts Law Review* 265. As well as offering the possibilities for licensed broadcasting considered in this article, the ability to specify and track uses to generate payment could well support independent music and audiovisual distribution, as Butt and Bruns suggest, or new intermediaries: see Jane C Ginsburg, 'Copyright and Control over New Technologies of Dissemination' (2001) 101 *Columbia Law Review* 1613.

of DTV, a central point to take from the existing plans is that technologies like CPCM — perhaps more than simpler flag-based models — and the ways in which they suggest regulation of reception and transmission, might offer a path for free-to-air television to retain some of the policy benefits it has given Australian society and simultaneously to find a valuable place within an enlarged media environment.

[16]

YAHOO! CYBER-COLLISION OF CULTURES: WHO REGULATES?

*Horatia Muir Watt**

I. AN OVERVIEW OF INTERNATIONAL
   CYBERCONFLICTS ISSUES 675
   A. *Prescriptive Jurisdiction in the International Context* 675
   B. *How Internet Technologies Exacerbate
      Traditional Difficulties*... 677
   C. *A Paradox: Technology in Lieu of Enforcement*................ 678
   D. *Two-Way Relationship Between Law and Technology*........ 679
II. LESSONS FROM THE REAL WORLD: THE LEGITIMACY
    OF INTERNATIONAL "EFFECTS" JURISDICTION.......................... 680
    A. *Cyberspace as Ideology*... 681
       1. The "Safe Haven" Argument.. 681
       2. Technology as a Given 682
       3. Normative Implications of Filtering Technology 683
    B. *Legitimacy of Real-World Yardsticks for
       Prescriptive Jurisdiction*.. 684
       1. Effects and Targeting.. 684
       2. "Zoning" Limits Ubiquity, Negating the
          "Notice" Argument... 687
III. LESSONS FROM CYBERSPACE: WHEN PRESCRIPTION
     AND ENFORCEMENT COINCIDE 689
     A. *Creating a Coasean Space of Watertight Compliance*........ 689
        1. Absence of Real World Inefficiencies......................... 690
        2. Enhanced Need for Optimal Definition of
           Prescriptive Jurisdiction 691
     B. *Burden of Implementation* 692
CONCLUSION... 695

An interesting aspect of cyberspace is the role it is playing in reviving the conflict of laws in the international arena—long relegated to quasi-oblivion in the U.S. experience[1] and now, too, a dying species in Europe, where private international law is now largely devoted to the

---

* Professor at the University of Paris I (Panthéon-Sorbonne), Co-director, Institute of Comparative Law, Paris (UMR de droit comparé, Paris I – CNRS); Secretary General of the "Revue critique de droit international privé"; Regular visitor at the University of Texas at Austin.

1. An important exception concerns conflicts of economic regulation, such as in the fields of securities or antitrust, which are more properly considered issues of prescriptive jurisdiction. For the distinction between the conflict of laws and prescriptive jurisdiction, see *infra* note 11.

interpretation of Community instruments.[2] Few and far between are the cases in which European courts are called upon to determine the law applicable to truly international issues arising in the real world[3]—and even then, they tend to be confined to very specific categories of litigation, such as in the field of family law, where they are clearly linked to the continuing use of nationality as a connecting factor in countries which are now home to large immigrant populations.[4]

However, while some of the conflicts now arising in cyberspace bear a familiar aspect, such as those arising in the course of electronic commerce, and require little more than mere technical adjustment of rules or methods applicable in analogous real-world situations,[5] a growing number of conflicts involve clashing fundamental public values in the international arena. These are typically cases in which freedom of expression, particularly as protected by the First Amendment of the U.S. Constitution, collides with the protection of concurrent values in States where information deemed offensive is made available. Of course, regulatory conflicts involving the clash of public values also take place in the real world, where publications or broadcasts originating in a foreign jurisdiction may also be perceived to contain material offensive to fundamental values in the receiving State, which may then take defensive or retaliatory measures.[6] However, if the cultural stakes appear

---

2. Most European case law concerns the implementation of Council Regulation 44/2001 on Jurisdiction and the Recognition and Enforcement of Judgments in Civil and Commercial Matters, 2001 O.J. (L 12) 1. The 1980 Rome Convention on Jurisdiction and the Enforcement of Judgments in Civil and Commercial Matters, 1972 J.O. (L 299) 32, as amended by 1990 O.J. (C 189) 2, which gives rise, relatively infrequently, to issues of interpretation, is not (as yet) a Community instrument. The rise of international commercial arbitration has removed much international contract litigation from European courts.

3. The term "real world" is used here in opposition to cyberspace.

4. ' Case law concerns adoption of children issuing from States which either do not recognize or expressly prohibit adoption, although, in France at least, recent legislation seems to have put an end to litigation. See C. CIV. arts. 370-3 to 370-5 (Fr.). Another source of litigation relates to the effect to be given to Muslim unilateral marriage repudiations. In France, the impact of the European Convention on Human Rights remains uncertain on this point. Convention for the Protection of Human Rights and Fundamental Freedoms, Nov. 4, 1950, Europ. T.S. No. 5, 213 U.N.T.S. 221.

5. Such conflicts concern the validity of electronic signatures, consumer protection, or advertising practice. Traditional territorial connecting factors may require adjustment. See, e.g., Paul Lagarde, *La Loi du le Février 2001 Relative à l'Adoption Internationale: Une Opportune Clarification*, 2001 REVUE CRITIQUE DE DROIT INTERNATIONAL PRIVÉ 774, 776–77 (Proposals of the *Groupe européen de droit international privé* on the reform of article 9 of the Rome Convention on conflicts relating to requirements for formal validity, session of September 21–23, 2001).

6. For a rare example of a "real-world" transAtlantic conflict involving freedom of expression and defamation, see the decision of the highest French civil law court, Cass. 1e Civ., Jan. 14, 1997, *Soc. Gordon and Breach Science Publishers c. Association The American Institute of Physics*, 1997 REVUE CRITIQUE DE DROIT INTERNATIONAL PRIVÉ 504 (Jean-Marc Bischoff). Here, for instance, the court ordered the seizure of the publication in France.

infinitely higher in cyberspace—a conclusion supported by the violence of reactions which the *Yahoo!* decision generated on both sides of the Atlantic[7]—it may well be that these conflicts implicate an additional ideological dimension unparalleled outside the Internet. Indeed, the *Yahoo!* litigation seems to point to the limits of analogy between cyberconflicts and their real-world counterparts.

This Article furthers this comparison of cyberconflicts and the real world, attempting to ascertain what lessons, if any, can be drawn from it. Part I of the Article explores the interests at stake in cyberconflicts and the relationship between technology and the law. Part II uses the French *Yahoo!* court's decision to show that real-world conceptions of prescriptive jurisdiction retain their legitimacy in cyberspace. Finally, Part III notes that the prospect of near perfect compliance offered by Internet technology provides the opportunity to engineer mature, well-calibrated solutions to international regulatory conflicts, which might then even serve as a model in the real world.

I. AN OVERVIEW OF INTERNATIONAL CYBERCONFLICTS ISSUES

A. *Prescriptive Jurisdiction in the International Context*

Cases such as *Yahoo!*,[8] *CompuServe*,[9] or more recently the *Barron's*

---

7. For an example of the (needlessly) aggressive comments by Ben Laurie, see Ben Laurie, *An Expert's Apology* (Nov. 21, 2000), *available at* http://www.apache-ssl.org/apology.html (denouncing the French court's ruling as "half-assed and trivially avoidable"); *see also* Joel Reidenberg, *Yahoo and Democracy on the Internet*, 42 JURIMETRICS J. 261, 277–78 (2002) (critiquing Laurie's response to the ruling).

8. UEJF et LICRA v. Yahoo! Inc., Ordonnance Référé, T.G.I. Paris, Nov. 20, 2000, *available at* http://www.juriscom.net/txt/jurisfr/cti/tgiparis20001120.htm; *see also* Yahoo! Inc. v. La Ligue Contre Le Racisme et L'Antisémitisme, 145 F. Supp. 2d 1168, 1179 (N.D. Cal. 2001) (declaring the French judgment contrary to Yahoo!'s freedom of expression as protected by the First Amendment).

9. When faced with the threat of criminal prosecution, CompuServe eliminated all access to news groups that fell under Germany's antipornography laws. It then attempted to provide "filtering" software in the form of "installment mechanisms," which were designed to allow parents to prevent children from viewing indecent material. CompuServe intended this solution to demonstrate a willingness to comply with German law while committing to provide continued access to users elsewhere. German authorities, however, found the installment mechanisms insufficient because the statute outlawed the dissemination of pornography, whether distributed to adults or children. For an abundant literature in English documenting all these events, generally disapproving German regulatory claims as excessive, see Asaad Siddiqi, *Welcome to the City of Bytes? An Assessment of the Traditional Methods Employed in the International Application of Jurisdiction over Internet Activities—Including a Critique of Suggested Approaches*, 14 N.Y. INT'L L. REV. 43, 89–90 (2001); Steven M. Hanley, Comment, *International Internet Regulation: A Multinational Approach*, 16 J. MARSHALL J. COMPUTER & INFO. L. 997 (1998); Mark Konkel, Note, *Internet Indecency, International Censorship, and*

litigation,[10] exemplify the rapidly expanding category of specifically international conflicts, which, by reason of their public interest dimension, are more appropriately described in terms of prescriptive jurisdiction than in traditional "conflict of laws" terms.[11] National regulations that conflict over activities conducted on the Web express fundamental cultural values for each of the States concerned; indeed, the colliding values are very often embodied in constitutional texts, international instruments dealing with human rights, or penal legislation. Typically, an assertion of freedom of expression in the State in which the website is located clashes with restrictive legislation in the receiving State, designed to protect such values as the right of privacy, to restrict hate speech or libel, or to prohibit indecency or pornography. The free availability of information collides with the negative right of the receiving State to protect itself against outside interference, thus creating a "true" regulatory conflict: If the receiving State can prohibit the emission of information, this comes

---

Service Providers' Liability, 19 N.Y.L. SCH. J. INT'L & COMP. L. 453 (2000); Kim Rappaport, Note, *In the Wake of Reno v. ACLU: The Continued Struggle in Western Constitutional Democracies with Internet Censorship and Freedom of Speech Online*, 13 AM. U. INT'L L. REV. 765 (1998); Kristina M. Reed, Comment, *From the great Firewall of China to the Berlin Firewall: The Cost of Content Regulation on Internet Commerce*, 12 TRANSNAT'L LAW. 543 (1999); Amber Jene Sayle, Note, *Net Nation and the Digital Revolution: Regulation of Offensive Material For a New Community*, 18 WIS. INT'L L.J. 257 (2000).

10. The *Barron's* decision has just been handed down by the Australian High Court. *See* Patti Waldmeir, *Regulating Cyberspace*, FIN. TIMES, Dec. 16, 2002. It ruled that Australian courts had jurisdiction to entertain a libel claim brought by an Australian businessman against Dow Jones, the U.S. publisher of the allegedly libelous material, loaded onto a server in New Jersey. *Id.*

11. Prescriptive jurisdiction is expressed in unilateral terms, allowing no room for applying foreign law. *See, e.g.*, William S. Dodge, *Extraterritoraility and Conflict-of-Laws Theory: An Argument for Judicial Unilateralism*, 39 HARV. INT'L L.J. 101 (1998). This approach characterizes conflicts of public law, or perhaps more exactly (as far as the United States is concerned), the reach of federal legislation in the international arena. The Restatement (Third) of Foreign Relations defines prescriptive jurisdiction, leaving the conflict of laws (whether international or interstate) to the Restatement (Second) on the Conflict of Laws. *See* RESTATEMENT (THIRD) OF FOREIGN RELATIONS LAW OF THE UNITED STATES §§ 402–03 (1986) [hereinafter RESTATEMENT (THIRD) OF FOREIGN RELATIONS]; RESTATEMENT (SECOND) OF CONFLICT OF LAWS § 2 cmt. d (1969) (referring questions of public international law to the Restatement of Foreign Relations); *id.* § 3 cmt. c (defining "state"). Lea Brilmayer explains that the real distinction between the Restatement of Foreign Relations and the Restatement on the Conflict of Laws lies in the source of domestic law: the former deals with conflicts involving federal law while the latter concerns solely state law conflicts. Lea Brilmayer, *The Extraterritorial Application of American Law*, 50 LAW & CONTEMP. PROBS. 11, 12–13. Thus, some international conflicts are subject to choice of law under the Restatement of Conflicts, when they arise in a field such as tort, which is not subject to federal legislation. On the other hand, when a claim is governed by federal regulation, federal courts have subject matter jurisdiction, and approach conflict-of-laws situations in terms of "prescriptive jurisdiction." For the moment, at least, little thought has been given to the potential role of foreign law in the solution of regulatory conflicts in cyberspace; courts assert adjudicatory jurisdiction with a view to applying forum law.

close to interference in the regulation of activities covered by constitutional immunity in the State where the website is located; conversely, not to do so looks very much like allowing cultural expansionism. Either way, the regulatory claim of one State will appear pernicious or intrusive to the other: For example, the United States jurisdiction in which the website is located will object to any corrective action taken by the receiving State as curtailing fundamental freedom of expression, while the latter, in turn, has no reason to accept that First Amendment protection should extend to activities conducted within its virtual borders in violation of its own constitutional or criminal law. Thus, on the one hand, persons in the United States denounce European regulations restricting the content of public expression as extraterritorial meddling with democratic values;[12] on the other, the same values cause European observers to denounce the perverse race to the bottom generated by First Amendment liberalism, as neo-Nazi websites seeking safe haven relocate massively across the Atlantic.[13]

B. *How Internet Technologies Exacerbate Traditional Difficulties*

Although such conflicts can and do occur through the use of traditional media, new communication technologies have sharply exacerbated the difficulties encountered in the real world. Indeed, data circulate instantaneously over the Internet, making the damage caused by the harmful use of information potentially far greater and far more difficult to prevent than in cases of data traveling through more traditional channels. Conversely, given the ubiquity of such effects, there is a risk that multiple courts will assert jurisdiction simultaneously over activity conducted on the Web, with potentially devastating consequences in the form of overregulation and contradictory decisions. Observers frequently express fear that the mere "press of a button" suffices to subject a given activity to foreign extraterritorial jurisdiction without proper notice.[14]

---

12. Some Europeans share this reaction. *See, e.g.*, Ben Laurie, *supra* note 7. A French author recently described the French *Yahoo!* court's decision as "exorbitant." *See generally* Daniel Arthur Laprès, *L'exorbitante affaire Yahoo*, 4 JOURNAL DU DROIT INTERNATIONAL 975 (2002).

13. This is not to suggest that the flocking of Nazi websites to the United States is not also denounced in this country. *See* Lisa Guernsey, *Mainstream Sites Serve as Portals to Hate*, N.Y. TIMES, Nov. 30, 2000, at G1; Reidenberg, *supra* note 7, at 275.

14. *See, e.g.*, Robert M. Harkins, Jr., *The Legal World Wide Web: Electronic Personal Jurisdiction in Commercial Litigation, or How to Expose Yourself to Liability Anywhere in the World with the Press of a Button*, 25 PEPP. L. REV. 451 (1997). The courts themselves sometimes express similar ideas. *See, e.g.*, Am. Libraries Ass'n v. Pataki, 969 F. Supp. 160, 171 (S.D.N.Y. 1997) (concluding that "no user could avoid liability under the New York Act simply by directing his or her communications elsewhere, given that there is no feasible way to preclude New Yorkers from accessing a website, receiving a mail exploder message, or participating in a chat room"); Jack L. Goldsmith & Alan O. Sykes, *The Internet and the*

However, if the conflict is more acute in virtual space, it is not only because of the inherent ubiquity of information and the magnified spillover effects of corrective action, but also and primarily because of the philosophical premises on which the World Wide Web is actually perceived to function, at least in the United States. For many, the Web's very architecture, which favors the free flow of information, anonymity, and geographical indeterminacy, embodies the United States' values of free expression, of which it constitutes the technological projection. Subsequently, any foreign regulatory attempt to inhibit the flow of information is considered not only as a violation of First Amendment immunity, but as vitiating the democratic values embedded in the structure of the Web.[15] Typically, the French decision in *Yahoo!* drew criticism in the Unites States as a claim to "control thinking" in cyberspace.[16]

C. *A Paradox: Technology in Lieu of Enforcement*

At first glance, therefore, cyberconflicts might seem to have little to teach private international law in the real world, and as little to gain from recourse to traditional analytical tools. In view of the acute ideological charge of international conflicts involving fundamental freedoms, nothing appears to prevent litigation from escalating into primitive *Laker*-type judicial warfare,[17] where the winner is clearly the most effective enforcer.[18] It may, however, be time to stop and consider that the free-flowing architecture of the Web results from man-made software, whereas real world constraints are given or, at least, tend to be perceived as inexorable. To what extent does this difference shed any light on the way in which *Yahoo!*-type conflicts could be managed? Paradoxically,

---

Dormant Commerce Clause, 110 YALE L.J. 785, 790–93 (2001) (discussing the *Pataki* court's application of the Dormant Commerce Clause to state criminal laws concerning Internet transmissions of pornography to minors).

15. Reidenberg, *supra* note 7, at 272–75.
16. See *infra* text accompanying note 24.
17. The complex *Laker Airways* antitrust litigation presents a notorious example of transAtlantic judicial warfare, in which British and U.S. courts exchanged anti-suit and counter-anti-suit injunctions to protect prescriptive jurisdiction. Laker Airways, Ltd. v. Pan Am. World Airways, 235 U.S. App. D.C. 207 (1984); *see also* ANDREAS F. LOWENFELD, INTERNATIONAL LITIGATION AND THE QUEST FOR REASONABLENESS 5 (1996) (analyzing this "struggle over jurisdiction").
18. The respective strengths of the contenders would thus appear to be measured exclusively in real-world terms; enforcement will involve the seizure of the defendant's assets located within the regulating State, diverse forms of injunctive relief, or more troubling forms of pressure applied directly on network participants. On the real dangers of exerting pressure through censorship on network participants, see Reidenberg, *supra* note 7, at 277, warning against the danger of overrating the chilling effect of State regulation of Internet communications, when far more troubling avenues are available. These may include "denial-of-service" attacks with a view to shutting down foreign websites, creation of viruses to cripple foreign computers, and more generally deployment of cyberenforcement agencies. *Id.*

whereas conflicts involving the exercise of free expression over the Web initially might appear infinitely more difficult to resolve than their real-world counterparts, the converse is probably true. This Article shows that, if man-made technology shapes cyberspace, it makes achieving a balanced solution of international regulatory conflicts potentially far easier on the Web than in geographical space. This is simply because transnational compliance is clearly more attainable than in the real world, through the use of technology itself. A regulating State now has the means to prevent given data from being made accessible within its borders simply by ensuring that adequate gateway software is put into place;[19] technology readily bypasses slippage, cost, and all the familiar difficulties generally linked to international enforcement of legislative prescriptions or judicial decisions in the real world. This means, in turn, that it is all the more important that States assert prescriptive jurisdiction only when it is clearly reasonable to do so, since unjustified technological interference with the free flow of information in cyberspace would be both destructive and counterproductive. To the extent that technology lends greater credibility to regulatory claims over cyberspace than in the real world, great care should be taken to see that such claims are properly calibrated.

D. *Two-Way Relationship Between Law and Technology*

This is where the real world may have much to teach about the relationship between law and technology. It has been witness in recent times to the gradual common acceptance of the effects doctrine[20] as a legitimate basis for international prescriptive jurisdiction. Similarly, the exercise of regulatory jurisdiction based upon the effects suffered within the forum State seems eminently reasonable in cyberspace. Adopting this approach would optimally regulate cross-border flows of information by allowing restrictions only when the regulating State has a substantial interest in preventing the flow of data within its territory, and only to the extent necessary to implement the protective policy involved. Technology increases the credibility of regulatory claims, but it also allows a State asserting prescriptive jurisdiction to adjust the scope of such claims functionally, so as to allow only those restrictions strictly necessary to prevent harm within its borders. The French *Yahoo!* court fully understood this complex relationship between law and technology on the Web.

---

19. The difficult question of who should bear the burden is discussed in the text below. *See infra* Section III.B. Here, we focus on the technical possibility of ensuring near-perfect compliance.

20. The "effects" doctrine will be described below. *See infra* Section II.B.1. This doctrine allows the regulating State to exercise prescriptive jurisdiction over foreign conduct with impacts on interests located within its borders.

It asserted regulatory jurisdiction on the basis of offensive effects of the data accessible on the Yahoo! Inc. website within French territory, but ordered that the data be made unavailable only to French-based internauts. The constraints it imposed on the free flow of information in the name of the fundamental values of French society did not affect access to that website from any other territory. Virtual space thus evidences a two-way relationship between international jurisdiction and technology. On the one hand, technology can make the assertion of jurisdiction effective to an extent unattainable in the real world. Conversely, proper definition of the limits of prescriptive jurisdiction is crucial to the coherence of State regulation of cyberspace.

II. Lessons From the Real World: The Legitimacy of International "Effects" Jurisdiction

To show that real world yardsticks retain their legitimacy in cyberspace, this Part first examines the "separatist" claim that the use of a borderless medium in some way modifies the bases of regulatory jurisdiction as designed for the real world. Relayed by conventional wisdom about the structure of cyberspace, the separatist claim draws normative conclusions from the freedom with which data can circulate over the Web.[21] Because the Internet provides a technical medium for unfettered expression, restrictive regulation is made to appear illegitimate—a denial of the democratic values it embodies.[22] In the international arena, the perception of Internet architecture as a given also creates important implications for the solution of regulatory conflicts. Thus, the free flow of information similarly gives rise to implicit normative conclusions regarding the allocation of international prescriptive jurisdiction. Assertion of regulatory authority by States seeking to impose restrictions on freedom of expression is seen as incompatible with the very structure of cyberspace. But it will be shown that such a perception reverses the proper relationship between law and technology, allowing separatist values to dictate the scope of international jurisdiction. Indeed, Section A shows that, on closer scrutiny, the design of the Internet depends entirely on the ideological choices that dictate technological development. So, as Section B demonstrates, no plausible reason exists to displace the yardsticks of regulatory authority as defined in the real world.

---

21. *See* Goldsmith & Sykes, *supra* note 14 (discussing conventional wisdom about cyberspace found in the cases).
22. Reidenberg, *supra* note 7, at 273–74.

A. Cyberspace as Ideology

This Section shows that the separatist claim is sustainable only insofar as the borderless quality of the Internet is accepted as a given. As Lawrence Lessig has demonstrated, the development of filtering technology for purely commercial purposes belies this premise, exposing conventional wisdom about the Web as ideology, not fact.

1. The "Safe Haven" Argument

As reactions to *Yahoo!* and similar litigation illustrate, many believe that the very design of the Internet carries strong normative implications for solutions of regulatory conflicts. In a borderless medium, claims by a State to restrict the flow of data perceived to affect welfare within its territory appear to lose their real-world legitimacy. Thus, when activities covered by freedom of expression at the place the website is located are considered elsewhere to undermine concurrent fundamental values such as privacy or the prohibition of hate speech, the defendant systematically invokes the "safe haven" argument. As Joel Reidenberg explains, the fact that the Web, instead of some other, more traditional medium, carries the cross-border effects of the regulated activity would seem to modify accountability, as though activities in borderless space somehow surmount local laws.[23] Favoring the free flow of data, the Internet is seen as conferring on expression carried through its medium a status that remains mysteriously beyond the thrust of the laws of the States in which users access the information. Thus, the geographical indeterminacy of cyberspace seems to set aside the principles governing prescriptive jurisdiction in the real world.

The explanation resides in the fact that the technological architecture of the Web clearly embodies values expressed in the First Amendment, making the very idea that the free movement of data might encounter the regulatory claims of other States seem an anathema to the Web's ideological foundations. Thus Ben Laurie, computer expert consulted by the French *Yahoo!* court, states that "what is being fought over is literally what people think. No one should be able to control what I know or what I think ... The Internet is pure information."[24] When, in the more recent *Barron's* litigation, the Australian Supreme Court ruled that Australian courts had jurisdiction to hear a claim that information loaded on a New Jersey server was libelous under Australian law, the demise of the Internet as a democratic forum of free expression was widely predicted.[25]

---

23. *Id.* at 272–75.
24. Laurie, *supra* note 7.
25. Waldmeir, *supra* note 10.

This perception of the relationship between law and technology extends beyond the international sphere. Similar attitudes exist in domestic litigation within the United States over the thrust of the Dormant Commerce Clause and in First Amendment cases. In the Dormant Commerce Clause context, restrictive regulation is perceived as unduly burdening interstate electronic commerce.[26] Where freedom of expression is involved directly, it is perceived as preempting restrictive regulation based on concurrent values, such as the protection of minors from access to pornography.[27] Whereas such a claim would hardly seem credible in a real-world context, free expression guaranteed by the First Amendment appears to acquire a worldwide immunity, to the point of excluding the regulatory claims of the State in which harmful effects are suffered. What about the Web makes such an argument appear sustainable?

2. Technology as a Given

Conventional wisdom about the Internet tends to present geographical indeterminacy and the free flow of data as givens. As Jack Goldsmith and Alan Sykes have shown, in the context of Internet litigation within the United States, courts looking for "facts" about the Web tend to find that the ubiquity of information, and the corollary risk of overreaching countervailing measures, justify giving precedence to the freedom of expression.[28] Claims about the architecture of the Internet include universal availability of information, absolute indeterminacy of geographical location and other identity factors, and indefinite exposure to liability under restrictive regulation, whatever real links exist between the exposed activities and the regulating State.[29] Correlatively, regulation itself is perceived as illegitimate. In other words, since the Web knows no frontiers, data must circulate freely; as no natural frontiers exist, States may not erect them artificially. This perception of the architecture of the Web clearly impacts the exercise of prescriptive jurisdiction in the international arena. Those who believe that the Internet represents

---

26. *See, e.g.*, Am. Libraries Ass'n v. Pataki, 969 F. Supp. 160 (S.D.N.Y. 1997); *see also* Goldsmith & Sykes, *supra* note 14, at 790–94 (discussing relevant case law); *id.* at 802–08 (applying economic analysis of cross-border burdens to Internet communications).

27. *See* Reno v. Am. Civil Liberties Union, 521 U.S. 844 (1997) (invalidating two provisions of the federal Communications Decency Act of 1996 due to the First Amendment); Am. Civil Liberties Union v. Reno, 217 F. 3d 162 (3d Cir. 2000) (upholding order preliminarily enjoining enforcement of the federal Child Online Protection Act due to likelihood that the Act violated free speech guarantees).

28. Goldsmith & Sykes, *supra* note 14, at 788.

29. Reidenberg, *supra* note 7, at 272–75.

undifferentiated space, find any attempt to introduce "zoning"[30] within its confines intolerable.

As such, arguments that tie up man-made space and law are not so unusual. Thus, free markets have obvious normative implications for the legitimacy of regulatory claims; as shown by litigation involving the Dormant Commerce Clause in the United States or the market freedoms in the European Union, risks of double burdens or overregulation are frequently invoked to limit prescriptive jurisdiction in such a context.[31] No one doubts that deliberate policy shapes free markets or that creating an economic space for the unfettered movement of goods and services requires constraints on regulatory jurisdiction. Curiously, however, the architecture of the Internet is not generally seen as being the product of software. As Ben Laurie's statement shows, the Internet is perceived as a natural space; because the Web enables the free cross-border flow of data, such a state of affairs should, as the argument goes, be taken as a *fait accompli*—an inexorable fact dictating regulatory abstention in the international arena.

3. Normative Implications of Filtering Technology

Although the vision of cyberspace as a borderless natural space still appears to carry weight, commentators also increasingly perceive it as delusional.[32] The rapid development of filtering and "zoning" techniques, now used for purely commercial reasons such as targeting advertising to a particular public, provides clear evidence that geographical indeterminacy on the Internet is not inevitable, but results from ideological choice. As the current state of Internet technology demonstrates, the borderlessness of the World Wide Web does not represent an intractable given. Concluding otherwise allows technology to disguise policy choices. We should ultimately reject the "safe-harbor" defense as having no more relevance than in the real world, precisely because the design of the Web is what we make it; if information flows freely, it is because we allow it to do so. Much of the conventional wisdom about the functioning of the Web grew out of the initial state of the art, under which "zoning" techniques were inconceivable. Improved technology, designed to identify various categories of users, means that claims of the ubiquity of information accessible on the Web, whether due to the inherent nature of the medium or to its accidental evolution, no longer ring true. This entails a

---

30. Lawrence Lessig & Alan Resnick, *Zoning Speech on the Internet: A Legal and Technical Model*, 98 MICH. L. REV. 395 (1999) (coining the expression "zoning").
31. For the impact of the "double burden" argument in the European Union, see JUKKA SNELL, GOODS AND SERVICES IN EC LAW: A STUDY OF THE RELATIONSHIP BETWEEN THE FREEDOMS (2001).
32. *See* Lessig & Resnick, *supra* note 30; Reidenberg, *supra* note 7.

fundamental consequence regarding the allocation of regulatory authority in cyberspace. If the Internet is not naturally borderless, then real-world yardsticks for the exercise of prescriptive jurisdiction retain their legitimacy.

B. *Legitimacy of Real-World Yardsticks for Prescriptive Jurisdiction*

This Section demonstrates that the regulating State may legitimately impose international "zoning" to protect itself from the effects of information made available elsewhere and perceived to be harmful. This requires showing that effects-based jurisdiction does not necessarily entail conflicting regulatory burdens, which would arise if national courts simultaneously asserted prescriptive jurisdiction over the same conduct "at the press of a button." New filtering technologies lessen the risk of accidental spillover and increase the means for preventing much-feared overregulation.

1. Effects and Targeting

Since the new types of regulatory conflicts emerging in cyberspace involve public values,[33] courts have tended to reason in terms not of conflict of laws but of prescriptive jurisdiction, using criteria developed in real-world clashes of public economic regulation.[34] In such cases, contemporary practice on both sides of the Atlantic seems to have converged more or less from a "place-of-conduct" rule to an "effects" test.[35] Indeed, recent applications of economic analysis to the conflict of laws have shown that the "effects" test seems to make the best sense in terms of

---

33. Indeed, globalization seems to have given rise to a new taxonomy of international conflicts—whether through increased interconnectedness or the use of new technologies—which now include regulatory clashes with strong public law components. On this new category of international conflicts, hitherto identified with conflicts of economic regulation, see Jurgen Basedow, *Conflicts of Economic Regulation*, 2 AM. J. COMP. L. 423 (1994); *see also* Jurgen Basedow, *Souveraineté territoriale et globalisation des marchés: le domaine d'application des lois contre les restrictions de concurrence*, 264 RECUEIL DES COURS 9 (1997).

34. *See supra* note 14.

35. On the three different tests (conducts, effects, and balancing of interests) which appear in U.S. practice, and their relationship to tests used in choice of law, see William Dodge, *Extraterritoriality and Conflict-of-Laws Theory: An Argument for Judicial Unilateralism*, 39 HARV. INT'L L.J. 101 (1998). While the balancing test proposed by section 403 of the Restatement (Third) of Foreign Relations was not rejected in the Supreme Court's most recent ruling, Hartford Fire Ins. Co. v. California, 113 S.Ct. 2891 (1993), it is clear that the way in which the test was implemented in that case comes very close to reinstating the "effects" test. The latter seems to have been adopted in fact, if not explicitly, by the Court of Luxembourg in the *Woodpulp* case. Case 89/95, Woodpulp, 1988 E.C.R. 5193. For a discussion of section 403, see LOWENFELD, *supra* note 17, chs. 2, 3.

global welfare[36]—a consideration which should bear a particular weight in the present context of worldwide interplay of regulatory authority. Where conflicts arise in cyberspace, courts both in Europe and the United States have asserted personal jurisdiction on this basis, and have then proceeded to apply forum law.[37] "Substantial effects" within the regulating State generally justify prescriptive jurisdiction, whether they arise in the real world or cyberspace.[38] To establish personal jurisdiction over the defendant, however, deliberate targeting may be both necessary and sufficient.[39] Using targeting as a yardstick has enabled courts of various countries to exercise jurisdiction sufficient to incriminate hate speech, indecency, libel, invasions of privacy, and copyright violations.[40]

"Targeting" involves the difficult task of discriminating between active and passive websites,[41] requiring considerable thought as to the

---

36. *See, e.g.*, Joel Trachtmann, *Economic Analysis of Prescriptive Jurisidction and Choice of Law*, 42 VA. J. INT'L L. 1, 34–41 (2001). However, Andrew Guzman argues that the effects test as such will not guarantee global efficiency, since a State regulating on that basis will have taken into account exclusively local costs and benefits. As a result, a globally optimal transaction (i.e., a transaction which increases global welfare) may nevertheless be regulated restrictively by any State in which its harmful effects are in excess of its local benefits, irrespective of its positive impact elsewhere. Andrew Guzman, *Choice of Law: New Foundations*, 90 GEO. L.J. 883, 897 (2002). To be allowed under the effects test, argues Guzman, a given transaction must not only increase global welfare but be perceived as optimal in all of the States in which it generates effects. *Id.* at 906–08. Although this argument is convincing, it is also clear that a global calculus of costs and benefits could only be carried out within a cooperative framework. Failing that, the (second) best yardstick of prescriptive jurisdiction is still the one that allocates legislative authority to the States with the greatest incentive to allow or refuse a given transaction, even if incentive must rhyme here with self-interest.
37. For an analysis of the case law, see Reidenberg, *supra* note 7, at 269–71. In the *Yahoo!* case, French personal and prescriptive jurisdiction was justified either under the territorial yardstick of Code Pénal article 113-2 (because the infraction presumptively took place on French territory since the harm, an element of the infraction, took place there), or the personal criterion of the victims' French nationality under Code Pénal article 113-7. Both yardsticks endorse the "effects" test.
38. Under section 403(2)(a) of Restatement (Third) of Foreign Relations Law, one of the tests of reasonableness to be applied to international prescriptive jurisdiction lies in the "substantial, direct and foreseeable effect [of the activity] upon or in the territory." RESTATEMENT (THIRD) OF FOREIGN RELATIONS § 403(2)(a).
39. Courts seem to use a "sliding scale" which requires either interactivity or purposeful availment in order to establish the minimum contacts required for the assertion of personal jurisdiction. For a very complete analysis of the case law on this point, see Siddiqi, *supra* note 9, at 72.
40. For an analysis of U.S. cases, see Reidenberg, *supra* note 7, at 269–71 and see *infra*, the text accompanying notes 45–48. In addition, on the trend toward an "effects" test in cyberspace, see Michael Geist, *The Legal Implications of the Yahoo! Inc. Nazi Memorabilia Dispute*, JURISCOM.NET (Jan.-Mar. 2001), *at* http://www.juriscom.net/en/uni/doc/yahoo/geist.htm.
41. The distinction is not an easy one—any more than is demonstrating purposeful available for jurisdictional purposes in real-world situations. For example, in *Panavision International*, the court required "something more" than a passive website to show that activity

weight to be given to various factors such as language, which may or may not be significant, according to the specific circumstances of each case.[42] However difficult the courts' task in defining the effects which legitimate the international assertion of prescriptive jurisdiction, targeting means that data deliberately made accessible within the forum State can lead to criminal liability there, even if it receives legal protection in the place of conduct. The fact that it is protected "at home," is no more a valid jurisdictional defense in cyberspace than it would be in the real world. Although defendants characterize such assertions of prescriptive jurisdiction as "imperialism," it is hardly necessary to show that once the ideological arguments linked to the architecture of the Web are set aside, there is nothing "exorbitant" about extraterritorial regulation on the basis of conduct targeted into the forum territory.[43] In this respect, the French court's "extraterritorial" injunction in the *Yahoo!* case is by no means exceptional: in the United States, the *Playboy* court required a website located in Italy to make material published under the United States trademark, "Playmen," inaccessible to users in the United States;[44] in *Nat'l Football League v. TVRadioNow Corp.*, a Canadian website was preliminarily enjoined from transmitting copyrighted programming into the United States;[45] the *People v. World Interactive Gaming Corp.* court ordered a casino based in Antigua to cease offering gambling over the

---

is targeted at the forum state. 141 F. 3d at 1320–22 (quoting Cybersell Inc. v. Cybersell Inc., 130 F.3d 414 (9th Cir. 1997) and distinguishing it due to a lack of targeting). In *Cybersell*, two corporations, organized in different states, used identical trade names on the Internet without specifically intending to injure each other. In Zippo Mfg. v. Zippo Dot Com, 952 F. Supp. 1119, 1124 (W.D. Pa. 1997), Judge McLaughlin explains:

> At one end of the spectrum are situations where a defendant clearly does business over the Internet. If the defendant enters into contracts with residents of a foreign jurisdiction that involve the knowing and repeated transmission of computer files over the Internet, personal jurisdiction is proper. At the opposite end are situations where a defendant has simply posted information on an Internet Web site which is accessible to users in foreign jurisdictions. A passive website that does little more than make information available to those who are interested in it is not grounds for the exercise of personal jurisdiction. The middle ground is occupied by interactive Web sites where a user can exchange information with the host computer.

Id. It is of course the defining of the "middle ground" which creates difficulty. *See* Siddiqi, *supra* note 9, at 72. Courts seem to use a "sliding scale" which requires either interactivity or purposeful availment in order to establish the minimum contacts required for the assertion of personal jurisdiction.

42. The *Yahoo!* case itself illustrates this difficulty. That the website targeted a French-speaking public seems clear from the use of French-language advertisements. But if using French in California evidences the targeting of users in France, it hardly follows that the use of English necessarily targets, say, an Australian public.
43. At least one French author strongly disagrees. *See* Laprès, *supra* note 12, 993–95.
44. Playboy Enters. v. Chuckleberry Publ'g, 939 F. Supp. 1032 (1996).
45. 53 U.S.P.Q.2d 1831 (2000).

Web to New Yorkers;[46] and in *Panavision International LP v. Teoppen*, an Illinois resident was held to be subject to suit in California for registering a domain name in Illinois, when his activity was directed to the forum state.[47]

2. "Zoning" Limits Ubiquity, Negating the "Notice" Argument

As we have already seen, the objection immediately raised in Internet litigation is the "notice" argument, linked to the alleged ubiquity of information on the Web. Because given content may come under the definition of libel or hate speech in innumerable jurisdictions, there appears to be a danger of massive overregulation; although the risk of conflicting regulatory burdens also exists in the real world, conventional media do not create the same likelihood of widespread unintentional effects. Two objections show the fallacy of this notice argument, one normative, the other technological.

First, prescriptive jurisdiction carries the same limits in cyberspace and the real world—the State cannot legitimately exercise jurisdiction over activities on the basis of effects that either do not specifically target its territory or remain insubstantial.[48] The distinction, now gaining ground in court practice, between interactive and passive websites, responds to this idea, linking the legitimacy of regulatory claims to the fact that information has been made deliberately accessible in the forum State, as in the *Yahoo!* case.[49] As filtering technology improves, the risk of accidental spillover decreases: "zoning" techniques lessen the force of the argument that effects can accidentally arise anywhere. The flow of information can be mastered in cyberspace, in the same way that one can avoid sending publications via traditional media deliberately into another State. This is precisely the thrust of the French *Yahoo!* decision, which took pains to check the feasibility of limiting access in France of material loaded on the California website; if the offensive data is nevertheless made accessible in France, it cannot be the result of an accident. As the *Yahoo!* court itself recognized, this does not entirely rule out seepage, particularly as engineered by third parties. However, it has been pointed out that it would be fair to provide a "reasonable efforts" defense, to

---

46. 714 N.Y.S.2d 844 (N.Y. Sup. Ct. 1999).
47. 141 F.3d 1316 (9th Cir. 1998).
48. The targeting of the regulating State's "territory" is of course metaphorical. In many cases, the stigmatized activity attempts to affect the forum State's economic interest.
49. *See, e.g.,* Zippo Mfg. v. Zippo Dot Com, 952 F. Supp. 1119, 1123-24 (W.D. Pa. 1997); *see also supra* note 5. For an interesting critique of the interactivity yardstick, see Siddiqi, *supra* note 9, at 74, pointing out that the commercial value of a website is not necessarily dependent upon its interactivity.

protect service providers who have taken care to comply with legislative restrictions in targeted States.[50]

Secondly, using zoning techniques, while limiting accidental seepage of information, also allows courts to adjust the restrictions required by their local law to mitigate harmful effects without overreaching. Contrary to popular belief, overregulation can be mastered more easily in cyberspace than in the real world, which provides far less opportunity for the fine-tuning of regulatory jurisdiction. Courts can limit the restrictive effect of regulation and incriminations to activities that directly affect welfare within their own jurisdiction. Unnecessary regulatory spillover can be avoided if restrictions to the free flow of information, for example, can be limited to a given set of geographically located users. Thus, the French *Yahoo!* court ordered that the content of the contentious website should be prevented from being accessed in France, where it was illegal, without affecting its accessibility elsewhere. In other words, in exercising prescriptive jurisdiction to apply a penal statute to foreign conduct on the basis of harmful effects suffered in France, the *Yahoo!* court made sure that the impact of its own corrective action was exactly adjusted to those effects. As little as five years earlier, at the time of the *CompuServe* litigation, striking a similar balance proved less easy; responding to the threat of criminal prosecution, CompuServe eliminated access worldwide to the pornographic chat group illegal under German law before it was able to come up with software (nevertheless judged inadequate by the German courts) enabling parents to install blocking mechanisms for children.[51] But because Internet technology now makes "zoning" possible, no compelling reason exists to alter the "targeting"/"effects" test which justifies prescriptive jurisdiction in the real world. When deliberate or targeted, obnoxious consequences felt within the forum State can hardly be challenged as a valid basis for restrictive regulation. Similarly, restrictions designed to operate exclusively with respect to effects produced within the territory of the regulating State remain clearly within the bounds of international legitimacy. All in all, it is far easier, technically, to parcel out prescriptive jurisdiction optimally in cyberspace than in the real world. Technology allows courts to prescribe the least intrusive solution. At the same time, and because regulatory reaction can be fine-tuned to harmful effects, the risk of externalities in the form of overregulation can practically be eliminated. This is particularly so since zoning techniques available on the Web cause prescription and enforcement jurisdiction to coincide. Technology harnessed to the law—and not the reverse—provides the means of ensur-

---

50. Reidenberg, *supra* note 7, at 276.
51. *See supra* note 9.

ing perfect compliance with regulation. Given this premise, cyberspace has much to teach the real world.

III. LESSONS FROM CYBERSPACE: WHEN PRESCRIPTION AND ENFORCEMENT COINCIDE

So what's so different about regulatory conflicts on the Web? Part II showed that, as a product of technology, cyberspace should not modify real-world principles of accountability. In fact, as this part of the Article will demonstrate, the very technology that defines the structure of cyberspace provides means to ensure near-perfect correlation between the scope of regulatory authority and the power of enforcement unattainable in the real world. The real specificity of cyberconflicts lies in the potential for filtering or zoning techniques to create a Coasean space of costless compliance,[52] which could not be achieved through real-world enforcement processes. Section A shows that enhanced means for ensuring compliance should provide a correlative incentive to fine-tune prescriptive jurisdiction. Section B acknowledges, however, that despite the normative potential of technology, the difficult question as to who bears the burden of implementation remains unanswered.

A. Creating a Coasean Space of Watertight Compliance

This Section argues that, putting aside for the moment the issue of the burden of implementation, filtering technology has the potential of allowing a regulating State to render illegality technically impossible within its prescriptive sphere. This ability allows the State to eliminate inefficiencies stemming from the real-world differences between the scope of regulatory claims and its power to enforce, creating incentives to adjust its prescriptive jurisdiction to match those restrictions functionally necessary to bring about its protective regulatory goals. Cyberspace thus provides conditions for optimal regulatory coordination, which remains unattainable in the real world.

As Lessig and Resnick point out, rules can be inscribed into the software itself[53]—in the very same way that, conversely, Internet technology can give expression to the idea that the Web is comprised of a lawless space. In a purely domestic context, for example, filtering

---

52. *See* Lawrence Lessig, *Reading the Constitution in Cyberspace*, 45 EMORY L.J. 869, 900 (1996) ("Perfectly zoned, cyberspace could be that place where there are no collective action problems—the Coasean space required by Roberto Unger's vision of plasticity; the plasticity of Unger assumed in the Coasean world.").

53. *See* Lessig & Resnick, *supra* note 30 (discussing costs and benefits of using different architectures to regulate speech); *see also* Reidenberg, *supra* note 7.

techniques that allow, say, identification of users' ages can ensure immediate near-perfect compliance with legal rules prohibiting the communication of pornographic data to minors. Software, put into the direct service of State regulation, can cause illegality to become a technical impossibility. Of obvious relevance to the domestic context, where it allows perfectly calibrated balancing of interests between conflicting values (in our example the adult freedom of expression and protection of minors), the same technology holds important potential to solve the regulatory conflicts on the international scene. Developing filtering techniques that allow geographical identification of Internet users can ensure the exact correlation of prescription and enforcement while "zoning" technology can similarly allocate regulatory authority. The end result is a far cry from the lawless space conceived by Internet separatists. Harnessed to regulatory objectives, filtering technology could free courts from the need to rely on the less-than-perfect enforcement techniques of the real world, while providing greater security to service providers, who would be protected from unwanted accountability due to accidental transgression of restriction regulation.

1. Absence of Real World Inefficiencies

In the real world, discrepancy between prescription and enforcement traditionally causes various inefficiencies, including evasion of the law. Thus, a judgment awarded in the forum State on a perfectly legitimate jurisdictional basis may nevertheless remain internationally ineffective if the defendant has no assets within forum territory on which enforcement can take place locally.[54] Since enforcement abroad will always be subject to some form of scrutiny of the content of the forum judgment by the foreign courts, it is easy for a defendant to remove assets to a safe harbor in any jurisdiction which will refuse to recognize that judgment—on public policy grounds, for instance. While obviously a cause for concern in cases where the basis for prescriptive jurisdiction is legitimate, this discrepancy between prescription and enforcement also serves as a natural check on exorbitant regulatory claims. When the regulating State overreaches its legitimate sphere of prescriptive jurisdiction, any judgment awarded in such conditions is doomed to nonrecognition abroad.

The same discrepancy between prescription and enforcement has also had important consequences on the effectiveness of regulation in cyberspace. A service provider who wishes to enjoy immunity from a given State's restrictive regulation—even if its regulatory claim is

---

54. Extraterritorial freezing orders and other forms of injunctive relief in common law jurisdictions can remedy this difficulty. Parties' desires to avoid being in contempt of court allow judges to effectively employ these injunctive control mechanisms.

legitimate by reason of substantial effects on welfare within its borders—may do so simply by removing all potential enforcement leverage—essentially assets—from that State's territory. Conversely, this discrepancy may act as an important check on the risk of overregulation. Not all the States claiming to regulate activities on the Web necessarily have the correlative power to enforce the restrictions they impose, so that exorbitant regulatory claims may be ignored by service providers who have no connection with the regulating State in the form of assets or other real-world bases for direct or indirect enforcement. Thus, the real world provides natural adjustment techniques for correcting exorbitant regulatory claims.

However, in cyberspace, regulating States may ensure immediate compliance by inscribing rules into the available software, even in situations where no real-world means of enforcement exist. If a regulating State employs zoning technology to block access to data it deems offensive, a content provider contemplating trading in such data can no longer choose to risk liability or criminal sanctions in the hope that enforcement processes cannot reach it. Understandably, such a perspective of immediate compliance might give rise to concern.

When the regulating State does not have any legitimate basis to assert prescriptive jurisdiction, the fact that technology nevertheless provides the means to ensure mandatory compliance might seem to herald the death of free enterprise and expression in cyberspace. However, if the regulating State does lack reasonable grounds for exercising prescriptive jurisdiction, this can only mean that effects within its territory are insubstantial or that it has not, in fact, been targeted. The filtering of access to data cannot therefore be of great import either to its own population or to the author of the regulated activity.

2. Enhanced Need for Optimal Definition of Prescriptive Jurisdiction

Nevertheless, it is true that in a world of perfect correlation between prescription and enforcement, excessive regulation can no longer be counterbalanced by real-world evasion techniques. Therefore, while the possibility of writing the rules into the software and ensuring immediate compliance presents obvious advantages, the need for an optimal definition of prescriptive jurisdiction deserves special thought. Because prescriptions can be enforced with accuracy, courts should aim at perfectly calibrated solutions, avoiding the friction that exists in the real world due to regulatory overreaching and resistance on the part of the regulated service provider. The incentive to do so should derive from the increased interconnectedness of activities conducted over the Web and the growing interdependency of regulating States. In other words, one may hope that moderation will breed moderation in the assertion of regulatory claims, since given the varying interests of States across the board, all will stand to gain from cooperative attitudes. For example, the interest of the United States is not systematically in favor of unbridled expression in cyberspace, particularly when its interest in protecting intellectual property is involved. Conversely, these are instances where other jurisdictions will be happy to invoke the free flow of data, which they may reject when it threatens competing local values such as privacy, the prohibition of hate speech, etc. Subsequently, all should be ready to subscribe, ex ante, to a rule of reason, under which the benefit from being able to ensure protection of local policies should balance out the concessions made to other States' conflicting regulatory claims.

Indeed, many courts are fine-tuning technical solutions, striving to limit the thrust of restrictive regulation in cyberspace to cases where effects felt within the forum State are either substantial or, in the case of criminal sanctions, deliberate. Thus, the *Yahoo!* court tailored its injunction to limit its intrusiveness; it ordered access to be blocked in France, where the targeted data was considered harmful, without interfering with the other activities of the defendant with respect to the rest of the world.[55] Perfect tailoring provides both the means and the incentives for perfect compliance and for fine-tuning regulatory claims. As Lessig emphasizes, zoning on the Web has efficiency unmatched in the real world,[56] and this certainly holds true when applied to international prescriptive jurisdiction.

B. *Burden of Implementation*

This last Section evokes the remaining, and most difficult, issue, deliberately set aside in preceding developments: Who should bear the burden, including costs, of implementing the perfectly tailored solutions discussed above? Indeed, it may be easier to determine "who regulates?" than to decide who should assume the burden of filtering the data which the regulating State wishes—legitimately—to make unavailable within its borders. Although the most realistic solution probably would place the burden on the regulating State, real-world inequalities between States may well intrude upon the implementation of a perfect Coasean space.

Curiously, this issue is very often neglected. For instance, the French decision in the *Yahoo!* case generated violent criticism in the United

---

55. The U.S. cases cited *supra* note 40 also enjoined the targeting of illegal content into the forum State, but did not regulate the availability of information elsewhere.

56. Lessig, *supra* note 52, at 889 (noting that "[z]oning is coming to cyberspace, with an efficiency unmatched in real space").

States and among Internet separatists of France's regulatory claim, even if such a claim was specifically tailored to the effects suffered within its borders and can hardly be said to be unreasonable by real-world standards. Yet the fact that the court put the burden of implementing the filtering on Yahoo! attracted much less attention. Nevertheless, the real issue seems to be far less "Who regulates?" as "Who bears the burden of zoning?"

Was it right that the cost of putting into place the filtering technology should have fallen on Yahoo!? Firstly, such a solution is obviously realistic only insofar as the court's order was enforceable, in the event of noncompliance, on local assets.[57] Moreover, familiar objections arise, with a slightly different thrust. Should a service provider located in the United States, whose activity is protected by the First Amendment, have to bear the costs of restricting access to information in all the countries which object to its availability? The issue is not to deny the equal right of States affected by the data to protect what they perceive as fundamental values, but to distribute equitably the cost of establishing such protection.

The obvious answer might be to say that, to the extent the website actively attracts business from the regulating State, there is no reason why Yahoo! should not bear the costs of compliance. For instance, Yahoo! was making substantial revenue from its business contacts (such as advertising contracts) with France; requiring it to adapt its software to the regulatory requirements of the State where it is doing business does not seem particularly unreasonable. The real world provides all sorts of instances where the marketing of a product or service requires compliance with local regulations. Given, too, that the service provider is exporting offensive material into the regulating State, the "polluter pays" principle could be invoked to justify the same result.

However, more practical considerations of incentive and regulatory advantage do not necessarily support the "polluter pays" principle in this context. The State in which the effects are suffered obviously has a greater incentive to set up technology which will ensure watertight enforcement of its own restrictive regulation: it would certainly make more sense to leave it to filter the undesired data, to avoid the risk of underenforcement. The receiving State also has the greater regulatory advantage, because it can best decide the extent of the prohibition that fits its

---

57. Or on assets in a "friendly" foreign State (that is, a State ready to enforce the forum judgment). In the *Yahoo!* case, Yahoo! was quick to ensure that enforcement would not take place in California. *See* Yahoo! Inc. v. La Ligue Contre le Racisme et l'Antisémitisme, 169 F. Supp. 2d 1181, 1184–86 (N.D. Cal. 2001) (discussing case history and Yahoo!'s arguments seeking a declaratory judgment). The revenues generated from its activities in France were arguably sufficient to ensure local enforcement.

conception of public welfare and sits in a better position to implement it. It is far less easy, from the content provider's point of view, to ensure that no information reaches users located in States where it might be considered obnoxious. Furthermore, making the receiving State bear the burden of ensuring compliance could be an efficient means of counteracting the inevitable temptation to overregulate; legislative spillover might best be avoided by imposing the cost of regulation on the regulating State.

Moreover, it may be that analogies with the real world should not be pushed too far by forcing the content provider to pay for filtering. If cyberspace is to be an area in which the rule of reason functions effectively, cooperation between regulating States might be better encouraged by a concession to conventional wisdom, which suggests that it is more "fair" for the regulating State to pay for filtering. This notion derives from separatist ideology and is relayed by traditional reliance on place-of-conduct conceptions. It no doubt felt excessively burdensome to Yahoo! to implement French regulation while protected at home by the First Amendment, as it feels burdensome to non-U.S. firms to comply with federal copyright law when their activity is perfectly legal in the place of conduct. As there is a likelihood that costs of implementation of restrictive regulation in these various fields will ultimately cancel each other out, there may be no point in insisting on a counterintuitive solution by burdening the author of cross-border effects with the cost of compliance which will inevitably be perceived as unfair—however legitimate it may be to do so in theory.

Whatever the arguments in favor of associating the right to regulate and the burden of cost, however, this is obviously far from being an easy issue. The above considerations only become valid if one supposes that access to filtering technology is equally easy (or burdensome) for all concerned. However, some regulating States with legitimate reasons to filter data may lack the technological means or public resources to do so. Real world inequalities intrude, once again, on the ways in which States eliminate regulatory conflicts. As a result, it might appear more equitable to burden private service providers generating revenue from activities directed at the regulating State rather than on the population of the regulating State. Unfortunately, however, as seen above, compliance will depend, once again, on the availability of traditional enforcement procedures.

Before the costless solutions of perfect Coasean space can be achieved, States must overcome their limited technological and pecuniary resources. This prospect leaves obvious room for real-world cooperation between States. Meanwhile, concessions necessary to subject cyberspace to a regulatory rule of reason are certainly worthwhile, if they can prevent States that feel threatened by excessive freedom of information on the Internet from taking far more radical, aggressive initiatives to control the flows of data.[58] The specter of technological warfare as a corollary to prescriptive jurisdiction should not be dismissed lightly!

Conclusion

Calling attention to new issues of prescriptive jurisdiction in the international arena, regulatory conflicts in cyberspace are now frequently linked to the worldwide availability in cyberspace of data perceived to be harmful or offensive to fundamental values in the regulating State, while protected by constitutional freedom of expression in the State in which they are made accessible. Looking for appropriate means of managing conflicting regulatory claims, which conventional wisdom sees as either illegitimate or irreducible, has, first of all, afforded the opportunity to confirm the legitimacy of yardsticks used to measure prescriptive jurisdiction in the real world. Effects-based jurisdiction, increasingly supported in the context of conflicts of market regulation in the real world, seems entirely appropriate here, where the assertion of prescription jurisdiction is generally designed to protect fundamental social values shared by a community living within the borders of the regulating State. Indeed, there is no reason that the interests of the society in which the harmful effects of free-flowing data are suffered should subordinate themselves to the ideological claim that the use of a borderless medium in some way modifies accountability for activities conducted through it. Analysis of such a claim has shown that it reverses the proper relationship between law and technology. Technology being purely manmade and thus subject to ideological choice, should in no way dictate the way in which law manages conflicting interests arising through its medium. Rather, once harnessed to the law, technology can facilitate the exercise of prescriptive jurisdiction in the international arena, by providing the means to ensure perfect compliance with regulatory claims over cyberspace, by the use of filtering techniques. In turn, the substitution of technology for enforcement should create incentives for States to calibrate their regulatory claims so as to avoid counterproductive overregulation. However, the foregoing depiction of an ideal world of costless cross-border compliance leaves several difficult issues unsolved.

---

58. On the possible development of spy systems, cyberenforcement agencies and other more "troubling avenues," see Reidenberg, *supra* note 7, at 277.

In particular, the burden of compliance requires additional reflection; in the present state of the world, unequal conditions of access to technology leave some States more vulnerable than others to the violation of fundamental social policies or values through the free flow of data in cyberspace. Although it may appear, therefore, that a case such as *Yahoo!* has raised as many difficulties as it suggests solutions, it must also be emphasized that it provides excellent food for thought not only on the relationship between law and technology, but on the proper calibration of prescriptive jurisdiction both in cyberspace and the real world.

[17]

SPECTRUM AUCTIONS: YESTERDAY'S HERESY, TODAY'S ORTHODOXY, TOMORROW'S ANACHRONISM. TAKING THE NEXT STEP TO OPEN SPECTRUM ACCESS*

ELI NOAM
Columbia University

Abstract

The auction paradigm for spectrum allocation has moved from heresy to orthodoxy, but like its predecessors it will not be the end of history. A better alternative, not driven by the revenue needs of government, is license-free spectrum. Users would gain entry to frequency bands on a pay-as-you-go basis, instead of controlling a slice of the spectrum. They would transmit their content together with access tokens. These tokens are electronic money. Access prices would vary with congestion, set by automatic clearinghouses of spectrum users. Spot and futures markets for spectrum access would emerge. Once technology and economics can solve the interference problem in ways other than exclusivity, the question arises whether the right to use the spectrum for electronic speech is the government's to sell in the first place.

I. Spectrum Paradigms

A. Three Old Paradigms and a New One

It will not be long, historically speaking, before spectrum auctions may become technologically obsolete, economically inefficient, and legally unconstitutional.

And it may not be long before a new form of frequency allocation may emerge where spectrum use does not require any license; when information traverses the ether as flexibly as an airplane in the sky instead of being straightjacketed into a single frequency and routed like a train on a track; and where congestion is avoided not by the exclusivity of ownership but by

---

* I have benefited from the assistance of Po Yi, Bernard Bene, Mobeen Khan, Nemo Semret, and Tracy White: helpful comments by Dennis Carlton, Bruce Egan, Robert Frieden, Thomas Hazlett, Wayne Jett, Evan Kwerel, Terrence McGarty, Milton Mueller, Michael Noll, Jom Omura, Robert Pepper, Peter Pitsch, Ken Robinson, Aaron Rosston, Andrew Schwartzman, Nadine Strossen, John Williams, and an anonymous referee are gratefully acknowledged.

access charges that vary with congestion, with the information itself often paying for access with tokens it carries along.

For today, auctions and usage flexibility are still the best way to allocate new frequencies. Yet it is one thing to support them pragmatically, as I do, because they tend at present to be a better approach than the existing alternatives, and quite another thing to behold auctions in dogmatic awe, blind to their technological relativism. Change the technology, and the economics and the law of spectrum use must change, too.

This article suggests the direction that such change will take. It analyzes, in Section II, the inherent problems of auctions—in particular that they become a tool of revenue generation rather than resource allocation, and that they encourage oligopoly. The article then proposes, in Section III, an alternative for the future, a system of unlicensed spectrum use in which users do not control an exclusive slice of the spectrum but rather are free to access various frequency bands by buying access tokens at a market-clearing price that is based on the extent of congestion. These access tokens could travel with the information itself as a form of electronic money. In effect, the information would pay for its way as it proceeds over wire and air toward its destinations.

What we have had in spectrum allocation is a classic case of a paradigm shift, along the lines of Thomas Kuhn's famous essay "The Structure of Scientific Revolutions"[1] on the rise and fall of schools of thought.

For spectrum, we can distinguish three successive paradigms and an emerging fourth one. In the beginning, there was a brief idyllic stage of spectrum allocation, based on *occupancy.* Entry to the virginal ether was free, and a kind of electronic original state of nature prevailed. Early radio users did not think in terms of permits to spectrum access any more than the Wright brothers considered filing a flight plan at Kitty Hawk. Radio amateurs, early broadcasters, radio telegraph operators, and the U.S. Navy all congregated on the air. But with the transmission technology improving faster for distance than for separation, and with only a few bands under technological mastery, it was not surprising that transmissions soon collided on the unregulated ether.

This inevitable crisis in the occupancy model led to its replacement by the *administrative* paradigm. Frequencies were allocated by the state, of course after it had first taken care of itself generously. The sparse residual was then allotted to various civilian purposes and assigned to private firms based on a combination of first come, best connected, and most persuasive. In some countries, the reception of signals was also licensed. On the whole,

---

[1] Thomas S. Kuhn, The Structure of Scientific Revolutions (2d ed. 1970).

this was a system that benefited influence brokers (whether in government or out), bureaucrats who could gain off-budget degrees of freedom, politicians who gained some influence over content, equipment makers who gained economies of scale, and incumbent firms that liked state-administered scarcity values and barriers to new entry. This was the orthodoxy—prosperous, powerful, potent.

The only problem was that the system did not work very well. As the utilization of spectrum grew, so did the latter's value. Fights over new allocation became shrill and (of course) lawyer intensive.[2] Competitors were excluded.[3] Foreigners were barred.[4] New technologies were excluded or delayed.[5] Politics intervened ham-fistedly.[6] Some spectrum bands were as deserted as Nevada, others crowded like Times Square, with no usage transfers possible. Government hogged vast stretches. Scarce licenses became highly valued, and fortunes were made in the reselling of licenses from the well connected to the merely efficient.[7] Media firms chased monopoly rents, and politicians chased the firms.[8] Because of their value, some licenses were loaded with requirements for off-budget public services. Licenses were temporary in theory—discouraging investments—but permanent in practice—diluting the attached requirements.

The old administrative paradigm was in crisis, but it had a powerful hold over the benefited mass media and politicians. For a short while, it was substituted by license lotteries, a bizarre system that attracted in the United States almost half a million "applications" looking for a windfall. Yet out

---

[2] Former FCC Commissioner Glen Robinson described spectrum allocation as the FCC's version of medieval trial by ordeal. See John McMillan, Why Auction the Spectrum? 19 Telecomm. Pol'y 192 (1995).

[3] The applicant for a fourth British license was told in 1920 that "the ether is already full." 1 Asa Briggs, The History of Broadcasting in the United Kingdom: The Birth of Broadcasting in the United Kingdom 78 (1961).

[4] In Britain reception licenses were limited to radios made by "genuine British manufacturers employing British labor." Eli M. Noam, Television in Europe 116 (1991), citing Briggs, *supra* note 3, at 112.

[5] For the early history of FM radio, see Eli M. Noam, ed., Edwin Armstrong: A Man and His Inventions (unpublished manuscript, 1998).

[6] After the 1952 election, newspapers that had editorially endorsed Eisenhower had a chance at getting a TV license. Stevenson supporters were left out. In other cases, politicians served themselves directly. For Senator Lyndon B. Johnson's good personal fortune from television licenses, see Robert A. Caro, Years of Lyndon Johnson (1990). In France, the allocation of the third mobile license was decided by the then–Prime Minister Balladour personally.

[7] Aftermarkets exist as secondary markets for licenses, or for the firms holding them, or for assets tied to licenses.

[8] Thomas W. Hazlett, Assigning Property Rights to Radio Spectrum Users: Why Did FCC License Auction Take 67 Years? (Working Paper Series No. 768, Columbia Inst. for TeleInfo., Columbia Univ., 1995).

of crisis, predictably, a new paradigm was born. And indeed, a new idea, that of spectrum sales to the highest bidder, was advocated first by a law student with little to lose, Leo Herzel,[9] and then by academic intellectuals, Ronald Coase[10] and later Arthur De Vany and colleagues[11] and Harvey Levin.[12] The idea was first dismissed out of hand as too "academic," ridiculed as impractical by the Federal Communication Commission's (FCC) former chief economist, the noted Dallas Smythe, as "of the realm in which it is merely the fashion of economists to amuse themselves"[13] and ignored or fought off by the established broadcasters. Eventually, however, most economists adopted it. With the intellectual battle won, with the TV community now split between the broadcasters and the newly powerful cable casters, and with mobile technology leading to an explosion of demand for over-the-air capacity, change was in the air. It was then only a matter of time before the need of the state for more revenue overpowered its propensity to micromanage societal resource allocations administratively. Economic efficiency provided the good-government cover for the change.

Today, the advocates of this *auction* paradigm are in the driver's seat. They have become the new conventional wisdom. And they are the darlings of the political establishment, providing it with vast new resources that make otherwise painful spending cuts or tax increases unnecessary. This is a heady experience for the dismal profession. But, just as Kuhn would have predicted, the new orthodoxy, too, has become complacent. Like generals fighting the last war, many of its adherents often reflexively oppose a questioning of the auction paradigm as a defense of the administrative model or of its beneficiaries, because that is where its opponents traditionally came from. Deep down, they believe, as Kuhn would have predicted, that their paradigm is the end of history in this field and that there is no beyond. Any problems are viewed as mere aberrations, probably because the auction concept is not executed with sufficient purity, rather than because of a systemic weakness. In short, the auction has progressed from a better mousetrap to a belief system. This, too, is classic. And it is similarly classic that this will not endure, that a new paradigm will emerge in turn, and that its proponents will be ridiculed as impractical by yesterday's heretics.

The new paradigm is not based on exclusive use, the technological and

---

[9] Leo Herzel, "Public Interest" and the Market in Color Television Regulation, 18 U. Chi. L. Rev. 802 (1951).

[10] R. H. Coase, "The Federal Communication Commission," 2 J. Law & Econ. 1 (1959).

[11] Arthur S. De Vany et al., A Property System Approach to the Electromagnetic Spectrum: A Legal-Economic-Engineering Study, 21 Stan. L. Rev. 1499 (1969).

[12] Harvey J. Levin, The Invisible Resource (1971).

[13] Dallas W. Smythe, Facing Facts about the Broadcasting Business, 20 U. Chi. L. Rev. 100 (1952).

economic foundation of both the administrative and auction paradigms. Indeed, both of these stages have much more in common with each other than their proponents would like to admit. Both basically allocate exclusive slices of the spectrum rainbow and differ only in the early mechanics of that allocation. Seen thus, these two paradigms really collapse into a single one, that of *licensed exclusivity*.[14]

But now, new digital technologies, available or emerging, make new ways of thinking about spectrum use possible that were not possible in an analog world. These new ways can be more daring than the question whether to buy spectrum from the FCC initially rather than from Westinghouse later or whether GE can use its TV channel sideband also for data transmission. The new paradigm is that of *open access,* in which many users of various radio-based applications can enter spectrum bands without an exclusive license to any slice of spectrum, by buying access tickets whose price varies with congestion. These tickets could be carried by the information itself. This brings us back, in several ways, to the earliest stage of frequency use, where there were no licenses. It is possible to do so because soon we can solve in new ways the problem of interference that had doomed the occupancy model and led to the licensing system in the first place.

The rumblings against the auction paradigm emerged in the mid-1990s. On the technology side, Paul Baran, a pioneer of packet switching, and George Gilder, a noted technology guru, argued against auctioned exclusivity. Gilder noted that "[y]ou can no more lease electromagnetic waves than you can lease ocean waves. . . . You can use the spectrum as much as you want as long as you don't collide with anyone else or pollute it with high-powered noise or other nuisances."[15] Underlying Baran's and Gilder's argument is the hope that technology solves scarcity and spares much of the need to deal with allocation questions.[16] My own position, since 1995, has been to go one step further.[17] With open access, scarcity emerges, the resource needs to be allocated, and a price mechanism is required.[18] But this does not require exclusive control over a specific slice of the rainbow.

---

[14] Resale or flexible use do not negate exclusivity of control.

[15] Paul Baran, Is the UHF Frequency Shortage a Self Made Problem? (paper presented at the Marconi Centennial Symposium, Bologna, 1995); George Gilder, Auctioning the Airways, Forbes, April 11, 1994.

[16] Indeed, there is much new high-frequency spectrum to open up, and much old spectrum to use more efficiently.

[17] Eli M. Noam, Taking the Next Step beyond Spectrum Auctions: Open Spectrum Access, 33 IEEE Comm. Mag. 66 (1995).

[18] It is a similar problem of pricing necessity discussed for the presently "free" Internet system as it is experiencing congestion problems. Jeffrey K. Mackie-Mason & Hal Varian, Economic FAQs about the Internet, 8 J. Econ. Persp. 75 (1994).

Many economists and policy advocates have been prisoners to the analogy of spectrum to land. But spectrum access is traffic control, not real estate development. It is about flows, not stocks.

B. Whose Spectrum Is It Anyway?

The emergence of technologies that make it possible for multiple users of spectrum to cohabit and move around frequencies has profound effects. It is not just that it is arguably a more efficient system in terms of technology, economics, and policy. On these points one might disagree. But more important, it is *constitutionally* the stronger system. The argument is simple. Electronic speech is protected by the First Amendment's Free Speech Clause. Therefore the state may abridge it only in pursuance of a "compelling state interest" and through the "least restrictive means" that "must be carefully tailored to achieve such interest."[19] A licensing scheme, however the license is given out, is a serious restriction on speech. Not only does it foreclose the electronic speech of those without a license, but it also limits the electronic speech of those with such a license, if they must comply with its conditions. Until now, government licensing could be justified due to the basic assumption that spectrum is a scarce resource whose uses collided with each other. Some allocation scheme was therefore in order. But suppose that the underlying assumption becomes invalid, and technology can solve the problem of frequency interference. A less restrictive means of control then becomes available.[20] Would not the entire licensing scheme then be subject to question, in the same way that changing transmission technologies in cable TV and computer networking have led to much lower levels of constitutionally permissible restrictions than for the "scarce"[21] broadcasting?[22]

---

[19] Sable Communications of California Inc. v. FCC, 492 U.S. 115, 109 S. Ct. 2829, 106 L. Ed.2nd 93 (1989).

[20] In Denver Area Educational Telecommunications Consortium, Inc. v. FCC, 1996 U.S. LEXIS 4261, the Supreme Court struck down Section 10(b) of the Cable Television Consumer Protection and Competition Act of 1992, which required cable operators to segregate "patently offensive" programming on a leased access channel from viewer access and to unblock it within 30 days of a subscriber's written request. Applying heightened scrutiny standard, the Court considered the availability of technology to block out indecent programming (for example, V-chips and scrambling) that is less restrictive on free speech. Prior to *Denver,* in Carlin Communications Inc. v. FCC (Carlin III), the Second Circuit directed the FCC to reopen proceedings for determining affirmative defenses to prosecution under § 223(b) of the Communications Act if a less restrictive technology became available.

[21] Part of the scarcity is due to the artificially small allocation of such licenses by the state; hence, a certain circularity bootstraps the government's regulatory powers.

[22] Turner Broadcasting System Inc. v. FCC, 512 U.S., 129 L. Ed.2nd 297, 114 S. Ct. 2445 (1994); American Civil Liberties Union v. Reno, 1996 U.S. Dist. LEXIS 7919 (E.D. Penn. 1996).

Instead of loosening the barriers to free entry, the U.S. government is going in the opposite direction, by selling off the spectrum. But is the spectrum the government's to sell in the first place? It is one thing to be a traffic cop, keeping the different users from colliding into each other. But it is quite another matter to assert ownership rights (in effect, to retroactively nationalize the spectrum) and to sell them off. Could the state sell off the right to the color red? To the frequency high A-flat? Preventing interference is based on the Commerce Clause of the Constitution. But what is the basis of asserting ownership?[23] If electronic communications are an aspect of our fundamental free-speech rights, on what ground can these rights be sold to the highest bidder? Imagine the state auctioning off, for perfectly good public policy reasons, the right to travel (in order to prevent overpopulation in Los Angeles), to print books (to protect forests), or to practice medicine (to keep down the cost of health care). Imagine, too, that these auctions are driven by the revenue needs of the state. Regulatory powers do not convey the authority to government to appropriate the economic value from attractive commercial opportunities. Nevertheless, most free-market advocates seem willing to concede this profit to the state.[24]

II. THE FUTURE PROBLEMS WITH AUCTIONS

Today, almost anyone in Washington loves auctions: most liberals, because it makes business pay its way and generates government revenues; and most conservatives, because it substitutes market mechanisms for government controls. Auctions have also been used in New Zealand, the United Kingdom, Australia, and Hungary. Others will follow, no doubt.

The arguments for auctions are well known. An auction is better than a mindless lottery or than comparative administrative hearings with their inevitable legal maneuvering. It takes politics out of the process. It gets spectrum resources quickly into the hands of users that value them highest. It rationalizes the assignment process while recovering the value of the spectrum to the public. It creates certainty and incentives to invest. Private auctions already exist in the form of a resale market.

The counterarguments to auctions are also well known. They are either those of existing stakeholders, of potential entrants who feel better served by the political process than the market, or of those who view spectrum as a public sphere subject to public goals. Broadcasters, for example, argue that the auctions should not extend to them, because (*a*) they are required

---

[23] Wayne Jett, May God Save the Constitution (with Our Help) from Its Friends (unpublished manuscript, 1996).

[24] For an exception, see *id.*

to perform public service obligations, (*b*) they have usually already paid for the license once by buying it in the aftermarket, and (*c*) it would be unfair to make them bid retroactively for an asset whose value they have created by their investments.

Other objections are those of governmental users who fear that their hold over vast chunks of free spectrum might be reduced once its opportunity cost is more precisely known; of radio amateurs, who tend a nonprofit spectrum garden dedicated to technology experiments and public service in the midst of a commercial and governmental wilderness; and of those who believe that vesting ownership based on today's technology will complicate the speedy deployment of new technologies in the future and lead to inefficient allocation. Parts of the public interest community fear (*a*) that regulatory power over TV on behalf of public interest goals would decline if renewable licenses were replaced by permanent property rights, (*b*) that an allocation to the highest bidder would raise barriers to small entrants and reduce diversity, and (*c*) that auctions would squeeze out free public access and nonprofit educational activities.

On the whole, the arguments in favor of auctions are stronger than the arguments against, partly because most legitimate problems raised by the critics can be dealt with in other and often more efficient ways. But this does not make auctions necessarily the best approach for the future.

Surprisingly missing in a critical evaluation of auctions are the free-market and free-speech perspectives. Where market-oriented criticism has been voiced it has focused on the specifics of the FCC auction schemes, such as the duration and flexibility of the licenses involved, not on the concept itself. Indeed, having fought a long, hard, and successful fight for auctions, their advocates often seem incapable of viewing different approaches opened up by future technological options as anything but a prostate position.

A. *Auctions Inevitably Deteriorate into Revenue Tools*

The FCC auctions have been sophisticated in game-theoretical terms and well executed as an operation. The underlying objective for the auction "game" is to raise revenues for government. This is usually denied quite heatedly, and other considerations are cited, such as moving spectrum to the users valuing it most, and so on. But the political fact is that auctions were finally approved, after years of opposition to them by powerful congressional barons and the broadcast industry, as a measure to reduce the budget deficit and to avoid spending cuts and tax increases. Allocating spectrum resources efficiently was a secondary goal in the political process.

The maximizing function may have been constrained in several ways, such as by rules against monopoly control and in favor of diversity. But these additional policy considerations were only the fig leaf on the main reason, raising money for the empty coffers of the federal government. The rest is merely technique. Conceived in the original sin of budget politics rather than communications policy, spectrum auctions are doomed to serve as collection tools first and allocation mechanism second.

Several problems are inexorably tied to the budget-driven auction system. One is a spend-as-you-go approach.[25] It is one thing to sell assets (spectrum rights) and reinvest the proceeds. But ours is a situation of funding current consumption through the sale of long-term assets. Around the world, countries aim to advance the national infrastructure. In the United States, there seems to be a widespread agreement that this should be done without government money. But the spectrum sales end up as the opposite of making public investments. Through auctions, the United States has been taking money away from infrastructure-providing private firms and throwing it into the black hole of the budget deficit. For decades, America's telecommunications system was superior to that of other countries, often because these countries used telecommunications as a cash cow for general government expenses. Now we have embarked on the same road, just as other countries have left it at our urging.

In fairness, this is not due to the auctions per se but due to the way the revenues are being used by Congress and the Executive. Therefore, to maintain sectoral neutrality and avoid siphoning resources from the infrastructure into general public consumption one would have to complement auctions with a recycling policy that returns the revenues to the communications infrastructure and its applications. Yet such a policy is unlikely (outside of a few crumbs), given budget pressure and the efforts by heavy-hitting constituencies to get more for less. And furthermore, such earmarking creates its own dynamic. The 1996 Telecommunications Act created a Development Fund, aimed at small and minority businesses, to be funded from the interest on auction bids. Vice President Gore advocated the use of auction revenues to finance the wiring of schools for the Internet. Congressional Subcommittee Chairman Fields wanted to fund public television with them. And President Clinton proposed their use for school reha-

---

[25] The short-term orientation of auction gains is also manifest in its accounting. Net revenues raised tend to be exaggerated because there is a trade-off between short-term revenue collection and long-term reduced tax yields. License payments can be depreciated against corporate income and are also likely to reduce dividends. Under quite reasonable assumptions, each dollar of auction revenue tax is reduced by about 25 cents of reduced tax revenues in present value.

bilitation. As various programs are funded in such a fashion, stakeholder groups inevitably emerge that seek ongoing funding, and therefore ongoing auctions. Once a certain budgetary dependency on revenues from communications has been created, it will inevitably color substantive policy, such as resistance to new technologies if they threaten auction revenues.

When all is said and done, an auction is a tax on the communications sector and its users,[26] based on an artificially created scarcity. It may be an invisible tax on an invisible resource, but its impact on policy will be real. Auction advocates deny an impact on prices, arguing that consumer pricing depends on marginal rather than historic cost and that the auction charge does not necessarily mean higher end user prices if demand is highly elastic or if the rents have previously been squeezed by government in other ways. It may be useful to start with a reality check. How can one possibly deny that the many billions of dollars raised by an auction are taken out of the private sector and end up with the government? That, after all, is the congressionally mandated point to the whole exercise.

The argument is that an auction bid is a fixed, lump-sum cost and not part of short-term marginal cost, thereby not affecting price, and that all an auction does is reduce profits to a normal level. Only demand characteristics count. This view supposes that there are no alternative long-term uses for the spectrum and for capital. But since alternative uses for spectrum exist continuously, the supply of the service using the bid-for spectrum is not fixed and can expand and contract with its expected profitability. Similarly, alternative uses for capital exist. And greater indebtedness may mean higher cost of capital to a firm generally.[27] Firms may price temporarily without regard to fixed cost, but they could not survive doing so in the long run. Hence an auction payment will be reflected in prices, with its incidence on consumers and producers depending on the respective demand and supply elasticities.

And where is all this going to end? Like diamonds, budget pressures are forever. There is never enough money. This creates a dependence on still more auctions, especially ones of the up-front cash type rather than the pay-as-you-go type. Even if a given auction is designed to achieve an efficient allocation, its existence may be based purely on revenue needs.[28] In 1996,

---

[26] Concern with effects of auction or services prices was raised by the European Commission in a Green Paper. Commission of the European Union, DG XIII, Towards the Personal Communication Environment 26 (January 12, 1994).

[27] McMillan, *supra* note 2, at 196.

[28] The absence of auctions for some spectrum allocations (such as for digital TV) suggests that Congress does not maximize revenues but political support.

for example, both congressional commerce committees were instructed to raise another $15 billion of revenues. Spectrum auctions were the obvious way to go. It had little to do with communications policy considerations.

Since spectrum use is derivative of international allocations of both spectrum and orbital slots, international organizations will also get into auctions. For example, the International Telecommunication Union's former secretary general, Richard Butler, argued that the 1967 Outer Space Treaty excluded a country from appropriating the profits from space frequencies for itself. Such revenues would have to be shared with the rest of the world. This means that they might become the foundation for funding international organizations, by international spectrum auctions.

It has been argued that auctions put a foreign government's decision process into the open, away from influence peddling and corruption, and that auctions thus play a liberalizing role. This might be true in some cases, but the opposite to liberalization is just as likely. A revenue-strapped country is likely to sell off a monopoly license rather than competitive ones because this will fetch the highest bid price.[29] A government's determination of the appropriate market structure will therefore provide ample opportunities for manipulative interventions. And the nonpolitical nature of the auction can be easily undermined by various domestic preference systems,[30] such as requiring bidders to join up with favored local partners, or by requiring bidders to undergo an approval process. After all, even in America foreign bidders are limited to 20 or 25 percent (depending on their corporate structure) of any spectrum, and "designated entities" of women, minorities, and small businesses initially received FCC bidding preferences.[31]

B. Auctions Encourage Oligopoly

An auction payment that must be paid in advance is a barrier to entry, unless capital markets are perfect, which they are not. This especially affects new firms and unproven technologies that cannot find partners to share the risks. Therefore, an up-front payment will reduce the pool of entrants.

Advocates of auctions claim that they are neither a barrier to entry nor a tax, because they merely duplicate the past "private" auctions of the aftermarket. What they seem to have in mind as an alternative to an auction

---

[29] Ideally, it would first sell a monopoly license, and later reneg on the exclusivity by instituting spectrum flexibility for other allocations. Such possibility would lower the auction price.

[30] In Canada, a 10-percent national preference exists.

[31] The discount in the narrowband spectrum auction to deregulated entities was up to 40 percent, plus a preferential payment schedule. McMillan, *supra* note 2.

is a lottery system with an aftermarket, which indeed creates windfalls, transaction costs, and delay. But suppose the alternative were not such an inefficient (though unfortunately real) system, but a merit-based comparative selection (for example, based on an explicit scoring criteria and evaluated by an expert panel, as is a scientific grant proposal) coupled with a 10-year non-re-sale provision. (This is definitely not my recommended solution, but at least it is a more sensible comparative yardstick to the auction than the lottery and resale system, against which most alternatives look good.) Such a system would have lower entry costs since no bids would have to be paid for.

The highest potential auction bid would be the present value of monopoly rent. The winner's profits would be normal, but price would be at monopoly level. The FCC recognized this and auctioned off several Personal Communication Services (PCS) licenses, not just one. This was wise, as well as easy, but it is much harder (if not impossible) to bar oligopolistic bids. The highest bidders will be those who can organize an oligopoly. This is facilitated by bidding consortia of companies that would otherwise be each other's natural competitors and who collaborate under some rationale of synergy. Those firms presently already holding market power under, for example, the cellular duopoly would bid highest to maintain it and its profit. And if precluded from bidding in their own territory (as they are in a departure from the highest-value-user principle), they could try to do it by proxy or by mutual back-scratching with other firms similarly situated elsewhere.

Further, after the auction, the high bidders may collectively suffer from "winner's curse" (winning bids unsustained by adequate profits) and, after some shake-out period, will collaborate, because otherwise they might not be able to support their bid price's cost. "Sunk cost" leads to passive acceptance only in competitive markets, and after the fact. Oligopolists, on the contrary, will attempt to raise prices in order to recover their bid price and more. This does not require an explicit agreement, just commonality of interest, and is therefore difficult to identify or control. Even with multiple service providers left nationally, there would be pressures for concentration to take place, similar to the dominance by airlines of "their" hub cities.

Oligopoly can be attacked in several ways: by adding spectrum allocations, encouraging spectrum flexibility, imposing structural rules of ownership limitation, and using antitrust law in cases of collusion. This is indeed FCC policy. However, ownership limitations are regulatory in nature, may conflict with potential efficiencies of scale, and are at tension with the stated goal of moving spectrum to the highest-value user. In addition, such structural rules would limit the ability of exit by a spectrum holder from one usage to another, since such exit may well impermissibly concentrate the market in the departed service. Flexibility of entry, however, is an excellent way to protect against oligopoly. The present auctions do not permit such flexibility, though the FCC is seeking it. But it must be kept in mind that entry into B means exit from A and may reduce competition there.

There must also be enough spectrum auctioned off to attack oligopolistic tendencies and reduce opportunity cost. But here, government is conflicted. Release more spectrum and its price drops. Just as New York cab drivers have used politics to prevent the issuance of additional taxicab medallions since the Great Depression in order to protect their investment, so will existing spectrum holders be united in the desire to stave off new entrants that will not only compete with them for future business but also depress the value of their past investment. Government has a related revenue-based incentive to keep spectrum prices high by limiting supply. Thus, government could become the spectrum warehouser and protector of oligopoly, a function it has played historically.[32]

The other major way to deal with oligopoly is through antitrust law. But that brings government right back, through its role in prosecution, adjudication, and enforcement. Some people consider antitrust enforcement purer than regulation, despite its sledgehammer style. They seem to have forgotten the political involvements of the Justice Department and its Antitrust Division in virtually any administration of this century and the experience of judicial micromanagement of the AT&T antitrust decree.

III. A Better Alternative: Open Spectrum Access

The alternative to the present auctions is not to return to the wasteful lotteries or comparative administrative hearings of the past, but to take a further step forward, to full openness of entry, which becomes possible with fully digital communications. Auctions are mostly good for now, given the state of technology, but there is a better next step, a free-market alternative: an open-entry spectrum system. In those bands to which it would apply, nobody would control any particular frequency. In this system no oligopoly can survive because anyone can enter at any time. There is no license and no up-front spectrum auction. Instead, all users of those spectrum bands pay an access fee that is continuously and automatically determined by the demand and supply conditions at the time, that is, by the existing congestion in various frequency bands. The system is run by clearinghouses of users.

The underlying present auction system is premised on an analogy to land ownership (or long-term lease). This is based on a certain state of technol-

---

[32] Hazlett, *supra* note 8.

ogy. In the past and present, the fixed nature of a frequency usage had a stability that is indeed reminiscent to land. But that was based on the relatively simple state of technology, in which information was coded (modulated) onto a single carrier wave frequency or at most a narrow frequency range. To forestall interference with other information encoded on the same carrier wave, the spectrum was sliced up, allocated to different types of usages, and assigned to different users. It is as if a highway was divided into wide lanes for each type of usage—trucking, busing, touring, and so on—and then further into narrow lanes, one for each transportation company. Once one accepts this model for spectrum, one can argue about how to distribute the lanes, whether by economics, politics, chance, priority, diversity, and so on. But it is important not to take this model as given and focus one's attention on merely optimizing it. To stay with the example, why not intermingle the traffic of multiple users? And if the highway begins to fill up, charge a toll to every user? And make this toll depend on the congestion, so that it is higher at rush hour than at midnight?

Access rights are economically relevant only when there is scarcity. Whenever there is no scarcity, there is no need to allocate, and the price would be zero. Anybody could enter. But absence of scarcity is not the interesting or usual case. Nobody "owns" the air route Cleveland–San Jose, and anybody could enter. But if landing slots or airport gates are scarce, an allocation must take place. In spectrum usage there are times of day and parts of the country where spectrum usage is always low. But it is realistic to assume that if there are multiple potential users and no restrictions, congestion will happen.

To allocate access one need not grant permanent allocation rights, but rather charge an access fee that is set dynamically at a level where the available capacity is fully utilized. The access fee could be an "edge price" that gives any users of the spectrum the right to enter information into the spectrum "cloud," or it could provide more limited access. Because demand for transmission capacity varies, the access fee would also vary—a high fee where demand is high, and zero when there is excess capacity.

A. The Open-Access Model

Technologically, the proposed system is not presently available, though its component parts exist or are within reach. It is not my purpose to try to work out the details here. They will evolve with time, discussion, and technology. What is important is the concept. Herzel and Coase did not design a multiround simultaneous Vickrey auction, either.

Such an open-access system might look as follows: For packets of information to be transmittable, they would require to be accompanied by an access code. Such a code could be a specialized token, a general electronic cash coin. The token would enable its bearer to access a spectrum band (rather than to a specific frequency), to be retransmitted over physical network segments, and to be receivable in equipment. Price for the access codes would vary, depending on congestion, and be determined by an automatized clearinghouse of spectrum users. Assured access, at a price certain, could be obtained from a futures market.

For example, a mobile communications provider, A, might face heavy for its service during the post–Labor Day morning drive time. It would therefore buy access codes to that capacity from the desired band, to unlock spectrum usage in a network environment. The tokens are bought from an automatic clearinghouse market of all users. Firm A and its customers, when initiating transmissions, add the access token to blocks of their transmitted information. Without the access codes, information could not be passed on to other networks and might not be readable by their intended receivers, if user equipment requires these codes for activation or descrambling.

If A finds itself using less capacity than it needs, it can offer its excess access codes on the clearinghouse's instant spot market to users who experience shortages or who have no real-time needs. In addition, A can assure itself of a long-term supply by contracting in a future market the access codes with B, who then delivers these codes at the time contracted for.

The buyer of capacity does not own any particular slice of spectrum, but rather the right to send so many information blocks over a band. At transmission time its equipment scans for a free frequency before occupying it. This search can be restricted to a single or a small number of frequencies or be free to roam widely across a band or bands. A receiver, similarly, scans for information addressed to it. This is similar to the way computer local area networks (LANs) work over wire-line networks and now also over the air.

The clearinghouse could also auction off long-term access codes. In that case, it would approach the present auction and license system, except that no frequency exclusivity needs to exist, though that could also be instituted.

The access codes are, in effect, like tokens paid by drivers at toll. They could resemble, in concept, the tokens used in one major category of computer data LANs. In these "token ring" LANs, in order to avoid congestion and collision of information streams, only that user can transmit bits who possesses a token that circulates from user to user. The prices of the tokens vary according to congestion. The blocks of information carry these tokens with them, together with the address they seek, and pay (that is, transfer the tokens) at various toll gates and access points. The tokens are thus electronic coins that are transferred from user to carrier and the clearinghouse. They are like money. With electronic cash emerging in the economy, they could be general money, not specialized tokens. In effect, the information not only finds its own way (which packets already do); it *also carries its own money for transit,* picking among various over-the-air and wire-line transmission options depending on price and performance. This resembles a person navigating a transportation system, choosing routes and transit modes depending on price and performance, and paying along the way.

Does this system require carriers? For wireline services, the need is obvious for pathways to be maintained. But for over-the-air transmission, there is no roadway in the sky. Transmission firms resemble airlines or shipping companies rather than railroad companies. They provide transmission and reception facilities[33] accessible by the information packets at a price. These facilities need not permanently control any particular frequency any more than United Parcel Service and Federal Express control a highway or air route.

B. How to Implement an Open Spectrum System

Who would administer such an open-access system? The options are (*a*) the government; but this would create powers of control and administrative inefficiencies that are undesirable. (*b*) The private owner of the spectrum; this is discussed further below. Or (*c*) the users themselves, by way of a clearinghouse that functions like an exchange.

In practical terms, a clearinghouse would be a computer that sets access prices based on demand for the spectrum endowment that it controls. The potential user of spectrum would use some intelligent software agent to deal with the clearinghouse. If the spectrum user is willing to pay the price,[34] which outside of slack periods is unlikely to be zero, it will receive authorization through access codes. Multiple clearinghouses[35] for different bands are also possible and would provide competition. There could also be dif-

---

[33] Howard A. Shelanski & Peter W. Huber, Administrative Creation of Property Rights to Radio Spectrum, in this issue, at 581.

[34] Prices must be initially announced by a signal of spectrum price being sent out by the clearinghouse, based on supply and demand conditions. When capacity is underutilized at that price, the price drops, and an updated price signal is sent. The reverse holds true if there is excess demand. The adjustment of demand could be facilitated by some packets that are coded with a reservation price. Usage that does not require real time is thus likely to make room when demand spikes occur.

[35] The mechanism of a clearinghouse of providers has precedent. It is the way in which the FCC has dealt with relocation issues in the PCS bands and is a mainstay of electricity distribution.

ferent prices for different frequency bands, because their different propagation characteristics differentiate their attractiveness.

Each user could apply its own standards and protocols, within general technical parameters of signal strength, and so on, to avoid interference. Enforcement of the system is straightforward for those flows of information that are transferred across networks. Without authorization code, they could not flow. For nonnetwork usage, the presence of transmissions without access codes would be closely watched by their competitors, and violators would be sued or reported.

In some cases, a frequency would be entirely dedicated to a user or usage, based on special circumstances, for example, to protect nonprofit, educational, or governmental usage.[36] Alternatively, such users could receive a credit against which they could obtain access in the open-access system and that they could resell.[37]

Who gets the proceeds? That is a political decision of allocation. It could be the user-owners of the clearinghouse; or, alternatively, the Treasury (as in the auctions, and with a similar negative potential of use for current consumption); or some earmarked functions. But the revenue flow is smoothed with the high fixed costs of entry converted into variable costs of usage. It therefore has a stabilizing function, because prices based on marginal costs, without regard to sunk cost, encourage collusive pricing. Transaction costs in an open-access system may be larger than in a traditional spectrum-assignment system, but that is true for any open economic system. The offset is increased utilization and efficiency. And, similar transaction costs would exist if spectrum owners would resell frequencies in a private resale market.[38] There would be incentives to develop new technologies and applications—just as aircraft manufacturers and airlines do for the utilization of airspace—and to create various instruments of contractual rights for access—just as for financial derivatives.

C. Objections to Open Spectrum Access

The concept of buying spectrum access as an input rather than owning a spectrum license is unfamiliar and disturbing to users and policy makers

---

[36] Existing frequency licensees would still have the assured right to their spectrum, under the terms of their license. It might be possible for others to buy or rent the terms of their license.

[37] In addition, in situations of natural or man-made catastrophes, blocks of access codes would be set aside for emergency communications.

[38] Similarly, the setting of technical specifications would be no more complex in a clearinghouse setting than in an ownership model because a user could employ any technology subject only to general noncollision rules that are set by statute, common law, or agreement of the users. Such an agreement would have less collusive potential, given the transparency of a clearinghouse process open to all users.

alike, and a number of objections are made, on the grounds of practicality, uncertainty, and property.

1. *Technological Considerations.*—There are various building blocks for an open system, all of them subject to rapid technical change. The challenge to technologists and entrepreneurs is to put the various elements together.

a. Signal processing has made enormous progress, pointing to a near future in which radios become portable digital computers. "Software radio"[39] shifts the processing of the received signal by conducting all functions like demodulation, filtering, and detection in software-defined units rather than, as at present, through manipulations of the electronic signal within hardwired systems. Intensive research is underway on this concept, which would allow, for example, for a single handset to access multiple systems or for a single base station's equipment to carry multiple types of calls.[40]

b. Intelligent agents are software programs that could deal with the clearinghouse and search the spectrum for the best value.

c. Digital communications have now reached broadcasting too. Their extension to packet- or cell-based technology is used in packet radio.

d. Spread spectrum technology permits frequency changing and frequency sharing by multiple users. Civilian applications exist in mobile code division multiple access (CMA). Spread spectrum cordless phones are commercially available. There has also been much progress in the development of dynamic channel assignment and distributed control processes for wireless LANs and wireless Private Branch Exchanges. (Spectrum hopping is also possible without the spread spectrum technology.)

e. Expanded spectrum availability is the result of expansion to an operational range of up to 60 GHz, a much higher frequency range than in the past.[41] Laboratory usage has proceeded to 300 GHz; theoretical range goes still further.

f. Advanced antennas can cover an increasing range of bands. Spatial sig-

---

[39] Joe Mitola, The Software Radio Architecture, IEEE Comm. Mag., May 1995, at 26.

[40] First-generation products utilizing this technology are already on the market for wireless infrastructure equipment, where the battery power and size are not the limiting factors. It allows a service provider to use the same equipment for multiple channel types in the same frequency band, for example, both analog and digital channels for mobile service. The biggest challenge faced by such equipment is the present inadequacy of processing power required for the massive computations that are required to be performed in real time, for interactive communications like voice. But this processing bottleneck is being reduced rapidly and will, no doubt, be solved in time. Rupert Baines, The DSP Bottleneck, IEEE Comm. Mag., May 1995, at 46. The number of operations per second per chip is in the hundreds of millions now and is becoming sufficient and affordable for digital mobile radio applications; Mitola, *supra* note 39.

[41] Simon Forge, The Radio Spectrum and the Organization of the Future: Recapturing Radio for New Working Patterns and Lifestyles, 20 Telecomm. Pol'y 53 (1996).

nal separation by directional beams permits space division multiple access (SDMA).[42]

g. Signal compression uses algorithms to reduce necessary transmission needs, especially for video, thereby reducing required bandwidth.

h. Encryption has made enormous progress. It could be used for the access codes that permit transmission to be part of a network. It could also be used to charge end users for reception and to prevent transmission without access codes.

i. Electronic cash is related to encryption and has made similar rapid development in order to serve commerce on the Internet. It could be used for the access-code tokens, making these access codes, in effect, electronic coins used for toll gates.

On the regulatory front, some steps in the direction of openness have been taken by the FCC in 1985 in its Part 15 rules, which increased the unlicensed use of spectrum bands used by industrial, scientific, and medical (ISM) low-power applications (such as garage openers) to a higher transmissions strength of 1 watt, provided that spread spectrum technology was used. Examples for new uses were wireless LANs and wireless bar-code readers.

The concept was expanded in 1994 to unlicensed personal communications (U-PCS), open to all users of asynchronous data and isochronous time-division duplex voice. The dynamic real-time coordination of use accomplished by users following a "spectrum etiquette" agreed on by the industry and approved by the FCC. These rules are, basically, "listen-before-transmit" on a channel, "don't talk too long without listening again," and "don't talk too loudly," that is, limitation on transmission power. A potential user seeking transmission, when encountering a "busy" channel, either switches to another or awaits his turn. This etiquette is embedded in the device itself. The etiquette does not require interoperability between the various devices or exchange of information among them.

Coordination, including the relocation of existing users and definition of channels and geographical regions, is administered by a private nonprofit company, UTAM, owned by equipment manufacturers and supported by them in proportion to their U-PCS equipment sales. UTAM is basically a cooperative.

The next steps in this evolution was initiated by two petitions to the FCC in 1995, by WIN Forum for a limited range high-speed Shared Unlicensed Personal Radio Network (SUPERNet), and by Apple Computer for a National Information Infrastructure (NII) band. In 1997, the FCC[43] allocated

---
[42] *Id.*
[43] ET Docket No. 96-102, January 9, 1997.

300 MHz of spectrum in the 5 GHz band for unlicensed National Information Infrastructure (U-NII) devices. The FCC also opened the band above 40 GHz to unlicensed usage.[44]

The main weakness of the unlicensed access approach in its present stage is that it deals with scarcity and congestion by a technological "etiquette," which cannot ensure real-time access if demand is high. The best-working etiquette for the allocation of a scarce resource in our society is a market-clearing price. Without it one may reenact the rise and fall of citizens band (CB) radio. CB radio is the poor man's open access. CD radios are unlicensed, and their usage was tremendous, even though much of it proved to be a fad. The weakness of CB radio was the absence of congestion prices and of commercial incentives for content provision.

2. *Regulatory Considerations.*—Auction advocates tend to stress the rapidity of its allocation, in contrast to the messiness of market trading. But this focuses on the short term. It is true that efficient resource allocations are accelerated by auctions. But soon thereafter, given the dynamics of markets and technology, an aftermarket must take over anyway. Spectrum efficiency therefore depends more on a smooth aftermarket than on the initial allocation mechanism.[45] Since the auction-based allocation system may lead to a spectrum oligopoly due to potential oligopolists' ability to bid higher, such a system may well end up requiring more government intervention than presently hoped for, in order to maintain market competition. In contrast, a system of continuous open entry makes it harder to sustain oligopolistic prices. In such a system, the government's role is that of providing an initial endowment (the same function as in an auction) and assuring nondiscriminatory access to a clearinghouse. Establishing multiple and competitive clearinghouses for different spectrum bands would add still further openness. It is true that government could intervene, but selling full property rights in spectrum does not eliminate opportunities for regulation either, just as private use of land is often heavily regulated.

3. *Property Rights Consideration.*—Without secure long-term tenure there may be less investment. In the exploitation of frequencies, however, greater competition also spurs innovation and investment. One needs to balance certainty with contestability. Uncertainty exists in every business, and no firm can control every input. Spectrum is no different in that respect from a gas station that cannot be certain of the price of its vital input, wholesale gasoline, or of a bakery that needs to buy flour at varying prices. Similarly, employers do not "own" their employees and are not dispos-

---

[44] ET Docket No. 94-124, February 10, 1997.

[45] Arthur De Vany, Implementing a Market-Based Spectrum Policy, in this issue, at 627.

sessed by their departure to firms offering higher salaries. But when it comes to spectrum, much of private industry is so used to the concept of long-term control (whether by ownership or license) that it finds it hard to conceive of regularly buying spectrum access like another input. Of course, for some firms certainty will be considered necessary, and for that purpose futures markets for capacity will evolve.

Couching the discussion in the terms of property rights is not helpful.[46] Even the old license system was one of property rights, regardless of the 1934 Communications Act's declaration that it did not establish ownership right.[47] It is similarly argued that the FCC auctions are only for a long-term usage rights, not for full ownership. But this is a legal distinction without a real difference. The strong expectation is that the lease will be almost automatically renewed, just as it has been for TV broadcast licenses, where of more than 10,000 renewals between 1982 and 1989, less than 50 were challenged and fewer than a dozen were not renewed, usually because of some malfeasance. A postcard suffices to renew a license. In cable TV the nonrenewal of franchises is similarly rare. For all practical purposes, the auctions are for permanent occupancy, though the slight uncertainty will lower the prices a bit.

As Richard Posner observes, "In economic, though not in formal legal terms, then, there are property rights in broadcast frequencies. . . . Once obtained the right is transferable. . . . And it is for all practical purposes perpetual. The right-holder is subject to various regulatory constraints, but less so than a public utility, the principal assets of which are private property in the formal legal sense."[48]

Today, scrambling and encryption technologies permit producers of information to exclude unauthorized access to it. Holders of information can thus create "bottoms-up" property rights through access control, and markets evolve. This means that the protection against the unauthorized transmission need not be accomplished through licensing but can be left to market forces governing the transfer of the information in networks.[49]

---

[46] Just calling some rights property does not make them the base of an economic efficiency. Under feudalism and absolutism, many rights and privileges were property and for sale, such as military commands and titles of nobility. People could sell themselves into bondage or buy their freedom. Yet by no stretch could one describe these systems as efficient. It all depends on the context, which in economics means on the market structure. A property-rights system that has a built-in tendency to oligopoly, for example, would not be efficient.

[47] 47 U.S.C. 301.

[48] Richard A. Posner, Economic Analysis of Law (2d ed. 1977).

[49] Eli M. Noam, The Revolution in Access Control: Markets for Electronic Privacy (paper presented at the Aspen Summit '96: Cyberspace and the American Dream, Aspen, Colo. August 1996).

4. *Could an Auction Winner Administer an Open System Itself?*—An appealing alternative route to the unlicensed open-access system would be for private spectrum managers to conduct the resale of their capacity. This would require the spectrum ownership to be diverse, because if a firm has market power in spectrum, it would charge monopsony prices, discriminate in prices, and appropriate the efficiencies of rivals. It would be as if, in the predivestiture days of AT&T dominance, AT&T could have auctioned off the right to compete against itself. Under such a system, MCI would not have emerged. If a market could evolve with many wholesale spectrum band managers controlling a lot of spectrum to make resale transactions with many resale users practical, a substantial openness would indeed be achieved. But such a world seems unlikely; even if a government would license many spectrum owners, there would be consolidation, as has been argued, toward oligopoly. Furthermore, for meaningful access to be provided by a wholesaler, it would need to control a significant band, which is not likely to be affordable by any but the largest of telecommunications consortia. Imagine a firm buying half the VHF TV broadcast band for resale to broadcasters. As Robert Crandall points out in an article on the New Zealand experience with spectrums of management rights[50] (the only concrete example to date for an effort to institute a resale system), on the basis of recent auctions, a single nationwide gigahertz would be worth about $300 billion in the United States, 12 times the value of the giant RJR Nabisco leveraged buyout. "It is far from clear who would be able to 'bid' for such a franchise if the U.S. government were to offer it as a management right at an auction."[51] Milton Mueller, similarly, finds that in New Zealand "spectrum management rights can be acquired since 1990, but they have not been resold to others."[52] Only two local bidders showed up for the management auction in New Zealand, the previous monopolists in telecommunications and broadcasting. It is hard to imagine that their motivation is to encourage usage by competitors.

Alternatively, spectrum slices for wholesalers could be drawn narrowly, but then the spectrum agility of users' access moving around the spectrum would be curtailed, and the system would be the traditional "slice-and-dice" of spectrum licensing, whose consolidation and utilization would impose major transaction costs.

Advocates of resale markets need to explain the empirical fact that there was never any meaningful resale of nonadvertising time slots for spectrum

---

[50] Robert W. Crandall, New Zealand Spectrum Policy: A Model for the United States? in this issue, at 821.

[51] *Id.* at 825.

[52] M. Mueller, New Zealand's Revolution in Spectrum Management, 5 Info. Econ. Pol'y 159 (1993).

access by broadcasters, even in multistation markets (or by cable companies for their bandwidth). Partly this was due to FCC restrictions, but there did not seem to be major complaints against these rules, and one suspects that few TV stations would become time brokers or common carriers even if they could, as they now partly do. In telecommunications, to take another example, resale exists primarily due to legal common-carriage obligations and has been strenuously resisted by incumbents everywhere. The basic problem is the resistance to provide a competitor with a vital input at a price that permits entry.[53]

Some resale is taking place in satellite transmission. Here, the huge hardware and launch costs and the need for government backing in international bodies cause indivisibilities and entry barriers that lead to a limited number of capacity providers reselling transponders (channels) to large and stable tenants. Such a market is moving in the right direction as long as the need of the handful of firms to shield their huge investments does not lead to a significant anticompetitive cooperation. PCS licensees are also able to resell their spectrum. But it appears this will be done primarily by the "small business" winners of small regional bids (Basic Traffic Areas) who resell to larger nationwide firms (excluded from the small business auctions) that complement their own spectrum holdings. Thus, resale is taking place upward to large aggregative firms rather than downward to multiple users.

Resale is clearly a step toward open access. It should be encouraged. It is likely to exist in some limited fashion. But it is not likely to generate a widespread openness of access.

IV. Conclusion

The open-entry spectrum exchange will not solve every problem of today's auctions. New ones will emerge. Many of these problems may be resolvable once the technologists focus on them, but to do so requires first that we get out of the box of the exclusivity paradigm.

Even if the open-access system has some flaws, the constitutional issue must still be answered. Efficiency of resource allocation and lower transaction costs do not overcome the protection of fundamental rights of which free (electronic) speech is one. If an open-access system is less restrictive than an auction/ownership model without causing spectrum chaos, the granting of exclusive speech rights may not pass the test of constitutionality. Even some inefficiencies and transaction costs cannot defeat constitutional rights.

What are some of the policy implications? First, it is not to stop auctions,

---

[53] Eli M. Noam, Beyond Liberalization II: The Impending Doom of Common Carriage 18 Telecomm. Pol'y 435 (1994); Noam, *supra* note 49.

since in the present state of technology they are still usually the better solution. But it means to limit the duration of auctioned licenses, in order to preserve future flexibility for other approaches.

Second, resale and spectrum use flexibility should be permissible to facilitate resale markets. License holders should be able, in most cases, to slice up the spectrum and resell and sublet them to others for various applications.

Third, experimentation and innovation in spectrum usage schemes should be encouraged. This would include expanding the unlicensed spectrum concept and dedicating frequency bands to the open-access, access-price model. Better to approach spectrum use in a pragmatic and searching fashion than with an ideological mind-set that equates the free market with one and only one particular technique. We should be ready to take the next step. The tremendous success of the Internet should lead us to seek its openness in spectrum use too. The Internet, with its multiple-route system, is an example for an open-access model in the wire-line environment. Here, too, congestion charges are being considered. Open does not mean free or nonprofit.

It took Leo Herzel and Ronald Coase almost 50 years to see their auction paradigm implemented. Similarly, the proposed open-access paradigm is not likely to be accepted any time soon. But its time will surely come and fully bring the invisible hand to the invisible resource.

Bibliography

Baines, Rupert. "The DSP Bottleneck." *IEEE Communications Magazine* 33 (May 1995): 46–54.

Baran, Paul. "Is the UHF Frequency Shortage a Self Made Problem?" Paper presented at the Marconi Centennial Symposium, Bologna, 1995.

Borenstein, Severin. "On the Efficiency of Competitive Markets for Operating Licenses." *Quarterly Journal of Economics* 103 (May 1988): 357–85.

Briggs, Asa. *The History of Broadcasting in the United Kingdom.* Vol. 1, *The Birth of Broadcasting in the United Kingdom.* London: Oxford University Press, 1961.

Calhoun, George. *Digital Cellular Radio.* Boston: Artech House, 1988.

Caro, Robert A. *Years of Lyndon Johnson.* New York: Knopf, 1990.

Coase, R. H. "The Federal Communication Commission." *Journal of Law and Economics* 2 (1959): 1–40.

Commission of the European Union. *DG XIII, Towards the Personal Communication Environment.* Brussels, January 12, 1994.

Congressional Budget Office (C.B.O.). *Auctioning Radio Spectrum Licenses.* Washington, D.C.: Congress of the United States, March 1992.

Crandall, Robert W. "New Zealand Spectrum Policy: A Model for the United States?" *Journal of Law and Economics* 41 (1998): 821–40.

De Vany, Arthur S. "Implementing a Market-Based Spectrum Policy." *Journal of Law and Economics* 41 (1998): 627–46.

De Vany, Arthur S., et al. "A Property System Approach to the Electromagnetic Spectrum: A Legal-Economic-Engineering Study." *Stanford Law Review* 21 (June 1969): 1499–1561.

Dixon, Robert C. *Spread Spectrum Systems.* 3d ed. New York: John Wiley & Sons, 1994.

Federal Communications Commission (FCC). "Inquiry and Proposed Rulemaking: Deregulation of Radio." *Federal Register* 44, No. 195 (October 5, 1979): 57,636–723.

Federal Communications Commission (FCC). "New Television Networks: Entry, Jurisdiction, Ownership and Regulation." In *Final Report of the Network Inquiry Special Staff.* Vol. 1. Washington, D.C.: U.S. Government Printing Office, October 1980.

Forge, Simon. "The Radio Spectrum and the Organization of the Future: Recapturing Radio for New Working Patterns and Lifestyles." *Telecommunications Policy* 20, No. 1 (1996): 53–75.

Fowler, Mark S., and Brenner, Daniel L. "A Marketplace Approach to Broadcast Regulation," *Texas Law Review* 60 (1982): 207–57.

Geller, Henry. *1995–2005: Regulatory Reform for Principal Electronic Media.* Washington, D.C.: Annenberg Washington Program in Communications Policy Studies, Northwestern University, 1994.

Gilder, George. "Auctioning the Airways." *Forbes,* April 11, 1994.

Hazlett, Thomas W. "The Rationality of U.S. Regulation of the Broadcast Spectrum." *Journal of Law and Economics* 33 (April 1990): 133–75.

Hazlett, Thomas W. "Assigning Property Rights to Radio Spectrum Users: Why Did FCC License Auction Take 67 Years?" Working Paper Series No. 768. Columbia Institute for Tele-Information, Columbia University, 1995.

Herzel, Leo. "'Public Interest' and the Market in Color Television Regulation." *University of Chicago Law Review* 18 (1951): 802–16.

Jett, Wayne. "May God Save the Constitution (with Our Help) from Its Friends." Unpublished manuscript, 1996.

Kuhn, Thomas S. *The Structure of Scientific Revolutions.* 2d ed. Chicago: University of Chicago Press, 1970.

Kwerel, Evan R., and Filcher, Alex D. "Using Auctions to Select FCC Licensees." O.P.P. Working Paper No. 16. Washington, D.C.: Federal Communications Commission, 1985.

Levin, Harvey J. *The Invisible Resource—Use and Regulation of the Radio.* Baltimore: Johns Hopkins University Press, 1971.

MacKie-Mason, Jeffrey K., and Varian, Hal. "Economic FAQs about the Internet." *Journal of Economic Perspectives* 8, No. 3 (Summer 1994) 75–96.

McMillan, John. "Why Auction the Spectrum?" *Telecommunications Policy* 19, No. 3 (1995): 192–99.

Melody, William H. "Radio Spectrum Allocation: Role of the Market." *American Economic Review* 70 (May 1980): 393–97.

Mitola, Joe. "The Software Radio Architecture." *IEEE Communications Magazine* 33 (May 1995): 26–38.

Mueller, Milton. "New Zealand's Revolution in Spectrum Management." *Information Economics and Policy* 5 (1993): 159–77.

National Telecommunications and Information Administration (NTIA). *Spectrum Management Policy: An Agenda for the Future.* NTIA Publication No. 91-23. Washington, D.C.: Department of Commerce, February 1991.

Noam, Eli M. *Television in Europe.* New York: Oxford University Press, 1991.

Noam, Eli M. "Beyond Liberalization II: The Impending Doom of Common Carriage." *Telecommunications Policy* 18, No. 6 (August 1994): 435–52.

Noam, Eli M. "The Impact of Competition on Television's Public Interest Performance in the USA." In *Television Requires Responsibility,* ed. European Institute for the Media. Gutersloh, Germany: Bertelsmann Foundation Papers, 1995.

Noam, Eli M. "Taking the Next Step beyond Spectrum Auctions: Open Spectrum Access." *IEEE Communications Magazine* 33 (December 1995): 66–73.

Noam, Eli M. "The Revolution in Access Control: Markets for Electronic Privacy." Paper presented at the Aspen Summit '96: Cyberspace and the American Dream, Progress and Freedom Foundation, Aspen, Colo., August 1996.

Noam, Eli M., ed. "Edwin Armstrong: A Man and His Inventions." Unpublished manuscript, 1998.

Noam, Eli M. "Interconnecting the Network of Networks." Unpublished manuscript, 1998.

Noll, Roger; Peck, M. J.; and McGowan, John J. *Economic Aspects of Television Regulation.* Washington, D.C.: Brookings Institute, 1973.

Posner, Richard A. *Economic Analysis of Law.* 2d ed. Boston: Little, Brown & Co., 1977.

Shelanski, Howard A., and Huber, Peter W. "Administrative Creation of Property Rights to Radio Spectrum." *Journal of Law and Economics* 41 (1998): 581–607.

Simon, M. K., and Omura, J. K. *Spread Spectrum.* New York: McGraw-Hill, 1994.

Smythe, Dallas W. "Facing Facts about the Broadcasting Business." *University of Chicago Law Review* 20 (1952): 96–106.

Zupan, Mark. "The Efficacy of Franchise Bidding Schemes in the Case of Cable Television: Some Systematic Evidence." *Journal of Law and Economics* 32 (October 1989): 401–56.

[18]

SPECTRUM FLASH DANCE: ELI NOAM'S PROPOSAL FOR "OPEN ACCESS" TO RADIO WAVES

THOMAS W. HAZLETT
University of California, Davis

Abstract

As Ronald Coase posited in his famous article on the nature of the firm, there are situations in which decentralized markets are relatively efficient for coordinating economic activity, and situations in which they are not. With spectrum access, assigning property rights to clearly specified private owners is the socially efficient policy because the relevant transaction efficiencies will be internalized by competitive "spectrum owners" selecting to what degree rights should be subdivided and in what manner marketed. Where spot markets are optimal, an owner will maximize profits by using them; where long-term contracts are efficient, the owner will enter profitably into them. Hence, Eli Noam's solution—imposing open-access rules on bands of radio frequencies by government mandate—is a mistaken attempt to duplicate the efficiencies of markets by mandating a particular subset of market solutions. Such a policy predictably will result in underutilization of the spectrum resource.

I. Introduction

ELI NOAM's paper[1] elucidates a property rights regime for radio waves that is under discussion among some thoughtful and influential thinkers within the computer community. It is provocative and deserves to be analyzed by economists. Unfortunately, the article begins with a list of policy failures incorrectly ascribed to the license auctions recently initiated at the Federal Communications Commission. Moreover, the problems asserted are not solved by the "open access" spectrum regime Noam imagines. In sum, the argument confuses defects in the status quo policy regime for allocating radio spectrum with the innovation of competitive bidding to assign rights. Thus it fails to appreciate the efficiency gains that auctions have provided and falls short in attempting to define useful remedies for the problems extant.

First this comment shall deal with the substance of Noam's critique of

---

[1] Eli Noam, Spectrum Auctions: Yesterday's Heresy, Today's Orthodoxy, Tomorrow's Anachronism, in this issue, at 765.

current institutions, including "spectrum auctions." Then it will focus on the "open access" regime that Noam's article offers as the optimal policy reform.

II. Critiquing Spectrum Allocation

The inefficiencies of the present system for allocating radio spectrum among competing uses have been observed by economists for decades.[2] The violence that the regime of comparative hearings for assigning radio and TV licenses does to the First Amendment's free speech clause has been noted by legal experts for about as long.[3] To combine and condense, the essential critique is that central planning of the spectrum resource (since the 1927 Radio Act created the system of allocating spectrum, and then assigning licenses, in the "public interest") leads to the inefficiencies of socialist economic organization, on the one hand, and to the political favoritism of classic rent-seeking rivalry, on the other hand. Pointing out the deficiencies, in terms of either lost consumer welfare or encroachment on a free broadcast press, is not a difficult sport. A consensus has developed that the road to improvement is paved with reforms that allow large swatches of spectrum to be privately used in flexible ways, as dictated by consumer demand.[4] Noam's discussion misses the essence of this logic and takes a dangerous detour in its advocacy of an imposed pricing scheme based on what he construes a state-of-the-art technology—albeit, a state yet to arrive.

A. Historical Development of Property Rights

The article's discussion of the emergence of the present system inaccurately portrays both the role of property rights and the political rationale for the 1927 Radio Act. Noam, while correctly noting that priority-in-use rights first policed spectrum access, mistakenly asserts that such a system worked only due to the absence of scarcity. Hence, he cites the popular fiction that the pre-1927 radio market was chaotic, asserting that "the problem of interference . . . doomed the occupancy model."[5] This is demonstrably false:

---

[2] R. H. Coase, The Federal Communications Commission, 2 J. Law & Econ. 1 (1959).

[3] See Harry Kalven, Jr., Broadcasting, Public Policy and the First Amendment, 10 J. Law & Econ. 15 (1967); Ithiel de Sola Pool, Technologies of Freedom (1983); Lucas A. Powe, Jr., American Broadcasting and the First Amendment (1987).

[4] See Harvey Levin, The Invisible Resource (1971); Evan R. Kwerel & John R. Williams, Changing Channels: Voluntary Reallocation of UHF-TV Spectrum (OPP Working Paper No. 27, Federal Communications Commission, November 1992); Thomas G. Krattenmaker, The Telecommunications Act of 1996, 29 Conn. L. Rev. 123 (Fall 1996); Gregory L. Rosston & Jeffrey S. Steinberg, Using Market-Based Spectrum Policy to Promote the Public Interest, 50 Fed. Comm. L. J. 87 (December 1997).

[5] See Noam, *supra* note 1, at 769. Noam also states that "transmissions soon collided on the unregulated ether," creating "an inevitable crisis," which led directly to the imposition of administrative allocation of airwave rights. *Id.* at 766.

orderly development of the radio broadcasting industry proceeded during the 1920–26 period under priority-in-use rules borrowed from common law.[6] Such rules then were abandoned in the 1927 Radio Act for distributional reasons, namely, to allow incumbent broadcasters and political regulators the opportunity to foreclose entry, divvying up industry rents in a manner advantageous to both parties.[7]

The Noam article fails to appreciate the availability of general legal institutions for solving the potential tragedy of the commons in spectrum use, let alone their demonstrated success. The author of the 1927 Radio Act, Senator C. C. Dill (D-Wash.), noted explicitly that the common law had a mechanism handy to deal with a wireless technology such as radio broadcasting: "right of user."[8] Such rules were employed administratively by the Department of Commerce up to July 1926 and were beginning to be recognized in state courts (when one Illinois broadcaster, in November 1926, won an injunction against a rival station "trespassing" in its airspace). It was to preempt such rules from further development, which would constrain political controls over this emerging influential medium of public opinion, that Senator Dill believed a statutory solution promoting "public interest" licensing—including a waiver of all vested claims in the "ether"—was necessary.

What is crucial about a common-law approach to policing radio transmissions is that it allows social use of this beneficial resource (spectrum) in a generalized legal framework not specific to communications.[9] By moving decisions from regulators with wide discretion to judges and juries with narrower discretion and by allowing spectrum rules to develop in the domain of property and contract law, the opportunity for regulatory capture diminishes. And that, of course, is the aim of proconsumer public policies that move us to more competitive market allocations.[10]

B. *Spectrum Allocation Overwhelmed by Revenue Demands*

The Noam article relies heavily on the idea that political pressure to maximize revenue is the driving—and overwhelming—force unleashed by auc-

---

[6] Thomas W. Hazlett, The Rationality of U.S. Regulation of the Broadcast Spectrum, 33 J. Law & Econ. 133 (April 1990).

[7] Thomas W. Hazlett, Physical Scarcity, Rent Seeking and the First Amendment, 98 Colum. L. Rev. 905 (May 1997).

[8] See Clarence C. Dill, Radio Law (1938).

[9] On the viability of common law as a regulatory regime in telecommunications, see Peter Huber, Law and Disorder in Cyberspace (1997).

[10] See Howard A. Shelanski & Peter W. Huber, Administrative Creation of Property Rights to Radio Spectrum, in this issue, at 581.

tions.[11] This is a curious assertion. If policy makers are so anxious to raise government dollars, why did it take 67 years (1927 to 1994) to initiate FCC license auctions? Moreover, as auctions as yet have been used only for nonbroadcast licenses, and TV broadcasters have procured zero-priced licenses for digital television service even in the postauctions era,[12] the idea that revenue raising dominates spectrum policy is refuted directly.

Yet, even if revenue raising were to become dominant, it is not clear that auctions are a negative policy change at the margin. That is because there is a profoundly promonopoly bias preexisting within our spectrum allocation policies. This is driven by agency capture by a coalition of licensees (that is, wireless incumbents) and assorted cross-subsidy recipients. It is plausible that instituting auctions will lead to a breakdown in the protectionist policy bent, releasing additional spectrum rights into the competitive marketplace. The logic involves Ronald Coase's durable goods monopoly problem: how can the monopoly seller of a durable good convince buyers that, should they pay a high (monopoly) price for the asset this period, the seller won't increase supply (say to the competitive level) next period?[13] The seller searches for a mechanism to assure buyers that it will not have an incentive to "dump" additional units on the market in future periods, or else it cannot enjoy full monopoly pricing.

The solutions Coase finds are that the supplier (1) does not sell, but only rents, units; (2) reliably restricts future output via a mechanism akin to "public interest" regulation. It is striking that the FCC, which can be seen as a monopoly seller of spectrum access rights, traditionally has employed the latter method for credibly restricting spectrum access in future periods.[14]

---

[11] "The underlying objective for the auction 'game' is to raise revenues for government." Noam, *supra* note 1, at 772.

[12] Joel Brinkley, Did Broadcasters Hoodwink Congress with False Promises about HDTV? N.Y. Times, September 15, 1997, CyberTimes on-line edition. It is important to note the financial magnitude of the "advanced television" licenses awarded without charge by the FCC in 1997 (well into the era of auctions, which were authorized in 1993 and initiated in 1994). The Federal Communications Commission and the Congressional Budget Office estimated the TV licenses to be worth between $12.5 and $70 billion. Edmund L. Andrews, Digital TV, Dollars and Dissent, N.Y. Times, March 18, 1996, at D3. Even the lower bound of this range represents more than half of all FCC license auction bids tallied by the Commission to date, and probably over 100 percent of the actual monies collected. (Note that the more than $10 billion bid for licenses in the PCS-C block auction, ending in May 1996, represents nearly one-half of total auction receipts. Perhaps 90 percent of these C-block revenues are as yet uncollected due to [a] the 10-year credit terms extended to firms designated as small businesses, [b] a series of defaults and bankruptcies by winning bidders, and [c] a Commission rule making allowing licensees several options for easing their debt obligations to the FCC.)

[13] R. H. Coase, Durability and Monopoly, 15 J. Law & Econ. 143 (April 1972).

[14] While FCC licenses assigned via competitive bidding may appear to be issued on a rental basis because underlying control of the spectrum resource remains vested in the Commission, this is incorrect, economically speaking. Once a license is assigned at auction, li-

The commitment to parsimoniously issue new rights has been faithfully rewarded by dominant rent seekers. These interests reliably apply pressure for policies that maximize existing license values.

If policy makers were, alternatively, to be captured by government revenue maximizers, they would shift from protecting existing license values (which yield no receipts, or entail sunk obligations as per past auctions) to promoting license values sold this period. While this yet might involve a restriction of the supply of licenses, it necessarily would involve a larger supply than a policy of pure incumbent protection (which supports zero additional supply). Moreover, each successive period would find policy makers attempting to auction progressively more licenses, capturing additional revenues (presumably, the residual monopoly amount) under the demand curve for wireless licenses.

Of course, if auction bidders anticipated such policy behavior, bids for licenses would collapse long before policy makers were able to extract such rents. (This is analogous to the durable goods monopoly problem.) Such a result would be the predictable, if ironic, effect of the prorevenue bias in spectrum policy that Noam claims to observe. But it should be noted that the push to sell additional licenses and the price collapse in anticipation of such supply expansion are both clear indicators that revenue maximization would be an improvement on the prevailing policy goal undeniably in place until (at least) the advent of auctions: incumbent protection. As incumbent licensees argued against entry, Noam's champions of revenue[15] would provide politically dominant support for additional licenses.

C. *Auctions as Barriers to Entry*

The assertion that FCC license auctions limit entry is incorrect. Auctions simply capture whatever surplus is available to the entrant—in the estimation of the winning bidder.[16] In arguing that "[f]irms may price [only] temporarily without regard to fixed cost. . . . Hence an auction payment will be reflected in [output] prices,"[17] Noam's article employs the sunk cost fallacy. Auction prices for licenses will reflect anticipated output prices for services, not vice versa.

---

cense renewals take place on a noncompetitive basis. Hence, the payment for the license comes up front, as in a sale, rather than via a market rent adjusted over time.

[15] Noam writes: "Like diamonds, budget pressures are forever. There is never enough money. This creates a dependence on still more auctions. There will be auctions everywhere." Noam, *supra* note 1, at 774. "Auctions Inevitably Deteriorate into Revenue Tools." *Id.* at 772.

[16] Winner's curse is possible—but the curse is incurred prior to entry and will not thereafter affect the marketplace.

[17] Noam, *supra* note 1, at 774.

It is also false that "[t]hrough auctions, the United States government has been taking money away from the infrastructure-providing private firms" in the telecommunications sector.[18] Were licenses to be awarded to wireless service providers at a price of zero, investors would simply rearrange their portfolios to distribute these windfalls so as to optimally diversify. In reality, auctions present no change in the status quo. By distributing licenses via lotteries (as in the cellular telephone market, 1984–89), operating firms had to purchase FCC permits in secondary markets from lucky drawing winners. Those monies can also be said to have been "withdrawn" from the telecom sector.

Auctions directly promote efficiency by eliminating costly rent seeking for licenses. Lotteries and comparative hearings both encourage wasteful social activity and disadvantage the most efficient (if less lucky) service providers by forcing them to incur the bargaining and transactions costs associated with procuring licenses in secondary markets.[19] The "auction is a tax on the communications sector and its users," claims Noam,[20] but it is not. The winning bid at auction constitutes either a payment that would otherwise accrue to a lottery winner or a capital windfall that would be redeployed by enriched stockholders. Policy analysts seeking to understand how competitive bidding changes the economics of license assignments should focus on transactions costs (including rent seeking and license resales). One salient fact that jumps out is that when lotteries assigned "free" cellular licenses in the 1980s, deployment (licensing and build-out) took much longer than when similar wireless telephone licenses for personal communications services (PCS) A and B blocks were auctioned in 1995.[21]

D. Auctions and Oligopoly

The FCC's allocation policies have protected industry incumbents to the detriment of potential entrants and consumers. But this procartel bias long predates auctions; hence, Noam's claim that "[a]uctions [e]ncourage [o]li-

---

[18] *Id.* at 773.

[19] Pure lotteries would not directly incur rent-seeking expense, but the actual apparatus used by the FCC maintained the fiction that to apply for a license it was necessary to be an "actual" telecommunications provider. This inspired "application mills" to crank out documentation for investors (including those lacking previous experience or financial interest in the sector) proving, as per FCC guidelines, that they could successfully build and operate the licensed system applied for. See Thomas W. Hazlett & Robert J. Michaels, The Cost of Rent-Seeking: Evidence from Cellular Telephone License Lotteries, 59 S. Econ. J. 425 (January 1993).

[20] Noam, *supra* note 1, at 774.

[21] Congressional Budget Office, Where Do We Go from Here? The FCC Auctions and the Future of Radio Spectrum Management (April 1997).

gopoly"[22] is unconvincing. Auctions determine initial assignments for a given set of licenses; the underlying spectrum regulation, which defines the terms allowing licensees to use radio waves (including the number of permits to be issued per market) is formally distinct.[23] The operative margin along which regulatory capture occurs is the "public interest" allocation of radio spectrum.

Indeed, the crux of the complaint lodged by economists against the prevailing system of spectrum allocation is that it is a process of administrative control. For better or for worse, auctioning FCC licenses under the current allocation regime does not change that. Auctioning spectrum rights would—but that is a radical departure from the status quo, one that would involve a dramatic (albeit socially useful) expansion of private property rights.[24]

E. Property Rights as Free Speech Infringement

The attempt to characterize auctions of broadcasting licenses as impediments to free speech is similarly uncompelling.[25] To the extent that a speaker must purchase an input to be (better) heard, then the free ability to pay money for such an input is not a barrier to speech but a facilitator. Certainly, Noam is correct when he offers that "[a] licensing scheme, however the license is given out, is a serious restriction on speech."[26] Of course this entirely offsets the assertion that "[i]nstead of loosening the barriers to free entry, the U.S. government is going in the opposite direction, by selling off the spectrum."[27] It is the license that restricts speech; the opportunity to bid freely for such is not the encroachment. Indeed, to the extent that competi-

---

[22] *Id.* at 775.

[23] There may be feedback between these two analytically distinct regulatory actions; indeed, this forms a key aspect of my explanation of the reluctance of policy makers to use auctions. See Thomas W. Hazlett, Assigning Property Rights to Radio Spectrum Users: Why Did FCC License Auctions Take 67 Years? in this issue, at 529. But interactive links between the dual regulatory functions of license assignment and spectrum allocation are not explored by Noam. Moreover, broadcasting licenses—where such linkages have been observed—have been excluded from assignment by competitive bidding.

[24] The article consistently misconstrues this crucial point, even in its title. The term "spectrum auctions" is inappropriate to describe the current policy regime; what is auctioned is an operating permit (to provide a given wireless service) issued by the FCC. Noam, it should be said, is not alone in his use of this unfortunately common phraseology.

[25] Since policy makers pointedly elected to exclude radio and TV licenses from assignment via competitive bidding in the 1993 auction authorization, it is not clear why this discussion appears in the article.

[26] Noam, *supra* note 1, at 770.

[27] *Id.* at 771. The confusion between "selling off spectrum" and actual FCC policy is readily apparent here.

tive bidding replaces comparative hearings (the relevant consideration for broadcast licenses, as Congress has never authorized lotteries for such properties), financial currency replaces political currency. Freedom of speech is surely enhanced. One of the prime selling points for auctioning radio and TV licenses is that it makes license assignment an arms-length transaction, depoliticizing a process that should—given the First Amendment—be immune from government discretion.[28]

III. Open Spectrum Access as a Policy Alternative

Noam's idealized vision as to the most progressive outcome for spectrum access is not controversial. Indeed, pricing spectrum usage at its social opportunity cost has been the proconsumer goal since 1951, when University of Chicago law student Leo Herzel, after reading Abba Lerner's 1944 classic, *The Economics of Control,* formally introduced the auction approach to the literature.[29] But the normative path charted by economists and lawyers from that point travels far afield from the trail that Eli Noam attempts to blaze. The law and economics approach has been to liberalize the existing complement of rights afforded FCC wireless licensees, moving away from operating permits (narrowly defined as to both inputs and outputs) and toward actual "spectrum rights" (where flexibility is granted to a spectrum user, and rights delimited according to interference parameters).

Because sweeping technological changes have tended lately to shift political influence away from incumbent spectrum users and toward potential entrants (and, most important, their potential customers and suppliers), some liberalization has occurred. For instance, "voluntary reallocation" of incumbent microwave users in the PCS band was allowed formally in the rule making crafted during the Bush and Clinton administrations. The PCS permit really defines an "overlay right," giving the licensee (assigned at auction) significant latitude to manage the bandwidth allocated to the license. That is a notable departure from previous policy, where administrative edict typically has constituted the sole mechanism for spectrum allocation.[30] If continued, liberalization of the rights associated with the FCC license will lead to the competitive market outcome that Noam envisions.

---

[28] "The present system involves the state in licensing preferred broadcasters and censoring from the air those values it does not share, whereas markets allocate resources by a game that, like any game, is not always fair but is at least insulated from government. Indeed, a principal advantage claimed for the present regulated licensing system over a market is that it enables government to implement its goals and, specifically, to subsidize the kind of broadcasting it favors." Pool, *supra* note 3, at 139.

[29] See Leo Herzel, My 1951 Color Television Article, in this issue, at 523.

[30] That personal communications services are broadly defined and that PCS licensees are not mandated to use a specified technology are additional indicators of the implicit deregulation that occurred in the PCS rule making.

The "open access" or "clearinghouse" approach suggested by the Noam article, however, heads in quite a different direction. Rather than extending the current rights system forward to connect with the more traditional regime of property law, the plan is to impose a technically sophisticated superstructure on the market for radio spectrum. According to engineering specifications not entirely worked out, and employing machinery not yet available, the right to use the airwaves for specific instances will be assigned by competitive bidding. There are, no doubt, some imaginative algorithms that may be derived to supply an elegant allocational solution along these lines.[31] Indeed, there are an infinite number of such solutions. But pursuing just one of them—as suggested by Noam, although he does not know which one yet—takes us on a very costly detour in our path to spectrum liberalization. For it veers away from decentralized decision making and empowers policy makers to rig this particular market—top-down—with an elaborate and untested allocation mechanism. It may not be a public policy disaster waiting to happen. But it surely will keep us waiting. And, in consumer welfare terms, that itself is a disaster.

As the Noam article concedes, the basic spectrum policy question concerns property rights.[32] The need for rights is mandated by scarcity, and the existence of scarcity is revealed (in Noam's model) by the above-zero prices paid for spectrum access to the "clearinghouse." This is what makes Noam's own taxonomy so problematic. He argues that the OCCUPANCY model (which I call the property rights model) was replaced by an ADMINISTRATIVE allocation system (which I refer to as the status quo) and alleges that both are really quite similar because they share the concept of "licensed exclusivity."[33] This he sets apart from his OPEN ACCESS paradigm, which purportedly ushers in a new era of shared spectrum use. But the limits to "open access" are immediately apparent: rights still must be auctioned, trespassers yet will be prosecuted. Technologies that yield greater communications capacity over a given bandwidth do not yield unlimited communications capacities; the possibility

---

[31] Indeed, Noam appears to be unaware that a key feature of the packet-switching technology employed in the Ethernet was first employed in a wireless architecture. The Alohanet at the University of Hawaii developed the protocol for randomly resending information after a collision between packets, a method of transmission that turned out to efficiently utilize bandwidth. Katie Hafner & Matthew Lyon, Where Wizards Stay Up Late: The Origins of the INTERNET 236 (1996).

[32] The article makes an effort to deny this—"couching the discussion in the terms of property rights is not helpful" [Noam, *supra* note 1, at 785]—but the substance of the argument revolves around how to ration access to airwaves. That implicitly concedes that access rights are scarce. Rules determining the allocation of scarce resources are typically called "property rights."

[33] Noam, *supra* note 1, at 766–69.

of sharing space previously only large enough for one does not end the necessity of determining who the marginal communicator will be. Noam, in fact, proposes not to open access but rather to ration access by a vast number of state-sponsored auctions.

The rights auctioned by Noam differ from "licensed exclusivity" only in the sense that they are artificially constrained by law to be sliced very, very thinly. The analogy to land—an excellent analogy—would be a regime under which the ownership of long-term rights for large parcels of land were illegal, with a monopoly government landowner leasing only daily (nanosecond?) usage on tiny (square millimeter?) plots.[34]

It is quite true that, ignoring transactions costs, the market would reassemble the slivers of rights in efficient configurations. Noam cites "technical reasons" that make such a Coasean (zero transactions cost) optimum possible. In essence, Noam assumes that transactions costs in a digital spread spectrum world are trivial. This is a trick; alert policy analysts will not be fooled. Not only is the organization of real-time auctions in tiny spectrum parcels not a costless endeavor, to vest authority over rules in a regulatory body charged with carrying out "public interest, convenience, and necessity" is to set up a full-fledged rent-seeking game. The history of such rule making in this sector is little more than a litany of successful attempts by one set of service providers (typically incumbent licensees) to delay or block new competition via the manipulation—or debate over—"technical issues."[35]

The ultimate irony is that Noam's professed goal of low-cost entry into wireless telecommunications will be achieved by the property rights system his article disparages. Noam's argument is that creating a larger bundle of private property rights should not be the goal of policy makers; rather, they should strive to define "rights of way." But rights of way are property rights. The examples used to establish the alleged difference between the two demonstrate a confusion over scarcity. There is no scarcity when I gain the right to walk over your property to get somewhere and you give up next to nothing for allowing me to do so. But the situation of no-cost coordina-

---

[34] Noam states, "But spectrum access is traffic control, not real estate development. It is about flows, not stocks." Noam, *supra* note 1, at 770. But real estate is all "about flows"—the expected present value of a flow of services delivered to tenants and customers over future periods is identically equal to the market value of a financial asset such as real estate. A spectrum developer vested with private property rights to a band would have all the same economic incentives as a real estate developer to promote service demands specific to the owned asset, so as to maximize capital value.

[35] AM radio managed to effectively block FM's development for better than 2 decades. VHF television managed to thwart UHF, and then cable, for nearly as long. The rule making for cellular telephony took decades to be completed, depending upon when one starts the clock. The list goes on.

tion collapses instantly when scarcity appears. When the nonscarce sky-route becomes a scarce landing slot, rights definition becomes necessary for efficient allocation. Then the right of way becomes just another property right and gives an "owner" the right to exclusive use of a resource.

In fact, the article's argument-by-analogy confuses rather than clarifies, because the analogies are ill-suited to the resource allocation problem at hand. Apparently, this springs from the analogy that haunts the article, that of the Internet. The "network of networks" seemingly has offered a glimpse of the future, in that the abundant capacities provided by fiber optics and packet switching have linked computers at very high speed for very little cost. But rather than provide a model for spectrum allocation, the Internet architecture demonstrates the inherent limitations of insufficiently developed private markets in bandwidth. The protocols and network facilities determine what traffic the Internet will bear and at what speed it will be delivered; there is currently no reliable mechanism for measuring consumer demands for alternative bandwidth uses and deploying the resource optimally.

While many (including Noam and this author) are impressed by the technical agility of "spread spectrum" and other techniques to squeeze much more electronic communications out of any given bandwidth, it is simply not true that the tragedy of the commons has been solved by science. It is, was, and will continue to be solved by rules. Noam's confusion is seen in his citation of George Gilder: "You can use the spectrum as much as you want as long as you don't collide with anyone else."[36] The "as long" phrase simply assumes away the economic coordination problem: if scarcity exists, collision avoidance will be costly. Without an owner to establish primacy (so users may know who, precisely, is colliding with whom), we anticipate a costly race to establish rights. Hence, the efficient assignment of property nights can economize on such rent-dissipating activity.[37]

Noam's vision is one of spread spectrum taking us down the Information Superhighway, dodging other messages, steering our way à la the airplane using visual avoidance through the skies. Yet airplane (and ship) routes are not typically scarce. Avoiding others is (*a*) not costly, and (*b*) very (privately) important. It is efficient, given the transaction costs of the relevant alternatives, to let shippers reliably (out of self-interest) avoid others (at trivial cost, compared to the benefits). Once congestion arrives, however, this solution fails. It becomes costly to avoid other traffic, and some of

---

[36] Noam, *supra* note 1, at 769.

[37] Indeed, we often establish property rights even when no one collides (that is, marginal cost equals zero). Such is the case with privately produced public goods, such as trademarks, patents, or copyrightable material.

those costs (in the absence of clearly defined property rights) may be imposed on others.

The Information Superhighway (or Internet) example further clouds the picture by introducing a centrally administered network infrastructure. Spectrum use has no tendency to natural monopoly, and a policy that enables many competing wireless broadband highways (on different blocks) is easily feasible. Hence, the more appealing solution is to create the underlying conditions for robust competition and allow marginal cost pricing to emerge spontaneously rather than to be imposed from above.

A better analogy? Try: Competing Tollways.[38] If we were to award private rights to multiple wireless broadbands (say, bands of the carrying capacity of the cellular and PCS licenses), endow them with full flexibility, and turn them loose on the marketplace (constrained only by interference parameters and the traditional, non-FCC, legal regime), we would observe the spontaneous emergence of competitive airwave pricing. As no highway owner has the ability to restrict entry, no one may increase price above the social opportunity cost of spectrum. These toll road competitors will rival each other on nonprice margins, as well. By investing in innovative technology and offering consumer-friendly standards for access, they will enhance revenue streams. Profit maximization will force competitive band managers to devise better technical means to increase wireless communications traffic; indeed, such rivalry will produce the metering devices that Noam divines—or produce a compelling efficiency explanation as to why such devices are not useful to deploy.

Magical technology will not end the process wherein entrepreneurs seek to increase spectrum yields by more efficient transmission standards. Neither should it be relied on to start that process. Indeed, policies designed to provide "open access" by fiat predictably will become complicated and standard specific. Rule makings working out just such complex terms of trade have proven the devil's playground, where incumbent service providers booby-trap the administrative process with anticompetitive trip wires and interminable delays.

For comparison, the salient features of three policy alternatives—the "public interest" status quo (ADMINISTRATIVE in the Noam article), property rights (OCCUPANCY), and Eli Noam's "open access"—are compared in Table 1.

IV. Seeing Red: The Twin Fallacies of "Open Access"

The "open spectrum access" model proposed by Noam does not fly. Perhaps the chief irony within the article is that it purports to critique auc-

---

[38] A catchier phrase might be the late Ithiel de Sola Pool's "technologies of freedom." Plural and spontaneous are the important elements.

TABLE 1

Comparing Three Spectrum Access Rights Regimes

	Regime		
	Status Quo	Property Rights	Open Access
How spectrum allocated among alternative uses	FCC central planning	Owners decide, via "voluntary reallocation"	Owners decide, via "voluntary reallocation"
How rights initially assigned	Auction (nonbroadcast licenses)	Priority-in-use for exclusive claims, auctions for competing claims	Auction each millisecond of access for tiny bandwidths as per FCC rules
How rights reassigned by market	Sale of licensed apparatus	Spectrum flexibility will induce subleasing and joint ventures; spectrum band owners maximize revenues by facilitating traffic	Resellers aggregate rights and sell access "packages" to companies that sink capital in frequency-specific technologies
Market structure	Cartels: government polices outputs	Competition: Price of spectrum access driven to opportunity cost	Cartels: incumbent operators influence the clearinghouse standard-setting process; FCC debates "technical issues"
Consumer welfare outcome	Inefficient due to (1) misallocation of spectrum, (2) rent seeking in allocation process	Efficient statically (see above), and dynamically (as band owners invest in technology to harvest more services from given bandwidth)	Inefficient due to (1) rent seeking in the rule-making process, (2) transactions costs of reselling, (3) underinvestment in band-specific technologies

tions—and then proposes that regulators conduct a million times as many auctions. All the perverse political incentives that are ascribed to auctions—like maximizing revenue by limiting service competition—would likewise haunt the clearinghouse program.

Noam's normative prescription founders on twin fallacies. The first is to oppose "spectrum auctions" on the specious analogy that auctioning use of the color red would be ill-founded public policy. Of course it would: the color red is not scarce.[39] So irrelevant is the analogy to a public good that Noam abandons it himself; the "clearinghouse" he advocates would ration access by auctioning tiny airtime rights bundles.

The second fallacy arises just here, with Noam's assertion that technology has eliminated transactions costs (and rent seeking) in arranging frictionless spectrum access rights sales. Rather than allow competitive markets to discover the most efficient technical standards and market forms for spectrum input transactions, the "open access" model would impose a particular, politically determined (and, as yet, technically unavailable) bidding structure. In one sense, Noam is right to portray his model as a radical departure from both the "exclusive licensing" approach of property rights and FCC "public interest" spectrum allocation. It would wholly depart from existing legal and market institutions, rather than build upon them. And that is a serious flaw in the proposal.

When the government must define property so that first appropriation can occur, it should do so in a way that incorporates the logic of established property law. This will lead rights definition to rules that tend to (a) approximate the final market organization, so as to minimize transactions costs; (b) be as cheaply reassigned as possible so as to allow market forces to adjust the inevitable misallocations;[40] and (c) minimize the possibility of regulatory capture. Whereas a private Spectrum Band Owner could adopt— or invent—the clearinghouse solution were it efficient to do so, the "open access" model could conversely render spectrum worthless and uninhabitable if the transactions costs of aggregating spectrum equities prove substantial. Most fundamentally, the move imposes a de novo solution by policy makers, when the extension of traditional laws and regulatory terms are readily available to facilitate commerce. It is dangerous to mandate that a committee reinvent the wheel; one could likely end up with a trapezoid. And because this mandated solution is specific to the telecommunications sector, that form will inevitably be vulnerable to manipulation by influential interests that gain by proffering rules that box out entrants.

It is simply not true that digital technology obviates all technical and economic decision making, including standard setting. PCS operators are enmeshed today in a battle royal over competing digital wireless telephone standards: TDMA v. CDMA v. GSM. Within the regulations imposed on advanced television service by the FCC, broadcasters, TV set makers, and computer companies are currently pitched in a high-stakes battle over which digital television format(s) to adopt. Convergence to a single stan-

---

[39] It is also discovered and in use, so no entrepreneurial incentives are required to create it. Patents and copyrights may still be necessary to call forth new colors.

[40] Harold Demsetz, When Does the Rule of Liability Matter? 1 J. Legal Stud. 13 (January 1972).

dard across all bands would be a rather strong outcome; it is unlikely and probably unwise. Experimentation, embedded capital, and competition between standards likely will steer different bands to distinct technologies. Proconsumer public policies will ensure that, during this competitive market discovery process, spectrum managers internalize the externalities, coordinating optimal use between (*a*) equipment manufacturers, (*b*) customers, and (*c*) service providers. Getting all ends of the market tied together without long-term contracts for spectrum access would be like building a skyscraper on rented land. The firms that invest in specific capital to make a particular bandwidth more productive can be expropriated, just as the building owner could be. All parties will naturally seek to reassemble private property in spectrum.

The clearinghouse approach would likely overdivide the spectrum. The market would incur transaction costs (including those of regulation) to then reassemble Humpty Dumpty so as to create long-term rights. Noam concedes this when he notes that futures contracts and long-term dealing would, by necessity, aggregate the micro spectrum units his model would auction. Trumpeting this solution is not, as the article argues, an argument in favor of the model. It undermines the model by forcing two unnecessary layers of transaction costs on the market, the first being the rent-seeking round wherein the various interests thrash out the auction and access rules at the FCC, the second being the auction in secondary markets wherein private, de facto spectrum owners (assuming that the first round of rent seeking produces a policy permitting terms sufficiently affordable as to justify the investment in trading and repackaging) emerge. This solution is reached far more efficiently, with less risk, and more quickly (in real, political time) by extending the currently evolving property rights bundle.

Noam augurs that the "open access" model, although heretical today as the Herzel-Coase auction idea once recently was, has the future on its side. I rather suspect that it is nestled deep in history. Leo Herzel, let us not forget, pulled the auction paradigm out of Abba Lerner's tool kit for market socialism. The Noam model fully embraces this lineage, advocating an ambitious government plan of truly Lerneresque proportions—ubiquitous auctions of tiny spectrum access rights would be implemented by omniscient and impartial state auctioneers. Such institutional devices would aim to discover marginal costs, mimicking competitive market outcomes in the absence of private capital owners. This method was once a grand vision. The evidence suggests that its glory days are a half-century past.

Bibliography

Coase, Ronald H. "The Nature of the Firm." *Economica* 4 (1937): 386–405.
Coase, R. H. "The Federal Communications Commission." *Journal of Law and Economics* 2 (1959): 1–40.

Coase, R. H. "Durability and Monopoly." *Journal of Law and Economics* 15 (April 1972): 143–49.

Congressional Budget Office. *Where Do We Go from Here? The FCC Auctions and the Future of Radio Spectrum Management.* Washington, D.C.: Congress of the United States, April 1997.

Demsetz, Harold. "When Does the Rule of Liability Matter?" *Journal of Legal Studies* 1 (January 1972): 13–28.

Dill, Clarence C. *Radio Law.* Washington, D.C.: National Law Book Co., 1938.

Hafner, Katie, and Lyon, Matthew. *Where Wizards Stay Up Late: The Origins of the INTERNET.* New York: Simon & Schuster, 1996.

Hazlett, Thomas W. "The Rationality of U.S. Regulation of the Broadcast Spectrum." *Journal of Law and Economics* 33 (April 1990): 133–75.

Hazlett, Thomas W. "Assigning Property Rights to Radio Spectrum Users: Why Did FCC License Auctions Take 67 Years?" *Journal of Law and Economics* 41 (1998): 529–75.

Hazlett, Thomas W., and Michaels, Robert J. The Cost of Rent-Seeking: Evidence from Cellular Telephone License Lotteries. *Southern Economic Journal* 59 (January 1993): 425–35.

Huber, Peter. *Law and Disorder in Cyberspace.* New York: Oxford University Press, 1997.

Herzel, Leo. "My 1951 Color Television Article." *Journal of Law and Economics* 41 (1998): 523–27.

Kalven, Harry, Jr. "Broadcasting, Public Policy, and the First Amendment." *Journal of Law and Economics* 10 (October 1967): 15–49.

Krattenmaker, Thomas G. 1996. "The Telecommunications Act of 1996." *Connecticut Law Review* 29 (Fall 1996): 123–74.

Kwerel, Evan, and Williams, John. "Changing Channels: Voluntary Reallocation of UHF Television Spectrum." OPP Working Paper No. 27. Washington, D.C.: Federal Communications Commission, November 1992.

Levin, Harvey. *The Invisible Resource.* Washington, D.C.: Resources for the Future, 1971.

Pool, Ithiel de Sola. *Technologies of Freedom.* Cambridge, Mass.: Harvard University Press, 1983.

Powe, Lucas A., Jr. *American Broadcasting and the First Amendment.* Berkeley and Los Angeles: University of California Press, 1987.

Shelanski, Howard A., and Huber, Peter W. "Administrative Creation of Property Rights to Radio Spectrum." *Journal of Law and Economics* 41 (1998): 581–607.

Name Index

Adelstein, Jonathan 508
Agnew, Spiro 168
Alston, Philip 288
Armstrong, Gary M. 156–8
Ayres, Ian 136

Baker, C. Edwin xviii, 80, 391–420
Balkam, Stephen 172
Balkin, J.M. 173
Bangemann, Martin 7
Bar, François xix, 433–56
Baran, Paul 195, 557
Barendt, E.M. xvii, 44
Barnouw, Erik 147
Beck, Ulrich 5
Benkler, Yochai 403–4
Bentham, Jeremy 202–3
Berlusconi, Silvio 283
Biedenkopf, Kurt 317
Birnhack, Michael D. xv–xvi, 219–71
Bockholt, Andrea 3
Booz, Allen 17
Born, G. 45, 59
Braithwaite, John 136
Brandeis, Louis Justice 270–1
Braren, Warren 148–9
Brennan, Justice 416
Brenner, Daniel 143, 146
Brogan, Patrick 167–9
Bruns, Axel 527
Budnitz, Mark E. 191
Busby, Nicole 23
Bush, George H.W. 348
Butler, Richard 563
Butt, Danny 527

Calabresi, Guido 109
Campbell, Angela J. xiv, xvi, 131–91
Charren, Peggy 152
Chaucer, 412
Chen, J. 29
Clark, David 195
Clinton, Bill 82, 132, 561

Coase, Ronald 104, 556, 566, 576, 579, 582
Cook, Philip 117
Copps, Michael 278, 508, 516
Coughlin, Charles E. 141–2
Crandall, Robert 574
Craufurd Smith, Rachael xvii, 43, 275–95
Curran, J. 37
Currie, David Lord 55

De Vany, Arthur 556
Dewey, John 84
Dill, C.C. 581
Douglas, Stephen 383
Doyle, Gillian xvii, 23, 287

Edelman, Murray 305–6, 321
Epstein, Richard 90
Etzioni, Amitai 220, 222, 224, 225, 226, 231, 245–6, 255, 269

Fields, Jack 561
Flew, Terry 525
Fowler, Mark 73, 144, 391
Frank, Robert 117
Frankfurter, Felix 383, 401

Galperin, Hernan xix, 433–56
Gantz, Walter 158–9
Geradin, Damien xx–xxi, 469–95
Gibbons, Thomas xviii–xix, 56, 421–9
Gilder, George 557, 589
Given, Jock 501, 527
Godfrey, Laurence 252–3
Goldberg, D. xi
Goldsmith, Jack 538
Gore, Al 561
Groebel, Jo 11, 17
Grossman, Lawrence K. 6, 18

Habermas, Jürgen 418
Hazlett, Thomas W. xxii, 579–94
Hearst, William R. 394
Helberger, Natali xix–xx, 457–67

Herzel, Leo 556, 566, 576, 586, 593
Hitchens, L. xvii
Hitler, Adolf 394
Hoffman-Riem, Wolfgang xi–xii, 3–22, 305, 307
Holmes, Oliver Wendell 34
Hotelling, Howard 276, 355
Hugenberg, Alfred 394
Humphreys, Peter xvii, 297–325
Hutchins, Robert 38
Hutton, W. 60

Irving, Larry 134

Jackson, Andrew 383
Jefferson, Thomas 383
Johanson, Jon 216
Johnson, David 212

Kenyon, Andrew T. xxi, 497–528
Kessler, Martina 17
Kirch, Leo 309, 317, 319
Kirch, Thomas 309
Kohl, Helmut 324
Kohl, Herb 172
Kruse, Jörn 13
Kuhn, Thomas 554, 556
Kunkel, Dale 158–9

Lascoutx, Elizabeth 160
Laurie, Ben 537, 539
Lerner, Abba 586, 593
Lessig, Lawrence xv, 194, 195, 202–6 *passim*, 207, 212, 216, 537, 545
Levin, Harvey 556
Lewinsky, Monica 82
Lieberman, Joe 172
Lincoln, Abraham 383
Linton, Bruce 142–3, 146
Little, Gavin 23
Long, Huey 413

McNally, Lord 50
Madison, James 383
Marsden. C. xi
Marshall, Justice 416
Meiklejohn, Alexander 232–3
Melamed, A. Douglas 109
Michael, Douglas C. 139–40, 177–81
Mikva, Abner 384
Morland, Justice 252

Mueller, Milton 574
Murdoch, Rupert 32, 311, 394
Murray, Andrew xv, 193–218

Nixon, Richard 168
Noam, Eli xxii, 553–78, 579–93 *passim*
Noelle-Neumann, Elisabeth 4

Owen, Bruce M. 13

Posner, Richard 573
Post, David 197, 212
Powell, Michael 452
Prosser, T. xi, 45, 59
Puttnam, David 50, 51

Reagan, Ronald 348
Reed, David, 195
Rehbinder, Manfred 6
Reidenberg, Joel 537
Reith, John 43, 45
Resnick, Paul 545
Roosevelt, Franklin D. 141
Rowbottom, Jacob H. xv–xvi, 219–71
Ruck, Silke 9

Saltzer, Jerome 195
Santer, Jacques 324
Saxer, Ulrich 5
Scannell, P. 45
Schlosser, Horst-Dieter 3
Scott, Colin xv, 193–218
Sedley, Stephen Sir 34
Sen, Amartya 83
Shaffer Van Houweling, Molly 518
Simpson, O.J. 82
Sinclair, John 527
Smith, Adam 33
Somm, Felix 251–2
Spencer, Herbert 90, 109
Stadek, Michael 4
Stallman, Richard 195
Starck, Christian 11
Steiner, Peter 351–5
Stoiber, Edmund 317
Sunstein, Cass R. xiii, 65–130, 195
Swire, Peter 137
Sykes, Alan 538

Valenti, Jack 171

Verhulst, S xi
Vick, Douglas W. xii, 23–61

Watt, Horatio Muir xxi, 529–52
Webster, Daniel 383
Wegener, Claudia 9
Weiler, Joseph 288
Westphal, Dietrich 287
Whish, Richard 493
White E.B. 401, 404

Wilberforce, Lord 228–9
Winterhoff-Spurk, Peter 9
Wisebrod, Dov 6
Wössner, Mark 324
Wright, Robin xxi, 497–528

Yoo, Christopher S. xviii, 327–89, 391–2, 397–420 *passim*

Zerdick, Axel 307